The ENVIRONMENT

Science, Issues, and Solutions

The ENVIRONMENT

Science, Issues, and Solutions

Mohan K. Wali

Fatih Evrendilek

M. Siobhan Fennessy

CRC Press
Taylor & Francis Group
Boca Raton   London   New York

CRC Press is an imprint of the
Taylor & Francis Group, an **informa** business

Cover design courtesy of Tim Meko. In the background are North and South America, NASA's "true color image," by Reto Stöckli, courtesy of NASA.

CRC Press
Taylor & Francis Group
6000 Broken Sound Parkway NW, Suite 300
Boca Raton, FL 33487-2742

© 2010 by Taylor and Francis Group, LLC
CRC Press is an imprint of Taylor & Francis Group, an Informa business

No claim to original U.S. Government works

Printed in the United States of America on acid-free paper
10 9 8 7 6 5 4 3 2 1

International Standard Book Number: 978-0-8493-7387-9 (Hardback)

Library of Congress Cataloging-in-Publication Data

Wali, Mohan K.
 The environment: science, issues, and solutions / Mohan K. Wali, Fatih Evrendilek, M. Siobhan Fennessy.
  p. cm.
 Includes bibliographical references and index.
 ISBN 978-0-8493-7387-9 (hardcover : alk. paper)
 1. Environmental education. 2. Environmental sciences--Study and teaching (Higher) 3. Interdisciplinary research. I. Evrendilek, Fatih. II. Fennessy, M. Siobhan. III. Title.

GE70.W35 2009
363.7--dc22
                                                                                    2009004456

Visit the Taylor & Francis Web site at
http://www.taylorandfrancis.com

and the CRC Press Web site at
http://www.crcpress.com

*Dedicated to present and future students
of all ages who seek environmental literacy*

Table of Contents

SECTION A: Ecosystem Structure and Function

The study of our natural environment in institutions of higher learning, even in secondary schools, has finally found a foothold. Unlike traditional academic disciplines, however, the introductory courses in environment are taught by instructors who are trained in fields ranging from biology, chemistry, geography, and geology to ethics, philosophy, and political science. Thus, a dichotomy between environmental *science* and environmental *studies* that began in the 1970s has also taken hold. The net result has been that introductory courses on the environment present a widely-diverse coverage of natural processes. What is needed, we believe, is a deeper understanding of basic scientific facts and principles interwoven with the social, economic, and political implications of environmental decision-making.

We have taught environmental science topics for many years, and these opportunities have provided us with a first-hand view of both the technical and the human dimensions of environmental subject matter. It was the experience from 18 years of teaching an introductory course that led the first author to take the lead on writing a new text. Fellow ecologists, Drs. F. Evrendilek and S. Fennessy, terrestrial and aquatic specialists, respectively, kindly joined the effort as coauthors. From the beginning, it has been our goal to develop a fully integrated textbook that rigorously explores environmental issues and their possible solutions. To achieve this dual objective we emphasize the basics of ecology, use this foundation to build an understanding of major environmental problems, and explore methods that might mitigate what has been degraded or destroyed. In doing so, we have endeavored to include an in-depth selection of references, examples and data, case studies, and websites. With these tools, students can further explore topics of special interest.

But why a new text? There are several reasons. We want to share our experience with teaching introductory environmental science using an approach that is strongly grounded in science, the scientific method, and evidence. We also noted that students in our classes were not wholly satisfied with the textbooks we used. What we have also noticed over the years was that students are eager to learn about the environment and are actively looking for ways to focus their talents and make a difference. This desire is important to us as teachers, for our students are the world's future scientists and environmental professionals, decision-makers, writers, poets, and artists. We firmly believe that environmental literacy for all is a must. We also recognize that students welcome the challenge to learn, ask thoughtful questions, and generally crave current, in-depth information.

This book, intended for a beginning college-level environmental science course, uses a back-to-the-basics, building-block approach. The subject matter is divided into three major sections. In Section A, we introduce principles of ecology that can be used to understand how ecosystems respond to disturbance. In Section B, we deal with how human population growth, expanded technology, and unprecedented economic development have altered ecosystems and created serious local, regional, and global environmental problems. The final Section C makes a case for seeking long-term solutions through the prevention and mitigation of environmental problems in their inter-connected, interrelated, and thus, interdependent ways.

We have undertaken this project in the belief that students can attain an environmental literacy that "requires a fundamental understanding of the systems of the natural world, the relationships and interactions between the living and non-living environment, and the ability to deal sensibly with problems that involve scientific evidence, uncertainty, and economic, aesthetic, and ethical considerations."* Nonetheless, given the speed with which new data and syntheses are becoming available from diverse sources, no textbook on the environment can claim to have covered everything.

Our gauge of success depends on how well this book is received by students and instructors alike. From both, we look forward to receiving comments and suggestions with gratitude.

Mohan K. Wali
Fatih Evrendilek
M. Siobhan Fennessy

* Environmental Literacy Council, Washington, DC. ⟨http://www.envroliteracy.org⟩

We acknowledge our gratitude to Professor Emeritus Robert P. McIntosh of the University of Notre Dame and Jerry M. Bigham, professor-director, School of Environment and Natural Resources at the Ohio State University (OSU) for their comments on multiple chapters and especially their encouragement. Our gratitude is due as well to the following colleagues who read drafts of chapters (noted in parentheses) and provided many useful comments and materials. They include, from OSU, Virginie Bouchard (5), Craig Davis (9), Don Eckert (4), Earl Epstein (21), Stacey Fineran (1, 2), Robert Gates (8, 24), Dale Gnidovec (14), Randall Heiligmann (13), Fred Hitzhusen (22), Greg Hitzhusen (26), Elizabeth Marschall (8), James Metzger (3), Martin Quigley (8, 9), Amanda Rodewald (8), Allison Snow (8), Brian Slater (12), Brent Sohngen (22), Alan Stam (4, 8), Lonnie Thompson (19), Robert Vertrees (5), and Roger Williams (13); and to Martin Quigley (University of Central Florida) (8, 25), Maharaj Raina (1) (NCERT, New Delhi), Nirander M. Safaya (1, 21) (North Dakota Public Service Commission, Bismarck), and Alan Stam (8, 25) (Capital University, Columbus). We also appreciate the encouragement of L. H. Newcomb and the assistance of Dennis Hull, Jim Pojar, and Paul Rodewald.

Our friends and colleagues deserve our gratitude for making photographs and illustrations available from their collections; they include Donald Anderson and Judy Kleindinst (Woods Hole Oceanographic Institution), Meinrat O. Andreae (Max Planck Institute for Chemistry, Mainz, Germany); James E. Beuerlein, Ralph Boerner, David Hix, Karen Mancl, Fred Michel, and Fred Miller (OSU); Dan Button (United States Department of the Interior - Geological Survey); Young D. Choi (Purdue University at Calumet); Gene Hettel and H. S. Khush (International Rice Research Institute, Las Banos, Philippines); Louis R. Iverson and Steve Mathews (Forest Service-USDA, Delaware, Ohio); Walter Jetz (Biological Sciences, University of California at San Diego); Karel Klinka (Faculty of Forestry, University of British Columbia); Dennis Lenardic (Photovoltaic Applications and Technologies, Jesenice, Slovenia), Lee Leonard and Charlie Zimkus (*The Columbus Dispatch*); Richard Pemble (Biological Sciences, Moorhead State University, Moorhead, Minnesota); Hans-Joachim Schellnhuber and Hermann Held (Potsdam Institute for Climate Impact Research, Potsdam, Germany); and Vivian Stockman. Many thanks are due also to those authors and publishers who have allowed us to use materials from their published works, all are mentioned at appropriate places throughout the book.

We have tried our best to make sure that the materials used in this book are properly attributed to their appropriate sources. We shall appreciate knowing about any of our lapses which, we want to assure you, are oversights.

Many comments and useful advice were offered from students. These include, at OSU, graduate students Cara Bosco, John Bowzer, Ellen Crivella, Laura Jacobs, Laura Kearns, Sujith Kumar, and Heather Whitman, and undergraduates Joshua Griffin, Rachelle

Acknowledgments

Howe, Melissa Gray, Kurtis Meyer, and Philip Renner. The "60+ Program" at OSU encourages senior citizens to enroll in courses of their choice, and it was the choice of senior librarian, Susan Logan to take ENR 201: Introduction to Environmental Science, a course for which this book is intended. She was not only meticulous in reading an early draft, but, as a librarian, indispensable in locating books post-haste. At Kenyon College, the Environmental Studies class that met during the spring of 2008 read an early draft of the text and provided many useful comments, as did Sadettin Erdonmez from Abant Izzet Baysal University, Turkey. We were ably assisted correcting the proofs by Philip Renner, John Bowzer, and Laura Kearns. Much office and computer assistance came from Patricia Patterson, Patricia Polczynski, and Anthony Utz (OSU). Our grateful thanks are due to them all.

The illustrations, redrawn from original sources, are by Timothy Meko. He has our deepest appreciation and grateful thanks. He acknowledges the courtesies of many sources (such as University of Maryland's Integration and Application Network).

Our warmest thanks also go to our friends at CRC Press–Taylor & Francis: Acquisitions editor Joseph Clements and his assistant Andrea Dale; managing editor Suzanne Lassandro, project editor Judith Simon, copy editor Maureen Kurowsky; typesetter Mark Manofsky; copy writer Stephany Wilken; graphic designer Shayna Murray; and to Helena Redshaw who, before moving on as implementation director academic–Informa, saw this book started. The academic sales manager Susie Carlisle and her staff deserve our gratitude for making draft copies of the book available to students during the winter and spring academic quarters 2007–2009.

We conclude this manifesto of appreciation with deep indebtedness to our respective spouses, Sara, Gülsün, and Ted, with much more than words can say.

MOHAN K. WALI (PhD, U. British Columbia, ecology and soils) is professor in the School of Environment and Natural Resources, and John Glenn School of Public Affairs at the Ohio State University (OSU). At OSU, in addition to other courses, he has taught introductory environmental science for 18 years. He and his colleagues and students have conducted research in the Kashmir Himalayas, the Danish woodlands, Canadian boreal forests, ecosystems of the mid-continent of North America, and the eastern temperate deciduous forest. From his early research in understanding the dynamics of plant communities in relation to their environmental gradients, he and his coworkers have investigated the processes of ecosystem rehabilitation/restoration of drastically disturbed ecosystems to ensure their long-term biological productivity. Along with his graduate students, his most recent work is related to combining restoration of ecosystems with aspects of global climate change. He was a national lecturer of Sigma Xi–the Scientific Research Society for 2 years, and is a Fellow of the American Association for the Advancement of Science.

FATIH EVRENDILEK received his BSc in landscape architecture from Ege University (Turkey), MA in energy and environmental policy from the University of Delaware, and his PhD in environmental science from the Ohio State University. He is an associate professor of environmental science and vice dean of the faculty of engineering and architecture at Abant Izzet Baysal University in Bolu, Turkey. In 2005, Dr. Evrendilek was a NATO Fellow in the United States. Dr. Evrendilek currently teaches the introductory environmental science course. He and his coworkers and students are active in modeling biogeochemical cycles research and global climate change, environmental and energy policy, and ecosystem restoration. His active research projects include "Modeling Temporal and Spatial Changes in Net Primary Productivity of Terrestrial Ecosystems in Turkey," and the "GIS-Supported Integrated Water Resources Management System for the Eastern Mediterranean: A Regional Clean Water Action Plan for the Seyhan River."

M. SIOBHAN FENNESSY received her BS in botany and PhD in environmental science from the Ohio State University. An associate professor of biology and codirector of environmental studies at Kenyon College, she and her students study freshwater wetlands, biological assessment methods, restoration ecology, and the role of temperate wetlands in the global carbon cycle. Earlier, she served on the faculty of the Geography Department of University College, London and held a joint appointment at the Station Biologique du la Tour du Valat investigating the human impact on Mediterranean wetlands. Later, at the Ohio EPA, she founded Ohio's wetland program and wrote the current rules designed to protect wetlands. She is a member of the U.S. EPA's Biological Assessment of Wetlands Workgroup, and currently leads a project to test the use of rapid assessment methods to evaluate the ecological condition of wetlands on a watershed basis. She is coauthor of the best-selling book, *Wetland Plants: Biology and Ecology* (Lewis Publishers, 2001).

In this book, we use the International System of Units, or SI. These units are the accepted currency of most scientific journals throughout the world. In those cases where the unit conversion from and to those in use in the United States is cumbersome, we provide both units in the text for convenience.

International System of Units (SI)

To convert Column 1 into Column 2, multiply by	Column 1 SI unit	Column 2 non-SI units	To convert Column 2 into Column 1, multiply by
Length			
0.621	kilometer, km (10^3 m)	mile, mi	1.609
1.094	meter, m	yard, yd	0.914
3.28	meter, m	foot, ft	0.304
1.0	micrometer, μm (10^{-6}m)	micron, μ	1.0
3.94×10^{-2}	millimeter, mm (10^{-3} m)	inch, in	25.4
10	nanometer, nm (10^{-9} m)	angstrom, Å	0.1
Area			
2.47	hectare, ha	acre	0.405
247	square kilometer, km^2 (10^3 m)^2	acre	4.05×10^{-3}
0.386	square kilometer, km^2 (10^3 m)^2	square mile, mi^2	2.590
2.47×10^{-4}	square meter, m^2	acre	4.05×10^3
10.76	square meter, m^2	square foot, ft^2	9.29×10^{-2}
1.55×10^{-3}	square millimeter, mm^2 (10^{-3} m)^2	square inch, in^2	645
Volume			
9.73×10^{-3}	cubic meter, m^3	acre-inch	102.8
35.3	cubic meter, m^3	cubic foot, ft^3	2.83×10^{-2}
6.10×10^4	cubic meter, m^3	cubic inch, in^3	1.64×10^{-5}
2.84×10^{-2}	liter, L (10^{-3} m^3)	bushel, bu	35.24
1.057	liter, L (10^{-3} m^3)	quart (liquid), qt	0.946
3.53×10^{-2}	liter, L (10^{-3} m^3)	cubic foot, ft^3	28.3
0.265	liter, L (10^{-3} m^3)	gallon	3.78
33.78	liter, L (10^{-3} m^3)	ounce (fluid), oz	2.96×10^{-2}
2.11	liter, L (10^{-3} m^3)	pint (fluid), pt	0.473
Mass			
2.20×10^{-3}	gram, g (10^{-3} kg)	pound, lb	454
3.52×10^{-2}	gram, g (10^{-3} kg)	ounce (avdp), oz	28.4
2.205	kilogram, kg	pound, lb	0.454
0.01	kilogram, kg	quintal (metric), q	100
1.10×10^{-3}	kilogram, kg	ton (2000 lb), ton	907
1.102	megagram, Mg (tonne)	ton (U.S.), ton	0.907
1.102	tonne, t	ton (U.S.), ton	0.907
Yield and Rate			
0.893	kilogram per hectare, kg ha^{-1}	pound per acre, lb acre^{-1}	1.12
7.77×10^{-2}	kilogram per cubic meter, kg m^{-3}	pound per bushel, lb bu^{-1}	12.87
1.49×10^{-2}	kilogram per hectare, kg ha^{-1}	bushel per acre, 60 lb	67.19
1.59×10^{-2}	kilogram per hectare, kg ha^{-1}	bushel per acre, 56 lb	62.71
1.86×10^{-2}	kilogram per hectare, kg ha^{-1}	bushel per acre, 48 lb	53.75
0.107	liter per hectare, L ha^{-1}	gallon per acre	9.35
893	tonne per hectare, t ha^{-1}	pound per acre, lb acre^{-1}	1.12×10^{-3}
893	megagram per hectare, Mg ha^{-1}	pound per acre, lb acre^{-1}	1.12×10^{-3}
0.446	megagram per hectare, Mg ha^{-1}	ton (2,000 lb) per acre, ton acre^{-1}	2.24
2.24	meter per second, m s^{-1}	mile per hour	0.447

To convert Column 1 into Column 2, multiply by	Column 1 SI unit	Column 2 non-SI units	To convert Column 2 into Column 1, multiply by
Specific Surface			
10	square meter per kilogram, $m^2\ kg^{-1}$	square centimeter per gram, $cm^2\ g^{-1}$	0.1
1000	square meter per kilogram, $m^2\ kg^{-1}$	square millimeter per gram, $mm^2\ g^{-1}$	0.001
Pressure			
9.90	megapascal, MPa (10^6 Pa)	atmosphere	0.101
10	megapascal, MPa (10^6 Pa)	bar	0.1
2.09×10^{-2}	pascal, Pa	pound per square foot, $lb\ ft^{-2}$	47.9
1.45×10^{-4}	pascal, Pa	pound per square inch, $lb\ in^{-2}$	6.90×10^3
Temperature			
1.00 (K − 273)	kelvin, K	Celsius, °C	1.00 (°C +273)
(9/5 °C) + 32	Celsius, °C	Fahrenheit, °F	5/9 (°F − 32)
Energy, Work, Quantity of Heat			
9.52×10^{-4}	joule, J	British thermal unit, Btu	1.05×10^3
0.239	joule, J	calorie, cal	4.19
10^7	joule, J	erg	10^{-7}
0.735	joule, J	foot-pound	1.36
2.387×10^{-5}	joule per square meter, $J\ m^{-2}$	calorie per square centimeter (langley)	4.19×10^4
10^5	newton, N	dyne	10^{-5}
1.43×10^{-3}	watt per square meter, $W\ m^{-2}$	calorie per square centimeter minute (irradiance), $cal\ cm^{-2}\ min^{-1}$	698
Plane Angle			
57.3	radian, rad	degrees (angle), °	1.75×10^{-2}
Electrical Conductivity			
10	siemen per meter, $S\ m^{-1}$	millimho per centimeter, $mmho\ cm^{-1}$	0.1
Water Measurement			
9.73×10^{-3}	cubic meter, m^3	acre-inch, acre-in	102.8
9.81×10^{-3}	cubic meter per hour, $m^3\ h^{-1}$	cubic foot per second, $ft^3\ s^{-1}$	101.9
4.40	cubic meter per hour, $m^3\ h^{-1}$	U.S. gallon per minute, $gal\ min^{-1}$	0.227
8.11	hectare meter, ha m	acre-foot, acre-ft	0.123
97.28	hectare meter, ha m	acre-inch, acre-in	1.03×10^{-2}
8.1×10^{-2}	hectare centimeter, ha cm	acre-foot, acre-ft	12.33
Concentrations			
1	centimole per kilogram, $cmol\ kg^{-1}$	milliequivalent per 100 grams, $meq\ 100\ g^{-1}$	1
0.1	gram per kilogram, $g\ kg^{-1}$	percent, %	10
1	milligram per kilogram, $mg\ kg^{-1}$	parts per million, ppm	1
Radioactivity			
2.7×10^{-11}	becquerel, Bq	curie, Ci	3.7×10^{10}
2.7×10^{-2}	becquerel per kilogram, $Bq\ kg^{-1}$	picocurie per gram, $pCi\ g^{-1}$	37
100	gray, Gy (absorbed dose)	rad, rd	0.01
100	sievert, Sv (equivalent dose)	rem (roentgen equivalent man)	0.01

SECTION A

Ecosystem Structure and Function

- What is the structure of natural systems and how do they function?
- How do natural animal and plant populations and communities live?

Section A (Chapters 1–10) begins by illustrating how the recent and widespread public recognition of contemporary environmental issues has been followed by a virtual explosion in environmental research and an appreciation of the need for improved environmental policies and controls. Discussed here are the fundamental principles of ecology that control the structure and function of both aquatic and terrestrial ecosystems in space and in time. This understanding of the natural systems is as fundamental to contemporary environmental issues and in formulating strategies for mitigating them as is the knowledge of such fields as anatomy, physiology, genetics, and pharmacology to the practice of medicine. The emphasis in these chapters is on natural forces that have shaped and continue to shape the existence, abundance, distribution, and diversity of individuals, ecological populations, and communities. This knowledge, therefore, while providing answers to the two questions posed here, is indispensable in understanding changes that are brought about by natural and human activities, disturbances, and phenomena in space- and time-bound relations of natural populations and communities.

The Environment and the Nature of Scientific Enquiry

Meaning and Significance of the Environment

All life on our Earth in its diverse color, form, and splendor owes its existence to the unique biophysical nature of its environment. Although the quest to discover life on other planets continues, none of the space explorations have as yet revealed any form of life anywhere else. The physical factors of our planet's environment—collectively the **physical environment**—comprise the incoming radiant energy of the sun, air, water, and soil without which life simply would not be possible. The **biological environment** encompasses the physical resources, circumstances, conditions, and events that affect the existence and survival of an organism or groups of organisms and their relationships to each other, including humans.

The Earth's natural environmental conditions vary tremendously in space and time. Linked to this natural variability is the fact that some species or groups of species thrive better, or exclusively, in one kind of environment than another, demonstrating their adaptation to a particular environment. The relationship of organisms to their environment was aptly described by Mason and Langenheim (1957):

> "About all that the environmental phenomena have in common is that they impinge in some significant way upon the organism. … More significant … is the *way* in which each phenomenon impinges upon the organism, and the fact that each phenomenon impinges differently."

Thus, the term **environment** is understood to mean the specific physical and biological conditions in a given space (**spatial**) and time (**temporal**) that support the development, growth, and reproduction of a particular species or group of species. Any human-induced change in the environment of an organism that causes an irreversible, adverse effect on its survival must be recognized as an environmental problem. Further, it may be useful to point out that the terms "environment" and "environmental problems" are used at a hierarchy of spatial and temporal scales.

At the broadest scale, sunlight, water, air, and soil constitute the overarching environment for all life. A drastic change in any of these resources at a global level—for example, through an irreversible climate change—can endanger most life forms. But we also know that the terrestrial and oceanic components of our planet support many distinct systems with their own unique environments and biota. Any drastic changes in these local or regional environments can endanger the biota of that system, either as a whole or as its subcomponents. For example, a significant change in the temperature or chemical composition of the waters may render a coral reef lifeless and incapable of regeneration, triggering other changes in the associated biota of that system. But such adverse impacts may not be reflected in terrestrial or aquatic biota of the same region that have altogether different requirements and/or broader ranges of adaptability. Thus, the term "environment" can also be defined as the result of interactions of all living organisms and non-living components or factors in a given space and time, working together to support, maintain, and, indeed, sustain life.

Environmental science is the scientific study of the dynamic relationships and processes that occur among the biological and physical components of nature and their impact on economic and social systems at multiple spatial and temporal scales. Thus, environmental science is, by its nature, an interdisciplinary area of study, addressing some of the most pressing questions of modern times as well as some of the oldest questions about our relationship to the environment (Botkin 1990). Environmental management is the act of applying ecologically sound principles and practices to manage natural resources and solve environmental problems in order to assure the well-being of all life, including humans. It is these principles that we address first in this book. Their indispensability in environmental management will become obvious.

The **nature of environment** as presented previously reveals remarkable complexity. The American psychologist Abraham Maslow placed environment at the very base of his triangular conceptualization of the **hierarchy of human needs.**[1] This base represented the "biological and physiological needs" that include the basic ingredients of life such as air, water, and food; other needs came later. To manage the environment well and sustainably, we need a firm understanding of many basic and applied sciences, including biology, ecology, chemistry, geology, oceanography, hydrology, climatology, geography, economics, political science, statistics, and mathematics. The interdisciplinary approach is dictated as well by the fact that environmental analysis involves numerous physical and biological factors that are in constant flux and dynamic interaction with each other. This complexity becomes obvious in developing and applying proper solutions to prevent, mitigate, or, better yet, eliminate the environmental problems that we face now and will face in the future.

To appreciate the distinctions between global and local environments, their significance and complexity, we must look first at the planet Earth from afar and then examine it from a much closer proximity.

The Pale Blue Dot

In a time span of a mere 40 or so years, as the illustrations show (Figure 1.1a), we have seen some profound changes in the multiple faces of Earth. Its blue face captured by the lens of a camera from space in 1969 was perhaps one of the most illuminating gifts of twentieth century science to the world. Carl Sagan referred to this photograph of Earth as the "pale blue dot" and noted:

> [T]here is no sign of humans in this picture, not our re-working of the Earth's surface, not our machines, not ourselves: We are too small and our statecraft too feeble to be seen by a spacecraft between the Earth and the Moon. … On the scale of worlds—to say nothing of stars and galaxies—humans are inconsequential, a thin film of life on an obscure and solitary lump of rock and metal. (Sagan 1994)

This photograph displayed both our planet's singular beauty and its spatial limits. It also

(a)

(b)

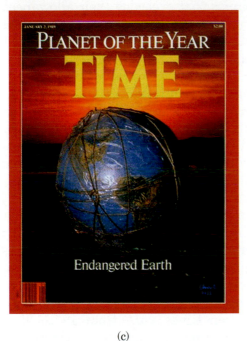

(c)

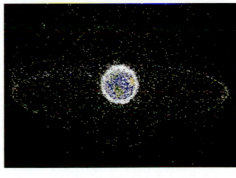

(d)

Figure 1.1 Faces of the Earth. (a) "The pale blue dot," earthrise, as captured from the moon by *Apollo 11* in 1969 (NASA); as visualized (b) in 1972 (*Newsweek* cover, June 12, 1972); (c) in 1989 (*Time* cover, January 2, 1989); (d) in 2002, depicting the space junk orbiting the Earth (*Science,* May 17, 2002, p. 1241).

highlighted why scientists and concerned citizens were alarmed by the widespread changes that were (and are) taking place in the global environment. Many newsmagazines and science journals depicted distress to our planet, from a mere dusty face of the Earth in June of 1972 (Figure 1.1b) to its falling apart in 1989 (Figure 1.1c). The diagrammatic representation of the Earth showing quantities of junk orbiting in space illustrates that human-caused problems exist even beyond the Earth's surface (Figure 1.1d). The last-mentioned fact is of no small consequence, for in 1997 a 500-pound piece of space junk—a discarded rocket engine—came disastrously close to smashing an ozone-measuring satellite; both were circling the Earth at 28,000 km (17,500 miles) per hour (Columbus *Dispatch* 1997).

Recently, the shuttle *Discovery* commander Eileen Collins, with her view from space, expressed great concern on the "widespread environmental destruction on Earth," noting that "the atmosphere almost looks like an egg-shell on an egg, it's so very thin. ... We know that we don't have much air, we need to protect what we have" (Franks 2005).

Many thoughtful people have remarked over and again that our future and the future of our children and their children is inextricably linked to natural resources and how well we take care of the environment. What we breathe, drink, eat, drive, drive on, live in, and live off are all linked. Therefore, the topics discussed in this book are quite important to all of us. Environmental awareness has now reached many segments of our society, including student bodies all over the world.

Stories in the News

The solutions to our past, present, and future environmental problems demand the serious and sustained attention of an environmentally literate citizenry. Two examples selected randomly from the mainstream media provide some optimism in this regard. The first is a special report in *Reader's Digest* that ranks the 50 most populated cities in the United States from the "cleanest" to the "dirtiest" (Burnett 2005). All five rating criteria were environmentally related: "air pollution, water pollution, toxic emissions, hazardous waste, and sanitation force." The report also provided data on the "background, problems, and solutions" required to enhance livability in these cities.

Understanding the second example may require a little explaining. "Dear Santa: Please bring me sulfur dioxide for Christmas" was a major headline spanning the full-page width of a major newspaper during a recent Christmas season (Edwards 1999). Sulfur dioxide! The article dealt with what is known as the sulfur dioxide allowance trading system. Sulfur dioxide is one of the gases produced by coal-burning power plants, and it is a major culprit in turning rain and snow so acidic that it is harmful to both aquatic and terrestrial systems (see Chapter 18 on air quality). The debate on how

best to trap this compound and stop it from entering the atmosphere has raged among power companies, environmental regulators, and decision makers in the United States, Canada, Western European nations, and other places for many years.

To strengthen the air pollution regulations in the United States, President George H. W. Bush signed the Clean Air Amendments on November 15, 1990. Under these provisions, each electrical power plant could reduce its allowable emission of sulfur dioxide into the atmosphere by using new technologies such as scrubbers that remove sulfur dioxide from the emissions from these plants. Any excess savings in allowable emissions could then be traded or sold to other companies that exceeded their allowances. The program is reported to have worked so well that, by 1995, none of the 445 utilities in the country exceeded, on an average, their allowance (U.S. CEQ 1997). The newspaper story refers to the buying of these allowances—not by utilities but, rather, by environmentally sympathetic individuals and groups—as Christmas gifts. This would deny the polluting industries an opportunity to buy them, thus making an "ultimate" gift to the nation and the society at large.

Less than a generation ago, such a headline in the mainstream press would have been unthinkable, given our acceptance of air pollution at the time. But what are the timescales of changes in our perception of environmental problems? What is the time span of a human generation? Our current understanding is that generation times vary from 18 to 35 years, with 28 years as an approximate length.[2] A human generation is long enough to witness a change in the lives of individuals, nations, and regions; however, for the development and establishment of the myriad plant and animal communities that enrich this world, the span of human generations does not qualify as units of great significance.

The timescale of interest varies with the issue at hand. For example, rain shower events from El Niño in the late 1990s breathed life into the dormant seeds of flowering plants in the deserts of the American Southwest; they flourished and bloomed 50 years or so after they were thought to be gone forever. The biological evolution of smaller organisms has taken from centuries to millennia, and for larger organisms, even hundreds of millennia. With the age of the Earth estimated at more than 4 billion years and the occurrence of the first known living organisms discovered from rocks more than 3 billion years old, a human generation does not count even remotely.

Stories related to the environment are now a common feature of most newspapers and magazines, even in *Sports Illustrated* (Boyle 1993). But where did the concern for the environment begin on a worldwide scale? It might be worthwhile to review that briefly.

Advent of Global Environmental Awakening

During the 1960s, a number of events that clearly pointed to environmental deterioration were being documented and reported. Examples of a few major events from the decade of the 1960s may help illustrate this ferment:

- In 1962, in her book, *Silent Spring*, Rachel Carson documented the ill effects of pesticides that persist in the environment for a long time. In a process known as biomagnification, these chemicals become concentrated as one moves up the food chain. For birds, this caused their eggshells to thin and break easily. As a result, the number of chicks that hatched was reduced and many populations dwindled. Because the specific insecticide of her discourse was DDT, a compound used widely throughout the world, this book ushered in recognition of pesticide effects worldwide.
- The "death" of Lake Erie: high loadings of phosphorus and other pollutants that led to the death of some economically important fish species and stimulated the growth of some noxious plant and bacterial species came to widespread attention during the late 1950s. It devastated the economic base of the lake by wiping out the economically important species like whitefish, leading

to the suggestion that the lake had "died." Those fish species that survived were contaminated with mercury and polychlorinated biphenyls (PCBs) and other toxic materials, so government advisories warned people against eating fish from the lake. Advisories for PCBs, mercury, and other chemicals remain in place for fish species such as common carp (*Cyprinus carpio*).

- In 1967, the U.S. Department of the Interior issued a report, "Surface Mining and Our Environment," which presented a pictorial review of the environmental and social costs of coal—"the silted streams, the acid-laden waters, and the wasteland left by surface mining." At the time of its writing, the report noted that 3.2 million acres had been disturbed by surface mining. Two years earlier, President Lyndon B. Johnson had sounded the clarion call for the environment in his address to the 89th Session of the U.S. Congress:

  > We see that we can corrupt and destroy our lands, our rivers, our forests, and the atmosphere itself … all in the name of progress. Such a course leads to a barren America, bereft of its beauty and shorn of its sustenance.
  >
  > We see that there is another course … more expensive today, more demanding. Down this course lies a natural America restored to her people. The promise is clear rivers, tall forests, and clean air … a sound environment for man.

- Caused by a platform fire, the Santa Barbara, California, oil spill of 1969 spilled 200,000 gallons of crude oil, which spread on the ocean surface. As a result, thousands of birds died, as did dolphins, seals, and other marine life.
- That same year, the Cuyahoga River in northeastern Ohio caught fire. This river was extremely polluted and carried with it oil, benzene, dead wood, tires, and such floating on its surface. The fire lasted about 20 minutes, burned down a railroad bridge spanning the river, and brought to worldwide attention how degraded our environment had become.

These are but a few examples of how environmental issues have caught the attention of the general public.

Focus on Environmental Problems Sharpens

These developments brought to the forefront numerous major and pressing environmental problems that had been simmering in the minds of many scientists and professionals. Although the environmental problems are indeed diverse, we next highlight some major examples that form the bulk of this book.

1. **Population growth.** The advances in modern science and technology that include breakthroughs in medical science have significantly decreased infant mortality and increased the human life span. The direct result is a burgeoning human population accompanied by a staggering demand for resources. In the last 70 years, the world population has tripled. The United Nations Human Population Studies estimated that, on November 18, 1999, the human population of the Earth reached 6 billion. Forecasts show that by 2050 this number will have grown to 8 billion *if* fertility rates stabilize. If they do not, human population may reach 12.5 billion. Although the general concern among demographers and some environmental scientists has largely centered on human population alone, what is equally important is the population size of domestic animals, such as cattle, sheep, goats, and horses. Collectively, their populations are estimated to number 3 billion, which we believe is an underestimation. We discuss this further in Chapter 11.

2. **Food and fodder production.** Advances in agricultural sciences have included mechanized farming in the developed countries, cultivation of high-yield and disease-resistant varieties of crops, enhanced growth of crops from the application of fertilizers and irrigation, and decreasing crop losses to pests by using biocides (pesticides, herbicides, and such). This has greatly enhanced agricultural production for 50 or more years in both developed and developing countries. But now many problems with agricultural systems are increasingly realized. These include the heavy reliance on fossil fuels, a decrease in the quantity and quality of water for irrigation, the persistence of pesticides, fertilizer runoff, an increasing loss of agricultural productivity of soils, and the shrinkage of arable land. These are discussed in Chapter 12.

3. **Forest resources.** Besides food, the prime resources of timber, pulp and paper, and biomass fuel come from the forests. Among environmental problems, deforestation was recognized early in the eighteenth century in Europe and in the nineteenth century in the United States. Forests have also been cleared for agriculture and for range lands. Even today, reports suggest that 13 million ha are cleared each year—25 ha per minute (Leape 2006). As important as these issues are, the vital roles of forests in the global water cycle and in climate change are also receiving widespread recognition. We discuss these topics in several chapters (e.g., Chapters 6, 13, 18, and 19).

4. **Mining for coal and minerals.** With increasing technological knowledge has come an increase in the myriad ways in which metals and minerals can be used. Thus, mineral use has increased several-fold. The extraction and processing of minerals require fuel energy. Thus, the mining of coal as a fuel energy source for mineral extraction and processing and for electric power generation has increased immensely. It has been estimated that from the dawn of human history to 1970, the cumulative production of coal was 140 billion tons; the cumulative production was approximately the same over the period 1970 to 2005, a mere 35 years. The role of coal in the economic growth of many nations is undisputed, but multiple environmental problems are involved in mining ores and coal, their processing, and generation of wastes; these are discussed in Chapters 14, 15, 18, and 19.

5. **Transportation and urbanization.** To meet the needs of a rapidly growing human population, extensive transportation systems such as roads, railroads, and air travel have become inevitable. The advent of automobiles and their convenience have resulted in an unprecedented increase in their numbers.

This has facilitated longer distance commutes between the workplace and home, making living farther and farther from cities possible. The result: unprecedented suburban growth and development, resulting in the phenomenon that we now refer to as urban sprawl. We discuss these aspects in Chapters 15–18.

6. **Quantity and quality of water.** Water is one of the basic necessities of life, yet more than half of the world's 6 billion persons suffer from a shortage of water. There are multiple reasons for this shortage; the primary one, of course, is that freshwater constitutes less than 3% of the world's water, while the remaining (salty) 97% is found in the oceans. The scarcity of freshwater has been exacerbated because we already use 70% of the supply, with water use rising sixfold in the last 70 years. Additionally, contaminants that result from what we do on land or send to the atmosphere often end up in water bodies. This degrades water quality, producing unsafe drinking water that contains waterborne diseases and claims the lives of millions of children each year. Water allocation is a prominent factor in disputes between countries that lack an adequate water supply. Chapter 16 is specifically devoted to water quantity and quality issues; however, the role of water is the backdrop of topics in several other chapters as well.

7. **Air quality and the loss of stratospheric ozone.** The quality of the air we breathe has a direct bearing on our health. A well-documented historical record exists of the degradation of air quality and the subsequent breathing stress experienced by both royalty and commoners of England dating as far back as the sixteenth century. Most of the air quality degradation then, as now, came from the burning of coal. Estimates are that more than a quarter of the world's population, or 1.5 billion people, breathes air that is unhealthy. Air contaminants also find their way into other biota and ecosystems, with deleterious effects. Considerable progress has been made in the last 30 years to improve the quality of air in the United States, although rapidly developing countries such as China suffer from serious air pollution problems. However, the use of certain chemicals called chlorofluorocarbons or CFCs (trade name *Freon*), a twentieth-century

phenomenon, has sent stable chlorine compounds up into the stratosphere, where, on a global basis, they degrade the ozone layer that otherwise keeps much of the sun's ultraviolet rays from penetrating to the Earth's surface. Details on air pollution, ozone degradation, and laws designed to improve air quality are discussed in Chapters 18, 19, and 21.

8. **The greenhouse effect and climate change.** Human activities lead to the emission of a number of gases to the atmosphere that have a long residence time and a warming potential. These gases, referred to collectively as greenhouse gases or GHGs, include carbon dioxide, methane, nitrous oxide, water vapor, and CFCs. They trap heat closer to the Earth's surface, thereby causing warming by what is known as the "greenhouse effect." Studies from a number of scientific disciplines now show conclusively that GHGs have considerably altered the temperature relations of the lower atmosphere. The changing temperature relations will have far-reaching impacts on nearly all regions of the world. Although Chapter 19 is devoted specifically to global climate change, the aspects of climate change appear in several other chapters as well.

9. **Loss of habitat and biodiversity.** The first five activities briefly mentioned earlier have transformed natural land areas in significant ways. The extent of such disturbance is truly staggering, and we know its magnitude only in an approximate way. In a widely cited report issued nearly 21 years ago, the Food and Agricultural Organization of the United Nations estimated the Earth's disturbed areas at 2,000 million ha, with five to 7 million ha added each year. An August 2003 report from the United Nations Convention to Combat Desertification notes "the loss of 250 million acres [more than 100 million ha] a year of fertile soil due to deforestation, drought, over-grazing and climate change." The result has been loss of habitats for organisms, rendering many species endangered or extinct. The loss of each such species means the potential loss of new sources for drugs, food, and fiber. Complicating this is the increasing invasion and survival success of many exotic species. Although these aspects are specifically discussed in Chapters 20 and 25, they also appear in other chapters about

topics such as ecosystem rehabilitation and restoration (Chapter 24).

10. **Resource loss and pollutant gain.** Collectively, changes in the environment from land transformation, overharvesting, and over-consumption have resulted in the degradation and loss of productivity in many areas. Such changes are not restricted to land areas alone; human activities have had significant impacts on aquatic resources as well. For example, fish catches globally have fallen, and populations of important species such as Atlantic cod, haddock, and herring have fallen considerably. At the same time, undesirable chemicals are present in nearly all environments, leading to global treaties on compounds known as POPs—persistent organic pollutants. Topics such as these and how to mitigate adverse impacts in the environment are discussed in chapters in Sections B and C.

Advances in Environmental Research and Regulation

The developments noted before showed that we must pay specific attention to the major environmental problems confronting us. The year 1969 witnessed a historic event in the United States: the passage of the National Environmental Policy Act (NEPA). This law, which requires all U.S. federal agencies to consider environmental impacts in their planning and decision-making, revolutionized environmental thinking in a legal–social sense. This landmark legislation not only paved the way for substantive legislation in many other environmental areas (see Chapter 21, "Environmental Policy and Law") but also provided the impetus for environmental legislation in many other countries.

Similar ferment was brewing on the international stage: In 1966, the United Nations (UN) proposed that a global discussion be conducted on "the problems of human environment." In 1968, the UN General Assembly passed a resolution that asked for a specific meeting to be held to examine such problems because there was

need for intensified action at the national, regional, and international level in order to limit and, where possible, to eliminate the impairment of the human environment and ... to protect and improve the natural surroundings in the interest of man ... and also to identify those aspects that can only, or best be solved through international cooperation and agreement. (Caldwell 1990)

Thus, in early June 1972, the first United Nations Conference on the Human Environment was held in Stockholm, Sweden. Governments of 114 countries, together with many nongovernmental organizations, participated. The conference enunciated a set of 26 principles referred to as the Stockholm Declaration (Box 1.1).

Since then, these global conferences have been held at 10-year intervals: in 1982 in Nairobi, Kenya; in 1992 in Rio de Janeiro; and in 2002 in Johannesburg, South Africa. Shortly after the 1972 Stockholm conference, the United Nations Environment Program was created. Public awareness of and demand for environmental policy have also resulted in a vigorous emergence of environmental research. What was once considered the domain of university researchers is now a part of governmental research at all levels—federal, state, county, and city. It has also been taken up by industry; by many laudable nonprofit, nongovernmental research centers (for example, Worldwatch Institute and World Resources Institute); by foundations (for example, Pew Charitable Trusts and Heinz Center for Science, Economics, and the Environment); and by many activist nongovernmental organizations (referred to as NGOs).

Most recently, United Nations Secretary-General Kofi Annan initiated the Millennium Ecosystem Assessment (MEA). The objective of the MEA, which took place between 2001 and 2005, was to assess "the consequences of ecosystem change on human well-being and the scientific basis for action needed to enhance the conservation and sustainable use of those systems and their contribution to human well-being" (MEA 2003). More than 1,360 experts worldwide contributed to the final report, which provides an evaluation of the status and trends of ecosystems and the services they provide (e.g., clean water, food, forest

BOX 1.1   Principles of the 1972 Stockholm Declaration on Human Environment (From Clarke and Timberlake 1982)

1. Human rights must be asserted, apartheid and colonialism condemned.
2. Natural resources must be safeguarded.
3. The Earth's capacity to produce renewable resources must be maintained.
4. Wildlife must be safeguarded.
5. Nonrenewable resources must be shared and not exhausted.
6. Pollution must not exceed the environment's capacity to clean itself.
7. Damaging oceanic pollution must be prevented.
8. Development is needed to improve the environment.
9. Developing countries therefore need assistance.
10. Developing countries need reasonable prices for exports to carry out environmental management.
11. Environment policy must not hamper development.
12. Developing countries need money to develop environmental safeguards.
13. Integrated development planning is needed.
14. Rational planning should resolve conflicts between environment and development.
15. Human settlements must be planned to eliminate environmental problems.
16. Governments should plan their own appropriate population policies.
17. National institutions must plan development of states' natural resources.
18. Science and technology must be used to improve the environment.
19. Environmental education is essential.
20. Environmental research must be promoted, particularly in developing countries.
21. States may exploit their resources as they wish but must not endanger others.
22. Compensation is due to states thus endangered.
23. Each nation must establish its own standards.
24. There must be cooperation on international issues.
25. International organizations should help to improve the environment.
26. Weapons of mass destruction must be eliminated.

products, flood control, and natural resources). The assessment also addresses options to increase the sustainable use of ecosystems.

Thus, in the last 40 or so years, many governmental and nongovernmental organizations and research centers of both national and international scope have emerged and produced much-needed environmental data and disseminated them in the form of papers, books, and reports. Almost daily, commentaries via radio, television, newsmagazines, and newspapers and coffee-room conversations have created a remarkable degree of environmental awareness in all of us. Put simply, the resources now available from these sources and on the Web sites of innumerable local, national, regional, and international organizations are phenomenal and captivating. Environmental literacy is rapidly catching on and will assume a high degree of importance in our daily lives.

The Scientific Method

Like all disciplines of science, environmental science shares a common and universally accepted code of rules that govern scientific inquiry (see Box 1.2). In addition, unlike other scientific endeavors, environmental science considers related fields such as ecological economics, environmental policy, and ethics. All inquiries, scientific and otherwise, must include what, in journalism, are referred to as

the five Ws: *who, what, where, when,* and *why.*
Of no less consequence is knowing *how.* This
book examines these questions in relation to
the structure and function of our living world.

At the heart of scientific inquiry are three
important attributes: **experiment, observation,** and **inference.** From a vast number of
experiments, observations, and inferences
emerges a body of knowledge that can be relied
upon. Conclusions based on observations that
cannot be verified by repeating experiments
are false and must be discarded. Through this
process, we arrive at fewer wrong, testable,
and supportable hypotheses. But this whole
process involves several steps that take place
in an orderly process, including:

- the process of induction that sets up a
  hypothesis that can be tested;
- reasoning, by a process of deduction,
  the likely outcome of a hypothesis; and
- designing experiments to verify a given
  hypothesis and then repeating them to
  establish the validity of their results.

Inductive reasoning has been emphasized
further in the concept of "strong inference" that
involves the application of the following steps:

- devising alternative hypotheses;
- devising creative experiments with alternative outcomes that will exclude one or
  more of the hypotheses;
- completing experiments with clear results;
  and
- repeating the procedures to exclude or
  refine other possibilities (Platt 1965).

Hogg (1999) noted, "Scientific hypotheses are
most securely 'validated' when (i) they make
successful predictions, (ii) there are conceiv-
able observations that could, in principle,
refute them, but have not, and (iii) there is a
comparably sensible competitor theory that is
faring worse."

The induction–deduction–verification approach is a continuous process, and each of the
steps feeds back to the others (Figure 1.2). It
is only after repeated experiments or observations that we establish the validity of phenomena from which scientists draw conclusions.

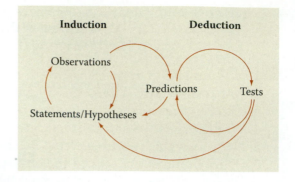

Figure 1.2 Scientific investigations are a continuous
process in which new hypotheses are proposed based
on new or revised inductions and deductions, inces-
sant observations and testing, and predictions based on
them. (Drawn by Brian P. Kraatz and David K. Smith
in Barnosky, A. D., and B. P. Kraatz. 2007. *BioScience*
57:523–532.)

Once these conclusions find widespread con-
firmation, a set of principles is established.
Well-substantiated principles become **theories**
or **laws.** The strength of science lies in carrying out all the mentioned steps systematically:
descriptions, explanations, predictions, claims
on knowledge, expansion and representation
of knowledge.

To summarize, the key words about the scientific method are **verification, validation,
repetition, prediction,** and **falsification.**
These fundamentals remain solidly in place
no matter how many scientists have defined
science (Box 1.2).

Some philosophers and scientists have
argued that scientific discovery may be less a
function of the inductive–deductive process but,
rather, may come more from **serendipity**—that
is, making discoveries by accident.[3] Others have
felt that discoveries come from **intuition,** "the
act or faculty of knowing without the use of
rational processes, a faculty of guessing accurately." Although both serendipity and intuition have played immense roles in scientific
discoveries, it must be emphasized that most
scientific investigations are tremendously labor
intensive, and the adage that science involves
1% inspiration and 99% perspiration is not too
far off the mark.

The scientific method, with its repeating
process of experimentation and verification,
allows continued progress in our understanding of environmental issues. As early as 1951,

BOX 1.2   Science: What It Is and What It Does

Science has been defined in many ways over the centuries. From more recent times, here are some examples of what science is and science does:

- Of Will Durant's eight elements of civilization, "the fifth element of civilization is science—clear seeing, exact recording, impartial testing, and the slow accumulation of a knowledge objective enough to generate prediction and control" (Durant 1935).
- The scientific way of forming concepts ... [lies] in the more precise definition of concepts and conclusions; more painstaking and systematic choice of experimental materials; and greater logical economy (Einstein 1940).
- Science is not so much about what is true but about being less and less wrong (O'Neill 1996).
- Science derives its authority from the fact that it is continually self-questioning and self-correcting and cannot venture beyond the limits of verifiability (or falsifiability) (de Duve 1999).
- [T]he systematic character of scientific knowledge [has] five features ...: how science describes, how science explains, how science establishes knowledge claims, how science expands knowledge, and how science represents knowledge (Hoyningen-Huene 1999).
- The task of science is twofold: to determine, as best we can, the empirical character of the natural world; and to ascertain why our world operates as it does (Gould 2001).
- To some, "science is producing objective and increasingly comprehensive descriptions of a largely invisible world." To others, "even our best scientific theories are only models, whose job is to generate accurate predictions, not to reveal hidden reality" (Lipton 2007).

Conant (1951) wrote that all of science is based on an "interconnected series of concepts or conceptual themes that have developed as a result of experimentation and observation and are fruitful for further experimentation and observation." The logical flow from observations → experimentation → inference invariably leads to the investigation of a cause and an effect. Several cause–effect relationships have been visualized in Figure 1.3.

The first two types of relationships (Figure 1.3a, b) are least common; ecological complexity arises because the relationships of environmental variables tend to follow the multiple cause–multiple effect and circular cause–effect pathways (Figure 1.3c, d). **Feedback loops** are inherent features of ecological systems and refer to a process whereby a change in the environment ultimately connects back to itself by triggering a chain of effects in a circular way or in a "loop" (Figure 1.3d). The designations "positive," "negative," and "neutral" for feedback loops indicate whether changes in

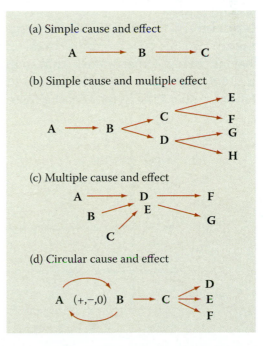

Figure 1.3 Types of cause–effect pathways. Causes are on the left and effects on the right. (Adapted from Grant, V. 1963. *The Origin of Adaptations*. New York: Columbia University Press.)

the system produce amplifying effects (i.e., the effects get bigger), dampening effects (the effects get smaller), or no effects, respectively. Feedback loops contain "delays," or "lags," in space and time when the effect of one variable on another takes time.

Types of Scientific Studies

Scientific investigations consist of observational, experimental, and modeling studies. **Observational approaches** aim at identifying, describing, and following the possible causes, effects, and controls associated with events or organisms for a given period of time in a given area without manipulating the system. Short-term or long-term observational studies can be (a) **retrospective,** where comparisons are sought between the past and the present; or (b) **prospective,** where comparisons are made between the present and the future. All such studies can range from molecular (e.g., microbial growth by electron microscope) to global (e.g., land use changes by remote sensing) spatial scales.

Experimental studies are designed to determine causal relationships and extract conclusions using experiments that manipulate a single factor or multiple factors (called **treatments**) about the hypothesis being tested.

There are two types of experimental approaches, whether short-term or long-term: **manipulative** and "**natural**" experiments (Figure 1.4). Laboratory or field manipulative experiments test educated guesses (**hypotheses**) by manipulating different environmental factors, letting everything else be constant (control groups) and evaluating associated responses. Natural experiments use natural gradients of variation in space (**toposequence**) and time (**chronosequence**) to gain insights into how that factor affects ecosystem behavior or processes (e.g., latitudinal and altitudinal transect- or gradient-driven studies of biological productivity).

Modeling studies are used extensively to understand the complexity of environmental systems. There are two basic types: empirical and process-based (or mechanistic) models. **Empirical models** rely on data in order to quantify the response of a system as a function of causal variables. These do not require the explanation of the underlying processes and structures that produced that system behavior. **Process-based ecosystem models** are used to mimic past, present, and future changes, based on the representation of components (structure) and their interactions (function) of a delineated environmental system. These provide a numerical solution over a given time period and spatial scale. In modeling terms, change through time and space means changes in the quantities of things (**state variables**), such as number, area, weight, concentration, abundance, and density.

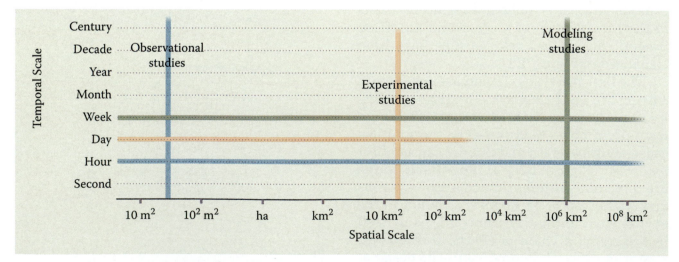

Figure 1.4 Time and space dimensions of the scientific studies (horizontal and vertical lines refer to the spatial and temporal scales, respectively).

State variables are expressed by differential equations and driven by forces or fluxes formulated as rates of processes (**rate variables**).

Overall benefits of process-based ecosystem models include:

- an increase in our understanding of the system;
- a quantitative description of environmental processes;
- the design of human interferences with environmental processes through preventive and mitigative measures, policies, and practices;
- the ability to predict the consequences of various courses of action;
- the ability to conduct experiments with ecosystems that would be impossible or unethical to carry out at regional to global scales (e.g., by the construction of "what-if" scenarios);
- establishment of interdisciplinary dialog and cooperation;
- identification of knowledge gaps and priority research areas; and
- testing the harmony among theory, experiment, and observation and their integration.

The Two Cultures and the Environment

Given the complex, dynamic nature of ecological–environmental science and the environmental problems we face, scientific methods must be augmented by philosophical and ethical precepts as well. Ever since the advent of science in human history, there has been tension between scientists and those who follow literary and cultural pursuits. This tension was explored in a brilliant series of lectures by the British science educator, C. P. Snow, in his book *The Two Cultures* (1959).

Snow made the following point: At one end are the literary intellectuals … at the other, scientists. The nonscientists have a rooted impression that the scientists are shallowly optimistic and unmindful of the human condition. On the other hand, the scientists believe that the literary intellectuals are totally lacking in foresight and ignorant of scientific experimentation and impact. In short, both groups show a tremendous intolerance of each other's views. Snow called intolerant people on either side Luddites.[4]

The viewpoint in *The Two Cultures* now pervades environmental issues as well. There are those who think that all environmental problems can be solved because we have plenty of resources (the "cornucopians") and all problems can be solved by human ingenuity and technology. Others think that resources of the Earth are limited and must be protected at all costs ("Malthusians," "neo-Malthusians," or "doomsayers").[5] But which of these two groups is scientifically correct? By presenting data and analyzing concepts, we shall examine the merits of both viewpoints throughout this book.

Contemporary historians note that "today's global forces for change are [moving us] into a remarkable new set of circumstances in which social organizations may be unequal to the challenges posed by overpopulation, environmental damage, and technology-driven revolutions and where the issue of winners and losers may to some degree become irrelevant" (Kennedy 1993).

Thus, like the variables of natural systems, economic systems and world trade have become interconnected to environmental issues, now more than ever.

The Urge to Live, Live Well, and Live Better

The mathematician–philosopher Alfred North Whitehead (1929) elegantly captured the place of scientific reasoning and living in the environment. He noted that the function of reason is to promote the art of life. Not only does environment affect organisms, but they, in turn, also modify the environment. The threefold urge of humans as they live in their environment is (1) to live, (2) to live well, and (3) to live better (as Whitehead noted for organisms in general). The fulfillment of that urge is the quest of people everywhere, especially in the less developed countries, where a staggering number of more than 2.85 billion people, by conservative estimates, go hungry every day.[6] It will all depend on how well we adopt scientific knowledge in making a sustained living for humans and our biotic and abiotic support systems.

Science provides the key to understanding the underpinnings of phenomena observed on our planet and in our universe. Therefore, in the following chapters, let us examine the scientific basis of our current knowledge of the environment, its processes, and its problems. We will follow this with discussions of how environmental science may guide us to a better future.

Notes

1. Abraham Maslow (1908–1970), noted as the father of humanistic psychology, proposed the "hierarchy of human needs," as seen in Figure 1.5.

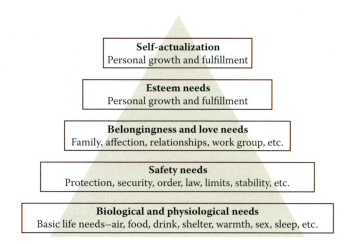

Figure 1.5   Maslow's hierarchy of needs.

2. *Strong's Exhaustive Concordance* suggests that the behavioral cycle is the natural tendency for humans to reproduce every 25 years; hence, the average length of a generation is about 25 years. In the generations table in Strauss and Howe's *Generations,* a time slice of 21 years is generous. The Boomer generational cohort birth years are generally taken as starting in 1943 and ending in 1960 (Strauss and Howe 1991, p. 299).
3. The word "serendipity" was coined by Horace Walpole after the characters in the fairy tale "The Three Princes of Serendip," who made such discoveries. "Intuition. Noun, the act or faculty of knowing without the use of rational processes; immediate cognition; a capacity for guessing accurately" (*The American Heritage Dictionary of the English Language*).
4. This little book, based on lectures that C. P. Snow gave at the University of Cambridge, was printed 18 times between 1959 and 1986; in 1964, Part II, "The Two Cultures and a Second Look," was added.
5. "Cornucopia" refers to the horn of plenty, implying inexhaustible resources; "doomsayers" refers to those who think that we are plundering our limited resources.
6. Professor Nicholas Negroponte of the Massachusetts Institute of Technology and founder of One Laptop per Child notes that of the 1.2 billion poor children in the world, half (600 million) do not have any education at the primary-school level (Public Broadcasting System's "News Hour," November 22, 2000). In many studies, literacy, particularly of women, has been strongly correlated with environmental consciousness and management.

Questions

1. How do our lives depend on the environment? How have past actions affected our situation today?
2. What constitutes scientific knowledge, and what are some ways in which science has been applied to solve problems?
3. Give some examples of positive and negative feedback relationships in your daily life.
4. Think of the three most important environmental issues where you live and what actions are being taken to improve the situation.
5. What are the roles of the media, the science community, and decision-makers in exploring and solving environmental issues?

Biosphere and the Science of Ecology

2

The Biosphere

The Earth has a total surface area of 509 million km^2; 359 million km^2 are water (the surface of the hydrosphere) and 149.6 million km^2 are land (surface of the lithosphere) (Table 2.1). In total, the land and water and the life they support are called the **biosphere.** The Russian geologist V. I. Vernadsky, who coined the term, described it thusly:

> The biosphere is the envelope of life, i.e., the area of existence of living matter … the biosphere can be regarded as the area of the Earth's crust occupied by transformers, which convert cosmic radiation into effective terrestrial energy: electric, chemical, mechanical, thermal, etc. (Vernadsky 1928)

The "biosphere is a cosmic phenomenon and a geological force" (Margulis and Lovelock 1989). It is the upper thin mantle on the Earth's crust that is enveloped by the atmosphere and supports life (Figure 2.1). Major climatic patterns and local weather phenomena affect both the living and the nonliving components. Above the

Table 2.1
The Earth's Surface Areas of Land and Water and Their Continental Distribution

Continent	Approximate Area		% of Earth's Land
	km²	mi²	
Total area	509,000,000	196,526,007	100.00
Oceans	359,000,000	138,610,680	71.00
Land	149,673,000	57,789,070	29.00
Asia	43,608,000	16,837,143	8.56
Africa	30,335,000	11,712,409	5.59
North and Central America	25,349,000	9,787,304	4.98
South America	17,611,000	6,799,645	3.46
Antarctica	13,340,000	5,150,603	2.26
Europe	10,498,000	4,053,300	2.06
Oceania	8,932,000	3,448,664	1.75

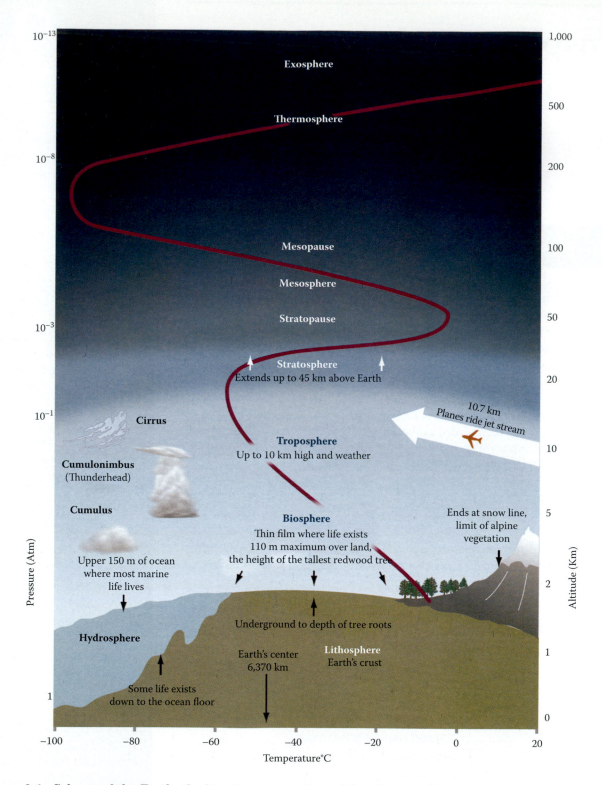

Figure 2.1 Spheres of the Earth, the biosphere, atmosphere, lithosphere, and hydrosphere, and the variation of atmospheric pressure and temperature (red line) with altitude above the Earth's surface. (Lower part based on a diagram by R. M. Chapin, Jr.)

biosphere, air temperatures cool as the altitude increases; this occurs at an average lapse rate of 6.5°C for every 1-km ascent in the troposphere (the lower atmosphere; see Chapter 3).

The biosphere is home to a splendid variety of plants (about 300,000 species reported so far) and of animals (1.4 million species described). However, scientists have a surprisingly poor understanding of just how many species the biosphere contains, with estimates ranging from at least 5 million to as many as 100 million species; the best estimates seem to hovering around 10 million (Wilson 2000a). Estimates change as new data are collected; for example, in a study of only 19 individual trees in Panama in the 1990s, fully 1,200 species of beetles were discovered, 80% (960) of which had never been described before (Thomas 1990). In order to document the rich diversity of life, an ambitious project was recently launched to create a central, online database of all 1.8 million described species; each will have its own Web page in this "Encyclopedia of Life" (Leslie 2007).

Some species are truly monumental. The most massive living thing on the Earth is the largest-known giant sequoia (*Sequoiadendron giganteum*), named the "General Sherman," which stands 84 m tall in California's Sequoia National Forest. It has a girth (or diameter at breast height) of 25 m and is an estimated 2,500 years old. It has been estimated to contain the equivalent of more than 500,000 m² of lumber. The single tree whose canopy covers the greatest area of ground is the Great Banyan tree (*Ficus benghalensis*) in the Indian Botanical Gardens in Calcutta, with 1,775 supporting roots and a circumference of 411 m. Overall, it covers some 1.4 ha. The tallest tree ever measured, at 145 m, was an Australian eucalyptus (*Eucalyptus regnans*) in Victoria, Australia. Other records have included a 120-m Douglas fir (*Pseudotsuga menziesii*) in Washington in 1905 and a 112-m coastal redwood, *Sequoia sempervirens,* in California in 1873.

The largest living land animal is the African bush elephant (*Loxodonta africana*); an average adult stands 3.2 m at the shoulder and weighs 6.3 tons. The tallest elephant ever recorded was a bull shot in Namibia on April 4, 1978. Its height was indicated at 4.2 m and an overall length of 10.4 m; its front foot alone had a circumference of 1.6 m. The giraffe (*Giraffa camelopardalis*), which is now found only in the dry savannas and semidesert areas of Africa south of the Sahara, is the tallest living animal. The tallest ever recorded was a 6-m tall Masai bull (*G.c. tippelskirchi*).

The gigantism of some organisms aside, the biosphere contains a diverse and rich plant and animal life. Most phenomenal are the microscopic living forms whose contributions to the functioning of the living world make them simply indispensable.

There is remarkable similarity between the changes that take place in the growth form of plants, as one proceeds from the equator to the poles, and changes that occur with increases in altitude. The lowest latitudes and altitudes display the greatest diversity of species, with a predominance of trees. As latitude or altitude increases, plant and animal communities show a diminution of species diversity, and tree forms give way to shrubs and perennial herbs until conditions become so harsh at high altitudes and far north and far south that no life exists.

Plants occupy nearly every available terrestrial habitat on Earth. For example, the yellow poppy (*Papaver radicatum*) and the arctic willow (*Salix arctica*) survive (the latter in an extremely stunted form) on the northernmost land at 83° N latitude, where temperatures are frigid. The highest altitude at which flowering plants have been found is at 6,400 m on Kamet Mountain (India), which rises to a full height of 7,756 m. The plants were members of the mustard family (*Ermania himalayensis*) and crowfoot family (*Ranunculus lobatus*). Another member of the mustard family (*Christolea himalayensis*) was found at 6,245 m, and a member of the Pink family (*Stellaria decumbens*) was discovered at 6,135 m in the Himalayas. Mosses and lichens have been recorded at 6,300–6,600 m and yeast and fungi have been recovered at elevations between 7,600 and 8,400 m. Salticid spiders have been collected at altitudes of 6,700 m (Swan 1992).

Lichens such as *Rhinodina frigida* were found in Antarctica's southernmost Moraine Canyon at 86°09′ S in 1965. The southernmost recorded flowering plants include the Antarctic hair grass (*Deschampsia antarctica*), which was found at latitude 68°21′ S on Refuge Island, Antarctica, and the carnation (*Colobanthus crassifolius*), which was found at latitude 67°15′ in Graham Island, Antarctica; both species live in the maritime Antarctic (Green, Schroeter, and Sancho 1999).

The Lithosphere

The lithosphere is the bedrock and other rocky material on which soils have developed. This includes the plant root zone, also known as the **rhizosphere.** Most plant roots are located in the upper 1 m of the soil. However, in dry regions where water is scarce, plants may extend their root systems to depths of 10–20 m below ground to reach the ground water table. In some cases, plant root systems may reach incredible dimensions. The greatest reported depth to which roots have penetrated is an estimated 122 m in the case of a wild fig tree in eastern Transvaal, South Africa. An elm tree root depth of at least 110 m was reported from England in the 1950s. It is also reported that the roots of a species of tropical acacia (probably *Acacia giraffa*) penetrated to an estimated depth of 46 m in southwestern Africa. The total length of all roots from a single plant can be surprisingly long. Scientists have reported a total root length in excess of 630 km on a single winter rye plant (*Secale cereale*); the surface area covered by the roots and root hairs was more than 630 m^{-2} in about a 0.5 m^{-2} plot of soil (Dittmer 1937).

The Hydrosphere

The hydrosphere is the area of the Earth covered by water, which makes up 70.9% of the Earth's total surface. The volume of the oceans is estimated to be 1.3 billion km^3, with an estimated mean depth of 3.7 km. Compare this with the relatively small volume of freshwater, which amounts to only 35 million km^3, or 2.7% of the total water on Earth. As in terrestrial environments, some huge animals occupy marine habitats. The longest and heaviest mammal in the world and the largest animal ever recorded is the blue or sulfur bottom whale (*Balaenoptera musculus,* also called Sibald's rorqual). The largest specimen ever recorded (1909) was a female that measured 34 m in length. Another female caught in 1947 measured 28 m and weighed 209 tons.

For humans, however, it is the interactions among the atmospheric, terrestrial, and aquatic processes in the biosphere that are perhaps of most interest. Between 1987 and 1991, a Texas businessman funded a $30 million project to duplicate the biosphere in miniature to ascertain if it could be self-supporting in terms of atmospheric, terrestrial, and aquatic processes. This grand-looking experiment, called Biosphere II, was more than a hectare of enclosed space (the largest closed system ever created) on a ranch in Oracle, northeast of Tucson, Arizona. The creation of Biosphere II involved a remarkably complex design (Figure 2.2), and the efforts and expense to bring together such a diverse collage of organisms were unprecedented. Key habitats included a coral reef, mangrove wetlands, savannah grasslands, and a farm for food production.

Four men and four women, called "biospherians," lived in the structure for 2 years and came out in December 1997. Many stories on this experiment appeared in newspapers, in newsmagazines, on radio and television, and much later in scientific journals. The major lesson from this experiment was that this supposedly self-contained enclosure was, in fact, not self-sustaining. Not enough oxygen could be generated by the plant life to support the animals and the eight biospherians (a captivating tale has been told by the biospherian Jane Poynter [2006]). Despite all the effort and money expended and the collective time of 16 man-and-woman years by the biospherians, it became more a tourist attraction than a scientific study. Because it was not properly designed as a scientific study, the series of experiments conducted in Biosphere II fell short of their goals and expectations. However, it did successfully reinforce the observation that we lack knowledge of the specific ecological processes operating at both small and large scales that support life on Earth.

The Science of Ecology

Understanding environmental issues, be they historical or contemporary, requires a firm

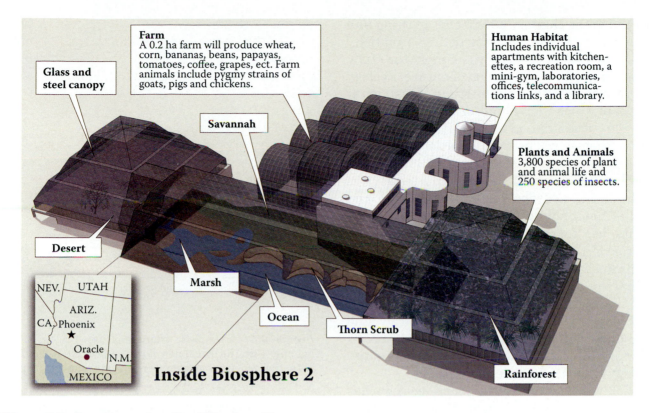

Farm
A 0.2 ha farm will produce wheat, corn, bananas, beans, papayas, tomatoes, coffee, grapes, ect. Farm animals include pygmy strains of goats, pigs and chickens.

Glass and steel canopy

Human Habitat
Includes individual apartments with kitchenettes, a recreation room, a mini-gym, laboratories, offices, telecommunications links, and a library.

Savannah

Plants and Animals
3,800 species of plant and animal life and 250 species of insects.

Desert

Marsh

Ocean

Thorn Scrub

Rainforest

NEV. UTAH
ARIZ.
CA. Phoenix
★
Oracle
● N.M.
MEXICO

Inside Biosphere 2

Figure 2.2   Location and profile of Biosphere II.

grasp of the science of ecology. It has been noted that ecology is to environmental sciences and resource management what anatomy, biochemistry, genetics, and physiology are to the practice of medicine (Wali 1992). Though ecology as a science was not formally named until 1866, the beginnings of ecological observations go back to Greek philosophers like Theophrastus, ca. 2300 b.p. (Theophrastus 300 b.c.). Since then, treatises on how the natural world and its inhabitants develop and change with time have been published from many parts of the world.

The word **ecology** is derived from "œkologie," a term first coined by the German zoologist, Ernst Haeckel; the term has the Greek roots *oikos,* meaning house, and *logos,* study of. Haeckel defined the content of his œkologie as "comprising the relation of the animal to its organic as well as its inorganic environment, particularly its friendly or hostile relations to those animals or plants that it comes in contact" (Haeckel 1866). The word *ecology* in its current Anglo-American usage seems to have been used first by the Scottish sociologist Patrick Geddes (Figure 2.3) (Mairet 1975; Wali 1999a). The first organized textbook on the

			Sociology
		Biology	Anthropology
	Physics and chemistry	Biochemistry	Ecology
Mathematics and logic	Mathematics (applied to physics)	Biometrics	Statistics

Figure 2.3   The visualization by Patrick Geddes in 1880 of "the hierarchy of the sciences" (in blue) and the placement of ecology in "the sciences—fields of combined application" (in red). Geddes appears to be the first to use ecology in its Anglo-American currency, but he placed sociology much higher in the hierarchy! (See Wali, M. K. 1999a. *Nature & Resources* 35(2): 38–50.)

subject was written in German by the Danish botanist Eugene Warming and appeared in 1898; its English translation with the word œcology in its title, *Œcology of Plants,* was published in 1905 (Warming 1905).

But even earlier, in 1807, von Humboldt and Bonpland (1807), writing on plant geography, stated the following ecological principle:

"In the great chains and effects, no thing and no activity should be regarded in isolation." To Odum (1959), ecology was simply the study of the structure and function of nature. Ecology has also been defined as the study of systems at a level in which individuals or whole organisms may be considered as elements of interaction, either themselves or with a loosely organized environmental matrix (Margalef 1968).

Haeckel observed in 1866 that ecology is both (1) the household of the organism, and (2) the household of nature. These two statements led to establishment of two distinct approaches to the study of ecology: autecology, the branch that dealt with the ecology of individual species, and synecology, the ecology of communities. Eugene Odum (1959) suggested that we drop these two terms entirely and introduce four divisions instead: (1) species ecology, (2) population ecology, (3) community ecology, and (4) ecosystem ecology.

For the sake of convenience, ecology has also been subdivided in several other ways: by habitat types, such as terrestrial or aquatic ecology (including freshwater ecology, marine ecology, estuarine ecology); by vegetation types, such as forest ecology, grassland ecology, and alpine ecology; or in conjunction with other disciplines, such as physiological ecology, statistical ecology, mathematical ecology, and molecular ecology. Specific types of organisms are also the object of study and include specialties such as algal ecology, microbial ecology, and mammalian ecology.

Terms bearing the words *eco, ecology,* and *ecological* have proliferated in recent years. Although in some cases the use of these words as prefixes or suffixes may make sense, the relevance of other uses is open to question (Wali 1999a). Nevertheless, as a science, ecology is fundamental—encompassing heredity (evolution), environment, and an organism's response to the environment (adaptation; Figure 2.4). It involves the diversity of an organism's activities and its successes and failures. In sum, ecology deals with the adaptations of organisms to their environment over time, how and how many organisms live in a given space at a given time, and how they obtain food, occupy space, and reproduce. The considerations of individuals, populations, and

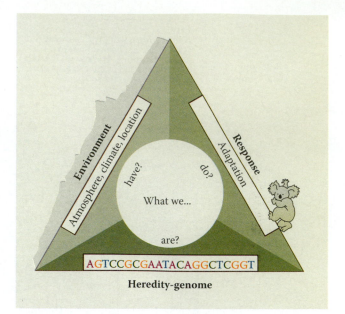

Figure 2.4 A heuristic triangle of life. Ecology deals not only with the attributes of the environment, but also with the genetic and behavioral adaptations of organisms to the environment.

groupings of different populations are all vital ecological considerations. Given its centrality, the following question will always be relevant: Why do populations exist, prosper, and proliferate where they do?

Although the roots of ecology have traditionally been grounded in the field of biology, its subject matter deals with every other fundamental and applied natural science: physics, chemistry, geology, mathematics, agriculture, soil science, and medicine. Ecology is interdisciplinary and multidisciplinary, but its strength is determined not only by how many fields it can cross, but also by how many disciplines it can integrate well to explain the complex workings of nature. The solutions of contemporary environmental issues lie in those integrations and syntheses.

Ecology is central to study and understanding of the environment. When we speak of the environment, we mean the totality of everything that surrounds us and with which we come in contact. Climatic variables (such as temperature, precipitation, wind, and hurricanes), atmospheric variables (such as intensity of light, the nature of gases, and types of pollutants), and the nature and composition of soils are all part of the environment. The environment also encompasses relations of living organisms—whether of one's own kind or with individuals belonging to other species. Thus, the environment includes the entire interactive complex of physical, chemical, and biological factors encountered by organisms.

The place where an organism lives and its surroundings, both living and nonliving, are its **habitat.** This is the so-called residential address that will vary for a muskrat, a tiger, and a buckeye tree. Habitats vary over large geographic regions (see Chapter 10) but also over small distances when **microhabitats** are created by features such as small topographical variations or the growth of other organisms (e.g., trees). These can bring about small but distinct differences in habitat conditions. Tree growth, for example, can bring about differences in the light and heat received on the forest floor, creating a number of **microhabitats.** Different microhabitats support different species. Thus, as the number of microhabitats increases, species diversity tends to increase as well.

The functional aspect of each species is described by its **niche** (Grinnell 1917). When first used in ecology by Grinnell, the word *niche* was viewed as a subdivision of the habitat, or the space where conditions are suitable for that species. Later, Elton (1927) defined niche as the role of the species in the community—"How does an organism function in the community?" Contemporary ecologists make a clear distinction between habitat and niche; the latter is viewed in the Eltonian sense as a summary of the tolerances and requirements of the species. Further elaboration of the niche concept by Hutchinson (1957b) recognized the niche as a "multidimensional space" whose limits are set by the tolerance of the species to differing environmental variables (i.e., the sum of conditions under which that species can survive, grow, and reproduce). Hutchinson's definition also took into account the effects of competition for resources with other species. For example, the niche that a species occupies when there is no competition from other organisms (called the **fundamental niche**) can be contrasted to the smaller niche space that a species occupies when a competitor is present.

Under pressure from competitors, organisms typically reduce the size of their niche in order to minimize competition. This smaller niche space is called the **realized niche.**

Our understanding of the laws that govern ecological systems has been established through years of observations and experiments; here, we shall examine a few of them that are related to the topics discussed in this book. A general law, first stated with respect to agricultural systems, applies universally to all ecological systems: the "law of minimum" (van der Ploeg, Bohm, and Kirkham 1999). In simple terms, the law states that each organism has a number of minimal resource requirements that it needs to survive, function, grow, and reproduce (Figure 2.5). If any one of the required factors is in a small quantity relative to the needs of the organism, then the growth or reproduction of an organism will be limited by the factor in shortest supply. In some cases, one environmental factor or condition may compensate for the minimal supply of the other—for example, cases where a warmer temperature may compensate for low light. But universally, some environmental factors (such as water or specific nutrient elements like calcium and iron that are required for growth) can be neither substituted for nor replaced.

A corollary of the law of minimum, sometimes stated as the "law of limiting factors" in plant physiology, is the "law of tolerance" (Shelford 1911), which states that too much of something may be as bad as too little. This concept was succinctly stated by Blackman (1909):

> [I]n treating physiological phenomena, assimilation, respiration, growth, and the like, which have a varying magnitude under varying external conditions of temperature, light, supply of nutrients, etc., it is customary to speak of three cardinal points, the *minimal* condition below which the phenomenon ceases altogether, the *optimal* condition at which it is exhibited to its highest observed degree, and the *maximal* condition above which it ceases again.

In this case, an environmental factor that is required in small quantities for growth may be present in so large a quantity (salt, for example) that it inhibits or stops growth.

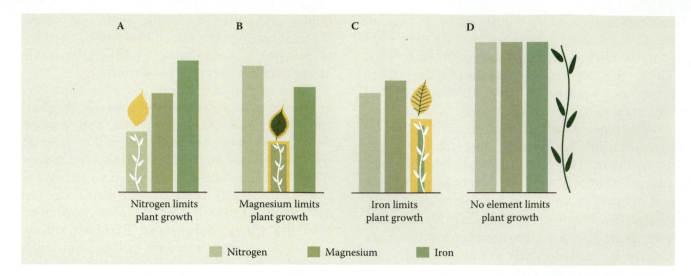

Figure 2.5 A simple illustration of the Sprengel–Liebig law of minimum (Patterned after Donahue, R. L. et al. 1971. *An Introduction to Soils and Plant Growth*. Englewood Cliffs, NJ: Prentice Hall.) showing limitation of essential nutrients on plant growth. (A) Nitrogen is in short supply, the leaves turn yellow because of lack of chlorophyll formation (chlorosis), and growth is stunted. (B) Magnesium is the limiting nutrient producing typical chlorosis around the edges of leaves. (C) Short supply of iron limits growth and produces the typical interveinal deficiency symptoms in the leaf, as shown. (D) All nutrients are in adequate supply and the growth of plants is optimal.

The Ecosystem Concept

As early as 1833, H. C. Watson, a British naturalist, visualized the importance of the interconnectedness of conditions that affect the growth of plants (see Gorham 1954). Watson noted that the distribution of vegetation could be explained using an array of factors: (1) temperature, (2) moisture, (3) the relative degree of shelter and exposure, (4) the physical and chemical properties of the soil, and (5) the physical and chemical properties of the underlying rocks. He emphasized that it was the "combined influence" of these factors that determined the flora and vegetation of countries. However, under local conditions, some factors might have a greater influence than others.

The concept of interconnectedness of ecological processes occupies a premier place in ecology and is expressed in the term **ecosystem.** Originally proposed by the British ecologist Sir Arthur Tansley, the term ecosystem has gained currency in the common lexicon as well. He visualized an *ecosystem* as an interacting group of organisms with "the whole complex of physical factors" and emphasized that although our interest may be in living organisms, "we cannot separate them from

their special environment, with which they form one system." Ecosystems are open to inflows and outflows, come in various kinds and sizes, and function in a relatively stable equilibrium. "It is the systems so formed … that are the basic units of nature on the earth" (Tansley 1935).

Many modifications of this definition are now a seminal part of the ecological literature. Of these, two are worthy of note here. An ecosystem was defined as "a system composed of physical–chemical–biological processes active within *space–time* units of any magnitude (i.e., the biotic community plus its abiotic environment) (Lindeman 1942). The consideration of both space and time (also referred to as spatiotemporal relations) was a major refinement of the definition, for no ecological process can be understood or complete without the consideration of these two components. A further refinement came when an ecosystem was defined as "any area of nature that includes living organisms and nonliving substances interacting to produce an *exchange of materials* between the living and nonliving parts" (Odum 1959) (emphasis added in both definitions). Both space and time relationships and the exchange of materials are crucial to understanding ecological systems.

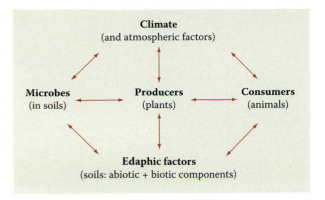

Figure 2.6 Functional and interacting components of a terrestrial ecosystem.

Functionally, an ecosystem comprises the abiotic components such as climate, water, and soils; the **producer** (or autotrophic) components, including the green plants, algae, and chemosynthetic bacteria; the **consumer** (heterotrophic) components of herbivores, carnivores, and omnivores; and the **detritivore** (saprophytic) components of the decomposer community, dominated by the microbes that break down dead, organic matter and are fundamental to the cycling of nutrients (Figure 2.6). From an ecological standpoint, the "producers" on planet Earth are the green plants, algae, and chemosynthetic bacteria (also known as the autotrophs or "self-feeders"). All animals (including humans) are heterotrophic consumers.

Structurally, a terrestrial ecosystem has a large number of components (Figure 2.7). The predominant work in ecology has focused on the structural elements of ecosystems, with work on individuals, groups, and communities of plants, animals, and microbes, singly or in combination, and it provides the richest base of information. Although illustrated here for terrestrial ecosystems, the same components apply to aquatic systems and the watershed with which they interact. Though ecosystem structure (the living and nonliving elements) and the function (the cycling of chemicals and energy flow) cannot be isolated from one another, it is function that is paramount in understanding contemporary environmental issues.

The *ecosystem* concept was ranked at the top of the 50 most important concepts in a survey of ecologists (Cherrett 1989). The ecosystem concept is now prevalent in scientific literature and in nonscientific writing. Though the term "ecosystem" originated in Europe, much of its development occurred in the United States, as a contemporary ecologist noted: "The ecosystem story is largely the American tale … the ecosystem concept appeared to be modern and up to date. It concerned systems, involved information theory, and used computers and modeling. In short, it was a machine theory applied to nature" (Golley 1993). Furthermore, visualizing of ecosystems as a set of interacting compartments makes it amenable to mathematical modeling.

The usefulness of the ecosystem concept is, in our opinion, overwhelming because of the following attributes (Kimmins 2004). These will be discussed in more detail in later chapters.

- **Structure.** Ecosystems are made of abiotic and biotic subcomponents. At the very least, a terrestrial ecosystem must have green plants, a substrate, and an atmosphere; in most ecosystems, there must be an appropriate mixture of plants, animals, and microbes if the ecosystem is to function. Terrestrial ecosystems normally consist of a complex biotic community, together with soil and atmosphere, a source of energy (the sun), and a supply of water.
- **Function.** The constant exchange of matter and energy between the physical environment and the living community constitutes function. Ecosystems at all times have organic materials that are alive or dead in various stages of decomposition. Thus, there are considerable advantages in looking at systems in biological–physical–chemical terms.
- **Complexity.** Each ecosystem has a high level of biological integration. As the number of abiotic and biotic components grows, so do the interactions between them and so does the complexity of the system. As ecosystem complexity increases, it becomes more difficult to predict how a system will behave—for instance, how it will respond to a disturbance.
- **Interaction and interdependence.** So complete is the interconnectedness

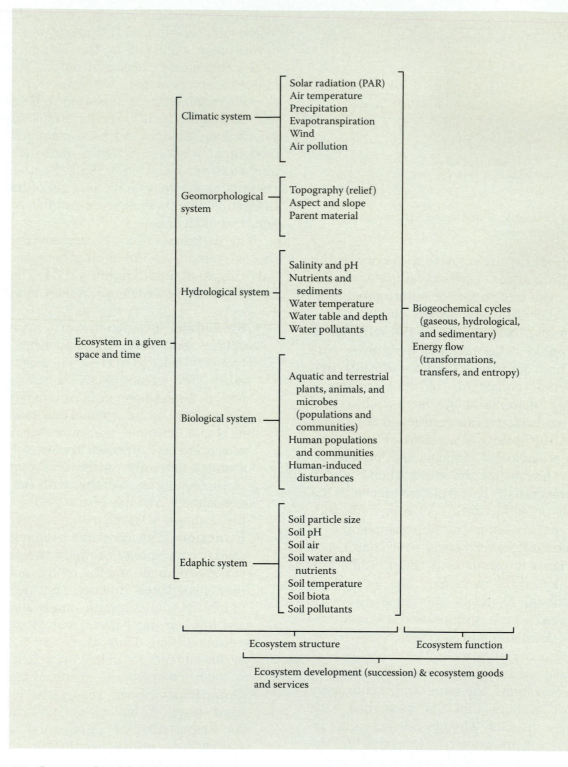

Figure 2.7 Structural and functional components of an ecosystem.

of various living and nonliving components that a change in any one will result in consequent changes in almost all the others.

- **Spatial relationships.** All populations of animals and plants (a flock of birds, a buckeye forest) occupy a given space. Ecologists study populations in relation to their abundance in a given space. The ecosystem per se does not restrict a researcher or observer to be constrained by spatial limits.

- **Temporal change.** The entire structure and function of populations and communities in nature undergo change over time. This temporal change (sometimes called ecological succession) is universal and is a very significant ecological process in assessing methods to restore disturbed systems.

Recently, these attributes have been cast in a thermodynamic perspective, noting that ecosystems (1) require an input of energy and are open to the flows of matter and information, (2) are so complex as to make accurate prediction about their function difficult, (3) are organized hierarchically such that one has to understand systems at lower levels in order to understand them at higher levels, and (4) have a complex response to disturbance (Jørgensen et al. 2007). In sum, the elegance of the ecosystem concept is that it emphasizes all the vital attributes of natural systems: interconnectedness of ecological variables in both space and time, structure, function, complexity, interaction and interdependence, temporal change, and flexible spatial dimensions.

Most recently, the United Nations embraced the ecosystem concept in its assessment of human impacts on ecosystems and the services they provide. Ecosystem services are the benefits that humans obtain from ecosystems, such as the supply of food, fiber, clean air and water, and climate regulation. This project, called the global Millennium Ecosystem Assessment (MEA 2003), was carried out between 2001 and 2005. The assessment provides information on the links between human well-being and the degree to which ecosystems have become degraded, thus establishing the scientific basis needed to promote their sustainable use. The number-one finding of the MEA confirms the extent of human impacts on the biosphere:

Over the past 50 years, humans have changed ecosystems more rapidly and extensively than in any other comparable period of time in human history,

largely to meet rapidly growing demands for food, fresh water, timber, fiber and fuel. This has resulted in a substantial and largely irreversible loss in the diversity of life on earth.

Worldwide, without exception, the consequences of human demands on ecosystems are governed by all ecological principles. This was realized by ecologists early, and from this recognition came the field of applied ecology. But as environmental problems became increasingly well documented, many others, professional and lay people alike, joined in. Despite the fact that scientific societies of ecology have been organized in many countries (the United Kingdom and the United States in the lead), a new genre of groups championed environmental science and environmental studies. Thus, ecology became the domain of professionally trained ecologists with their own journals,[1] or *eco* (e.g., eco-music, eco-watches, eco-anarchists) and *ecological* as prefixes (e.g., ecological sensitivity, ecological spirituality) or *ecology* as a suffix (e.g., clinical ecology, critical ecology, romantic ecology, theology of ecology) came into the public domain as ideology, multiple metaphor, allegory, myth, and even gospel (Wali 1999a).

The fields of environmental sciences and environmental studies have, therefore, become established. If ecology is a field of science, then environmental issues are considered in the broad context of human activities such as economic and social systems (Figure 2.8). Appropriately, in this book we discuss the principles of ecology first.

It seems remarkable that poets, without any formal background in ecology, visualized the interconnectedness of ecosystems well before many scientists did:

> All things by immortal power
> Near or far
> Hiddenly
> To each other linked are
> That thou canst not stir a flower
> Without troubling a star.

Francis Thompson, ca. 1897

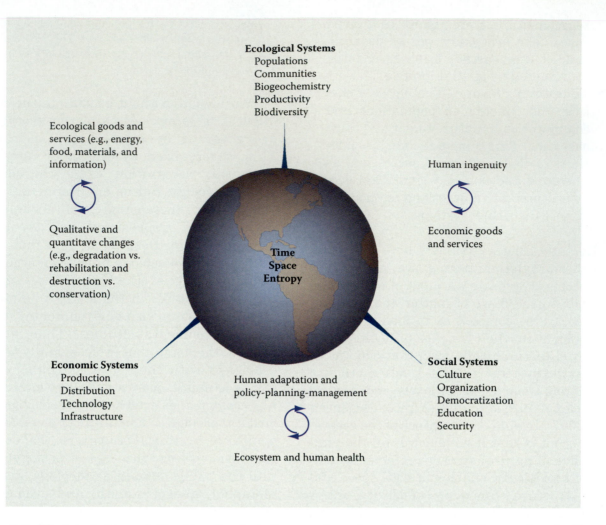

Figure 2.8 The interlocking relationships of ecological, economic, and social systems of the human domain. Note the centrality of time, space, and energy relations of all systems and fields of principle formation in each respective system. (Modified from Wali et al. 1999b.)

Note

1. The oldest ecological society, the British Ecological Society, was founded by Arthur Tansley in 1914. The society publishes several journals: *Journal of Ecology, Journal of Animal Ecology, Journal of Applied Ecology,* and *Functional Ecology.* The Ecological Society of America (ESA), now much larger than the British Ecological Society in membership, began in 1915. It publishes *Ecology, Ecological Monographs, Ecological Applications,* and *Frontiers of Ecology and the Environment.* Several noteworthy and comprehensive issue papers have been published by the ESA, and we shall refer to them in the appropriate chapters.

Questions

1. Given that in order to repair something you must have a basic understanding of how it works, what kinds of scientific advances have occurred that allow humans to understand and thus properly manage the biosphere?

2. What components make up the basic structure of an ecosystem?
3. What do we mean by "ecosystem processes"?
4. How do the time and space scales of environmental processes differ, and what effect does scale have on interactions between systems?
5. How is the comparison of renewable and nonrenewable natural resources related to the timescales at which natural processes operate?

Atmosphere, Climate, and Organisms

Atmosphere

The biosphere, hydrosphere, and lithosphere are all enveloped by an invisible gaseous mantle—the Earth's atmosphere. Life exists only near the zone of contact between the atmosphere and the Earth's land and water surfaces. Thus, living things are greatly influenced by changes in the atmosphere.

Based on distinctive temperature gradients, atmospheric scientists have recognized several vertical regions of the atmosphere (see Figure 2.1). The layer of atmosphere closest to the Earth is the **troposphere,** which extends from the Earth's surface up to a height of about 10 km. Above it is the **stratosphere,** a region that extends from 10 to 45 km with calmer and clear air.

The lower region of the stratosphere is the ozonosphere, or the **ozone layer,** which is vital to life because it absorbs much of the sun's damaging ultraviolet radiation. It is here that ozone (O_3) is both produced and destroyed. The ozone layer and how it is altered by human-made chemicals have been shown to be of great significance to life and are discussed in Chapter 21.

The layer between 45 and 80 km is the **mesosphere,** characterized by the lowest temperatures in the atmosphere (as low as −138°C). Above it, from 80 to 500 km, is the **thermosphere,** a zone of very high temperatures that may approach 900°C. This zone is also referred to as the ionization zone because, in this zone, sunlight reduces molecules to individually charged particles called ions. The uppermost atmospheric region is the exosphere, which begins about 500 km above the Earth's surface and converges with interplanetary space.

Temperatures within the atmosphere vary with altitude; in the troposphere, the decrease in temperature with altitude is referred to as the **adiabatic lapse rate.** The temperature of the troposphere decreases with increasing altitude at an average lapse rate of −6.5°C for every kilometer (3.5°F for every 1,000 ft) (Figure 2.1). Likewise, atmospheric pressure is highest at the Earth's surface and decreases rapidly with increasing altitude by a factor of about two for every 5 km.

Although the stratosphere and troposphere are the most important layers of the atmosphere for organisms, the lower troposphere is of the most immediate consequence to the living world. This zone consists of a mixture of gases—some with constant and some with variable concentrations—as well as suspended solid and liquid particles. About 99.9% of the air by volume consists of three major gases: nitrogen, oxygen, and argon.

The variable gases collectively account for only a tiny proportion of the air; they include carbon dioxide, water vapor, oxides of nitrogen and sulfur, a more specialized group of compounds containing chlorine and bromine, and, of course, tropospheric ozone. Also present are specialized groups of compounds called photochemical oxidants, which are formed when sunlight strikes some organic molecules, and traces of other variable gases. Suspended particles, both solid and liquid particles of natural and anthropogenic origin, are **aerosols** and include dust, smoke, soot, and sulfuric acid. The impacts of such gases and aerosols on the structure and function of ecosystems are presented in Section B.

Climate and Weather

Although this book does not deal with climatology (the study of climates), the impact of climate is so vital to the growth, distribution, and abundance of organisms that a brief description of climate as it affects the environment is imperative. The principal difference between the terms "climate" and "weather" is the timescale of each. **Climate** is an average condition derived from long-term periods of observation of day-to-day weather conditions. **Weather,** on the other hand, refers to the short-term conditions of the atmosphere at any time (e.g., daily, monthly, short-term annual) or place; it is expressed by a combination of:

   a. temperature;
   b. precipitation (rain, hail, snow, fog) and
      humidity;
   c. air pressure; and
   d. winds.

Thus, the weather of a place is the sum of its atmospheric conditions for a short period of time. It should be noted at the very outset that there is an underlying randomness in climatic phenomena, much of which is not understood at present.

For ecological purposes, the average values of any weather variable over a given time period are not as important as its variations over time—yearly, seasonally, and daily. Although most climatic data in the past have been obtained from airports and other open locations, plants and animals do not generally live in such open situations. Habitats occupied by both plants and animals are greatly influenced not only by atmospheric weather patterns but also by modifications brought about by the organisms themselves. For example, forest trees can intercept light and rain and even absorb light differentially. As a consequence, those plants and animals that inhabit the forest floor must have the necessary adaptations to live under the modified conditions that are typical of the ground floor.

Ecologists use the term **microclimate** to denote the small-scale atmospheric conditions that are actually encountered by growing organisms. Within each climatic pattern, a mosaic of microclimates exists. We shall expound on these characteristics later in the chapter, but first we provide an overview of the general links between the climate and the atmosphere.

Global Climate and General Circulation

Several features of the Earth control the climate. These include:

- the angle of the sun as it strikes the Earth's surface as influenced by latitude and season;
- ocean currents;
- relative placement of land masses and bodies of water;
- semipermanent low- and high-pressure cells in the atmosphere;

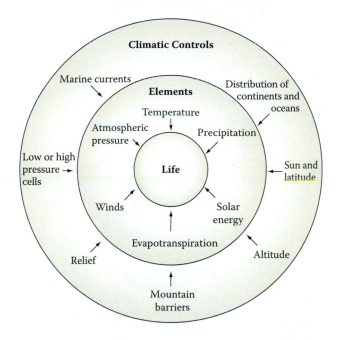

Figure 3.1 Climatic controls and elements acting upon ecosystems. (Synthesized from Finch and Trewartha 1942; Dansereau, P. 1957. *Biogeography: An Ecological Perspective.* New York: Ronald Press Company.)

- air masses, winds and storms, and the jet stream;
- altitude; and
- mountain barriers.

These climatic controls help create the elements of climate such as temperature, precipitation, humidity, air pressure, and winds at any given place. The climatic elements, in turn, produce day-to-day variations in weather (Figure 3.1).

Global Circulation System

The Earth's rotation on its axis and its revolution around the sun cause different amounts of net solar radiation to be received on different areas of the Earth's surface; this has huge consequences for the climate of different regions. For example, as one moves from the equator toward the poles, the incoming solar energy strikes the Earth at a more oblique angle and passes through a longer distance of the atmosphere. As this occurs, the solar energy becomes less intense, producing lower temperatures over a given area. The inclination of the Earth's axis (23.5° from the vertical) results in seasonal changes in temperature as the Earth

rotates around the sun; the tilt results in the Northern Hemisphere's being more directly oriented toward the sun from March 21 to September 22 (summer) and away from the sun from September 22 to March 21 (winter) (Figure 3.2). These differences result in differences in day length, as well as in the distance the light must pass through the atmosphere before hitting the surface of the Earth.

Temporal and spatial variations in solar energy result from several additional processes. Atmospheric conditions such as the presence of clouds or differences in pollutants can affect the intensity and character of solar radiation striking the Earth's surface. Finally, at the surface of the Earth, latitude, altitude, land use and cover, vegetation, slope, and aspect (direction relative to due north) influence the distribution of solar energy at the local level. This uneven distribution of solar energy received at varying points on the Earth's surface creates and drives the vertical (e.g., precipitation and evapotranspiration) and horizontal (e.g., winds) motions of the climatic system of the Earth.

The Earth's rotation from west to east causes the **Coriolis effect,** a large-scale deflection of the direction of wind and water to the right in the Northern Hemisphere and to the left in the Southern Hemisphere (Figure 3.3). Differences in net solar radiation produce variations in air temperature and, consequently, horizontal and vertical air pressure gradients. The interaction of four forces—the Coriolis effect, the centripetal (center-seeking) forces of air pressure gradients, friction, and gravity—initiates and then controls the movement of air away from areas of high air pressure toward areas of low air pressure in the form of winds. The wind is stronger where the air pressure gradient is steep and can form extremely fast-moving currents of air, or **jet streams,** at high altitudes where friction is less than at the Earth's surface. The resultant winds can flow either east or west, depending on latitude.

The large-scale movement of air and ocean waters redistributes heat and moisture across the Earth's surface; it is estimated that air circulation accounts for 87% of this heat distribution. Year-round heating in the equatorial regions causes a thermal low-pressure belt

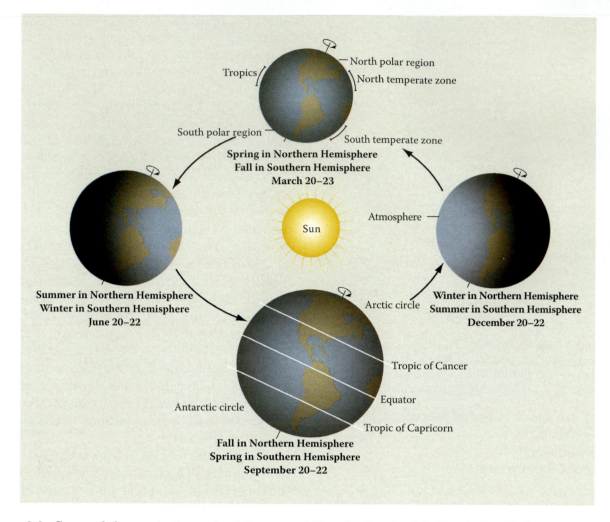

Figure 3.2   Seasonal changes in the angle of the sun and the orbital path of the Earth around the sun (Gould et al. 1996. *Biological Science*, 6th ed. New York: W. W. Norton & Company.)

in that zone known as the **intertropical convergence zone** (**ITCZ**), or equatorial low. As moisture-laden, warm air in the ITCZ rises and cools, it releases moisture as precipitation, giving rise to the tropical rain forests on the land areas around the equator. This cooler and drier air then moves poleward in both hemispheres and descends around latitudes 30° N and 30° S, creating an area of high pressure known as the subtropical highs.

At 30° N and 30° S, this cool air descends. As it does so, it warms and absorbs moisture, reducing the amount of precipitation that falls in this zone. This creates the arid conditions under which the world's major deserts, including the Sahara of North Africa and the Sonora of Mexico and southwestern United States, are located. **Trade winds** blow out of the high-pressure area around the subtropical latitudes

(30° N and 30° S) toward the low-pressure area around the equator (Figure 3.3). These were used by captains of sailing vessels for hundreds of years to move their ships across the oceans. By changing the latitude at which they sailed, they could pick up favorable winds and currents to move either east or west. The polar easterlies (dense, cold air masses) flowing out of large cells of high pressure centered over each pole (called polar highs) toward the equator meet and override the westerlies (milder, lighter air masses) flowing out of the subtropical highs along a narrow zone of transition called the polar front.

This global pattern of prevailing winds drives the cell-like circulation pattern of ocean currents called **gyres** (Figure 3.4). Most currents are called "drifts" because they lag far behind the average speeds of the surface winds, averaging only about 8 km hr^{-1}. Variations in

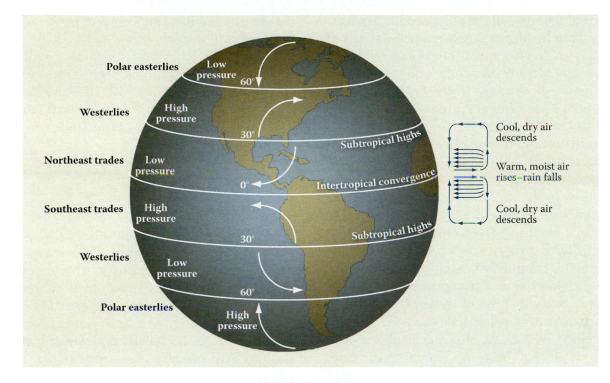

Figure 3.3 Global atmospheric circulation pattern and the Coriolis deflection of surface winds.

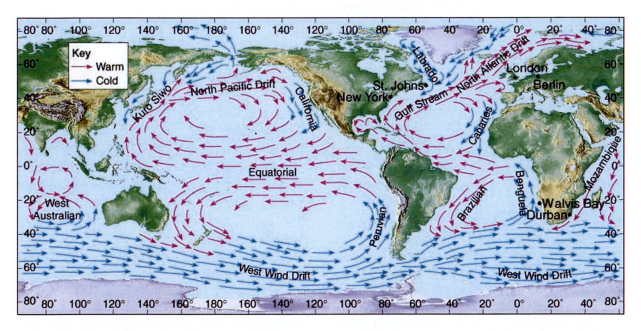

Figure 3.4 Global distribution pattern of ocean currents. (From de Bilj, H. J., and P. O. Muller. 2004. *Physical Geography of the Global Environment.* New York: Oxford University Press.)

water temperature, density, and salinity are another major source of oceanic circulation. For example, increases in salinity or decreases in temperature make water denser and cause it to sink. Cold currents of the polar areas sink and flow toward the equator, while warm currents from the equator flow toward the higher

latitudes transporting heat. **Upwelling** involves the rising of cold, nutrient-rich bottom water to the surface. This is associated with a decline in surface air temperature and evaporation, as well as the production of such arid coastal areas as the coast of Chile. Because of their nutrient-rich status, areas of upwelling along coastal zones produce some of the most productive fisheries on Earth, supporting large populations of sardines, anchovies, and other ocean fish.

The interaction of winds and ocean currents can cause changes in the typical pattern of ocean surface temperatures. This has been shown to have an immense impact on weather patterns throughout the world. Under normal conditions, trade winds blow westward across the Pacific Ocean, causing a mass of warm water to pile up in the western Pacific near Australia (see Figure 3.5A, a). This, in turn, allows an upwelling of cold water along the west coast of South America, leading to an area of high marine productivity, supporting a major anchovy fishery (a major commercial fishery). However, every 2–7 years (possibly related to the occurrence of solar flares), a large-scale disturbance occurs in this atmospheric and ocean circulation pattern: Warm water does not move westward as it normally does, and the cold-water upwelling off the coast of Peru is halted. This is called the **El Niño/southern oscillation** (**ENSO**).

Under El Niño conditions, weakening westerly winds allow the surface water along the South and North American coasts to warm (3–4°C above normal) and suppress the upwelling of cold, nutrient-rich bottom water (see Figure 3.5A, b). This creates a lack of nutrients, which in turn reduces marine primary productivity, including a collapse of the Peruvian anchovy fishery. The eastward displacement of warm water causes a shift in global atmospheric circulation. For example, precipitation that typically falls on the oceans makes it to shore under El Niño events, causing coastal floods or landslides in California and Peru. Ultimately, El Niño results in more rain in the United States and western South America (e.g., Peru) and drought in Australia and eastern Africa. Similar patterns can be illustrated for the Andes (Figure 3.5B).

The other cycle of the southern oscillation, the **La Niña/southern oscillation,** is characterized by strong easterly trade winds that, in this case, strengthen the normal pattern of cold upwelling, leading to sea surface temperatures as much as −14°C below normal. La Niña

essentially creates conditions opposite to those of El Niño. The world has witnessed the effects of several El Niño and La Niña oscillations in the last 25 years, including strong El Niño events in 1982, 1983, 1986, 1987, 1992–1994, 1997, 1998, and, to a milder extent, in 2003 and 2005. La Niña was pronounced in years 1975, 1988, and 1995.

Regional and Local Circulation Systems

Surface wind belts at more localized scales create mesoscale and microscale circulation patterns such as breezes, monsoons, thunderstorms, tornadoes, hurricanes, urban heat islands, and temperature inversions (see Chapter 17). In general, winds, along with solar energy, enhance water loss from both soils and water bodies by the physical process of **evaporation.** Plants lose water by **transpiration,** the process by which water is taken up from the soil and lost through the leaves, thus dissipating excess heat and aiding gas exchange. Many times the two processes are referred to together as **evapotranspiration.** Land surfaces warm up during the day and cool at night more quickly than water surfaces do. This leads to local onshore breezes (wind moves landward) during the day and local offshore (waterward) breezes at night. Moist sea and ocean winds moving inland lose their moisture in the form of precipitation (called **orographic precipitation**) as they ascend along the windward sides of mountains (Figure 3.6) because the decreasing atmospheric pressure with increasing elevation causes the air to expand and thus cool.

Cool air holds less water than warm air does, so as the air cools, the excess water condenses and falls as precipitation. On the leeward side of mountain ranges (i.e., the side away from the prevailing winds), the air warms as it descends. Warming air absorbs water, increasing the loss of water from vegetation, animals, and soils. This phenomenon is known as the **rain shadow,** or a zone of low water availability on the lee sides of mountains. In the state of Washington, orographic precipitation can be

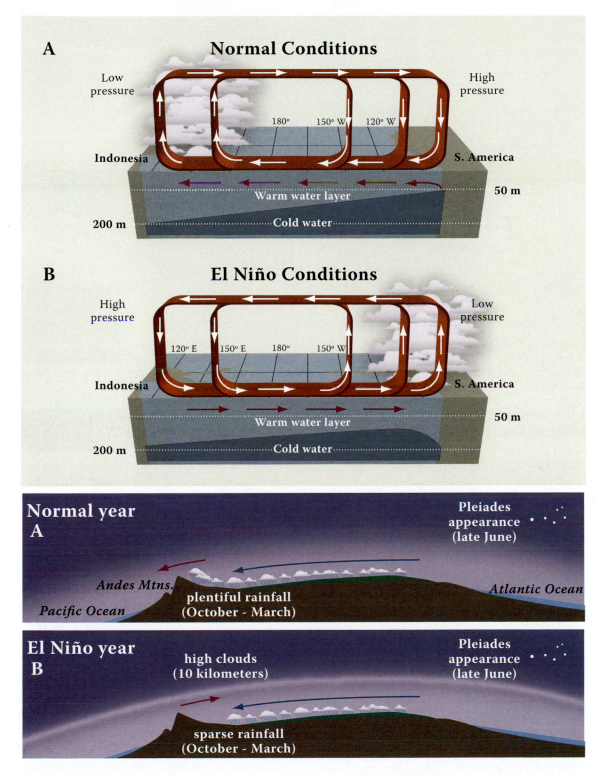

Figure 3.5 Top panel, (A) Equatorial Pacific waters during normal and (B) El Niño/southern oscillation conditions. (Redrawn from C. S. Ramage. 1986. El Nino, *Scientific American* 254(6) p. 76.) Bottom panel, The flow of high-level summer winds from east to west (red arrow) during a normal year over the Andes (A), and a reversal of the winds from west to east during an El Niño year (B). (Orlove, B., J. Chiang, and M. Cane. 2002. *American Scientist* 90:428–435.)

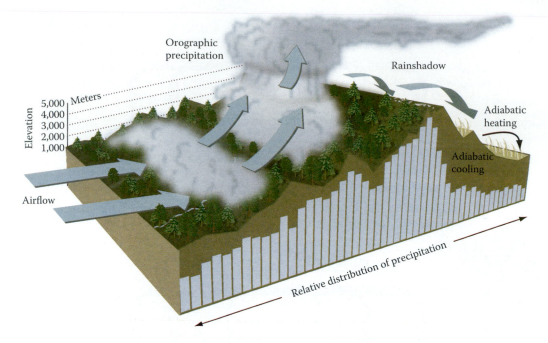

Figure 3.6 Orographic precipitation and rain shadow effect; note the influence of moisture-laden clouds and green vegetation (left) and desert conditions (right). (Redrawn from Marsh, W. M. 1987. *Earthscape: A Physical Geography,* p. 99, Figure 6.8. New York: John Wiley & Sons.)

seen on the western slopes of the Olympic and Cascade Mountains, giving rise to a lush temperate rain forest. This leaves a pronounced rain shadow in eastern Washington, particularly when the air is forced over both mountain ranges. The high deserts of Washington and Oregon result. The same pattern occurs from the combination of the coastal ranges and Sierra Mountains in California (Figure 3.6).

Water loss from plants through transpiration increases with the square root of wind velocity. Above wind speeds of 6 km hr^{-1}, however, transpiration is mainly determined by the drying power of the air. The drying or evaporative power of the air is the difference between the actual water vapor pressure and the saturation vapor pressure at the same temperature. Wind chill increases the loss of heat from ecosystems by evapotranspiration and by convection. The local topography also produces different wind patterns in the valley, such as upslope and downslope winds due to daytime warming and nighttime cooling, respectively. In urban–industrial ecosystems, however, these processes can cause air pollutants to remain trapped in high concentrations in the air near the ground because it cools faster than the air at higher levels, causing **thermal inversions.**

Solar Radiation and Light

All organisms are dependent upon the availability of sunlight. Nuclear reactions within the core of the sun produce **radiant energy,** which arrives at the top of the atmosphere in the form of **electromagnetic radiation.** In keeping with the second law of thermodynamics, energy always flows in one direction, from high-quality to lower-quality, less useful energy. The change from ordered energy to disordered energy represents an increase in entropy, sometimes called "time's arrow." Thus, the sun is the sole driving variable that provides life on the Earth with a continuous flow of high-quality, low-entropy energy, at a rate of 5.6×10^{24} J yr^{-1}.

Radiant energy from the sun includes ultraviolet, visible, and infrared wavelengths, and it provides light and heat. In addition, this radiation undergoes transformations to kinetic (motion), potential, and chemical forms of energy. These forms of energy drive biological, physical, and chemical—or **biogeochemical**—changes in the environment. Some of these energies can be harnessed as **renewable energy sources,**

Table 3.1
Interactions between Plants and Light

Type of Activity	Phenomenon or Response	Organism or Example
Production of biomass	Photosynthesis	Green plants and bacteria
Orientation in space	Movements in response to light	Motile algae and bacteria; nonmotile plants and fungi
Orientation in time	Rhythm timing	Many metabolic activities; cell division and growth; stomatal opening; directional movements (phototropism)
	Photoperiodism	Flowering, dormancy induction, leaf abscission, and production of tubers, corms, bulbs and runners
Determination of form (photomorphogenesis)	Greening	Pigment synthesis and chloroplast development
	Effects on growth	Stem growth, leaf expansion, branching pattern, and root growth

Source: From Hart, J. W. 1988. *Light and Plant Growth.* London: Unwin Hyman.

including solar, hydro, wind, tidal, and geothermal energy, and even biomass. Short-wavelength radiation striking the top of the atmosphere is known as incident (or direct) solar energy, or **insolation** (**in**coming **sol**ar radi**ation**). Passing down through the atmosphere, some of the insolation is reflected by clouds and the Earth's surface, scattered by atmospheric gas molecules, and absorbed by water vapor, ozone, carbon dioxide, and the Earth's surface.

Albedo is a measure of the reflectance of the Earth's surface or its atmosphere. Snow and ice, for example, reflect the most light and thus have the largest albedo. Of the incoming sunlight that strikes snow and ice, as much as 70% is reflected back to space; only 30% is absorbed as heat. By contrast, when sunlight hits water (low albedo), only about 6% is reflected and 94% is absorbed as heat. Forest canopies have the lowest albedo. Ultraviolet wavelengths are largely absorbed by the layer of ozone (O_3) in the stratosphere. This protects aquatic and terrestrial life against the damaging effects of these wavelengths.

Several gases naturally present in the atmosphere have an important role in maintaining the heat balance of the Earth. Water vapor is the most abundant gas, followed by carbon dioxide (CO_2), methane (CH_4), nitrous oxide (N_2O), and ozone (O_3). These are collectively called the **greenhouse gases** (**GHGs**). These GHGs are transparent to incoming short-wavelength solar radiation but absorb outgoing long-wavelength terrestrial radiation emitted by the Earth's surface and atmosphere. When the outgoing long-wavelength terrestrial radiation is trapped by the GHGs, heat is radiated back toward the Earth's surface. The effect is similar to the way in which glass panes trap heat in a greenhouse. Their heat-trapping ability creates the **greenhouse effect,** a concept first introduced by Joseph Fourier in 1824. In the absence of these greenhouse gases, the Earth's average surface temperature would be about 33°C colder. As will be discussed later, the concentration of many of these gases has increased over the past 150 years due to human activities, raising concerns about climate change and global warming.

The solar energy absorbed by the Earth's surface heats the atmosphere and drives terrestrial and aquatic ecosystems and the water cycle. This transformed energy is eventually transferred from the Earth's surface back to the atmosphere and, finally, to space in the form of outgoing long-wavelength terrestrial radiation.

Light

The energy from the sun that penetrates through the atmosphere—together with the green pigments in the chloroplasts of plants and a series of enzymes and other chemicals—powers **photosynthesis,** the process responsible for producing the food and fiber that all organisms use in one form or another.

Table 3.2
General Sensitivities of Biological Processes to Different Intensities of Natural Light

Light Intensity (W m^{-2})	Equivalent Conditions	Relative Sensitivity of Biological Processes
10^3	Sunlight, noon, summer	Photosynthesis (saturates at 200–300 W m^{-2})
10^2	Daylight, cloudy	
10		Photosynthesis (compensation point)
1		Flowering
10^{-1}	Twilight	Seed germination
10^{-2}	Moonlight	
10^{-3}		Movement in response to light (algae); color vision (man)
10^{-4}		Plant greening
10^{-5}		Movement in response to light (coleoptile)
10^{-6}		Movement in response to light (many fungi)
10^{-7}		Black and white vision (man)
10^{-8}		
10^{-9}	Starlight	
10^{-10}		Growth inhibition (dark-grown oats)
10^{-12}		Detection limit of human eye
10^{-19}		Detection limit of largest telescope

Source: Modified from Bjorn 1976.

In addition to the production of biomass by photosynthesis, the competition for light between plants produces different growth forms and orientation of plant parts in time and space (Table 3.1).

Photosynthesis may be represented in a generalized way by the following equation:

$$\text{sunlight}$$
$$6CO_2 + 12H_2O \rightarrow C_6H_{12}O_6 + 6H_2O + 6O_2$$
$$\text{chlorophyll}$$

That is, carbon dioxide is chemically reduced to form sugar, while water is *oxidized* to form free oxygen. It is an energy-requiring biochemical reaction, and that energy comes from the sun. The chemical energy stored in sugar (or protein or fats) can be oxidized later to carry out various synthetic reactions, growth, or even movement. The anatomy of the photosynthetic apparatus in algae is relatively simple, but higher plants have evolved three distinct chemical **photosynthetic pathways** (C$_3$, C$_4$, and crassulacean acid metabolism [CAM]), which confer adaptation for their respective environments. These pathways are discussed in detail in Chapter 7; here we shall simply examine this process in relation to three of light's characteristics: intensity, quality, and duration.

Light intensity is the radiant energy received per unit area per unit time (expressed in units of watts per square meter, or megajoules per square meter per day). As sunlight travels to the Earth, it is lost (attenuated) in the atmosphere in several ways. It can be absorbed by ozone (O$_3$), oxygen (O$_2$), or water, or it can be scattered by molecules or aerosols. The oxygen, ozone, and water absorption produces the spectrum of direct sunlight, and the scattering processes result in the production of diffuse skylight. Together, direct sunlight and diffuse skylight constitute daylight (Table 3.2).

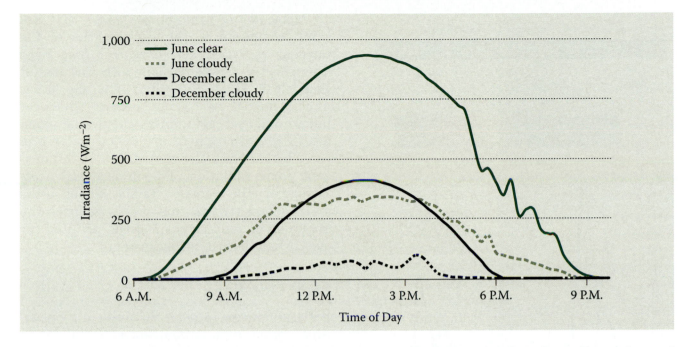

Figure 3.7 Daily variation in light intensity (horizontal incidence) at Columbus, Ohio (latitude 40° N), on a clear and a cloudy day in June, and on a clear and cloudy day in December. (Figure courtesy of James Metzger.)

Daily and seasonal variations in light intensity affect plant growth (Figure 3.7); at least 1% intensity of full sunlight is necessary to trigger photosynthesis. Plants are said to have reached a **compensation point** when the light intensity is enough to produce an equivalent amount of energy by photosynthesis that is being used in respiration. From there, photosynthesis increases linearly, up to between 10 and 20% of full sunlight, when light saturation may be reached. Beyond the saturation point, light intensity may actually impair photosynthesis by causing photooxidation of recently produced photosynthetic products or by increasing the breakdown rate of chlorophyll beyond its renewal rate. Inadequate light intensity limits the radiant energy available for photosynthesis. Only 1–2% of total radiant energy is absorbed and converted into chemical energy through photosynthesis.

Light Quality

Solar radiation is characterized by its wave motion and wavelengths. The wavelengths of the electromagnetic spectrum range between 10 and 100,000 nanometers (nm). A single pulse of electromagnetic energy is called a quantum or a **photon** when it is in the visible spectrum. Energy wavelengths shorter than visible light are known as **ultraviolet** (**UV**), while the longer, lower energy wavelengths are known as **infrared** (Figure 3.8). If long wavelengths (i.e., those above 1,000 nm) are absorbed, they are wholly converted to heat.

The number of photons or photon flux is often measured in units of 10^{-6} mol absorbed per unit time per area. Spectral wavelengths below 10 nm constitute the high-frequency shortwave radiation; of these, **x-rays** and **gamma rays** are capable of penetrating matter and producing ionization that has very harmful effects (for example, damage to organisms and mutations to genes). Between 100 and 390 nm is the ultraviolet range. Ultraviolet A (UVA between 315 and 400 nm) is the least detrimental; UVB (280–315 nm) is detrimental to most plants, and UVC (<280 nm) kills plants rapidly.

The visible electromagnetic spectrum lies between 390 and 760 nm and consists of the violet, blue, blue-green, green, yellow, orange, and red spectral bands (i.e., the rainbow) (Figure 3.8). The rates of photosynthesis, growth, and development by plants depend upon the visible wavelength range of solar radiation

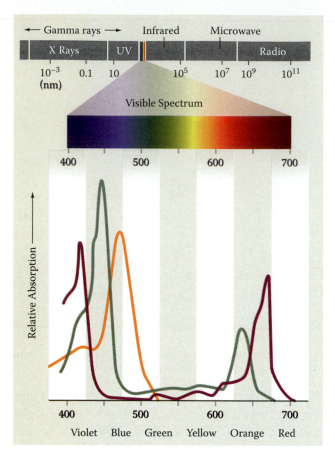

Figure 3.8 The spectral composition of photosynthetic pigments (chlorophyll *a*, chlorophyll *b*, and a carotenoid); note the photosynthetic peaks in relation to the spectra.

(about 50% of the incoming solar radiation); this is known as **photosynthetically active radiation** (**PAR**). Chlorophylls and accessory carotenoid pigments in plants and algae are adapted to maximize their absorption of light in specific wavelengths in the blue and red regions of the spectrum. High reflectance and transmittance of PAR in the green wavelengths cause most algae and higher plants to look green, although about 20% green light is absorbed on a photon basis.

All algae and land plants and most photosynthetic bacteria have chlorophyll *a*, but others also have chlorophyll *b*, *c*, or *e*, which absorb different wavelengths. Light absorption by photosynthetic pigments will greatly reduce photon flux in aquatic systems or under a canopy of terrestrial plants. In these environments, the competition for light can be fierce.

Molecules of atmospheric gases that are small compared to the wavelengths of the incident radiation act to scatter blue light (the shorter wavelengths), with the result that clear skies look blue. On the other hand, aerosol scattering occurs whenever human-induced pollutants (chemicals) present in the atmosphere are of a similar size as or bigger than the wavelength of the incident solar radiation. The preferential scattering of the longer visible wavelengths (orange and red) in this process causes the brownish or reddish color of the polluted air over urban–industrial ecosystems.

Light Duration

With a few exceptions, virtually all organisms—plant and animal—need exposure to light for a given number of hours per day for a variety of growth and reproductive functions. This day length is called the **photoperiod** and the phenomenon is called **photoperiodism.** In other words, photoperiodism refers to the variations in the duration of daylight and darkness that stimulate responses in the **biological clocks** of organisms. The change in photoperiod or day length is a function of latitude (Figure 3.9).

Since the early twentieth century, photoperiodic responses of a large number of plants and animals have been studied. In plants, bud dormancy in woody plants, tuber and bulb formation, and flowering all seem to be related to photoperiodic responses. Photoperiodism is a transducer—a mechanism that supplies a signal from the environment. For example, under an appropriate photoperiod, plants switch from a growth phase to a reproductive phase, responding to the synthesis of a flowering hormone and its subsequent translocation to buds (Bartholomew 1977).

Mammalian responses such as the induction of estrus, spermatogenic activity, and accelerated hair growth are also closely related to light duration. The behavior of many animals and plants follows a diurnal–nocturnal pattern of approximately 24 hours. This is called the **circadian rhythm.** Feeding and drinking, singing and calling, leaf movement,

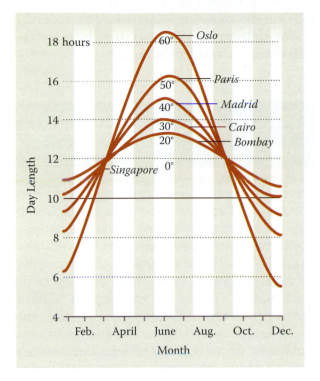

Figure 3.9 Seasonal changes in day length at different latitudes. (From Hart, J. W. 1988. *Light and Plant Growth*. London: Unwin Hyman.)

and flower opening all recur over a 24-hour period. These activities are controlled by many environmental factors (e.g., light, temperature, and humidity), but light seems to be the principal factor. Photoperiodic responses have also been documented in humans. For example, extensive work has been done on the causes of the symptoms of depression associated with seasonal affective disorder (SAD), which has been "estimated to afflict 10 million to 11 million American adults, with mild cases occurring in 25 million more, ... the onset of symptoms is clearly related to lack of sunlight" (personal communication: Rosette Kolotkin, clinical psychologist, Duke University Diet and Fitness Center, Durham, North Carolina).

Heat and Air Temperature

Heat is the total kinetic energy of the atoms or molecules of a substance, while **temperature** is the average kinetic energy of the atoms or molecules of a substance. In every transfer and transformation process of energy, heat energy is released. Transfer of heat energy occurs by three processes: conduction, convection, and electromagnetic radiation. **Conduction** is the transfer of heat that occurs when two objects are in physical contact. **Convection** is the mass transfer of heat from a solid object to a fluid, either air or water. Electromagnetic radiation can transfer heat between substances without any contact between them. For example, electromagnetic radiation transfers heat energy between the sun and Earth through the void of space. Heat always flows along a temperature gradient from where it is warmer to where it is colder, and its flow rate is determined by the features of the objects involved and the steepness of the temperature gradient. The ability of a substance to conduct heat decreases from solids to liquids and gases, respectively.

Air temperature, in conjunction with soil temperature, initiates and controls the rate of chemical and biological processes. Within the tolerance range of living tissues, temperature influences in all organisms follow the general rule that for every 10°C increase in temperature, the rate of a chemical process is doubled, or $Q_{10} = 2$ (ranging from 1.5 to 2.5). Conversely, with a decrease in temperature of 10°C, the rates of biochemical reactions decrease by half or more. At a certain threshold of temperature, development and growth are optimal. Generally, for land plants, the rate of net photosynthesis increases from zero to an optimum value at temperatures of 20–35°C and decreases to zero at higher temperatures (Fitter and Hay 1987).

Organisms grow and reproduce under a remarkable range of temperatures: Snow algae grow at 0°C or below and algae in hot springs at 70°C; living bacteria have been collected at 77.5°C (Ruttner 1953). However, for all growth and reproductive processes, it is the daily, seasonal, and yearly variations that are important. The variation in mean monthly temperatures increases as one moves from the equator toward the poles. The range of an environmental gradient along which an organism can thrive is denoted as its ecological amplitude.

To illustrate how ecological amplitude influences aquatic communities, consider the spatial and temporal distribution of five North American frog species along a temperature gradient (Wallace and Srb 1964). Wood frogs, restricted to Alaska and Labrador, breed in late March when water temperature is 10°C, and they lay their eggs in water as cold as 2.5°C. Their young take 60 days for metamorphosis. Meadow frogs, found in southern Canada, breed in late April when water temperatures are around 15°C, and they take 90 days to develop. Green frogs, a predominantly southern species, do not breed until June, when water temperatures are about 25°C; their eggs develop at 33°C. Bullfrogs prefer warmer temperatures and are completely southern in their distribution. Thus, temperature differences play a role in separating amphibian reproduction in time. This provides an excellent example of how temperature differences can bring about species isolation.

Both plants and animals have adaptations that allow them to live when temperature conditions are not conducive or congenial to growth. Thus, plants may produce seeds, buds, underground bulbs, or rhizomes to tide them over a period of cold. In temperate climates, when temperatures start to drop, deciduous plants shed their leaves in autumn and remain leafless in winter. In tropical areas, however, the leaf fall occurs when the temperature increases in summer or when rainfall ceases.

Animals have many adaptations to their environments. Animals display two distinct adaptive patterns: Those that do not have an internal mechanism for thermal regulation—hence, their temperatures vary as the air temperatures do—are referred to as **poikilotherms** ("cold blooded"). Such organisms become inactive when temperatures are too cold (below 8°C) or too warm (above 42°C). Organisms (including humans) that have the ability to regulate their body temperature are **homeotherms** ("warm blooded"). In the latter, metabolic energy is used to maintain a constant body temperature. Other adaptations such as hair growth and fat deposits facilitate protection from low temperatures and help with the retention of water when temperatures are high. However, if conditions become too harsh, some animals of both categories undergo **hibernation** (winter dormancy) or **estivation** (summer dormancy).

Precipitation, Humidity, and Wind

Precipitation refers to rain, snow, hail, sleet, and fog, all forms of water that reach the surface of the Earth through the processes of **condensation** (vapor → liquid) and deposition (vapor → ice). Water returns to the atmosphere through **evaporation** and **transpiration** from plants (liquid → vapor) and sublimation (ice → vapor). Condensation of the water vapor into tiny droplets with decreasing temperatures forms clouds or fog. The temperature at which condensation occurs, called the **dew point,** is a function of the **humidity,** or the amount of water vapor in a given volume of air. Tiny suspended particles (aerosols) on which water droplets can collect are called condensation nuclei and must be present in air for precipitation to occur.

The reason for the suspension of clouds in the atmosphere is that a typical cloud droplet, grown of water with a diameter of about 10 μm, can fall only a few centimeters in dry air before evaporating (Toon 2000a). A typical raindrop a few millimeters in size can fall many kilometers in dry air before evaporating, but about 10^6 cloud droplets must collide and coalesce to form a precipitation-sized drop. Ice generation requires that the radius of cloud droplets be above the threshold of 12 μm; when this occurs below the freezing point, ice particles grow rapidly by adding vapor from surrounding liquid droplets. These large ice crystals then coalesce with liquid droplets to form precipitation. Research by Rosenfeld, Rudich, and Lahav (2001) has shown that desert dust in the atmosphere prevents formation of droplets as large as this, thus preventing precipitation and contributing to desertification, which in turn produces more dust, further decreasing precipitation.

Daily, seasonal, and yearly variations in temperatures and the amount of precipitation in a given area have produced a seemingly

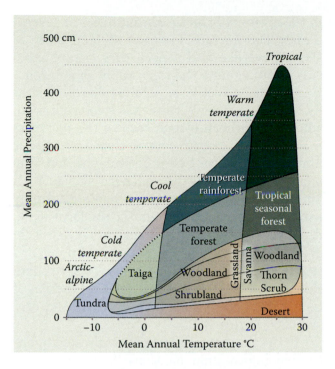

Figure 3.10 The distribution of the Earth's major vegetation as a function of the interaction of mean annual precipitation and temperature. (Redrawn from Whittaker, R. H. 1975. *Communities and Ecosystems,* 2nd ed. New York: Macmillan Publishing Company.)

infinite variety of diverse ecosystems. How much and how sustained the biological production of a terrestrial ecosystem depends largely on the interaction of these two factors. Thus, it is obvious why all climatic classifications consider these primary factors (Figure 3.10; also see Chapter 10).

Humidity refers to the content of water vapor in the atmosphere that ranges from practically 0 to 4% (4 g of water in every 100 g of air). It can be expressed as absolute humidity (density of water vapor, grams per cubic meter) and as relative humidity (ratio of actual vapor pressure to saturation vapor pressure, expressed as a percentage). Warm air holds more water vapor than cold air, and maximum absolute humidity (normally 12 g m^{-3}) can increase up to 40 g m^{-3} at high temperatures. Humidity is an important determinant of the amount and rate of evapotranspiration; hence, it is a critical climatic factor in the rate of moisture and heat loss by plants and animals. The more arid the air is (i.e., the lower its humidity), the higher is the water pressure

gradient, thus creating conditions for greater evaporation and transpiration.

Like air temperatures and other factors, the mean annual precipitation of an area is only part of the story. It is the yearly and seasonal distribution pattern of precipitation that determines the diversity of plant communities and, as a consequence, the animal communities.

Wind

Wind contributes to the dispersal of pollen, seeds, and reproductive propagules of most plants over considerable distances, as well as to the transport of small animals. The cooling and drying effects of wind can contribute to growth reduction (dwarfing) and desiccation damage (water deficits) in plants, and they can influence the feeding behaviors of animals. Windblown salt or sand near marine ecosystems can kill the foliage and buds of vegetation and result in flag-shaped crowns or shrub-like growth form of trees (known as a krumholz formation). Soil erosion by wind contributes to the degradation and loss of productive lands. Strong winds can also adversely affect water quality in aquatic systems by increasing mortality, turbidity, nutrient loading, and dissolved organic carbon. In areas like the western United States, wind may be a constant feature, leading to the following well-known refrain from Lerner and Loewe's Broadway play *Paint Your Wagon:*

> Way out West, they got a name for
>   rain and wind and fire;
> The rain is Tess, the fire is Joe, and
>   they call the wind Mariah.
> Mariah blows the stars around and
>   sets the clouds a-flyin';
> Mariah makes the mountains sound
>   like folks up there were dyin'.

In forests, the high wind velocities typical of storms (typhoons, tornadoes, and hurricanes) result in windbreaks (tree trunks snapped off) and windthrows (roots pulled out). Intense storms such as hurricanes have a major impact on ecosystems; the effects of one such event, Hurricane Andrew, have been well documented (Pimm et al. 1994). The hurricane made landfall on the east coast of Florida on August 24, 1992, with sustained winds of 242 km/h. It cut a swath across South Florida and the Everglades that was approximately 100 km long and 50 km wide. The hurricane changed the water quality near shore, scoured the tops of some reefs, uprooted between 20 and 30% of the trees, and defoliated most large hardwood trees. The storm affected a diversity of ecosystem types, including marine resources, upland and freshwater swamp forests, freshwater marshes, and mangroves. Applied ecologists have great interest in understanding the course that the ecosystems will take to recover from such events. Similar studies of the long-term effects of Hurricanes Katrina, Rita, and Wilma, which occurred in 2005, not only have focused on the human suffering these storms caused but also have centered on their impacts on the coastal wetlands of North America.

Influence of Topography

The physical features of the landscape have a significant influence on a number of climatic factors, and this influence is directly reflected in the diversity of biotic communities at several scales. The surface features of land such as slope, elevation, and aspect (direction of exposure to sun) define the topography of an area.

In the Northern Hemisphere, north-facing slopes receive the least solar radiation, whereas southern slopes receive the most. As a consequence, north-facing slopes are cooler and lose less water by evapotranspiration. The opposite is true for the southern slopes (Wali 1999b). This is one example of how topography can modify the environment at the local scale. As discussed earlier, variations in altitude greatly influence the distribution of precipitation by the formation of rain shadows. In turn, this also affects the structure and function of terrestrial populations (Figure 3.11) and ecosystems (Figure 3.12).

In the chapters that follow, we present quantitative information on these factors in their relation to the processes of ecosystem function, the cycling of materials, the productivity and development of ecosystems, and the global and regional patterns of plant and animal community distribution. Many early ecologists marveled at the role of climate in producing an amazing diversity of terrestrial plant and animal communities in different parts of the world. This will be discussed in Chapter 10. Recent

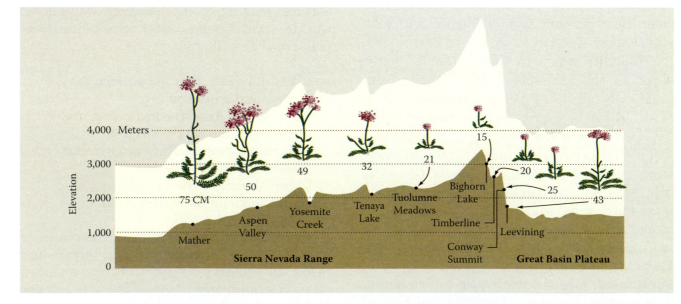

Figure 3.11 The size variations of the herb yarrow (*Achillea lanulosa*) as governed by the atmospheric and soil conditions along an altitudinal gradient. (Based on Clausen, J., D. D. Heck, and W. M. Hiesey. 1948. *Carnegie Institution of Washington Publication* 581:1–129. (Modified from Gould, J. L. et al. 1996. *Biological Science,* 6th ed. New York: W. W. Norton & Company.)

Figure 3.12 The similarity in the pattern of vegetation height along a gradient of (A) increasing aridity from moist forest in the Appalachian Mountains westward to desert in the United States; (B) increasing aridity from rain forest to desert in South America; (C) elevation up a tropical mountain in South America, from tropical rain forest to the alpine zone; (D) temperature from tropical seasonal forest northward to the arctic tundra. (Redrawn from Whittaker, R. H. 1975. *Communities and Ecosystems,* 2nd ed. New York: Macmillan Publishing Company.)

research on environmental problems has shown that atmospheric pollutants have the potential of modifying animal and plant growth processes considerably and rapidly. We discuss these aspects in Section B, where chapters address contemporary environmental issues.

Climate, Weather, and Humans

The congeniality of some climates—that is, those with comfortable temperature and adequate precipitation patterns—is to a large extent the reason why, historically, human populations have shown a clumped distribution. The limitations of the homeothermal nature of humans have been somewhat overcome by our ability to use technology to make our dwellings livable in the face of widely fluctuating temperatures. This has resulted in a great expansion of our distribution.

However, humans cannot alter the interconnectedness of the atmospheric conditions and other factors, so they must contend with the forces of nature. For example, weather patterns affect skin temperature, which, in turn, sends stimuli to other parts of the body, particularly the hypothalamus. These stimuli generate a neurological response, making conditions such as serotonin irritation syndrome more common. Thus, behavioral changes become obvious; for example, with temperature extremes, conditions such as traffic accidents, aggression, sexual assaults, and suicides increase (Persinger 1980).

Questions

1. How is "weather" different from "climate"?
2. How do climate factors control the distribution of plants and animals?
3. What factors determine the distribution of a species on a local scale?
4. On a global scale, what forces provide the basis for climate regimes and ocean current patterns?
5. Explain the relationship between the global solar energy balance and the greenhouse effect.

The Soil Environment

Soil Defined

Soil is the sustenance of life and water its elixir. An old saying, attributed to the Chinese, sums it all up: "Man—despite his artistic pretensions, his sophistication, and his many accomplishments—owes his existence to a six-inch layer of topsoil and the fact that it rains." This truism is as old as human knowledge of the use of soils. For more than 150 years and from many continents, studies have documented the interplay of geologic substrates, climate and organisms, and their interactions that produce a rich diversity of soils that, in turn, support the mosaic of life.

The definitions of soils are numerous, but those presented in the following discussion capture both the essence and the progression in scientific thinking. Plant growth was central to the definition of soils nearly a hundred years ago when soils were defined as "the surface stratum of earthy material, as far as the roots of plants reach" (Hillgard 1911). It continues to be so in the latest glossary of soil science, where **soil** is "the unconsolidated mineral or organic material on the immediate surface of the Earth that serves as a natural medium for the growth of land plants" (Soil Science Society of America 1997).

As more scientific knowledge has accumulated over the past century and a half, it has also become clear that many potentially arable soils brought under cultivation throughout the world have been degraded by management practices. Soils have also been used as a repository for waste products that human society generates. These often exceed the capacity of soils to assimilate such wastes, resulting in the accumulation of toxins and the pollution of ground waters and adjacent ecosystems. These issues will be discussed in Section B.

The following, more recent definition of soil incorporates the classical and the modern, and the use and abuse dilemma:

> The thin skin at the Earth's surface known as soil is the primary reason that life is possible on this planet; this is the life-sustaining pedosphere. This biologically active, porous, and structural medium serves as:
>
> - an effective integrator and dissipater of the mass flux of elements and energy;
> - the medium through which plant productivity (biomass) is sustained;
> - the foundation upon which structures are built;

- the fabric to which organisms are anchored and housed (habitat);
- the repository of solid and liquid wastes; and
- the living filter for bioremediation of waste products and water supplies.

Soils are the long-term capital on which a nation builds and grows. It is a basic component of ecosystems and is one of the most vulnerable to degradation through mismanagement. (Wilding 1994)

Soils form the base and basis of life. All soils have three components: solid, liquid, and gaseous. Each of these phases affects the others, and their interactions result in the structure and function of soils. Plant productivity depends on the fertility, root environment, and stability of soils; this, in turn, defines the sustenance of and welfare for all consumers, including humans. Soils, therefore, must be viewed in a holistic way, noting the interconnectedness of their atmosphere, biosphere, lithosphere, and hydrosphere (Figure 4.1). The integration of these variables as a soil ecosystem is at the

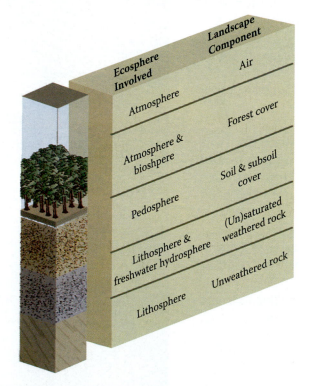

Figure 4.1 An ecological prism showing the relationships between the ecosphere involved and the landscape components in a forest ecosystem. (Modified from Fortescue, J. A. C. 1980. *Environmental Geochemistry: A Holistic Approach*. New York: Springer–Verlag.)

core of regional, national, and global sustainability issues and will be an important topic of discussion throughout this text.

Origin of Soils

All soils develop in parent materials that are derived from rocks. There are three kinds of rocks:

Igneous rocks are those formed from molten magma and may be of two types: volcanic or extrusive rocks (e.g., basalt) if lava cools quickly above ground after eruption, or plutonic or intrusive rocks (e.g., granite, diorite) if the cooling is underground and slow.

Sedimentary rocks are those formed from the deposition and recementing of weathering products of other rocks modified by the action of climatic variables and water (e.g., sandstone, shale).

Metamorphic rocks are products of the first two rock types that have been subjected to tremendous pressure and high temperature and are transformed (hence metamorphosed), for example, from igneous granite to form gneiss or from sedimentary shale to form slate.

Over time, a complex group of processes, collectively termed **weathering,** brings about a breakdown and disintegration of any rock type. The immediate products of weathering are termed **parent materials.** When found where they are formed, parent materials are referred to as residual. However, they may be carried large distances by natural forces and are then called transported (Figure 4.2.). The natural forces of transport are so important to the formation of soils that the methods of transport are used to characterize them. Thus, when ice is the agent of transport, the parent materials are glacial, and when wind is the agent, they are aeolian. Parent materials transported by water and deposited as sediments in streams are called alluvial, in lakes they are called lacustrine, and those in seas

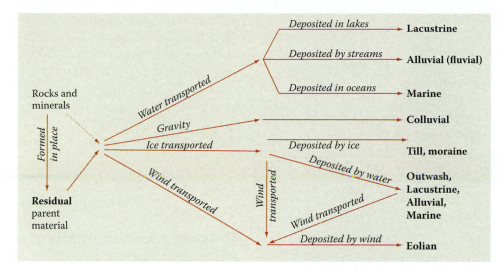

Figure 4.2 The weathering, transport, and deposition of parent materials. (Brady, N. C., and R. F. Weil. 2008. *The Nature and Properties of Soils*, 14th ed. Upper Saddle River, NJ: Prentice Hall.)

and oceans are called marine (Figure 4.2). It is with these parent materials that the processes of soil formation begin. Soil formation is discussed toward the end of this chapter.

Weathering includes both **physical forces** (such as freezing and thawing of water and abrasive action of water, wind, and snow) and **chemical processes** (such as hydration, hydrolysis, carbonation, and oxidation). Although physical and chemical weathering invariably go together, it is possible to see that one may be dominant over the other. For example, forces of physical weathering dominate in the cold tundra regions, while chemical weathering dominates in the tropical rain forest. Temperature, the quantity of precipitation, and the type of vegetation in a given area will determine how quickly soil formation can proceed. Biological activity significantly accelerates weathering, even at moderate temperatures (Figure 4.3).

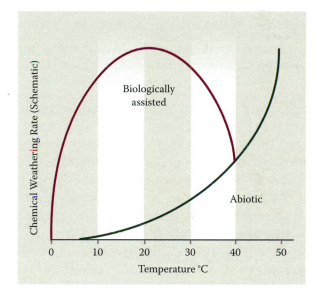

Figure 4.3 Idealized representation of the effect of biological processes and temperature change on the rates of chemical weathering of rocks. (Kump, L. R., and J. E. Lovelock. 1995. In *Future Climates of the World; a Modeling Perspective,* ed. A. Henserson-Sellers, 537–553. Amsterdam: Elsevier Science B.V.)

or 50–2 μm), and clay (<0.002 mm or 2 μm). The relative percentages of sand, silt, and clay define the soil's particle size distribution or **soil texture** (Figure 4.4). The textural triangle aids scientists in grouping soils into convenient "textural" categories that indicate their physical and chemical properties.

Soil **structure,** which may be described as platy, blocky, granular, columnar, or prismatic, is the aggregation of soil particles under

Soil Physical Characteristics

Particle Size and Structure

The primary mineral constituents of soil, arranged by their particle size, are sand (very coarse to very fine particles, ranging from 2.00–0.05 mm in diameter), silt (0.05–0.002 mm

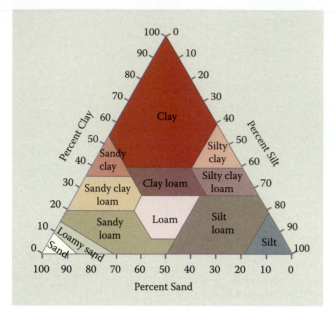

Figure 4.4 Textural triangle showing the grouping of soils by particle size and hence the relative contribution of sand, silt, and clay to their composition. (USDA. 1994. *Keys to Soil Classification,* 6th ed. Washington, D.C.: Soil Conservation Service, Government Printing Office.)

field conditions into primary and secondary structural units. Micropores (small spaces or voids) are formed when the primary particles clump together into primary aggregates, leaving spaces between them. Further, as primary aggregates form larger secondary aggregates, the spaces or voids between structural units become larger, forming larger macropores. Together, the two kinds of spaces or voids (micropores and macropores) constitute the total soil porosity.

Soil porosity is intimately related to the mass per unit volume of soils, referred to as their bulk density and typically expressed in units of grams per cubic centimeter. The lower the bulk density is, the lighter the soil is and the greater the void space to accommodate roots. Bulk density provides a good estimate of the physical resistance that roots will encounter as they penetrate into the soil. The average bulk density of agricultural soils is in the range of 1.1–1.5 g cm^{-3}; for organic soils, bulk densities are generally below 1.0 g cm^{-3}. For soils that are compacted, either naturally or by human activities (e.g., by cultivation, machinery or foot traffic, off-road vehicles), the bulk density may be 1.5 g cm^{-3} or greater. Conversely, natural

phenomena, such as freeze–thaw and the activities of soil animals such as ants, termites, and earthworms, may considerably increase the soil porosity and, in doing so, decrease bulk density. Actual pore space may be estimated by considering both bulk density and the true specific gravity of soil particles (called particle density) and is calculated as follows:

$$1 - (\text{bulk density/particle density}) \times 100$$

$$= \% \text{ total porosity.}$$

The average particle density of mineral soils is commonly taken as 2.65 g/cm^3.

Together, soil texture, structure, and the amount of organic matter present in the soil determine gas exchange rates, the soil's ability to retain water, nutrient availability, temperature relations, and the distribution and abundance of soil biota. These properties are routinely used to evaluate soil fertility and the potential productivity for the growth of agricultural crops, rangeland grasses, and forest trees. For example, ecological studies in many parts of the world have demonstrated a relationship between soil texture and the productivity of forest trees. This is measured by the "site index," which is the total height to which dominant trees will grow on a given site at some index age, usually 25 or 50 years.

Soil Water

The amount of water remaining in the soil after a precipitation event (e.g., rain, snow, or hail) after the excess water has drained away determines the potential for growth, abundance, and distribution of plant and animal life, fungi, and bacteria. When water is in short supply, as in the arid regions of the world, vegetation is sparse. When human practices accelerate the loss of vegetation in such arid areas, soils become unproductive, resulting in temporary or permanent desert conditions. This process is termed **desertification.** In high-rainfall areas, exposed soils and their nutrients are prone to erosion. In many tropical rain forest regions, vegetation is cut down, burned, and cleared away to make land available for farming. This practice is known as "slash-and-burn" agriculture, and these fields often lose their fertility in 5–10 years through a combination of crop removal (the harvested crops take nutrients with them), soil erosion, and nutrient leaching processes. Some of these aspects are discussed further in Chapter 14.

The role of water in producing distinctive plant communities led early ecologists to produce qualitative groupings to indicate the distribution and abundance of plants. Thus, dry habitat conditions in a given area have been designated as **xeric;** plants that grow on such habitats are **xerophytes.** These plants (such as cacti) have morphological and physiological characteristics that allow them to withstand drought conditions. They possess a sturdy and extensive root system, and roots may reach to great depths in search of water. These plants also have a reduced or modified leaf area to cut down on water loss through transpiration. Many plants (like sagebrush) obtain water from deep underground water tables; these are the **phreatophytes.**

On the other hand, aquatic habitats are **hydric.** Plant species growing there, such as cattails, are **hydrophytes** that do not need well-developed root systems or adaptations to cut down on water loss. Their leaves have large spaces among their cells to trap gases. Intermediate conditions are **mesic;** plants there (such as maples) are **mesophytes.** These plants lack adaptations that would allow them to live in either xeric or hydric environments. Although these terms are in wide use in the ecological literature, it should be underscored that they are qualitative and reflect relative conditions in specific areas: Mesic conditions, for example, in temperate Ohio are not the same as the mesic conditions of parts of California with a Mediterranean climate.

The movement of water from the soil through the plant and to the atmosphere represents a more or less continuous stream of water; this is referred to as the soil–plant–atmosphere continuum. Much of our understanding of this transfer of water has come from soil scientists. The continuum concept offers a holistic model of how the components function as a system rather than considering each phase on its own (i.e., the plants or soil in isolation from the other components).

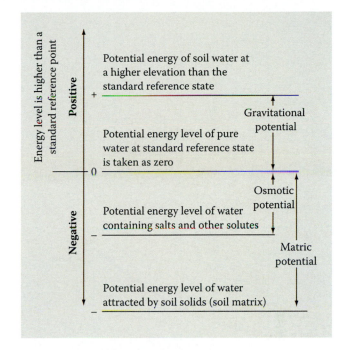

Figure 4.5 Factors making soil water more available (positive: elevation and gravity) or less available (negative: osmotic potential and matric forces) to plants relative to pure water. (From Brady, N. C., and R. F. Weil. 2008. _The Nature and Properties of Soils_, 14th ed. Upper Saddle River, NJ: Prentice Hall.)

The availability of soil water depends on the quantity of water in the soil, soil particle and pore size, bulk density (because heavier soils present more mechanical resistance to roots), and the chemical composition of the soils. The availability of soil water for plant uptake is influenced by four forces: gravitational potential, pressure potential, osmotic potential, and matric potential (Figure 4.5).

The available soil water capacity of soil is closely related to soil particle size and is defined as the difference between the soil water content at **field capacity** (water remaining in soil after drainage due to gravity has occurred) and the **wilting point** (when soil is too dry to allow further water extraction by plants) (Figure 4.6). Notice that the soil particle sizes identified on the x-axis of Figure 4.6 correspond to those identified in the soil textural triangle in Figure 4.4; that is, the relative contribution of clay, sand, and silt in a soil helps determine its available soil water capacity. The range of available water seen in Figure 4.6 can be used to approximate the plant species adapted to grow under these soil moisture conditions with

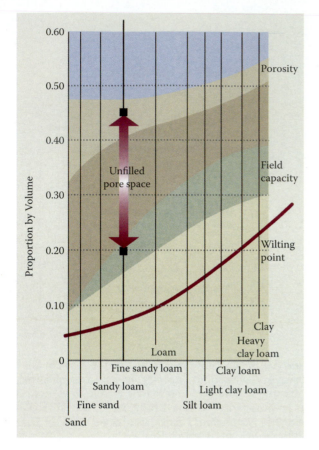

Figure 4.6 General relationship between water-holding properties of soils and combinations of different particle sizes. For a fine sandy loam, the approximate difference between porosity, 0.45, and field capacity, 0.20, is 0.25, meaning that the unfilled pore space is 0.25 times the soil volume. The difference between field capacity and wilting point is a measure of unfilled pore space. (From FISWRG 2003, based on USDA. 1955. Yearbook. *Water.* Washington, D.C.: Government Printing Office, and Dunne and Leopold 1978.)

the caveat that field capacity and permanent wilting point are not set values but, rather, are gross generalizations of soil water availability. For example, not all plants wilt at the same soil matric potential.

The ways in which vegetation distribution varies as a function of water availability have been illustrated for a number of ecosystems (e.g., Wali and Krajina 1973; Wikum and Wali 1974). However, soil water characterization has limitations when used alone in ecological studies. For example, the direct interception of precipitation by tree crowns and the deposition of dew may increase water availability to plants independently of soil water reserves. In several studies, soil water measurements

have provided good correlations with plant distribution, but only when used in combination with internal water stress measurements (Waring and Cleary 1968; Ruess and Wali 1980). A newly proposed geographic information system (GIS)-derived integrated moisture index predicts species composition in forests and holds promise in elucidating ecological relationships of natural communities (Iverson et al. 1997).

The importance of soil water in human welfare is beyond question, and recent work on a broad regional scale in the Middle East emphasizes the importance of soil water in the rise and fall of civilizations (Hillel 1994). Regardless of the region of the world, soil–water relations play a critical role in all ecological processes; we shall explore this topic further in Chapter 16 on water quality and quantity.

Soil Air

The rich biological diversity that soils support includes organisms adapted to both aerobic and anaerobic conditions. The **aerobic** organisms require oxygen. Small fauna—ants and termites, for example—can increase the pore space substantially by tunneling and churning the soil, increasing soil aeration. Soil air dynamics (or aeration) bear an intimate relationship with water. As water infiltrates soil and percolates into and through soil voids, it drives out gases, including oxygen. Once saturation occurs and the available oxygen is used up, chemical processes begin to take place under **anaerobic** conditions. Such conditions lead to changes in the reduction–oxidation potential and alter the forms in which ions occur (Table 4.1). The relative amounts of the different oxidized and reduced forms of these elements will vary with concentration, pH, and temperature in any given situation.

Thermal Relations

Like all chemical and biochemical processes, soil processes are intimately related to energy transfers and hence temperature. In some European ecological literature, temperature relations are referred to as **heat economy** or **heat budgets.** In general, soil temperatures

Table 4.1
Predominant Chemical States[a] of Essential Nutrients in Aerobic and Anaerobic Conditions and Approximate Redox Potential for Transformation

Element	Oxidized Form	Reduced Forms	Approximate Redox Potential for Transformation (millivolts)
Nitrogen	NO_3^- (nitrate)	N_2O, NO_2^-, NH_4^+	250
Manganese	Mn^{4+} (manganic)	Mn^{2+} (manganous)	225
Iron	Fe^{3+} (ferric)	Fe^{2+} (ferrous)	+100 to −100
Sulfur	SO_4 (sulfate)	S (sulfide)	−100 to −200
Carbon	CO_2 (carbon dioxide)	CH_4 (methane)	Below −200

Source: After Mitsch, W. J., and J. G. Gosselink. 2007. *Wetlands,* 4th ed. New York: Van Nostrand Reinhold.
[a] Oxidized and reduced forms.

fluctuate with air temperatures. However, a major distinction is that there is a daily (diurnal) time lag in the temperature change across the atmosphere–soil interface. Diurnal temperature extremes are thus dampened, and plant roots and microbial life forms are subject to narrower temperature fluctuations. Temperature dampening effects increase with soil depth to the point where daily temperatures are constant. At even greater depths, the soils have constant annual temperatures, modified only by the temperature changes in the groundwater and bedrock.

Soil temperature is related to soil particle size, soil color, and water content. Soils with dark-colored surfaces absorb heat readily and are warmed quickly by the sun. Soils with high water content (e.g., wetlands) require greater energy to warm up because water requires about three times more energy to warm than does soil. Although sandy soils are highly reflective, especially when they are low in organic matter, their surface temperature is high during the day. At night, they lose heat quickly. Because they heat quickly during the day and cool quickly at night, these soils show a wide diurnal temperature range. This is in part because sandy soils tend to lose water quickly; without the moderating effects of water, they warm rapidly during the day. In agricultural fields, sandy soils allow early planting compared to heavier textured (silts, clay) or poorly drained soils that require much more solar energy to warm.

Soil Chemical Relations

Ion Exchange and Plant Nutrition

Soils are a vast storehouse of chemical elements and compounds. Most dissolved chemical elements are found as **cations** (positively charged ions) or as **anions** (negatively charged ions). These are the ingredients with which plants grow and add to their biological mass, or biomass. How plants acquire these ions from soils is one of the most fascinating marvels of nature.

Through their sensitive surface membranes, plant roots exchange hydrogen ions (H^+) for the nutrient ions that are adsorbed to the charged surfaces of clay particles and organic colloids. There are several types of clay minerals, including kaolinites, illites, smectites, vermiculites, and various oxides of iron, aluminum, and manganese. Plant uptake of nutrients from soils is not simply a function of ion concentration but is also influenced by valence and the size of the ion. **Ion exchange** between the two surfaces (soil and root) depends on the types and concentrations of ions, their charge density and strength, and the surface area available for exchange. The soil's ability to adsorb ions is measured as the **cation exchange capacity (CEC)** or **anion exchange capacity (AEC)**; these are extremely useful measures of soil fertility.

The larger the number of small particles making up the soil mass is, the greater is the surface area per unit mass or volume

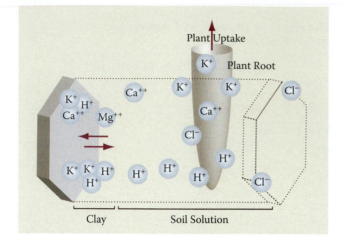

Figure 4.7 The exchange of soil calcium and hydrogen ions between soil colloid and a plant root. (From Thien and Graveel 1997.) Obviously, the types of soil colloids and clays present are critical, as is the pH of the soil. The smaller the soil particle size is, the larger is the number of surfaces per unit area for ion exchange. As an example, a 1-mm cube has a volume of 1 mm^3 with six surfaces and a surface area of 6 mm^2; however, 1,000 0.1-mm cubes also have a total volume of 1 mm^3, but they have a total surface area of 60 mm^2.

(Figure 4.7). Coarse mineral particles have low surface area and correspondingly low cation exchange capacities. The surface area per unit mass of soil shows a marked increase in the initial stages of weathering and then a gradual reduction toward the final stages of weathering. Thus, the exchange of ions between plant roots and soils is related to surface chemistry and by clay mineralogy. The ability to retain cations (CEC) usually dominates in temperate soils, while anion-holding power (AEC) may be more important in many tropical soils.

Physiological experiments with a number of plant species have shown that some specific ions are essential for plant growth; these are termed essential elements. Depending upon the relative quantities of each element required by plants, essential elements have been divided into two groups: **macronutrients,** which are needed in relatively large quantities by plants and animals, and **micronutrients,** which are needed in very small quantities (see Appendix 4.1). A number of elements have been shown to be beneficial to specific plant and animal groups, some of which are beneficial to humans (see the periodic table of elements in Figure 4.8).

Within an ecosystem, the dominant pathway of nutrient uptake is from soils to plants and then from plants to herbivores (or vegetarians): soils → plants → herbivores → carnivores. Nutrients play many roles in growth: (1) as the basis for building tissues, (2) as enzymes that have a catalytic or regulatory role, (3) as parts of organic molecules associated with enzyme function, called cofactors, and (4) as messengers in nerve and muscle cells called signal transduction. Some elements, like calcium, may have multiple roles. Although the specific roles of some elements and their ionic forms are still unknown (see Appendix 4.1), the elements used by plants show phenomenal diversity. Some elements, such as silicon, may also be taken up by plants through passive water flow without playing a specific metabolic role (Figure 4.8).

In agriculture, forestry, and range management, several chemical properties are used to assess the fertility of soils. These include soil pH, which refers to the hydrogen ion activity (or concentration) in solution, expressed as $-\log[H^+]$. In any circumstance, a small fraction of water dissociates into its two ions, H^+ and OH^-. At pH 7, the H^+ ions equal the number of OH^- ions, with a concentration of 10^{-7} *M*. This is the "neutral" condition. Below pH 7, soil solutions are **acid** and above pH 7 they are **alkaline** (Figure 4.9). Another widely used property is **electrical conductivity** (**EC**). EC is a measure of the solution's ability to conduct electricity. As the concentration of ions in a solution increases, that solution is more able to conduct an electrical current. Thus, EC is an indirect measure of the concentration of dissolved salts, measured in units of siemens (S). When ECs exceed 0.4 S m^{-1} (4 millimho/cm), the soil solution is too salty for the growth of most plant species, and the soil is classified as **saline.** Such conditions are most common in arid and semiarid regions.

The role of pH as a diagnostic tool is beautifully illustrated in the example of *Hydrangea,* which produces blue flowers in acid soils and pink flowers in neutral and alkaline soils (Figure 4.10). The mechanisms for such color change, however, are complex. In this case, they are related to the availability of aluminum, which increases at low pH.

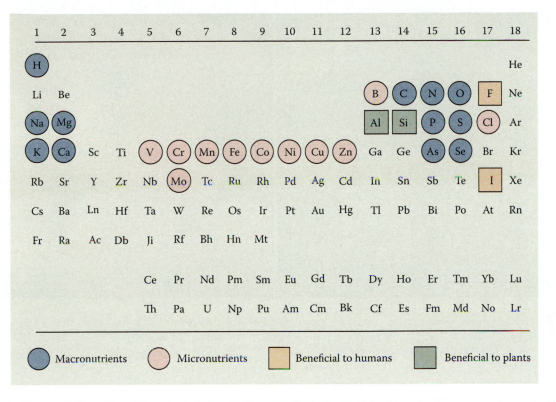

Figure 4.8 The periodic table of chemical elements showing the macronutrients and micronutrients needed by living organisms; also shown are the elements that may be beneficial to some plants and humans.

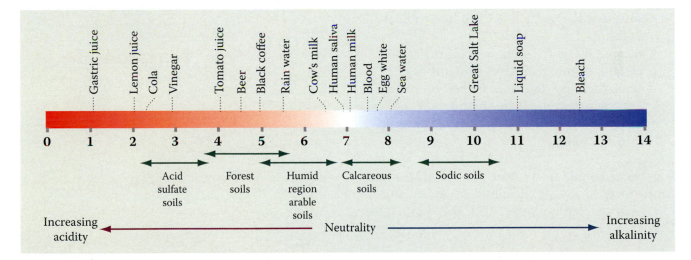

Figure 4.9 Comparable pH ranges for commonly used products and soils: The scale is a logarithmic (not linear) expression of hydrogen ion concentration or activity. Each pH unit represents a 10-fold change in H^+ ion concentration, but there are 100 times more H^+ ions than OH^- ions present at pH 6 than at pH 7. pH 8 has 10,000 times more OH^- ions than H^+; hence, all 1-unit pH changes result in 100-fold changes in the H^+/OH^- ratio.

Nutrients in soils have been classified into four groups: (1) those dissolved in soil water, (2) those adsorbed in exchangeable form, (3) those bound in nonexchangeable form by mineral particles and organic matter (OM), and (4) those present in soluble and insoluble forms in the OM.

Several processes help determine the flux of nutrients into and out of the soil, namely:

- atmospheric precipitation that carries nutrients along;
- nutrient inputs from the weathering of minerals;

Figure 4.10 *Hydrangea macrophylla* produces blue flowers at acid pH levels (left) and pink flowers at alkaline pH (right). Studies have shown that the blue color is induced by the availability of aluminum in acid soils. (Photo courtesy of The Ohio State University Extension Service.)

- the dry deposition of nutrients from the atmosphere;
- adsorption to and desorption from soil particles;
- nutrient losses by leaching and volatilization;
- decomposition and mineralization of soil organic matter (the process of transforming organic forms of nutrients to inorganic forms); and
- nutrient cycling by plants, animals, and microbes.

Electrical conductivity and pH are important indicators of the availability of many nutrient ions for both plants and soil biota (see Figure 4.11).

Plant Roots

Scavenging of water and nutrients by roots is indispensable to plant nutrition. Depending upon genetic and physiological traits, roots may be "surface feeders"—a condition in which roots are most dense just under the soil surface—or they may penetrate deep into the water table. Some plants that inhabit arid and semiarid environments, like the grasses, have intimate contact with soil particles; ecologists and soil scientists often find it difficult

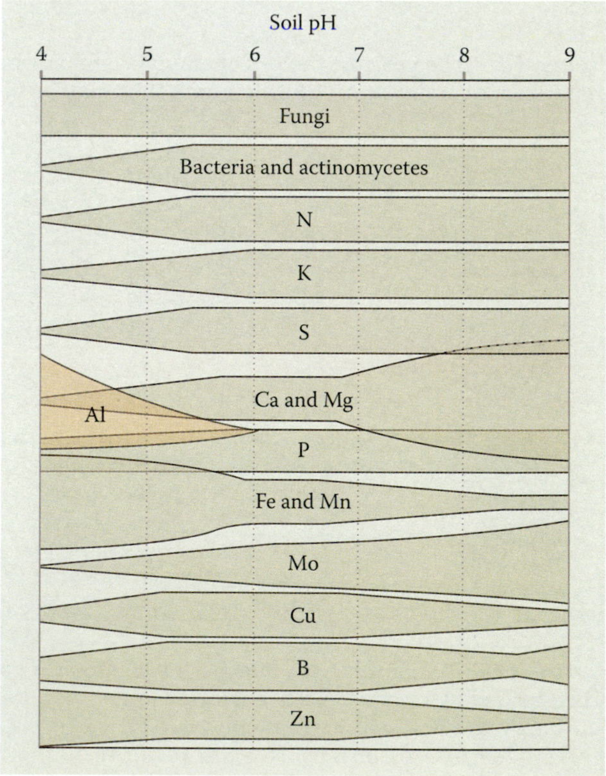

Figure 4.11 Relationships of pH, microbial activity, and nutrient availability (Brad, N. C., and R. F. Weil. 2008. *The Nature and Properties of Soil*, 14th ed. Upper Saddle River, NJ: Prentice Hall.) The bandwidth implies the relative availability of elements across the pH range indicated. Fungi seem to function equally well across the pH ranges shown, whereas bacteria and actinomycetes are favored by neutral to alkaline conditions.

to separate soils from roots because roots and soil bacteria produce mucilaginous sheaths that extend into the immediate soil matrix. Formation of the sheaths (also known as soil sheaths or rhizosheaths; McCulley and Canny 1988) requires water allocated by the plants to the roots. The energy expenditure required to send water to the roots is repaid because rhizosheaths help keep plants from desiccating and help them obtain nutrients under dry soil conditions.

Studies on the role of micronutrients and other "trace elements" in plant growth have focused primarily on agricultural crops and horticultural varieties (Table 4.2). The role of micronutrients in the distribution of natural communities is relatively recent and has involved research on the chemical forms these elements take, their availability for plant uptake, and their movement within soils.

Although trace element uptake varies with such factors as soil pH, light, heat, and rainfall, the relative abundance of trace elements in soils, known as a trace element profile, can be used to "fingerprint" soils and the crops that grow on them. Because the trace element profile of soils in a given region is unique and plants tend to take up and store trace elements in their tissues in proportion to their concentration in the soil, once the trace element profile of a given region is known, an analysis of plant tissues can confirm that the plants were indeed grown on that soil. In this way, tissue analysis of the plant material growing in different areas allows the U.S. Department of Agriculture to identify the true place of origin for different crops. These include coffee, fruits, and other produce on which importers might otherwise try to avoid paying import duties of millions of dollars.

Soil Biota

Soil organisms belong to several major taxonomic groups and contribute to ecosystem processes through a myriad of functions (Table 4.2), including the production and processing of organic matter (Figure 4.12). The

staggering numbers of both identified and unidentified (estimated) species constituting the soil biota contribute to the breakdown of organic matter in several ways (Edwards, Reichle, and Crossley 1970):

- fragmenting plant and animal tissues and making them more easily invaded by microorganisms;
- selectively decomposing and chemically changing organic residues;
- transforming plant residues into humic substances;
- increasing the surface area available for bacterial and fungal action;
- forming complex aggregates of organic matter with the mineral fraction of soil; and
- mixing the organic matter thoroughly into the upper layers of soil.

The biomass of soil organisms varies with the type of ecosystem (Figure 4.13); it also varies spatially within the soil; for example, the abundance of organisms in the rhizosphere is often dramatically higher than in the root-free soil environment (Tables 4.2 and 4.3). This is the soil equivalent of a "hotspot" of biological activity.

Bacteria

Soil microbes are dominated by bacteria and fungi, along with some algae, cyanobacteria, and minute animals. Important bacterial activities include the decomposition of organic matter and nutrient transformations such as nitrogen fixation (conversion of N_2 gas to ammonium, NH_4) and denitrification (conversion of nitrate, NO_3^-, ion to N_2 gas). In addition to these processes, nitrifying bacteria alter the availability of other nitrogen compounds by converting ammonium ions to nitrite (NO_2^-) and then to nitrate, the form preferred for uptake by many plants.

One type of bacteria, *Rhizobium*, is responsible for nitrogen fixation (see Figure 6.7). Rhizobium reside in nodules—gall-like structures on the roots of legumes (e.g., peas, beans, clovers, and alfalfa). The result is a highly beneficial **symbiotic** relationship between

Table 4.2
Functions of Soil Organisms

Type of Soil Organism		Major Functions
Photosynthesizers	Plants	Capture energy
	Algae	Use solar energy to fix CO_2
	Bacteria	Add organic matter to soil (biomass such as dead cells, plant litter, and secondary metabolites)
Decomposers	Bacteria	Break down residue
	Fungi	Immobilize (retain) nutrients in their biomass
		Create new organic compounds (cell constituents, waste products) that are sources of energy and nutrients for other organisms
		Produce compounds that help bind soil into aggregates
		Bind soil aggregates with fungal hyphae
		Nitrifying and denitrifying bacteria convert forms of nitrogen
		Compete with or inhibit disease-causing organisms
Mutualists	Bacteria	Enhance plant growth
	Fungi	Protect plant roots from disease-causing organisms
		Some bacteria fix N_2
		Some fungi form mycorrhizal associations with roots and deliver nutrients (such as P) and water to the plant
Pathogens	Bacteria	Promote disease
Parasites	Fungi	Consume roots and other plant parts, causing disease
	Nematodes	Parasitize nematodes or insects, including disease-causing organisms
	Microarthropods	
Root feeders	Nematodes	Consume plant roots
	Macroarthropods (e.g., cutworm, weevil larvae, and symphylans)	Potentially cause significant crop yield losses
Bacterial feeders	Protozoa	Graze
	Nematodes	Release plant available nitrogen (NH_4^+) and other nutrients when feeding on bacteria
		Control many root-feeding or disease-causing pests
		Stimulate and control the activity of bacterial populations
Fungal feeders	Protozoa	Graze
	Microarthropods	Release plant available nitrogen (NH_4^+) and other nutrients when feeding on bacteria
		Control many root-feeding or disease-causing pests
		Stimulate and control the activity of bacterial populations
Shredders	Earthworms	Break down residue and enhance soil structure
	Macroarthropods	Shred plant litter as they feed on bacteria and fungi
		Provide habitat for bacteria in their guts and fecal pellets
		Enhance soil structure as they produce fecal pellets and burrow through soil

Table 4.2 (continued)
Functions of Soil Organisms

Type of Soil Organism		Major Functions
Higher-level predators	Nematode-feeding nematodes	Control the populations of lower trophic-level predators
	Larger arthropods, mice, voles, shrews, birds, other aboveground animals	Larger organisms improve soil structure by burrowing and by passing soil through their guts Larger organisms carry smaller organisms long distances

Source: From USDA-NRCS 1999.

the plants and bacteria; the bacteria supply the plants with nitrogen for growth, and the plants provide a food supply for the bacteria. Wherever they grow, legumes increase soil nitrogen levels considerably, a property that is the basis for crop rotations to maintain high-productivity agriculture. One well-known rotation in the United States alternates growing corn in one year and soybeans in the next. Synthetic nitrogen fertilizers augment or have supplanted this practice in most industrialized countries.

So successful was the practice of legume-based crop rotations in agriculture that it prompted several scientists to suggest similar legume-based rotational systems for forest ecosystems where soils were nitrogen deficient for tree growth. In addition to legumes, plants that have bacterial associations that allow for N fixation include alders (*Alnus* spp.) in the cool regions of the Northern Hemisphere; California lilac (*Ceanothus*) and buffalo berry (*Shepherdia*) in North America; Russian olive (*Elaeagnus*) in Asia, Europe, and North America; bayberry (*Myrica*) in the temperate regions of both hemispheres; *Coriaria* in Japan, New Zealand, Central and South America, and the Mediterranean region; and *Casuarina* in the tropics and subtropics from East Africa to the Indian Archipelago, Pacific Islands, and Australia. On the other hand, the nitrogen-fixing ability of these taxa has resulted in their success as invasive plants and noxious weeds, particularly in regions with nitrogen-poor soils. In the tropics, particularly in areas where rice is cultivated in flooded paddies, substantial nitrogen fixation comes from cyanobacteria, particularly by the taxa *Anabaena* and *Nostoc*.

Fungi

Recent estimates indicate that more than 95% of all terrestrial plants are dependent on special types of fungi known as **mycorrhizae.** Mycorrhizal fungi form symbiotic relationships with most plant species by "infecting" their roots. Ectomycorrhizae are fungi that form a mat on the outer surface of the plant root and also penetrate into the root cortex (Marx 1991a). There are more than 2,100 species of ectomycorrhizal fungi that form associates with forest trees in North America. Because trees with ectomycorrhizae have a larger surface area for the adsorption of nutrients and water, they are able to accumulate N, P, K, and Ca more rapidly than trees without ectomycorrhizae. The result is faster growth and reproduction rates for trees that have mycorrhizal infection. Ectomycorrhizae also appear to increase the tolerance of trees to drought, high soil temperatures, organic and inorganic soil toxins, and extremes of soil acidity associated with high levels of S, Mn, or Al (Marx 1991b).

The most common mycorrhizal symbionts are endomycorrhizae or vesicular–arbuscular mycorrhizae (often abbreviated VAM). These invade the root itself and live within the root tissues. Almost all angiosperm families and some ferns, fern allies, and mosses form VAM relationships with this small group of primitive fungi related to bread molds. As a consequence of the differences in the diversity between the

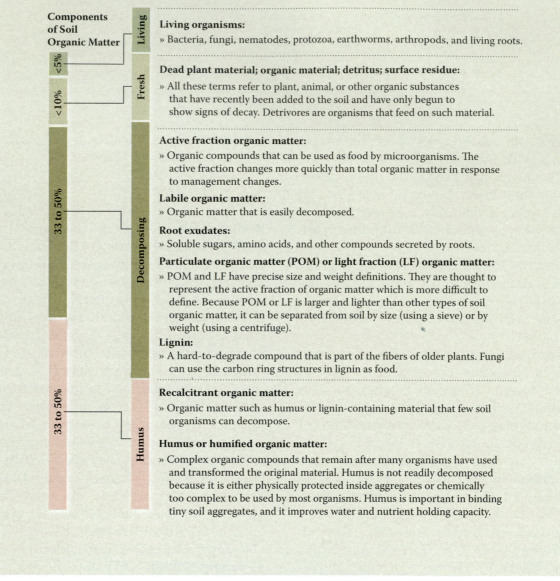

Food Sources for Soil Organisms
"Soil organic matter" includes all the organic substances in or on the soil. Here are terms used to describe different types of organic matter.

Components of Soil Organic Matter

Living

Living organisms:
» Bacteria, fungi, nematodes, protozoa, earthworms, arthropods, and living roots.

Fresh

Dead plant material; organic material; detritus; surface residue:
» All these terms refer to plant, animal, or other organic substances that have recently been added to the soil and have only begun to show signs of decay. Detrivores are organisms that feed on such material.

Decomposing

Active fraction organic matter:
» Organic compounds that can be used as food by microorganisms. The active fraction changes more quickly than total organic matter in response to management changes.

Labile organic matter:
» Organic matter that is easily decomposed.

Root exudates:
» Soluble sugars, amino acids, and other compounds secreted by roots.

Particulate organic matter (POM) or light fraction (LF) organic matter:
» POM and LF have precise size and weight definitions. They are thought to represent the active fraction of organic matter which is more difficult to define. Because POM or LF is larger and lighter than other types of soil organic matter, it can be separated from soil by size (using a sieve) or by weight (using a centrifuge).

Lignin:
» A hard-to-degrade compound that is part of the fibers of older plants. Fungi can use the carbon ring structures in lignin as food.

Humus

Recalcitrant organic matter:
» Organic matter such as humus or lignin-containing material that few soil organisms can decompose.

Humus or humified organic matter:
» Complex organic compounds that remain after many organisms have used and transformed the original material. Humus is not readily decomposed because it is either physically protected inside aggregates or chemically too complex to be used by most organisms. Humus is important in binding tiny soil aggregates, and it improves water and nutrient holding capacity.

Figure 4.12 Characterization of some of the major components of soil organic matter (USDA-NRCS 1999).

plant hosts (250,000–300,000 described species) and fungal symbionts (400–500 species in five genera), there is little specificity between the fungi and their plant hosts. Fossils of the earliest land plants show strong VAM colonization of their roots, indicating that this relationship developed very early in the evolutionary history of land plants (Allen 1991).

Soil fauna are diverse and generally classified on the basis of size into three groups (Coleman and Crossley 1995):

• *Microfauna* are generally less than 200 µm and include mostly protozoa and a few small mites, nematodes, and rotifers.

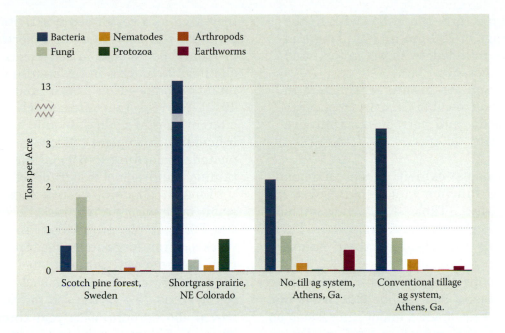

Figure 4.13 Comparison of the biomass of soil organisms in two natural and two agricultural soil ecosystems. (From USDA-NRCS 1999.)

Table 4.3
Relative numbers and biomass of biota commonly found in surface soil horizons (from Brady and Weil (2008).

Microflora and earthworms dominate the life of most soils.

Organisms	Number[a]		Biomass[b]	
	per m^2	per gram	kg/ha	g/m^2
Microflora				
Bacteria and Archaea[c]	10^{14}–10^{15}	10^9–10^{10}	400–5000	40–500
Actinomycetes	10^{12}–10^{13}	10^7–10^8	400–5000	40–500
Fungi	10^6–10^8m	10–10^3m	1000–15,000	100–1500
Algae	10^9–10^{10}	10^4–10^5	10–500	1–50
Fauna				
Protozoa	10^7–10^{11}	10^2–10^6	20–300	2–30
Nematodes	10^5–10^7	1–10^2	10–300	1–30
Mites	10^3–10^6	1–10	2–500	0.2–5
Collembola	10^3–10^6	1–10	2–500	0.2–5
Earthworms	10–10^3		100–4000	10–400
Other fauna	10^2–10^4		10–100	1–10

[a] For fungi the individual is hard to discern, so meters of hyphal length is given as a measure of abundance.

[b] Biomass values are on a liveweight basis. Dry weights are about 20 to 25% of these values.

[c] Estimated numbers of Bacteria and Archaea from Torsvik et al. (2002); others from many sources.

- *Mesofauna* are generally within the size range of 200 µm to 1 cm long and include millipedes, most mites, and most nematodes.
- *Macrofauna* are several centimeters in length and include earthworms, large arthropods, and vertebrates that live in the soils.

These organisms exhibit much taxonomic diversity and function in soil–plant systems in diverse ways (Table 4.2). They include detritivores, herbivores, and carnivores. Their diversity is as important as their numbers and biomass.

Three groups of organisms may illustrate the importance of soil fauna. **Earthworms,** whose role in soils made a major impression on Charles Darwin (Hartenstein 1986), perform a multitude of functions: soil turnover; incorporation of organic matter; improvement of soil aeration; the formation of complex organic, or *humic,* compounds; conversion of organic nitrogen and phosphorus into plant-available forms; stimulation of respiratory and enzyme activities of soil; destruction of phytopathogenic fungi; increase in yield of fruit, roots, grass, and other crops; preservation of soil structure through humification; and increase in nitrogen-fixing bacteria. Earthworms are also essential in organic waste management, soil amelioration, and land reclamation, as well as bioindicators of environmental contamination.

Ants are also abundant in some soils, and ant mounds or hills are dispersed over large areas in many parts of the world. By one estimate, the presence of 20 anthills per hectare will result in a pileup of more than 3,800 kg of earth on the soil surface. In the Northern Great Plains, 1,000 mounds were found in a 4-ha area; there may commonly be as many as 50 colonies per hectare, with a total of more than 13,800 individual ants per hectare. By tunneling and moving the earth, ant activity was shown to increase the porosity of soil by more than 30%, thereby considerably altering water transport and nutrient mobility.

Two contrasting situations that show that ant activity has effects on soil chemistry have been reported in Europe and North America. In Poland, the low pH values of pine soils increased as a result of ant activities, whereas the high soil pH values in a Northern Plains grassland decreased; in both cases, there were significant changes in the resulting plant community composition (Wali and Kannowski 1975). The role of termites, which occur on even a larger spatial scale than ants, is particularly influential in tropical soils, with some termite mounds exceeding 2–3 m in height and containing hundreds of kilograms of soil.

Finally, gophers, prairie dogs, and other small, burrowing vertebrates affect soil properties. For example, prairie dogs and other herbivores consume large quantities of plant material; they then deposit unassimilated materials in the form of urine and feces back onto the soil. As much as 75% of the soil nutrients may be moved aboveground each year through a combination of plant uptake and herbivory by prairie dogs. As a result, higher nitrogen content is found in areas colonized by prairie dogs than in those not colonized. Relatively recent studies of the impact of large vertebrates on soil properties and vegetation composition via field studies and simulation modeling have documented the effect of large mammals on African ecosystems, particularly through impacts on nitrogen cycling. However, from the dominance of microbial biomass compared to other faunal groups, it is evident that microbial activity exceeds that of every other faunal component: Microbial biomass is estimated to be 100 g m^{-2}, in contrast to that of invertebrates (0.1–1.0 g m^{-2}), small vertebrates (0.2–2 g m^{-2}), and large vertebrates (1–5 g m^{-2}) (Clark 1975).

Soils and Plant Distribution

Soils have a strong influence on the diversity, abundance, and distribution of plants. For example, the requirements of plants for, or their tolerance to, different micronutrients have been directly implicated in population and community differences at several levels and have been referred to as **edaphic specialization.** The concentrations of particular elements in soils and the ability of plants to take up these elements are important factors that define edaphic specialization. Many studies have shown that some plant species can accumulate high concentrations of metals in their tissues without ill effect (e.g., Al, Cu, Mo, Ni, and Se). These accumulators are referred to as **indicator plants.** They have been used in geochemical prospecting, in geological and soil mapping, and in groundwater surveys (Cannon 1971).

From species with small, localized populations to those with wide distributions, plants show many adaptations to soil conditions. Thus, plants provide excellent evidence of edaphic specialization at many levels. Some examples include:

- *Evolution in closely adjacent populations.* The selection pressures associated with soils containing high metal concentrations can produce genetically based physiological **ecotypes** (i.e., populations that have evolved to be tolerant of those metals). These often evolve near populations of the same species that do not have such tolerance. For example, the ecotypic differentiation in sweet vernalgrass (*Anthoxanthum odoratum*) and sheep's fescue grass (*Festuca ovina)* evolved in some areas in response to high Zn concentrations in soil (Antonovics, Bradshaw, and Turner 1977). A fern (*Pteris vitatta*) was recently shown to accumulate more than 14,000 parts per million (ppm) of arsenic in its leaf tissue without showing signs of toxicity (Ma et al. 2001). This provides an example of how plants might be used in the "phytoremediation," or cleanup, of contaminated sites.
- *Single-species–single-nutrient response.* Some species have a strong preference for specific soil nutrients. At the Rothamsted Experiment Station, England, areas with high K applications had very high dandelion infestations, whereas other plots were relatively free of dandelions. Subsequent research found that dandelion had much higher concentrations of K in these tissues than several grass species. As a practical application, it was suggested that the exclusion of K from

fertilizer could be an effective control of dandelions (Tilman et al. 1999).
- *Single-species populations occupying specific habitats.* California manzanita (*Arctostaphylos myrtifolia*) has been shown to grow in dense stands in the foothills of the Sierra Nevada Mountains in California, where parent materials are low in cations and fertility, have low pH values of 2–3.95, and contain a high concentration of soluble Al (Gankin and Major 1964). This specialization has been termed **edaphically controlled endemism.**
- *Metal accumulations in individual genera.* Cattails (*Typha* sp.) have been found to be accumulators of Hg, Se, and Al (Sundberg-Jones and Hassan 2007). In a wetland designed to treat wastewater, plant tissue concentrations of Hg, Se, and As were up to 1,900, 10,980, and 4,900 above background levels, respectively. Because of their ability to bioconcentrate metals, this species is sometimes used in bioremediation of metal-contaminated water and sediments.
- *Calcicole ("calcium-loving" species) and calcifuge ("calcium-hating" species) plants.* Much work has been done to determine the mechanisms by which plants are able to inhabit limestone areas where soil calcium levels are very high. Those plants that have the ability to take up high amounts of Ca (calcicoles) have a need for as well as a tolerance of the element. Studies suggest that these species have adaptations that help them to take up essential nutrients such as P and Fe, which can be lacking in such environments. In contrast, calcifuge species show strong toxic effects from high levels of Ca. Because they are unable to solubilize and take up P and Fe on calcareous soils, they suffer from nutrient deficiencies in these habitats and do not persist (Zohlen and Tyler 2000).
- *The effects of serpentine soils.* In many regions of the world, serpentine [$Mg_6Si_4O_{10}(OH_8)$] soils, derived from high-Mg ultrabasic rocks, have low pH

values and low Ca contents but high Ni, Mg, Cr, and Fe contents; they support specialized plant taxa, resulting in "edaphic endemics" (Kruckeberg 1984). So distinct is the flora of these unusual areas that it has been suggested that "serpentine vegetation is not a seral [intermediate successional] stage in the vegetation of other soils but a distinct climax of its own distinctive successions" (Whittaker 1954). Many of these plants attain extremely high concentrations of metals in their tissues.

• *Permafrost and soil moisture.* Plants that are **diploid** (contain two copies of each chromosome in their cells) and those that are **polyploid** (contain more than two copies of each chromosome) are distributed closely in relation to soil environmental gradients that result from glaciation. Research has shown that polyploid plants are common in the Arctic, and evidence suggests that they are more successful in colonizing soils as glaciers recede than are diploid species (Brochman et al. 2002). In this case, polyploidy is presumably an adaptation to prevent a rapid genetic response to temporary changes in the weather patterns. If a plant has eight copies of every gene, its rate of evolutionary change will be slow!

Soil Formation

Relative to human timescales, soil formation is a very slow process. Indeed, soils have been characterized as the least renewable of our natural resources. Building on the original Russian soil science concepts (particularly those of Dokuchaev), Jenny (1980) proposed an "environmental formula of soil forming factors" or the state factor approach. Specifically, the rate of soil formation $s = f(cl, o, r, p, t)$, where soil formation (s) is a function (f) of the interactions of climate (cl), organisms (o), relief or topographic features (r), parent materials (p), and time (t).

Soil varies in its horizontal and its vertical dimensions. The dynamic and interactive nature of the soil-forming processes results in general patterns that characterize soils throughout the world. In the vertical dimension, water percolates down through the soil, carrying with it soluble nutrients. In this process, clay is also removed from the upper horizons and accumulates in the deeper horizons. Two other major processes that contribute to soil formation are biocycling (plants act as nutrient pumps, continually bringing nutrients toward the surface from deeper soils' depths) and the churning of soil by animal activity (a process known as pedoturbation). Removals, additions, biocycling, and pedoturbation are universal in all ecosystems; their relative dominance depends on the geographic area. Under a given regional climate and the indigenous flora and fauna, these four processes result in soils that are characteristic of a region. Developed soils show distinct horizons (see, for example, a well-developed soil in a northern coniferous forest in Figure 4.14). We illustrate several examples of other soils in association with their dominant vegetation in Chapter 10.

Major Soils of the World

Over time, scientists have organized many attributes of soils not only to understand the genesis of soil systems, but also to decipher general patterns that could be used to better manage soils for agriculture and forestry. Thus, **soil classification** schemes have evolved and incorporated the prevailing understanding and knowledge of soil systems. Such classification systems exist in a number of countries; their form depends upon the tradition in which the science of soils developed. These include the United States, Russia, France, Canada, and a system in development at the United Nations Food and Agriculture Organization.

The U.S. system currently in use was introduced in 1960 and improved since then (USDA 1994). Previously, the U.S. taxonomic system was biased in favor of both Russian terminology and concepts of soil genesis. At the broadest

Figure 4.14 A spodosol developed under a coniferous forest in a northern boreal forest climate. At left is a soil profile showing organic horizon (O), mineral horizon A showing the leaching/eluviation (E), mineral horizon B showing deposition with red color from iron and aluminum oxides, unconsolidated mineral horizon C, and the parent material (R). Right is the minimonolith showing the morphology of soil horizons. (Photos courtesy of M. K. Wali.)

level of integration, the new system is divided into 12 soil orders. Worldwide, these orders have 55 suborders, 238 great groups, 1,243 subgroups, 7,504 families, and 18,807 series. A simplified key (Figure 4.15) shows the distinctions among the orders (Brady and Weil 1999).

The formation and development of different soil orders result from the general climatic and vegetation conditions in an area. Temperature, evapotranspiration, and the balance between precipitation and evapotranspiration are particularly important factors in soil genesis; when precipitation is greater than evapotranspiration, salts are leached from soils. Conversely, in regions where evapotranspiration is greater than precipitation, salts remain in the upper soil layer, and salt-tolerant plants dominate (Ahmad et al. 1983) (Figure 4.16). The cumulative effect of all of these processes acting on the indigenous rocks and minerals is a remarkably diverse distribution of soil types (Figure 4.17) (USDA 1997). As will be seen in more detail in Chapter 10, soils profoundly influence the distribution of major plant groups and the organisms that depend upon them for food and cover.

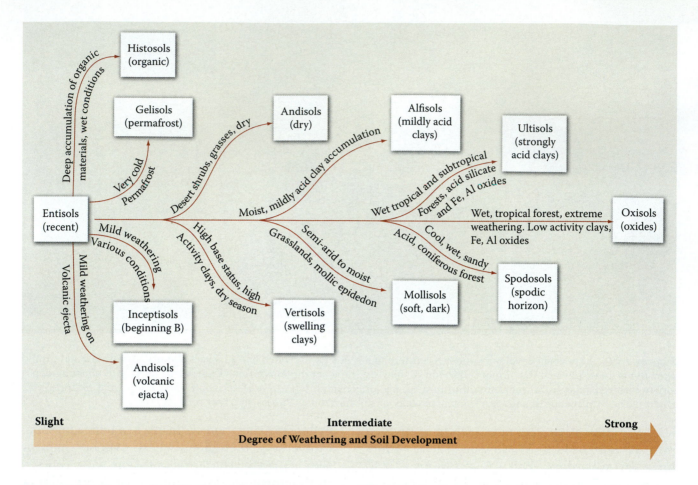

Figure 4.15 Degree of weathering and soil development in different soil orders under general climatic and vegetation conditions. (Brady, N. C., and R. F. Weil. 2008. *The Nature and properties of Soils*, 14th ed. Upper Saddle River, NJ: Prentice Hall.)

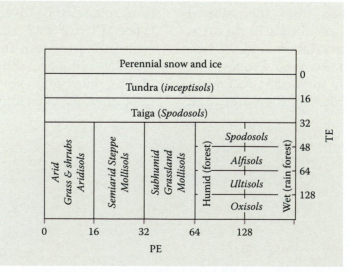

Figure 4.16 Distribution of soil orders in the major geographic regions of the world (Ruhe, R. V. 1975. *Catena* 2: 309–320). PE refers to precipitation—evaporation ratio—and TE to temperature—evaporation ratio.

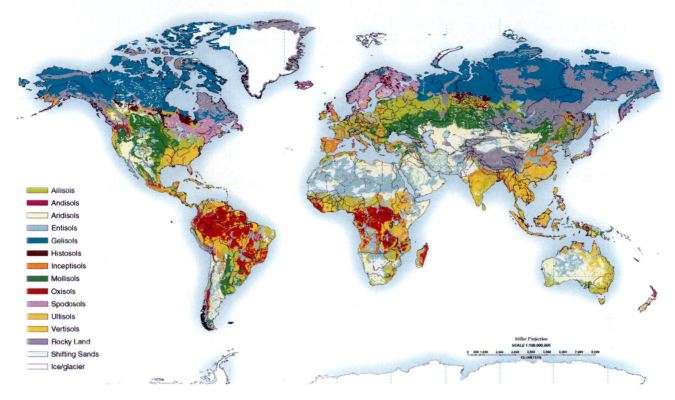

Figure 4.17   Global soil regions. (USDA-NRCS–Soil Science Division, World Soil Resources.)

Table 4.4
Some characteristics of the 12 Orders in the U.S. Classification of Soils[1]

Order	Salient Features	%Global Extent[2]
Alfisols	Soils with thin surface horizons and clay-rich subsoils; well-defined horizons and moderate native fertility; mainly formed under forest vegetation (e.g., in north-central U.S.A.)	9.6
Andisols	Soils formed from volcanic ash or other volcanic ejecta (e.g., in northwest U.S.A.)	0.7
Aridisols	Low organic matter soils that are dry for long periods, often containing calcium carbonate, gypsum, or more soluble salts (e.g., desert soils)	12.7
Entisols	Young, weakly developed soils with no horizon differentiation, dominated by mineral matter (e.g., river banks)	16.3
Gelisols	Soils with irregular or broken horizons, characterized by the presence of permafrost within 200 cm of soil surface (arctic and some alpine areas)	8.6
Histosols	Soils with 20-30% organic matter by weight in the upper 80 cm but no permafrost (e.g., bogs)	1.2
Inceptisols	Soils with horizons that are weakly developed; found in a wide variety of climates, landscapes and ecological settings (e.g., mountainous forested areas)	9.9
Mollisols	Soils with thick, dark surface horizons characterized by soft, crumb structure and high organic matter content; high native fertility (e.g., grasslands)	6.9
Oxisols	Highly weathered soils rich in iron and aluminum oxides (e.g., tropical and subtropical regions)	7.6
Spodosols	Acid soils with strongly leached (ash colored) surface horizons overlying reddish subsoils; usually sandy textured (e.g., northern conifer forests)	3.6
Ultisols	Soils of warm, humid climates with clay-rich subsoils; strongly leached with low native fertility (e.g., forest soils of southeastern U.S.A.)	8.5
Vertisols	Soils with high content of swelling clays that develop deep, wide cracks during dry periods (e.g., dry lake beds)	2.4

Sources: From [1]USDA-NRCS. 1999. *Soil Taxonomy, A Basic System of Soil Classification for Making and Interpreting Soil surveys.* 2nd ed. Washington, DC: Government Printing Office; [2]Brady and Weil (2008).

Appendix 4.1: Role of Chemical Elements
Some roles of elements essential for growth, function, and reproduction of plants, animals, and humans are described; elements beneficial to each group specifically are also shown.

Elements	Plants	Animals and Humans
Essential Macronutrients[a]		
Nitrogen (N)	Photosynthesis (chlorophyll)	Reproduction and protein synthesis (DNA, RNA)
	Reproduction and protein synthesis (DNA, RNA); enzymes, amines, structural proteins	Enzymes, amines, structural proteins
Phosphorus (P)	Energy metabolism and transfer (ATP, ADP); reproduction and protein synthesis (DNA, RNA); membrane structure (phospholipids)	Energy metabolism and transfer (ATP, ADP); reproduction and protein synthesis (DNA, RNA); membrane structure (phospholipids); bones and teeth (calcium phosphates)
Potassium (K)	Osmoregulation; enzyme cofactor	Osmoregulation, enzyme cofactor, neural transmission
Calcium	Cell wall structure (calcium pectate) and development; membrane stabilization	Bones and teeth (calcium phosphates)
Magnesium	Photosynthesis (chlorophyll); enzyme cofactor	Enzyme cofactor
Selenium	Not essential	Enzyme cofactor
Sodium	Beneficial[c] for enhanced osmoregulation in some species	Essential for neural transmission
Sulfur	Protein structure and activity of many enzymes and coenzymes	Protein structure and activity of many enzymes and coenzymes
Essential Micronutrients[b]		
Boron	Cell wall formation and cell elongation; tissue differentiation; membrane function	Low quantities (0.3 ppm) stimulated growth in chicks
Chlorine	Water splitting in photosynthesis; osmoregulation and stomatal function	Osmoregulation; Na–K–Cl transport
Cobalt	Metal component of vitamin B12 needed by N-fixing microorganisms (not known to be essential for higher plants)	Metal component of vitamin B12, essential for nonruminants
Copper	Electron transport in respiration and photosynthesis	Electron transport in respiration
Iron	Redox reactions in respiration and photosynthesis (cytochromes); chlorophyll synthesis	Functional component of hemoglobin; redox reactions in respiration
Manganese	Water splitting in photosynthesis; enzyme cofactor	Metalloenzyme cofactor
Molybdenum	Metal component of nitrogenase (dinitrogen fixation) and nitrate reductase (nitrate reduction) enzymes	Molybdoenzymes, xanthien oxidase; cofactor; unusual amino acid S-sulfocysteine depressed by deficiency of molybdoenzyme
Nickel	Metallic component of urease enzyme	Deficiency causes depressed growth, swollen joints in rats
Zinc	Protein and auxin synthesis; membrane integrity	Component of carbonic anhydrase
Beneficial Nutrients[c]		
Aluminum	Generally toxic, but enhances growth in some tolerant species and gives blue color to flowers and fruits	Not known

Appendix 4.1 (continued): Role of Chemical Elements
Some roles of elements essential for growth, function, and reproduction of plants, animals, and humans are described; elements beneficial to each group specifically are also shown.

Elements	Plants	Animals and Humans
Silicon	Increases dry matter yield, cell wall elasticity, and tolerance for aluminum, iron, and manganese, particularly in some grasses	Maximal bone formation; suggested metabolic and structural roles in connective tissue

[a] Essential nutrients needed in large quantities by all organisms.
[b] Essential nutrients needed by all organisms in small quantities.
[c] Nutrients that are beneficial in small quantities. Some elements have been found beneficial for specific plant and animal groups and humans. In the latter, for example, iodine prevents goiter (hence, iodized table salt), fluoride prevents dental caries (hence, fluoride added to toothpaste and drinking water). Several elements are toxic, and some of these are mentioned in Section B. (See McDowell 2003; O'Dell and Sunde 1997. Our grateful thanks to Don Eckert for his help with this table.)

Note

1. This chapter draws from "The Soil Ecosystem" by M. K. Wali and F. P. Miller in the *Encyclopedia of Environmental Analysis and Remediation* (see references).

Questions

1. Why do soils vary by region?
2. What is the role of soil microorganisms in soil development?
3. How do plants obtain nutrients from soils? Which nutrients are typically limiting for plant growth?
4. What causes the formation of soil horizons?
5. How is soil organic matter formed?

The Aquatic Environment

Characteristics of Water

Water is the elixir of life, and the Earth seemingly has plenty of it. Indeed, the surface area of the oceans (salt water) covers 71% of the Earth, but only 2.72% (by volume) of the Earth's total is freshwater. Most of the latter occurs as glaciers, polar ice caps, and deep groundwater, so the remaining quantity of freshwater in lakes, reservoirs, and river channels that is accessible for human use is less than 1% of the total freshwater on the Earth (Table 5.1). The study of freshwater ecosystems, including inland waters, lakes, reservoirs, rivers, streams and wetlands, is the science of **limnology** (derived from *limnos,* the Greek word for lake).

By their sheer size, area, and volume (Tables 5.2 and 5.3), water bodies (particularly the oceans) affect all climatic and weather systems of the Earth (Chapter 3). This water undergoes transport around the Earth and transforms from gas to liquid to ice and back again, making up the hydrological cycle discussed in Chapter 6.

Water is a universal solvent with unique chemical and physical properties (Table 5.4). It affects every facet of life, from rocks and soils to biota, but its distribution is uneven. Based on riverine drainage basin area and flow, Brazil possesses the largest renewable freshwater resources in the world, followed by the nations of the former Soviet Union collectively, and then China and Canada (Table 5.3). Australia and the many arid regions of the world have the least. The availability and apportionment of water (especially freshwater) already ranks as one of the top causes for disputes among regions and political units (we discuss this topic in Chapter 18). In this chapter, we provide a narrative on the structure of aquatic ecosystems and its bearing on the function of freshwater and salt water systems.

Hydrological Cycle

Precipitation, evaporation, transpiration, cloud formation, and subsequent runoff and percolation are all variables of the water cycle (the cycle is discussed in Chapter 6). The water cycle is the circulation of the Earth's fixed supply of water among land, freshwater

Table 5.1
Water Resources on the Earth

Water Reserves	Area (10³ km²)	Volume (10³ km³)	Percent of Total water	Percent of Freshwater
Ocean	361,300	1,338,000	96.5	—
Groundwater	134,800	23,400	1.7	—
Freshwater		10,530	0.76	30.1
Soil moisture		16.5	0.001	0.05
Glaciers and ice caps	16,227	24,064	1.74	68.7
Antarctic	13,980	21,600	1.56	61.7
Greenland	1802	2340	0.17	6.68
Arctic islands	226	83.5	0.006	0.24
Mountainous regions	224	40.6	0.003	0.12
Ground ice/permafrost	21,000	300	0.022	0.86
Water reserves in lakes	2058.7	176.4	0.013	—
Fresh	1236.5	91	0.007	0.26
Saline	822.3	85.4	0.006	—
Swamp water	2682.6	11.47	0.0008	0.03
River flows	148,800	2.12	0.0002	0.006
Biological water	510,000	1.12	0.0001	0.003
Atmospheric water	510,000	12.9	0.001	0.04
Total water reserves	510,000	1,385,984	100	—
Total freshwater reserves	148,800	35,029	2.53	100

Source: From Shiklomanov, I. A. 1993. In *Water in Crisis: A Guide to the World's Fresh Water Resources,* ed. P. H. Gleick. New York: Oxford University Press.

Table 5.2
Major Oceans and Seas of the World

Oceans and Seas	Surface Area (km²)
Pacific Ocean	165,721,000
Atlantic Ocean	81,660,000
Indian Ocean	73,445,000
Arctic Ocean	14,351,000
Mediterranean Sea	2,966,000
South China Sea	2,318,000
Bering Sea	2,274,000
Caribbean Sea	1,943,000
Gulf of Mexico	1,813,000
Okhotsk Sea	1,589,700

Source: From Craig, J. R., D. V. Vaughan, and B. J. Skinner. 1996. *Resources of the Earth,* 2nd ed. Upper Saddle River, NJ: Prentice Hall.

lakes and rivers, the seas and oceans, and the atmosphere. This cycle collects, purifies, and distributes water around the world. Water on the land and in the oceans is evaporated into atmospheric water vapor by solar energy. Masses of water vapor are moved around the Earth by winds and then condensed into water droplets that form clouds or fog. Through precipitation, water returns to the Earth in the form of dew, rain, hail, or snow.

Water that falls on the land is absorbed by the roots of plants, passed through their stems and other structures, and released from stomata in their leaves into the atmosphere as water vapor by transpiration. Through runoff, water moves from the land to water bodies and into the ground, where it is stored as groundwater. From there, it eventually returns to the surface, lakes, streams, or oceans. Human activities can adversely affect the quality and

Table 5.3
Ten Largest Total Annual Flows of Rivers of the World

River	Average Runoff (km³ yr⁻¹)	Area of Basin (10³ km²)	Length (km)	Continent
Amazon	6930	6915	6280	South America
Congo	1460	3820	4370	Africa
Ganges (with Brahmaputra)	1400	1730	3000	Asia
Yangtze	995	1800	5520	Asia
Orinoco	914	1000	2740	South America
Paraná	725	2970	4700	South America
Yenisei	610	2580	3490	Asia
Mississippi	580	3220	5985	North America
Lena	532	2490	4400	Asia
Mekong	510	810	4500	Asia

Source: From Shiklomanov, I. A. 1993. In *Water in Crisis: A Guide to the World's Fresh Water Resources,* ed. P. H. Gleick. New York: Oxford University Press.

quantity of water. Water quality is degraded by agricultural runoff or by overloading a water supply with wastes, such as from wastewater treatment plants or factories. Water quantity to different regions may be affected if global precipitation patterns are altered through climate change (see Section B).

Streams and Rivers

The flowing and running water ecosystems such as rivers and streams are referred to as **lotic** or **fluvial** (flowing) systems. They owe their origin to precipitation, springs, and melting of snow and ice. As precipitation comes down from the atmosphere, part of it is evaporated and transpired, part of it infiltrates through the soil, and the rest is the surface runoff that forms much of the water in streams and rivers (Figure 5.1). Five environmental conditions govern the amount and quality of runoff (Wetzel 2001):

- the nature of the soil (porosity and solubility);
- the degree of surface slope;

- the development stage and type of vegetation;
- local climatic conditions of temperature, wind, and humidity; and
- the volume and intensity of a precipitation event.

Upon close examination, even seemingly flat land shows relief and microrelief features that all contribute to drainage and the hydrological cycle. The ecological unit that includes a river system and the area of land that catches the precipitation that drains to that river are known as a **watershed** or **catchment area.** Obviously, the density of streams and their flow rates are strongly affected by precipitation, with humid and semiarid areas varying greatly in the distribution of surface flow (Figure 5.2). All streams and rivers, irrespective of their size, carry soil particles collectively called **sediments,** as well as dissolved chemical elements. The quantity of sediment loads depends partially upon how precipitation strikes the land surface and on flow velocity. Lands devoid of plant cover lose more sand, silt, clay, and associated chemical elements to erosion, with much of these materials entering the streams and rivers.

The amount of sediment load that can be carried by water is proportional to the fourth

Table 5.4
Physical and Chemical Properties and Aesthetic Quality of Water

Chemical structure. Water has a very simple atomic structure that consists of two hydrogen (H) atoms bound to one oxygen (O) atom. The hydrogen atoms are not on opposite sides of the oxygen atom, but are closer together so that their H:O bonds are at an angle of 105°. This atomic structure causes molecules of water to have unique electrochemical properties. The hydrogen side of the water molecule has a slight positive charge, and the oxygen side of the molecule has a negative charge. The resulting molecular polarity of charge causes liquid water to be a powerful solvent and to have strong surface tension. Water's strong surface tension makes it adhesive and elastic, and it tends to aggregate in drops rather than spread out over a surface as a thin film. Hence, water forms droplets and waves, allowing plants to move water (and dissolved nutrients) from their roots to their leaves and blood to move through tiny vessels in the bodies of many animals.

Superior solvent. More materials can be dissolved in water than in any other solvent. Water can permeate living cell membranes; dissolved materials also diffuse through the membranes or are "pumped" through using energy-requiring active transport.

Strong attractive force. The strong attractive force between water molecules allows them to be transported through the spaces in soil, into roots, and through the conducting tissue in plants and then to be transpired through leaves and needles back into the atmosphere.

State changes. In the hydrological cycle, the sun supplies the large amount of energy (540 cal g^{-1}) required to vaporize water from the liquid state to a gaseous state. When water vapor condenses to a liquid, the same amount of energy is given off, helping to distribute the sun's energy across the Earth. Similarly, freezing of water involves the loss of 60 cal g^{-1}, and melting requires an equal input of energy.

High specific heat. It takes substantial amounts of energy (1 cal g^{-1}) to raise the temperature of a body of water, and, conversely, it is slow to cool. In contrast, the average rock requires only 0.2 cal g^{-1} to raise its temperature 1°C. This property protects aquatic environments against rapid temperature changes. The different heating and cooling rates of bodies of water and adjacent land masses help to circulate air, create winds, and establish weather patterns.

Expansion at freezing. Water is the only common substance to expand when it freezes such that its density is 0.917 g cm^{-3} at 0°C. Ice floats! Other substances contract when they freeze.

Density at freezing. Pure water reaches its maximum density (1.0 g cm^{-3}) at 3.98°C (39°F), which is heavier than when it is at its freezing point (0°C). As water approaches freezing (<4°C) by losing heat to the atmosphere, it becomes lighter and floats on the surface. When it loses enough heat to freeze (60 cal g^{-1}), it is lighter yet and floats on the surface of the water as ice. Thus, streams, lakes, and ponds freeze from the top down, helping protect aquatic environments from freezing solid in the winter. Sea water, on the other hand, freezes before it reaches its greatest density at ~−2°C.

Oxygen- and nutrient-holding capacities. More oxygen dissolves in cold water than in warm water, but warm water generally holds more nutrients. The largest populations of aquatic organisms are found where nutrient-rich warm waters and oxygen-rich cold waters mix.

Flow. Water flows downhill over land toward streams, rivers, lakes, ponds, and oceans. Water also percolates through the ground, sometimes collecting in huge, slowly moving underground aquifers, some of which resurface as springs and seeps.

Aesthetic quality. Flowing water, shimmering water, cascading water, waves, small drops of dew collected on a spider's web, rainwater pattering on a roof or running down a windowpane, puddles, a single drop of water from a pond teeming with life—each affects the human psyche in ways science is unable to explain.

Sources: Modified from Kaufman, D. G., and C. M. Franz. 2000. *Biosphere 2000: Protecting Our Global Environment.* Dubuque, IA: Kendall/Hunt Publishing Company; Wetzel, R. G. 2001. *Limnology—Lake and River Ecosystems,* 3rd ed. New York: Academic Press.

power of the velocity with which water flows, which in turn is affected by the steepness of the terrain through which it passes. The largest rocks in a mountain stream are thus found associated with the stream's steepest regions. Conversely, the smaller rocks and sediments transported from the steeper slopes (often at the highest altitudes) upstream are deposited as the water movement slows in downstream regions that have shallower slopes. When received repeatedly in specific areas over a period of time, such depositions form **deltas** in rivers, lakes, or the ocean. Examples include the Mississippi River delta (Louisiana) and the Mackenzie River delta (Northwest Territories, Canada). In many river basins, water overflows the river's banks periodically, and sediments are deposited in what are known as

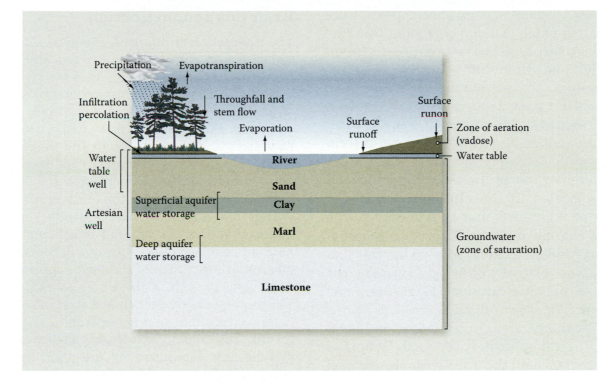

Figure 5.1 Interactions and vertical and horizontal transfers of water among the components of a riverine ecosystem. (U.S. Geological Survey.)

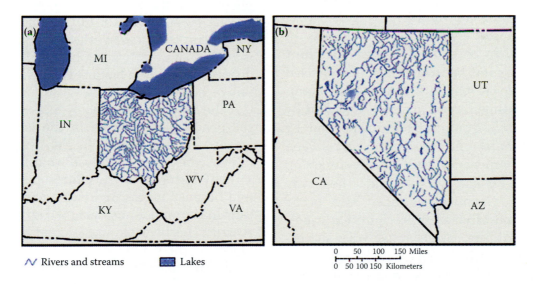

Figure 5.2 Two contrasting drainage network systems in (a) Ohio, a temperate climate (precipitation-to-evapotranspiration ratio ranging from 1.5:1 to 1:1), with a drainage density of 784 m km^{-2} of land, and (b) Nevada, a desert climate (precipitation-to-evapotranspiration ratio of 0.5:1), with a drainage density of 676 m km^{-2} of land. Ohio's rainfall on average exceeds the water evaporated by plants, whereas in Nevada precipitation replaces only one-half of that released by plants. Converting drainage density to U.S. units, Ohio has about 1 mile of stream for every square mile of land, whereas Nevada has only 0.86 mi/mi^2. Note the scale of the two maps: Ohio is drawn at one-half the scale of Nevada. (Drawing courtesy of Dan Button, U.S. Geological Survey.)

the **floodplains,** which have unique physical, chemical, and biological characteristics quite distinct from those of the stream or the adjacent upland.

Not only do the slopes of rivers vary from their headwaters to their junction with the sea, but their channel size (it becomes wider) and their discharge (volume of water increases) also vary. A small stream that appears from a spring or patch of damp earth and that has no tributaries is called a first-order stream (Cushing and Allan 2001). In this classification system, the stream is considered first order until it joins another stream to make a larger, second-order stream. It will, in turn, join with another second-order stream to form a third-order stream, and so on. First-order streams may be overarched by trees and shrubs located in the **riparian zone**—the border of vegetation that runs alongside the stream channel and is influenced by seasonal flooding and saturated soils. The riparian zone leaves relatively little light reaching the water, so the water can run cold (and dark). As a stream flows downhill and gains flow from a larger drainage area, it gets wider and more light penetrates because the trees' branches no longer reach all the way across the channel (Figure 5.3). Rivers are open systems, and thus the chemistry of river waters varies as stream order changes and the river flows through different watershed environments.

The **nutrient cycles** we will discuss in Chapter 6 have been described as **nutrient spirals** in streams because materials released to the water do not stay only in the water column above the sediments but are also adsorbed by benthic (bottom-dwelling) organisms and by particulate matter. As nutrients cycle between ecosystem components, they are also moving downstream—hence, they "spiral" their way downstream. Chemical changes also occur seasonally and after each precipitation event. A stream's salinity can decrease with successive dilution as tributaries enter the river as it moves downstream, or it can increase if evaporation occurs faster than new water is added.

Rivers are highly diverse, supporting a diversity of microbes, plants, and animal life that includes the insects, crustaceans, mussels, snails, fish, and amphibians that feed on materials falling in from the riparian zone—material that is produced in the stream itself—and on each other. Stream biota are highly specialized to live in flowing waters.

River Continuum Concept

The interaction of physical, chemical, and biological factors that occur in streams and rivers from the headwaters to high-order rivers at their mouths has been summarized by Vannote et al. (1980) (Figure 5.3). Their model, called the river continuum concept (RCC), suggests how stream biota respond to the predictable changes in physical–chemical conditions that occur as stream order increases. It suggests that the food sources for the biota in a stream switch from allochthonous materials—organic material such as leaves and twigs that originate outside the stream—to autochthonous production, where much of the food supply comes from algal growth in the water column itself. This represents the switch in a stream's energy balance from that of a heterotrophic condition to an autotrophic condition. Heterotrophic systems are dominated by the allochthonous energy sources in which the organic matter consumed by the bacteria, fungi, and animals in the stream exceeds that produced by photosynthesis within the stream (i.e., photosynthesis/respiration < 1.0). Under autotrophic conditions, the photosynthesis/respiration ratio is >1.0. As we discuss in Chapter 16, the consequences for stream biota of removing the riparian vegetation are severe.

The allochthonous plant material entering the headwaters is often in the form of whole leaves and twigs and is relatively indigestible because the cellulose that helps make up leaf tissue is an insoluble polysaccharide and because leaves and twigs usually contain tannins—compounds designed to prevent terrestrial herbivores (e.g., caterpillars) from digesting them. One group of organisms collectively known as "shredders" feeds on relatively large chunks or coarse particulate organic matter (CPOM; >1-mm pieces). The shredders break up the CPOM, gaining the easily digestible materials such as simple sugars and proteins, and egest the remainder of the material as relatively small chunks, or fine particulate organic matter (FPOM; <1-mm pieces).

The FPOM has a greater surface area/volume ratio because it is smaller and is more easily attacked by bacteria and fungi that have enzymes (cellulases) that can break down cellulose into simple sugars that the bacteria and fungi use as an energy source. When the FPOM is eaten again by filter feeders or collectors (organisms that collect small particles of organic matter as their food source), the bacteria and fungi that have grown on the FPOM are digested. The bacteria and fungi contain proteins, lipids, nucleic acids, etc. and thus are a more balanced meal than was the starchy cellulose. The collectors release the remaining undigested cellulosic material again as FPOM, and the process continues. The relative balance of these feeding groups changes as one moves downstream; interestingly, it is the predators that maintain a constant presence throughout the stream orders (Figure 5.3).

More recent comments on the generality of the RCC point to the fact that headwater streams in semiarid locales may not have trees to shade the stream or shed leaves into it and that the relative size of inflowing tributaries to the river may alter the "continuum" through inputs of silt that interfere with photosynthesis (Benda et al. 2004). Others have pointed out that dams and natural barriers to water flow alter the sequences, too. Nevertheless, the concept neatly emphasizes how the pattern of organisms found in a stream varies from the headwaters to its mouth, integrating flow, light penetration, depth, allochthonous inputs of organic matter, autochthonous production, the digestibility of these items as food, the methods required to capture or collect food items, the roles of bacteria and fungi, and their impact on important water-quality parameters such as oxygen concentration.

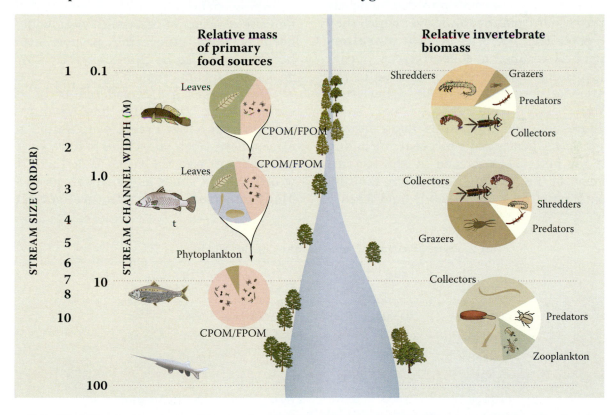

Figure 5.3 Variation in stream trophic structure with increasing stream order as summarized by the river continuum concept (RCC) of Vannote et al. (1980). The functional feeding groups (feeding guilds) found among stream macroinvertebrates switch from dominance by the shredders in the headwaters (organisms that break up large pieces of leaves and other organic matter, or CPOM, to dominance by collectors eating FPOM and grazers and rock scrapers, e.g., snails, eating autochthonous plant materials in the intermediate river sites). The lower reaches of a river are dominated by the collectors found feeding on FPOM and algae suspended in the water column. Insects and fish that prey on these organisms can be found throughout the stream continuum. (Vannote, R. L. et al. 1980. *Canadian Journal of Fisheries and Aquatic Science* 37: 130–137.)

Freshwater Lakes

Standing water bodies, also called **lentic** (calm) systems, are principally lakes and ponds. These vary considerably in size, from the prairie potholes of less than a hectare in size to the Caspian Sea, which covers 374,000 km², to the five Great Lakes of North America, which include Lake Superior, covering more than 82,000 km² (Table 5.5).

Most of the world's lakes have been formed through glacial activity (e.g., the Great Lakes of North America). The Earth has undergone several glaciations with intervening interglacial periods. Once glaciers retreat and melt, they leave behind a vast number of basins and depressions that accumulate water. Movement of glaciers transports vast quantities of debris that they scour from the lands over which they move, depositing them as **glacial moraines** (Figure 5.4). In marine coastal areas, glacial overdeepening of river valleys results in the formation of fjords that can be either fresh or saline.

Figure 5.4 also illustrates lakes formed through other geological processes, including one of the lake types formed by the tectonic movement of the Earth's crust. Here, faulting and shifting of rocks provides a land feature called a **graben** that accumulates water. A spectacularly large example is Lake Baikal in Siberia, Russia. Other lakes have been formed from the geological uplifting of mountains (e.g., the lakes in the Himalayan region). Lakes may also be formed by volcanic activity. In this case, the core of a volcano collapses down into the space vacated by its vented magma, creating large basins (**calderas**) (e.g., Crater Lake, Oregon) (Figure 5.4b), or flows of molten lava block existing valleys and back up water, much like a reservoir (e.g., Snag Lake, California).

In addition to those formed by these mighty geological agents, lakes are also formed by stream action, by the movement and deposition of materials by winds, by driftwood, and by animal activities (e.g., dams built by beavers [*Castor canadensis*]). However, human construction of artificial dams and levies, the changing of the courses of streams and rivers, and the creation of reservoirs worldwide have been of such staggering proportions that humans can *also* be characterized as mighty geological agents (Table 5.6).

Many lakes are considered to be closed systems that have water flowing into them but have no outlet except evaporation to the atmosphere. Such lakes tend to have very high salt content (e.g., the Great Salt Lake, Utah, which has six times as much salt as the sea). Many important physical features govern both

Table 5.5
Ten Largest Naturally Formed Lakes of the World

Lake	Area (km²)	Volume (km³)	Salinity	Continent
Caspian	374,000	78,200	Mostly marine	Asia
Superior	82,100	12,230	Fresh	North America
Victoria	68,460	2,700	Fresh	Africa
Aral	64,100	1,020	Mostly marine	Asia
Huron	59,500	3,537	Fresh	North America
Michigan	57,750	4,920	Fresh	North America
Tanganyika	32,900	18,900	Mostly fresh	Africa
Baikal	31,500	22,995	Fresh	Asia
Great Bear	31,326	3,381	Fresh	North America
Great Slave	28,568	2,088	Fresh	North America

Source: From Horne, A. J., and C. R. Goldman. 1994. *Limnology,* 2nd ed. New York: McGraw–Hill.

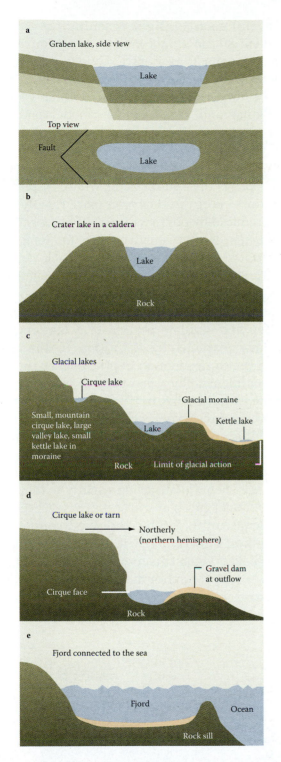

Figure 5.4 The origins and present situations of the most common types of lakes: (a) tectonic movements involving the sinking of a fault block; (b) caldera in an extinct volcano; (c) and (d) glacial processes scouring out basins, depositing outwash material in moraines, and "kettle" lakes formed from subsequent melting of blocks of ice buried in glacial outwash plains; (e) fjords formed by glacial deepening of coastal valleys that are connected to the ocean. (From Horne, A. J., and C. R. Goldman. 1994. *Limnology,* 2nd ed. New York: McGraw–Hill.)

Table 5.6
Ten Largest Reservoirs of the World Formed by Dams

Dam	Location	Volume (× 10³ m³)	Year Completed
Three Gorges	China	39,300,000	UCᵃ
Syncrude Tailings	Canada	540,000	UC
Chapetón	Argentina	296,200	UC
Patie	Argentina	238,180	UC
New Cornelia Tailings	United States	209,500	1973
Tarbela	Pakistan	121,720	1976
Kambaratinsk	Kyrgyzstan	112,200	UC
Fort Peck	Montana	96,049	1940
Lower Usuma	Nigeria	93,000	1990
Cipasang	Indonesia	90,000	UC

Source: http://www.infoplease.com.ipa/A0001334.html
[a] UC = under construction.

the structure and the functional processes of lakes. Called morphometric or geomorphologic features, these include geological origin, shape, area, volume, and the length of the shoreline.

Lake Stratification

Lakes and ponds also exhibit thermal and chemical stratification or zonation brought about by light (heat) and other environmental factors (e.g., salts) that in turn govern the growth and abundance of organisms. For example, light has a profound influence on photosynthetic activity. Light (both blue and red wavelengths) is used in photosynthesis; however, as light penetrates the deeper layers, it changes in its intensity and its quality. Clear waters have the deepest light penetration (Figure 5.5a), lakes with moderate nutrient concentrations have less (Figure 5.5b), and nutrient-rich waters have the least penetration (Figure 5.5c). Blue light is least absorbed by water and most easily scattered by fine particulate matter, so lakes with the clearest water appear blue.

The clearer the water is, the deeper the sunlight can penetrate. Suspended particles, algae, microbes, and chemicals that impart a color to water decrease the penetration of light, thus

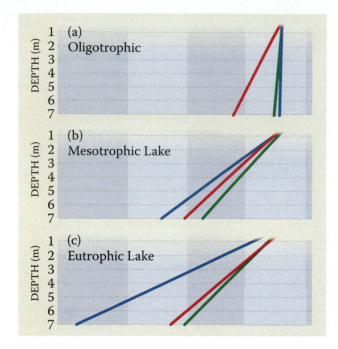

Figure 5.5 Light penetration and absorbance of different colors of light in (a) an oligotrophic (nutrient-poor) lake; (b) a mesotrophic lake; and (c) a eutrophic (nutrient-rich) lake where light penetrates only to shallow depths and is absorbed more quickly with depth. Oligotrophic lakes such as Crater Lake, Oregon, as well as the open ocean, show the light penetration pattern seen in panel (a). (From Dodds, W. K. 2002. *Freshwater Ecology: Concepts & Environmental Applications.* San Diego, CA: Academic Press.)

increasing its **turbidity.** The depth at which light intensity is just sufficient for the rate of photosynthesis to equal the rate of respiration is called the **compensation depth.** This corresponds to a light intensity of approximately

1% of full noon sunlight reaching the surface (or about 30 μmol photons m^{-2} s^{-1}) (Figure 5.6). The compensation depth divides the water column in two; above is the **photic** (light) zone and below is the **aphotic** (dark) zone.

Plants and algae in the photic zone produce a net photosynthetic gain during daytime; that is, their photosynthetic rate is higher than their rate of respiration. Below the compensation depth is the aphotic zone, where respiration consumes energy and oxygen more rapidly than it is produced by photosynthesis. Photosynthetic organisms cannot survive here for long, and essentially all food energy available in the aphotic zone has been manufactured somewhere in the photic zone. These zones are so important to the plant distribution that the **littoral zone,** the top and shallow portion of a lake nearest the land, extends by definition to the point where the compensation depth reaches the bottom sediments. The open body of water beyond this depth is the **limnetic zone.** The depth of these boundaries varies throughout the season, depending on the effective light penetration.

Incoming sunlight warms the upper waters of the lake most quickly because infrared radiation (i.e., heat) is absorbed rapidly with depth. As the upper layer of water warms, it becomes less dense until, finally, the wind cannot mix the lighter, warm water with the colder, heavier water below; when this happens, the lake is said to have undergone **thermal stratification.** The **epilimnion** represents the uppermost layer of

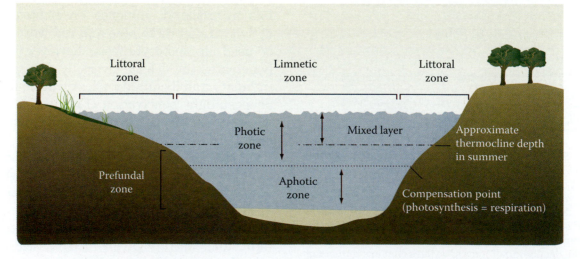

Figure 5.6 Lake zonation as a function of light penetration. (From Horne, A. J., and C. R. Goldman. 1994. *Limnology,* 2nd ed. New York: McGraw–Hill.)

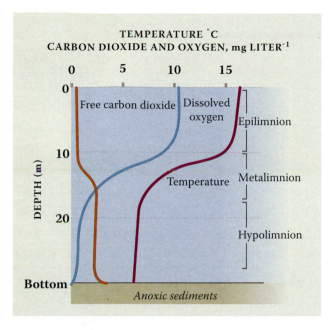

TEMPERATURE °C
CARBON DIOXIDE AND OXYGEN, mg LITER⁻¹

Figure 5.7 Variation in dissolved oxygen, carbon dioxide, and temperature with depth for midsummer in a typical productive (eutrophic) temperate zone lake that has become stratified. (From Horne, A. J., and C. R. Goldman. 1994. *Limnology,* 2nd ed. New York: McGraw–Hill.)

water in direct contact with the atmosphere. Below the epilimnion is the metalimnion, a layer characterized by a rapid decrease in temperature. The depth of steepest temperature decline is referred to as the thermocline. The deep, cold region of water at the bottom of the lake is the hypolimnion (Figure 5.7). Oxygen and carbon dioxide availability vary with depth as well; oxygen is high in the epilimnion, where algae and other primary producers release oxygen in the process of photosynthesis. The hypolimnion often becomes completely lacking in oxygen, or anoxic, in nutrient-rich lakes as microbes decompose the heavy load of organic matter produced in the epilimnion that has

fallen to the bottom, consuming the oxygen contained in the water. Carbon dioxide follows the opposite pattern as it is consumed in photosynthesis in the epilimnion and released in respiration near the bottom (Figure 5.7).

Generally, lakes in temperate regions (and in high mountains in subtropical regions) are directly stratified in summer and *inversely* stratified in winter; that is, in the winter, the warmest water (at 4°C) is at the bottom because water is densest near 4°C (Figure 5.8). These lakes can undergo a complete mixing of water twice a year, in autumn and in spring, and are thus called **dimictic** (Figure 5.8). Lakes in warm, oceanic climates and in low mountains of subtropical latitudes, where surface layers do not freeze, mix only once a year through the winter and are thus **monomictic.** Their water, which is never below 4°C at any depth, cools sufficiently to allow complete water circulation only during the coldest time of the year. Conversely, lakes in the polar and arctic regions may rarely rise above 4°C, and circulation occurs only in summer. In equatorial regions, where there is little change in seasonal temperature but extreme differences in daytime and nighttime temperatures occur, lake waters may circulate frequently at temperatures above 4°C; hence, they are **polymictic** (Hutchinson 1957a; Wetzel 2001).

Water Chemistry

Lakes vary widely in the composition of the chemicals dissolved in their water (Table 5.7). Lakes Superior and Baikal lie in drainage basins with relatively insoluble (granitic) rocks, whereas water in Lake Ontario has flowed from Lake Superior through the somewhat

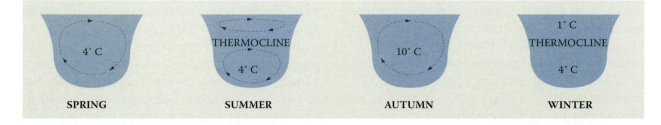

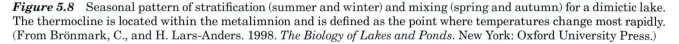

Figure 5.8   Seasonal pattern of stratification (summer and winter) and mixing (spring and autumn) for a dimictic lake. The thermocline is located within the metalimnion and is defined as the point where temperatures change most rapidly. (From Brönmark, C., and H. Lars-Anders. 1998. *The Biology of Lakes and Ponds.* New York: Oxford University Press.)

Table 5.7
Composition of Dissolved Salts[a] in Various Lakes from Dilute to Very Salty

Lake	Ca^{++}	Mg^{++}	Na^+	K^+	Cl^-	$SO_4^=$	HCO_3^-	$CO_3^=$	Total
Superior, Minnesota	13	3	3	1	1	2	28	0	60
Ontario, New York	32	7	6	1	8	12	62	0	135
Baikal, Russia	15	4	4	2	1	5	64	0	98
Fayetteville Green, NY	419	7	17	3	34	1,119	200	0	1,864
Pelican, Oregon	45	52	580	71	158	438	0	612	1,983
Redberry, Saskatchewan	72	1,586	1,548	110	142	9,093	0	333	12,898
The Sea	396	1,221	10,098	330	18,150	2,541	0	132	33,000
Great Salt, Utah	346	5,616	67,498	3378	112,896	13,593	0	183	203,490
Dead Sea	9876	30,781	25,176	5469	149,838	633	0	0	226,000

Sources: Data from multiple sources; the bottom five sets of data are from Hutchinson, G. E. 1957a. *A Treatise on Limnology*, vol. I. *Geography, Physics and Chemistry.* New York: John Wiley & Sons.
[a] Milligrams per liter.

more soluble limestone rocks in the Lake Erie drainage basin before reaching Lake Ontario (note the total dissolved salts shown in the last column of Table 5.7). Fayetteville Green Lake has glacial till rich in limestone cobbles in its drainage basin; the calcium bicarbonate input is so great that limestone (calcium carbonate) precipitates every summer in the lake, raining down in tiny white crystals.

Saline Lakes

Saline lakes such as the Great Salt Lake of Utah; Pelican Lake, Oregon; Redberry Lake, Saskatchewan; and the Dead Sea have a high pH and an abundant concentration of salts. However, they vary in their relative concentrations of carbonate, sulfate, and chloride ions (Table 5.7), thus reflecting the chemical composition of the rocks, sediments, and soils in their drainage basins and the relative solubility of potassium, sodium, magnesium, and calcium ions in combination with bicarbonate (HCO_3^-), carbonate (CO_3^{2-}), sulfate (SO_4^{2-}), and chloride and the pH of their waters. Pelican and Redberry Lakes have waters high in carbonate and sulfate, respectively. The remaining saline bodies of water have such high concentrations

of salts that the most soluble ions, sodium, potassium, and chloride dominate. Note that the last two lakes are saltier than the sea.

Salt ponds with a thin layer of freshwater on top (solar ponds) capture heat like a greenhouse and can allow the salty bottom layer temperature to be as high as 80°C. Recent research seeks to exploit this heat source to generate electricity or to produce freshwater from saltwater by reverse osmosis.

The biological characteristics of saline lakes and ponds can differ greatly from freshwater lakes, in large part because few fish can survive in waters much saltier than the sea, but many algae and higher plants can. As a result, saline lakes and ponds often support extremely high densities of insects and large crustacean organisms such as fairy shrimp (*Artemia salina*) that would be eaten by fish elsewhere. In Utah's Great Salt Lake, fairy shrimp are so abundant that their resting cysts form large rafts floating on the surface; they are harvested commercially, canned, and sold all over the world as fish food. With appropriate exposure to aerated saltwater, the planktonic larvae of *Artemia* will hatch out of the cysts, providing living prey for small fish. One problem with managing solar ponds (see earlier discussion) is that algae accumulate at the freshwater–saltwater interface, blocking light penetration into the saline layer. It would seem that stocking fairy shrimp would be helpful here!

Lake Biology

Lakes also show biological stratification comparable to their physical and chemical zonation, resulting in distinct subhabitats (Figure 5.9). The littoral zone, or the shallow zone around the edge of the lake, is generally occupied by floating plants, rooted plants (emergent, floating-leaved, and submerged forms), and attached algae (Cronk and Fennessy 2001) (Figure 5.10). The open water or limnetic zone supports planktonic plants (**phytoplankton**) and animals (**zooplankton**); these are small organisms, often not visible to the naked eye, whose movements are more or less dependent on water currents, although both groups have taxa that can swim. Organisms, like the fishes and amphibians, that are strong swimmers and are able to navigate through all regions of the habitat are referred to as nekton. The lake bottom–water interface forms the benthic zone, and organisms growing here are collectively called benthic organisms or the benthos. Included are the periphyton—the organisms, both plant and animal, that attach to or cling to stems, leaves, rocks, or other surfaces projecting above the bottom. This is why things taken off the bottom of a lake or pond often feel slippery. Other benthic animals include a diversity of taxa, including insects, clams, crayfish, worms, etc. The dissolved oxygen distribution in the lake (Figure 5.9) is influenced by biological activity (e.g., respiration) and by temperature and stratification. This greatly influences the distribution of the biota.

The biological growth that a lake can support depends not only on light, temperature, and oxygen concentration but also on the concentration of required nutrients. Essential nutrients include nitrogen (N), phosphorus (P), iron (Fe) and, to a lesser extent, calcium (Ca), carbon (C), sulfur (S), and silica (Si). N and P are typically the nutrients that, due to their low supplies, most limit the growth of primary producers in freshwater systems. A number of micronutrients can also be limiting, such as cobalt (Co) and molybdenum (Mo). Nutrient-poor systems (i.e., those low in N and

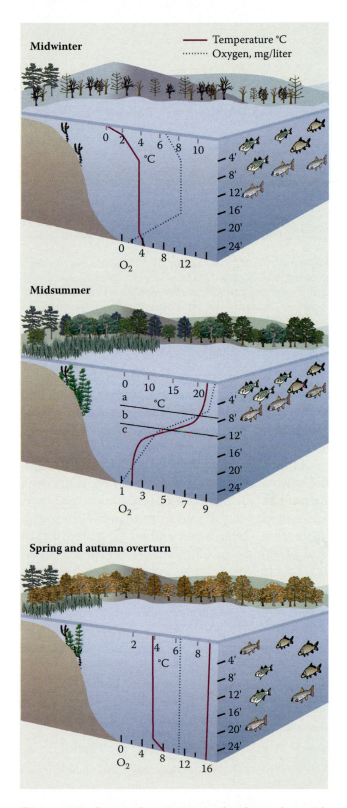

Figure 5.9 Seasonal variation in the thermal stratification and oxygen availability regulate the distribution of fishes in a temperate zone lake. Stratification in midsummer can lead to low oxygen levels in the hypolimnion, thus restricting fish to the upper, well-oxygenated epilimnion. (Modified from Smith, R. L. 1990. *Ecology and Field Biology*, 4th ed. New York: Harper Collins.)

Figure 5.10 A cross-section of a littoral zone of a lake showing the typical plant zonation. Emergent plants like *Typha* (cattails) and *Phragmites* (reed) grow close to shore, floating leaved plants such as water lilies (*Nuphar*) grow in deeper water, and where the water depth is too great, submersed and free-floating species dominate. (From Brönmark and Hansson 1999.)

P) that do not support abundant biological growth are referred to as **oligotrophic** lakes (Table 5.8). Productive aquatic systems, called **eutrophic,** are considered "well nourished" (i.e., with high N and P concentrations).

The susceptibility of a lake to eutrophication (becoming nutrient enriched and thus eutrophic) is a function of basin morphometry, watershed geology, and human activities in the watershed. Given a watershed area identical to that of the deep, oligotrophic lakes discussed earlier, shallow lake systems tend to be richer in organic matter and nutrients and therefore support more luxuriant plant growth simply because the nutrient input is discharged into a smaller volume of water. Therefore, the depth and volume of lakes tend to be inversely related to their nutrient status (e.g., oligotrophic lakes tend to be 20–120 m deep; mesotrophic lakes, 5–40 m deep; and eutrophic lakes 3–20 m deep). In the natural course of time, deep lakes fill in with silt, clay, and sand from the drainage basin and organic matter produced in the lake, so the volumetric loading of nutrients (grams per cubic meter per year) becomes higher with time. This so-called natural eutrophication can be accelerated by disturbance, particularly human activities such as agriculture or urban or suburban development in the watershed. Known as cultural eutrophication, this increases the rate of filling due to the discharge of nutrients into the lake.

Whether oligotrophic, mesotrophic, or eutrophic, the photosynthesis in the photic zone of a lake is dominated by microscopic phytoplankton algae in the open water of the lake. These vary from filamentous forms a few millimeters long to spherical algae only 2 μm (2×10^{-6} m) in diameter. Because P is often in short supply, its availability (in the form of the orthophosphate ion, PO_4^{3-}) greatly influences the growth of algae and the types of algae dominating the algal community (Table 5.8).

Table 5.8
Characteristics of Oligotrophic and Eutrophic Lakes

Oligotrophic Lakes	Eutrophic Lakes
Low nutrient content (especially P and N)	High nutrient content (especially P and N)
Deep, steeply banked	Shallow with broad littoral zone
Highly transparent	Limited transparency
Water blue or green	Water green to yellow or brownish-green
Oxygen abundant through all levels at all times	Oxygen depleted in hypolimnion in summer
Not much phytoplankton; green algae and diatoms dominant	Abundant phytoplankton; cyanobacteria dominant
Abundant aerobic decomposers in profundal zone	Limited diversity in profundal zone; anaerobic species favored
Low biomass in profundal zone	High biomass in profundal zone
Fish community dominated by salmon, trout, and other deep-water fishes intolerant of low dissolved oxygen	No fishes in hypolimnion that cannot tolerate low oxygen

Source: Modified from Cole, G. A. 1994. *Textbook of Limnology,* 4th ed. Prospect Heights, IL: Waveland Press.

Wetlands

Wetland ecosystems (e.g., bogs, fens, swamps, and marshes) are areas where shallow water covers the land or saturates the soils for at least a portion of the growing season (Table 5.9). Estimates of the global extent of wetlands vary, depending on the methods used to formulate the estimate, ranging from 5.3 to 12.8 million km², or approximately 4–6% of the Earth's land surface. Wetlands are found in every climate, from the tropics to the boreal tundra (Figure 5.11). More than 90% of wetlands in the United States are freshwater inland wetlands, with coastal systems making up the remainder. Russia and South America have the largest coverage (1.51×10^6 km²), followed by Canada, Alaska, the United States mainland, Africa, and Southeast Asia (1.27, 0.69, 0.42, 0.36, and 0.24×10^6 km², respectively).

Numerous definitions of what constitutes a wetland have been advanced. The reason for this definitional diversity is that both scientists and regulatory agencies need a workable definition upon which decisions on management can be based. Keddy (2000) describes wetlands as ecosystems that "arise when inundation by water produces soils dominated by anaerobic processes and forces the biota, particularly rooted plants, to exhibit adaptations to tolerate flooding." The U.S. Army Corps of Engineers has adopted a technical definition designed to allow wetlands to be identified on the ground. For this purpose, they are described as shallow systems (less than 2 m) that are saturated by surface- or groundwater at a frequency and duration sufficient to support a prevalence of vegetation typically adapted for life in saturated soil conditions (Mitsch and Gosselink 2007). Common diagnostic features under this definition are the presence of hydric soils (those that have formed under the influence of anaerobic conditions) and hydrophytic vegetation (plants adapted to life in saturated soils).

Wetlands include a diversity of ecosystem types—for example, marshes, bogs, swamps, tidal flats, coastal mangroves, and the very shallow depressions (prairie potholes) found in grassland regions (Figures 5.12 and 5.13). Wetlands of the north temperate regions include marshes, bogs, and fens. Marshes are wetlands that are generally inundated and dominated by emergent (nonwoody) vegetation. Swamps (sometimes called deepwater swamps) are dominated by trees and shrubs with little else in the understory. Bogs are peat-accumulating wetlands that are strongly acidic (low to very low pH)—low-nutrient ecosystems that

Table 5.9
Ten Largest Wetlands of the World

Wetland	Area (× 10³ km²)	Description	Continent
West Siberian lowlands	780–1,000	Peat bogs, boggy forests, meadows	Eurasia
Amazon River	>800	Large and small river floodplain	South America
Hudson Bay lowlands	>200	Peatlands	North America
Pantanal	140–200	Marsh, swamp, floodplain	South America
Upper Nile swamps (the Sudd)	>90	Swamps, floodplain	Africa
Chari-Logone	90	Seasonal floodplain	Africa
Papua–New Guinea	69	Swamp, bog	Eurasia
Zaire–Congo system	40–80	Riverine swamps, floodplain	Africa
Upper Mackenzie River	60	Marsh, fen, floodplain	North America
Chilean fjordlands	55	Estuarine, swamp, floodplain	South America

Source: Keddy 2000.

Note: Although wetlands have been widely drained for agriculture and filled for construction of homes, factories, and roads, many regions still have huge areas of wetlands.

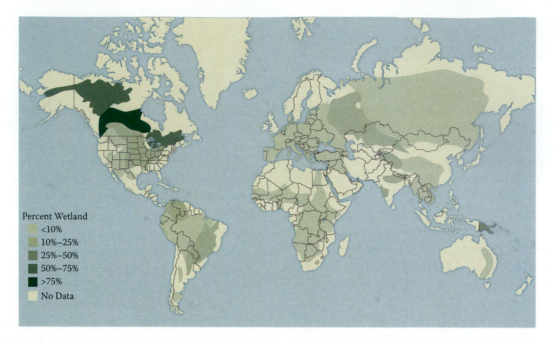

Figure 5.11 Map showing global wetland distribution by percent of land area. (From World Resources Institute. 2000. World Resources Report. *People and Ecosystems: The Fraying Web of Life.* Washington, D.C.: the World Bank, the United Nations Environment Program, and the World Resources Institute.)

receive their water and nutrients from precipitation. They are often dominated by growth of the acidophilic moss *Sphagnum* spp., which can float on the surface of the water at the edge of a lake forming a "quaking bog" that one can walk on (with care!). *Sphagnum* even increases the acidity of the water by excreting hydrogen ions. As the moss grows, the lower layers of the mat die, forming peat and eventually even filling in the lake. Fens also accumulate peat; however, they are alkaline (pH 7–8) systems rich in calcium and magnesium and receive most of these ions from the surrounding area—for example, from glacial till containing limestone.

Oceans, Seas, and Estuaries

These ecosystems are represented by marine, coastal, and estuarine areas. Marine resources support profitable business industries (e.g., commercial fishing) in a number of regions in the world, including the United States (Pacific and Atlantic coasts), Britain, India, Norway, Japan, Australia, New Zealand, Russia, South

Africa, Canada, and a number of other countries, particularly in Scandinavia.

A variety of animal and plant resources is harvested from saltwater systems, including seaweeds (kelps), which are estimated to yield 3–20 million tons annually. Agar (made from species of the genera *Gracilari* and *Gelidium*) is used as a therapeutic agent. Carrageenin (*Chondrus*) is used as a gelling agent for desserts and as a stabilizer in a number of products: chocolate milk, cheese, ice cream, salad dressings, and fruit syrups. Algin (from various seaweeds) is used as nonflammable camouflage material. Industrial chemicals come from species of *Pterosiphonia* and *Polysiphonia*. Fertilizers and carbohydrate products come from *Sargassum* and *Laminaria*. Animal resources are primarily for food, including shrimp, lobsters, clams, oysters, squid, fish, seals, and even whales.

Seas and oceans, like deep freshwater lakes, show both horizontal and vertical zonation in light, temperature, as well as other factors (Figure 5.14). The coastal zone extends from the high-tide mark on land through the intertidal zone (between the low- and high-tide marks) to the sloping, relatively shallow, submerged plain at the edge of the continent

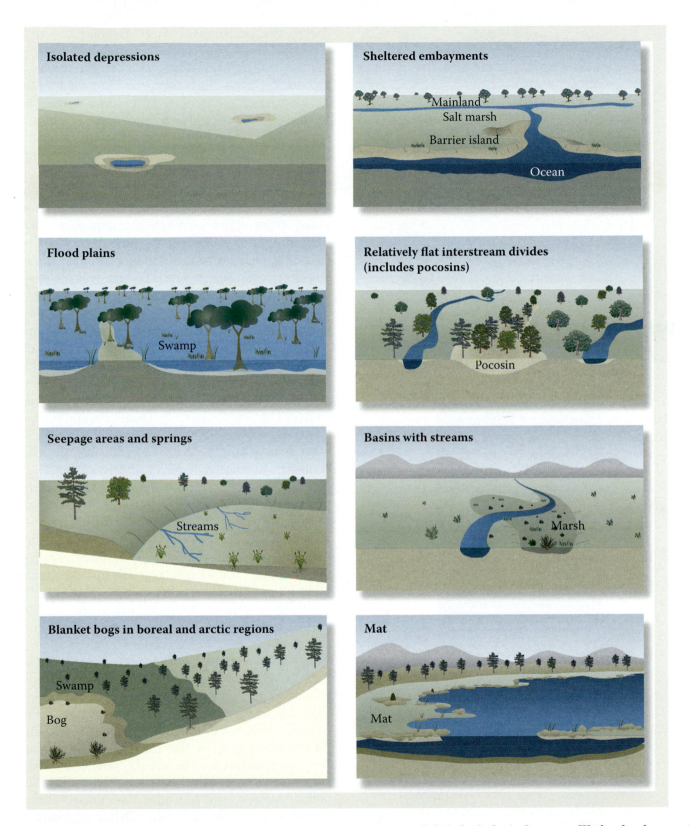

Isolated depressions

Sheltered embayments

Mainland

Salt marsh

Barrier island

Ocean

Flood plains

Swamp

Relatively flat interstream divides (includes pocosins)

Pocosin

Seepage areas and springs

Streams

Basins with streams

Marsh

Blanket bogs in boreal and arctic regions

Swamp

Bog

Mat

Mat

Figure 5.12 Comparison of different wetland types showing some of their hydrologic features. Wetlands always involve saturated or pooled water for part of the growing season. Here, saturation of water in the soils associated with prairie potholes, estuaries, flood plains, springs, river meanders (oxbows), thin soils over rock substrate, and the growth of a *Sphagnum* mat in an acid lake has resulted in a diversity of wetland communities. Pocosins are acidic swamps with evergreen trees such as pines. (From Tiner, R. W. 1996. In *National Water Summary on Wetland Resources*, chap. 2, water supply paper 2425. Denver, CO: U.S. Geological Survey.)

Figure 5.13 Examples of different wetland types: (a) a freshwater marsh near the Lake Erie shore; (b) a tidal marsh along the southern coast of England showing exposed mudflats at low tide; (c) a swamp forest in Ohio; and (d) a coastal mangrove forest in northeastern Australia. (Photo by S. Fennessy.)

called the continental shelf. Beyond the continental shelf, the bottom drops off steeply as the continental slope. It extends through the bathyl zone that may exhibit geologically active trenches, canyons, and ridges subject to underwater erosion and avalanches. The continental slope then levels off somewhat to a continental rise before dropping to the deeper but relatively level abyssal plain.

Life in oceans extends to all depths. Like the deep freshwater lakes, the compensation point from light penetration separates the euphotic zone (producing region) from what in this system is a vastly thicker aphotic zone. The forms of life are plankton(ic), free-floating phytoplankton, and zooplankton; the nektonic forms such as the squids, fishes, and whales that have a good power of locomotion; and the benthic or bottom-dwelling organisms such as worms and clams. In the bathyl and abyssal zones, all energy must come from outside the zone; that is, it depends on the organic matter that rains down from the euphotic zone ("marine snow"). The oxygen that these organisms use to break down organic matter also comes from photosynthesis at the surface but reaches the deeper zones by major currents in the oceanic circulation. Bottom decomposers are numerous and include more than just worms. Echinoderms such as brittle stars, for example, may be found as deep as 1,000–2,000 m; some of their sea cucumber relatives (Echinodermata: Holothuria) have been collected at 10,000 m!

In the dimly lit (twilight) and dark regions of the continental slope and rise are found many living marine organisms that can produce light by bioluminescence (e.g., dinoflagellate algae, many bacteria and fungi). Other organisms, including fish, produce light by harboring these organisms in their bodies. Bioluminescence requires the expenditure of energy (in the form of ATP) via an enzyme-catalyzed chemical reaction that generates very little heat, generates light of various colors, and is under the control of the nervous system in animals. Other organisms produce light through phosphorescence, which is light produced by some chemicals that have stored energy from previous illumination.

All the oceans are hydrologically connected, and the chief barriers between them

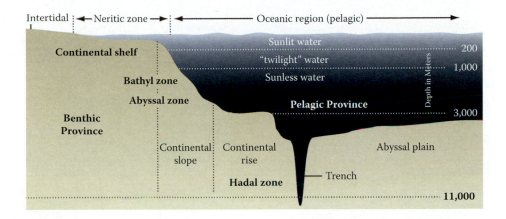

Figure 5.14 Major zones in an ocean. (After Enger and Bradley 1992.)

Figure 5.15 Estuarine section of the Connecticut River. Note the transition from freshwater marsh to brackish water marsh to salt marsh, which depends upon the successive mixture of salty marine water from Long Island Sound with the river water under the influence of river flow, wind-driven surges, the tides, and the greater density of salt water than freshwater. (Google Earth, digitalglobe/terrametrics 2007.)

Figure 5.16 A wood sandpiper (*Calidurus mauri*) foraging among the emergent plants in an estuarine salt marsh along the coast of Alaska. The sandpiper is hunting insects, small crustaceans, and other estuarine organisms that are more exposed at low tide. (Photo by Vernon Byrd, courtesy of USDI–Fish and Wildlife Service.)

total, followed by magnesium sulfate, calcium bicarbonate and carbonate, and potassium salts (Table 5.7). The essential nutrients N, P, and Fe can be very limiting.

Relatively warm, nutrient-rich coastal ecosystems are important ecologically and economically. As rivers flow into oceans and seas, they form highly productive, distinct, transitional zones (ecotones) called estuaries, where freshwater and seawater mix. The confluence of the two waters produces a different mix of temperatures, suspended particulate matter, chemical ions, and thus density. The organisms inhabiting this area must be able to cope with conditions in a constant state of flux, which favors the growth of unique assemblages of organisms (Figures 5.15 and 5.16). Similar organisms can be found at the ocean's edge in sandy and rocky habitats (Figures 5.17 and 5.18).

are not only continents but also density differences caused by differences in temperature and salinity. An average value for salinity (salt content in weight per volume) is about 3.5% (35 g L^{-1}), whereas the salinity of freshwater is less than 0.05% (0.5 g L^{-1}). Sodium chloride salts dominate at about 27% of the

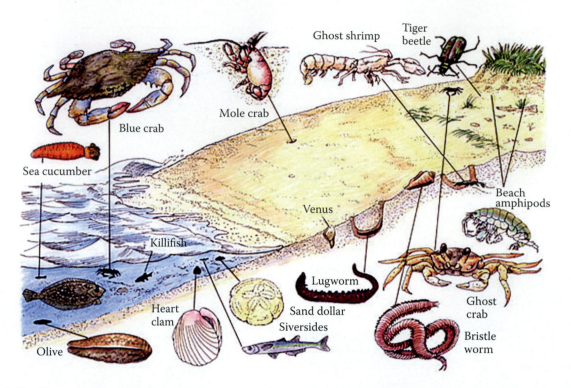

Figure 5.17 Characteristic organisms of the intertidal zone of a marine sandy shore. These organisms obtain their food by filtering phytoplankton from the ocean or by processing organic matter filtered out from waves breaking over the sand, or they are predators. (From Enger, E., and B. F. Smith. 1995. *Environmental Science—A Study of Interrelationships.* Dubuque, IA: W. C. Brown.)

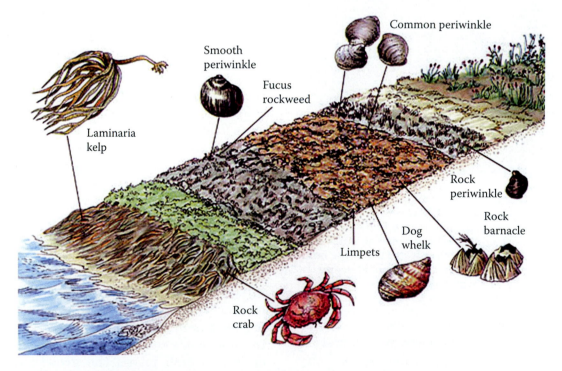

Figure 5.18 Characteristic organisms of the intertidal zone of a rocky shore at low tide. The kelps and rockweeds are photosynthetic brown algae and, with phytoplankton from the water, provide the primary energy sources for the system. The barnacles remove fine particles from the water, whereas the other animals illustrated are all predators (Enger and Smith 1997). (From Enger, E., and B. F. Smith. 1995. *Environmental Science—A Study of Interrelationships.* Dubuque, IA: W. C. Brown.)

Questions

1. What are the different sources of water and how are they replenished?
2. What determines water quantity and quality in a given region?
3. What processes drive lake stratification?
4. What are the functions of wetlands? Why are these important?
5. How can a water resource be depleted?

Biogeochemical Cycling of Materials

Biogeochemical Cycles

Biogeochemical cycles and energy flows establish not only the links among ecosystem components (biosphere, atmosphere, lithosphere, and hydrosphere) but also the link among ecosystems at local, regional, and global scales. Unlike energy that flows in one direction only, in nature, materials (including water) are cycled and reused within and among ecosystems over and over again. But together, the energy flow (energetics) and material cycling are the two fundamental and complementary models of ecosystems (Reiners 1986).

Within the biosphere, biogeochemical cycles describe how an element moves through the biotic and abiotic portion of an ecosystem. In doing so, both the transformation and transport of elements occur. The transport of elements can occur by the movement of wind, water, or animals. In some cases, cycling involves little movement—only a conversion from one form of a chemical to another (e.g., the uptake of nitrate by a plant and its conversion to an amino acid) or by a chemical reaction (e.g., the oxidation of methane in a lake to carbon dioxide and water by a bacterium). These relationships are often summarized in compartmental models, or "box and arrow" diagrams. These consist of compartments (or pools) representing the mass of a given element in a particular chemical form and location. Compartments are connected to one another with links reflecting the movement of elements as they are transported or transformed between them.

Because the cycles of materials involve biological, geological, and chemical components of an ecosystem, they are called **biogeochemical cycles.** In many cases, the cycles of elements are also referred to as nutrient cycles; however, all elements (including toxic elements) cycle in nature. Appropriately, **nutrient cycles** include only those macro- and microelements that are essential for biotic growth (see Chapters 4 and 5).

Three groups of biogeochemical cycles are recognized: the hydrological (or water) cycle; gaseous cycles, which have gas, liquid, and solid phases (e.g., those of carbon, nitrogen, oxygen, and sulfur); and sedimentary or mineral cycles (e.g., iron, calcium, magnesium, and phosphorus), where relatively insoluble elements are taken up from the soil or water by plants, passed on to the herbivores and carnivores, and, through excretion, death, and decay, are returned to the soil or water. These have no atmospheric component.

Drivers of Biogeochemical Cycles

The flux of solar energy is the engine that powers the hydrologic cycle, drives the Earth's climate patterns, and energizes the biogeochemical cycles. These cycles are crucial to the functioning and maintenance of the ecological processes that are requisite for the sustenance of life. As the name implies, the fluxes of biogeochemical cycles are determined by several factors: climatic variables—particularly temperature and precipitation, the nature of the element (for example, its solubility or the different forms it can take), and types and chemical compositions of soils, plants, and animals. Nutrients can be lost from and carried into ecosystems by wind, precipitation, flowing water, and migrating animals, as well as by various forms of pollution, management practices, and land uses.

All these biotic and abiotic factors (including human activities) affect four ecological processes that drive biogeochemical cycles: (a) the removal of elements from a compartment—for example, by leaching, emissions, harvesting, and wind and water erosion; (b) the addition of elements to a compartment through processes such as the chemical weathering of rocks, biological fixation, and atmospheric deposition; (c) the transfer of elements between pools— for example, the uptake of nutrients; and (d) transformation—for example, chemical reactions in which the chemical form of the element changes (e.g., atmospheric nitrogen, N_2, to ammonia, NH_3). Thus, the sizes of each ecosystem pool, the fluxes between the pools, and the residence time of materials in the pools vary widely. **Ecosystem pools** or "stocks" refer to those phases of the cycle where a material is held in varying quantities in an ecosystem compartment such as in the atmosphere,

water, soil, animals, or vegetation. The materials in some reservoirs (e.g., phosphate in rock) may not be available for use by humans or other organisms without physical, chemical, or biological modification.

Biogeochemical cycles are often thought of as being in equilibrium or "steady state"; that is, the inflows and outflows to a given compartment are balanced. However, if the reservoir size in an ecosystem component decreases (a greater magnitude of outflows compared to inflows, such as in vegetation, soils, geological formations, or oceans), then it has become a source for materials. If the reservoir size increases (greater inflows than outflows), then it is referred to as a sink for materials; that is, **sequestration** occurs. **Fluxes,** flows, or transfer rates refer to how much material passes through a reservoir over a given period. The **residence time** of a material in a particular pool is the average length of time a molecule of that material stays in that pool before it moves on to another pool (Table 6.1). Residence time is meaningful for a reservoir that is in steady state (i.e., the total amount of material in the pool stays constant) and is expressed as

Residence time = amount in reservoir/outflow

(or inflow) at steady state

In contrast, **turnover time** describes how long it takes for all the material in a pool to be replaced (i.e., how frequently the pool "turns over"). Carbon in the atmosphere has a turnover time of 5 years (i.e., it takes, on average, 5 years for all the C atoms in the atmosphere to be replaced), while the average C in the deep ocean organic pool turns over every 5,000 years. Turnover time is calculated as the reciprocal of residence time and can be expressed as

Turnover time = 1/residence time

The amounts of the various elements on the Earth and the different compartments in which these elements may be found (e.g., there is 1,000 times as much C as N on the Earth, but 5 million times as much N as C in the atmosphere) (Table 6.1) indicate how the flux

Table 6.1
Quantity and Residence Time of Major Elements

Element	Quantity	Turnover Time	Element	Quantity	Turnover Time
Carbon (10^{15} g C)			Nitrogen (10^{12} g N) (continued)		
Atmosphere	750	5 years	Marine biomass	470	?
Terrestrial biomass	550–680	50 years	Soils	300,000	2,000 years
Marine biomass	2	0.1–1 years	Sediments	$50,000 \times 10^4$	10^7 years
Soils	1,500	<10–10^5 years	Ocean (dissolved N_2)	$22,000 \times 10^3$	1,000 years
Sediments, rocks	$77,000 \times 10^3$	>>10^6 years	Ocean (inorganic)	600,000	
Surface ocean	1,000	Decades	Sulfur (10^{12} g S)		
Surface sediments	150	0.1–1,000 years	Atmosphere	4.8	8–25 days
Deep ocean (inorganic)	38,000	2,000 years	Soils	300,000	1,000 years
Deep ocean (organic)	700	5,000 years	Lithosphere	$20,000 \times 10^6$	10^8 years
Oxygen (10^{15} mol O_2)			Sediments	$30,000 \times 10^4$	10^6 years
Atmosphere	37,000	3×10^6 years	Marine biota	30	1 year
Long-lived biota	180	1,000 years	Ocean	$30,000 \times 10^5$	10^6 years
Biota	11	50 years	Lakes	300	3 years
Sedimentary rocks		10^6 years	Phosphorus (10^{12} g P)		
Surface ocean	6	22 days	Atmosphere		0.028 days
Ocean	219	500 years	Terrestrial biota	3,000	~50 years
Nitrogen (10^{12} g N)			Land	200,000	2,000 years
Atmosphere (N_2)	$40,000 \times 10^5$	10^7 years	Sediments	$40,000 \times 10^5$	2×10^8 years
Atmosphere (N_2O)	14,000	100 years	Surface ocean	2,700	2.6 years
Terrestrial biomass	13,000	50 years	Deep ocean	87,000	1,500 years

Source: From Reeburgh, W. S. 1997. *Bulletin of the Ecology Society of America* 78 (4):260–267.

of each element through each pool varies widely, as does its turnover time.

Hydrological or Water Cycle

For each material cycle, the amount in each pool, the residence time in a particular pool, and the rate of transfer (flux) from one reservoir to another vary dramatically with the chemical character of the material. Characteristics such as its solubility, volatility, and reactivity all have a bearing on elemental transfers. For example, water, which occurs in all three states (gaseous, liquid, and solid) within a relatively narrow and common temperature range,

can be cycled in large volumes rather quickly (Table 6.2; Figure 6.1). Calcium and silicon, on the other hand, are chemical elements that are commonly tied up in sediments and rocks for long periods. Thus, it may take many years, even millennia, for them to be solubilized or physically moved (by erosive water, wind, glacial action, or volcanic eruption) from one pool or location to another.

Carbon, nitrogen, and oxygen are major elemental building blocks of the biosphere and constitute major components of the atmosphere. They can cycle quickly through the biosphere, atmosphere, and hydrosphere as well as become lodged in sediments as part of the lithosphere. Finally, the turnover times of some pools (e.g., phosphate ions in lakes) can be *extremely* short (on the order of minutes or less) because organisms

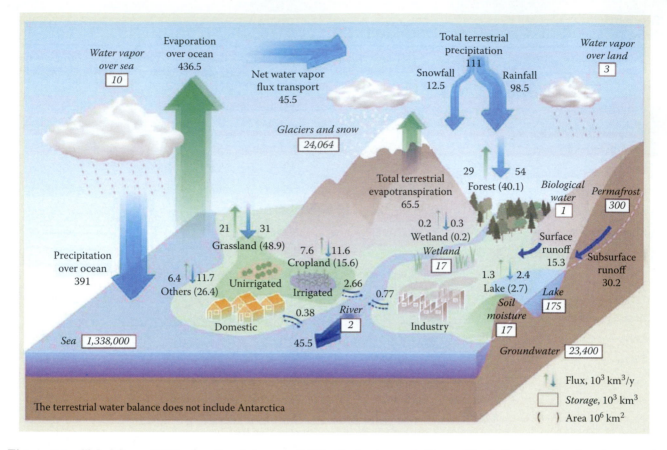

Figure 6.1 Global fluxes (10^3 km^3 yr^{-1}) and storages (10^3 km^3) of water are influenced by both natural and anthropogenic forces. The big vertical arrows show total annual precipitation and evaporation over land and ocean (10^3 km^3 yr^{-1}), which include annual precipitation and evapotranspiration in major landscapes (10^3 km^3 yr^{-1}) presented by small vertical arrows; parentheses indicate area (10^6 km^2). The direct groundwater discharge, which is estimated to be about 10% of total river discharge globally, is included in the river discharge. (From Oki, and S. Kanae. 2006. *Science* 313: 1068–1072.)

invest considerable energy in actively taking up and chemically transforming these compounds. In contrast, the turnover time of phosphorus in sea sediments (often as the calcium salt, apatite) is estimated as 200 million years.

Water makes up most of life because it is the major compound in living biomass. Its cycle moves immense quantities of water. The hydrologic cycle (Figure 6.1) has a profound influence on the cycle of many other compounds because it carries lots of other compounds along with it (discussed later). It also provides a good illustration of the complexity of global cycles: It is a compound that has three phases (gas, liquid, and solid), has vastly varying pool sizes and turnover rates, is influenced by global climate, and is extensively altered by human activity. Note that in Figure 6.1, the sizes of pools (referred to as "storage") are shown in boxes, the area covered by individual pools on land is shown in parentheses, and the annual

fluxes are shown as numbers associated with each arrow.

Water is continuously circulated through the processes of evaporation, condensation, precipitation, runoff, infiltration, transpiration, snowmelt, and groundwater flow. The cycle is driven by solar energy, where the main "pump" of water into the cycle is the evaporation of ocean water. The oceans hold 97% of the world's water. Although the oceans are the main source of water to the atmosphere (supplying about 86% of total water evaporated each year), the turnover time of the water held in the oceans is on the order of 37,000 years (Table 6.2). Other sources of water to the atmosphere include evaporation from freshwaters and transpiration from vegetation (combined, these two processes are referred to as evapotranspiration); these pools cycle much more rapidly. Precipitation on land is the driving force for transporting chemicals

Table 6.2
Turnover Times of Water Resources on the Earth

Hydrosphere	Turnover Time
Oceans	37,000 years
Polar ice, glacier	16,000 years
Ground ice of the permafrost zone	10,000 years
Mountain glaciers	1,600 years
Groundwater (actively exchanged)	300 years
Saline lakes	10–10,000 years
Freshwater lakes	1–100 years
Bogs	5 years
Soil moisture	280 days
Rivers	12–20 days
Channel network	16 days
Atmospheric water vapor	9 days
Biological water	Several hours

Source: From Reeburgh, W. S. 1997. *Bulletin of the Ecology Society of America* 78 (4):260–267.

and sediments, weathering rocks (ultimately changing the shape of the landscape), and filling freshwater ecosystems such as lakes, rivers, and wetlands. Given long enough, all this water makes its way downhill to rejoin the oceans (Smil 2007).

From an ecological standpoint, it is important to note that the global distribution of water is uneven, resulting in a surplus of water in some regions (precipitation > evapotranspiration) and a deficit in others (precipitation < evapotranspiration). This leaves a substantial proportion of the Earth's human population without an adequate supply of water, a topic that will be discussed in Chapter 16.

Gaseous Cycles

Oxygen (O_2) Cycle

Oxygen, carbon, sulfur, and nitrogen are four additional elements whose biogeochemical cycles involve gaseous phases. Oxygen is the second most abundant element of the atmosphere (about 21% by volume) (Table 6.1), but all of the oxygen in the atmosphere (or the hydrosphere) has come from the splitting of water in the process of photosynthesis either on land or in the oceans, lakes, streams, wetlands, and rivers. The Earth's original atmosphere consisted of only the gases released by volcanoes (primarily H_2O, N_2, NH_3, H_2, H_2S, SO_2, CO_2, and CO); a group of photosynthetic bacteria known as the Cyanobacteria are responsible for generating atmospheric O_2 beginning 3.5 billion years ago, making the evolution of life as we know it possible. They are sometimes referred to as the "architects of the Earth's atmosphere." Generally, the oxygen produced through photosynthesis is taken up by both autotrophs and heterotrophs during cellular respiration.

Photosynthesis only occurs in the presence of light of proper intensity and quality, but plants continue to respire 24 hours per day. Among the most reactive of all the elements, oxygen is also removed from the atmosphere by chemical reactions with reduced abiotic or biotic materials in a process known as **oxidation.** Through oxidation, fossil fuels such as coal, oil, and natural gas can power many of our machines. The recycling of living and dead organic materials occurs by microbial decomposition and fires; both are forms of oxidation—the first controlled and the second uncontrolled. The carbon and oxygen cycles are thus intimately coupled through the processes of photosynthesis and respiration. Schlesinger (1997) accordingly modeled the global biogeochemistry of oxygen with pools (in boxes) of free oxygen in the atmosphere balanced by reduced organic and inorganic carbon compounds in the land surface (and oceans) and in the crust, with the processes (in circles) of primary production (photosynthesis) contributing to the accumulation of surface organic matter and atmospheric oxygen (Figure 6.2).

Respiration and decay do the opposite (Figure 6.2). Much lower rates of burial and weathering of organic matter and crustal pools round out the model. Schlesinger points out that small variations in the balance between net primary production and respiration/decay will result in the accumulation of organic matter on land or in the oceans, lakes, and wetlands,

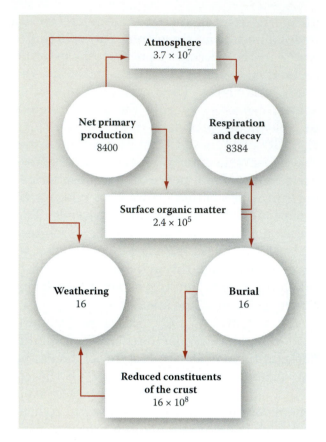

Figure 6.2   A model of the global oxygen cycle; units are in 10^{12} mol yr^{-1} of O_2 or the equivalent amount of reduced compounds. (Schlesinger, W. H. 1997. *Biogeochemistry—An Analysis of Global Change*, 2nd ed. San Diego, CA: Academic Press.)

and accumulation of oxygen in the atmosphere. Burning of fossil fuels will do the opposite.

Carbon (C) Cycle

The relatively small amount of carbon in the biosphere belies its central importance in both biogeochemistry and building living tissues. Like oxygen, carbon has gaseous forms that reside in the atmosphere. It is driven by four central biochemical processes involving coupled oxidation and reduction (or redox) reactions: (1) photosynthesis, (2) respiration, (3) aerobic oxidation (C compounds are oxidized when oxygen is present), and (4) anaerobic oxidation (C compounds are oxidized during respiration when oxygen is absent). In carrying out these redox reactions, organisms also transform other key elements such as N, P, and S. The biotic processes that dominate the carbon cycle include photosynthesis and respiration

by organisms, decomposition and mineralization of organic matter by microbes, and human activities such as deforestation and burning of fossil fuels (Figure 6.3).

Carbon dioxide (CO_2) is present in the atmosphere (0.038%) and dissolved in water, where it can also occur in the forms of bicarbonate (HCO_3^-) and carbonate ($CO_3^=$) ions; the latter constitute an important abiotic component of the carbon cycle (Table 6.1) because carbonates can precipitate with calcium and magnesium to produce limestone and dolomite and because carbon dioxide is removed from the atmosphere and the water by photosynthesis in plants and stored as plant organic matter (**biomass**).

Two pathways dominate the C cycle: the annual flux of carbon between the atmosphere and the biota and the CO_2 flux between the atmosphere and its dissolution in the oceans. Carbon dioxide in the atmosphere is the only source of carbon needed for photosynthesis; it is used directly by terrestrial plants and, once it is dissolved in water, used by aquatic plants (Smil 2007). The carbon derived from photosynthesis provides the energy for all other organisms to acquire the resources they need for growth and development. Carbon is returned as carbon dioxide to the atmosphere by respiration of the living terrestrial and aquatic biota. The transfer of carbon from organisms (plant, animal, and microbes) to the soil occurs through shedding (e.g., litter fall), mortality, and secretion of soluble organic compounds by roots into the soil (root exudation).

These carbon transfers from plants to soil eventually give rise to soil organic matter (SOM). The C in organic matter is returned by **decomposition** (breakdown of coarser biotic residues into finer ones) and **mineralization** (conversion of organic materials into inorganic ones) of organic matter. Plants release carbon through several pathways in addition to respiration. There is the much slower process of subduction of carbonate rocks and organic matter under the continental plates and release of the carbon through volcanoes as carbon dioxide, carbon monoxide, and methane (not shown in Figure 6.3).

The largest inorganic carbon stocks in the biosphere are sedimentary carbonates—especially calcium carbonate (limestone) and

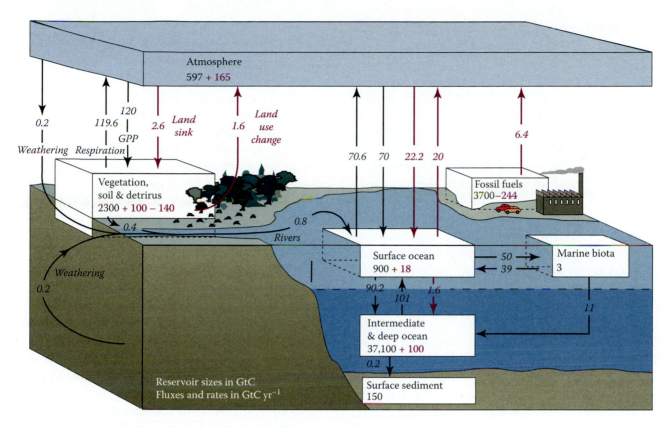

Figure 6.3   The global carbon cycle shows the carbon reservoirs in Gt (gigatonne = 1 billion t) C and fluxes in Gt C yr⁻¹. The indicated figures are annual averages over the period 1980–1989. The component cycles are simplified and the figures present average values. The riverine flux, particularly the anthropogenic portion, is currently very poorly quantified and is not shown here. Evidence is accumulating that many of the fluxes vary significantly from year to year. In contrast to the static view conveyed in figures like this one, the carbon system is dynamic and coupled to the climate system on seasonal, interannual, and decadal timescales. (From Denman, K. L., G. Brasseur, A. Chidthaisong et al. 2007, p. 515.)

magnesium carbonates (dolomite) formed from the shells of small (often microscopic) marine organisms that grew in ancient seas—and kerogen, the remains of buried biomass held in calcareous and oil shales, followed by marine dissolved organic carbon, soils, and surface sediments (Table 6.1). The long-term burial and storage of organic carbon in deep-sea sediments effectively removes it from the C cycle. Over geologic time, this led to the formation of fossil carbon (aka fossil fuels) and aided in the accumulation of oxygen, the by-product of photosynthesis, in the atmosphere. The latter came about because the carbon fixed in photosynthesis was not released through respiration, which consumes oxygen. Natural weathering of uplifted marine shales and human-induced combustion of fossil fuels turn these major long-term sinks for carbon into sources for atmospheric carbon dioxide. The atmospheric carbon reservoir is presently increasing quickly

due to human activities that are accelerating the release of carbon dioxide, a very important topic discussed in Chapter 19.

Carbon sequestration in an ecosystem occurs when the sum of carbon fixed through photosynthesis, including that stored by the formation of soil organic matter (humus) and calcium carbonates in aquatic systems, exceeds the sum of carbon losses by plant, animal, and microbial respiration; soil erosion; and combustion of biomass in wildfires or of fossil fuels. Carbon sequestration in vegetation and soils is reversible and can be altered by natural and human-induced disturbances of ecosystems. The carbon reservoir in trees is obviously much greater than that in grasses and crops. The size of the carbon reservoir in soil depends on the balance between the annual additions of plant, animal, and microbial biomass to the soil and the annual losses by microbial soil respiration and soil erosion. Soils are a major reservoir of organic matter in the carbon cycle. The soils of the world contain more carbon than the total amount occurring in vegetation and the atmosphere combined.

Just like in plants, the amount of carbon stored in soils shows no change when an ecosystem reaches a steady state (i.e., litter inputs into the soil and microbial soil respiration are in balance). The size of carbon reservoirs in ecosystems varies in response to climatic, geological, soil, and biotic conditions. For example, in wetlands, where soil respiration rates are slow, the amount of soil organic matter accumulation is large. However, in semiarid and arid areas, where soil respiration is rapid and the biological productivity of vegetation is low, the inputs of biotic residues into the soils are small, resulting in a small reservoir of soil organic matter. In general, organic carbon in the soil is concentrated in the top 1 m of the soil profile; on a global basis, this stores approximately twice as much carbon as the atmosphere, amounting to approximately 1,500 Pg (petagrams) C. Much of this pool is held in the soils of the tundra and boreal forest ecosystems. These ecosystems are among the most vulnerable to global climate change, and increasing temperatures are predicted to lead to increased rates of their soil respiration, causing the release of soil C to the atmosphere as CO_2.

Currently, the world's oceans store about 40,000 Gt of inorganic and organic carbon and serve as both a sink and a source for atmospheric CO_2. Carbon dioxide dissolves readily in water through the physical exchange across the air–water interface. The oceans can serve as a source or a sink of carbon depending on whether the surface seawater concentrations are above or below saturation with respect to CO_2 solubility. As atmospheric CO_2 increases the equilibrium, CO_2 concentrations in ocean water increase accordingly. The main route by which CO_2 enters the deep oceans takes place at high latitudes where surface water cools before sinking. Because the solubility of CO_2 increases as temperatures decrease, these cooler waters absorb atmospheric CO_2. Inorganic carbon is partitioned among three main chemical species in the oceans: dissolved CO_2, bicarbonate ions (HCO_3^-), and carbonate

ions (CO_3^{2-}). About 98% of ocean inorganic carbon is present as bicarbonate ions; carbonate ion is present as 1% and CO_2 as 1%. This cycling mechanism is known as the "solubility pump." Conversely, deep waters upwelling to the surface in tropical regions release CO_2 to the atmosphere.

The transfer of carbon dioxide from the surface waters of the oceans to the deep water is brought about by photosynthesis. Through the process of the oceans' **biological pumps,** about a fifth of the atmospheric carbon fixed by algae in the euphotic zone (the upper layer of ocean water where light is available for photosynthesis) is lost from the surface layer to the deeper ocean through the sedimentation of dead algal cells. Thus, the total concentration of carbon dioxide decreases in the surface ocean and the atmosphere and increases in the deep ocean. Over longer timescales (e.g., centuries), a large fraction of organic carbon accumulated in the deep ocean is mineralized by bacteria into dissolved inorganic form and recirculated to the upper zones again; the remainder is buried in the sediments. The equilibration time of the transfer of water between the surface and deep oceans takes several thousand years, much longer than the life span of humans. Sequestration of anthropogenic CO_2 by the deliberate fertilization of the ocean to stimulate algal growth can only work if the solubility and the biological pumps are increased.

Methane (CH_4) Cycle

Methane is produced by specialized anaerobic bacteria that are found in natural wetlands and rice paddies, in the anoxic rumen of cattle, and in the gut of termites and other wood-consuming insects. These bacteria use carbon dioxide in respiration, reducing it to methane. Accordingly, the anthropogenic sources for emissions of methane to the atmosphere include burning of fossil fuels and biomass, animal husbandry, rice agriculture, and waste disposal and landfills. Wetlands, both natural (contributing 115 Tg CH_4 yr^{-1}) and agricultural (i.e., rice paddies, which contribute 100 Tg yr^{-1}), are the biggest contributors of methane to the atmosphere, followed by

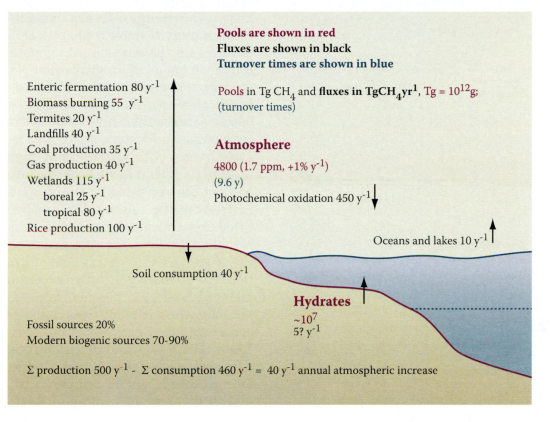

Figure 6.4 The global methane cycle, showing pool sizes in units of teragrams (Tg = 10^{12} g), fluxes in Tg y^{-1}, and turnover times (in parentheses) in years. (Reeburgh, W. S. *Bulletin of the Ecology Society of America* 78(4): 260–267.)

methane production by cows (known as enteric fermentation; 80 Tg yr^{-1}) (Figure 6.4). The latter occurs because the anaerobic rumen, one compartment of a cow's stomach, is filled with methane-producing bacteria. Hence, anthropogenic activities have dramatically increased the load of methane to the atmosphere. There are three major methane reservoirs: natural gas (a fossil fuel), the atmosphere, and a compound known as methane hydrates (or clathrates) (Figure 6.4). Methane hydrates are a crystalline combination of natural gas and water frozen into a substance that looks like ice but burns if ignited. It is found under the ocean floor and in permafrost.

Although methane concentrations in the atmosphere are quite low (1.8 ppm compared to 380 ppm for CO_2), they are increasing more quickly than any other greenhouse gas— on the order of a 1% increase per year. The methane molecule is a long-lived, radiatively active atmospheric trace gas that traps about 25 times more heat than a molecule of CO_2. The major atmospheric sink for methane is photochemical oxidation by hydroxyl radical (OH) in the troposphere. In contrast, atmospheric methane uptake by well-drained soils through the oxidation of atmospheric CH_4 constitutes a relatively small sink.

Nitrogen (N) Cycle

Nitrogen is extremely important in the biology of all organisms because it is a component of all amino acids and nucleic acids (DNA, RNA). Although biota require N in relatively small quantities, it is a scarce commodity, so the availability of N (along with P) is the most common factor that limits the growth of both autotrophic and heterotrophic organisms. The ultimate source of nitrogen in ecosystems is the atmospheric, elementary nitrogen gas (N_2, or dinitrogen gas). This makes up the largest component of the air—a full 78% by volume—and is the largest global pool of nitrogen (3.9×10^{21} g; Table 6.1), on the order of a million times the amount found in terrestrial biomass and 100 million times that found in marine biomass.

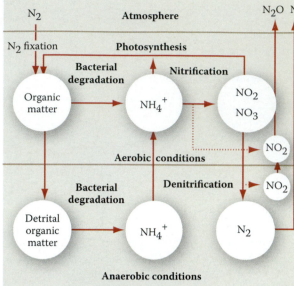

Figure 6.5 Microbiological transformations in the nitrogen cycle. (From Wollast, R. 1981. In *Some Perspectives of the Major Biogeochemical Cycles*, ed. G. E. Likens. New York: John Wiley.)

Though extremely plentiful, the nonreactive gaseous form of nitrogen in the atmosphere is not available for use by organisms. The conversion of N_2 into a form that is usable by organisms is very difficult due to the extremely high stability of nitrogen gas (this molecular nitrogen consists of two nitrogen atoms joined by a triple bond). The conversion of gaseous nitrogen to more widely bioavailable forms (ammonia, NH_3, or nitrate, NO_3^-) is called **nitrogen fixation** and can occur by biotic or abiotic processes (Figure 6.5).

The N cycle is unique among the nutrient cycles in the extent to which it is driven by microbes that transform N from one form to another. Nitrogen transformations in the biosphere take place in three main components: the atmosphere, habitats where aerobic conditions predominate, and habitats where anaerobic conditions predominate. Most of the N transformations shown in Figure 6.5 occur in both aquatic and terrestrial environments (Figure 6.6). Furthermore, the fact that many of these compounds occur as gases that are highly soluble in water is extremely important to their cycling and availability to higher plants, algae, bacteria, and fungi.

Only certain types of bacteria (heterotrophic and photosynthetic types) and fungi,

known as **nitrogen fixers,** have the capacity to use energy to drive a complex series of reactions (using the enzyme nitrogenase) to break the triple bonds of atmospheric N_2 and reduce the nitrogen chemically to NH_4^+ (ammonium), which they use for their own growth. These bacteria are either free-living forms in soils, sediments, and waters or they live in symbiotic associations with the roots of higher plants (Figure 6.7). That is, no higher plant can fix N_2 without the assistance of these organisms, and their symbiotic relationship involves sophisticated anatomical, physiological, and biochemical adaptations in the root nodules to support the bacteria that carry out nitrogen fixation. An estimated 1 kg N is fixed per square meter of land area per year around the globe, although the distribution is not spread evenly across different ecosystem types (Schlesinger 1997). Nitrogen fixation also occurs during lightning strikes, when momentary conditions of high pressure and temperature enable N_2 and O_2 to combine, forming nitrate (NO_3^-).

Industrial nitrogen fixation has had an extraordinary impact on the global N cycle, primarily through the processes of fertilizer production in which chemical manufacturers chemically convert N_2 into ammonium nitrate fertilizer. The effects that the extra N circulating around the globe is having on air and water quality and global climate are discussed in Chapters 18 and 19.

In aquatic systems, bacteria fix nitrogen in the sediments, and many taxa of photosynthetic bacteria, the Cyanobacteria, will also fix nitrogen if other N sources, such as ammonia or nitrate, are too low. Algal growth in the marine environment is often limited by low ammonia and nitrate concentrations, so nitrogen fixation is an important process there. Recently, it has been shown that the Cyanobacteria genus *Trichodesmium* may be responsible for contributing half of the fixed nitrogen used for growth by photosynthetic organisms in the sea (Figure 6.8). This genus is so abundant that huge rafts of it can be seen from space.

The most significant source of fixed nitrogen for terrestrial plants is from nitrogen-fixing bacteria (*Rhizobium* spp.) that inhabit nodules in roots of plants known as legumes (such as beans, peas, and alfalfa)

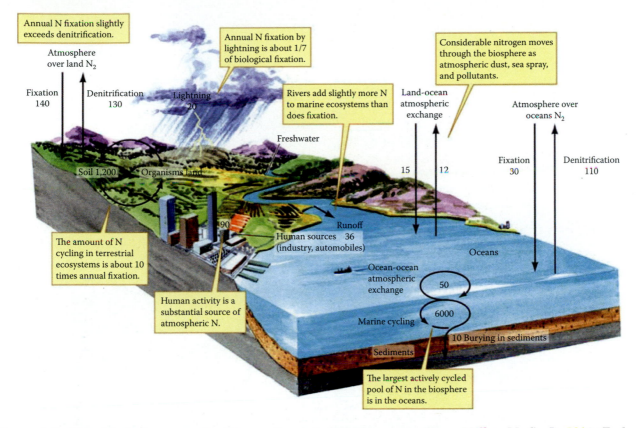

Annual N fixation slightly exceeds denitrification.

Annual N fixation by lightning is about 1/7 of biological fixation.

Considerable nitrogen moves through the biosphere as atmospheric dust, sea spray, and pollutants.

Atmosphere over land N$_2$

Fixation 140

Denitrification 130

Lightning 20

Rivers add slightly more N to marine ecosystems than does fixation.

Freshwater

Land-ocean atmospheric exchange

Atmosphere over oceans N$_2$

15   12

Fixation 30

Denitrification 110

Soil 1,200   Organisms land

The amount of N cycling in terrestrial ecosystems is about 10 times annual fixation.

90

Runoff

Human sources 36 (industry, automobiles)

Human activity is a substantial source of atmospheric N.

Oceans

Ocean-ocean atmospheric exchange

50

Marine cycling

6000

10 Burying in sediments

Sediments

The largest actively cycled pool of N in the biosphere is in the oceans.

Figure 6.6 The global nitrogen cycle; data are fluxes in 10^{12} N per year. (From Molles, M. C., Jr. 2008. *Ecology: Concepts, Applications*. New York: MacGraw–Hill Higher Education. Based on data of Schlesinger 1991 and Söderlund and Rosswall 1982.)

Figure 6.7 Soybean root nodules due to *Rhizobium* symbiosis. Each nodule contains billions of *Rhizobium* bacteria. (Photo courtesy of J. E. Beuerlein.)

(see Chapter 4). It has been recognized for centuries that rotation of certain crops, such as clovers, alfalfa, peas, beans, and other such species (N-fixation is present in all legumes), is a natural means of enriching the nitrogen content of cropland soils.

The process of atmospheric deposition also adds inorganic nitrogen to terrestrial and aquatic ecosystems through three main ways: wet deposition, dry deposition, and cloud water deposition. Wet deposition occurs when nutrients such as nitrate and ammonium ions are dissolved in precipitation and fall as rain or snow. Nitrous oxides can mix with atmospheric water and form nitric acid, HNO_3, which falls to earth. Dry deposition involves the deposition of particles such as dust by sedimentation, horizontal deposition, or direct absorption of gases such as nitric acid vapor. Cloud water deposition involves the deposition of nutrients in water droplets onto plant surfaces that are immersed in fog. These mechanisms and their role in acidifying ecosystems are discussed in Chapter 18.

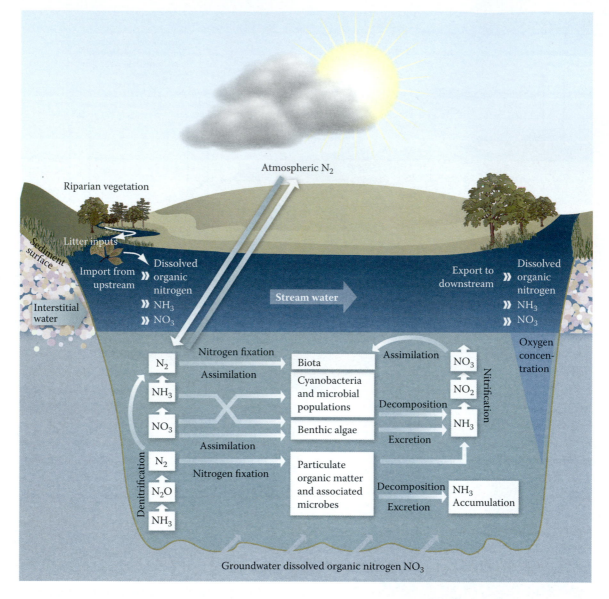

Figure 6.8 Dynamics and transformations of nitrogen in a stream ecosystem. Nutrient transformations occur with changes in nutrient inputs, as well as temperature and oxygen availability. All of this occurs as water moves downstream, displacing the nitrogen "cycle" into a nitrogen "spiral." (FISWRG 2001.)

The pool of inorganic nitrogen available to biota is made up of NH_4^+ and NO_3^-. Nitrate is lost from the soil by the processes of leaching and denitrification, while NH_4^+ is lost through volatilization. Nitrate is extremely soluble and travels easily with water movement. **Leaching** refers to the movement of nitrate into the groundwater below the root zone, where it can no longer be utilized by terrestrial plants, or its movement into surface waters, where it can contribute to downstream eutrophication of aquatic systems. Under anaerobic (lack of oxygen) conditions, some bacteria are able to use nitrate as the terminal electron acceptor in respiration in place of oxygen ("N-breathing"). In this way they obtain energy by using organic matter in respiration with the associated reduction of nitrate to nitrous oxide (N_2O) and then to dinitrogen gas, N_2. In this process, known as **denitrification,** the N_2 produced is released back to the atmosphere, thus completing the cycle that began with nitrogen fixation. Ammonium ions tend to be helped by soil particles, but at pH values above 9, ammonium (NH_4^+) can be converted to free ammonia gas (NH_3), which may be lost to the atmosphere in a process called **volatilization.**

In habitats where oxygen is plentiful, nitrogen is mobilized in a series of steps. This process begins with the decomposition of organic matter. Microbes break down the complex proteins in plant and animal waste to yield simpler amino acids (aminization). Eventually, these amino acids are converted into free ammonia and ammonium ions (**ammonification**) and fatty acids. Specialized microbes (nitrifying bacteria such as *Nitrosomonas*) oxidize ammonium (NH_4^+) to nitrite (NO_2^-), which is rapidly converted to nitrate (NO_3^-). The overall process is called **nitrification:**

$$NH_4^+ \;\rightarrow\; NO_2^- \;\rightarrow\; NO_3^-$$
$$\textit{Nitrosommonas} \quad \textit{Nitrobacter}$$

These reactions release the energy stored in the NH_4^+ molecule that the microbes use to fuel their cellular processes; as such, both steps of this reaction require oxygen. Because a majority of terrestrial plants investigated prefer to take up nitrogen in the form of nitrate, the oxidation steps mediated by nitrifying bacteria are crucial. These bacteria are aerobic and prefer a circumneutral pH. Plants in wetland areas, where conditions are anaerobic, have the ability to utilize nitrogen in the form of ammonium because the process of nitrification is not as prevalent. Algae, on the other hand, have been shown to prefer ammonia as a fixed nitrogen source. When they use nitrate, they first convert nitrate to ammonia using the iron- and molybdenum-containing enzymes nitrate reductase and nitrite reductase.

Carbon and nitrogen are both intimately related to biological activity and thus are indicators not only of soil fertility but also of ecosystem productivity. Interactions between (1) biotic (biological) factors, such as the growth efficiency of decomposer organisms, and (2) abiotic (physical and chemical) factors, such as the quantity and quality of the substrate, control the decomposition and mineralization of soil organic matter. The ratio of carbon to nitrogen (C:N) concentration in microbes ranges from 5 to 10 for bacteria and from 8 to 15 for fungi. During the breakdown of organic matter, microbes generally incorporate about 40% of the carbon into their microbial biomass and release the remaining 60% of the carbon back to the atmosphere as CO_2 (40% growth efficiency).

Sulfur (S) Cycle

Sulfur is an important element in cellular chemistry because it is responsible for the three-dimensional folding pattern of proteins that is necessary for them to function. It is estimated that for every 1,000 atoms of carbon fixed in photosynthesis, one atom of sulfur is taken up by plants (Smil 2007). Like nitrogen and carbon, the sulfur cycle undergoes oxidation-reduction reactions and has both gaseous and mineral components (Figure 6.9). Weathering of the sulfide components of rocks is the primary natural avenue for inputs of sulfur to most ecosystems—for example, pyrite (FeS_2), in which oxidation produces iron hydroxide from the iron and sulfuric acid from the sulfur. The atmosphere receives most of its sulfur compounds from burning of biomass and fossil fuels, decomposition of organic matter, volcano eruptions, drainage of wetlands, sulfate-containing dust in arid regions, and sea salt over the oceans (as aerosols). Most sulfur is added to the atmosphere as SO_2 by human use of sulfur-rich fossil fuels, particularly coal. The SO_2 subsequently dissolves in cloud water droplets to produce strong sulfuric acid (H_2SO_3), forming acid precipitation.

The most important trace sulfur gases emitted to the atmosphere are hydrogen sulfide (H_2S), dimethyl sulfide (DMS), methyl mercaptan (CH_3SH), carbon disulfide (CS_2), and carbon sulfide (COS). As ocean temperatures increase, phytoplankton activity increases, releasing DMS that is subsequently oxidized to SO_2 and then to $SO_4^=$ in the atmosphere. These sulfate aerosols increase the Earth's albedo (reflectivity) by acting as cloud droplet condensation nuclei. As cloud cover increases, more solar energy is reflected back to space, contributing to global cooling. Large amounts of sulfur emissions from the volcanic eruption of Mount Pinatubo in the Philippines in June 1991 (the largest input of stratospheric SO_2 ever recorded) are believed to have resulted in the average global cooling observed in the Earth's atmosphere over the next year. Sulfur is responsible for various important biogeochemical processes, such

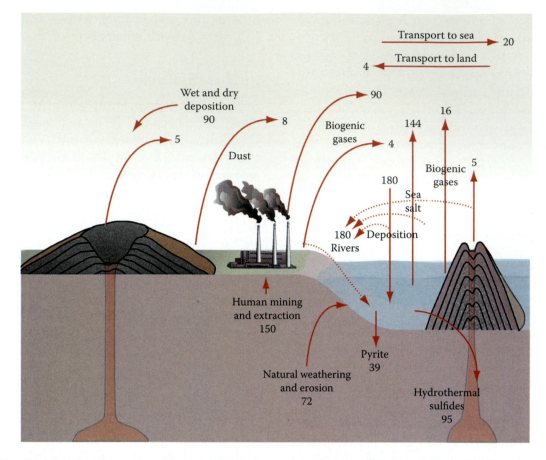

Figure 6.9 The global sulfur cycle with annual flux shown in units of 10^{12} g S yr^{-1}. (From Schlesinger, W. H. 1997. *Biogeochemistry—an Analysis of Global Change,* 2nd ed. San Diego, CA: Academic Press.) Volcanic sulfur is primarily in the form of H_2S. DMS and oxides of sulfur (SO_2, SO_3) are important in human and marine sources of atmospheric sulfur.

as sulfate reduction, pyrite (FeS_2) formation, metal cycling, salt-marsh ecosystem energetics, and atmospheric sulfur emissions. The important role of sulfur compounds in the formation of acid precipitation is discussed in Chapter 18.

Mineral or Sedimentary Cycles

For mineral or sedimentary cycles, snow, rain, wind, and now pollution can act to increase nutrient loading into ecosystems. The quantities of nutrients that return to the soil or surface waters depend upon a number of factors: (1) type of vegetation, (2) type of habitat, and (3) climatic factors. Sedimentary cycles include important nutrients such as calcium, magnesium, phosphorus, and a number of micronutrients like iron, manganese, and copper. We discuss the P cycle here as an example.

Phosphorus (P) Cycle

Unlike the cycles discussed earlier, phosphorus does not undergo oxidation-reduction reactions in soils or water and has no major gas phases or atmospheric components (Figure 6.10) (although dust storms in desert areas can transport large amounts of phosphorus). Many of the reactions that control phosphorus availability are geochemical rather than biological in nature, although the biota *strongly* influence the rapid cycling of the small amounts of orthophosphate ion (PO_4^{-3}) present in aquatic systems. Except for decomposing plant materials, animal excreta, and phosphorus fertilizers applied by man, phosphorus is made available to the biotic component of terrestrial ecosystems only as it is released (solubilized) from the rocks found in soil; this limits its availability compared with many other nutrients. Because phosphorus is quite rare in the Earth's crust relative to the other elements needed by plants, it often

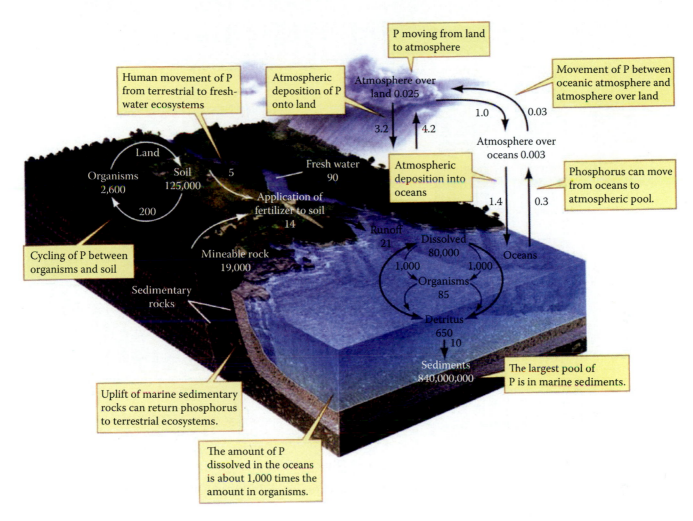

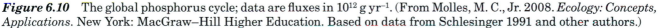

Figure 6.10   The global phosphorus cycle; data are fluxes in 10^{12} g yr^{-1}. (From Molles, M. C., Jr. 2008. *Ecology: Concepts, Applications*. New York: MacGraw–Hill Higher Education. Based on data from Schlesinger 1991 and other authors.)

can be limiting to plant growth. The form of P most available to organisms is orthophosphate, H_3PO_4. This molecule dissociates as the pH of the environment changes—for example, losing an H^+ ion to become the less highly charged form of phosphate, $H_2PO_4^-$. This form is even more mobile in soil and therefore more available to plants, illustrating the importance of soil pH to plant nutrition.

Much of the phosphorus in the environment that is precipitated in iron, aluminum, calcium, and organic compounds is essentially unavailable to plants and is referred to as **occluded phosphorus.** The major global sink for phosphorus is burial of these same precipitates in marine sediments (Figure 6.10). Because PO_4^{-3} is bound very tightly to organic matter or to soil minerals in most soils, 90% of the phosphorus loss from agricultural fields

occurs through surface runoff and erosion of phosphorus bound to soils rather than through leaching to groundwater. Very little P enters groundwater; of that which does, two-thirds is in an organic form and is therefore less reactive with soil minerals.

The phosphorus transported to the sea precipitates out of solution, forming calcium phosphate deposits. This is a long-term sink for P. One way that it returns to "circulation" is when humans mine exposed ancient ocean sediments as phosphate ores to make laundry detergents and inorganic fertilizers. The flux of P in rivers has thus been enhanced by human-induced erosion, pollution, and fertilizer runoff. The structure and function of aquatic ecosystems are particularly sensitive to accelerated additions of phosphorus by human activities; this is discussed in Chapter 16.

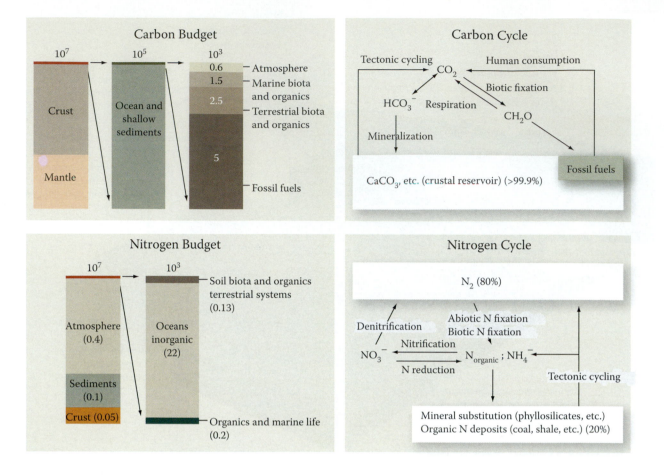

Figure 6.11   Comparative carbon and nitrogen cycles on Earth (units are in 10^{15} g). On the left are Earth's major reservoirs of C and N and, on the right, major redox changes catalyzed by biota. (From Capone, D. G. et al. 2006. *Science* 312:708–709.)

Interactions of Biogeochemical Cycles

It should be obvious by now that the net primary production of organic C by terrestrial and aquatic plants is related to the relative availability of many nutrients, including C, N, S, O, and P, in suitable chemical forms. Stoichiometry describes the relative quantities of the reactants and products in a chemical reaction. Stoichiometric relationships have been determined to describe the ratios of C, O, N, P, and S in organisms such as plants, animals, and microbes. This enables one to estimate the stocks or fluxes of other nutrients based on any known nutrient stock or flux. For example, carbon taken up by phytoplankton is released during the mineralization of organic matter in proportion

to the other major nutrients that were part of the original phytoplankton, based on an average molar ratio of $C_{organic}$:N:P:O of about 103:16:1:72 (Takahashi, Broecker, and Langer 1985). This means that, for every 103 atoms of C bound in the organic matter, 16 atoms of N, 1 of P, and 72 of O go along. This ratio is referred to as the Redfield ratio after the scientist who first pointed out this fundamental stoichiometric relationship. The field of ecological stoichiometry has recently been elaborated by Sterner and Elser (2002).[1]

The availability of C and N indicate the fertility and productivity of ecosystems (see, for example, Figure 6.11). The element most limiting to the growth of plants will be the one that does not meet these relative needs. Ironically, phosphorus, the element needed in the lowest relative amount, is the one that usually controls plant growth in freshwater aquatic systems, whereas nitrogen is usually limiting in terrestrial and marine systems. Examples of the interdependence of biogeochemical cycles include a high demand for P by organisms that are able to fix N, the adsorption of P to iron minerals in marine sediments caused by high levels of O_2, and the use of O_2 produced by photosynthesis in the oxidation of NH_4 (nitrification) and in the decomposition and mineralization of dead organic matter.

Human activities have dramatically altered the interactions of biogeochemical cycles. The C and N cycles are closely linked such that the natural scarcity of N often limits the growth of plants, in effect slowing their photosynthesis rates. In this way, the availability of N is said to regulate the fixation of C, so these cycles are said to be "coupled." One consequence of humans' ability to fix nitrogen on an industrial scale for fertilizer production is that N is much more plentiful in the biosphere than it has ever been. The amount of biologically available, or fixed, N in terrestrial ecosystems has doubled since 1960 (MEA 2005). As a result, N has become less limiting to growth; when N no longer regulates plant productivity, the N and C cycle is said to be "uncoupled." Research is under way to help us understand the implications of this large-scale biogeochemical change.

Note

1. Chemical stoichiometry is the study of the quantitative aspects of chemical relations. Its fundamentals are based on the law of conservation of mass, the law of definite proportions (i.e., the law of constant composition), and the law of multiple proportions. In the present context, it refers to a definite ratio of chemicals combining in a chemical reaction. Its importance came from Alfred C. Redfield, who demonstrated that the molecular ratio of carbon, nitrogen, and phosphorus in phytoplankton shows a stoichiometric ratio of C:N:P = 106:16:1. Since then it has been referred to as the Redfield ratio. A seminal paper by Reiners (1986) and the recent work by Sterner and Elser (2002) make ecological stoichiometry an exciting field of specialization.

Questions

1. How do biogeochemical cycles in an ecosystem differ from energy flows?
2. Why are biogeochemical cycles important?
3. What do we mean when we say "stocks and flows" in describing biogeochemical cycles?
4. How are nutrient cycles with atmospheric stocks different from those without gaseous components?
5. How can understanding biogeochemical cycles benefit our attempts to manage natural resources?

Energy Flows and Ecosystem Productivity

7

Solar Energy—the Driving Variable

Thus far, we have studied the ingredients of life and where and under what conditions they operate. We now turn our attention to what drives the systems that produce the amazing diversity of life as we see it. The sun, directly and indirectly, sustains all life systems and drives all life processes. Those who model ecological systems refer to this energy as the "driving variable." The energy reaching the Earth from sunlight, or **solar energy,** is captured and converted into chemical energy (photochemical energy) or into heat (thermal energy). The photochemical energy is captured by green plants, which use it to convert simple molecules (carbon dioxide, water, minerals) into complex ones such as carbohydrates, proteins, fats, and other compounds. The thermal energy warms up the earth, drives the water cycle, and provides currents of air and water. Regardless of the circumstances, all forms of energy continually move from something warmer (high-quality energy) to something cooler (low-quality energy)—never the reverse.

The sunlight, or radiation, is thus continuously captured, converted, and reconverted, with changes brought about by a flow of energy. The science of energy transfer and transformation is **thermodynamics.** Thermodynamically, there are three types of systems: (1) isolated (or adiabatic) systems that can exchange neither energy nor matter with their surroundings, (2) closed systems that can exchange energy but not matter, and (3) open systems that can exchange both energy and matter (von Bertalanffy 1950). The Earth is a closed system (see Figure 7.1), but its ecosystems are open systems.

To illustrate the sealed nature of the Earth, Philip and Phyllis Morrison (1987) compared it to the features of a sealed glass sphere that is bounded by a wall. The Earth, though lacking walls, is held together by gravitational forces (Figure 7.1). They noted that, like the glass sphere,

> Earth too has fixed material balance and a constant weight. We have not measured it precisely, but we firmly believe its weight remains all but constant. Everything that happens to fix our fate, certainly over millions of years, is

but continual rearrangement, the constant ebb and flow of all the materials that pass ceaselessly from creature to creature, from place to place, from form to form. The internal balance is guaranteed by constancy of weight through change.

For material movement to occur, there must be a high-quality energy source out of which energy flows and a low-quality energy sink into which the energy flows. This **energy flux** makes mechanical work possible. Energy is recognized in two forms: **potential energy,** which is not associated with motion and gets stored in an object (e.g., a candy bar), and **kinetic energy,** which is any form of energy associated with motion, such as that found in a moving car. An estimate of the Earth's global energy budget is illustrated in Figure 7.2. Not all of the energy that reaches the Earth is captured; even photosynthesis harvests only a small part of it (Figure 7.3).

Energy is measured in several units; for convenience, ecologists often use the calorie (cal), defined as the amount of heat energy necessary to raise the temperature of 1 g (or mL)

Figure 7.1 Two closed ecosystems: the Earth (ecosphere) sealed by gravity and the small sphere sealed by fusing the glass. (Morrison, P., and P. Morrison. 1987. *The Ring of Truth: An Inquiry into How We Know What We Know*. New York: Random House.)

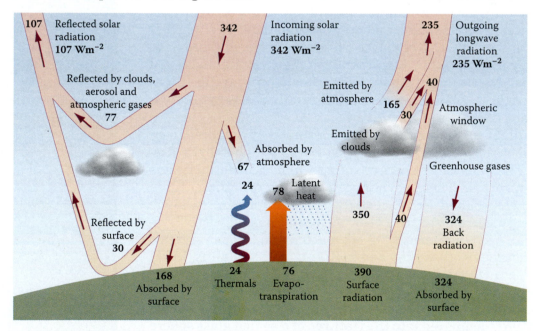

Figure 7.2 Estimate of the Earth's annual and global mean energy balance. Over the long term, the amount of incoming solar radiation absorbed by the Earth and the atmosphere is balanced by the Earth and atmosphere releasing the same amount of long-wave radiation. About half of the incoming solar radiation is absorbed by the Earth's surface. The energy is transferred to the atmosphere by warming the air in contact with the surface (thermals), by evapotranspiration, and by long-wave radiation that is absorbed by clouds and greenhouse gases. The atmosphere in turn radiates long-wave energy back to Earth as well as out to space. (From IPCC. 2007. *The Physical Science Basis*, 93–127. Working Group I Report. Frequently asked questions. New York: Cambridge University Press. Based on Kiehl, J. T., and K. E. Trenberth. 1997. *Bulletin of the American Meteorology Society* 78:197–208.)

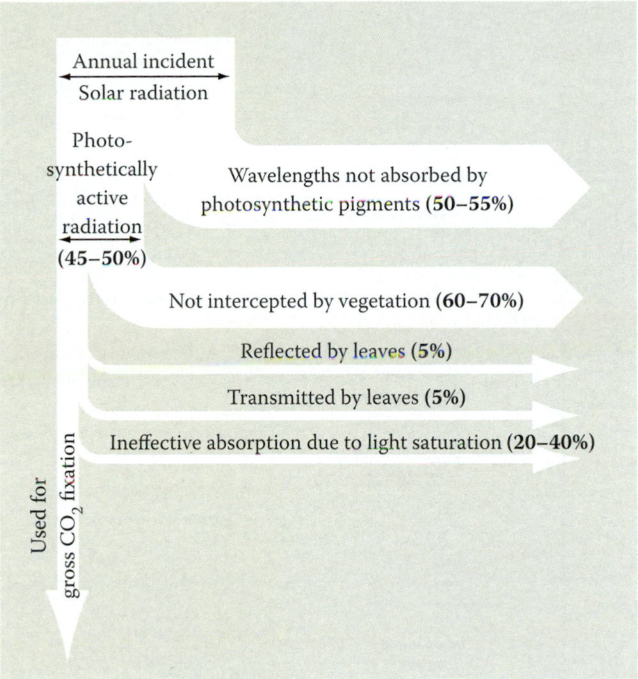

Figure 7.3 Average partitioning of solar radiation reaching photosynthesizing terrestrial plants (Smil 1999). After several forms of inefficient capture, only 7–14% can be used for photosynthetic processes. Of that, only one-third is eventually stored as starch; the rest is given off as heat.

of water at 4°C by 1°C. When larger quantities of energy are involved, the kilocalorie (kcal) is used. Calories of energy listed on food labels are also in kcal; hence, the word "calorie" in the designation of a 140-Calorie soft drink is capitalized to symbolize kcal. Ecologists are more commonly concerned with energy flow than content, so they use kilocalories per space per time (e.g., kilocalories per square meter per day) or any metric conversion thereof (e.g., watt seconds, joules) (Figure 7.4).

Laws of Thermodynamics

Although four laws of thermodynamics are recognized in somewhat differing formulations, the first and the second laws are recognized for their application to biological (living) systems and ecological (living and nonliving) systems. Conversion of energy—that is, energy transfers and transformations—follows the **first law of thermodynamics,** which states that energy can be converted from one form to another but can neither be destroyed nor created during

the process. For example, light energy from the sun can be converted to heat, warming the Earth, or it can be captured by plants and converted to the potential energy stored in food. In neither case, however, is energy destroyed.

To demonstrate this law, the Morrisons and their staff devised a complex maze of interconnected devices attached to a magnet that, once released, were capable of setting up a domino-like chain of events. This entire maze was then sealed and placed on a crude balance (scale) (Figure 7.5). Once the magnet was released, the chain reaction caused combustion of every component in the sealed cube. "[A]ll those mechanical, chemical, electrical, visual changes in position and state, nothing changed the balance." Thus, no matter what the energy transfers are, there is no gain or loss in energy.

"You cannot win." In the conversion of potential energy to kinetic energy, some reactions release (dissipate) a lot of heat. Such reactions are **exothermic** (heat producing). Other reactions, however, need an input of energy or heat to progress and are thus **endothermic**. In photosynthesis, an endothermic process, energy comes from the sun. Although energy

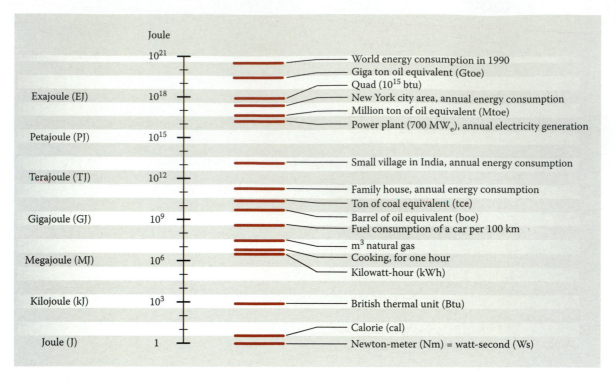

Figure 7.4   The ranges of various human energy needs converted to a standard unit (joules) on a logarithmic scale (IPCC). Recall that 1 kcal = 1 cal = 1,000 cal = 4,184 J. 1 W = 1 J s^{-1}; 1 kW-hour = 3,600,000 J.

is neither lost nor gained (i.e., the first law of thermodynamics is obeyed), much of the energy captured in this process is degraded and thus incapable of doing any further work. This energy ends up as **heat,** serving to disorganize or randomly disperse the molecules involved and making them useless for work.

"You cannot even break even." The preceding result illustrates the **second law of thermodynamics,** which states that whenever energy is transferred or transformed, part of the energy always assumes a form that cannot be passed on any further; energy is always degraded. No process that involves an energy transformation can occur spontaneously without a degradation of energy from a concentrated form into a dispersed form. This can be illustrated in several ways. The irreversible degradation of energy as it is transferred means that energy quality (usable energy) becomes increasingly unavailable for producing work; it becomes waste heat. The measure of relative disorder is termed **entropy,** which is a gauge of the inability of energy to do work, Table 7.1. For example, energy resulting from photosynthesis stored as fossilized plant and animal tissues is converted, over millions of

years, into **fossil fuels** (coal, oil, and natural gas). When we burn these energy resources to meet the energy demands for our economy, we convert the potential energy stored in the fuels to kinetic energy to run our vehicles. We do this with relatively low efficiency, dissipating a large amount of that energy as waste heat into the environment. No transfer of energy, however, can be 100% efficient.

The second law of thermodynamics also explains the limits to the number of links that can be present in a food chain. With each successive transfer of energy to a higher trophic level, energy is lost (Figure 7.6). This can be thought of in several ways; as one moves up the food chain, a smaller and smaller proportion of the sun's energy remains. For instance, only about 0.05% (1/20 of 1%) of the original solar energy makes it to the top levels (Figure 7.6a). This also represents a concentration of energy in the sense that, for each calorie that is made (as biomass, for example) at the top trophic levels, the number of solar calories received at the Earth's surface needed to generate that biomass increases. In this way the ratio of solar calories arriving on Earth to the number

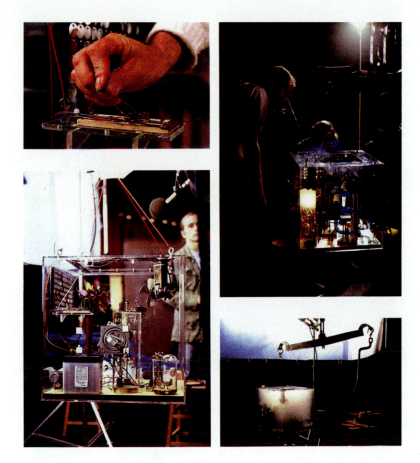

Figure 7.5   The well-sealed box of changes encloses a variety of events (e.g., unwinding, battery discharge, motion, flame, light, sound, explosion, falling, spinning), and the scale with its load of bricks does not change by even so much as the weight of one twenty-five cent piece. (Morrison, P., and P. Morrison. 1987. *The Rings of Truth: An Inquiry into How We Know What We Know.* New York: Random House.)

Table 7.1
Examples of Ordering and Dissipative Processes in Ecosystems[a]

Ordering Processes (Decrease in Entropy)	Dissipative Processes (Increase in Entropy)
Biological	
Photosynthesis in plants	Respiration
Growth of all living organisms	Senescence
Formation of humus in soil	Decomposition and mineralization of humus
Physical	
Water flow (profile development)	Water flow (erosion, leaching)
Flocculation	Dispersion
Aggregation	Disaggregation
Development of structure	Breakdown of structure
Larger units	Smaller units
Fewer of them	More of them
More ordered	Less ordered

Source: Modified from Addiscott, T. M. 1995. *European Journal of Soil Science* 46:161–168.

[a] Ordering processes consume energy, whereas dissipative processes release it.

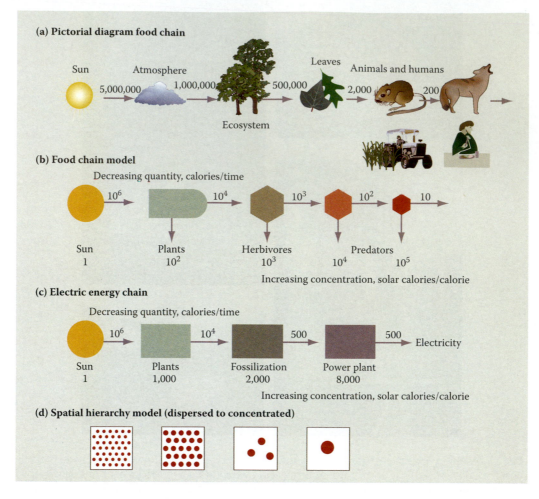

Figure 7.6 Comparison of the decreasing energy quantity and increasing energy concentration (quality) that accompany progression up food chains and the process of electric energy generation. Both use energy originating from the sun (in kilocalories per square meter). (a) Only 20% of the sun's energy passes through the atmosphere, and of this 0.5% ends up as leaves edible by herbivores (rabbits or humans). Only 0.05% of the sun's energy is finally passed on to the fox or human that eats the rabbit. (b) The diagrammatic model demonstrates that as the quantity of original energy declines, the proportion of solar calories involved per calories passed up the chain increases. (c) Inefficiencies in energy transfer are also seen in a power plant, with only 125 kcal of the original 10^6 kcal m^{-2} originally coming from the sun ending up as electrical power. (d) Energy from the sun arriving as photons produces far fewer, but larger plants, which support fewer herbivores and thence only one carnivore. (From Odum, E. P., and G. W. Barrett. 2005. *Fundamentals of Ecology,* 5th ed. New York: Thompson/Brooks–Cole.)

of calories present at a given trophic level increases with trophic levels (Figure 7.6b–d).

How much energy can be lost during biological activities? Bike riders perform a large amount of physical work pedaling against wind and friction while transporting themselves and their bikes long distances. The Morrisons tabulated an authentic menu of a bicycle rider in a race (Table 7.2). They converted all the food into JD units (i.e., jelly doughnut units! 1 JD unit = 250 nutritional cal). The racer's menu: 30–32 JD units, or 7,500–8,000 calories per day. Most of this energy is burnt and lost as heat (Figure 7.7).

Ecological Pyramids

The second law of thermodynamics has particular relevance in ecosystems because they progressively lose energy with chemical transformations. This means that chemical energy in the form of food leaves progressively less energy as it flows from one organism to the next. Energy fixed by plants (**primary producers**) is consumed by **primary consumers** (herbivores), which are eaten by **secondary consumers** (carnivores) and **tertiary consumers** (also carnivores), and so on, forming a food chain. In nature, however, organisms

Table 7.2
Day's Menu of a Hard-Working Cyclist[a]

Breakfast	Later
Large glass of orange juice	1 apple
2 large glasses of milk	1 candy bar
Double stack of pancakes with lots of butter and syrup	**Supper**
3 strips of bacon	Large piece of roast beef
Snack while riding:	2 baked potatoes with butter
2 candy bars	3 slices of bread and butter
1 banana	Salad
4 pint bottles of slightly flat Coke	2 servings of pudding
After the race:	**Later**
Lunch	1 candy bar
Barbecued chicken	1 malt shake
2 large servings of potato salad	1 hamburger
4 or 5 pieces of bread and butter	But the racer woke during the night a little hungry and got a snack at 2:00 a.m.
2 glasses of milk	2/3 quart of ice cream
2 pieces of pie	6 Oreo cookies

Source: From Morrison, P., and P. Morrison. 1987. *The Ring of Truth: An Inquiry into How We Know What We Know.* New York: Random House.

[a] The high-energy input illustrated here is typical of that taken in by Tour de France bicycle racers.

Figure 7.7 Left: 32 jelly doughnuts are equivalent to the caloric content of the cyclist's menu listed in Table 7.2. Their high-energy content is illustrated by the energy lost as waste heat and light. Right: from their combustion in a fire. (Morrison, P., and P. Morrison. 1987. *The Rings of Truth: An Inquiry into How We Know What We Know.* New York: Random House.)

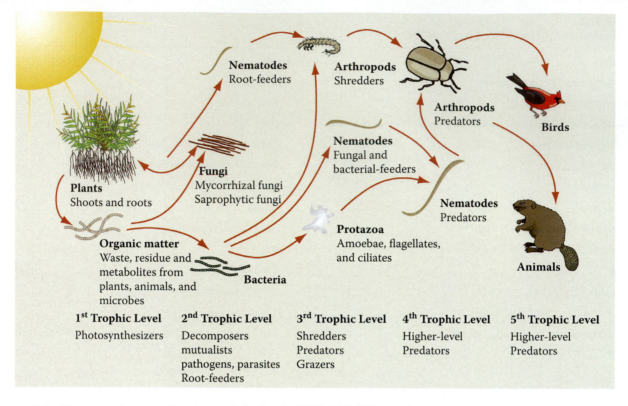

Figure 7.8 Diagram of energy flow in a soil food web. (USDA-NRCS 1999.)

at higher consumer levels (e.g., humans) can obtain food from multiple sources and levels, making them omnivores and generating a complex network of food acquisition, forming a food web (see one simplified example in Figure 7.8) rather than a food chain. In the transfer of energy from one trophic level to another, on average, 90% of usable energy is lost as heat. As one goes up the food chain, the available energy left at the top is considerably less than at the starting point, thus forming an **ecological pyramid of energy** (Figure 7.9). These **trophic structures** are also referred to as Eltonian pyramids (so named after the well-known British ecologist, Charles Elton).

Three types of ecological pyramids can be illustrated: 1) the pyramid of numbers, which shows the number of organisms at each level; 2) the pyramid of biomass, in which the total dry weight or caloric value of the living material is shown; and 3) the pyramid of energy, which shows the rate of energy flow, or productivity of each trophic level. The absolute amount of biomass or the number of organisms at each level varies from one ecosystem to another, depending on the species involved, climatic factors,

etc. Within an ecosystem, the loss of usable energy with each transfer between trophic levels explains why top carnivores in natural ecosystems are low in number: There is simply not enough energy moving to the top trophic level to support dense populations (Figure 7.9). One ecologist titled a book after this phenomenon, calling it *Why Big Fierce Animals Are Rare* (Colinvaux 1993). It also explains why top carnivores are more susceptible to the effects of ecosystem degradation and the loss of plant productivity. All things being equal, a given area of land can support more omnivores (eaters of plants and animals) eating at lower trophic levels than if they were eating at higher trophic levels. This is true for human populations as well; vegetarians eat more efficiently in terms of energy use than do omnivores.

A **pyramid of numbers** represents the number of individuals at each successive trophic level. The assumption is that the number of producers will be high. The herbivores will be larger in size but smaller in number. In turn, carnivores will be even larger in size but fewer yet in number. Such a relationship holds well in some ecosystems (e.g., oceans and grasslands),

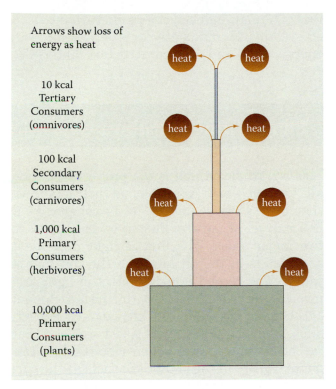

Figure 7.9 An ecological pyramid illustrating energy stored and energy lost in successive trophic levels.

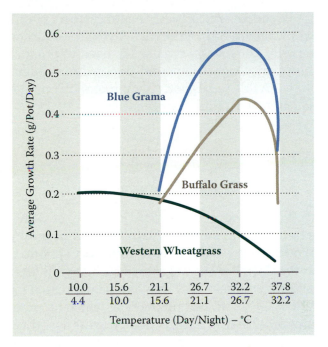

Figure 7.10 Average growth rate of C3 western wheatgrass (*Pascopyrum smithii*) and C4 blue grama grass (*Bouteloua gracilis*) and buffalo grass. (*Buchloe dactyloides*) grown at different day/night temperatures. (Knievel and Schmer 1971.)

but not in others (e.g., in forest ecosystems where trees may be quite large in size but few in number).

A better illustration than the pyramid of numbers is provided by the **pyramid of biomass** showing the amount of dry matter found at each trophic level (e.g., kilograms per square meter). In this depiction, the relative biomass of consumers reflects their numbers. In all three cases, the trophic pyramids of energy remain fundamental to understanding ecological systems.

Terrestrial Primary Productivity

Primary producers are autotrophs (literally, self-nourishing organisms) and include photosynthetic plants, photosynthetic Cyanobacteria, and eukaryotic algae, as well as chemosynthetic organisms (e.g., sulfur-oxidizing bacteria) that oxidize reduced inorganic compounds to obtain energy. Green plants are the only "real"

producers that are widespread on the face of the Earth in the sense that the survival of all other organisms (including humans) depends upon them. Photosynthesis takes place in the chloroplasts of plant cells and consists of light-dependent and light-independent reactions.

In the initial, light-dependent phase of photosynthesis, light-absorbing green pigments in plants' leaves (chlorophylls) absorb the photosynthetically active radiation (PAR) portion of visible light and convert it to chemical energy. In the final light-independent phase (dark reactions, or the Calvin cycle) of photosynthesis, plants use this biochemical energy to fix carbon from atmospheric carbon dioxide (carboxylation). Thus, through the utilization of the sun's radiant energy, plus carbon dioxide, water, and nutrients, plants produce carbohydrates ($C_6H_{12}O_6$) and release oxygen to the atmosphere as a waste product:

Photosynthesis:

$$6CO_2 + 12H_2O + \text{light energy}$$

$$\rightarrow C_6H_{12}O_6 + 6H_2O + 6O_2$$

In the reverse process (**respiration**), plants oxidize about half of the carbohydrate to carbon dioxide. The energy released is used to produce the energy-rich compound adenosine triphosphate (ATP) to support their metabolic activities of growth and maintenance:

Respiration:

$$C_6H_{12}O_6 + 6O_2$$

$$\rightarrow 6CO_2 + 6H_2O + \text{released energy}$$

The photosynthetic fixation rate of atmospheric carbon dioxide is called **gross primary productivity (GPP)**. **Net primary productivity (NPP)** is the net rate of carbon gain after respiration losses (R). This can be expressed as: NPP = GPP – R. Net production, which accumulates as biological mass in plant organs over time, is referred to as **biomass** and can be expressed in terms of dry weight (g) or energy content (kcal). The overall photosynthetic efficiency of plants, measured as the amount of solar energy available to them that they actually covert to biomass, is typically about 1% or less.

Net primary productivity in land plants is often measured as the change in biomass over time, giving a value for aboveground NPP in kilograms of carbon per hectare per year. In contrast, measuring root (belowground) biomass is cumbersome and time consuming. A large proportion of net primary productivity, particularly in temperate zones, is returned to the soil as litter dead root material or leaf litter. This is subsequently mineralized by decomposer organisms through heterotrophic respiration and the nutrients released become available again for the growth of the plant. This results in the release of roughly another 45% of the original GPP. Overall net ecosystem productivity (NEP) is the accumulation rate of all biomass of both primary and secondary producers.

Net terrestrial ecosystem productivity (the remaining 5% of GPP) forms the fraction of slowly decomposable soil organic material known as **humus.** The biological production of most terrestrial and aquatic ecosystems is affected directly and indirectly by human activities (including agriculture), so net ecosystem productivity also reflects the magnitude and rate of human-induced effects (disturbances) on ecosystems. Humans use parts of NEP for food, energy, fiber, and even construction materials.

Photosynthetic Pathways

Photosynthesis, which is *the* process by which all food becomes available, is not a simple chemical reaction; indeed, in its complexity it is a marvel of nature. Photosynthesis involves two major steps: The first is light dependent and results in the photolysis or the splitting of water.[1] The second is independent of light and involves the reduction of CO_2 (by the attachment of H-compounds), a process known as carbon fixation. About 60 years ago, it was discovered that carbon fixation occurs when ribulose bisphosphate (RuBP) in the plant tissue acts as the carbon acceptor; the first product of photosynthesis is a three-carbon, or C_3, compound. This process is catalyzed by the enzyme ribulose bisphosphate carboxylase-oxygenase, or RuBisCO. More than 80% of plants have this pathway of photosynthesis and are referred to as **C_3 plants.**

Many years later it was discovered that plants that grow in semiarid environments, where conditions are high in light and low in water availability, use another enzyme for fixation of carbon. Here, phosphoenolpyruvate, or "PEP" carboxylase, fixes carbon from atmospheric carbon dioxide during photosynthesis and forms a four-carbon product. Hence, the plants are called **C_4 plants.** PEP carboxylase has a higher affinity for CO_2 than does RuBisCO in the C_3 pathway, enabling it to reduce the concentration of carbon dioxide inside the leaf to a lower concentration (Table 7.3) making C_4 plants more efficient.

There is yet a third pathway called **crassulacean acid metabolism (CAM)**, which is found primarily in plants growing in deserts. Here, water is at a premium and its loss must be avoided by strictly controlling the time periods when the stomata (the pores in leaves through which gas is exchanged) can be open. These plants shut their stomata in the day, during times of high light intensity, and open them during the night. During the night, the leaves

Table 7.3
Anatomical and Physiological Characteristics of Vascular Plants with C_3, C_4, and CAM Modes of Photosynthesis

Characteristic Leaf Structure	C_3 Plants (Chloroplasts Distributed Throughout the Leaf)	C_4 Plants (Chloroplasts Arranged Around Bundle Sheaths)	CAM Plants (Mesophyll Cells with Large Vacuoles)
Primary CO_2 acceptor	RuBP (substrate = CO_2)	PEP (substrate = HCO_3^-)	In light: RuBP; in dark: PEP
First product of photosynthesis	C_3 acids (PGA)	C_4 acids (oxaloacetate, malate, aspartate)	In light: PGA; in dark: malate
Photorespiration in presence of light and O_2	Yes	No	Yes
CO_2 release in the light	Yes	No	No
CO_2 compensation concentration at optimal temperature, $\mu L/L$	30–50	<10	In light: 0–200; in dark: <5
Temperature of optimal photosynthesis, °C	15–25	35–40	35

Source: After Larcher, W. 1995. *Physiological Plant Ecology. Ecophysical and Stress Physiology of Functional Groups,* 3rd ed. Heidelberg, Germany: Springer–Verlag.

take in CO_2 using PEP carboxylase. During the day, when the stomata are closed, the CO_2 is released into the air spaces inside the leaf itself and C_3 photosynthesis using RuBisCO takes place (see Table 7.3 for anatomical and physiological adaptation of plants).

If the weight of all the biological enzymes in the world could be tallied up, the weight of RuBisCO would be the greatest; it is the most common enzyme on Earth. Interestingly, it is also not terribly efficient in the photosynthetic process. Not only does it have the ability to bind CO_2 (though it acts as a carboxylase), but it also has the ability to bind O_2 (its actions act as an oxygenase). When CO_2 concentrations are low, particularly in high temperatures, RuBisCO preferentially binds O_2 and, through a series of reactions, liberates CO_2. This undoes the work of photosynthesis. Although O_2 is consumed and CO_2 is produced as in respiration, because light is driving the process, it is known as photorespiration. Photorespiration makes photosynthesis inefficient, releasing 20–40% of carbon just fixed back to the atmosphere.

C_4 photosynthesis is an adaptation that minimizes the loss of fixed carbon to photorespiration. C_4 plants have a number of other adaptive advantages in environments with high light intensity, high temperature, and low water availability (Table 7.3; Figure 7.10).

Although the large majority of plants are C_3 plants, the C_4 photosynthetic pathway is found primarily in the grasses, in the saltmarsh grass *Spartina patens,* and in other herbaceous plants that are abundant in arid, hot environments. Several important crop species, such as sugar cane, corn, and sorghum, are also C_4 plants. Most woody species are C_3 plants.

Ecological Controls on Primary Productivity

Atmospheric and climatic factors (light, temperature, and wind, discussed in Chapter 3), factors of soil (water, nutrients, temperature, air, and pollutants, discussed in Chapter 4), and biotic factors (of competition, disease, herbivory, and human activities) all affect primary productivity.

Light

As we learned in Chapter 3, the interception and absorption of incident light by plants' leaves and canopies vary due to changes in sun angle, variation in cloudiness, and changing patterns of sun flecks (patches of direct sunlight that penetrate to leaves within a plant canopy). Net primary productivity is strongly related to intercepted and absorbed PAR (megajoules per square meter per month). This relationship gave rise to the concept of radiation (light) use efficiency (RUE) as the average amount of carbon gained per unit of PAR intercepted or absorbed (grams of carbon per megajoule of PAR) (Monteith 1972). The product of the absorbed PAR and RUE sets an upper limit to potential (maximum possible) NPP. As the following equation states, light is greatest at the top of the canopy and decreases exponentially within a plant canopy:

$$I = I_o\, e^{-kL}$$

where

   I is the light intensity beneath the canopy;
   I_o is the light intensity at the top of the canopy;
   k is the extinction coefficient; and
   L is the leaf area index (LAI), or the leaf area per unit of ground area.

LAI varies widely among ecosystems, ranging from 0 to 12 m^2 of leaf per square meter of ground, but typically has values of 1–3 for ecosystems with a closed canopy.

The extinction coefficient, k, is a constant that describes the exponential decrease in light intensity as it passes through a canopy. The extinction coefficient is low (i.e., light penetration is high) for vertically inclined or small leaves (e.g., 0.3–0.5 for grasses), but high for near-horizontal leaves (0.7–0.8). At very low light intensities—for instance, at the ground surface of a forest—only 1–2% of the light intensity present at the top of the canopy remains. Under these conditions, respiration completely offsets photosynthetic carbon gain and net photosynthesis becomes zero. (This is the same as the light compensation point as it occurs in lakes and oceans.)

Even within the same tree, leaves, which can be thought of as "solar panels," have differing adaptations to maximize their efficiency in light capture. On the same tree, "sun leaves" develop under high light, have more cell layers, are thicker, and therefore have greater photosynthetic capacity per unit leaf area than do "shade leaves," which are produced under low light. Shade leaves are produced in the shade of other leaves and are typically larger and thinner, with higher concentrations of chlorophyll per unit mass so that they are able to capture light more efficiently. This can be observed by comparing a leaf from the top and one from the middle (inside) of a broadleaf shrub.

Light also plays an important role in governing the distribution (allocation) pattern of plant biomass among plant parts such as leaves (to maximize carbon gain) and roots (to maximize acquisition of belowground resources). Plants allocate new biomass preferentially to roots when water or nutrients limit growth and to shoots when light is limiting. Thus, the ratio of root production to shoot production is greatest in low-productivity ecosystems where low water or nutrient availability limits plant nutrient supply (Vogt et al. 1997).

Carbon Dioxide (CO_2)

Unlike C_4 and CAM, the photosynthetic rate of C_3 plants is not CO_2 saturated at current atmospheric concentrations, and therefore photosynthetic rates increase with increased atmospheric concentration of CO_2. On the other hand, all plants can close their stomates more often when CO_2 levels are elevated, leading to a decline in transpiration and increasing water use efficiency of all plants. Effects of increasing atmospheric CO_2 concentration and associated increased air temperature are further discussed in Chapter 21.

Air Temperature

Temperature generally follows the same latitudinal and seasonal patterns as light, typically with a lag time of a few days to several weeks. Thus, temperatures are high and equitable in the tropics and fluctuate more with increasing latitude (see Chapter 3). Because they are immobile, all plant species, including crops, are adapted to a limited range of temperatures and are extremely sensitive to temperature extremes. Most biological metabolic activity takes place within the range of 0–50°C, above or below which there is little activity. The optimal range of air temperatures for plant productivity generally falls between 15 and 25°C, although this varies widely (e.g., 15°C for a moss from a boreal forest and 44°C for a desert shrub). Temperature influences both photosynthetic and respiration rates.

Many living systems that do not maintain constant temperatures exhibit a two- to threefold increase in respiration with every 10°C increase in temperature. Gross primary productivity generally increases with temperature up to an optimum. Respiration rates, however, continue to increase with increasing temperature. Net primary productivity shows an optimum range, above and below which it decreases. At temperatures above the optimum, photorespiration increases, photosynthesis declines, and photosynthetic pigments are destroyed. An increase in air and leaf temperature also increases the rate of transpirational water loss. When water supply is limited, stomates close to reduce this water loss, which in turn reduces the photosynthetic rate. Conversely, all of these processes slow down with decreases in temperature.

Biological productivity of most ecosystems is hindered by periods when it is too cold or too dry for significant photosynthesis to occur. The length of time when climatic conditions are suitable for photosynthetic activity of plants is referred to as the **growing season** (Figure 7.11). At temperatures higher or lower than those at which photosynthetic capacity can sustain plant metabolic activity, plants die (annuals), lose their leaves (deciduous plants), or become physiologically inactive. In extreme

environments that are consistently cold, like the arctic and alpine, or hot, like deserts, plants acclimate and adapt to their environment chiefly by morphological and physiological adaptations, and by the reallocation of available nutrients and water to those organs in most need.

Water and Nutrients

Water and soil nutrient availability are closely coupled in determining plant productivity. Leaves and stems in the canopy can directly intercept rain, fog, or snow. Some intercepted precipitation can be absorbed directly by the leaves. The remaining contributes to the soil water by dripping to the ground as **throughfall** or running down stems to the ground as **stemflow.** Soil water affects net primary productivity not only because it is needed for growth, but also because it regulates the rate of decomposition of organic matter, which in turn liberates nutrients, adding to the nutrient supply of the soil (see Chapter 4). Soil nutrient uptake by organisms depends on the concentration gradient between the exchange sites of nutrient uptake and nutrient supply and occurs through the processes of diffusion, mass flow, and root interception.

The process of diffusion refers to the movement of molecules or ions along a concentration gradient to the root surface. It is driven by the fact that nutrient uptake by roots reduces nutrient concentration at the root surface, thus increasing the relative concentration elsewhere in the soil. The concentration gradient that is established induces nutrients to move toward the root. Roots also take up nutrients in a process known as "mass flow," which refers to the movement of nutrients dissolved in soil water flowing to the root. The movement of the soil water is primarily driven by transpirational water loss through leaf surfaces; water moves to the root, into the plant and to the leaves, and through the stomates to the atmosphere. Root interception refers to elongation of roots into new soil to intercept available nutrients in this unoccupied soil and to produce new root surface area to which nutrients can move by diffusion and mass flow. In semiarid and arid regions, there is a linear increase in plant productivity

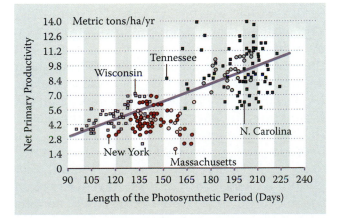

Figure 7.11 The effect of growing season length on net primary productivity of land plants. (From Lieth 1975.)

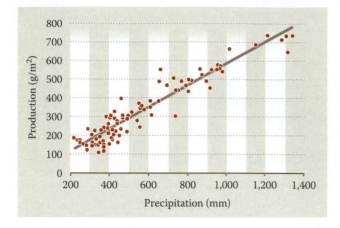

Figure 7.12 The effect of precipitation on primary productivity of land plants. (From Lieth 1975.)

with increased water availability (Figure 7.12). In more humid regions, plant productivity begins to level off at higher levels of precipitation because other factors, including light and nutrient availability, become limiting.

Topography

In northern latitudes, topographic features have a significant effect on both community composition (Wikum and Wali 1974) and productivity. The north- and east-facing slopes are cooler and hence lose less soil water; the south- and west-facing slopes are warmer and thus water loss through evapotranspiration is higher. As a result, north-facing and east-facing slopes have been found to have higher primary productivity, litter production, and development of soils than south- and west-facing slopes (Killingbeck and Wali 1978).

Herbivory

Consumption of plants by herbivores is sometimes a major avenue of biomass and nutrient loss from plants, ranging from a relatively small (2–10%) to a large (25%) proportion of plant production, as is found in productive grasslands. In addition to quantitative plant biomass losses, herbivory affects plant nutrient budgets not only by preventing retranslocation of nutrients among plant parts but also by consuming plant tissues that are high in nitrogen. Because of this, chemical and morphological defenses against herbivores and

pathogens are best developed in long-lived tissues and in environments where the supply of nutrients is inadequate to readily replace nutrients lost to herbivores.

Anthropogenic Disturbances

Human activities increasingly affect the primary productivity of ecosystems through land use and land cover changes, overgrazing, introduction of invasive species, loss of biological diversity, nutrient enrichment, stratospheric ozone depletion, global climate change, acid rain, and other types of pollution. Increases also result from agricultural nutrient applications (especially N and P). These environmental issues are discussed in Section B.

Aquatic Primary Productivity

The foundation of the biological productivity of the oceans is single-celled, short-lived algae, called phytoplankton, that drift freely near the surface of the water and range in size from 0.02 to 2 mm in length. Phytoplankton, also known as microalgae, are C_3 species and include a wide range of photosynthetic organisms such as diatoms, dinoflagellates, coccolithophores, and Cyanobacteria (sometimes called blue-green algae). A substantial portion of the ocean's total primary productivity comes from a class of tiny plankton, the picoplankton, whose size ranges from 0.2 to 5.0 μm (or 0.0002–0.005 mm in length) (Figure 7.13). The picoplankton do not produce a large crop of standing biomass at any one time, but their rate of C fixation is high, making them an important food source for higher trophic levels. Higher trophic levels also depend on the C generated by bacterial growth in the water, the so-called microbial loop (Figure 7.13). The bacteria grow on dissolved organic compounds excreted by other organisms, and small zooplankton (or heterotrophic plankton) graze on the bacteria. The zooplankton then become part of the autotrophic food web.

In certain aquatic conditions where light intensity is insufficient, such as in the deep

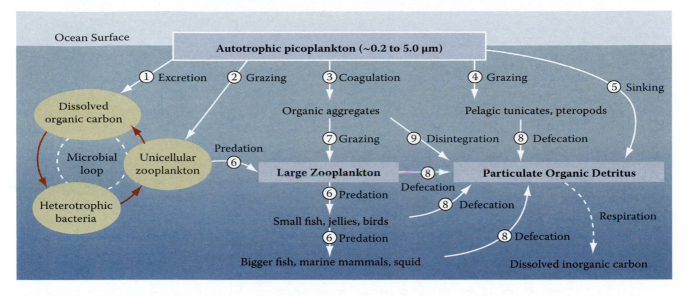

Figure 7.13 An oceanic food web based on the primary productivity of picoplankton shows the pathways of carbon flux. On the left the "microbial loop" is shown. The light-blue boxes are large zooplankton and particulate organic matter, two carbon pools that use the C fixed by picolankton. (From Barber, R. T. 2007. *Science* 315: 777–778.)

ocean and in anaerobic sediments, chemosynthetic bacteria can harvest the energy for carbon fixation by oxidizing reduced sulfur compounds, methane, hydrogen, and ammonia. Photosynthesis in aquatic ecosystems, as in terrestrial ecosystems, is controlled by the interaction of light intensity, carbon dioxide, nutrients, and water. Some oceanic Cyanobacteria plankton can convert N_2 gas dissolved in the ocean into ammonium ion (NH_4^+), but this process is primarily limited by the intensity of light, the supply of phosphorus due to slow weathering rates of rocks and subsequent inflow rates of phosphorus into the sea, and iron and molybdenum needed to form the nitrogenase enzyme involved in N fixation.

Primary productivity differs between terrestrial and aquatic ecosystems in several ways, including the rate at which the intensity of light diminishes with an increasing water depth, in what form the carbon is available, the fact that water can be limiting in terrestrial systems, and what particular nutrients are in short supply. The rate of marine primary production is mainly controlled by two factors: the light intensity in the **photic** zone and the supply of nutrients. As sunlight passes through water, selective absorption of various wavelengths causes changes in its intensity and quality, both of which are very important in controlling the primary productivity of aquatic ecosystems.

A greater diversity of pigment types is found in the marine primary producers (red, green, and brown algae) than on land. Again, this is associated with the complex light environment of water. The depletion of light through water follows the same exponential decrease with depth given earlier for penetration of light in terrestrial plant canopies. Aquatic ecosystems also benefit from the fact that chemical forms of carbon dioxide other than gaseous CO_2 are available for photosynthesis, such as calcium bicarbonate ($Ca(HCO_3)_2$), calcium carbonate ($CaCO_3$), and carbonic acid (H_2CO_3). Although aquatic plants preferentially use dissolved CO_2, some Cyanobacteria and green macroalgae like sea lettuce (*Ulva* sp.) can use bicarbonate ions under certain conditions.

Although the oceans cover 70% of the surface area of the Earth, they are responsible for only one-third of total annual net primary production (about 172.5 Gt dry weight biomass per year = 78 Gt C yr^{-1}) of the Earth (Table 7.4). On average, annual net primary productivity per unit area of land is about five times greater than that of the ocean, and dry weight biomass per given area is about 1,200 times greater on land than in the ocean (Table 7.4). Small, short-lived phytoplankton are predominant primary producers in aquatic ecosystems, whereas most primary producers in terrestrial ecosystems are relatively large and long lived.

Table 7.4
Net Primary and Secondary Productions by Ecosystem Types of the Earth

Ecosystem Type	Net primary Production (dry matter)				Biomass (dry matter)						
	Area (10^{12} m²)	Normal Range (g/m² yr)	Mean (g/m³ yr)	Total (10^{15} g/yr)	Normal Range (10^3 g/m²)	Mean (10^3 g/m²)	Total (10^{15} g)	Litter Mass (10^{15} g)	Animal Consumption (10^{12} g/yr)	Animal Production (10^{12} g/yr)	Animal Biomass (10^{12} g)
Tropical rain forest	17.0	1000–3500	2200	37.4	6–80	45	76.5	3.4	2,600	260	330
Tropical seasonal forest	7.5	1000–2500	1600	12.0	6–60	35	260	3.8	720	72	90
Temperate evergreen forest	5.0	600–2500	1300	6.5	6–200	35	175	15.0	260	26	50
Temperate deciduous forest	7.0	600–2500	1200	8.4	6–60	30	210	14.0	420	42	110
Boreal forest	12.0	400–2000	800	9.6	6–40	20	240	48.0	380	38	57
Woodland and shrubland	8.5	250–1200	700	6.0	2–20	6	50	5.1	300	30	40
Savanna	15.0	200–2000	900	13.5	0.2–15	4	60	3.0	2,000	300	220
Temperate grassland	9.0	200–1500	600	5.4	0.2–5	1.6	14	3.6	540	80	60
Tundra and alpine	8.0	10–400	140	1.1	0.1–3	0.6	5	8.0	33	3	3.5
Desert and semidesert scrub	18.0	10–250	90	1.6	0.1–4	0.7	13	0.36	48	7	8
Extreme desert (rock, sand, ice)	24.0	0–10	3	0.07	0–0.2	0.02	0.5	0.03	0.2	0.02	0.02
Cultivated land	14.0	100–4000	650	9.1	0.4–12	1	14	1.4	90	9	6
Swamp and marsh	2.0	800–6000	3000	6.0	3–50	15	30	5.0	320	32	20
Lake and stream	2.0	100–1500	400	0.8	0–0.1	0.02	0.05	—	100	10	10
Total continental	149		782	117.5		12.2	1837	111	7,810	909	1005
Open ocean	332.0	2–400	125	41.5	0–0.005	0.003	1.0	—	16,600	2,500	800
Upwelling zones	0.4	400–1000	500	0.2	0.005–0.1	0.02	0.008	—	70	11	4
Continental shelf	26.6	200–600	360	9.6	0.001–0.04	0.001	0.27	—	3,000	430	160
Algal beds and reefs	0.6	500–4000	2500	1.6	0.04–4	2	1.2	—	240	36	12
Estuaries (excluding marsh)	1.4	200–4000	1500	2.1	0.01–4	1	1.4	—	320	48	21
Total marine	361	—	155	55.0	—	0.01	3.9	—	20,230	3,025	997
Full total	510	—	336	172.5	—	3.6	1841	—	28,040	3,934	2002

Source: From Whittaker, R. H., and G. E. Likens. 1973. Carbon in the biota. In *Carbon and the Biosphere*, ed. G. M. Woodwell and E. V. Pecan, 281–302. Brookhaven Symposium of Biology. 24. Springfield, VA: National Technology Information Services; and Whittaker, R. H. 1975. *Communities and Ecosystems*, 2nd ed. New York: Macmillan Publishing Company.

Thus, the mean residence time of carbon in plant biota (the ratio of the stock of living biomass to the flux of net primary productivity) is much shorter in the oceans than on land. The open oceans account for about 80% of the total net primary production of about 55 Gt C yr^{-1} as a result of their large area, with the coastal zone accounting for 20% (Table 7.4).

There are two fates for the organic matter, and the nutrients it contains, produced in marine primary production: nutrient recycling within the sunlit (photic) zone of the portion that decomposes and "export production" of organic matter that sinks out of the photic zone before the nutrients can be recycled. This is a "biological pump" that moves carbon and nutrients from the ocean surface to deeper waters. In coastal regions, primary productivity is enhanced by the inflow of nutrient-rich estuarine water and land runoff. Also, upwelling and deep winter mixing enhance primary productivity by bringing the nutrient-rich deep ocean water to the surface. Human effects on atmospheric deposition and land runoff on biological productivity of aquatic ecosystems are profound and are discussed in Chapter 18.

Secondary Productivity

The accumulation rate of biomass by consumers or **heterotrophic organisms,** including herbivores, carnivores, omnivores, and detritivores, is referred to as **secondary productivity** and is expressed as animal growth in biomass or calories per unit area and time. Just as in primary production, secondary production can be divided into gross and net productions. Gross secondary production refers to the amount of biomass consumed by the consumers. However, of the total amount ingested by consumers, only some is digested and assimilated; the rest is excreted as feces and urine or used in growth and respiration (Figure 7.14). Net secondary production is about three times higher in aquatic ecosystems than in terrestrial ecosystems because a larger fraction of the net primary productivity is consumed in the oceans than on land (Table 7.4).

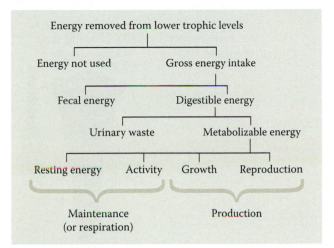

Figure 7.14 This diagram illustrates the successive loss of energy by a herbivore or carnivore. (After Krebs, C. J. 1985. *Ecology: The Experimental Analysis of Distribution and Abundance,* 3rd ed. New York: Harper & Row, Publishers.)

Much of the unconsumed terrestrial production is in relatively indigestible woody stems and trunks of trees.

Consumption efficiency is a measure of how much of the total energy available is actually consumed by the next higher trophic level. Consumption efficiency is 25–40% in the oceans and 5–10% in terrestrial ecosystems. **Assimilation efficiency** is the ratio of the amount of energy assimilated to the amount of energy ingested. Secondary productivity by herbivores is also constrained by two major ecological factors: the inability of herbivores to consume the majority of primary production consistently and the nutritional value of primary production ingested. Carnivores experience a benefit from the fact that herbivores' muscles are relatively easy to digest.

Bioaccumulation through Food Webs

Whereas energy, as discussed previously, is lost as it moves from one trophic level to the next, compounds that are nonbiodegradable, such as metals or some complex organic compounds, follow an opposite course; that is, their concentration increases with successive trophic levels. This phenomenon is referred to

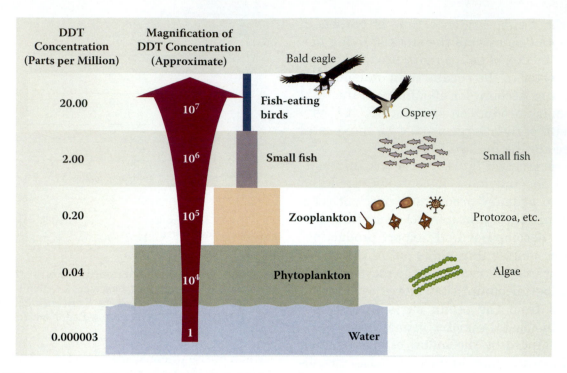

DDT Concentration (Parts per Million)	Magnification of DDT Concentration (Approximate)		
20.00	10^7	Fish-eating birds	Bald eagle / Osprey
2.00	10^6	Small fish	Small fish
0.20	10^5	Zooplankton	Protozoa, etc.
0.04	10^4	Phytoplankton	Algae
0.000003	1	Water	

Figure 7.15 Right: a trophic pyramid of energy with decreasing energy content from phytoplankton (highest) to bald eagle (lowest). Left: lowest DDT content in water and highest in the bald eagle.

as bioaccumulation, bioconcentration, or bio-magnification (Figure 7.15).

Although energy dissipates going up the food chain, the insecticide DDT (**d**ichloro-**d**iphenyl-**t**richloroethane) accumulates. This finding was historic: It was the story that Rachel Carson told in her book, *Silent Spring,* in 1962. In it, she wrote of the bird eggs of some top-predator species, such as eagles, that were made brittle by insecticide, resulting in dramatic population declines. She made a strong case against the use of such chemicals and their impact on environmental and human health. Although people were initially incredulous about her ideas, over time the indiscriminate use of DDT and some compounds like it was banned.

The experimental data to support Carson's observations and hypothesis were provided by an ecologist, George M. Woodwell, just a few years after her death.[2] His work on an east coast estuary, an extensive salt marsh on the south shore of Long Island, New York, showed the accumulation of DDT from water and up through several trophic levels (Woodwell, Wurster, and Isaacson 1967) (Figure 7.16). Many years later, he wrote an elegant essay,

titled "Broken Eggshells," on the impact of Carson's book and its contribution to the global environmental awakening; this essay was part of the American Association for Advancement of Science's "20 Discoveries That Shaped Our Lives" (1984).

Trophic Cascades

The interactions of different trophic levels in an ecosystem mean that changes in one level can affect all the other levels. This occurs when ecological processes or the sizes of different populations are altered. In this sense, the effects of such changes travel, or "cascade," through the system. This is essentially a predator-driven effect, occurring when predator populations are added to or removed from a community. When predators are introduced, they act to suppress their prey populations. Lower prey populations, in turn, mean that the food supply of the prey is released from predation, so songbird numbers will increase. In this way the effects alternate across trophic levels, until the level of the primary producers is reached. Some of the earliest demonstrations of this came from studies of freshwater lakes (Carpenter and Kitchell

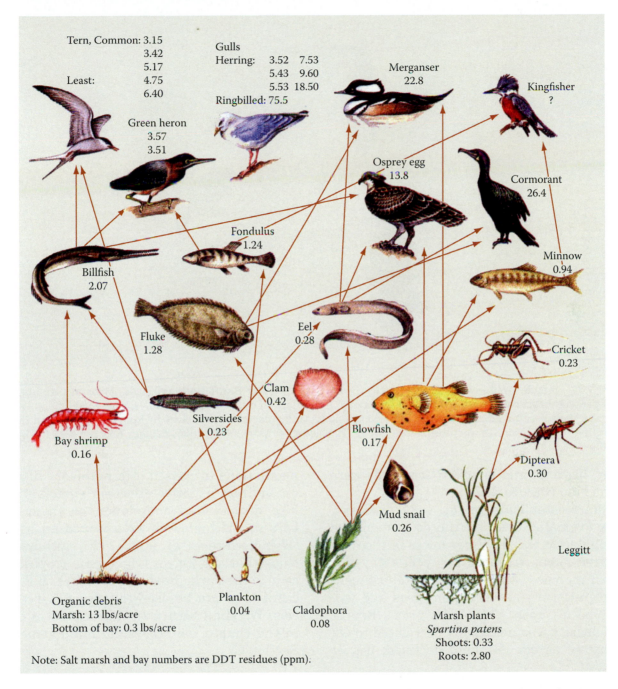

Figure 7.16 The bioaccumulation of DDT in a salt marsh food web, from plankton, algae (*Cladophora*) to marsh plants to insects, fishes, and birds. (From Enger, E., and B. F. Smith. 1995. *Environmental Science—a Study of Interrelationships*. Dubuque, IA: W. C. Brown.)

1993), although there is some debate about how universal this phenomenon is.

Based on their ability to alter the trophic structure of a community, trophic cascades have been used as a management tool in freshwater lakes (to manage algal blooms and fish diversity) and in terrestrial ecosystems (to protect wildlife species). One of these

studies documented the effect of reintroducing wolves, a top predator, to Banff National Park, Canada. In what the authors call a "serendipitous natural experiment," a portion of the park was colonized by wolves (high wolf area), while another portion was not (low wolf area) (Hebblewhite et al. 2005). The impacts of wolves on the population density of elk, aspen,

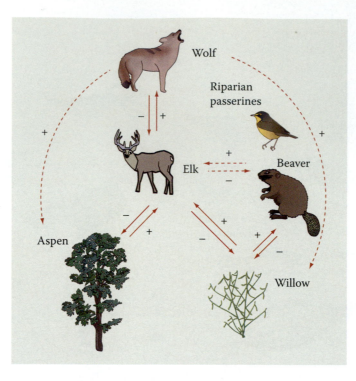

Figure 7.17 Simple trophic interactions model of the Bow Valley of Banff National Park, Alberta, Canada. Solid lines represent direct consumer–resource interactions: + for effect on upper trophic level, – for negative effect. Indirect effects are shown with dashed lines and represent an indirect displacement of wolves by human activity and indirect (exploitative) competition between herbivore levels. (Hebblewhite, M. C. et al. 2005. *Ecology* 85:2135–2144; Figure 1, p. 2138.)

willow, beavers, and riparian songbirds were documented (Figure 7.17).

The reestablishment of wolf predation had a dramatic effect on the ecosystem; the elk population was 10 times lower in the high wolf area compared to the low wolf area. This in turn increased the seedling establishment and population sizes of aspen trees and willows, species upon which elks graze. Beaver lodge density was negatively correlated to elk density, so as elk numbers declined, beaver populations increased. Elk herbivory also has an indirect negative effect on riparian songbird density and abundance, so their numbers in the high wolf area increased as well (Hebblewhite et al. 2005). The authors conclude that the wolves triggered a trophic cascade in the park, providing evidence that large carnivores can be used to help conserve species. We shall return to this aspect in detail in the chapter on ecosystem management in Section C.

Notes

1. Much research is under way to understand how plants "break" water into hydrogen and oxygen. It may provide most beneficial clues to what is known as the "hydrogen economy"—that is, getting supplies of fuel energy from hydrogen. We discuss this more in Chapter 17 on fuel energy in Section B.

2. Several papers by George Woodwell are given in the references. The paper "Broken Eggshells" is highly recommended. We will return to the insecticide–pesticide topic in Section B.

Questions

1. Explain how terrestrial and aquatic ecosystems make approximately equal contributions to total global photosynthesis despite great differences in the biomass of their primary producers.
2. How are nutrients brought from the deep ocean to the surface? Why is this important?
3. What factors determine variations in terrestrial and aquatic ecosystem productivity?
4. Explain why we cannot use, convert, transfer, or store energy with 100% efficiency.
5. How is the energy gained by autotrophs during photosynthesis lost at the higher trophic levels?

Population Ecology

8

Populations—the Central Units in Ecology and Evolution

As we know from the fossil record and direct observation, living organisms can (and do) change over time. Some groups have become extinct, while others have become more diverse. Population changes do not come about because an individual changes genetically in response to its environment. Rather, it is a population that changes as the relative frequency of different genes (or different forms of genes, alleles) changes within the population. Any genetic change that confers an advantage in survival and reproduction is more likely to be passed on. If, over generations, this leads to a change in gene frequency, then evolution has occurred. However, environmental factors often act randomly and do not necessarily lead to change. A rodent that has a greater range of diet and is capable of producing healthier offspring might just as easily be killed by a predator or a storm as her littermate that lacks these characteristics. The environment is composed of a complex interplay of factors that may confer advantages and disadvantages to a species and to individual organisms.

For centuries, humans have used their knowledge of "selection" to select for certain desirable traits in countless domestic plants and animals. However, despite a long history of breeding by humans into endless variations from original forms, we can still recognize all dogs, all apples, and all horses as such. Biologists refer to these as species. The branch of biology that covers the study and naming of different kinds of species is called **systematics.** The specialists in this field (called systematists) generally agree that a species—whether it is a fungus, protist, bacterium, plant, or animal—consists of a group of organisms that possesses characteristic common morphological, anatomical, behavioral, and physiological features and can interbreed and produce fertile offspring (or have the potential to do so). Each species has a particular common ancestry and descent and is reproductively isolated from other species.

The species concept, therefore, has been central to the field of biology in studying genetic composition and evolution and now is very useful in protecting the species diversity of the world, particularly those that have become rare and endangered. Thus, the species concept has also a legal component, discussed in Chapter 22. Worldwide, biologists have given a scientific name to (or "described") at least 1.5–1.8 million species of plants and animals (Wilson 2000a).

For many ecologists, however, the population is the central focus of study. A **population** comprises the interbreeding members of a species found in a given habitat, and these individuals collectively express genetic, morphological, physiological, and behavioral variations. For this reason, populations are the functional ecological and evolutionary units. In the definition of populations, there should be separation or discontinuity (also called disjunction) from other populations in one of the following three characteristics: (a) spatial location, (b) gene flow, or (c) demographic structure (Wells and Richmond 1995). In other words, population A is considered distinct when it is disjunct from population B by at least one of these features.

The number of organisms of a species found in a given area at a given time is the **population density.** Not all individuals of a population at any given time are of the same age, size, or sex. Some are young, some maturing, and others old. New individuals are produced while others die, thus forming a definite age structure for the population. Based on the abiotic and biotic characteristics of a particular habitat, as well as their adaptations through evolutionary time, populations exhibit a characteristic dispersion (pattern of distribution) in both space and time.

The rates at which individual organisms are added to and lost from a given population due to births, deaths, emigration and immigration, the age structure of the population, and variations in population size over time (year to year) all constitute **population dynamics.** Populations gain and lose individuals continually and yet maintain a dynamic balance that is influenced by a host of environmental factors and characteristics intrinsic to the species. For example, competition among individuals will increase as a resource becomes scarce. This, in turn, may lead to decreased survival and reproduction within the population. As a consequence, population growth will slow or may even become negative. Such populations could be regarded as self-regulating systems.

Historically, nearly all of the population ecology concepts have come from the studies conducted by animal ecologists. It was relatively recently that these concepts have been applied to terrestrial plant ecology, which will be discussed in Chapter 9.

Population Growth

All organisms, without exception, have a propensity to produce as many offspring as they can; if they do so, this leads to the intrinsic rate of population growth, also known as the "biotic potential" of the species. Thus, if environmental resources and other ecological conditions do not constrain population growth, exponential growth will occur. For example, if a single cell of baker's yeast (*Saccharomyces cervisiae*) were to reproduce unconstrained, it is estimated that in just 2 weeks it could form a layer around the Earth that would be 3 m deep (Hofman 1992). A pair of fruit flies may in one year produce a mind-boggling 10^{41} descendants (Evans 1995). Similarly, if the biotic potential of a mating pair of houseflies were actually reached, the number of flies in the seventh generation would be 5.7 trillion (Table 8.1) (Kormondy 1996).

These stupendous numbers illustrate the power of exponential growth (See Box 8.1). If a population were to achieve exponential growth with unlimited availability of resources such

Table 8.1
Biotic Potential of House Flies (*Musca domestica*) **in 1 Year**[a]

Generation	Total Population If All Survive 1 Year But Reproduce Only Once
1	120
2	7,320
3	439,520
4	26,359,320
5	1,581,559,320
6	94,893559,320
7	5,693,623,559,320

Source: From Kormondy, E. J. 1996. *Concepts of Ecology,* 4th ed. Upper Saddle River, NJ: Prentice Hall.

[a] Based on the reasonable assumptions that a female lays 120 eggs per generation, that half of these eggs develop into females, and that there are seven generations per year (and, of course, that none of them die).

BOX 8.1   Population Growth Equations

In the following equations, N is a population density, N_0 is an initial population density, N_1 is density after a unit time ($t = 1$) of population growth, and N_t is density after any time period, with growth at a constant rate of increase.

A. Growth without a limit

Geometric formulae:

$$N_1/N_0 = R \qquad\qquad \text{rate of increase per unit time} \qquad (1)$$

$$N_t = N_0 R^t \qquad\qquad \text{density after time units } t \qquad (2)$$

Exponential formulae:

$$dN/dt = N_0 r \qquad\qquad \text{rate of increase} \qquad (3)$$

$$Nt = N_0 e^{rt} \qquad\qquad \text{density after time } t \qquad (4)$$

B. Growth with limits

Logistic formulae:

$$dN/dt = rN\,(1 - N/K) \qquad\qquad \text{rate of increase} \qquad (5)$$

$$N_t = K/1 + [(K - N_0)/N_0]e^{-rt} \qquad\qquad \text{density after time } t \qquad (6)$$

K is the carrying capacity or limiting density

e is the base of natural logarithms, 2.718

C. Competition among two species

$$dN_1/dt = r_1 N_1 . (K_1 - N_1 - \alpha N_2 / K_1)$$

$$dN_2/dt = r_2 N_2 . (K_2 - N_2 - \beta N_1 / K_2)$$

N_1 and N_2 are populations of species A and B at a given time t.
r_1 and r_2 are their intrinsic rates of population increase of A and B.
K_1 and K_2 are the environmental resource limits or carrying capacity of the environment in the absence of one another.
α and β are competition coefficients that express through the effects of population levels of the competing species.

These equations imply that, for most relationships, one species will increase while the other will decrease until, at equilibrium, the latter is extinct. There is, however, another possibility: in the case where $\alpha < K_1/K_2$ and $\beta < K_2/K_1$, one species can survive in the presence of the other. This point can be expressed in two ways: (1) the two species have adjusted their mutual strategies, or (2) the species have divided the environmental resources regulating their own growth. (Whittaker 1975; for C, two-species equations, see Hutchinson 1965.)

as food and space, it would realize its biotic potential as in the previous examples of exponential growth. Assuming that there are no environmental constraints, an exponential rate of population growth (a *J*-shaped curve) can be represented by the equation

$$dN/dt = rN,$$

where *dN/dt* is the change in the number of organisms over time, *N* is the number of organisms, *t* is time, and *r* is the intrinsic rate of increase (Figure 8.1). This equation can also be written as

$$N_t = N_0 e^{rt}.$$

In this model, *r* represents the difference between the per-capita birth rate (average number of births per individual per unit time) and per-capita death rate (average number of deaths per individual per unit time). If the value of *r* is positive, it means that the population size is increasing; if *r* is negative, then the population size is decreasing. An important assumption of this model is that *r* is constant over time, which can only happen if resources are not limiting. This is obviously an unrealistic assumption.

Under natural conditions, populations of any species never achieve their biotic potential because environmental conditions impose strong constraints on indefinite population growth. These constraints include the physical availability of resources (such as food and space) and such biotic influences as competition for limited resources, predation, and diseases. When environmental constraints are imposed by limited resources, they determine the **carrying capacity** of the environment (*K*) or simply how large a population a given habitat can support. Population growth will slow as it approaches carrying capacity due to increased death rates, decreased birth rates, or both.

Carrying capacity, sometimes also called environmental resistance to growth, describes the maximum sustainable population size that a given environment can support. Thus, it is what determines the actual population size in a given environment. As population size increases, a typical growth curve shows four phases: (1) accelerating, (2) logarithmic, (3) decelerating, and (4) equilibrium (steady state) (Figure 8.1). The equilibrium, also called the asymptote, can oscillate around the mean carrying capacity (Figure 8.2). This curve is generally *S*-shaped and referred to as a sigmoidal, or logistic, curve.

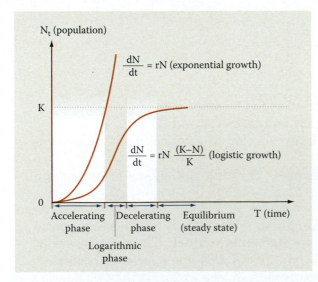

Figure 8.1 Exponential and logistic growth curves of a population. *r* = intrinsic rate of population increase; *K* = population size at carrying capacity; *N* = number of individuals at a particular time; *dN/dt* = rate of population increase.

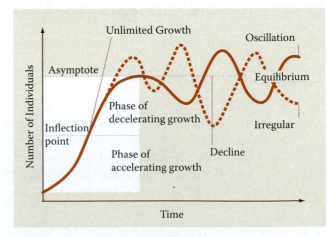

Figure 8.2 Oscillation and irregular forms of dynamic equilibrium. These are two forms of logistic-growth population size over time. The dotted line represented by "decline" is associated with the "unlimited growth" characteristic of the exponential growth curve. This illustrates the fact that the population of a species with exponential growth will ultimately overexploit available resources, and the population will dramatically decline.

The carrying capacity of the environment (*K*) will set a limit to the number of organisms in a population; hence, the logistic equation takes the form:

$$dN/dt = rN \frac{(K - N)}{K}$$

where (*K* − *N*) indicates the amount of unused resources or the "space" available for additional individuals in the environment. *K* is not a constant value; rather it varies over space (i.e., where a particular population is located and the resources available there) and through time (e.g., food availability in a habitat may vary from year to year).

The logistic equation states, in very simple terms, the relationship between *K* and the population size, *N*. For example, if *K* is much larger than *N*, resources will be abundant, and population size may increase. Mathematically, the term (*K* − *N*)/*K* denotes that as the number of individuals (*N*) approaches *K*, resources become more limiting. For example, if the carrying capacity of a habitat is 100 (meaning 100 individuals can be supported, or *K* = 100) and the actual population size is 10 (*N* = 10), then (*K* − *N*)/*K* = (100 − 10)/100 = 0.9 (90%), and population growth (*rN*) proceeds at 90% of its maximum level (*rN*). If, on the other hand, *N* = 90, then (*K* − *N*)/*K* = (100 − 90)/100 = 10/100 = 0.1 (10%). In this case, the population will increase at only 10% of its maximum level. When the population reaches the carrying capacity, *K* = *N* and the population is said to have stabilized with respect to the carrying capacity. A prediction of the logistic equation is that the rate of population growth will first rise and then fall until it reaches equilibrium and stops, as shown in Figure 8.3. An outstanding question is whether or not human populations have reached or will reach carrying capacity. This will be discussed in Chapter 11.

The concept of carrying capacity arose in livestock management to describe the interactions for food, space, and shelter between members of the same population and with members of other populations. The emphasis in this context was on obtaining the maximum yield of animals. But now the concept of

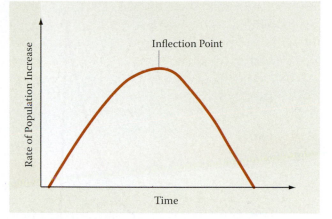

Figure 8.3 Parabolic curve showing the rate of population growth over time for populations undergoing logistic growth.

carrying capacity is used in a much wider context, ranging from the management of wildlife populations to the sustenance of humans in a given area. Thus, it has been rightly argued that carrying capacity should be scientifically investigated to reflect each different situation (MacNab 1985).

Fluctuations in environmental characteristics, such as climate, can increase or decrease the carrying capacity of a habitat, thereby affecting population size. Consumption of resources (e.g., food, nutrients) by a species can also impact the carrying capacity. The rates of resource consumption and resource regeneration within the habitat are important with regard to the sustainability of populations.

If a population does increase above its *K*, then overexploitation of resources will occur, adversely affecting the capacity of the environment to regenerate those resources and meet demand. This leads to a decrease in the carrying capacity, and the population will decrease—possibly quite dramatically. An example of what can happen when a population exceeds its resource base is shown by the growth of a reindeer population that became established on a small island off the coast of Alaska. Originally, 29 individuals were released. Grazing on the abundant lichens of the island, the population grew exponentially over the next 20 years until the herd reached approximately 6,000 individuals. The high reindeer numbers were an overshoot of the carrying capacity of the island and led

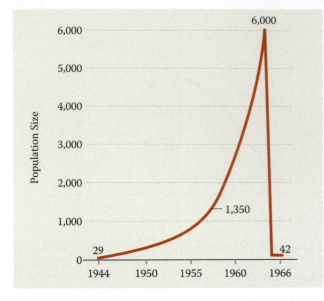

Figure 8.4 Population size of reindeer on St. Matthew's Island, Alaska, showing that, following the release of 29 individuals, the population increased exponentially, peaking at more than 6,000. Overgrazing finally led to a population crash, with only 42 animals surviving. (Klein, D. R. 1959. Special Science Report Wildlife no. 43, Washington, D.C.: U.S. Department of the Interior, Fish and Wildlife Service. Image at http://dieoff.org/page80.htm.)

to severe overgrazing of lichens. With the destruction of their food supply, the reindeer population crashed, leaving only 42 animals surviving the following winter (Figure 8.4). More generally, when a population overshoots the carrying capacity and then declines, the result can be that population numbers oscillate around *K* without ever arriving at equilibrium (Figure 8.2) (Gotelli 2001).

A population can decrease the carrying capacity of its habitat not only through overconsumption of resources (such as the overgrazing just described) but also by the production of toxic waste products. Hence, consumption of resources and the production of wastes should only occur at levels that the environment can accommodate without reducing its long-term carrying capacity. This is important for any species, be it a plant, a fungus, an animal, or a human.

The population density of a habitat is influenced not only by extrinsic factors such as food availability, disease, or climate but also by factors intrinsic to the species, which include fecundity (number of offspring produced by an organism) and age to sexual maturity. Another intrinsic factor is the body size of the

organism. Using reasonably large data sets (which should always be a requirement of ecological studies) of 400 different benthic stream invertebrates from two separate stream communities in Austria and Wales, researchers recently found that population density is inversely proportional to body size. Body size influences several other aspects of population dynamics, such as the energetic requirements of an organism, the way the population exploits its potential resources, and its susceptibility to predation (Schmid, Tokeshi, and Schmid-Arya 2000).

Population Survivorship Patterns

The members of a population that are the same age constitute a **cohort.** Following the addition of a new cohort to a population through births, the cohort's numbers will decrease over time through deaths or emigration. Typically, for any species, there will be certain ages during which the mortality rate of the cohort is high and other ages when it is lower. This pattern of age-specific mortality on the survivorship of cohort members is represented by **survivorship curves** (Figure 8.5).

There is tremendous variation in age-specific death rates of individuals among species. Consequently, survivorship curves showing

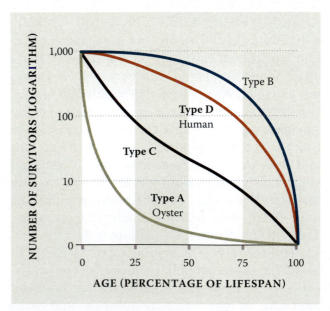

Figure 8.5 Possible types of survivorship; in the curves, x-axis shows increasing age from left to right and y-axis the number of organisms on a logarithmic scale.

the relative number of survivors as a function of time can assume several forms (Figure 8.4). A concave-shaped curve results when a large proportion of offspring die young and only a few survive to old age (Figure 8.5, curve A). This is the most common survivorship pattern found when many offspring are produced but few reach maturity (e.g., oysters, amphibians, and oak seedlings) (Odum 1971)). A less common situation occurs when juvenile mortality in the population is low, but a large number of individuals are lost as the population matures, resulting in a convex survival curve (Figure 8.5, curve B). A few large mammals display this survivorship pattern, particularly when extensive parental care protects offspring, but mortality is higher for aging adults.

The survivorship of the current human population closely approaches that represented by type "D," where, thanks to the availability of food and medicine, infant mortality has been reduced considerably and the longevity of life enhanced. An intermediate type of survivorship would appear to be more or less linear (Figure 8.5, curve "C"): Individuals of a population are lost at a constant rate throughout the potential life cycle of the organism. This is the least common survivorship pattern seen in nature. Given the different mortality pressures experienced by juvenile and adult organisms, it is not surprising that few populations in nature show such a constant rate of mortality.

Regulation of Population Size

Density-Dependent Regulation

The increase of a population to the carrying capacity of an environment invariably leads to density-dependent (intrinsic) effects that act to regulate population size. As described in the logistic equation, when population numbers are high, the impacts of disease or predation may become more severe, and the competition for resources such as nesting sites, food, nutrients, or water will intensify. In this negative

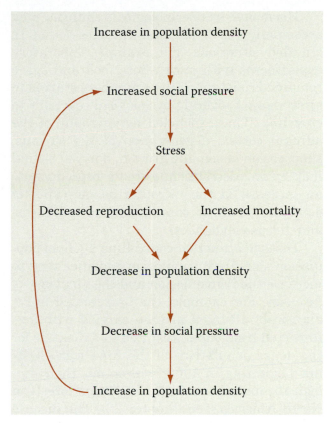

Figure 8.6   Stress hypothesis of populations illustrated by the processes driving the cycles in population abundance of lemmings in Canada's Northwest Territories: Increased social pressure results in increased physiological stress. Stress, in turn, adversely affects fecundity, increased mortality, and eventually population growth rate. As population density decreases, so does physiological stress, and the population increases. (Krebs, C. J. 1964. Technical paper no. 15, Arctic Institute of North America.) The stress of defending territories and crowding are strong inducements for individuals to emigrate.

feedback loop, competition increases, resources become more limiting, and population growth slows. For prey populations, predation pressure may also become more intense as their population size increases. This will be discussed in a later section on interspecific interactions.

An increase in population density can be detrimental to the population as a whole, perhaps bringing about behavioral changes that in turn can trigger stress, a decrease in reproduction, and an increase in death rates (Figure 8.6). In this way density regulates population numbers and the severity of density affects increases with population size. Depending on the life span of the organism, recovery of the population may be protracted.

Hormonal (endocrinological) changes may accompany the behavioral changes found in crowded situations. For example, laboratory experiments with mice whose food and space remained constant as their numbers started to grow showed several physiological reactions to crowding. These included hypertrophy of the adrenal cortex, degeneration of the thymus, suppression of somatic growth, delayed sexual maturation, decreased desire for reproduction, an increase in the intrauterine mortality of embryos, death or cannibalism of newborns, and disease epidemics.

Disease is another controlling agent of population size. High population densities tend to increase the transmission and the virulence of diseases. For example, the unintended introduction of a fungal disease carried with logs imported from Japan resulted in a disaster for the dense populations of the American chestnut beginning in 1905, wiping out their ecologically and economically important forests in North America (see Chapter 22). The disease constricts growth in this species in such a way that individuals cannot attain a normal tree-like form. Regrowth from the stump of the dead tree assumes a shrub-like form with multiple shoots. By 1950, the abundant chestnut forests had been wiped out, and, despite all the recent advances in genetics and biotechnology, the species has still not recovered.

Density-Independent Regulation

Population densities are also strongly influenced by density-independent (extrinsic) factors. In the case of density-independent factors, the size of the population has no influence on the magnitude of the effect. We owe much of our understanding of density-dependent factors to mammalian ecologists and of density-independent factors to insect ecologists. Many insect populations cycle from periods of extreme rarity to superabundance; their numbers never show signs of coming under density-dependent controls (Figure 8.7). These boom and bust cycles are often due to weather; when conditions are favorable, population numbers rise exponentially and then crash when conditions change.

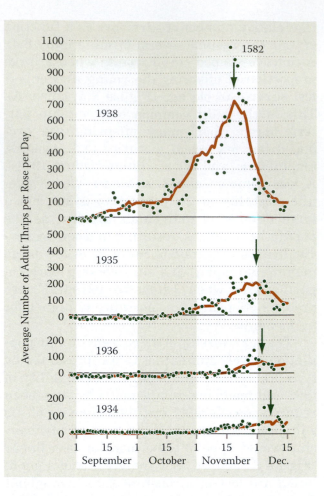

Figure 8.7 Annual population cycles of thrip irruptions over 4 successive years, illustrating the variability of year-to-year irruptions and density-independent controls. (Andrewartha, H. G., and L. Birch. 1954. *The Distribution and Abundance of Animals.* Chicago, IL: University of Chicago Press.)

A classic example involves the Australian thrip species *Thrips imaginis,* tiny insects that produce many generations of offspring per year. As adults they live in flowers, so their success depends on the number of flowers available, which in turn depends on weather. Populations can grow to huge numbers during wet seasons. In the dry season, however, populations quickly crash as the flower supply becomes limiting (Andrewartha and Birch 1954). No density-dependent population control is evident in this case, although some researchers contend that density-dependent factors set the baseline (lower limit) of the population. Population "explosions," or irruptions, are possible from this low point when resources become temporarily abundant. In any case, a short generation time is necessary to take advantage of this temporary resource glut.

Weather patterns are the most commonly cited density-independent factor that alters the rise, fall, and abundance of populations. Extreme occurrences, such as storms, floods, freezes, or droughts, can have particularly large effects on population numbers. These are sometimes called stochastic events because they occur by chance, regardless of the size of a population. Recent typhoons in Asia, estimated to have killed tens of thousands of people, are an example of density-independent impacts on human populations.

Human Impacts

Of all the factors that affect plant and animal populations, human activity is perhaps the most important. The selective introduction and elimination of some species; introduction of harmful chemicals (such as pesticides) into air, water, and soil; and the loss of habitats all have a detrimental impact on plant and animal populations. The loss of stratospheric ozone, increase in ultraviolet-B radiation, and air pollution have had (and will continue to have) differential but deleterious impacts on populations. These are discussed in detail in Section B of this text.

The alterations and modifications of the landscape that result in habitat destruction have a major impact on population densities. The alteration of some natural ecological factors—particularly disturbance events (e.g., floods and fire)—can greatly affect the composition, abundance, and species richness of plants and animals. Elimination of fires over extended periods in the western United States has resulted in the excessive accumulation of organic (plant and animal) debris that frequent small fires would have removed. This organic matter served as a fuel source to sustain record-setting large fires in several western U.S. states in the summers of 2000 and 2002, and the adverse effect on populations was immense. Although many of the populations will return over time, the large changes in population densities confirm that fire is a major factor in controlling population numbers.

Spatial Patterns of Population Abundance

Many studies have demonstrated that populations have aggregated or clumped distributions in space. If the populations were dispersed uniformly or randomly, it would mean less competition for resources among organisms, but such distributions are uncommon for several reasons. First, environmental resources occur neither uniformly nor randomly; organisms congregate where the resources they require are most abundant. The clumped nature of human populations, both across the world and in specific regions of individual countries, favors this kind of distribution due to both resource availability (e.g., a constant water supply) and social connections.

In addition to environmental resources, both behavioral and reproductive patterns of populations favor clumped distributions. Indeed, several characteristic behavioral patterns help them share environmental resources, including those illustrated later. A new dimension has been added recently to the clumping patterns, now also referred to as the "swarm theory." The central tenet of this theory is that groups or "swarms" of organisms, including humans, have greater collective power in negotiating situations than do individuals.[1] However, in the case of humans, "urban aggregation" also presents very specific problems (see Chapter 17).

Territoriality and Home Range

Animals tend to inhabit a given space—their **home range**—for most of their lives and may even aggressively defend it against encroachment by others. These animals use the area to obtain their food (e.g., lions and tigers hunting for prey). The defended portion of the home range is called a **territory.** An individual defends territory, establishing an area whose resources are protected. This is a means of ordering competition for essential resources.

Defending a reproductive territory is frequently described for birds but has also been well documented in ungulates and is achieved in several ways (Owen-Smith 1971). For example, two-thirds of the adult male white rhinoceroses in South Africa were found to use a series of four behaviors to defend their territories. First, the bulls practiced range exclusiveness: They prevented other bulls from entering their non-overlapping ranges, forming a mosaic of territories of about 2 km² in area. Bulls seldom leave this territory except to obtain water in dry seasons. Second, the bulls defended their territories through ritualized encounters that take place whenever two bulls approach a common line between their territories. The encounters are more of the horn-touching type than serious clashes. Third, the cows in estrus were confined in a territory by gentle squealing or blocking. But the bulls do not pursue if the cows move beyond 200–330 m of the territory boundary. Lastly, urination and defecation were used for scent-marking of territories. Owen-Smith notes that this territorial system allows courtship and mating to proceed without interference from other males.

Although large animals may maintain territorial boundaries behaviorally, many small animals and plants use chemical means (Schmid et al. 2000). The advantages of territoriality include keeping population density well below the carrying capacity of a range, increasing the sustainability of the habitat's resources, and, in the case of reproductive territories, limiting reproduction and slow population growth.

Metapopulations

A typical landscape contains a variety of habitats. For example, in many areas, one might see wetland ecosystems distributed across the landscape, with areas of upland (forest or agricultural land) separating them. Each wetland may support a population of ambystomid salamanders that use the wetlands for reproduction in the early spring. Following the development of larva and their eventual metamorphosis, the adult salamanders will disperse, with some proportion relocating to nearby wetlands that also contain salamanders. In dry years, some small wetlands will dry out before metamorphosis occurs, leading to population crash in those sites. This scenario, in which individuals move between discrete habitat patches—each of which supports an extinction-prone population—is described by the concept of the **metapopulation.**

A metapopulation, defined as a "population of populations," consists of several local populations that are linked by immigration and emigration (Levins 1969). Generally, three basic criteria are used to define a metapopulation: The populations must be spatially subdivided, they must be linked by dispersal and gene flow, and each population must have its own independent probability of extinction and colonization. Metapopulation dynamics are important to the persistence and survival of the overall populations involved. If a subpopulation becomes extinct in one patch, that patch can be recolonized by immigrants from nearby subpopulations through their powers of dispersal.

In a metapopulation, individual local populations that occupy a specific habitat are separated from one another by less suitable areas, producing a patchy distribution of the species over the landscape. For example, the metapopulation of Southern California spotted owls consists of populations of owls that occur in 22 distinct patches over the San Bernardino Mountains landscape (Figure 8.8) (Lahaye, Gutiérrez, and Akçakaya 1994). Within an individual habitat patch, the members of a local owl population will likely interact in a variety of ways, including competing for resources and mates. Interactions between members of the local populations in separate patches within the metapopulation will become more limited as distances between patches and connections between patches decrease. This isolation of local populations can lead to increased genetic diversity over time within the metapopulation, and it can also isolate local populations from disease and other environmental conditions such as fire.

When connections between local populations are present, the movements of individuals between patches can be very important in stabilizing local population levels. Emigration (the movement of individuals out of a local population) will decrease intraspecific competition and minimize overexploitation of the habitat,

Figure 8.8 Metapopulation of the Southern California spotted owl in the San Bernardino Mountains. The numbers in parentheses are estimated carrying capacities for each patch. (From Lahaye, W. S. et al. 1994. *Journal of Animal Ecology* 63:775–785.)

especially when the population originally had a high density. Habitat patches with declining populations may have their populations supplemented by the immigration (movement into the habitat) of individuals from over-populated patches, thus decreasing the likelihood of extinction in underpopulated patches of habitat. It is easy to see why the concept of metapopulations is of great importance in the design and maintenance of wildlife preserves and parks (Hanski 1998).

Social Hierarchies and Societies

Individual organisms within populations tend to self-organize into loosely accepted hierarchies or highly ordered societies, providing another mechanism for successfully adjusting to living in a common habitat (See Box 8.2). For example, in a flock of hens, a social hierarchy of dominance–subordination develops through the establishment of a "pecking order" wherein dominant individuals bring about a hierarchy in the flock by physically pecking the subordinates, who in turn peck others.

In contrast, the societies of hymenopteran insects such as ants, bees, and wasps are much more ordered and based on a division of labor. Queens, soldiers, and workers retain their functional roles rigidly, as reflected in their anatomical diversification. For example, within complex societies of ants, the queen's role is solely that of reproduction. The workers and soldiers will function to protect the queen and the developing ants (eggs, larvae, and pupae) and provide food, while maintaining optimal living conditions within the nest. This cooperative arrangement enhances the success and perpetuation of the colony. Some human societies also provide good examples of hierarchies based upon a division of labor. In the historical caste system in India, members of the society were designated thousands of years ago into several distinct groups—teachers, farmers, warriors, and cleaners—that became so rigidly codified that today it still influences societal structure and the types of employment available to individuals born into particular castes.

BOX 8.2   Social Organization of Elephants (from Meredith 2001)

"Modern biologists have discovered that [elephants] possess one of the most advanced and harmonious social organizations in the world of mammals.

"The [Scottish biologist Iain Douglas-Hamilton writes that the] basis of the elephant society is the family unit, led by a matriarch and consisting of sisters and cousins, with their various offspring. Family members are bound by strong ties of loyalty, which lasted for life. Cows were devoted to the well-being and protection of their offspring, exercising parental care well into their [calves'] early teenage years. In times of distress and danger they held fast to their family ties and swiftly combined to ward off threats by predators. Family units in turn were linked to wide 'kinship' groups, usually related [to] whose company they often shared.

"They form close relationships with older brothers and sisters, spending hours in playful activity (p. 152). In the event of trouble, elephants act readily in mutual help out of loyalty and compassion. When alarmed, they quickly bunch together, with the matriarch taking a prominent position while calves are protected within the phalanx (p. 153). Beyond their immediate circle, family groups maintain strong ties with other elephant families. ... Family reunions invariably inspire displays of affection. Even after short periods of separation, while feeding apart, family members greet one another with loud rumbles, head postures and ears spread out. Families meeting after days apart make a special occasion of it (Meredith, M., *Elephant Destiny*, New York: Public Affairs.)" (p. 154).

Life Histories of Organisms

The life history of an organism refers to the pattern of growth and reproduction over its lifetime. This includes such characteristics as changes in morphology during development, age at sexual maturity, fecundity, modes of dispersal, and life span. Life history strategies have been postulated to have different modes, characterized by either *r*- or *K*-selection. (The variables r and K are named for the terms in the logistic equation: r refers to exponential growth rate of a population and K to the carrying capacity.) The *r*-selected species are characterized by an early age of reproduction, high reproductive rates, and low survival (Table 8.2). Species with *K*-selected traits are just the opposite.

Pianka (1970) suggests that insects, which have wide population fluctuations, are *r*-selected, while mammals, with less fluctuation, are *K*-selected. Though some aspects of this hypothesis are still debated, the recognition of these strategies is useful, for it may provide insight into the trade-offs species make in order to complete their life cycle. For instance, individuals must allocate the energy available

to them into growth, maintenance, and reproduction. Not all life functions can be maximized at the same time, so trade-offs must be made. One such trade-off is between the number of offspring produced versus the amount of parental care each receives. One can invest a great deal of resources and parental care in very few young (*r*) or, conversely, invest little or no care in many offspring (*K*). Humans typically have one offspring at a time, with each receiving extended parental care, while many species of fish, which can lay hundreds of eggs during each reproductive event, invest no parental care to helping with their survival. In addition to growth, development, and reproductive biology, the body size of a species is one of the important determinants of the potential pace of population growth and the vulnerability of a species to extinction (Bonner 1965) (Figure 8.9).

In general, the time required for the growth of members of a species to full size and reproductive maturity increases as body size increases. Correspondingly, the likelihood of an individual's death before it reproduces becomes greater with increasing time to maturity. In addition, large animals (for example,

Table 8.2
Some Attributes of *r*- and *K*-Selection

	r-Selection	*K*-Selection
Environment	Unpredictable	More stable
Mortality	High juvenile mortality	Low juvenile mortality rate
Population size	Variable, below and above carrying capacity	Constant, close to carrying capacity
Intra-and interspecific Competition	Variable, often weak	Usually strong
Selection favors	Rapid development	Slow development
	High-resource thresholds[a]	Low-resource thresholds
	Early reproduction	Delayed reproduction
	Small body size	Large body size
	No parental care/small seed size	Extensive parental care/large seed size
Length of life	Usually shorter	Usually longer
Leads to	High productivity	Higher efficiency in reproductive output

Source: Modified from Pianka, E. R. 1970. *American Naturalist* 104:592–597.
[a] High-resource threshold means that these species can survive in wide ranges of conditions, in contrast to *K*-strategists that have narrow tolerance ranges.

whales, rhinoceroses, and elephants) typically give birth to a small number of offspring at any one time. Consequently, the population growth rates for these *K*-selected species are low. If death rates are also high—for example, due to overharvesting—and births are low, then there will be a decline in the population toward extinction. Some other aspects of species' *r*- and *K*-strategies are discussed in relation to ecological succession in the next chapter.

Interspecific Interactions and Population Dynamics

The interactions between organisms belonging to different species are referred to as **interspecific interactions.** The long-term persistence, size, and density of a population are commonly influenced by its interactions with other species. Such relationships may benefit or harm the two species in different ways, so several categories of interaction are recognized (Table 8.3). The totality of the relationships between populations forms the web of interactions that organize a community (Chapter 9).

In **commensalism** between two species, one benefits while the other neither benefits nor is harmed. Benefits of commensalism range from physical support to receiving food, shelter, or transport. Examples include the nesting of birds in trees (the birds benefit, while the tree is indifferent) and climbing plants that use other plants for support. Many lichens grow on the bark of trees without harming the tree. Examples also include some fish species (e.g., clownfish) and sea anemones. The fishes derive shelter and protection from predators, while the sea anemones are not harmed in any way. In fact, the relationship between the two species is so well established that the clown fish have adapted, over evolutionary time, to survive the sea anemones' toxins.

In **mutualism,** populations of both species receive benefits from each other. Species interactions of this kind have evolved over long periods and have tremendous evolutionary significance. Mutualism is **facultative** if both species are able to lead independent lives in the absence of each other; it is **obligate** if they cannot do so. Because it is such an important interaction in the evolution of species, we give several examples here.

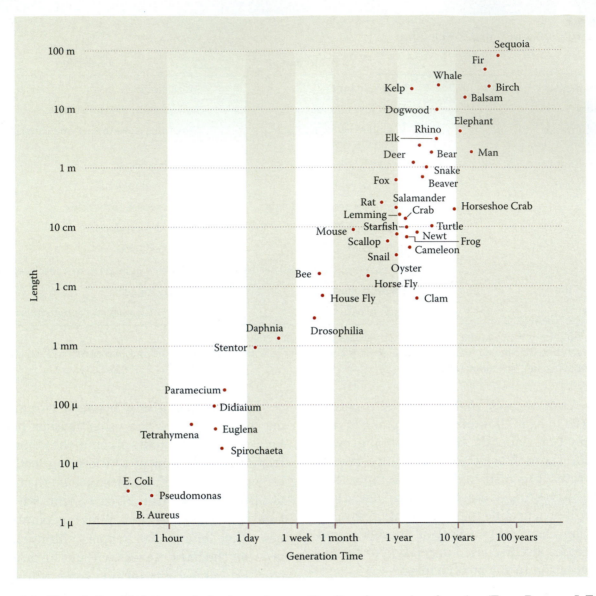

Figure 8.9 The relationship between body size and generation time for a series of species. (From Bonner, J. T. 1965. *Size and Cycle.* Princeton, NJ: Princeton University Press.)

Table 8.3
Interspecific Relations between Populations of Two Species

Relationship	Benefit/Harm	
	Species A	Species B
Commensalism	+	0
Mutualism	+	+
Predation/ parasitism	+	−
Neutralism	0	0
Amensalism	−	0
Competition	−	−

Notes: + = positive; − = negative; and 0 = neutral.

Lichens are organisms that result from an intimate association between an alga and a fungus. Worldwide, lichens grow on a remarkable diversity of habitats, including bare rocks, boulders, stone buildings, the bark of trees, and soil. Lichens have a remarkable resistance to drought, cold, and desiccation, which is afforded by the fungal walls, whereas the photosynthetic alga living inside the fungi provide food to the fungus.

A remarkable case of mutualism occurs between ants and trees. Janzen (1966) found that acacia trees in places such as Panama and Nicaragua avoid being browsed by herbivores by a mutualistic association with ants.

The trees provide food and shelter for the ants; ants can collect sugar in nectar-containing structures outside the flowers and oil and protein in structures contained on modified leaf tips, and the ants live in large hollow thorns grown by the tree. In turn, the ants attack any herbivores that seek to eat the leaves and kill any plants that sprout within several meters of the acacia.

Mutualistic "root fungi" called mycorrhizae have received much attention because of their importance in the nutrition of plants. These fungi live on root surfaces (ectomycorrhizae) or within the root cells themselves (endomycorrhizae) and are associated with most temperate and tropical plants. Recent work shows that the multiplying fungal hyphae greatly increase the surface area of roots; therefore, mycorrhizal associations greatly enhance the capabilities of plants to obtain water and nutrients from the soil. Making phosphorus available to plants is a particularly significant contribution of these fungi.

Perhaps the most remarkable mutualistic relationships are to be found in the pollination of flowers, dispersal of seeds, and their pollinating and dispersing agents. The coloration of flowers, the chemicals that plants synthesize to produce scents, and the production of nectar all cater to the specialized mouth or other body parts that have coevolved in insects, birds, and even bats. Thus, for example, plants that are only pollinated by bees cannot produce seeds in their absence (unless the plants can self-pollinate). Pollination by bees, beetles, birds, and other animals has been a fascinating study in **coevolution** (i.e., reciprocal evolution of morphological and chemical traits that have arisen in two species to provide ecological advantages to both), which is measured as increased reproductive success.

A noteworthy case of obligate mutualism was described more than a hundred years ago showing that Mojave yucca (*Yucca shidigera*) is pollinated exclusively by the yucca moth (*Tegeticula maculata*). This moth takes the sticky pollen of the yucca to other plants for the small price of the moth larvae feeding exclusively on yucca seeds (Riley 1892; Powell and Mackie 1966). This example of coevolution shows that mutualistic relationships can evolve to the extent that the survival of each species depends on that of the other. Their dependence on one another also points to the fragility of the system. If environmental changes were to wipe out one species, the other would be lost as well. For example, date palms have lost their coevolved pollinator and now have to depend on very inefficient wind pollination or, much more commonly, on agile humans with ladders.

Predation

Predator–prey relations are a form of exploitive species interactions in which one organism benefits, while the other is decidedly harmed. Predation includes carnivory, where one animal hunts and eats one or more prey species, and herbivory, where animals eat one or more kinds of plants. Predators such as lions, tigers, and some birds, who kill their prey and thus remove individuals from the population, are also called true predators. Grazing by animals, although still a form of predation, may not kill the plants altogether. In some cases, grazing has been shown to stimulate plant growth.

Parasitism is a type of predation. In true predation the prey is killed, but with parasitism, the host (prey) is not killed, at least not in the short term. The principle here is that parasites obtain nourishment from their "host" without immediately killing it. They may live on the surface of a plant or animal body (lice, for example) as ectoparasites; if the parasites live within the prey's tissues (such as fungi in plants or worms in animals), they are endoparasites. Host–parasite relationships are often complex, such as when different life stages of a parasite are associated with two or more specific hosts (e.g., the malarial parasite *Plasmodium*). Parasites may be facultative—that is, they can live even if the host is not available—or obligate if they die off without access to their host.

Some insects lay eggs within the bodies of their host. This type of parasite is known as a parasitoid; in this case, the parasitoid eggs are laid in the bodies of the larval stage of the host. The larvae develop using the host tissues as food, eventually emerging as adults and killing their hosts in the process. These parasitoids keep some insect populations under control. The latter ability is being explored as a biological mechanism in place of the need for using pesticides on some crops (see integrated pest management in Chapter 14). Ample evidence suggests that parasitic populations coevolve with their hosts.

Competition

Interspecific competition—that is, competition between populations of different species—is one of the most important regulators of population density. In competition, both species involved are negatively affected by the presence of the other, and the population growth rates of both are reduced.

The extent to which competition occurs between two species depends on how much their niches overlap. A **niche** is defined as the sum of the environmental conditions (e.g., temperature) and resources (food or water availability) that a species can tolerate or that it needs to survive (see Chapter 2). When competition is absent and the full spectrum of resources is available, a population is said to be using its fundamental (full) niche. Competition with another species for limiting resources can restrict a population to use only a subset of its fundamental niche, or the realized niche. The realized niche describes the more limited range of conditions that allow a population to persist in the face of competition.

In some cases, niche overlap between two species may be high. A series of classical experiments to test the limits of niche overlap was conducted by Gause (1934) with one-celled organisms. He introduced two species of *Paramecium*—*P. aurelia* and *P. caudatum*—into tubes containing a known supply of bacterial food (Figure 8.10). When grown in separate tubes, each grew to the carrying capacity of the tube. However, when both species were placed in the same tube, *P. aurelia* increased in numbers while *P. caudatum,* a weaker competitor, died out.

When the experiment was repeated with *P. aurelia* and *P. bursaria*, the results differed markedly. When these two species were grown together, they were able to coexist because *P. bursaria* was confined to eating bacteria on

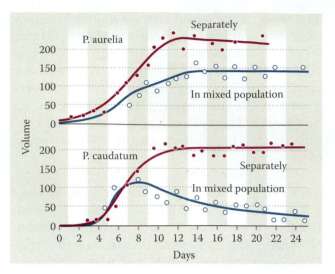

Figure 8.10 Competition experiments with two species of *Paramecium* and resultant competitive exclusion. (Gause, G. F. 1934. *The Struggle for Existence.* Baltimore, MD: Williams and Wilkins.) The red lines show the growth of each species when it was grown in separate tubes, while the blue lines show growth when the two species were grown together in the same tube (mixed population). In the latter case, *P. caudatum* was competitively excluded from the tube.

the bottom, while *P. aurelia* ate the bacteria suspended in solution. By dividing the space, the two populations avoided direct competition for food. Experiments like these have been repeated by other researchers using beetles and other animals. From these experiments came a now universally accepted principle: the **competitive exclusion principle,** which states that two species with identical ecological niches cannot coexist (sometimes paraphrased as "one species, one niche"). In the first case, *P. aurelia* and *P. caudatum* occupied the same niche, leading to the exclusion of the weaker competitor.

Competition can also lead to an evolutionary response by the species involved in ways that allow them to live harmoniously. An excellent example is the manner in which plant roots obtain water and nutrients by growing spatially at different depths and by adjusting the spread of their roots (Figure 8.11). Thus, "resource partitioning" avoids intense competition among populations for resources. Resource partitioning is a common means to mute the potential for competition. MacArthur demonstrated this for a group of warblers that appear to have very similar niches; they live in the same species of trees, hunt the same food,

Figure 8.11   Root systems of several prairie plants growing in a dry, well-aerated soil show how resources are partitioned by differences in the depth and spread of roots systems: *h,* hawkweed (*Hieracium scouleri*); *k,* hairgrass (*Koeleria cristata*); *b,* touch-me-not (*Balsamina sagittata*); *f,* fescue (*Festuca ovina ingrata*); *g,* cranesbill (*Geranium viscosissimum*); *p,* meadowgrass (*Poa sandebergii*); *ho,* hoorebekia (*Hoorebekia racemosa*); *po,* cinquefoil (*Potentilla blaschkeana*). (From Kramer, P. J. 1983. *Water Relations of Plants.* New York: Academic Press.)

and tolerate the same climate. Why is competitive exclusion not evident here? In this classic study, he showed that the warblers inhabit different zones of the tree; that is, the populations minimize competition by partitioning the resource of space (Figure 8.12) (MacArthur 1958). This allows the populations to coexist.

Forms of Competition

Competition has been recognized to take the following different forms, each of which may occur singly or in combination:

- consumptive competition involving the use of resources;
- preemptive competition that deals with the occupation of space (e.g., plants or barnacles);
- overgrowth competition, where greater numbers of one population overtake the other;
- chemical competition through the production of chemical compounds that deter the growth of others;
- territorial competition involving the defense of territories; and
- encounter competition, which may include fights, harm to others, and subversive tactics (Schoener 1983).

Many of these concepts apply equally to human population interactions (see Chapter 13).

A profound example of competition can be seen in the production of a variety of chemical compounds synthesized by plants. These compounds, by themselves and through their degradation products, can have major impacts on adjacent plants. More than 75 years ago it was shown that juglone, a compound produced by walnut trees, inhibited the growth of plants underneath these trees (Massey 1925). This phenomenon, whereby the metabolic products of one species are injurious to others, is known as **allelopathy,** and the compounds causing them are called allelochemicals.

Many studies on allelopathy are available on both the compounds synthesized and their effects. Allelochemicals can cause effects such as reduced permeability of cell membranes, inhibition of cell division, reduction in enzyme production and protein synthesis, and interference in the uptake of minerals. It appears that in some instances compounds from the same species may be self-inflicting (also called autoallelopathy), whereby detrimental effects may be felt by the plant (or its offspring) producing the compounds (Iverson and Wali 1982). The most logical explanation is that these compounds are synthesized as a defensive mechanism against herbivores.

Figure 8.12 Niche differentiation in birds is exemplified by the division of space within each tree by warblers. (From Molles, M. C., Jr. 2008. *Ecology: Concepts, Applications.* New York: MacGraw–Hill Higher Education. Based on MacArthur, R. H. 1958. *Ecology* 39:599–619.)

Competition and Keystone Predators

The effect of predation on the outcome of competition in a community is sometimes mediated by the presence of a keystone predator. As an illustration, consider the intra- and interspecific competition between animals for space in the rocky intertidal zone of marine shores. In a classic experiment, Paine (1966) found an intertidal community of 15 species of invertebrates on the coast of Washington state, all of which were consumed from time to time by the predaceous sea star, *Pisaster*. The regular grazing by *Pisaster* opens up patches of rock by removing the attached filter-feeding barnacles (acorn and goose-necked barnacles), attached mussels, predators (anemones and snails), and browsers (chitons, limpets, and snails) present in the community. When Paine regularly removed *Pisaster* from one section of the rocks and threw them into the very deep water adjacent (it takes them a long time to crawl back), the bare patch previously cleared by the starfish was quickly colonized by the larval form of the acorn barnacle *Balanus*.

However, *Balanus* was soon forced out by the more rapidly growing mussels (*Mytilus*) and goose-necked barnacles (*Mitella*) that also colonized the patch and came to dominate the community. All three taxa filter-feed on phytoplankton, but their real competitive interaction is for space.

The net result was that when *Pisaster* was removed from the community, its overall diversity fell from 15 to only 8 species, a reduction in diversity of nearly 50%. Paine called *Pisaster* a **keystone predator,** the presence or absence of which determines the outcome of competition and eventually determines the diversity and taxonomic composition of the entire community. When *Pisaster* was present, the prey populations were kept low, limiting the competition between them and preventing competitive exclusion from occurring. Higher community diversity was the result.

Although ecological relationships between two species are easy to visualize, in nature it is often the multiple species relations that have a profound influence on population dynamics, and a recent example may illustrate this. This example comes from a three-species relationship involving leaf-cutting ants (species of *Acromyrmex* and *Atta*) on whose leaf surfaces they grow colonies ("gardens") of fungi that are harvested by the ants for food. In addition, there are actinomycetes (a type of bacteria) that produce antibiotics that help maintain the fungus in a healthy state. Researchers believe that this obligate mutualism may be as old as 65 million years and is one of the few examples of an animal species developing an agricultural system. The ants tend the fungus to keep it healthy and productive. However, if the fungi become infected—for example, with a virulent parasite (*Escovopis*)—the ants will treat the infection using a third symbiotic partner, an actinomycetes bacterium that grows on the bodies of the ants. Investigating how ants overcome these infections, Currie, Bot, and Boomsma (2003) found that the infection induced an accelerated growth of the actinomycetes that, in turn, secrete larger quantities of antimicrobial compounds that the ants transport to the fungi to treat the infection.

The authors write: "Our finding of a mutualistic bacterium protecting its symbiont from a pathogen might be a common but vastly overlooked benefit within mutualisms in general" (Currie et al. 2003).

Human populations are not exempt from biological pressures and evolutionary change. Most of the concepts of species interaction, whether negative (competition) or positive (mutualism), apply equally well to human populations, though many people erroneously consider humans to be independent of such natural population processes. But what is warfare but competition for space, resources, or identity (niche)? What is charity or altruism but the sharing of resources among members of a species? We even talk about some people being "parasites" on their families or on society at large. Although humans are organisms most able to alter the environment, we are by no means immune to the harshest consequences of population dynamics and interactions with other species. These aspects are discussed in Chapter 11.

Note

1. "A single ant or bee isn't smart, but their colonies are. The study of swarm intelligence is providing insights that can help humans manage complex systems, from truck routing to military robots," writes Peter Miller (2007) in his article on swarm theory (*National Geographic*, volume 21, July 2007). This concept is elaborated in the book by James Surowieck, *The Wisdom of Crowds—Why the Many Are Smarter Than the Few and How Collective Wisdom Shapes Business, Economics, Societies, and Nations* (2004). Recently, also, medical science has been studying the human health effects of staying together or "clumping." Hear, for example, the program on "aging gracefully" with David Freudberg (program #103, December 2, 2007) on Public Radio International (PRI, humanmedia.org).

Questions

1. What has happened to deer populations in the eastern United States without large predators or hunting?
2. How do populations change over time in response to biotic (density-dependent) and abiotic (density-independent) factors? Give some examples of density-dependent and density-independent factors.
3. Discuss whether or not the concept of carrying capacity can be applied to human populations.
4. What are the effects of competition, mutualism, predation, parasitism, herbivory, and commensalism on population dynamics?
5. What feedback mechanisms affect populations as they approach carrying capacity?

Community Ecology and Succession

9

Ecological Communities

Unlike animals, terrestrial plants do not possess autonomous movements; that is, they cannot move from one place to another at will. Over evolutionary time, communities of plants and the animals and microorganisms that accompany them have developed, linked by their adaptations to the climate and soil types that support them. These communities show characteristic groupings of plants and are often named after the dominant species, such as oaks in an oak forest. The mere mention of an oak or a pine forest, a prairie, or a desert conjures up images of what that forest, grassland, or desert community will look like. Animal communities (including microbiota) are inextricably coupled to specific plant communities for their food and shelter; thus, animal and plant populations are equally characteristic features of communities.

A **community** is a group of coexisting populations of different species (plants, animals, microorganisms) living in the same place at the same time. Essentially, it is the living part of an ecosystem, consisting of all forms of life, their structure, and their interactions (Bazzaz and Sombroek 1996). The study of such communities is termed **community ecology.** The interspecific interactions that were described in Chapter 8—competition, mutualism, commensalisms, and predation—are responsible for linking species together in a complex web of direct interactions (e.g., a wolf eating an elk) or indirect interactions (e.g., the primary productivity of aspen trees increasing as wolf predation reduces elk population size, thus reducing elk herbivory).

Ultimately, the type of community found in a given place is a function of its environmental conditions, climate, soils, and the type of disturbance it typically receives. The members of a given community must be adapted to the same general set of conditions in order to persist; when environmental conditions are similar, the same types of communities may appear, even on different continents. Thus, the northern forests of Canada have a remarkable similarity to the forests of northern Scandinavia and Siberia; forest communities of the eastern United States are remarkably similar to the communities found in some parts of northern Europe and China (see Chapter 10 on biomes). In sum, communities exhibit patterns that are striking and often unique and yet show some kind of geographical regularity.

Because these abiotic and biotic components function as a whole, communities and ecosystems have properties in addition to the properties of their component parts. These are known as **emergent properties** (i.e., properties that are apparent when the group is viewed as a whole that are not apparent, or even predictable, when looking at the individual species). Because emergent properties are characteristic of the system behaving as a unit, they do not reside in any individual species. Rather, they have meaning only with reference to the community and ecosystem levels of organization and include properties such as coevolution, trophic structure, biodiversity, and stability. Even when the constituent populations of a community change over time and space, the emergent properties of the community are maintained.

Plant community ecology has had a long and sometimes contentious history. One issue stems from the recognition of early European ecologists that plants can serve as indicators of environmental conditions. They were so impressed by how consistently certain groups of species were associated with each other—"bound together in mutual dependence"—that they named the study of communities **plant sociology,** reflecting the "social organization" of plants (Crawley 1997). Although recent work has shown little evidence to support the idea that plants organize themselves into discrete units (Huston 1994), the degree of similarity between communities can provide an indication of the similarities and differences among habitats in the following ways:

- Each kind of community is characterized by a group of species that show distinctive patterns of relative abundance over the landscape.
- Any large unit of vegetation is a mosaic of plant communities, the distribution of which is governed by a corresponding mosaic of habitats.
- The more heterogeneous the environment of an area is, the more numerous

are the kinds of plant communities that it supports. Communities differentiate by virtue of the interaction of the environmental tolerances of organisms and the heterogeneity of the environment (Daubenmire 1968).

Although "no two bits of vegetation have precisely identical floristic composition," vegetation can nonetheless be organized into communities with similar combinations of species called **associations** (Braun-Blanquet 1932). An association was formally recognized as "a plant community of definite floristic composition, uniform site conditions, and uniform physiognomy" in 1910 (Crawley 1997). In the American ecological literature, an association is typically called a **community type.** An impressive body of plant ecological studies characterizing the patterns of species that comprise a community has accumulated from Europe and elsewhere. The resulting understanding of community structure has led ecologists to define the following attributes as useful in characterizing a community:

- **abundance:** the number of individual plants per unit area;
- **cover:** how much of an area is covered by individual species;
- **dominance:** the extent to which species dominate in a given area;
- **frequency:** how frequently different species are in a given area; and
- **biomass:** how much is the dry weight mass of a species per unit area.

Studies of community structure also include qualitative characteristics such as:

- **Stratification** or layering. All vegetation can be assigned to one of the following strata: trees, shrubs, herbs (nonwoody green plants, including vines), mosses, and epiphytes.
- **Phenology** or seasonal periodicity. This includes germination, establishment, break of buds, and production of foliage, flowering, and fruiting.

Two aspects of community ecology in the United States have generated particularly strong debate. The first was based on the definition of the plant community itself, as proposed by the American ecologist Frederic Clements. He saw communities as closely integrated systems with numerous emergent properties that could be considered analogous to a "super-organism" (Clements 1916). As in the definition of an association, members of a community in this definition were tightly bound, self-organizing groups that will always co-occur. If the "organism" is harmed in any way—for example, by a disturbance such as a flood or fire—it will "heal" itself, returning to the same state as before the disturbance. Sir Arthur Tansley, a British ecologist whose views differed, argued that a community has enough collective properties to qualify in some way as unified but is more analogous to a "quasi-organism" than a super-organism (Tansley 1935).

But some American ecologists, foremost of whom was the noted botanist Henry Gleason, saw plant communities more as assemblages of adapted species that did not exhibit any integrated properties (Gleason 1926). Gleason's view, also called the "individualistic hypothesis," states that the species that coexist in a community do so because of similarities in their niches—that is, because they have similar environmental requirements or tolerances. In this view, communities are groupings of species adapted to local conditions, and changes in the community are brought about by the response of each individual species to changes in the environment. Evolutionary processes favor this view; it is difficult to imagine circumstances under which a group of species would evolve as a unit. Instead, individual species have evolved to best meet the challenges of their particular physical and biological conditions. Although it took until the 1950s, modern community ecology has abandoned the organism analogy and incorporated the adaptations of individual species in explaining community structure and function (Ricklefs 2007).

The second debate centers on trying to understand how the *association* concept became such a fundamental part of community ecology, particularly given the evidence that supports Gleason's individualistic hypothesis.

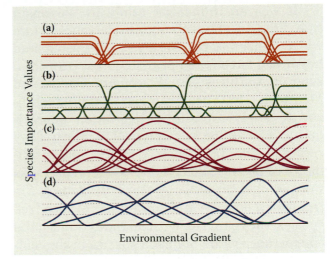

Figure 9.1 Four hypothetical illustrations of how species are distributed in communities along environmental gradients: (a–c) represent sharp breaks in species distributions, and the transitional zones or ecotones are small. In (d), species distributions are more or less continuous with wide ecotones. (From Whittaker, R. H. 1975. *Communities and Ecosystems,* 2nd ed. New York: Macmillan Publishing Company.)

The answer to this has to do with ecological conditions that vary spatially and that affect the distribution and abundance of organisms. These are referred to as **environmental gradients.** When environmental gradients are sharp—for instance, when average soil moisture changes suddenly—the plant communities also show a sharp spatial delineation (Figure 9.1a–c). Interestingly, Clements did most of his work in regions where environmental gradients are sharp.

When environmental gradients are diffuse, the communities are more or less continuous and grade from one community into another with no distinct boundary; this represents a continuum (Figure 9.1d). This continuous gradation in the distribution of species is the basis for the continuum concept, which states that, within major community types such as forests or grasslands, individual species replace one another continuously along environmental gradients. This lends support to the individualistic hypothesis. Transition zones or **ecotones** occur between major communities and are often characterized by greater species diversity as a consequence of mixture of edge species and tolerant species from the adjacent communities (**edge effect**) (see Chapter 10).

It was an influential study in Wisconsin that brought the individualistic concept into wider recognition. Working in the upland hardwood forests in a prairie–forest transition zone of southwestern Wisconsin, Curtis and McIntosh (1951) used random sampling methods to study an undisturbed area approximately 256 by 368 km (160 by 230 miles). They recorded the frequency, density, and dominance of 80 trees selected as random pairs, as well as data on other species. By adding the frequency, density, and dominance of each tree species, they came up with an index dubbed the "importance value" for each species in the community. The relative positions of four tree species—black oak, white oak, red oak, and sugar maple—were then plotted along a "continuum index" to illustrate their position in the study area (Figure 9.2). The researchers noted that the plants formed a "vegetational continuum," with each species distributed independently. The authors also assigned each species a ranking to indicate its position from early to later stages of community development. The gradient (Figure 9.2) from left to right moves from dry conditions with low pH and exchangeable calcium to more moist conditions with higher pH and exchangeable calcium.

Using the European ecological methodology (Wali and Krajina 1973), a similar community gradient was found in a north-central British Columbia area dominated by lodgepole pine, aspen, black spruce, white spruce, and subalpine fir. A number of environmental gradients were rigorously quantified in 80 forest stands, and special attention was given to the types of undergrowth species found in the dominant communities. Pines occupied the lowest positions along the available water gradient due to their ability to grow in dry areas with low nutrient concentrations, whereas river alder–subalpine fir–white spruce communities dominated the wet portion of the gradient. The middle portion was dominated by white spruce (Figure 9.3a). A gradient of silt, clay, and nutrient content of soils (Figure 9.3b) showed similar species and

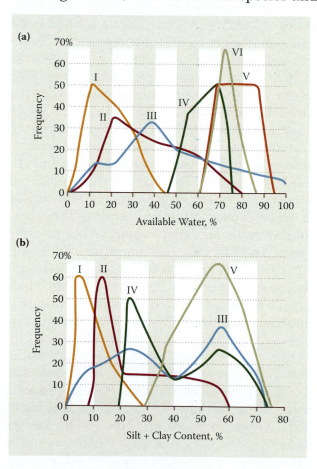

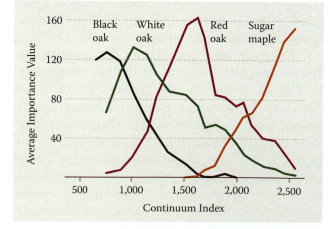

Figure 9.2 Relative importance values of four major species of trees in forest communities of southwestern Wisconsin arranged along a continuum index. The environmental gradients of soil moisture, exchangeable calcium, and pH increase to the right on the continuum index. (Curtis, J. T., and R. P. McIntosh. 1951. *Ecology* 32: 476–496.)

Figure 9.3 Distribution of communities along gradients of (a) available water (%) and (b) silt and clay content (%). Community I is a xeric lichen–pine, II is mesic moss–white spruce, III is more mesic moss–spruce–fir, IV is a fern–devil's club–subalpine fir–white spruce, V is a sphagnum moss–black spruce, and V and VI (in both a and b) are dead nettle–fern–river alder. (Wali, M.K., and V. J. Krajina. 1973. *Vegetation* 26: 237–381.)

community distribution as the available water gradient. A measurement of light intensity under these forest canopies, however, showed the opposite trend. Pines occupied sites with high light conditions; sites with white spruce and subalpine fir had progressively less light. Light quality and spectral composition have important implications for species tolerance or intolerance of shade.

Community Structure

Naturally, the methods and measures we use in studying communities are of great significance. The late Robert H. Whittaker, one of the most illustrious ecologists of the last century, emphasized that the following four features characterize communities and should be measured in the field when characterizing the community structure (Whittaker 1975):

- **Growth form and structure.** Ecological communities comprise several growth forms, such as trees, shrubs, herbs, mosses, and lichens. These occupy different strata and each of these groups of organisms has definite growth form and structure. As was emphasized by ecologists before him, Whittaker concluded that each of these attributes must be carefully recorded and described. Rigorous field natural history is the backbone of ecology.
- **Species diversity.** Generally, the number of different species of plants and animals found in a community is of interest—what we refer to as species richness. Patterns of species richness vary widely from community to community and geographically; on the global scale, it is typical to find more species per area as one moves toward the equator from the poles.
- **Dominance and abundance.** "Only a few species within a community are dominant, either by virtue of their size, numbers, or activities. Dominance can be expressed as absolute abundance (density), relative abundance, or biotic productivity." Dominant species exert the greatest influence on the community, although most dominant species are generally found with one or two other species as codominants.
- **Trophic structure.** The flow of energy and the transfer of materials from plants to herbivores to carnivores (the food webs), as discussed in Chapter 7, are important determinants of the presence, abundance, and spatial and temporal dispersion of members of the community. As noted in Chapter 7, recent work illustrates the theoretical and practical importance of trophic cascades.

The debate about the definition of a "community" continues, not only because it has an important place in understanding ecological systems but also because it is now of great importance in ecosystem management, in understanding the evolution of biological diversity, and for the conservation of natural areas. These issues are continually under discussion by scientists and decision-makers.

Ecological Succession

Change is the order of nature. This universal process of gradual (progressive or retrogressive) changes over time in vegetation—namely, the composition and relative abundance of species in a community, as well as the associated biota—is called **ecological succession** (Huston 1994). Naturalists have observed the changes in communities for many centuries. As a result, there is a rich documentation of community changes in bogs, forests, and other ecosystems since the time of Theophrastus (Table 9.1). To the casual observer, the process of succession may be imperceptible if disturbance has not occurred or if climatic fluctuations are not pronounced (e.g., in tropical rain forests). Successional changes may also elude observation if the ecosystems are in a relatively steady-state condition—that is, where rates of changes in climate, soil, flora, and

Table 9.1
Pre-1900 Development of the Concept of Ecological Succession

Year	Author	Context in Which the Concept was Used
ca. 2300 b.p.	Theophrastus	Changes in vegetation of river floodplains
a.d. 1685	William King	Changes in bog vegetation; inferred past changes from plant remains in peat
1714	G. M. Lancisi	Successional changes of seashore vegetation near Rome
1735	C. Linnaeus	Succession in bog communities
1742	Georges Louis Leclerc, Comte de Buffon	Earliest clear record of changes in predominant species in forests
1749	I. J. Beiberg	Pioneering role of mosses and lichens, development of bogs
1789	Gilbert White	Natural history of Selborne, U.K.
1842	G. W. Francis	Succession from lichens to mosses, ferns, grasses, flowering plants, etc.
1847	John W. Dawson	Vegetation changes after fire, agriculture
1850	Alexander von Humboldt	Vegetation changes
1851, 1857	C. Vaupell	Reconstructing postglacial vegetation changes by studying macrofossils in peat deposits
1860	Henry David Thoreau	Coined "forest succession" after logging
1895	E. Warming	Formulated "laws of succession"
1899	H. C. Cowles	Succession on the sand dunes of Lake Michigan
1916	F. E. Clements	Climate in sere development; monoclimax
1926	H. A. Gleason	Individualistic successional patterns
1977	J. H. Connell and R. O. Slatyer	Roles of herbivores, predators, and pathogens

Sources: From Wali, M. K. 1980; Modified from Wali, M. K. 1999a. *Nature & Resources* 35(2): 38–50.

fauna are very slow compared to the life span of humans.

A survey of the members of the British Ecological Society on the 20 most "important" concepts in ecology showed that most respondents ranked succession as second only to the ecosystem concept (see Chapter 1; Cherrett 1989). Both concepts—ecosystems and succession—are in fact remarkably intertwined. It has been emphasized that the ecosystem concept could serve to integrate the understanding of succession with other ecological knowledge. In turn, long-term studies of succession could strengthen the ecosystem concept (Tansley 1951). Other than these two, few ecological concepts have accumulated such a wealth of information from diverse ecosystem types.

The types of successional changes in a community vary with the temporal scale involved (Table 9.2). Succession begins the moment that

autotrophs gain a foothold in the ground with their attendant physiological processes; the consumer populations follow. As we shall see in the following discussion, the systems are simple when young and become increasingly complex with the passage of time.

Ecologists have many reasons for studying succession. First, as the phenomenon of community changes over time, succession is universal and observable. Second, each ecogeographic region presents its own patterns of developmental changes in soils, vegetation, and associated biota, all of which bear fascinating ecological details. Third, based on the first two reasons, succession provides an excellent framework for seeking common patterns in the operation of ecosystems and for deciphering principles that may be widely applicable. It is in the latter that a diversity of opinion arises. Studies on succession are also indispensable in ecosystem

Table 9.2
Kinds and Timescales of Vegetation Change over Time

Duration (Years)	Kind of Change
<0.01	Daily rhythms of the component species (e.g., transpiration, photosynthesis)
1.0	Seasonal: related to the ontogeny of particular species of the community, changes in tolerances in different growth stages
ca. 4	Small rodent cycles possibly related to cycling of necessary minerals from animal to plant populations
Few tens (decades)	Weather cycles producing changes in productivity, seed supply, cover, abundance of annuals
Several tens	Changes in species' vitality with age, developmental changes in perennials
Several to many tens	Replacement cycles within the community, pattern, and process cycles related to limited life spans of species
Hundreds (centuries)	Changes leading to development of a new community from the former one; directional and therefore predictable as the preceding shorter changes that are not; the independent factors of the environment are constant and therefore differ from the following allogenic changes
Several hundreds	Change caused by man's activities in some factor of the environment producing changes in the vegetation: burning or fire control, mowing, forestry (as distinguished from timber harvesting)
Thousands (millennia)	Natural changes in factors of the environment: historical changes related to changes in climate, physiography, flora, or fauna; slow modifications of the soil parent material by geological processes
Many thousands	Evolution of communities related to changes in flora by migration, extinction, or evolution

Source: Modified from Major, J. 1974. In *Handbook of Vegetation Science,* vol. 8, 392–412. *Vegetation Dynamics.* The Hague: Dr. W. Junk, Publishers.

management and ecosystem restoration (see Chapter 26).

Let us look at some of the basics of the concept. Succession is **primary** when it begins in an area where no plants have grown before and, as such, the area has had no biological activity or accumulation of soil organic matter, also called a biological legacy. Examples are bare rock surfaces, sandy outwashes, lava flows and other volcanic landscapes, and abandoned surface mines after the extraction of coal and minerals. Succession is **secondary** when it begins in areas from which the original biotic community has been removed due to a disturbance (e.g., crop fields left fallow after harvesting, logged areas, forests after fires, and areas of heavy grazing). Under U.S. environmental laws, abandoned mined lands are required to have topsoil replaced on them to provide a substrate, including soil and organic matter, to provide a boost to processes akin to secondary succession (see Chapter 24 on restoration of ecosystems).

Succession that starts primarily in dry habitats is referred to as **xerarch** (Figure 9.4). Water is scarce in these habitats, and the plants inhabiting such areas are **xerophytes** (some attributes of xerophytic plants were given in Chapter 4). Some of the driest habitats where succession begins are the rock substrates, such as those in arctic and alpine areas. **Pioneer plants** (first arrivals) on the rocks are lichens. They live there because they can withstand drought and take advantage of short intervals of wetness. Lichens secrete acids that corrode rock surfaces. Dead lichens and dust provide anchorage for the mosses, and mosses grow in clumps and collect more debris, beginning soil development. These will be followed by other herbaceous plants, and so on.

Succession that starts primarily at the land–water interface, such as around the margins of lakes and ponds, is called **hydrarch** succession (Figure 9.4). Over time, modification of both dry and wet types of habitats alters them greatly such that dry sites, with

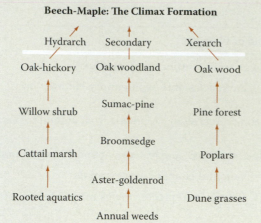

Beech-Maple: The Climax Formation

Hydrarch   Secondary   Xerarch

Oak-hickory   Oak woodland   Oak wood

Willow shrub   Sumac-pine   Pine forest

Cattail marsh   Broomsedge   Poplars

Rooted aquatics   Aster-goldenrod   Dune grasses

Annual weeds

Figure 9.4 An oversimplified course of succession attributed to Clements: dry-xerarch, wet-hydrarch, and secondary succession types; each successive stage is longer lived toward the climax formation. (From Colinvaux, P. A. 1993. *Ecology 2*, 371–372. New York: John Wiley & Sons.)

the accumulation of organic matter in the soil, gain the ability to conserve water. Similarly, the wet areas tend to fill, becoming less wet. In both cases, communities attain a mesic character—that is, not so dry and not so wet. In nature, multiple processes mediate these transitions; none of these processes is simple and they do not generally occur over short periods or decades. A directional change from a less to a more complex community may be considered a **progression**, and a change from a more to a less complex community is a **retrogression.**

When habitat conditions are modified by the biota that grow there such that conditions become suitable for colonization by other species, the changes in succession are referred to as **autogenic** (changes in species are driven by changing conditions coming from within the community). When the changes are imposed by factors external to the biotic community, such as fire, grazing, and human activity, the succession is **allogenic** (external to the community). The distinction of these two types of changes is hazy, however, because autogenic and allogenic processes occur simultaneously in all communities. Overall, the differentiation is from simple one-layered communities to complex and multilayered ones (Figure 9.5). Also, as plant species change in the communities, so do the animals (Figure 9.6).

Regardless of whether the factors driving successional change are internal (intrinsic) or external (extrinsic), each succeeding community of organisms, called **seral stages** or **seres,** tends to last for a longer period than the preceding one. Each community in the course of progressive succession becomes relatively more stable in the face of disturbances. All successional stages tend to proceed toward a relatively steady endpoint originally designated as the **climax community.** The climax community is the final, self-perpetuating sere originally envisioned as the community best adapted to the ecological conditions where it is found. In this view, there is a single climax community for each area. It will persist (or replace itself) until the system is reset by disturbance. However, because communities and ecosystems are open—that is, they exchange energy, materials, and species with one another—and constantly changing (dynamic), a mature ecosystem is considered to vary in its composition in relationship to environmental gradients. It can be said to be in a condition of **dynamic equilibrium** (see Chapter 7 for the discussion on nonequilibrium thermodynamics). How long this relatively stable state will last has been a matter of debate for many years.

The credit for developing a cohesive set of concepts for the development of succession goes to Clements; at the very beginning of the last century, he laid the foundations of community change in his classic work on plant succession (Clements 1916). He recognized six processes as fundamental:

- creation of a bare or partially bare substrate by a disturbance event (**nudation**);
- the arrival of seeds and other plant propagules at a site (**migration**);
- the establishment of the new arrivals (pioneers) at a site (**ecesis**);
- intraspecific and interspecific interactions (**competition**);
- the collective modification of the habitat by these organisms and their relative success in dominating the site (reaction) (as we note later, this is now referred to as **facilitation**); and

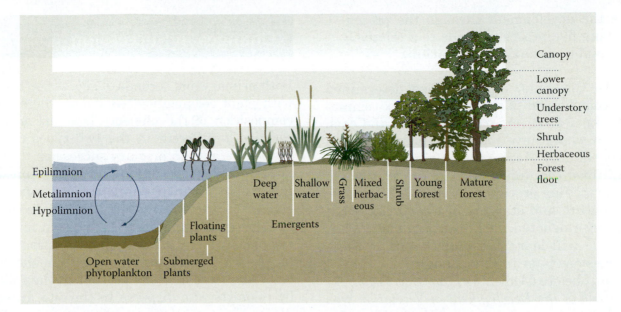

Figure 9.5 The sequence of stages from aquatic to terrestrial communities: Stratification and complexity of the community become greater along the gradient from aquatic to terrestrial communities. (From Smith, R. L. 1990. *Ecology and Field Biology*, 4th ed. New York: Harper Collins.)

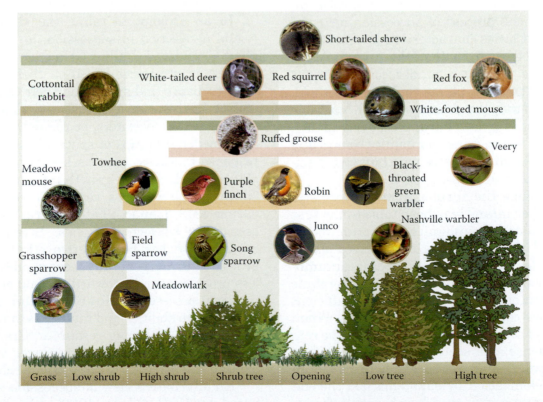

Figure 9.6 Concomitant changes in plant and animal species in the course of succession. (Based on Odum, E. P. 1969. *Science* 164:262–270; and Odum, E. P., and G. W. Barrett. 2005. *Fundamentals of Ecology*, 5th ed. New York: Thompson/Brooks–Cole. Animal representations courtesy of multiple resources.)

- the development through the preceding five processes leads to a relatively stable climax (**stabilization**).

To Clements, climate alone determined the composition of the stable climax community for a given ecographic region, and each region would have just one climax (hence, a monoclimax). Many studies since then have confirmed the universality of the first five of the preceding processes. It is only the sixth—the nature and stability of the climax—that has been debated, modified, and elaborated, and we will learn the main points of these discussions later in this chapter.

Just before Clements's *Plant Succession: An Analysis of the Development of Vegetation* was published, his contemporary, Henry Chandler Cowles, published what is now a classic treatise on the succession of plants on sand dunes adjacent to southern Lake Michigan (Cowles 1899). As glaciers retreated from the Great Lakes region, several beach and dune systems were left on the edges of Lake Michigan. As the lake water level receded somewhat, the beaches and dunes became more expansive. Plant colonization began the process of succession near the beach, and Cowles characterized several stages as he proceeded from the lake inland. Areas nearest the lake were under constant wave action, and thus few plants could get a foothold. Further inland, the dunes were still unstable but supported a few hardy succulent plants, followed by a fore dune community comprising some sand-binding grasses with burrowing spiders, beetles, and grasshoppers.

Then came the shrubs. Some surface stability farther inland was hospitable to the first tree-dominated community of cottonwoods. These and other deciduous broad-leaved plants returned leaves to the ground each autumn, which, upon decomposition, built up the organic matter content of the soils. With sufficient light, pines could colonize other dunes. When the cover of cottonwoods and pines became reasonably dense, oak seeds started germinating because oaks require some shade to get established. Colonization by oaks enriched the soil further, giving way to a beech–maple forest, which Cowles called the climax.

Changes in Succession

Both primary and secondary succession begins with plant colonization. As noted, the species composition of the pioneers that are able to colonize is different in the two types of succession because in primary succession there is no true soil, so no organic matter is present in the substrate. Therefore, the water-holding capacity of the substrate is low and its essential nutrient composition is such that few species can survive. Secondary succession often begins with seeds germinating from the seed back (the store of seeds present in the soil) or by colonists who disperse to the site. In both cases, the early **pioneers** are *r*-strategists; that is, these species are fast growing, have high powers of dispersal, generally have a small size, and produce a large number of seeds. A characteristic of pioneers is that they are able to exploit areas where conditions are harsh, due, for example, to their low nutrient requirements (the traits of *r*- and *K*-selected populations were discussed in Chapter 8).

In an attempt to refine the strategies of *r*- and *K*-selection in succession, an influential plant ecologist, J. Phillip Grime, defined three life history strategies based on their tolerance to disturbance and stress and their competitive ability. In this model, species are classified as ruderal, stress tolerators, or competitors. These are functional groups of species based on their response to stress, such as low nutrient availability (which reduces growth rates) and disturbance (which removes vegetation or biomass) (Grime 1979; Tilman 1990). Ruderal species are essentially *r*-selected species. Stress-tolerant species have low reproductive effort and low growth rates and generally occur in undisturbed, less productive areas toward the end of the successional process. Finally, competitors have low reproductive abilities and high growth rates and specialize in capturing resources (Grace 1991). The interaction of these three over time can explain the patterns of successional change.

An alternative view of succession was proposed by E. P. Odum (1969), who took into account the development of the whole ecosystem

Table 9.3
Trends in Community Attributes during Succession

Ecosystem Attributes	Developmental Stages	Mature Stages
Community Energetics		
Gross production/community respiration (P/R ratio)	Greater or less than 1; high	Approaches 1; low
Gross production/standing crop biomass (P/B ratio)	Low	High
Biomass supported/unit energy flow (B/E ratio)	High	Low
Net community production	Linear, predominantly	Web-like
Food chains	Grazing	Predominantly detritus
Community Structure		
Total organic matter	Small	Large
Inorganic nutrients	Extrabiotic	Intrabiotic
Species diversity—variety component	Low	High
Species diversity—equitability component	Low	High
Biochemical diversity	Low	High
Stratification and spatial heterogeneity	Poorly organized	Well organized
Life History		
Niche specialization	Broad	Narrow
Size of organisms	Small	Large
Life cycles	Short, simple	Long, complex
Nutrient Cycling		
Mineral cycles	Open	Closed
Nutrient exchange rate between organisms and environment	Rapid	Slow
Role of detritus in nutrient regeneration	Unimportant	Important
Selection Pressure		
Growth form	For rapid growth	For feedback control
Production	(r-selection) quantity	(K-selection) quality
Overall Homeostasis		
Internal symbiosis	Undeveloped	Developed
Nutrient conservation	Poor	Good
Stability (resistance to external perturbations)	Poor	Good

Source: From Odum, E. P. 1969. *Science* 164:262–270.

rather than the replacement of species over time. His comparison of ecological processes in the early and later stages of succession provides useful information on trends expected during the developmental process (Table 9.3), with an emphasis on ecosystem functions such as rates of ecosystem productivity and respiration.

The physiological and life history attributes of species in the early and late successional stages are distinct (Table 9.4). Species composition changes continuously during succession, and the change is more rapid in early stages than later ones. Total number of species increases initially and then becomes more or less

Table 9.4
Physiological and Life History Characteristics of Early and Late Successional Plants

Attribute	Early Succession	Late Succession
Photosynthesis		
Photosynthetic rates	High	Low
Light saturation intensity	High	Low
Light compensation point	High	Low
Efficiency at low light	Low	High
Dark respiration rates	High	Low
Water Use Efficiency		
Transpiration rates	High	Low
Stomatal and mesophyll tolerances	Low	High
Resistance to water transport	Low	High
Seeds		
Seed dispersal in time	Well dispersed	Poorly dispersed
Seed germination enhanced by light	Yes	No
Fluctuating temperatures	Yes	No
High NO_3^-	Yes	No
Seed germination inhibited by far-red light	Yes	No
High CO_2 concentration	Yes	No?
Resource acquisition rates	Fast	Slow?
Recovery from resource limitation	Fast	Slow
Root/shoot ratio	Low	High
Mature size	Small	Large
Structural strength	Low	High
Growth rate	Rapid	Slow
Maximum life span	Short	Long

Source: From Bazzaz. 1979.

stabilized. As a consequence, the total biomass in the ecosystem and the amount of nonliving organic matter in the soil increase as well. New organic matter is produced or synthesized at a rapid rate by the producers in the initial stages and slows down in the later stages. However, the quantity of organic matter produced in the later stages is much greater, given the increased numbers of species and their growth forms. The increasing diversity of species, biomass production, and stability enhance the complexity of food webs. In the later stages of succession, the relationships among producers, consumers, and decomposers become more complex, and there is

also an increasing specialization of niches among the species.

Controls on Succession

Depending on their frequency and intensity, both natural forces and human activities have impacts on succession. Natural disturbances—fire, flooding, snow- and windstorms, hurricanes, typhoons, and volcanic activity—have an impact on succession. However, unlike human activities, these are periodic, rather than pervasive and chronic. The result is that ecosystems under natural disturbances are able to recover and follow a successional course in a relatively shorter period of time than when ecosystems are drastically disturbed. While we devote a whole section to human-caused disturbances and their restoration (Chapter 24), some of the effects of disturbances on succession are presented next.

- Cultivation involves the complete removal of existing vegetation to make land areas available for the growth of crops. Cultivation includes a thorough working of the upper 15–30 cm of topsoil and seeding with desired species and some effort to control the growth of naturally occurring species (often referred to as "weeds"). Areas such as these show no succession. If cultivated fields are abandoned, secondary succession will commence. The effects of cultivation on ecosystems are discussed in Chapter 14.
- Logging is the removal of trees for timber, pulp, and firewood. The extraction process entails the use of logging equipment and heavy trucks. All have an impact on the ground flora and soils. The rate of secondary succession depends on the extent and magnitude of disturbances during extraction. The impacts of deforestation and setting fires to clear forests are discussed in Chapter 15.
- Fire. Both natural and human-induced fires set succession back to an earlier seral stage. Low-temperature ground fires can be beneficial in burning the dead materials that accumulate in forests, breaking the dormancy of seeds, killing wood-rotting fungi, and making nutrients readily available. Intense or "crown" fires are destructive not only of soil organic matter but also of seed sources. The destruction of soil organic matter makes it vulnerable to wind and water erosion through loss of soil particle aggregation, leading to topsoil and nutrient losses.

  Ever since learning to make fire, humans have used it extensively throughout the world for agriculture and other purposes. Humans appear to have reached the New World 30,000–40,000 years before the present day by way of the Bering Strait. Their migration through the arctic region in an interglacial period makes it reasonable to conclude that they had knowledge of the use of fire. Later, fire was used by indigenous people as a tool to deter the encroachment of trees into the prairies. Under natural conditions, fires maintain many forest communities in the United States and Canada, such as those of lodgepole pine, ponderosa pine, jack pine, and black spruce.
- Overgrazing retards succession. Although light grazing may be beneficial in some cases, heavy grazing destroys native vegetation and organisms dependent on it. As the vitality of grasses is diminished, their competitive ability is lowered; shrubs and other undesirable species will invade. In extreme grazing, soils are laid bare and are prone to loss by wind and water erosion. The loss of indigenous species because of grazing and subsequent invasion by exotic species is discussed in Chapter 20.
- Pollutants. Numerous air, water, and soil pollutants resulting from industrial and other human activities have differential effects on species abundance and distribution—indeed, even their survival. They may also have a detrimental effect on succession. Some of these impacts are presented in Chapters 18–20.
- The greenhouse effect. The excessive quantities of gases emitted into the atmosphere by human activities are altering the chemical composition of the atmosphere and raising the average air temperature. Associated changes in precipitation patterns are also likely. Sophisticated models have predicted the impacts that climate changes are likely to have on distributions of trees and other species as well as the responses of plants and animals in the ecosystems of the world. These are presented in Chapter 21.

The degree of vulnerability or stability of ecosystems affects both the disturbance and productivity regimes of ecosystems. **Stability,** a term introduced into ecology nearly a century ago, can be used in two different ways. The equilibrium point is considered stable if the state of the system returns to the equilibrium point after a disturbance or if the system does not return to the equilibrium but oscillates around it with certain amplitude.

Ecological stability has two main constituents: resistance and resilience. **Resistance** refers to a measure of ecosystem capacity to absorb disturbances before a change in the state of an ecosystem occurs (Figure 9.7). The higher the resistance of an ecosystem is, the harder it is to change a given state of the ecosystem in the face of disturbances. **Resilience** is a measure of ecosystem capacity to recover from disturbances. The more quickly the disturbed ecosystem reaches a steady state from its initial displacement, the shorter its recovery time (RT) and the greater its resilience (1/RT).

Mechanisms of Succession

Succession of communities is not a simple process. Depending on the habitats and the various influences that may intervene during the development of communities, several pathways are possible. Indeed, several "mini" successions may be found along the formation of a "final" community. For example, whenever an opening or gap is available (e.g., in the northern hardwood communities of beech in New Hampshire), the pioneering yellow birch, which grows rapidly, may fill the gap. When the yellow birch dies, shade-tolerant sugar maple seedlings grow up to take its place. Beech seedlings and saplings come into the community as the sugar maple dies. All these changes take place while beech trees form the tree canopy. Indeed, it has been noted that in forests with lower diversity, sugar maple tends to replace beech, and beech replaces sugar maple (Figure 9.7) (Forcier 1975).

A similar type of succession may be observed in grassland communities. For example, in a

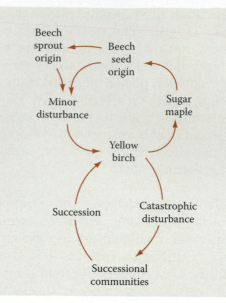

Figure 9.7 Cyclic succession in a climax hardwood forest in New Hampshire. (From Forcier, L. K. 1975. *Science* 189:808–810.)

Michigan field, cyclic changes were observed in the growth and development of herbaceous perennial species. When soils are exposed through animal activities such as the nest building of ants or burrowing of ground squirrels, panic grass is able to germinate and establish itself. This plant dominates for 15–20 years, after which it degenerates. The degenerated community provides an opportunity for the development of a new clump. This process may take place for several generations (Figure 9.8) (Evans 1975).

A related type of succession may be observed in dry-land communities. For example, in a review of trends in ecosystem development, predictable changes were observed in species diversity and the growth and development of herbaceous species over time (Figure 9.9). In this case, species diversity was highest during the period when a site transitioned from being highly degraded due to disturbance to one that has recovered (undisturbed). At this midpoint, both early colonizing annual and later colonizing perennial species are present, leading to high species diversity and productivity. In the later stages of succession, biomass continues to increase but diversity declines as many species are eliminated from the community due to competitive exclusion. This example illustrates how

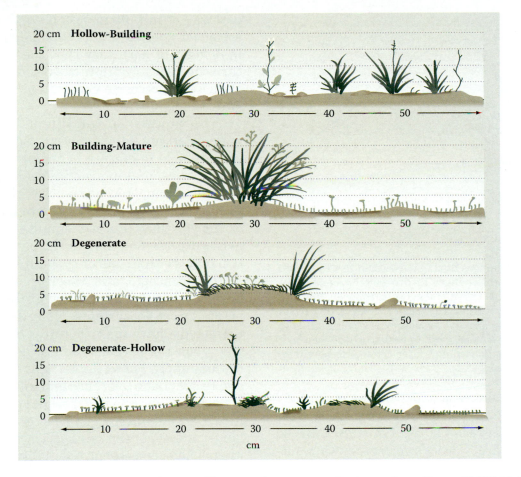

Figure 9.8 Successional phases in panic grass (*Panicum depauperatum*) microcycle on Evans Old Field, E. S. George Reserve. (From Evans, F. C. 1975. In *Prairie: A Multiple View,* ed. M. K. Wali, 27–51. Grand Forks: University of North Dakota Press.) Local erosion pavements provide a suitable germination bed for panic grass seeds (hollow phase). Successful establishment (building phase) leads to development of an elevated mound or tussock with considerable dead material at its base (mature phase). With senescence, the central portion of the clump dies back and is covered by a mat of lichens (species of *Cladonia*), leaving a ring of lateral shoots at the periphery (degenerate phase). Final dissolution of the clump results in new erosion pavements; completion of this cycle requires 15–20 years.

the "reset" provided by disturbance is important in maintaining diversity (Guo 2005).

Changes in wetland communities are often driven by the extent and duration of flooding. Wet–dry cycles and the activities of herbivores such as muskrats lead to vegetation changes over time as the seeds of different species (stored in the seed bank) germinate when conditions become favorable. Studies by van der Valk and Davis (1978) showed that cyclic hydrology leads to a seed bank with two types of seeds: those produced by plants adapted to flooding and those produced by plants adapted to dry conditions. The seed bank in the prairie wetlands studied in this example was immense: 42,000/m² seeds in the top 5 cm and 225,000/m² in the top 35 cm of the soil. The germination

and establishment of the plants depended on where seeds were buried, wet and dry cycles, temperature, and salinity (Figure 9.10).

It should be clear that the course of succession is not a simple unidirectional process. The pathways and "final" communities can indeed be several. Olson, who studied the original beech and dune sites used by Cowles along Lake Michigan, provided an elegant exposition of the phenomena involved. He began by building a time series up to nearly 12,000 years by radiocarbon-dating the dunes and documenting both the plant communities and the chemical composition of the soil underneath them. Several different dominant, mature ("climax") communities were indeed possible, as was documented by Olson (1958). Other possible

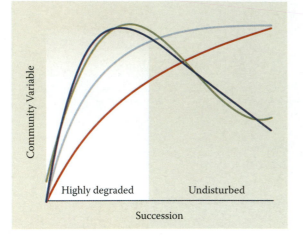

Figure 9.9 Changes during succession in grasslands showing temporal changes in diversity (green curve), productivity (blue curve), biomass (turquoise), and stability (red). (From a literature review by Guo 2005.) Over successional time (x-axis), species diversity is highest in the transition from disturbed to undisturbed when both early colonizing annual and later colonizing perennial species are present.

pathways on succession, depending on species' adaptations and life history strategies, were provided (Figure 9.11) (Marsh 1987). Based on this evidence, Cowles's contention that succession on the dunes would always end in a beech–hickory forest was not supported. In this case, the final sere on the dunes tended to be dominated by black-oak forests, even after 12,000 years of succession time. But that should in no way lessen the great impact of Cowles's work.

A useful tautology of a three-pathway approach in visualizing the processes of succession of plant communities has been suggested (Figure 9.12). These are the facilitation, inhibition, and tolerance pathways. In **facilitation,** an initial group of organisms inhabiting a given area modifies so much that, over time, the habitat becomes more hospitable for the establishment of other succeeding species. Facilitation results from various types of changes, including modifications in microclimatic conditions, changes in soil physical and chemical properties, and changes in the relative competitive abilities of species.

Inhibition involves modifications of the environment by pioneer species that inhibit subsequent recruitment of other species until the pioneer species eventually die. A different

pathway of inhibition was observed in which a pioneer species produced enough chemicals in defense of herbivory that the chemicals in turn considerably diminished the growth form of its own successive generations (Iverson and Wali 1982).

The process of **tolerance** occurs when more tolerant, late-succession species invade and are able to mature in the presence of pioneer species, making conditions less favorable for subsequent recruitment of the pioneers. These successional processes are not mutually exclusive; they proceed simultaneously, and most successional changes involve a combination of more than one or even all three.

The concepts and theories of succession have developed primarily from studies of temperate forest ecosystems. However, grassland ecosystems have many features that set them apart from temperate forests, so not all generalizations apply. For example, light requirements and hence shade tolerance and intolerance play a much smaller role in grassland communities. Fire is a constant allogenic factor in the grasslands, and generally soil nutrients are not limiting. The ever-present possibility of droughts makes the progression of grassland communities more a "fits-and-starts" cycle than the smooth progression presented for the temperate forest ecosystem succession (Figure 9.13) (Woodmansee 1978).

Rangeland Succession

Rangeland ecologists and managers have developed somewhat different models of succession. Herbaceous species are considered to be decreasers, increasers, and invaders in response to grazing pressure and the differential palatability of the plants. Decreasers are those species that are highly palatable to grazing animals and that decrease in vigor and abundance due to differential grazing pressure. Increasers initially increase in abundance when grazing occurs but ultimately decrease under excessive grazing pressure. Invaders are usually not present in the climax community but dominate under excessive grazing pressure

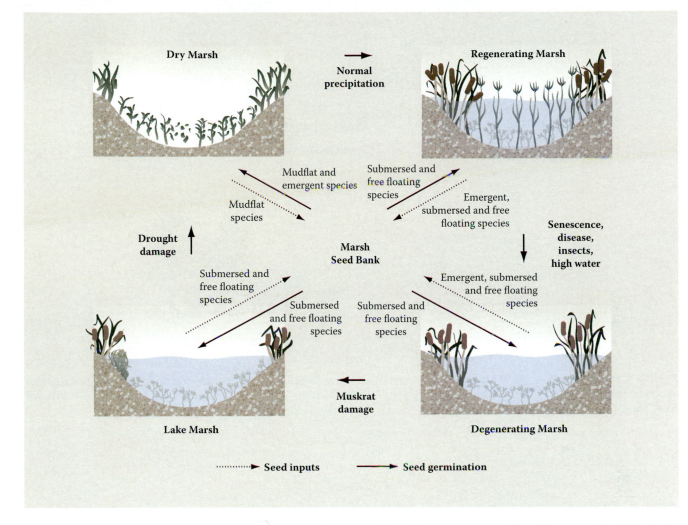

Figure 9.10 Seedbank dynamics and vegetation cycles in North American prairie marshes. (From van der Valk, A. G., and C. B. Davis. 1978. *Ecology* 59:322–335.)

and may be short-lived species such as annual grasses and forbs or long-lived species such as shrubs, many of which will have antigrazing defenses such as thorns and leathery leaves. These successional changes in species composition are considered to be reversible by altering the timing and intensity of grazing.

Although the classical model of succession works well in high-moisture rangelands such as the tall-grass prairie, the **state-and-transition** model has been presented as an alternative model for arid and semiarid rangelands (Dyksterhuis 1949; Woodmansee 1978; Westoby, Walker, and Noy-Meir 1989). In this model, vegetation dynamics in a given area are described by a set of discrete states, with defined transitions between states. Natural and anthropogenic disturbance events such as drought, fire, grazing, and fertilization are

the triggers for transitions between states. In contrast to the classical understanding of succession—where successional change is reversible—transitions from one state to another may not be reversible when a threshold is crossed. For instance, conversion from grass dominance to shrub dominance or from perennial grasses to annual grasses may not be reversible, even with considerable management inputs (see Laycock 1991 for examples). Finally, there is no single endpoint or even multiple endpoints to succession under the state-and-transition model. The state of a community at any given time is only relative to its previous states and the transition events that have occurred.

From the preceding discussion, it should be clear that succession does not flow in a single pathway or even in only a few pathways. That would assume a tremendous uniformity of

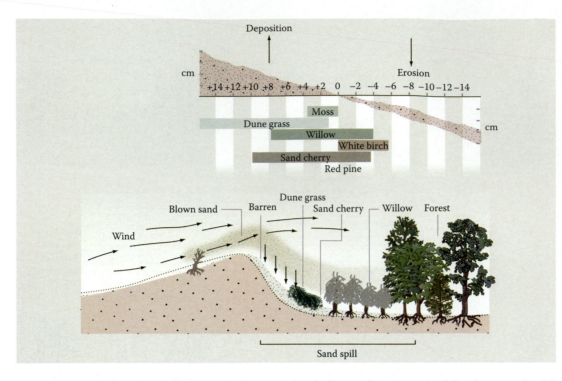

Figure 9.11 The impact of wind and blowing sand in creating habitats that are suited for the growth of dune grass, sand cherry, willows, and other tree species. (Redrawn from Marsh, W. M. 1987. *Earthscape: A Physical Geography.* New York: John Wiley & Sons.)

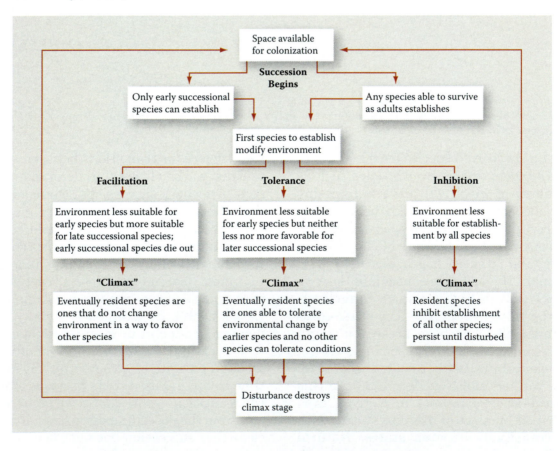

Figure 9.12 The three-pathway approach to understanding processes of succession in natural plant communities. (From Connell, J. H., and R. O. Slatyer. 1977. *American Naturalist* 111:1119–1144.)

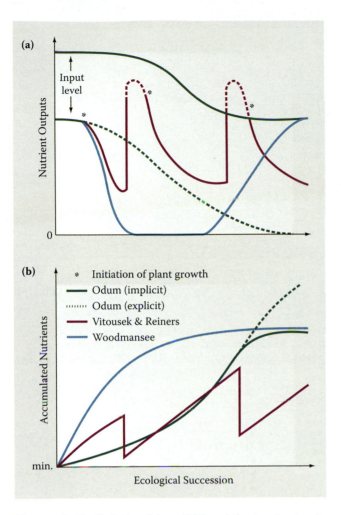

(a)

Input level

Nutrient Outputs

0

(b)

* Initiation of plant growth
— Odum (implicit)
······· Odum (explicit)
— Vitousek & Reiners
— Woodmansee

Accumulated Nutrients

min.

Ecological Succession

Figure 9.13 Relationships of (a) nutrient output rate to the successional status of ecosystems and (b) nutrient accumulation to ecological succession. (The curves are from Odum, E. P. 1969. *Science* 164:262–270; Vitousek and Reiners (1975); and Woodmansee, R. G. 1978. *BioScience* 28:448–453.) (From Woodmansee, R. G. 1978. *BioScience* 28:448–453.)

habitat conditions and just a few strategies to colonize, inhabit, and succeed in a given environment. Given the enormity of habitat and biological diversity, it is to be expected that the paths of succession will be multiple (as they are) and will vary from region to region, giving rise to cyclic and retrogressive–progressive pathways (Figure 9.14). Succession can also be very slow and seemingly imperceptible when environmental conditions dictate; hence, some may think that the concept is untenable and should be discarded.

To conclude, one more point must be made. Most discussions of succession have largely been based on plant communities, and not enough has been documented in the ecological

literature on the effect of other trophic levels in altering the paths of succession (e.g., Figure 9.15). It is known that the effects of herbivory on successional processes (particularly by insects) can be multiple and the adaptive response by plants of deterring them by the synthesis of chemicals can be significant (Iverson and Wali 1982). We discuss this later (Chapter 24 on restoration).

Concept of Climax, Old-Growth, or Mature Communities

As noted earlier, the proposal by Clements that each location on Earth has a single climax has generated much discussion. One of its foremost detractors was Robert H. Whittaker. He argued that the vegetation of a landscape does not represent a single type of environment or community or habitat. Rather, the environment represents a mosaic of environmental gradients along which species are continuously replaced according to their adaptations to that location. Among these climax community types, one is usually most widespread in the landscape. Whittaker termed his interpretation of vegetation distribution and development the **climax pattern hypothesis** (Whittaker 1974).

Does the climax concept have some utility? It seems so, particularly in both ecosystem management and ecosystem restoration. Despite his strong criticism of the monoclimax, Whittaker believed that the concept had definite usefulness, an exposition that we find concise and compelling. He elaborated (1974, pp. 153–154):

The climax concept has these continued values:

(1) It expresses a real and important (though relative) difference in stability between successional and self-maintaining communities.
(2) It offers by this means a basis for reducing research variables. Effects of climate, topography, parent material, etc. on community composition and function may be compared on the basis of climaxes, without complication of the comparison by different successional stages.

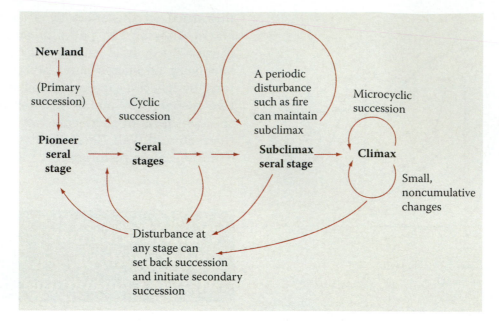

Figure 9.14 Pathways of different types of succession: primary, secondary, and cyclic. The climax stage is in a state of dynamic equilibrium. (Barbour, M. G. et al. 1999. *Terrestrial Plant Ecology*, 3rd ed., p. 270, Figure 11-1. San Francisco, CA: Benjamin/Cummings.)

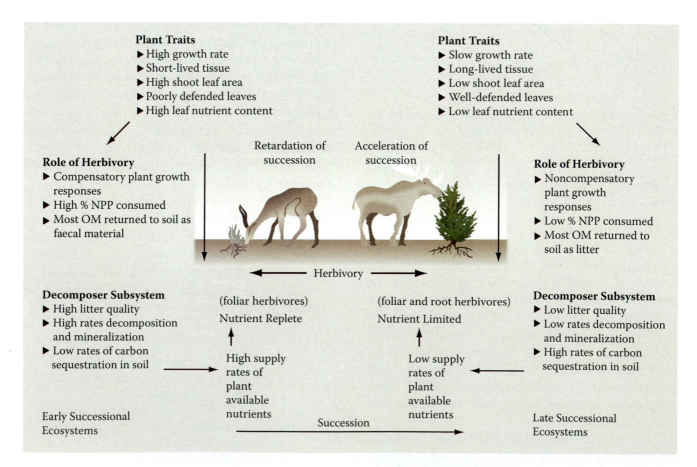

Figure 9.15 The effect of herbivores on the decomposer subsystem at a plant community level, thus altering successional pathways. (From Bardgett, R. D., and D. A. Wardle. 2003. *Ecology* 84:2258–2268.)

(3) It provides a standard of comparison for the study of the characteristics, function, and practical management of successional communities.

(4) It provides a means of ordering and understanding in relation to one another, the diverse stable and unstable communities of an area.

(5) It thus makes possible comprehension of the vegetation of a landscape through a climax pattern (or mosaic) and the successional communities related to parts of this.

(6) It provides a standard of comparison by which may be measured the increasingly widespread effects of disturbance and pollution by (hu) mans that cause communities to retrogress …

(7) It contributes through concepts of regional climaxes to understanding of the broad patterns of climatic adaptations of the vegetation of the world. The climax concept remains an essential means to the interpretation of the complexity of vegetation in space and the flux of vegetation in time, in the world around us.

The nature of community organization[1] and the myriad processes involved in succession are of fundamental importance in understanding ecological systems. But only a few ecosystems in the world have been studied rigorously, so for most ecosystems, information and understanding remain inadequate. Yet, given the pressing nature of finding sustainable solutions to environmental problems and proposing adaptive strategies for environmental management, a generalized and generalizable pattern of community distribution is a must. Thus, the study and recognition of a climax pattern in a given region are very useful. For example, the association of old-growth forests, the "climax" community of the Pacific Northwest, with the northern spotted owl provides useful information on the critical habitat characteristics of an endangered species. Similarly, the delineation of ecoregions and major habitat types based on climax-type communities is of great value in the conservation of ecosystems (see Chapters 22 and 26).

Note

1. The fundamental concept of plant community has recently been re-examined by Callaway (2007). He provides a comprehensive synthesis of recent literature that shows that the facilitation of one plant by another (or one group by another) is not merely a random or individualistic accident but indeed a well-evolved mechanism of how the natural plant world operates. Callaway's facilitation concept, supported by his elegant work, more than brings to life the concepts of "constancy", sociability," and "fidelity" that were in vogue for more than 70 years from the early part of the last century (Braun-Blanquet 1932).

   The evidence presented by Callaway includes an impressive range of examples from the ability of plant roots to "recognize" each other (Callaway and Mahall 2007) to much broader aspects of community structure and function. He begins by asserting that "facilitation affects plant community structure and diversity in very different ways than competition. . . . [P]ositive interactions suggest that some interactions among plants *expand* niches . . . and directly *promote* coexistence and community diversity (pp. 11–12; emphasis in original)." He concludes that facilitation effects are "powerful and ubiquitous" (p. 333; Callaway 2007).

Questions

1. How is a community different from a population?
2. Explain the process of ecological succession. What is the link between succession and disturbance?
3. What are "emergent properties" and why are they important?
4. Discuss the differences between climax and steady state, and between resistance and resilience.
5. What factors determine the time required for forest regeneration?

Biomes—the Major Ecosystems of the World

10

Introduction

For many years, ecologists have sought to understand how geography, climate, and other factors generate global patterns of species distribution. European scientists in the nineteenth century stressed the roles of precipitation and temperature. In the United States, C. Hart Merriam (1894) stressed how temperature controls the geographic distribution of terrestrial animals and plants and formulated the "life zones." The life zone characteristics were also the basis for delineating the crop zones of the United States and are still in use. In the field of ecology, the recognition of "climax" (or old-growth) communities, discussed in the last chapter, led to the concept of biomes.

Biomes

The following three definitions capture the essence of our current understanding of the biome concept:

> Regional climates interact with regional biota and substrate to produce large, easily recognizable community units, called biomes. The biome is the largest land community unit [that] it is convenient to recognize. (Odum 1971, p. 378)

> A biome is a grouping of terrestrial ecosystems on a given continent that are similar in vegetation structure or physiognomy, in the major features of the environment to which this structure is a response, and in some characteristics of their animal communities. (Whittaker 1975, p. 135)

> A biome includes all animals and plants adapted to common climate, which thus share habitats and must accommodate to each other. The animals of a biome are further constrained by the physical structures of plant design … Thus a biome is community in which the living things offer solutions to a common problem; they are environmentally linked. (Colinvaux 1993, pp. 371–372)

Of the biomes recognized by scientists, we will discuss 10 major types here (Table 10.1), including their global geographic positions

Table 10.1
Relative Areas of the Major Biomes on the Earth

Biome	Area (10^6 km^2)
Arctic and alpine tundra	25
Taiga	12
Temperate deciduous forest	7
Grasslands (including temperate savannas)	24
Temperate coniferous forest	5
Temperate rain forest	0.5
Desert	32
Tropical forest	24.5
Mediterranean shrublands	1.8
Tropical savannas	15
Others—woodlands, scrublands, and shrublands	8.5

Sources: After Whittaker, R. H., and G. E. Likens. 1973. In *Carbon and the Biosphere,* ed. G. M. Woodwell and E. V. Pecan, 281–302. Brookhaven Symposium of Biology. 24. Springfield, VA: National Technology Information Services; and Archibold, O. W. 1995. *Ecology of World Vegetation.* London: Chapman & Hall.

and their characteristic climates, topographies, soil types, vegetation, and animal life. Because all are influenced by this same list of attributes, organisms in these communities share common types of adaptations—physiological, morphological, and behavioral; thus, organisms' adaptations to specific environmental conditions will be discussed with each biome.

The discussion included here draws from the work of Archibold (1995), who has provided good comparative information to our understanding of the terrestrial biomes. Our review will help to

- illustrate how edaphic features structure vegetation communities, and hence ecosystems, in terrestrial biomes;
- identify characteristics that determine what trees, shrubs, and animals humans can use, and where;
- understand climax communities in biomes so that we better understand the factors that affect successional processes;
- identify biome changes associated with glaciations and, conversely, with global warming; and

- understand the role of disturbances such as fire.

When summarizing the distribution of the various biomes on the Earth, however, we should never lose sight of the effect of the spatial scales of our sampling. In other words, an illustration showing the distribution of the world's biomes (Figure 10.1) will generalize over huge areas, showing relatively few biomes per 1,000 km^2 and relatively smooth boundaries for each. If the biomes in the United States were illustrated in a similar space, much more spatial heterogeneity of biome distribution would be evident (Figure 10.2); even more heterogeneity would be seen in a representation of the biomes in a single state (e.g., Ohio; Figure 10.3).

It is also important to bear in mind that the boundary lines between biomes are not sharp; rather, each biome grades into the next with a swath or belt of intermediate character. These transitional communities are called **ecotones.** The geological processes, soils, and plant and animal species composition of the ecotones reflect the nearest regions of the biomes on either side. There is some elaboration of ecotones toward the end of this chapter.

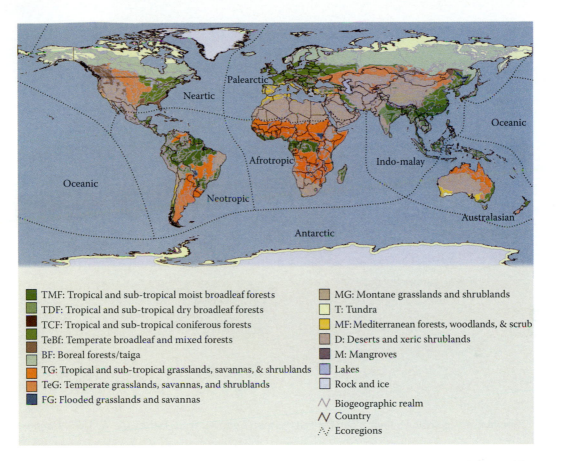

Figure 10.1 Distribution of major biomes in the world. (Olson, D. S. Terrestrial ecoregions of the world: a new map of life on earth. *Bioscience* 51: 933–938.)

Tundra and Alpine

The word *tundra* comes from the Finnish word *tunturi,* meaning treeless plain, and the Siberian word meaning north of the timberline. Globally, tundra covers about 25 million km² in far northern and southern latitudes, including polar deserts, alpine tundra (high-altitude mountaintops), and permanent ice caps (i.e., Greenland and Antarctica) (Archibold 1995).

Arctic Tundra

Tundra and polar desert in North America range from about 55° N along Hudson Bay to 83° N at northern Ellesmere Island and Greenland. Tundra is the most circumpolar of biomes, interrupted only in two places in the northern hemisphere: the Bering Sea and the North Atlantic.

In most arctic tundra, temperatures seldom rise above 0°C for 6 or more winter months of the year, and annual precipitation is below 25 cm. In summers, incoming solar radiation produces long-day photoperiods exceeding 18 hours. However, the angle of incidence of solar radiation is such that it has low thermal energy and thus does not warm the landscape significantly above 10°C (Figure 10.4; see also Figure 3.2). The result is a short growing season between first snow melt in spring and freeze-up in the autumn—from 1.5 to 2.5 months in the high arctic and 3 to 4 months in the low arctic (Table 10.2).

The tundra landscape is diverse, with mountains, shield areas, and many areas of low-elevation lands with the appearance of vast undulating plains. Frost action, freeze and thaw cycles, and the presence of a permanently frozen layer—the permafrost—give tundra its unique geological character. A symmetrical pattern characteristic of the tundra known as patterned ground results from frost action. The fine soil materials and clays, which hold more water than the coarser

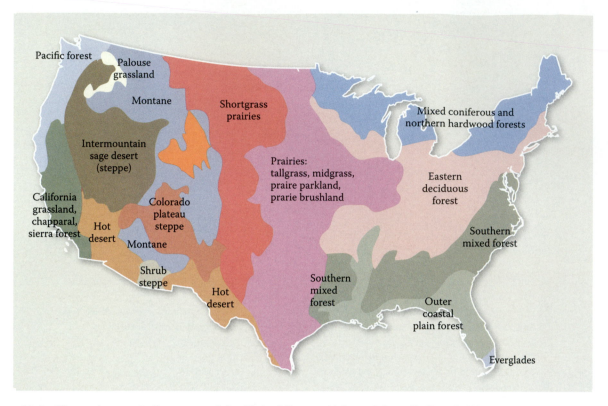

Figure 10.2 The major vegetation zones of the United States. (Adapted from Bailey, A. W. 1976. *Rangeman's Journal* 3:44–46.)

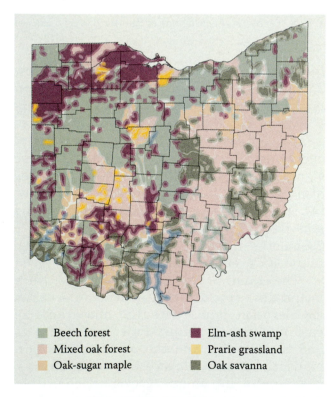

▨ Beech forest	▨ Elm-ash swamp
▨ Mixed oak forest	▨ Prarie grassland
▨ Oak-sugar maple	▨ Oak savanna

Figure 10.3 The vegetation zones of Ohio. (Modified from Hutchins, E. F. 1979. The Ohio country. In *Ohio's Natural Heritage*. ed. M. B. Lafferty, 4–15. Columbus, OH: Ohio Academy of Science.)

materials, expand while freezing and contract upon thawing. This action tends to push the larger material upward and outward from the mass. Depending on the extrusion and sorting, patterned ground results in many shapes, and scientists have designated them accordingly as rock polygons, rock circles, garlands, and stripes (Figure 10.5).

Among other dominant processes is **solifluction,** whereby the surface layers of the soil and ice melt during the summer, resulting in saturated soil that flows downslope over the soil that remains frozen underneath (Figure 10.6). The process of solifluction produces a characteristic undulating landscape of dry and wet soils. Freeze–thaw cycles also produce vertical ice wedges that later increase in width and depth, forming low-centered polygons that may turn into shallow ponds.

Soils

The presence of permafrost and continual freezing and thawing cycles govern both the formation of soils and the development of vegetation that these soils support, creating

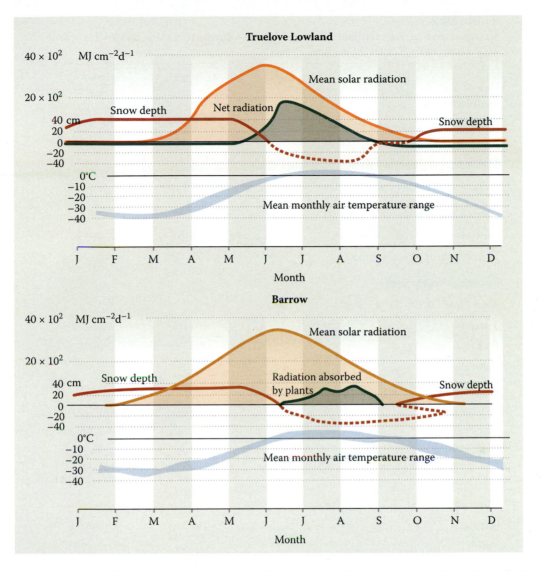

Figure 10.4 A comparison of environmental features of two sites in the Arctic; note the solar radiation received, mean minimum and maximum temperatures, and the depth of active layer. (Redrawn from Chapin, F. S., III, and G. R. Shaver. 1985. In *Physiological Ecology of North American Plant Communities*, ed. B. F. Chabot and H. A. Mooney, 16–40. New York: Chapman & Hall; and Bliss, L. C. 2000. In *North American Terrestrial Vegetation*, ed. M. G. Barbour and W. D. Billings, 1–40. New York: Cambridge University Press.)

conditions unique to the tundra area. In short summers, thawing may penetrate from a few centimeters to two-thirds of a meter. But the deeper ground has permafrost that is impenetrable to water and plant roots. It is the thin layer of water above the permafrost, called the active layer, that allows life to exist during the summer months (Table 10.2; Figure 10.6).

The northern, high arctic tundra (Figure 10.7a) has mostly herb–cryptogram–graminoid plants. The **low arctic tundra** has communities that include cotton grass on permafrost soils (Figure 10.7b–c). South to the low arctic are shrub communities of willows and birches.

An excellent paper by Duke ecologist Dwight Billings (1974) compared the characteristics and adaptations of tundra and alpine communities. Plants that grow in tundra possess characteristics that allow them to grow in cold, harsh, and windy environments. These include a root system capable of living near frozen ground, low growth forms, and a capability for vegetative (rather than sexual) reproduction. The short growing season reduces the opportunity for sexual reproduction and seed formation. Thus, lichens, nonvascular plants like mosses, grasses, and grass-like plants (also called graminoids) dominate the tundra. The biome is

Table 10.2
Comparison of Environmental and Biotic Characteristics of the Low and High Arctic in North America

Characteristics	Low Arctic	High Arctic
Environmental		
Length of growing season (months)	3–4	1.5–2.5
Mean July temperature (°C)	8–12	3–6
Mean July soil temperature at –10 cm (°C)	5–8	2–5
Annual degree days above 0°C	600–1400	150–600
Active-Layer Depth (cm)		
Fine-textured soils	30–50	30–50
Coarse-textured soils	100–300	7–150
Botanical/Vegetational		
Total Plant Cover (%)		
Tundra	80–100	80–100
Polar semidesert	20–80	20–80
Polar desert	1–5	1–5
Vascular plant flora (species)	700	350
Bryophytes	Common	Abundant
Lichens	Common	Common
Growth form types	Woody and graminoid common	Graminoid, rosette, and common cushion
Plant Height (cm)		
Shrubs	10–500	5–100
Forbs	5–30	2–10
Sedges	10–50	5–20

Source: From Bliss, L. C. 2000. In *North American Terrestrial Vegetation,* ed. M. G. Barbour and W. D. Billings, 1–40. New York: Cambridge University Press.
Note: Graminoid refers to a grass-like growth form.

not rich in species (numeric diversity), but the population size of the species present is large.

Animal Life

The species richness of tundra animals, like the plants, is low. Although technically the species of the North American and European tundra differ, they are remarkably similar functionally. Dominant herbivores are the North American barren ground caribou (*Rangifer arcticus*), European reindeer (*Rangifer tarandus*), and musk ox (*Ovibos muschatus*) (primarily on Banks Island in the Canadian arctic). Musk oxen are intensive grazers with limited herd movements, whereas caribou are extensive grazers, migrating across large areas of the tundra. The arctic hare (*Lepus arcticus*), lemming (*Lemmus trimucronatus*), and ground squirrel (*Spermophilus parryii*) are common small mammals. Carnivores include the brown bear (*Ursus arctos*) and polar bear (*Ursus maritimus*), wolves (*Canis lupus*), and arctic foxes (*Alopex lagopus*). Migratory birds common to the tundra during the summer include snowy owls (*Nyctea scandiaca*), snow geese (*Chen caerulescens*), arctic terns (*Sterna paradisaea*), and tundra swans (*Cygnus columbianus*).

Figure 10.5 Left: freeze-and-thaw cycles extrude rocks to form the patterned ground—in this case, raised rings of stones surrounding circular fine-grained domains. The slope determines the shape, forming polygons, circles, streams, stripes, and garlands. (Redrawn from Kessler, M. A., and B. T. Werner. 2003. *Science* 299:380–383.) Right: different forms of patterned ground. (Modified from Zwinger, A. H., and B. Willard. 1972. *Land above the Trees, a Guide to American Alpine Tundra.* New York: Harper & Row Publishers.)

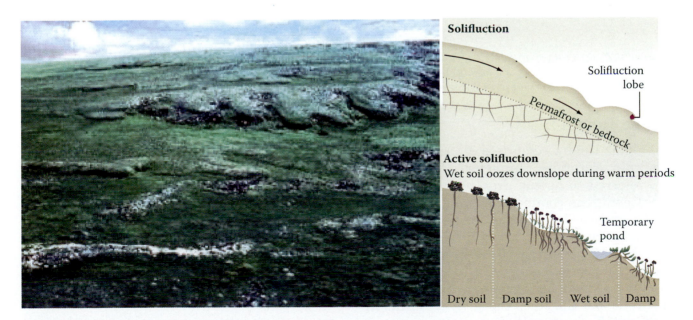

Figure 10.6 Left: solifluction lobes and terraces, Lewis Hills, Gros Morne National Park, Newfoundland. (Photo courtesy of Geological Survey of Canada.) Right: solifluction—the movement of thawed saturated soil over rock or permafrost—seeps down a slope as (upper part) an active process and (lower part) as stabilized slope. (Modified from Zwinger, A. H., and B. Willard. 1972. *Land above the Trees, A Guide to American Alpine Tundra.* New York: Harper & Row Publishers.)

(a)

(b)   (c)

Figure 10.7   (a) Tundra sparsely vegetated with saxifrage and arctic willows (high arctic). (b) Cotton grass tundra (low arctic). (Photos courtesy of National Land and Water Information Service, Agriculture Canada.) (c) Typical tundra soils, the Gelisols have permafrost within 200 cm of the soil surface. (Photo courtesy of USDA-NRCS.)

Alpine Tundra

The alpine tundra comprises all areas of high mountain elevations throughout the world— also referred to as the area beyond the tree line. Conditions common to arctic and alpine tundra systems include the short growing season and snow cover for most of the year; however, there is no permafrost in most alpine regions. Precipitation, especially snowfall and humidity, is higher in the alpine areas than in the arctic, but the steep topography results in rapid runoff of water (Billings 1974).

For the alpine tundra, three types of environmental spatial gradients have been recognized: macrogradients, mesogradients, and microgradients (Billings 1974). In alpine environments, the macrogradients extend from timberline to the upper limits of plant growth on a particular mountain range. The mesogradients in alpine environments are on a topographic scale and are superimposed upon the altitudinal and latitudinal gradients. Because of the interaction of slope angle and direction, wind, blowing snow, and long-lasting snowdrifts, they are much more important in influencing plant growth and distribution in mid- to high-latitude mountains than they are in the tropics. The microgradients in alpine environments may be caused by the presence of small features such as a rock (Figure 10.8).

(a)   (b)

(c)

Figure 10.8 (a) in a seemingly bare alpine area, a closeup reveals the microgradient created by rocks that shelter (b) *Silene* sp. and (c) *Saxifraga* sp. from wind. (Photos by M. K. Wali.)

On its windward side, the vegetation is blasted by snow, whereas in its lee it is protected by a miniature snowdrift. Soil instability and differences in soil particle size are also important components of microgradients.

Species adaptations that are morphological, physiological, reproductive, and ecological contribute to the survival of the whole plant, and thus of the local population, under alpine conditions. Among the morphological characteristics are the small plant size, perennial life cycle, and the ability to grow immediately after the snowmelt. The physiological adaptations to short alpine growing season and low temperatures include quick regrowth from vegetative parts and the channeling of first year's photosynthate into the development of a root system. This serves as a carbohydrate bank and as insurance against drought. For both arctic and alpine tundra, flower primordia are produced at least the year before flowering and sometimes 2–3 years before. These preformed flower buds are ubiquitous adaptations in arctic–alpine, ensuring that there is no delay in flowering after snowmelt (10–20 days) and that there is time for seed-set, assuming that all other environmental conditions are met.

Wildlife in the alpine tundra is quite variable depending on the geographic location. In North America, some common mammals found in the alpine tundra include pikas (*Ochotona princeps*), marmots (*Marmota* spp.), mountain goats (*Oreamnos americanus*), and mountain sheep (*Ovis* spp.).

Taiga

Geographic Position

South of the tundra and circumpolar in extent (Figure 10.9) lies the taiga, or the boreal forest. Estimates of its extent vary, ranging from 12.0 million km² (Whittaker and Likens 1973) to 15.8 million km² (Archibold 1995). Like the tundra, it forms a circumpolar belt extending from Alaska and Canada to Scandinavia, Russia, and Siberia.

Climate

The taiga has a continental climate with cold, frigid winters. Though the summer is longer and warmer than it is in the tundra,

Figure 10.9 Left: an overview of taiga or boreal vegetation showing a gradation from white spruce (top) to black spruce bogs (bottom). Right: the dominant soils are Spodosols with pronounced leached (white) layers. (Photos by M. K. Wali.)

temperatures seldom exceed 25°C. The area lies between the July temperature isotherm of 13°C in the north and 18°C in the south (Barbour and Billings 1988). Annual precipitation in this biome mostly ranges below 50 cm, most of it received during the summer (Archibold 1995). The length of the frost-free season generally ranges from 50 to 100 days. Within this macroclimate, however, the microclimates vary considerably, depending upon the nature of the soils, topography, plant abundance, and community structure.

Landscape and Topography

Both tundra and taiga are young regions because they were the last to have been exposed by the retreat of the Paleocene ice sheets. Permafrost is discontinuous, generally found in the northern end of the biome and at higher elevations, and totally absent in the south. Generally, like the tundra, this landscape is also level to undulating. The taiga has an abundance of lakes, ponds, and bogs, which have an important effect on local microclimates and plant and animal diversity.

Soils

The dominant soil order of the taiga is Spodosols, with some Alfisols and Inceptisols, and in bogs and other wetland areas, Histosols (Archibold 1995). **Spodosols** (Figure 10.9,

right) are relatively infertile soils with low pH, a leached (eluviated) surface (A-horizon) layer, and a subsoil zone characterized by iron and aluminum accumulation. **Alfisols** are soils that have developed primarily on glacial till in mountainous areas and have a layer of clay accumulation. Soils without well-developed organic layers (e.g., **Inceptisols**), such as soils developed from volcanic deposits, are often found under young pine communities. **Histosols** are soils high in organic matter comprising poorly decomposed organic matter and are found in poorly drained areas throughout the taiga.

Vegetation

The boreal vegetation (Figure 10.9, left) is represented by upland forest, bogs, and aquatic vegetation of the lakes. The dominant tree species of the boreal region show a remarkable similarity across the longitudinal gradient from Alaska and western Canada to eastern Siberia (Table 10.3). White spruce is generally regarded as the climatic climax species of the North American taiga, but it is more or less restricted to finer textured soils and moist situations and not often found in pure stands.

On younger fine-textured soils, even aged aspen stands are found frequently; a fire history in such stands can be discerned from the presence of charcoal in surface soil layers.

Table 10.3
Major Species of the Taiga Biome by Longitudinal Sector

Genus	North America, 55° W–160° W	Northern Europe, 5° E–60° E	Western Siberia, 60° E–120 °E	Eastern Siberia, 120° E–170° W
Conifers				
Abies (fir)	*balsamea lasiocarpa*	*sibirica*	*sibirica*	*sibirica*
Larix (larch)	*laricina*	*sukachzewski*	*sibirica*	*dahurica*
Picea (spruce)	*glauca mariana*	*abies obovata*	*obovata*	*obovata*
Pinus (pine)	*banksiana contorta*	*silvestris*	*sibirica silvestris*	*pumila silvestris*
Hardwoods				
Alnus (alder)	*crispa tenuifolia*	*incana*	*fruticosa*	*fruticosa*
Betula (birch)	*glandulosa papyrifera*	*pendula pubescens*	*pendula pubescens*	*ermanii*
Populus (poplar)	*balsamifera tremuloides*	*tremula*	*tremula*	*tremula*

Source: Modified from Hare, F. K., and J. D. Ritchie. 1972. *The Geographical Review* 62:333–365.

Low-intensity fires seem to promote aspen and exclude white spruce (Rowe 1952). Aspen and white spruce have differential effects on the development of understory vegetation. In older aspen stands, seedlings of white spruce are frequent, so spruce replaces aspen over time. On dry, coarse glacial outwashes, jack pine communities are common in the north-western and eastern range and lodgepole pine in the southwestern range; the two pine species are ecological equivalents in their respective regions.

Communities of pine, aspen, and spruce all have their distinctive lichen, moss, herb, and, where present, shrub components (Wali and Krajina 1973). Bogs in the boreal regions are widespread and dominated by black spruce, Labrador tea, and several species of the moss *Sphagnum*. Black spruce is regarded as the edaphic climax, maintained as such by soil conditions, poor drainage, and periodic fires. However, sedge-dominated fens are also found. Along the freshwater lakeshores, communities of river alder are found.

Animal Life

The dominant large animals of the North American taiga are the moose (*Alces alces*) and woodland caribou (*Rangifer tarandus*). Common large carnivores include black bears (*Ursus americanus*) and, less frequently, brown bears (*Ursus arctos*) in the west; Canada lynx (*Lynx canadensis*); wolves (*Canis lupus*); and red foxes (*Vulpes vulpes*). Smaller mammals include snowshoe hares (*Lepus americanus*), red squirrels (*Tamiasciurus hudsonicus*), beaver (*Castor canadensis*), and muskrat (*Ondatra zibethicus*). Mosquitoes (species of *Anopheles*, *Culex*, and *Aedes*) and black flies (species of *Prosimulium* and *Simulium*) are pervasive; indeed, biting insects have been characterized as the best known group of boreal invertebrates (Pruitt 1978). Common birds of the North American taiga include year-round species such as the boreal owl (*Aegolius funereus*), Canada jay (*Garralus canadensis*), and raven (*Corvus corax*), as well as migratory species such as the Canada goose (*Anser canadensis*), common loon (*Gavia immer*), and various ducks and passerine species.

Temperate Deciduous Forest

Geographic Position

Occupying 7 million km² (Whittaker and Likens 1973), these forests originally covered eastern North America, all of Europe, eastern Asia, parts of Japan, portions of Australia and New Zealand, and the southern part of South America (Figure 10.10). Of all the biomes, this one has perhaps seen the greatest transformation through human impacts.

Climate

The north temperate deciduous forest biome is characterized by moderate temperatures and longer summers, with the growing season ranging from 120 days in the north to 250 days in the south (Archibold 1995). Precipitation ranges from 75 to 150 cm and is fairly evenly distributed, although a slightly greater part of it comes down as rainfall during the summer months.

Landscape and Topography

The landscape in the biome ranges from flat areas to hills and mountain slopes. Greller (1988) suggests that topography in deciduous forest is described by two major types of landforms: platform areas that are flat or show minor relief and foldbelts that show pronounced relief (up to 1600 m) and other geomorphological variations. He appropriately emphasizes that local topographic and geological variations are more important than major features in determining forest composition.

Soils

Alfisols are the predominant soil order of the temperate deciduous forest (Figure 10.10, right), along with Inceptisols and Ultisols (Archibold 1995). **Inceptisols** are found primarily in the Appalachian Mountains and have developed on weathered shales and sandstones. They have a deep, fertile surface horizon maintained by rapid decomposition of organic matter and typically support a diverse flora. **Ultisols** have developed on igneous and metamorphic

Figure 10.10 Left: vegetation of temperate deciduous forest, shown is a sugar maple stand. (Photo courtesy of D. M. Hix.) Right: dominant soils are Alfisols. (Photo courtesy of USDA-NRCS.)

substrates and are primarily found in the hilly piedmont region, as well as terraces throughout the region, and the coastal plains. Ultisols are relatively infertile due to excessive leaching and a low cation exchange capability and have a well-developed layer of clay accumulation.

Vegetation

Brockman (1966) listed 67 trees and shrubs in North America whose distributions are centered on or restricted to the deciduous forest (Greller 1988) and several climax communities. Braun (1950) recognized nine major North American forest regions. The hemlock–white pine–northern hardwoods region is located from the Atlantic coast to Minnesota and just south of the taiga. Sugar maple (*Acer saccharum*), beech (*Fagus grandifolia*), basswood (*Tilia americana*), yellow birch (*Betula alleghaniensis*), eastern hemlock (*Tsuga canadensis*), white pine (*Pinus strobus*), and red spruce (*Picea rubens*) in the east, and white and red pine (*Pinus resinosa*) along with the deciduous species toward the west, dominate this region.

South and east of this forest, the oak–chestnut forest occupies most of the ridge and valley and Blue Ridge provinces and extends across the Piedmont onto the Atlantic coastal plain from northern Virginia northward and into glaciated southern New England. Because of the chestnut blight (*Endothia parasitica*), this area now has only small sprouts of chestnut (*Castanea dentata*) in the understory, so the overstory is dominated by various species of oaks (*Quercus* spp.), maples (*Acer* spp.), and hickories (*Carya* spp.). West of the oak–chestnut forest and centrally located in the biome is the mixed mesophytic forest of the unglaciated Appalachian Plateau. Inhabiting moist and well-drained soils are beech, tulip tree (*Liriodendron tulipifera*), sugar maple, chestnut sprouts, red oak (*Quercus rubra*), white oak (*Quercus alba*), and several species of basswood (*Tilia* spp.).

The western mesophytic forest, located primarily in the states of Kentucky and Tennessee, includes a mosaic of communities growing on a variety of soils. This forest includes oak–hickory to the west and oak–tulip tree and beech–chestnut to the east. Further west are the oak–hickory forests of the Ozark and Ouachita Mountains, extending northward and northeastward onto glaciated areas. The maple–basswood forest is found in the northwestern region of the deciduous forest and extends into south-central Minnesota. Dominant species are American basswood and sugar maple, usually with red oak intermixed.

The oak–pine forest region occupies most of the Piedmont Plateau from Virginia southward and stretching across the Gulf States. Although oaks (particularly white oak) and hickory are considered the dominant deciduous species, this forest region is significant for a dominance of loblolly pine (*Pinus taeda*) and yellow pine (*Pinus echinata*). Longleaf pine (*Pinus palustris*) dominates this area, with deciduous forests to the north and evergreen magnolias (*Magnolia* spp.) in the south.

Animal Life

The most common large mammal is the white-tailed deer (*Odocoileus virginianus*). Carnivores include black bear, gray fox (*Urocyon cinereoargenteus*), and bobcat (*Lynx rufus*). Small mammals include gray squirrel (*Sciurus carolinensis*) and fox squirrel (*Sciurus niger*). Common birds include wild turkey (*Meleagris gallopavo*), red-eyed vireo (*Vireo olivaceus*), wood thrush (*Hylocichla mustelina*), tufted titmouse (*Baeolophus bicolor*), and several woodpecker species.

Grassland

Geographic Position

Temperate grasslands cover an area of approximately 9 million km², whereas savannas (rolling grasslands scattered with shrubs and isolated trees) cover 15.0 million km² (Whittaker and Likens 1973). Grasslands originally covered an area as large as 46 million km² (Shantz 1954, cited by Sims and Risser 2000). However, much of the grassland biome has been converted to cropland in order to grow grain crops such as wheat (*Triticum aestivum*) and maize/corn (*Zea mays*). The areas remaining as natural grassland tend to be those least suited to cultivation because of climate, soils, and/or topography.

Climate

The grasslands characteristically are low-rainfall areas (semiarid and arid climates). Temperate grasslands have a mean annual temperature between 6 and 12°C, but have wide climatic fluctuations from year to year; thus, averages are poor representatives of the thermal character of the region. The average total precipitation ranges from 20 to 100 cm per year (most receive less than 50 cm per year), but this, too, may vary considerably from extremely wet to extremely dry years (Coupland 1958). Buol et al. (1997) summarized Borchert (1950), who described the climate as "severe dry winters with much wind and relatively slight accumulations of snow, relatively moist springs in most years, droughty summers with some thunderstorms and tornadoes." In North America, temperatures get warmer and precipitation more uneven as one proceeds from east to west and southwest. The constant winds are a major ecological factor because they affect both evaporation and transpirational stress of plants in summer and, lacking a good snow cover, may cause severe wind erosion of topsoil.

Landscape and Topography

The grassland landscapes are fairly flat and, to a lesser extent, somewhat undulating. Driving through a grassland landscape, one might get the impression that these areas have no relief or microtopographical features. However, they do exhibit considerable microtopography. Indeed, a better view of a grassland area may be obtained from an aerial view, when the differences in soil color and vegetation communities reveal a corresponding mosaic of vegetation types (Figure 10.11, left).

Soils

Grassland soils exhibit considerable variation. Typical, however, are humus-rich, dark, and deep soils, the **Mollisols** (Figure 10.11, right). Melanization, the dominant soil-forming process of grasslands, is the darkening of soils by decaying organic matter, which occurs by a combination of the following processes: (1) extensive and deep spread of grass roots (as much as

Figure 10.11   Left: Stands of big and little bluestem in grassland biome of midcontinent North America. (Photo by M. K. Wali.) Right: a Mollisol. (Photo courtesy of USDA-NRCS.)

80% of the total biomass in many grasslands is in the roots); (2) decay of organic matter, producing dark, stable compounds; (3) churning of the soils by soil biota; and (4) formation of residues of organic matter resistant to decomposition that give these soils a black color even after cultivation for decades (Buol et al. 1980).

However, grasslands do have a variety of other soils, particularly those that have an enrichment of limestone (calcium carbonate). The concentration of salts (especially sodium chloride) exacerbates problems with soil permeability and hence impedes deeper infiltration of water. Entisols, Inceptisols, and carbonate- and sodium-rich members of these soil orders are also fairly common.

Vegetation

In North America, the French word *prairie* is used for grasslands, in both common usage and scientific literature. On an east–west gradient, several prairie types are recognized (Table 10.4):

- the tallgrass prairie dominated by big (*Andropogon gerardii*) and little (*Schizachyrium scoparium*) bluestem, Indiangrass (*Sorghastrum nutans*), and switchgrass (*Panicum virgatum*);

- mixed or midgrass prairie dominated by western wheatgrass (*Pascopyrum smithii*) and needlegrasses (*Stipa* spp.);
- shortgrass prairie, dominated by blue grama (*Bouteloua gracilis*) and buffalo grass (*Buchloe dactyloides*);
- desert grasslands of the southwest, dominated by black grama (*Bouteloua eriopoda*), three-awn grass (*Aristida* spp.), creosote bush (*Larrea tridentata*), and cacti (e.g., *Opuntia* spp.);
- the Palouse Prairie of eastern Washington and northwestern Idaho, dominated by bluebunch wheatgrass (*Pseudoroegneria spicata*) and Idaho fescue (*Festuca idahoensis*); and
- the California annual grasslands characterized by bunchgrasses (*Stipa* spp.) and wild oats (*Avena fatua*) among many others.

Grasses possess unique ecological features that allow them to inhabit a wide range of habitat conditions. Their physiological adaptations to a combination of widely fluctuating temperature and water availability, as well as their ability to withstand high winds, frequent drought cycles, and fire, allow them to occupy continuous ranges in many areas of the world. They have evolved to match different temperature regimens by utilizing the C_3

Table 10.4
Major Features of the Grassland Biome in North America

Attribute	Shortgrass	Mixed Grass	Tallgrass
Vegetation types (Sensu Küchler)[a]	64–65	66–70	74–78, 81–82, 86
Presettlement area (km^2)[b]	507,970	510,316	745,680
Remaining area (km^2)[b]	32,910	264,070	198,710
Remaining area (%)[b]	63.6	51.7	26.6
Annual precipitation (cm)[c]	25–50	40–75	50–100
Potential evapotranspiration (P.E.T.) (cm yr^{-1})[c]	85–195	71–160	62–139
Precipitation/P.E.T. ratio (during growing season)[c]	0.3–0.5	0.4–0.7	0.6–1.0
Species richness, spp./km^2	30–60	50–200	150–300
Canopy height, cm	15–60	40–120	80–180
Fire periodicity, years[d]	5–10	2–7	1–3
Dominant grasses[e]	Aristida purpurea	Agropyron spp.	Andropogon gerardii
	Bouteloua gracilis	Aristida longiseta	Elymus canadensis
	Buchloe dactyloides	Bouteloua curtipendula	Panicum virgatum
	Hilaria mutica	Koeleria cristata	Sorghastrum nutans
		Schizachyrium scoparium	Spartina pectinata
		Stipa spp.	
Dominant forbs[e]	Artemisia frigida	Chrysopsis villosa	Amorpha canescens
	Opuntiu spp.	Gutierrezia sarothrae	Echinacea pallida
	Phlox hoodii	Haplopappus spinulosus	Eryngynium yuccifolia
	Yucca sp.	Psoralea spp.	Liatris punctata
		Ratibida columnifera	Phlox pilosus
		Solidago missouriensis	Ratibida pinnata
			Silphium spp.
			Solidago rigida

Sources: After Burton et al. 1988; modified by Iverson, L. R., and M. K. Wali. 1992. In *Ecosystem Rehabilitation: Preamble to Sustainable Development,* vol. 2, 85–129. *Ecosystem Analysis and Synthesis,* ed. M. K. Wali. The Hague: SPB Academic Publishing.
[a] Principal source: Küchler. 1964.
[b] Principal source: Klopatek et al. 1979.
[c] Principal source: United States Geological Survey. 1970.
[d] Principal sources: Sauer. 1950; Wright, H. A., and A. W. Bailey. 1982. *Fire Ecology.* New York: John Wiley & Sons.
[e] Principal sources: Weaver. 1954; Weaver and Albertson. 1956.

and C_4 pathways of photosynthesis discussed in Chapter 7.

The C_4 photosynthetic process results in increased water use efficiency, but it is relatively inefficient at low temperatures and low light intensities. Thus, C_3 grasses are generally considered to be cool-season grasses, whereas C_4 grasses are considered to be warm-season grasses.

The grasses have also evolved xeromorphic features and mechanisms by which they are able to withstand frequent droughts. They have extensive and deep root systems to scavenge for more water and nutrients and, when conditions are unfavorable, develop rhizosheaths. Mechanisms also include the ability to reduce water loss by stomatal closure while maintaining gas exchange. If unfavorable conditions persist, grasses become dormant, shutting down all growth processes. Grasses also have the ability to take up nutrients and water rapidly during short periods of water availability and to store and recycle nutrients internally.

Because grasses burn easily and grasslands have a large reserve of dry stubble and organic matter—all acting as fuel—they are prone to frequent natural fires. At least as importantly, throughout history, humans have used fire as a tool for converting forests to open areas for cultivation and grazing. Thus, grasslands, droughts, and fire have been regarded as "inseparables." Borchert (1950) noted that "grassland climates favor fire just as they favor grasses whether there are fires or not." But fire has been shown to increase the productivity of these ecosystems by increasing the release of nutrients and stimulating tiller, flower, and seed production. Fires also deter the growth and encroachment of woody vegetation into grasslands. The adaptability of grasses comes from protected meristematic tissue and rapid regrowth after fires. Overall, fires generally are considered beneficial to growth in grasslands. Note that many of the attributes differ profoundly and that the tallgrass prairie has the greatest precipitation, potential evapotranspiration, and the most frequent fires.

Another common feature of grasslands is that they support large populations of herbivores and hence are regularly grazed. Light to moderate grazing has been shown to stimulate grass growth. Regular grazing increases penetration of light and reduces mutual shading. This in turn increases the energy translocation to young leaves, which increases photosynthetic response. However, plants have developed several mechanisms against grazing, such as the production of toxic chemicals and coarse stalks to make plants unpalatable. Heavy grazing is detrimental because it denudes the vegetative cover, exposing the soils to high temperatures and erosion by wind and water.

Animal Life

The animal most associated with North American grasslands is the bison (*Bison bison*). However, bison were nearly extirpated during settlement, and domestic cattle (*Bos taurus*) and sheep (*Ovis aries*) have taken their place. Other common large wildlife include pronghorn (*Antilocapra americana*) and mule deer (*Odocoileus hemionus*). Carnivores include coyotes (*Canis latrans*) (having replaced the mostly extirpated wolves), bobcats (*Lynx rufus*), and badgers (*Taxidea taxus*). Small mammals associated with grasslands include ground squirrels (*Spermophilus* spp.), jack rabbits (*Lepus* spp.), and prairie dogs (*Cynomys* spp.), all serving to dig in the soil and graze on plants and their seeds.

Temperate Coniferous Forest

Geographic Position

Estimates of the extent of the temperate coniferous forest range from 3.3 (Archibold 1995) to 5 million km² (Whittaker and Likens 1973). Globally, the primary locations of this biome are western North America and the Alps of Europe. In North America, this includes the montane forests of the Rocky Mountains, Cascade Mountains, and Sierra Mountains, as well as the forested land west of the Cascade Mountains. The temperate coniferous forest merges into the boreal forest in a broad ecotone in northern British Columbia and Alberta. It is delineated into dry-mesic forest and rain forest based on whether annual rainfall is less or greater than 140 cm.

Dry-Mesic Forest

Geographic Position

Approximately 80–90% of the temperate coniferous forest may be classified as dry-mesic forest because it lies in the rain shadows of the western mountain ranges in North America, such as the Rocky Mountains, the interior of British Columbia, the Cascade Mountains west to the coastal mountains, and the Sierra Nevada of California.

Climate

Climate is quite variable in the dry-mesic portion, with great differences from its boundary with the grasslands in the east to its boundary with the rain forest in the west. To the north and south, it is bounded by the boreal forest and the desert, respectively. Annual precipitation ranges from 40 to 140 cm and is heavily influenced by orographic effects. Precipitation increases with increasing elevation and is highest on the windward sides of mountains with a rain shadow on the leeward sides. The Rocky Mountains receive the bulk of the precipitation in the summer, whereas the Cascade Mountains and the Sierras receive the majority of their precipitation in the winter. Heavy snowfall and deep snow packs during the winter are typical, especially in the north and at high elevations.

Landscape and Topography

The majority of the dry-mesic coniferous forest is found in mountain areas. Thus, the topography is rough and varied, with common steep slopes. However, the area between the Cascade and coast mountains in Oregon and Washington is relatively flat-to-rolling topography, with only moderate slopes.

Soils

Because of the broad geographic extent and great climatic variation, soils are also quite variable. Spodosols are common in the wetter and cooler areas. Alfisols are widespread in the warmer and drier montane areas, and Andisols (volcanic soils) are found in the Cascade Mountains.

Vegetation

In the drier and warmer areas, various species of pines are dominant. Ponderosa pine (*Pinus ponderosa*) and aspen (*Populus tremuloides*) are signature trees of much of the Rocky Mountain region. As elevation increases, dominant trees species change from Douglas fir (*Pseudotsuga menzeisii*) and ponderosa pine

to lodgepole pine (*Pinus contorta*) and then to Englemann spruce (*Picea engelmanii*) and subalpine fir (*Abies lasiocarpa*) (Archibold 1995). In the Sierra Nevadas, ponderosa pine is also common, as well as Jeffrey pine (*Pinus jeffreyi*) and sugar pine (*Pinus lambertiana*) (Barbour and Billings 1988). The south-central Sierra Nevadas are also the habitat for the one of the most massive tree species in the world, the giant sequoia (*Sequoiadendron giganteum*), which may reach heights in excess of 80 m, trunk volumes greater than 1,000 m^3, and trunk diameters larger than 800 cm. The western slopes of the Cascade Mountains and the broad valleys between the Cascades and coast mountains are dominated by Douglas fir, along with western hemlock (*Tsuga heterophylla*) and western red cedar (*Thuja plicata*) (Archibold 1995). The drier habitat on the eastern side of the Cascades supports ponderosa pine, lodgepole pine, and Douglas fir. High elevations in the Cascades have stands of Engelmann spruce and subalpine fir, with white spruce also common in the north.

Animal Life

Common large herbivores are elk (*Cervus elaphus*) and mule deer in the Rocky Mountains, moose in the north, mule deer in the eastern Cascades, and black-tailed deer (*Odocoileus hemionus*) in the western Cascades and western valleys. Common carnivores include mountain lions (*Felis concolor*), coyotes, and black bears throughout, and wolves and brown bears in the north. The northern spotted owl (*Strix occidentalis caurina*) is an important endangered bird species in the Cascades and West.

Temperate Rain Forest

Geographic Position

The temperate rain forest covers approximately 0.5 million (Archibold 1995) to 1.0 million km^2 globally (Steiner 1998) (Figure 10.12). The geographic extent of temperate rain forest

Figure 10. 12   Left: A view of the coastal western hemlock forests of British Columbia, Canada. Right: growing on a Spodosol. (Photos courtesy of Karel Klinka.)

in North America covers a latitudinal range of 38–61° N (Alaback 1996) and includes the Pacific Northwest west of the coastal mountains from southeastern Alaska, through British Columbia and the Olympic peninsula of Washington, to the coastal redwoods of southern Oregon and northern California. It is also found in southern Chile, western Tasmania (Australia), and on the west coast of New Zealand's south island.

Climate

The temperate rain forest has a maritime climate characterized by cool annual temperatures with relatively small seasonal variation and high annual precipitation (Alaback 1996). Annual precipitation in the temperate rain forest exceeds 140 cm, with the least amount occurring during the summer. Interception of fog by vegetation and subsequent dripping of collected moisture onto the ground may also be an important element of the hydrologic budget

(Harr 1982). Snowfall is common in the winter in the northern range of North America, but it is absent in the south and in Chile, Tasmania, and New Zealand. Mean annual temperatures range from 4 to 12°C, with individual annual temperature ranges no more than 15°C.

According to Alaback (1996), there are four primary climatic zones of temperate rain forest in North America: (1) subpolar (59–61° N), with significant summer precipitation (greater than 20% of annual), cool summer temperatures, and heavy snowfall that accumulates on the ground; (2) perhumid (50–58° N), with moderate summer precipitation (greater than 10% of annual), cool summers, and moderate snowfall that generally melts fairly quickly; (3) seasonal (43–50° N), with light summer precipitation (less than 10% of annual) leading to common droughts and subsequent fires; and (4) warm (38–43° N), with little summer precipitation (less than 5% of annual) and where droughts and fires may occur during any season of the year.

Landscape and Topography

Topography of the temperate rain forest is very rugged and mountainous. Steep terrain and landscapes heavily influenced by glaciation are common. The coastal mountains of North America reach elevations greater than 5,000 m in the north, although rain forest is restricted to the more equable climates of lower elevations.

Soils

Spodosols are the most widespread soil type of the temperate rain forests; however, Ultisols are also found in the warmer portions and Histosols in extensive areas of bogs (Figure 10.12, right) (Archibold 1995). In the Spodosols, the iron content of the B-horizon is typically lower than in the boreal forest. The very high levels of organic matter on the soil surface lead to humus formation and dark-colored surface horizons. Soil pH is very acid, with pH in the surface horizons ranging from 3.5 to 5. The high rainfall leads to leaching of soil nutrients from the mineral soil, leaving most of the nutrients concentrated in surface organic matter. Because soils are generally shallow, tree rooting depths are also shallow, making them very susceptible to uprooting by strong winds.

Vegetation

The temperate rain forests of North America contain the longest-lived and largest species of most dominant conifer genera, including *Abies, Chamaecyparis, Larix, Libocedrus, Picea, Pseudotsuga, Sequoia, Thuja,* and *Tsuga* (Figure 10.12, left) (Alaback 1996). The most dominant and widespread tree species in the North American temperate rain forest are Sitka spruce (*Picea sitchensis*), Douglas fir (*Pseudotsuga menziesii*), western red cedar, and western hemlock (*Tsuga heterophylla*). However, in southern Oregon and northern California, the coast redwood (*Sequoia sempervirens*), which may grow to heights greater than 100 m, is the dominant species. The Tasmanian rain forest contains the tallest hardwood tree species in the world—mountain ash (*Eucalyptus regnans*)—which may also reach heights in excess of 100 m. Lichens, bryophytes (mosses and liverworts), and ferns are also prevalent throughout the temperate rain forest.

Animal Life

Five species of Pacific salmon—chum (*Oncorhynchus keta*), coho (*O. kisutch*), chinook (*O. tshawytscha*), pink (*O. gorbuscha*), and sockeye (*O. nerka*)—spawn each fall in North American temperate rain forest streams and rivers, providing a major source of food for brown bears in British Columbia and Alaska, as well as American bald eagles (*Haliaeetus leucocephalus*) throughout. Other carnivores include populations of black bears, mountain lions, and wolves. Large herbivores include black-tailed deer and Roosevelt elk (*Cervus elaphus roosevelti*). The temperate rain forest is also notable for being a primary habitat for the endangered marbled murrelet (*Brachyramphus marmoratus*).

Desert

Geographic Position

Found on all continents, deserts are present in both cold and hot climates—hence the terms cold or polar deserts and hot or arid deserts, respectively. The extreme deserts of ice, rock, and sand comprise about 24 million km², the deserts that bear some life occupy about 18 million km² (Whittaker and Likens 1973). Much of the area around latitudes of 30° N and S are deserts associated with high pressure related to downward air currents of air (e.g., Sahara in Africa and Australian deserts; see Chapter 3). Other deserts are associated with coastal areas at latitudes of 20–30° N (or S), where prevailing winds blow from land to sea (e.g., southern California and northwestern Mexico; Atacama Desert, Chile; and the Namib Desert, Namibia). Deserts associated with rain shadows in the lee of mountain ranges may be found in many parts of the world and at many latitudes (e.g., intermountain region of North America), and these may also be associated with being a great distance from oceans (e.g., Gobi Desert in China and Mongolia).

Climate

The arid regions of the world are characterized by high temperatures, high ranges of diurnal temperature fluctuation, and the lack of water (Figure 10.13). Maximum daily temperatures for extended periods of time range between 35 and 40°C, but temperature extremes of 48–58°C in the shade and ground surface temperature maxima of 69–82.5°C have been recorded. Temperatures drop considerably during clear nights, producing a large diurnal range; although daily ranges of 17–22°C are considered normal, ranges of 35–42°C have been recorded (Goudie and Wilkinson 1977). The mean annual precipitation is low and of uneven distribution, ranging from near 0 cm (Atacama Desert) to as much as 60 cm on mountain slopes, with most of the desert receiving less than 30 cm. Warm deserts seldom see snow, whereas cold deserts may receive a large proportion of their total annual precipitation in the form of snow. Throughout all of the deserts, potential evapotranspiration greatly exceeds precipitation.

Landscape and Topography

Much of the desert consists of gently rolling plains interspersed with small mountain ranges, steep canyons, buttes, and rock outcrops. Extensive sand dunes may also be found in many deserts, especially the Sahara.

Soils

Desert soils tend to have a high pH, have high concentrations of soluble salts, and be low in organic matter (Figure 10.13, left) (Archibold 1995). Aridisols are the primary soil order of the deserts; however, Entisols are also found in the widely distributed sand dunes. Older landscapes with Aridisols that developed under more humid conditions typically have subsurface layers of clay accumulation, whereas the younger soils on alluvial fans do not. The soil surface in the latter is often gravelly as a result of wind erosion that has removed the finer soil particles.

Vegetation

Desert vegetation is dominated by drought-resistant shrubs and grasses, succulent perennials such as cacti, and ephemeral species (Figure 10.13, right). The vegetation of North American deserts can be classified based on

Figure 10.13   Left: desert vegetation. (Photo courtesy of Arizona State University.) Right: the dominant soils here are Aridisols. (Photo courtesy of USDA-NRCS.)

the four major deserts: the Mojave, Sonoran, Chihuahuan, and Great Basin. The most distinctive plant species of the Mojave Desert is the Joshua tree (*Yucca brevifolia*), but creosote bush (*Larrea tridentata*) is more widespread along with bursage (*Ambrosia dumosa*) and many winter-active ephemeral species (Archibold 1995). The Sonoran Desert is especially known for a wide variety of cacti such as the Saguaro (*Cereus giganteus*) and organ-pipe (*Cereus thurberi*) cacti. Small, thorny trees characteristic of the Sonoran desert include mesquite (*Prosopis glandulosa*), ironwood (*Olneya tesota*), and palo verde (*Cercidium microphyllum*). Creosote bush, bursage, and winter-active ephemerals are also common. In the Chihuahuan Desert (Mexico), agaves (*Agave* spp.) and yuccas (*Yucca* spp.) are the most distinctive species, whereas creosote bush is the most common species, along with tarbush (*Flourensia cernua*) and mesquite. In contrast to the Mojave and Sonoran deserts, the ephemerals of the Chihuahuan Desert tend to be active in the summer rather than the winter, reflecting the season of greatest precipitation.

The Great Basin Desert is a cold desert and therefore the species must be capable of surviving hot and dry summers as well as cold winters. Big sagebrush (*Artemisia tridentata*) is the most common shrub species, often associated with antelope bitterbrush (*Purshia tridentata*); however, winterfat (*Ceratoides lanata*), shadscale (*Atriplex confertifolia*), and bud sage (*Artemisia spinescens*) are also common in many of the saline and dry areas. Characteristic perennial grasses include blue-bunch wheatgrass, Sandberg's bluegrass (*Poa secunda*), Indian ricegrass (*Oryzopsis hymenoides*), basin wildrye (*Leymus cinereus*), and bottlebrush squirreltail (*Elymus elymoides*); however, the exotic annual cheat grass (*Bromus tectorum*) has come to dominate large areas.

Animal Life

Globally, the animal most associated with the desert is the camel (*Camelus* spp.), well known for its ability to go for long periods of time without water. Lizards and snakes are also widespread, including many venomous species such as rattlesnakes (*Crotalus* spp.). In North America, characteristic mammals include jack rabbits (*Lepus* spp.), ground squirrels (*Spermophilus* spp.), kangaroo rats (*Dipodomys* spp.), coyotes, kit foxes (*Vulpes velox*), and bobcats. Animals of the desert have developed strategies to deal with heat and low water availability, such as nocturnal activity, rest in shade or underground burrows during the day, ability to withstand elevated blood temperatures, and use of enlarged organs such as ears (i.e., jack rabbits) as blood-cooling radiators. The kangaroo rat can even obtain all of its needed water from metabolizing dry seeds.

Tropical Rain Forest

Geographic Position

This biome is an important area for keeping carbon sequestered and hence for global climatic stability. Tropical forests occur in the humid tropics around the equator, as well as up and down to latitudes 23° 27′ north and south. This zone contains about 40% of the Earth's land surface and corresponds to the area where the sun is directly overhead at some time during the year. Approximately 17 million km² of evergreen tropical rain forest and 7.5 million km² of seasonal tropical rain forest may be found globally (Whittaker and Likens 1973). The largest contiguous rain forest, accounting for about 50% of the total, is found in the Amazon basin in South America (Archibold 1995). The Congo basin in Africa is next largest, followed by the southeastern Asia mainland and islands. Tropical rain forest may also be found in smaller units in various places in Africa, Madagascar, South and Central America, and Australia.

Climate

There is great climatic variation in the tropics due to the specific distribution of continents and oceans and the circulation of air masses and sea currents. The geographic equator is less important than the climatic equator—the line of maximum uniformity of humidity and temperature. Seasonality in the tropics is governed by the annual march of the convergence zone, which follows the position of the sun in the zenith. Seasonality increases with distance from the climatic equator, although displacement of fronts may carry tropical air masses north (India, Mexico, Florida) or south (East Africa, Madagascar) and thus extend tropical climatic conditions beyond the geographic tropics.

At low altitudes, the climate is governed by a potentially high level of incident solar radiation due to the shorter path of the sun's rays through the atmosphere. Still more important is the variation in rainfall and air humidity, which forms the basis of widely used classification of Köppen climatic types: (1) permanently wet rain forest (all months have sufficient precipitation), (2) seasonally humid or subhumid evergreen rain forest (a few months have arid characteristics), and (3) areas with dry periods in the winter of the corresponding hemisphere—subhumid or xeromorphic forests, savanna woodlands, and savannas.

Critchfield (1966) found that typical regions with a rainy tropical climate included the Amazon basin, the windward coast of Central America, the Congo basin, the eastern coast of Madagascar, and much of tropical Southeast Asia. These regions have rainfall totals surpassing 200–300 cm per year, which are distributed more or less equally over the year. Much of this area has the potential to be covered by tropical rain forest, with almost all trees evergreen. The regions with a monsoon tropical climate include the western coasts of India and Myanmar (Burma), a few parts of Southeast Asia, the coastlands of West Africa, the northern coast of South America, small portions of northeastern Australia, and some of the Pacific Islands. Although these areas may not differ greatly in annual total rainfall from the rainy tropics, the year is divided into seasons of unequal precipitation, humidity, and temperatures. The potential vegetation type of these regions ranges from tropical evergreen seasonal forests to seasonal deciduous forests in response to precipitation patterns.

The mean annual air temperature in regions covered by tropical forests is often about 27°C, and the temperature never falls below freezing. Monthly means lie generally between 24 and 28°C, so the seasonal range is less than the diurnal fluctuations of 8–10°C. Maximum temperatures recorded in the rainy and monsoon tropics rarely exceed 38°C.

Landscape and Topography

Topography is varied in the tropical forest. The large basins in South America and Africa tend to have relatively flat topography. However, abrupt and rocky hills are common in tropical plains (Thomas 1974). In contrast, the Andes of South America, the Central American mountains, and the volcanic islands of Southeast Asia all have very rough and mountainous topography. Large areas of the river basins have low enough relief that they may experience significant flooding during rainy seasons over a very large geographic area.

Soils

Soils in the tropics are heavily leached (Figure 10.14). "Next to climate," noted Buol et al. (1997), "soils are the most powerful factor controlling both the distribution and composition of tropical forests." Dominant soils in the tropics are Entisols, Inceptisols, Spodosols, Ultisols, Oxisols, and Histosols. Tropical soil development is controlled by strong weathering, leaching, biological activity, and input of elements by bulk precipitation. Because of warm temperatures and high precipitation, strong chemical weathering leads to deep soils that are somewhat uniform in clay composition and overall chemistry. The most characteristic soils are **Oxisols,** which are highly weathered soils with clay mineralogy dominated by koalinite and sesquioxides; they comprise 22.5% of the tropics (Van Wambeke 1991). Oxisols are characterized by a low cation exchange capacity (CEC) and resultant low inherent fertility, relatively indistinct horizon boundaries, typically low pH, and low water-holding capacity.

Vegetation

Profound differences in forest communities create a number of microenvironmental conditions with characteristic vertical stratification and immense species diversity. Longman

TROPICAL RAIN FOREST

TEMPERATE
DECIDUOUS FOREST

Figure 10.14 Top: a comparison of the canopy stratification of (left) 20 × 30 m forest plot from Trois Sauts, French Guiana (located on the northern coast of South America) and (right) similar temperate red oak–maple forest at Tom Swamp Harvard Forest, Massachusetts. (Redrawn from Halle, F., R. A. A. Oldeman, and P. B. Tomlinson. 1978. *Tropical Trees and Forests.* Berlin: Springer–Verlag.) Bottom: characteristic soils of the two forest types: Oxisol (left) and Alfisol (right) (USDA-NRCS.)

and Jeník (1987) note that four points are characteristic in describing the stratification of these forests:

- maximum height achieved by the canopy and emergent trees;

- aboveground layering of shoots and dependent organisms;
- underground layering of roots, rhizomes, and dependent soil organisms; and
- the maximum depth reached by roots.

Table 10.5
Cauliflory and Its Subdivisions

Growth Form	Description
Cauliflory	Flowers on the trunk; leafless twigs and roots
Ramiflory	Flowers on large branches; leafless twigs, but absent from the trunk
Trunciflory	Flowers on the trunk, but not on the branches and twigs
Phylloflory	Inflorescences are borne on the midrib of assimilating leaves
Basiflory	Flowers at the base of the trunk
Flagelliflory	Flowers on pendulous twigs spreading down from the lower trunk to the ground surface
Rhizoflory	Flowers on the roots

Source: From Longman, K. A., and J. Jeník. 1987. *Tropical Forest and Its Environment.* Essex, England: Longman Scientific & Technical.

The characteristic feature of the stratification of the aboveground vegetation presents a point of distinction with the temperate deciduous forest (Figure 10.14).

The well-known tropical ecologist Paul Richards (1996), who has described these communities in detail, finds few canopies exceeding 50 m; only the exceptional ones may reach 70 or 80 m. The tallest forests have distinct wet and dry periods, rainfall cover, and well-drained soils. As he notes, growth is limited by high rainfall, lack of seasonality, poor drainage, winds, and salt-laden ocean sprays.

A peculiar characteristic found only in the tropical rain forest is the production of flowers on twigs, leafless branches, and roots—a feature known as **cauliflory.** Richards (1996) notes that "to a visitor from temperate climates, cauliflory seems one of the strangest features of tropical trees" (Table 10.5).

Animal Life

The animal life in the tropics is as diverse as or more diverse than the plant life. Invertebrate life has the greatest species number as well as diversity. These two features account for quick degradation of organic matter on the forest floors. Large groups of arthropods represent several trophic levels: decomposer, herbivore, and carnivores. Earthworms are an important component, as are the numerous species of ants. The group of leaf-cutting ants alone, note Longman and Jeník (1987), has been "estimated to harvest approximately 0.2% of gross primary production. Termites build large and conspicuous mounds on the forest floor." Species of reptiles, birds, and mammals are diverse as well.

Tropical Savannas

Geographic Position

Tropical savannas are typically located around the edges of the tropical forests and eventually grade into the tropical deserts. They cover an area of approximately 15.0 million km², mostly in Australia, India, Africa, and South America.

Landscape and Topography

The topography of the savannas tends to be plateaus, level plains, and rolling hills, with common rock outcrops.

Soils

Areas that supported forests in the past may be underlain by Oxisols. Most of the savanna, however, is underlain by Alfisols. The Alfisols tend to have high base saturations but are relatively shallow and subject to erosion. Other areas are underlain by Entisols and Vertisols. **Vertisols** are "self-churning" soils that have a high proportion of montmorillinitic clays that shrink and swell during wet–dry cycles. They are found in extensive areas of India and eastern Australia.

Vegetation

Savannas are characterized by widely spaced small trees interspersed among a grassland matrix. The perennial grasses tend to use the C_4 photosynthetic pathway in response to regular drought stress, high temperatures, and high solar radiation. However, annual grasses are also common, especially in drier areas. Trees tend to be evergreen with sclerophylous leaves, although drought deciduous trees are also common, especially in Africa. They usually have very extensive root systems leading to large root-to-shoot ratios. Fire is also a common component of savannas during the dry season. The dried, dormant grasses provide large amounts of fine fuels and result in fast-moving fires whose main influence is to kill young tree saplings. Fires are thus necessary to keep trees from encroaching into the grass-dominated areas and reducing the dominance of grasses.

Animal Life

Large ungulates and predators are characteristic animals of the African savannas. Birds can be quite diverse, particularly at midlatitudes, decreasing in the temperate zone and arctic.

Other Vegetation Types

There is a variety of vegetation types that do not fit well within the major biomes. These are often called woodlands, scrublands, and shrublands. They may account for approximately 8.5 million km^2 globally.

Ecotones

As noted in the beginning of this chapter, biomes do not have the sharp boundaries that may be depicted for convenience in maps. They grade into one another over relatively shorter ranges if the environmental gradients change rapidly or over longer distances if the environmental changes are more gradual, as is shown by the ecotone from the boreal to the deciduous forest vegetation in North America (Table 10.6). Transitions occur at all spatial scales, and a very useful hierarchy of ecotones is illustrated in Table 10.7. At the highest level of generalization—the biomes—the main interactions considered are those of climate and topography. At smaller spatial scales, many more interactions must be considered. Hence, broad-scale studies have a value of generality but less detail, while just the opposite is true at narrower scales.

Ecoregions

Theoretical discussions in ecology notwithstanding, the concept of biomes has been

Table 10.6
Elevation of Transition from Boreal to Deciduous Forest Vegetation in Eastern North America

Location	Elevation
Mount Katahdin, Maine	152 m
Cape Breton, Nova Scotia	213 m
Mount Washington, New Hampshire	762 m
White Mountains, New Hampshire, and Green Mountains, Vermont	
Northern portion	732 m
Southern portion	914 m
Adirondack Mountains, northern New York	914 m
Catskill Mountains, southern New York	1067 m
Great Smokey Mountains, Tennessee	1524 m

Courtesy of R. E. Boerner.

Table 10.7
Ecotone Hierarchy for a Biome Transition Area

Ecotone Hierarchy	Probable Constraints
Biome ecotone	Climate (weather) × topography
Landscape ecotone (mosaic pattern)	Weather × topography × soil characteristics
Patch ecotone	Soil characteristics × biological vectors × species interactions × microtopography × microclimatology
Population ecotone (plant pattern)	Interspecies interactions × intraspecies interactions × physiological controls × population genetics × microtopography × microclimatology
Plant ecotone	Interspecies interactions × intraspecies interactions × physiological controls × population genetics × microtopography × microclimatology × soil chemistry × soil fauna × soil microflora

Source: From Gosz, J. R. 1993. *Ecological Applications* 3:369–376.
Notes: Each level in the ecotone hierarchy has a range of constraints and interactions between the constraints; × symbolizes interactions between the constraints. The primary constraints vary with the scale of the ecotone, with an increase in the number of possible constraints at finer scales.

useful in meeting the needs for ecological information to describe easily generalizable units for ecosystem conservation and management. An ecoregion, which encompasses the biome concept, is defined as a relatively large area of land or water that contains a geographically distinct assemblage of natural communities. These communities (1) share a large majority of their species, dynamics, and environmental conditions; and (2) function together effectively as a conservation unit at global and continental scales (Dinerstein, Graham, and Olson 1995; Ricketts et al. 1999). Dinerstein et al. recognize 10 major habitat types (MHTs), which "are not geographically defined units; rather, they refer to the dynamics of ecological systems and to the broad vegetative structures and patterns of species diversity that define them. In this way they are roughly equivalent to the biomes." These authors further use two "discriminators" for each MHT: the biological distinctiveness index and conservation status index. All in all, this is a major advance.

Questions

1. Why are there different biomes around the world, and what factors determine their locations?
2. Name the biome you live in and describe its climate, soil, vegetation, and animal life characteristics.
3. Why do hot and cold deserts support a much lower NPP than do tropical and temperate forests and grasslands?
4. How are tropical and temperate rain forests different in terms of species richness and species abundance? Does this imply anything about the stability of either ecosystem?
5. What is the significance of ecotones?

SECTION B

Contemporary Environmental Issues

- How have human impacts altered the structure, function, and organization of ecosystems?

In Chapters 1–10, the emphasis was on learning how the inter-related and interdependent factors of atmosphere, soil, water, and biological processes act upon living organisms, their populations, and their communities. In this section (Chapters 11–20), attention is focused on how the vast increase in our human population and its environmental influences have impacted original ecosystems and altered both community structure and function. This human force, greatly aided and abetted by advances in technology, has been characterized as "the mighty geological agent."

All discourses in the forthcoming chapters bear intimate relationships with myriad human activities on land—a mere 29% of the Earth's surface. All such activities determine the extent, magnitude, and nature of materials that finally find their way into the atmosphere and into our water bodies. Equally important, the resources that humans harvest from seas and oceans and materials are all related to land use. We therefore provide a brief overview of land use as a preamble to Section B.

Land Use Changes—Widespread and Worldwide

In his seminal historical work on world ecological degradation, Chew (2001) captures two fundamentals of human actions across the historical gradient of 5,000 years. One fundamental is "that the history of civilizations, kingdoms, empires, and states is also the history of ecological degradation" (p. 1). The second fundamental is that "ecological relation is as primary as the economic relation in the self-expansionary process of societal systems" (p. 2). Land use changes, without exception, are a phenomenon of worldwide concern,

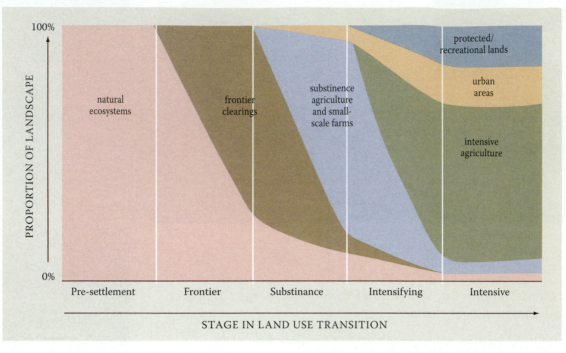

Figure B.1   Land use transitions. (From Foley, J. A. et al. 2005. *Science* 309:570–574; Figure 1, p. 571.)

for as human and livestock populations are going up in numbers, our productive resources are shrinking.

The impact of this human force on both terrestrial and aquatic systems has been phenomenal and continues to be widespread and chronic. Removals of beneficial resources from the Earth are matched in their rapidity only by the additions of pollutants to the land, air, and water. Most recently, the magnitude of these processes has been illustrated by the remarkable effects of greenhouse gases on the global climate. The transitions of land use over time in a given region (Figure B.1) are aptly summarized in the following paragraph by Foley et al. (2005, p. 571):

As with demographic and economic transitions, societies appear also to follow a sequence of different land use regimes: from presettlement natural vegetation to frontier clearing, then to subsistence agriculture and small farms, and finally to intensive agriculture, urban areas, and protected recreational areas. Different parts of the world are in different transition stages, depending on their history, social and economic conditions, and ecological context. Furthermore, not all parts of the world move linearly through these transitions. Rather, some places remain in one stage for a long period of time, while others move rapidly between stages.

Expansionary Land Use Changes

Indeed, the nature, extent, and magnitude of these disturbances are staggering. Some scientists estimate that between one-third and one-half of the Earth's land surface has been transformed by human action (Vitousek et al. 1997). Recent reports note the current loss of 13 million ha of forests each year—25 ha a minute (Leape 2006). The United Nations Environment Program (UNEP) commissioned a study 23 years ago specifically to quantify the extent of disturbance and soil degradation from agricultural systems on a worldwide basis. The study, which is widely cited, found that approximately 2,000 million ha had been disturbed by 1983 and the rate of disturbance was 5–7 million ha per year (Oldeman 1994). In dryland zones alone, 3,600 million ha were found degraded; 2,600 million ha exhibited degraded rangeland vegetation without soil degradation; and 1,000 million ha had experienced degradation of both soil and vegetation (Dregne and Chou 1992).

Information available for biomes of the world (Table B.1; Hannah et al. 1995), by land cover classes (Table B.2), and for regions and

Table B.1
Human Disturbance in Biomes

Biome	Total Area (10^6 km²)	Undisturbed Habitat (%)	Human Dominated (%)
Temperate broadleaved forests	9.5	6.1	81.9
Chaparral and thorn scrub	6.6	6.4	67.8
Temperate grasslands	12.1	27.6	40.4
Temperate rain forests	4.2	33.0	46.1
Tropical dry forests	19.5	30.5	45.9
Mixed mountain systems	12.1	29.3	25.6
Mixed island systems	3.2	46.6	41.8
Cold deserts/semideserts	10.9	45.4	8.5
Warm deserts/semideserts	29.2	55.8	12.2
Most tropical forests	11.8	63.2	24.9
Tropical grasslands	4.8	74.0	4.7
Temperate conifer forests	18.8	81.7	11.8
Tundra and arctic desert	20.6	99.7	0.3

Source: After Hannah, L., J. L. Carr, and Lankerani. 1995. *Biodiversity and Conservation* 4:128–155.

Note: When undisturbed and human-dominated areas do not add up to 100%, the difference represents partially disturbed lands.

specific countries of the world is none too comforting (Figure B.2). Reports indicate that 1 ha of Canadian boreal forests was being destroyed in less than 30 s and that the Russian boreal forests were being destroyed at twice the rate of tropical rain forests in Brazil (Acharya 1995). This has some well-known scientists lamenting that North American and European ecologists have been preoccupied with the impacts of disturbance effects in South America, whereas impacts equal to or greater than those effects on boreal ecosystems in North America and Europe have been ignored (Schindler 1998).

For the second most populous country in the world, India reports that of the total land area of 329 million ha, more than 175 million ha are prone to severe erosional loss and that 129 million ha could already be classified as "waste land" (Khoshoo 1992). These staggering estimates on disturbance, though highly variable, illustrate how urgent the need is for a systematic collection of quantitative data on the extent of total terrestrial lands disturbed around the world.

The soil degradation in South, Central, and North America reveals similar trends in the extent and magnitude and the causative factors (Table B.3). The extent of disturbance, however, becomes abundantly clear when one studies the grassland biome of North America. The conversion of natural grasslands to croplands in many parts of this biome in the United States and Canada exceeds 99% (Table B.4). Indeed, the enviable productivity of North American croplands is solely due to the centuries of root growth of grasses that has immensely enriched the organic matter content of these soils. High erosional losses begun during the late 1920s and early 1930s, though greatly lessened, continue to take a heavy toll on the productivity potential of these ecosystems.

Disturbance Defined

When a land area is converted from a forest to rangeland, the trees and shrubs are cut and the fodder grasses are promoted to grow. When the forests or natural grasslands, on the other

Table B.2
Current Land Cover of the Americas as Depicted by DISCover[a]

Land Cover Classes	Current Vegetation (km^2)	Potential Vegetation (km^2)[b]	Human-Induced Change (km^2)[b]
Evergreen forest	10,036,268	14,200,623	−4,164,355
Deciduous forest	957,683	2,748,511	−1,790,828
Mixed forest	3,333,949	4,427,083	−1,093,134
Woody savanna	2,578,185	638,819	+1,939,366
Savanna	1,350,552	4,772,712	+3,422,160
Shrubland	6,724,171	6,286,924	−437,247
Grassland	2,795,509	2,968,184	−172,675
Desert	2,711,947	2,008,580	+703,367
Cropland	2,970,387	0	+2,970,387
Cropland mosaic	4,488,301	0	+4,488,301
Urban and built up	104,484	0	+104,484
Wetlands	314,998	314,998	NA
Snow and ice	3,918,725	3,918,725	NA
Region total	42,285,159	42,285,159	42,285,159

Sources: From Loveland, T. R., and A. S. Belward. 1997. The International Geosphere-Biosphere Program Data and Information System Global Land Cover Data Set (DISCover). *Acta Astronautica* 41: 681–689. Mathews, W. 1983. Global vegetation and land use: New high-resolution databases for determining externality costs. *Environmental Science and Technology* 34: 1390–1395.

[a] The Data and Information System data set (DISCover) was initiated by the International Geosphere Biosphere Program and implemented through collaboration of many agencies (in particular, the U.S. Geological Survey Earth Resources Observation System Data Center) because of the need for global land-cover data with known classification accuracy (Loveland and Belward 1997). The 1-km resolution of DISCover captures the heterogeneity missed by the coarser resolution of past remote sensing estimates and is the first to utilize this resolution on a global scale (Loveland et al. 2000). Greenness classes were defined by monthly AVHRR NDVI composites from images taken between April 1992 and March 1993 (Loveland and Belward 1997, slightly modified by Gibbs).

[b] Data on potential vegetation (Mathews 1983) and human-induced change synthesized by Holly Gibbs (personal communication).

hand, are converted to croplands, not only is the aboveground vegetation removed but the soils are also turned over. In both cases, the natural ecosystems have been disturbed—more so in the latter case than in the former. Given the extent and magnitude of disturbance as detailed earlier, we need firm foundations for stating a problem as well as for seeking mitigative measures. Thus, it may be useful to define **disturbance.**

Among many definitions that have been advanced, the following two capture the essence of the problem. A disturbance is "any discrete event in time that disrupts ecosystem, community, or population structure, changes resources, availability of substratum, or the physical environment" (White and Pickett 1985). Following the attributes of the ecosystem concept, "an ecosystem disturbance may be defined as an event or a series of events that results in the altering of relationships of organisms and their habitats from their natural state both spatially and temporally" (Wali 1987).

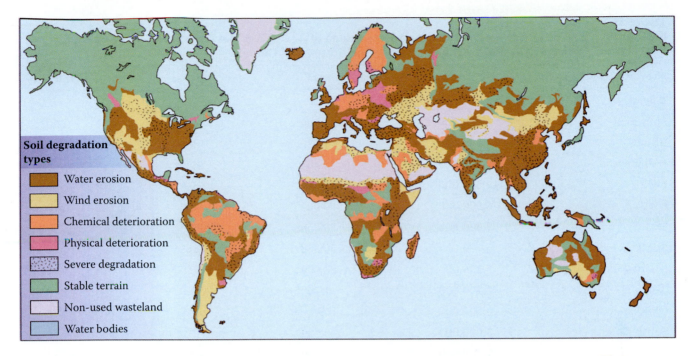

Figure B.2   The causative agents for human-induced soil degradation. (Oldeman, L. R., R. T. A. Hakkeling, and W. G. Sombroek. 1990 (Revised 1991). World Map of the Status of Human-Induced Soil Degradation. Waginingen: ISRIC and Nairobi:UNEP. United Nations Environment Program.)

Table B.3
Soil Degradation of the Americas

	South America (Mha)	Central America (Mha)	North America (Mha)
Extent			
Water erosion	123	46	60
Wind erosion	42	5	35
Chemical	70	7	+
Physical	8	5	1
Total	243	63	96
Causative Factors			
Deforestation	100	14	4
Overexploitation	12	11	–
Overgrazing	68	9	29
Agricultural activities	64	28	63

Source: Based on Oldeman, L. R. 1994. In *Soil Resilience and Sustainable Land Use,* ed. D. J. Greenland and I. Szabolcs, 99–118. Wallingford, Oxford, U.K.: CAB International Publishers.

Table B.4
Estimated Current and Historic Areas and Percentage of Area That Have Declined and Are under Protection in Major Grassland Areas of the United States and Canada

Province/State	Historical	Current	Decline	Protected
Tallgrass				
Manitoba	600,000	300	99.9	N/A
Illinois	8,900,000	930	99.9	<0.01
Indiana	2,800,000	404	99.9	<0.01
Iowa	12,500,000	12,140	99.9	<0.01
Kansas	6,900,000	1,200,000	82.6	N/A
Minnesota	7,300,000	30,350	99.6	<0.01
Missouri	5,700,000	30,350	99.5	<0.01
Nebraska	6,100,000	123,000	98.0	<0.01
North Dakota	1,200,000	1,200	99.9	N/A
Oklahoma	5,200,000	N/A	N/A	N/A
South Dakota	3,000,000	449,000	85.0	N/A
Texas	7,200,000	720,000	90.0	N/A
Wisconsin	971,000	4,000	99.9	N/A
Mixed Grass				
Alberta	8,700,000	3,400,000	61.0	<0.01
Manitoba	600,000	300	99.9	<0.01
Saskatchewan	13,400,000	2,500,000	81.3	<0.01
Nebraska	7,700,000	1,900,000	77.1	N/A
North Dakota	13,900,000	3,900,000	71.9	N/A
Oklahoma	2,500,000	N/A	N/A	N/A
South Dakota	1,600,000	N/A	N/A	N/A
Texas	14,100,000	9,800,000	30.0	N/A
Shortgrass				
Saskatchewan	5,900,000	840,000	85.8	N/A
Oklahoma	1,300,000	N/A	N/A	N/A
South Dakota	179,000	N/A	N/A	N/A
Texas	7,800,000	1,600,000	80.0	N/A
Wyoming	3,000,000	2,400,000	20.0	N/A

Source: Sims, P. L., and P. G. Risser. 2000. In *North American Terrestrial Vegetation,* 2nd ed., ed. M. G. Barbour and W. D. Billings, 323–356 (Table 9.5, p. 346). New York: Cambridge University Press.

Note: Estimates of current and historical prairie areas are based on information from The Nature Conservancy's Heritage Program; U.S. Department of Agriculture Forest Service; Canadian Wildlife Service; Provinces of Alberta, Manitoba, and Saskatchewan; and state conservation agencies.

Agents of Disturbances Are Many

Ecosystem disturbances result from several major activities (list of Bazzaz 1983, modified):

- extensive clearing of natural vegetation for growing crops, for pasture, for roads and other transportation corridors, for housing developments, and for other purposes;
- selective harvesting of desirable species and introduction of alien ones;
- abandonment of unproductive agricultural and range land;
- mining for coal and other minerals;
- draining of wetlands for agriculture and human settlements;
- introduction of chemicals into the environment;
- the impacts of war that include bombing, defoliation by chemicals, and the movement of personnel and matériel; and
- introduction of greenhouse gases into the atmosphere and climate change.

Many of these disturbances are local or regional and their collective impact on the loss of productive ecosystems is enormous. None of the data and figures (even those included here) present a complete picture of disturbance. These land use changes have already contributed to global climate change, adding 156 Pg C to the atmosphere between 1850 and 2000 (Houghton 2003). As land use changes continue, they will further diminish considerably the multitude of services provided by ecosystems. This progressive diminution of ecosystem services assumes profound importance, particularly when one considers that human population numbers are increasing concomitantly. Thus, during the last three or so decades, scientists have studied ecosystem disturbances in many diverse regions of the world; strategies for their rehabilitation or restoration have been treated in hundreds of books and thousands of research reports.

A useful metric that has entered the environmental lexicon is based on the concept of **ecological footprint.** The ecological footprint of a given population or a nation or, collectively, of all humans living on Earth is defined as "the total area of productive land and water required on a continuous basis to produce the resources consumed, and to assimilate the wastes produced, by that population, wherever on Earth the land (and water) is located" (Rees 1997). Although concept is discussed in great detail in Chapter 22, we mention it here to emphasize that growing crops and trees, using minerals and water, changing land use, and pouring emissions into the atmosphere all have their unique ecological footprints.

Because many productive ecosystems have deteriorated considerably—some irreversibly for human or animal use—meeting natural resource demands in the future will depend largely on how well disturbed areas can be rehabilitated and restored. These aspects of ecosystem restoration are presented in Chapter 24.

Human Population Growth

11

Introduction

The factors, phenomena, and principles that govern the numbers of all naturally occurring animal and plant populations, despite their many manifestations, apply to human numbers as well. Biotic potential, equilibration with the carrying capacity of the environment, density-dependent and density-independent relationships, and the competitive phenomena discussed in Chapter 8 apply to the human population as well.

To begin, it may be worthwhile to consider a discourse on the potential of burgeoning human numbers relative to constrained environmental resources as visualized by the Reverend Thomas Malthus in 1798, now a landmark in environment literature. Malthus wrote:

> "Taking the earth as a whole, … the human species would increase as the numbers, 1, 2, 4, 8, 16, 32, 64, 128, 256, and subsistence as 1, 2, 3, 4, 5, 6, 7, 8, 9." Further, "1. Population is necessarily limited by the means of subsistence; 2. Population invariably increases where the means of subsistence increase, unless prevented by some very powerful and obvious checks."

As noted in Chapter 1, these postulates—an exponential increase in population numbers, but a linear growth in available resources—are referred to as *Malthusian*. The consequences of resource constraints encountered by vastly expanding animal and plant populations are now well documented in population ecology. Despite that, it is intriguing that this essay—now over 200 years old—continues to evoke a strong denunciation (see, for example, Trewavas 2002). His hypotheses on famines and such may well be debated but, as Patel (2007) notes, "Malthus' ghost haunts." Even the "father of the Green Revolution," Norman Borlaug, noted in his 1970 Nobel Peace Prize Lecture that "most people still fail to comprehend the magnitude and menace of the 'Population Monster.'" Why? In this chapter, we discuss the temporal trends of increase in human population and their implications for the welfare of humans and other biota.[1]

Forces Driving Human Population Growth

Demographic Causes

The scientific study of the size, composition, and rate of change of human populations (demography) shows that they have the potential to grow like all other populations in nature—namely, exponentially—and are subject to all the ecological principles discussed in Chapter 8. Change in population size over a given time and space in a closed system, such as the world, is the product of its current population size and the rate of change, or growth rate (the birth rate minus the death rate). It is generally expressed as the number of births and deaths per 1,000 people per year:

Future population size

= present population size

× (birth rate – death rate)

Growth rate (r) of the human population at the global scale (usually expressed as percent per year) can undergo a natural increase or a natural decrease, or it can remain stable, a

condition known as **zero population growth,** where the birth rates equal the death rates:

Annual growth rate (%)

= [(birth rate – death rate)/1,000 persons] × 100

When population dynamics (changes over time and space) are considered at the local, regional, and international levels, **migration** (immigration and emigration) is included as a factor in the estimation of population change. **Immigration** and **emigration** are the migration of people into or out of a population from another area. With the inclusion of the migration factors, population change can be estimated as follows:

Annual growth rate (%)

= [(births + immigration)

– (deaths + emigration)]/1,000 persons

Exponential growth is a system behavior that is exhibited when a component of a system feeds upon itself by a **positive feedback loop.** As an example, consider the world's human population, in which the birth of humans increases the initial human population size, which further increases the number of humans who are born (Figure 11.1). The

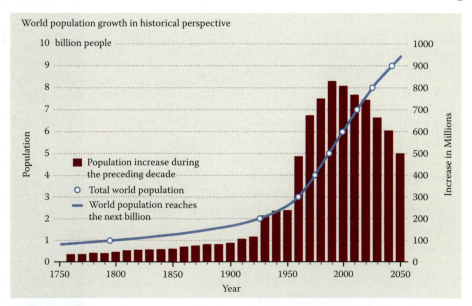

Figure 11.1 Exponential growth of human population and estimated population by 2050. (McDevitt, T. M. 1999. *World population profile: 1998*. Washington, D.C.: U.S. Census Bureau.)

BOX 11.1   The Power of Exponential Growth:
The Story of a King and a Mathematician (Bartlett 1978)

The growth in numbers may be illustrated by the doubling of numbers associated with the board used in the game of chess. Having performed a service for the king, the mathematician, the story goes, was asked how he could be rewarded. The mathematician told the king that he would like to be paid a wage in grains of wheat. He asked that the king place a grain of wheat on the first square of a chessboard and double the number of grains thereafter on each subsequent square. Note the number of grains the mathematician would win when he reached the 64th square (see following table). How much wheat is 2^{64} grains, 1 grain doubled 63 times? Simple arithmetic calculation shows that it is approximately 500 times the 1976 annual worldwide wheat harvest! This quantity is probably larger than all the wheat that has been harvested by humans in the history of the Earth.

Filling Squares on the Chessboard with Wheat Grains

Square Number	Grains on Square	Total Grains Thus Far
1	1	1
2	2	3
3	4	7
4	8	15
5	16	31
6	32	63
7	64	127
.	.	.
.	.	.
.	.	.
64	2^{63}	$2^{64}-1$

larger the human population gets, the faster is its growth rate (see Box 11.1 for power of exponential growth and Box 11.2 for doubling times in human population growth). When the numbers of people on Earth are plotted over time, the form of exponential growth takes a characteristic "**J**" **shape** (Figure 11.1). A useful measure of population growth rate is the **doubling time** (Td) for the population size, which is calculated from the annual percent growth rate (r) as follows[2]:

$$Td = 70/r$$

The world population of ca. 500 million in 1650 increased to ca. 4 billion (10^9) by doubling in 1800, 1930, and 1975 (Anderson 1981;

Table 11.1
Population Growth over the Past 200 Years

A Progression		
Population	Period	Time Required for Each 1 Billion Increase in Population (Years)
1 billion	1800s	200,000?
2 billion	1930	130
3 billion	1960	30
4 billion	1975	15
5 billion	1987	12
6 billion	1999	12

Source: From Scheidel, W. 2003. In *Encyclopedia of Population,* vol. 1, ed. P. Demeny and J. McNicoll, 44–48. New York: Macmillan Reference.

BOX 11.2   The World Ushers in 6 Billion People

On Tuesday, October 12, 1999, the United Nations announced the arrival of the 6 billionth person on Earth. Actually, no one knows the real population of the world. Even with the availability of the best technologies, the 1990 and 2000 censuses in the United States have been reported to be off in their counts by 4–8%. Be that as it may, the 6 billion mark has been a quotable landmark.

On Monday, October 16, 2006, Stephen Ohelmacher of the *Associated Press* wrote, "America's population is on track to hit 300 million on Tuesday [October 17] morning, and it's causing a stir among environmentalists." It is expected to be 400 million in the United States in 2043.

It is worthwhile to note that, by his calculations, Edward S. Deevey, Jr., an ecologist at Yale University, had predicted the 6-billion human population growth mark 40 years before the actual event (see the following table based on his data).

Year	Population	Doubling Time (Years)
1,000,000 B.C.	125,000	
300,000 B.C.	1,000,000	230,000
25,000 B.C.	3,340,000	160,000
8000 B.C.	5,320,000	22,000
4000 B.C.	86,500,000	1,000
A.D. 0	133,000,000	6,400
1650	545,000,000	830
1750	728,000,000	240
1800	906,000,000	160
1900	1,610,000,000	120
1950	2,400,000,000	87
2000 (est)	6,270,000,000	36

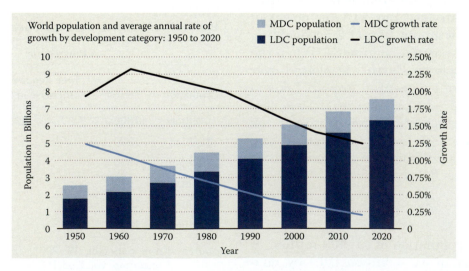

Figure 11.2   Annual growth rate for the world and major development groups, 1950–2020. (McDevitt, T. M. 1999. *World population profile: 1998.* Washington, D.C.: U.S. Census Bureau.)

Greep 1998) (Table 11.1). The current world population of over 6 billion is projected to rise to 9.3 billion by 2054 (PRB 2007) (Figure 11.1). However, the annual rate of population growth in both developing and developed countries is expected to continue to decline. Changes in the population growth rate of developing countries lag behind those of developed countries by about 30 years (Figure 11.2). It is important here to emphasize the relationship between rates and stocks. What is declining is the annual growth rate of the world's human population, which means that population size (stock) still continues to increase (albeit more slowly) until population stability is reached globally. Regional variations in the growth rates and current stock size of the 20 most populous countries will shift the ranks of their population sizes between 2000 and 2050 (Table 11.2).

Human birth and death rates depend on many interacting internal and external factors, such as fertility rates, the proportion of men and women of reproductive age (age structure), infant mortality, life expectancy, religious and cultural norms, level of education and affluence, and quantity and quality of natural resources. There are two types of fertility rates: replacement-level fertility and total fertility rate. **Replacement-level fertility (RLF)** is the number of children that a couple must have to replace themselves. **Total fertility rate (TFR)** is the average number of children born to each woman during her lifetime (Figure 11.3). A TFR must be slightly

Table 11.2
The World's 20 Most Populous Countries, 2000 and 2050

2000		2050	
Country	Population (Million)	Country	Population (Million)
China	1275	India	1531
India	1017	China	1395
United States	285	United States	409
Indonesia	212	Pakistan	349
Brazil	172	Indonesia	294
Russia	146	Nigeria	258
Pakistan	143	Bangladesh	255
Bangladesh	138	Brazil	233
Japan	127	Ethiopia	171
Nigeria	115	Dem. Rep. of Congo	152
Mexico	99	Mexico	140
Germany	82	Egypt	127
Vietnam	78	Philippines	127
Philippines	76	Vietnam	118
Egypt	68	Japan	110
Turkey	68	Iran	105
Iran	66	Uganda	103
Ethiopia	66	Russia	102
Thailand	61	Turkey	98
United Kingdom	59	Yemen	84

Source: From Brown, L. R. 2005. *Outgrowing the Earth: The Food Security Challenge in an Age of Falling Water Tables and Rising Temperatures,* Table 2.1, p. 25. New York: W. W. Norton & Company.

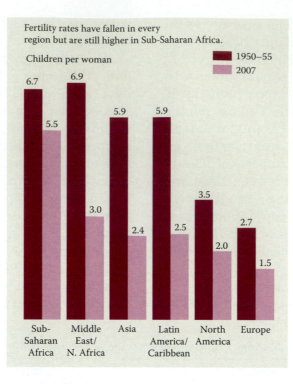

Figure 11.3 Global fertility rates have fallen worldwide, but the drop in Sub-Saharan region has not been so pronounced. (Population Reference Bureau. 2007. *2007 World Population Datasheet*. Accessed February 4, 2009: http://www.prb.org/pdf07/07WPDS_Eng.pdf)

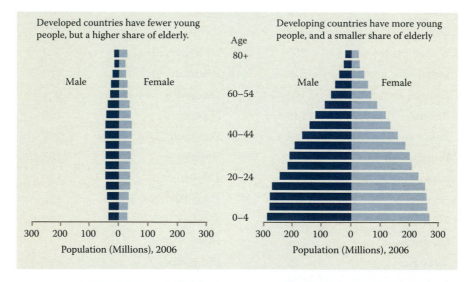

Figure 11.4 Age structure of developed and developing countries, males and females (in millions). (From United Nations. 2007. *World Population Prospects: The 2006 Revision*.)

above two children per woman in order to reach replacement level (2.1 in countries with low mortality) because some female children die before reaching reproductive age.

Average RLF in developing countries with high infant mortality is as high as 2.7. If TFR remains low for a prolonged period, then populations experience a natural decrease. Global average TFR dropped from 5.0 in 1950–1955 to 2.9 in 1990–1995. In China, the process of reducing TFR from about 6 to 2.4 children per woman took 20 years. On the other hand, the high fertility period between the 1950s and early 1970s in many developing countries has resulted in the current high reproductive **age structure** (Figures 11.4 and 11.5). The high

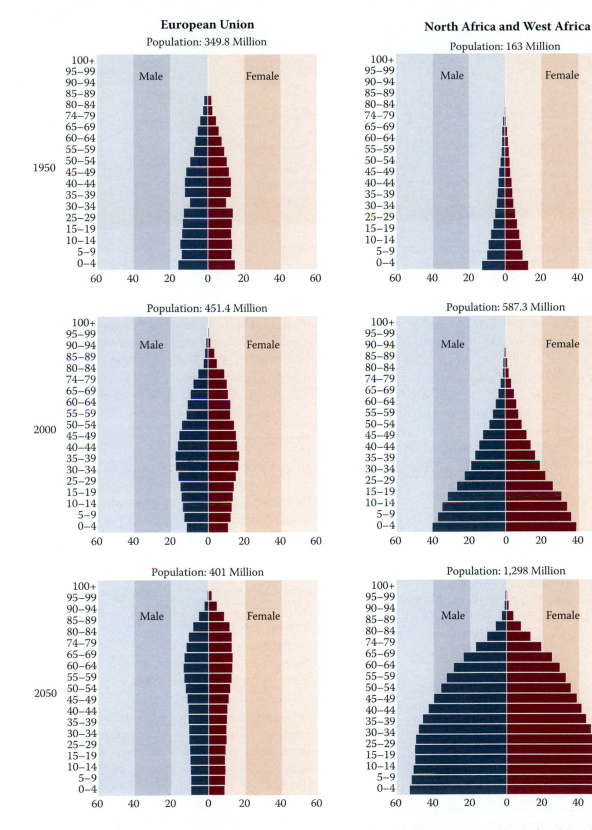

Figure 11.5 Population size and age distribution for 1950, 2000, and 2050 in an anticipated enlarged European Union of 25 countries and in 25 countries of northern Africa and western Asia between India's western border and the Atlantic Ocean, excluding countries of central Asia that were part of the former Soviet Union, those of Muslim black Africa, and Israel. Horizontal scale gives million persons separately by sex; vertical scale gives age groups in increments of 5 years. (Cohen, J. 2003. Human population: the next half century, *Science* 302: 1172–1275.)

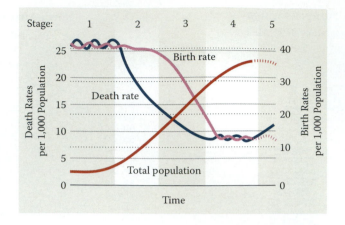

Figure 11.6 Classic stages of demographic transition. *Human Population: Fundamentals of Growth and Change* is an update of *World Population: Fundamentals of Growth and World Population: Toward the Next Century*. Revisions and additions were made by Cheryl Lynn Stauffer, November 2000.

proportion of individuals of reproductive age in a population causes what is called **population momentum,** one of the most important factors creating a time lag in the stabilization of world population growth.

Demographic models generally distinguish between four stages of **demographic transition** that lead to stable population growth in more highly developed countries. This occurs when there is a transition from high birth and death rates to relatively low birth and death rates in stages (Figure 11.6):

- In the **preindustrial stage** (**stage 1**), both birth and death rates are high.
- In the **transitional stage** (**stage 2**), the population grows rapidly because death rates fall as standards of hygiene and the availability of medical treatments increase (leading to an increase in life expectancy), but birth rates remain high.
- In the **industrial stage** (**stage 3**), the growth rate of the population declines as birth rates begin to fall because the incentives for having large families decrease.
- In the **postindustrial stage** (**stage 4**), the population stabilizes at a relatively low level and may even undergo a natural decrease. In this stage populations are said to have undergone a full demographic transition.

Two intimately coupled indicators of overall public health and environmental quality in a country are life expectancy and infant mortality rate (IMR). **Life expectancy** is the average number of years that a newborn infant can be expected to live. **Infant mortality rate** is the number of babies per 1,000 newborns each year who die before their first birthday.

Social and Economic Causes

In addition to demographic factors, social and economic factors play an important role in the growth rate of the human population. These include migrations, poverty, adult literacy, socioeconomic (decision-making) status of women, family planning, religious beliefs, and cultural norms. Generally, life expectancy has increased in all regions of the world (Figure 11.7). People living in poverty can see having more children as improving their hope for future security and well-being because they depend on their children for household, industrial, or agricultural work and for support in old age. Globally, about 87% of men and 77% of women are literate (Figures 11.8 and 11.9). Factors causing lower enrollment of women than men in schools include poverty, gender bias, early marriage and childbearing, and limited job opportunities for women in many economic sectors. However, the second half of the twentieth century saw an expanding access for women to education and reproductive health services as well as the development of family planning methods, which cumulatively have been called the "reproductive revolution."

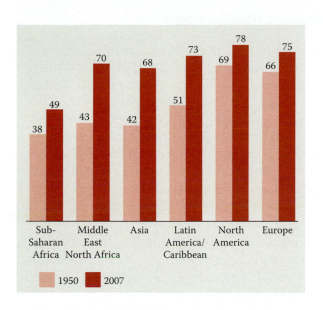

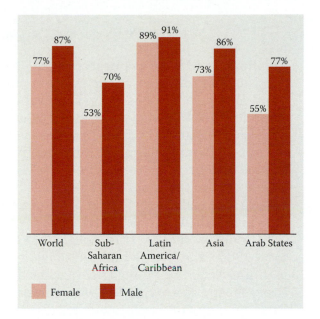

Figure 11.7 Life expectancy by world regions. (Population Reference Bureau. 2007. *2007 World Population Datasheet*. Retrieved February 4, 2009 from http://www.prb.org/pdf07/07WPDS_Eng.pdf)

Figure 11.8 World relative literacy rates for adult females and males in selected regions, 2000–2004. (Source: UNESCO Institute for Statistics. Accessed online at www.uis.unesco.org/TEMPLATE/html/Exceltables/education/Literacy_Regional_April 2006 on May 21, 2006.)

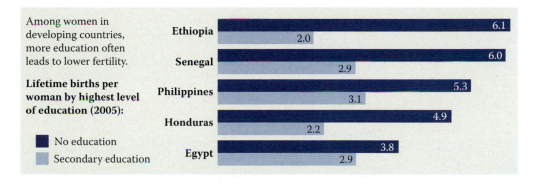

Figure 11.9 Fertility rate (children per woman) as a function of educational level in selected countries (2005). (From: *Demographic and Health Surveys,* 2007.)

Population growth is also affected by the economic and political status of women, which in turn affects women's choices about the number and spacing of births. Some good examples of how countries can slow their population growth, as described by Lester Brown (2005), come from two countries with very different climates, cultures, and economies: Thailand and Iran (see Box 11.3).

Reproductive health characteristics of a society are closely coupled with the development status. For example, prevalence of HIV/AIDS infection is highest in southern Africa, although these infections are seen in both poor and rich countries (Figure 11.10). Similarly, maternal mortality is determined by interactions of biological, socioeconomic, and cultural factors and is considered to be an indicator of societal development (Figure 11.11), as are high fertility rate and child mortality (Figure 11.12).

Immigration policies and poverty are important socioeconomic factors affecting population growth independent of the status of women. Immigration policies can be used to regulate population growth: For example, encouragement of immigration can increase population of countries despite their decreased birth rates. Conversely, restriction of immigration or encouragement of emigration relieves crowding and unemployment.

BOX 11.3   Two Success Stories (Brown 2005)

Thailand and Iran have been remarkably successful in slowing their population growth despite their differences in culture, climate, and economy. Predominantly Buddhist Thailand has a rice-based agricultural economy under a humid and subtropical climate, while Muslim Iran has a wheat-based one under a semiarid and temperate climate. The main contributor to Thailand's success is an individual, Mechai Viravaidya, who promoted the topics of family planning, reproductive health, and contraception throughout the country and mobilized the resources of the Thai government to that effect during the 1970s and was elected to the senate by the people of Thailand. Today, Thailand's annual population growth rate is 0.8% instead of 3%. Its current population of 63 million is projected to stop growing at about 77 million by 2050.

In nearly a decade, Iran reduced its annual population growth rate from 4.4% in the early 1980s, the world's highest, to just over 1%. The country's leadership started a family planning program by recognizing that its record population growth rate was burdening the economy, destroying the environment, and overwhelming schools. The program mobilized not only family planners but also the ministries of education and culture, radio and television broadcasts, and religious leaders. The population growth rate in Iran was cut in half from 1987 to 1994; the only other two countries succeeding in the same effort were Japan and China. In 2004, the population of Iran was growing only slightly faster than that of the United States.

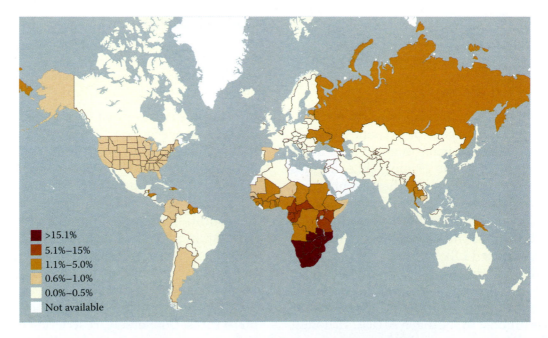

Figure 11.10   HIV prevalence is highest in southern Africa. (Population Reference Bureau. 2007. *2007 World Population Datasheet.* Retrieved February 4, 2009 from http://www.prb.org/pdf07/07WPDS_Eng.pdf)

Rapid Growth in Numbers Means Greater Consumption

Rapidly growing population and the accompanying rapid industrialization/urbanization have diminished the world's forests and woodlands by 1.2 billion ha and grasslands and pastures by 580 million ha since 1700 (Richards 1990) (Figure 11.13). Humans have increased the area of arable land by 1.2 billion ha since 1700 in proportion to increased human needs for food and fiber (Richards 1990; Evans 1998). Croplands expanded by 466% from 1700 to 1980, mostly at the expense of forests, grasslands, and biodiversity. Since 1960, however, the rate of cropland expansion has slowed considerably in proportion to population growth such that the arable land total is now about 1.3 or 1.4 billion ha. At the same time, global average cereal yields have increased at a ratio of 1 ton ha^{-1} for every 2 billion people due to increased uses of high-yielding crops, fertilizers, biocides, and irrigation water in a process called the Green Revolution (Evans 1998; Postel 1998) (Figure 11.13). The Green Revolution in the 1960s enabled the doubling of global food production in the past 35 years without a large increase in arable land (Figure 11.13).

Nevertheless, per-capita grain production has declined by 8% (0.5% yr^{-1}) worldwide since 1984 because of shortages of productive land and water, and the diminishing returns from expensive fertilizers, pesticides, and fossil fuels (FAO 1998). The agricultural intensification of the past 35 years has had major detrimental impacts on both terrestrial and aquatic ecosystems as a result of a 6.87-fold increase in nitrogen fertilizer, a 3.48-fold increase in phosphorus fertilizer, and a 1.68-fold increase in the amount of irrigated croplands (Naylor 1996; Tilman 1999).

As humans became the dominant species, human activities have considerably homogenized and simplified species composition, diversity and abundance of terrestrial and aquatic ecosystems, and their natural disturbance regimes. With each passing day, an estimated 150 species are eliminated because of increasing human-induced disturbances, including deforestation; pollution of soil, water, and air; biocide use; urbanization; and industrialization (Reid and Miller 1989).

Agricultural monocultures have caused four once-rare plants (barley, maize, rice, and wheat) to become the dominant plants globally. In combination, these now cover 39.8% of global croplands and, because they are monocultures, they have a higher natural

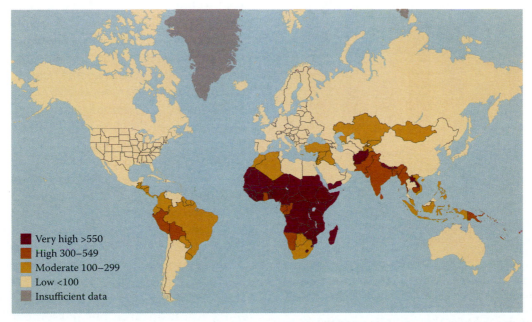

Very high >550
High 300–549
Moderate 100–299
Low <100
Insufficient data

Figure 11.11 World maternal mortality is highest in low-income countries. (Population Reference Bureau. 2007. *2007 World Population Datasheet*. Retrieved February 4, 2009 from http://www.prb.org/pdf07/07WPDS_Eng.pdf)

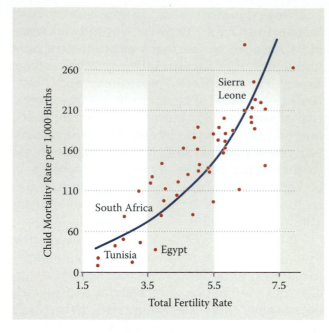

Figure 11.12 High fertility rate and high child mortality rate are highly associated. (From United Kingdom All Party Parliamentary Group on Population, Development and Reproductive Health. 2007. *Return of the Population Growth Factor: Its Impact upon the Millennium Development Goals.* London: House of Commons.)

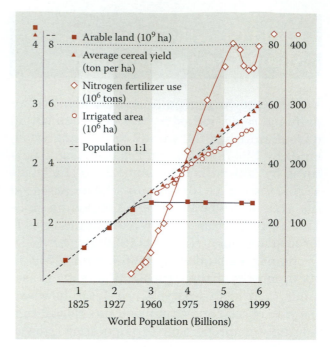

Figure 11.13 The relationship among world population, arable land, average cereal yield, nitrogen fertilizer use, and irrigated area. (Evans, L. T. 1998. *Feeding the Ten Billion: Plants and Population Growth.* Cambridge University Press. New York, NY.)

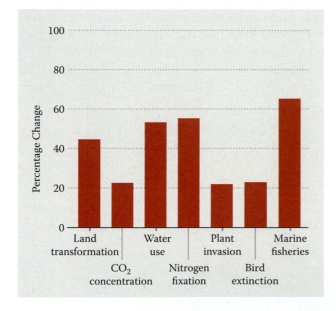

Figure 11.14 Several major components of the Earth system altered and dominated by human beings, expressed as percentage (from left to right) of the land surface transformed, the current atmospheric CO_2 concentration, accessible surface freshwater used, terrestrial nitrogen fixation, invasive plant species in Canada, globally extinct bird species in the past two millennia, and major marine fisheries that are fully exploited, over-exploited, or depleted. (From Vitousek, P. M. et al. 1997. *Science* 277:494–499).

susceptibility to pathogens and pests (Tilman 1999). Approximately one-third of the world's food supply depends on the pollination and biocontrol services provided by insects, birds, and mammals that live in neighboring natural ecosystems (O'Toole 1993). Thus, such monocultures result in about 40% of total potential crop production lost to pests (such as southern corn blight, western corn rootworm, and wheat head rust) despite the annual use of 2.5 million tons of biocides (Pimentel 1997).

About 40% of global grain production goes to animal feed (70% in the United States and 24% in Asia), and current total meat consumption has been predicted to increase from 211 million tons in 1997 to 513 million tons in 2050. This will further increase pressures on environmental quality and food security (WRI 1998; Brown, Gardner, and Halweil 1999; Socolow 1999). Global land area in grain expanded from 590 million ha in 1950 to 670 million ha in 2004, with its historical peak of 730 million ha in 1981 (Brown 2005). The grain land area per person declined from 0.23 ha in 1950 to 0.11 ha in 2000 and is projected to shrink to 0.07 ha in 2050 (Brown

2005). A serious decline in food self-sufficiency already exists in developing countries. Cereal imports of 20 million tons in 1969–1971 by developing countries as a whole increased to about 112 million tons in 2000 (Sadik 1991). The overwhelming bulk of these deficits has so far been met by surpluses in North America, thus making world food security highly susceptible to any variation in the performance of farmers and climatic conditions in that single region or to the use of crops for nonfood products such as ethanol.

Biophysical Controls over Population Growth

Generally, demographic projections for countries do not consider the controls over them imposed by local and global biophysical limitations. Exponential growth generated by positive feedbacks in a system cannot continue indefinitely. In other words, there are always limiting factors and negative feedbacks to prevent the exponential growth of a system from going on forever; these are also known as **limits to growth** or the **carrying capacity.** For example, the presence of limited food, water, energy, and land will eventually constrain the well-being of social, natural, economic, and political systems so as to limit the human population. The size of a population takes the shape of "S" (**logistic growth**) over a longer period of time when the **biotic potential** (a maximum growth rate under ideal conditions) of that population is regulated by the limiting factors and the carrying capacity in a given area.

Natural capital (resources) contains all the stocks of ecosystem goods and all the flows of ecosystem services upon which the survival and health of plants, animals, and humans depend (Table 11.3). Ecosystem goods and services provide productive, regulative, protective, and cognitive functions for the well-being of humans and their surrounding natural and managed ecosystems. Natural resources can be mainly classified into two groups: renewable and nonrenewable natural resources.

Renewable resources are **rate limited** (available in fixed flows) by virtue of rates of their self-regenerative capacity at less than a human time span. **Nonrenewable resources** are **stock limited** (available in fixed quantities) due to their very slow renewal rates. Global human carrying capacity can be overshot if human resource use and waste generation exceed rate- and stock-limited ecosystem goods and services at the local and regional scales. This, in turn, progressively damages the productivity, integrity, biodiversity, and stability of the Earth. The Earth's natural capital is finite and unsubstitutable, which eventually limits the indefinite growth of human population and consumption.

Three reasons explain how humans have achieved the numbers they have now. First, they expanded into new territories and new habitats. Second, by resource importation and increases in productivity due to technological advances, the carrying capacities of environments already occupied were expanded. Finally, a number of means masked and delayed the workings of the limiting factors. For example, trades, technological advances, and market failures may defer the ecological limits on increasing human demands for natural resources and services at local, regional, and global scales.

Market failures refer to the inability of market prices to reflect in a timely and accurate way the social value of nonmarketed and nontradable ecosystem goods and services. Market failures include:

- externalities (e.g., inadequate valuation of nonmarketed ecosystem goods and services in the process of decision-making);
- inequitable income distribution for the collective welfare of the society;
- lack of long-term environmental analysis and indicators for the well-being of current and future generations; and
- lack of signals that guide decisions in the presence of common natural resources and services (e.g., atmosphere, ozone layer, greenhouse effect) to which it is impossible to assign property rights.

Many researchers attribute the degradation and depletion of natural capital to a range

Table 11.3
Ecosystem Goods and Services

Ecosystem	Goods	Services
Agroecosystems	Food crops Fiber crops Crop genetic resources	Maintain limited watershed functions (infiltration, flow control, partial soil protection) Provide habitat for birds, pollinators, soil organisms important to agriculture Build soil organic matter Sequester atmospheric carbon Provide employment
Forest ecosystems	Timber Fuelwood Drinking and irrigation water Fodder Nontimber products (vines, bamboos, leaves, etc.) Food (honey, mushrooms, fruit, and other edible plants; game) Genetic resources	Remove air pollutants, emit oxygen Cycle nutrients Maintain array of watershed functions (infiltration, purification, flow control, soil stabilization) Maintain biodiversity Sequester atmospheric carbon Moderate weather extremes and impacts Generate soil Provide employment Provide human and wildlife habitat Contribute aesthetic beauty and provide recreation
Freshwater ecosystems	Drinking and irrigation water Fish Hydroelectric Genetic resources	Buffer water flow (control timing and volume) Dilute and carry away wastes Cycle nutrients Maintain biodiversity Provide aquatic habitat Provide transportation corridor Provide employment Contribute aesthetic beauty and provide recreation
Grassland ecosystems	Livestock (food, fiber, game, hides) Drinking and irrigation water Genetic resources	Maintain array of watershed functions (infiltration, purification, flow control, soil stabilization) Cycle nutrients Maintain biodiversity Generate soil Sequester atmospheric carbon Provide human and wildlife habitat Provide employment Contribute aesthetic beauty and provide recreation
Coastal ecosystems	Fish and shellfish Fishmeal (animal meal) Seaweeds (for food and industrial use) Salt Genetic resources	Moderate storm impacts (mangroves, barrier islands) Provide wildlife Maintain biodiversity Dilute and treat waste Provide harbors and transportation routes Provide human and wildlife habitat Provide employment Contribute aesthetic beauty and provide recreation

Source: From WRI (World Resources Institute). 2000. *People and Ecosystems: The Fraying Web of Life.* Washington, D.C.: the World Bank, the United Nations Environment Program, and the World Resources Institute.

Table 11.4
World's Land Area, Uses of Land, and Population Density in 2002

Region	Land Area	Land Use (× 10⁶ ha)				Population Density (People ha⁻¹)
		Cropland	Permanent Pasture Grassland	Forest–Woodland	Other	
Africa	2,962	184	900	685	1198	28
Asia	3,097	511	1110	532	931	123
North and Central America	2,143	257	349	710	794	23
South America	1,753	112	515	829	316	20
Oceania	849	50	412	157	207	3
Europe	2,260	287	182	157	94	32
World	13,066	1404	3471	3898	4345	48

Source: FAO (Food and Agriculture Organization of the United Nations). 1998. FAOSTAT. Rome, Italy.

of major causes and regard them as **underlying driving forces** or **causes.** These include the rapid growth of human population and consumption, use of environmentally incompatible technologies, lack of appreciation and valuation of ecosystem services, complexities in the marketplace, and distributive injustice of wealth and power. Detection and recognition of these underlying causes are essential to effective, long-term solutions of environmental problems. A demographic measure of human population impact on environmental quality was first proposed by Cloud (1969) as follows:

Impact

= total resources available/population density

× per-capita consumption

The following measure, proposed by Ehrlich and Holdren (1974) and cited widely, calculates the impact (*I*) of human population on the environment as the product of population (*P*), level of affluence (*A*), and the damage done by the particular technologies (*T*) that support that affluence (per-capita consumption):

$$I = P \times A \times T$$

The relationship among ecosystem productivity, human population size, and consumption growth has also been quantified by others. One such recent analysis receiving the

Table 11.5
Resources Used and/or Available[a] in the United States, China, and the World to Supply Basic Needs

Resources	U.S.	China	World
Land (ha)	0.71	0.08	0.27
Cropland	0.91	0.33	0.57
Pasture	1.00	0.11	0.75
Forest	3.49	0.52	1.59
Total			
Water (× 10⁶ L)	1.7	0.46	0.64
Fossil fuel (oil-equivalent liters)	8740	700	1570
Forest products (kg)	1091	40	70

Source: From Pimentel, D. et al. 1999. *Environment, Development and Sustainability* 1:19–39.
[a] Per person per year.

most attention is the **ecological footprint analysis** (Wackernagel and Rees 1996), which is discussed in some detail in Chapter 22 on economic systems, growth, and development.

Increasing Consumption and Decreasing Natural Resources

The growing gap between the quality and quantity of the natural resource base and the human population and its consumption and use of environmentally destructive technology reduces the Earth's carrying capacity and thus its ability to provide for the welfare and health of current and future human and non-human constituents. The total ice-free area of the Earth's terrestrial ecosystems is about 13.1×10^9 ha (Table 11.4). Ecologically productive areas constitute only 8.9×10^9 ha of the world's land surface. The assumption that at least 1.5×10^9 ha must be left intact for the protection of life-supporting ecosystems leaves an unrealistically available productive land area of only 7.4×10^9 ha for human uses. The disparity of resources available per-capita presents distinct contrasts (Table 11.5).

As the number of people per unit space at a given time (**population density**) increases, the world's limited land availability shrinks, a fact first recognized by a British economist, Thomas Malthus, in 1798.[3–5] The 300% increase in global population size from 1.5 billion in 1890 to 6 billion in 1999 has diminished the global land supply per person, which must provide all the productive and protective functions necessary for the survival and well-being of an individual, by 291%—from 90×10^3 m^2 in 1890 to 23×10^3 m^2 in 1999

(Figure 11.13). Human alteration of the Earth system has been increasingly felt in all the ecological components such as the atmosphere, hydrosphere, biosphere, pedosphere, and lithosphere (Figure 11.14). The quantities of fossil fuels used have increased collectively in an unprecedented way on the global level (see Chapters 14 and 17). This has also created a huge disparity between the developed and the developing countries (Figure 11.15).

Thus, cumulative and synergistic (interacting) impacts of these local and regional human-induced disturbances have led not only to the transboundary pollution of air, water, and soil at the regional scale but also to environmental and climate change at the global scale (Figure 11.14). Global average figures as to population size and the availability and quality of land, energy, and water resources conceal great regional and local variations in the types, severity, magnitude, and rate of human pressures and the vulnerability of ecosystems to them. For example, this analysis may obscure the danger of population pressure in regions with semiarid or arid climates on the total amount of available water in aquifers and water courses (**demographic water scarcity**) and overexploitation of freshwater resources

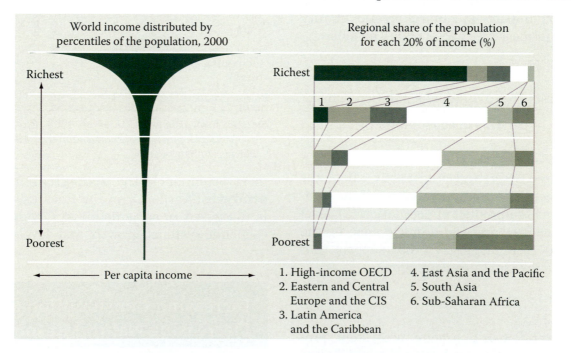

Figure 11.15 Distribution of world economic income. One-fifth of the world, living mainly in industrialized countries, owns the vast bulk of global wealth. (Dikhanov, Y. 2005. Background note for *Human Development Report 2005*. New York.)

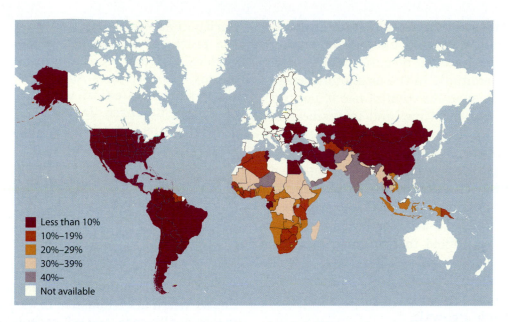

Less than 10%
10%–19%
20%–29%
30%–39%
40%–
Not available

Figure 11.16 In low-income and middle-income countries, 30% of children under age 5 are underweight. (Population Reference Bureau. 2007. *2007 World Population Datasheet.* Retrieved February 4, 2009 from http://www.prb.org/pdf07/07WPDS_Eng.pdf)

(**technical water scarcity**) (see Chapter 16) (Shiklomanov 1996; Falkenmark 1997; Penning de Vries, Rabbinge, and Groot 1997).

The fraction of the Earth's resources appropriated by humans (i.e., **human appropriation of net primary production—HANPP**) is among the widely used indicators of the "human domination of Earth's ecosystems" (Haberl et al. 2002). Mankind is already appropriating 40% of terrestrial NPP (Vitousek et al. 1997) and 30% of rich coastal marine NPP (Pauly and Christensen 1995). Globally, human activities already appropriate 54% of readily accessible renewable freshwater (12,500 km³) (Postel, Daily, and Ehrlich 1996). Human-induced soil degradation has adversely affected 15% (1,966 million ha) of the Earth's land surface (Oldeman 1994). The global extent of the soil degradation is 38% for agricultural land, 21% for permanent pasture, and 18% for forest and woodland (FAO 1990). The global loss rate of land to urbanization and transportation ranges from 10 to 35 million ha yr^{-1} (ca. 1%), with half of this coming from croplands (Döös 1994). From 1970 to 1995, the energy-use rate of 2.5% (doubling every 30 years) exceeded the population growth rate of 1.7% (doubling in about 40 years).

"The Tragedy of the Commons" by Garrett Hardin (1968) relates how the growing human population and its consumption threatens its own livelihood and the sustainability of the quality and quantity of publicly owned ecosystem services upon which the global human community depends (such as atmospheric, hydrologic, and sedimentary cycles; biodiversity; and protective ozone layer). Environmental consequences of growing population and consumption that compete for a finite resource base include shortages in land- and marine-based food production, energy, and water; deforestation; losses of biodiversity and productive habitats; climate change; and depletion of the protective ozone layer. Social and economic consequences include poverty, environmental refugees, overcrowded urban areas, increased health problems and crime, gridlocked transportation difficulties, and an undermined sense of community, solidarity, and attachment to nature.

Uneven Access to Natural Capital Threatens Global Security

Another vital factor for the resolution of environmental problems in the face of uneven and unsustainable use of scarce natural resources

is **distributive justice of wealth and power.**
In proportion to human population size, human
uses of scarce natural resources have been
growing exponentially in the form of increased
expansion (the spatial extent of use of natural
resources in a given time) and intensification
(the intensity of use of natural resources in a
given space and time). Globally, the rapidly growing human population has been accompanied
by growing consumption: a 50-fold increase in
industrial production, a 30-fold increase in fossil
fuel consumption, a 20-fold increase in the world
economy and a 10-fold increase in water use
since 1900 (Walker and Steffen 1996; Biswas
1998). The cumulative effect of the growth in
these per-capita multipliers is often referred to
as **demophoric growth.**

Despite increasing economic production, the
distribution of global economic income is still
so distorted that 80% of the world's population
has an income less than the average. The average income of the top 20% of the world's population is about 50 times the average income of
the bottom 20% (Dikhanov 2005). World income
share as divided into five population segments
(20% in each) resembles a tornado, with the
poorest individuals at the bottom and the richest at the top (Figure 11.15). The richest 20%
of the population receive three-quarters of the
total global income, whereas the poorest 20%
hold just 1.5%. The poorest 40% consists (more
or less) of 2 billion people living on an average of less than $2 a day. Currently, the global
income distribution is such that one in every
two people in sub-Saharan Africa, one in every
four people in South Asia, one in every five
people in East Asia, and one in every twelve
people in Latin America are located in the poorest 20%, with nine in every ten people in rich
countries located in the top 20% (Dikhanov
2005; UNDP 2005).

Uneven and unsustainable use of scarce
natural resources threatens distributive justice
of wealth and power, which in turn impairs common security of global community. **Distributive
justice of wealth** refers to the degree of
unequal economic growth (or the gap between
poverty and overconsumption); **distributive
justice of power** refers to associated dis-

parity in access of the poor versus the rich to
the processes of decision- and policy-making
(Tables 11.5 and 11.6). For example, average
energy use estimates among nations suggest
that a single U.S. citizen uses the same amount
of energy as three Japanese, six Mexicans,
13 Brazilians, 14 Chinese, 38 Indians, 168
Bangladeshis, 280 Nepalis, or 531 Ethiopians.
About three-quarters of a billion people (one
person in eight) are believed to be chronically
hungry due to poverty—not only in developing countries but also in developed countries
(Table 11.6; Figure 11.16).

Even if hunger decreases in response to
increased food availability and international
trades of food, malnutrition may not decrease
due to distributive injustice in sanitation, clean
water, health care, and education. Watson et al.
(1998) state:

> Today more than 1.3 million people live on less than
> $1 per day and 3 billion people live on less than
> $2 per day, 800 million people are malnourished,
> 1.3 billion people live without clean water, 2 bil-
> lion people live without sanitation, 2 billion people
> lack electricity, and 1.4 billion people are exposed
> to dangerous levels of outdoor air pollution. These
> unfulfilled needs for a clean and healthy environ-
> ment cause millions of people to die prematurely
> each year.

Environmental Refugees

Rapid population growth in combination
with environmental degradation and limited biophysical capacity of natural capital
undermines food, energy, land, and financial
security, especially in developing countries,
thus leading to externally and internally displaced people called **environmental refugees**
(Figure 11.17). This term was coined by
El-Hinnawi in 1985 and described people who
have been forced to leave their traditional habitats because of environmental disturbances
that jeopardized their existence and the quality of their life (Ramlogan 1996).

Environmental refugees can result mainly
from two causal agents: natural disturbances
(earthquakes, volcanic eruptions, landslides,

Table 11.6
Poverty Trends in Developing Economies

Region	Incidence of Poverty (%)[a]		Number of Poor (× 10⁶)	
	1985	2000	1985	2000
South Asia	50.9	26.0	525	365
India	55.0	25.4	420	255
Sub-Saharan Africa	46.8	43.1	180	265
Middle East, North America, Western Europe	31.0	22.6	60	60
East Asia	20.4	4.0	280	70
China	20.0	2.9	210	35
Latin America, the Caribbean	19.1	11.4	75	60
Eastern Europe	7.8	7.9	5	5
Total	32.7	18.0	1125	825

Source: From *World Food Security: Prospects for the future.* World Bank. Washington, D.C.
[a] The incidence of poverty is the share of the population below the poverty line, which is set at an annual income of $370.

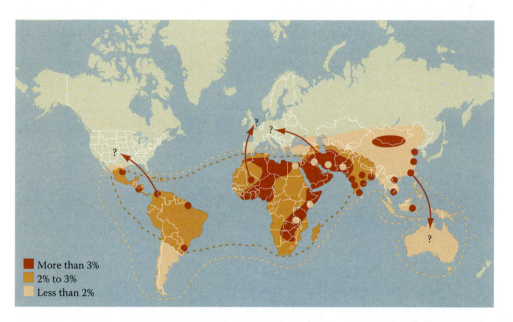

More than 3%
2% to 3%
Less than 2%

Figure 11.17 Possible scenarios of large-scale migrations of environmental refugees caused mainly by rapid population growth and insufficient local food production. The dots indicate regions where the population will have increased by 100 million prior to 2025. (From Döös, B. R. 1994. *Ambio* 23 (2): 124–130.)

avalanches, cyclones, storms, floods, and droughts) and human-induced disturbances (climate change; loss of prime farmlands; desertification; deforestation; salinization; depletion of aquifers and fish stocks; pollution of air, water, and soil; dam and road constructions; industrial accidents; and wars). Current environmental refugees are estimated to number 25 million, including internally displaced people;

the number is projected to rise to as many as 100 million by 2050 (Myers 1995).

Major examples of environmental refugees emerging from human-induced disturbances include temporarily or permanently dislocated people due to deforestation (e.g., in Haiti), air pollution (e.g., in the region known as "the Black Triangle," lying in southwestern Poland, the northwestern part of the Czech Republic, and the southeastern part of Germany), improper disposal of chemicals (e.g., the Love Canal incident in New York), radioactive emissions (e.g., the Chernobyl incident in the former Soviet Union), emissions of poisonous gases (e.g., the Bhopal incident in India), and dam constructions (e.g., in Pakistan and China) (Ramlogan 1996).

Overall Assessment of Increasing Human Numbers

Although many professionals, individually and collectively, have stressed the disconnect between increasing numbers of people and environmental costs, proposed solutions have had only limited success. We sum up our discussion by the notable conclusion in the report of the United Kingdom All Party Parliamentary Group on Population, Development and Reproductive Health (2007): *Return of the Population*

Growth Factor: Its Impact upon the Millennium Development Goals, which declares:

The Millennium Development Goals (MDGS) are difficult or impossible to achieve with current levels of population growth in the least developed countries and regions. …

Population growth is not the only cause, or even the leading cause, in the success or failure of the MDGs. It is, instead, a critical factor. Population growth and high fertility rates are strongly linked. In fact, in many regions, the MDGs are simply not attainable without a greater focus on slowing population growth, through making voluntary family planning universally accessible. …

Reversing the loss of environmental resources cannot be achieved in the context of rapid or even moderate population growth. …It is clear that without addressing the issue of population growth and high fertility in the poorest regions of the world, these regions have little chance of achieving.

It is surprising that, although all professionals agree with the criticality of the growing human population, a majority seem to shy away from facing the reality.

Notes

1. Two books recommended on the topic of this chapter are P. R. Ehrlich's *The Population Bomb* and J. L. Cohen's *How Many People Can the Earth Support?* (see references).
2. When a quantity such as the rate $r(t)$ of consumption of a resource grows a fixed percent per year, the growth is exponential: $r(t) = r_0 e^{kt} = r_0 2^{t/T}(1)$, where r_0 is the current rate of consumption at $t = 0$; e is the base of natural logarithms; k is the fractional growth per year; and t is the time in years. The growing quantity will increase to twice its initial size in the doubling time T_2, where: T_2 (yr) = (ln 2)/k » 70/P(2).
3. "The cause to which I allude is the constant tendency in all animated life to increase beyond the nourishment prepared for it. (Malthus 1914, p. 1)
"Taking the earth as a whole,…the human species would increase as the numbers, 1, 2, 4, 8, 16, 32, 64, 128, 256, and subsistence as 1, 2, 3, 4, 5, 6, 7, 8, 9. In two centuries the population would be to the means of subsistence as 256 to 9; in three centuries as 4096 to 13, and in two thousand years the difference would be almost incalculable.
"In this supposition no limits whatever are placed to the produce of the earth. It may increase for ever and be greater than any assignable quantity; yet still the power of population being in every period so much superior, the increase of human species can only be kept down to the level of the means of subsistence by the constant operation of the strong law of necessity, as a check upon the greater power" (Malthus 1914, pp. 10–11).
4. That "Malthus was foiled over and over again" by the invention of agricultural tools; the understanding of nutritional requirements of plants; discovery of the Haber process for the manufacture of ammonia, thus making it possible the widespread use of fertilizers; and the breeding of new crop varieties that resulted in Green Revolution is the view of Trewavas

(2002). The author notes, however, that there is "no room for complacency," given the population size that will need food.

5. Lutz, Testa, and Penn (2006) find "a consistent and significant relationship between human fertility and population density. Also individual fertility preferences decline with population density." They used fixed effects models on the time series of 145 countries for key social and economic factors.

Questions

1. What forces are responsible for human population growth?
2. What biophysical controls exist to counter population growth?
3. Discuss the potential social, economic, and environmental consequences of a decrease in the quantity and quality of natural resources available in your region that could result from an increase in human population density.
4. Discuss the role of demographic, social, and economic conditions in encouraging or curtailing human population growth.
5. How have economic growth and innovation supported human population growth? Can consumption of natural resources and rapid economic growth continue indefinitely?
6. What relationship (feedback) causes the logistic growth population model to differ from the exponential growth curve?

12

Land Use— Agriculture

Agriculture in Retrospect

To meet its need for food and fiber, the human population depends on primary and secondary production. Archaeological evidence suggests that a transition from gathering and hunting to agricultural society evolved independently in several regions around the world about 10,000–12,000 years ago as a result of **plant** and **animal domestication** (Figure 12.1). It is thus the oldest transformational activity of land. During their history, humans have cultivated or gathered 7,000 plant species for food, but today they depend on only 20 plant species to provide more than 90% of the world's food; just three species—maize, wheat, and rice—contribute more than half of this food (Wilson 2000a).

The selection and growth of only a few desirable plant species have resulted in homogenous systems with little natural genetic diversity. The spread of these domesticated/cultivated varieties (or cultivars) from their centers of origin (and adaptation) to regions with different growing conditions has required multiple modifications of the environment—tillage, irrigation, fertilizers, biocides, and sowing dates—and the manipulation of genetic makeup of plant species. Globally, on a dry matter basis, only 16 crops account for most (92%) of the food consumed by humans today (Table 12.1); about 40% of cereal grains are fed to domestic animals.

Traditional and Industrialized Agriculture

There are two major types of agriculture: traditional and industrialized. **Traditional agricultural systems** (mostly practiced in developing countries) are more labor- and land intensive and consist of four main types: polyculture, subsistence agriculture, shifting agriculture, and nomadic herding. **Polyculture,** where several mutually nurturing crops (e.g., nitrogen-fixing plants) or those that show mutually compatible uses of resources (e.g., the root systems) are grown simultaneously, is found mostly in traditional agricultural

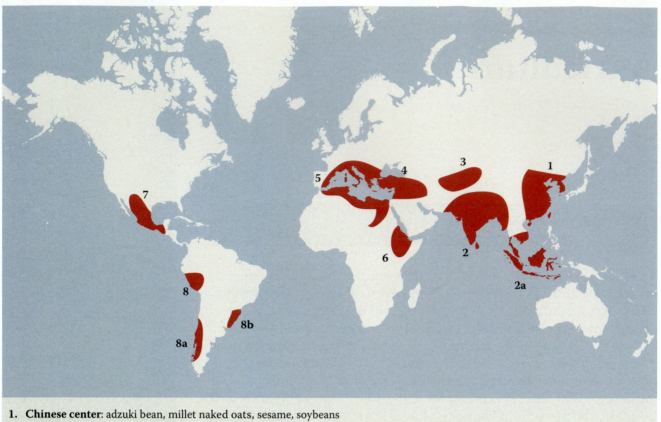

1. **Chinese center**: adzuki bean, millet naked oats, sesame, soybeans
2. **Indian center**: rice bean, chickpea, aboreum cotton, jute, finger millet, mungbeen, rice, sugarcane, taro, yam
2a. **Indomalayan center**: banana, coconut, yam, sugarcane
3. **Central Asiatic center**: chickpea, flax, lentil, pea, rye, safflower, sesame, bread, wheat
4. **Near Eastern center**: alfalfa, barley, chickpea, lax, lentil, meion, red oats, pea, rye, sesame
5. **Mediterranean center**: broad bean, cabbage, lettuce, hulled oats, durum wheat
6. **Ethiopian (formerly Abyssinian) center**: barley, chickpea, flax, lentil, finger millet, pea, sesame, teff, tetraploid wheat
7. **South Mexican and Central American center**: common bean, corn, upland cotton, cucurbits (gourd, squash, pumpin), sisal, hemp
8. **South American (Peruvian–Ecuadorian–Bolivian) center**: lima bean, sea-island cotton, potato, sweet potato, tobacco, tomato
8a. **Chilean center**: potato
8b. **Brazilian–Paraguayan center**: cacao, manioc, peanut, pineapple, rubber tree

Figure 12.1 World map of centers of origin for a number of crop plants proposed by N. I. Vavilov. (Poehlman, J. M., and D. A. Sleper. 1995. *Breeding Field Crops*. 4th ed. Ames, IO: Iowa State University Press.)

systems. **Subsistence agriculture** refers to the production of enough food to feed the farming families with little left over to sell for income or hold in reserve for hard times.

Shifting agriculture (also called slash-and-burn agriculture) is common mostly in tropical forest areas. It involves the cutting and burning of forests, cropping the land for a few years until the soil fertility is exhausted, and then abandoning the area and moving on to clear another forest patch for agriculture. Shifting cultivation renders cleared and abandoned areas prone to extreme soil erosion, which eventually results in the loss of their regenerative capacity permanently. The

practice renders millions of people as "environmental refugees." **Nomadic herding** occurs where the land conditions are not favorable to support crops for livestock growth, and hence herders must continually search for fodder by moving their animals.

Industrialized agriculture relies on high inputs of fossil fuel energy, water, fertilizers, and pesticides to produce high yields of single crops (monoculture) per unit area of cropland. Industrialized agricultural systems range from hobby farms to commercial farms and occur mostly where well-to-do farmers can afford the expensive fossil fuel energy, fertilizers, pesticides, and seeds required.

Table 12.1
Trends in the World's Major Crops

Crop	Area Harvested ($\times 10^6$ ha)		Yield (kg ha^{-1})	
	2002	2007	2002	2007
Barley	55.266	56.609	2473	2406
Cassava	17.304	18.665	10804	12223
Maize	138.365	152.874	4358	4971
Millet	32.972	35.836	725	889
Nuts	.548	.583	1169	1244
Oats	12.452	11.952	2042	2175
Potatoes	19.078	19.322	16583	16647
Rice	147.298	156.953	3851	4152
Rye	9.102	6.892	2300	2285
Sorghum	41.344	43.295	1299	1475
Soybeans	78.983	94.899	2303	2278
Sugar beet	6.033	5.295	4262	46815
Sugar cane	20.508	21.977	64907	70878
Sweet potatoes	9.358	9.093	14460	13810
Wheat	213.703	217.433	2688	1792

Source: From FAO. 2008. FAOSTAT. Rome, Italy. This table was constructed from selected data given at http://faostat.fao.org/site/567/DesktopDefault.aspx?PageID=567#ancor via - http://faostat.fao.org/site/567/default.aspx. Retrieved January 23, 2009.

Food Bounty of the United States

Agricultural production in the United States has been the envy of the world, and U.S. citizens spend only a small proportion of their per-capita income on buying food compared with people in the rest of the world. It may therefore be worthwhile to provide a brief historical overview of the food production system that has led to this success.

Bountiful land, adequate precipitation, soils immensely rich in organic matter in many parts of the country, and a continual desire to enhance plant and livestock growth all have had a major role. Attention to agricultural production systems in a concerted way dates back to 1860, when the U.S. Department of Agriculture (USDA) was born and farmers comprised 58% of the labor force. Shortly thereafter, the Morrill Act (1862) was passed authorizing public land grants for teaching agriculture and mechanic arts (hence such terms as "Land Grant University" and Texas A&M [Agricultural and Mechanical] University). The country's overall population was small and the land areas vast, generating a constant urge for farm mechanization.

In the years since 1860, the goal has consistently been to realize the largest productivity from farms. Along the way, there were major problems such as many fungal diseases (rusts and smuts), corn borers, tobacco mosaic virus, foot-and-mouth disease of animals, and grasshopper infestations in the West, but worst of all were the droughts of the late 1920s and early 1930s (creating the "Dust Bowl"). Beginning in 1928 and lasting for several years, the worst drought in 1934 covered 75% of the country and billions of tons of precious topsoil blew away.

Fortunately, two legislative acts were in place: (1) to conduct research specific to agricultural problems, and (2) to disseminate the research results to farmers for implementation in the field. For the first goal, the Hatch Act (1887) provided grants for the establishment of agricultural experiment stations. For the second, the Smith–Lever Act (1914) authorized the establishment of the Agricultural Extension Service, which provides agricultural extension agents in every county. Together with other steps from time to time (e.g., the establishment of the Soil Erosion Service in 1933, now called Natural Resources Conservation Service), the goal of maximizing agricultural productivity has been enviably attained.

Historically, the land area under cultivation each year was at an all-time high during the 1950s—about 162 million ha.[1] This area has remained more or less constant for the last 50 years (in fact, it has decreased somewhat), despite an increase in population and a much greater world demand for food.[2] With both population growth and world demand, authorities contend, more land should have been brought under cultivation, but "technological advances"— namely, energy subsidies, mechanization, and improved cultivars—have saved an additional land area of about 97 million ha (Figure 12.2).

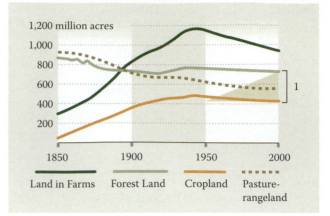

Figure 12.2 Quantitative changes in major land use and population in the United States, 1850–2000: Land in farms, forest land, croplands, and pasture–rangeland; "1" refers to land saved by technological advancement since 1950. (From Miller, F. P. 2000. "Land Grant Colleges of Agriculture: Preempting a Postmortem—Requisites for Renaissance." Invited lecture, School of Natural Resources, The Ohio State University, Columbus, April 6. 42 pp.)

farm as soon as researchers produced them. But, with all these in place, energy subsidies went up and human labor inputs plummeted (Figure 12.2; Table 12.2). The energy input into U.S. agriculture, from plowing and seeding to the table, is between 18 and 20% of the total national energy budget. Many of these processes are energy intensive, as can be seen from an energy budget for corn production (Table 12.2).

Energy Input into Agriculture

How was this success brought about? The answer lies in the application of multiple strategies, each with its own advantages and, as we know now, disadvantages. Farm mechanization made it possible to work on larger tracts of land than would have been possible with the labor of humans and horses. These included plowing, seeding, application of fertilizers and biocides, irrigation, and harvesting. Attention was also paid to food processing, packaging, and transportation so that the bounty of the fields was not wasted. Much-improved, high-yielding plant varieties were brought to the

The Green Revolution

The development and spread of high-yielding varieties and more nutritious content of crops during the 1960s are referred to as the **Green Revolution.** Sponsored by the Rockefeller and Ford foundations in 1943, the International Maize and Wheat Improvement Center in Mexico, under the leadership of agricultural scientist Norman Borlaug, developed "miracle strains," such as high-yielding varieties of wheat and rice with higher protein content and disease-resistant potatoes and beans. Borlaug was awarded the Nobel Prize in 1970.[3]

Table 12.2
The Energy Input for Corn Production, 1920–1985

Years	1920	1945	1954	1964	1975	1980	1985
Direct Energy Inputs ($\times 10^3$ kcal ha^{-1})							
Labor	65	31	23	15	10	7	6
Machinery	278	407	648	907	925	1,018	1,018
Draft animals	886	0	0	0	0	0	0
Fossil fuel	0	1,428	1,842	1,991	1,512	1,378	1,278
Indirect Energy Inputs ($\times 10^3$ kcal ha^{-1})							
Nitrogen	168	630	1,555	2,331	2,940	2,940	3,192
Phosphorus	0	50	82	227	353	409	365
Potassium	0	15	50	70	155	195	187
Lime	3	46	39	64	69	134	134
Seeds	44	161	421	520	520	520	520
Insecticides	0	0	13	27	50	60	60
Herbicides	0	0	7	40	300	300	350
Irrigation	—	125	250	625	2,000	2,125	2,250
Drying	0	9	15	145	458	640	760
Electricity	1	8	24	60	90	100	100
Transport	25	44	67	89	82	90	89
Total input (I)	1,302	2,492	4,111	5,935	8,855	9,916	10,309
Total yield (Y)	7,520	8,528	10,288	17,060	20,575	26,000	26,000
Energy efficiency (Y:I)	5.8	3.4	2.5	2.9	2.3	2.6	2.9

Source: From Pimentel et al. 1990. In *Agroecology,* ed. S. R. Gliessman, 305–321. New York: Springer.

Since Borlaug's Green Revolution, the research in all areas of crop growth has indeed been phenomenal. The work of Khush (2006) on rice at the International Rice Research Institute may be cited as just one clarion example of this trend and the following highlights are based on one of his lectures.

The share of calories consumed worldwide is 23% (rice), 17% (wheat), and 9% (maize). High-yielding rice has been transformed genetically and phenotypically to enhance grain-production plants (Figure 12.3). Rice development research continues on aspects such as:

- Nitrogen responsiveness. Conventional varieties showed an initial increase with N fertilization, followed by a major drop. Varieties were bred to show increased yield with increasing N applications.

Tall conventional plant   Improved high-yielding plant   Low-tillering ideotype (new plant type)

Figure 12.3 Improved rice types illustrating the role of genetic modification in the Green Revolution. Note the differences in rhizosphere, stalks, and inflorescences from the conventional to the new type. (Photo courtesy of International Rice Research Institute.)

- Photoperiodic response. Conventional varieties (e.g., Peta) mature in ~180 days. Varieties were bred to mature in 95 days, allowing for multiple harvests in a year.
- Disease resistance. Varieties have been bred to overcome (1) grassy stunt virus, (2) Tungro virus (brown plant-hopper nymphs—vectors of 1 above), (3) bacterial blight, and (4) stem borers.
- Transgenic modifications. Varieties have been bred that are resistant to stem borers, drought, submergence, and high-salinity soil.
- Transgenic varieties (genetic engineering). These have been developed to overcome human vitamin A deficiency by introducing beta-carotene for vitamin A synthesis in rice endosperm ("golden rice").
- Trials of varieties to overcome human iron deficiency have been conducted.
- Research is under way for converting rice from C_3 photosynthesis to C_4 mode.

Although the Green Revolution has boosted food production per hectare of cropland mostly in Asia and Latin American countries, the intensification of agricultural production there has led to increased use of fertilizers, pesticides, mechanized machinery, fossil fuels, expensive seeds, and an associated increase in the loss of genetic and species diversity, plus increased greenhouse gas emissions. The environmental and socioeconomic effects of the intensified agriculture will be explained in greater detail later in this chapter.

Land Use Conflicts

Expanding and competing human uses of land and water resources (such as urban–industrial activities, transportation, agricultural lands, grasslands, forests, fisheries, and nature conservation) can adversely affect the structure (biological diversity and species distribution and abundance) and function (exchanges of energy, materials, water, and species) of neighboring ecosystems; this is called land use conflicts. Land cover refers to the biophysical appearance and the vegetation type present on the land surface, such as forest, agriculture, grassland, and barren lands.

Land cover change describes differences in the area occupied by cover types as well as fragmentation (shifts in the spacing) of vegetation cover types across the landscape over time. **Land use** refers to the transformation of land from its original state to a new use (e.g., clearing of trees for agricultural or grazing purposes), or converting to farmlands to other uses (such as shopping malls). For example, a place with forest cover may be used for low-density housing, logging, or recreation, potentially causing conflict among different human user groups. The term **land use change** includes variations in human uses of the land over time.

Quantitative changes among land uses occur either **irreversibly** as the losses of ecosystems or **reversibly** as the conversions from one ecosystem type to another. For example, the global area of croplands is estimated to have expanded from about 320 million ha in 1850 to 1,360 million ha in 1990. In response to the world's growing population size, the expansion of the lands devoted to agriculture reached 4.87 billion ha in 1994, exceeding the physical limits of the ecologically suitable arable lands of 3.3 billion ha (FAO 1997; Lund and Iremonger 2000) (Figure 12.4). The continuation of the agricultural expansion beyond its physical limits has resulted in the conversions of forests, grasslands, wetlands, and nature conservation areas to croplands and the fragmentation of remaining intact habitats.

Over a 140-year period since 1860, about 730 million ha of croplands and pastures were cleared from forests (17% reduction) and woodlands, and almost 500 million ha of grasslands and other nonforest ecosystems were converted to croplands, with most of the loss occurring in the tropics (Houghton 1999). U.S. cropland area peaked at about 500 million acres in the 1940s—a 400% increase since 1850—and then declined to 400 million acres by 2000. The losses of prime (productive) farmlands to the expansion of residential, commercial, industrial, and transportation sectors pose important threats that destabilize the balance between food supplies and human demands at local, regional, and global scales (see Chapter 17). Conversely,

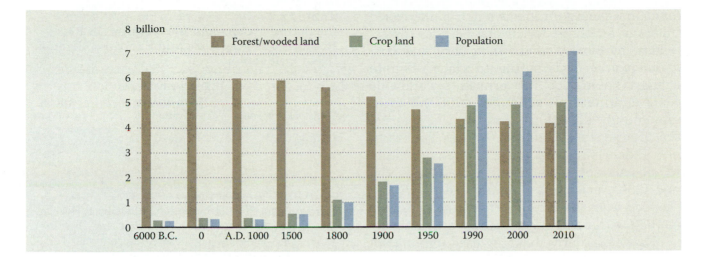

Figure 12.4 Historical and projected changes in land use (in hectares) and population; time sequence is not drawn to scale. (Lund, H. G., and Iremonger. 2000. *Forest Ecology and Management* 128: 3–10.)

the losses of wetlands to croplands and the conversions from grasslands, forests, and natural areas increase the loss of biodiversity and the release of greenhouse gases sequestered in the pools of soils and vegetation to the atmosphere, as well as modify the hydrological regimes and microclimate of an area (Chapters 13 and 19).

Qualitative changes occur when adjacent land uses and management practices within land uses are ecologically incompatible and thus have adverse off-site and on-site effects on the structure and function of ecosystems. For example, in many semiarid and arid regions of the world (e.g., western North America, Africa, the Middle East, Australia, and central Asia), water-intensive industries (e.g., mining and textile mills) can deplete or contaminate the water supplies of an agricultural area; conversely, an irrigated cropland can reduce the quality and availability of water necessary to support the biodiversity in the surrounding wildlife areas.

Growing Monocultures

Simplification and homogenization of agricultural ecosystems as monocultures cause the loss of biodiversity and, consequently, high incidences of diseases and pests because sustained productivity and stability of these ecosystems depend on biological and ecosystem diversity just as in other ecosystems. The World

Watch Institute states that an alarming rate of decline in the number of different varieties of crops grown has occurred around the world. For example, wheat varieties grown in China decreased from 10,000 in 1949 to 1,000 by the 1970s; more than 90% of the varieties of peas and cabbages and 81% of the varieties of tomatoes grown in the United States in 1904 are no longer grown, and no seeds of these cultivars have been stored for future use. Only 20% of the corn varieties grown in Mexico 60 years ago remain in cultivation. Not only are monocultures more susceptible to disease and pests, but also the loss of cultivars means their unique combinations of genes are unavailable for improving currently grown strains.

The clearing of vegetation for agriculture increases flooding because the new agricultural ecosystems retain less precipitation than do the forests. Agricultural expansion into infertile soils, where high-diversity ecosystems generally occur, and the selection of crops that do not match local growing conditions lead to increases in loss of biodiversity and irrigation, especially under semiarid and arid climates.

Sustaining Energy Subsidies

Agricultural intensification since the Green Revolution in the 1950s and 1960s has become fossil fuel (energy) intensive due to increased inputs of synthetic fertilizers, biocides (insecticides, herbicides, and fungicides), and irrigation water contributing to global climate change (Table 12.2). Governments worldwide could eliminate an estimated $700 billion in annual subsidies that stimulate the destruction of ecosystem goods and services. For example, Indonesia spent $150 million annually to subsidize pesticide use in rice fields in the mid-1980s, poisoning humans and wildlife in the food web; the country ended the subsidies in 1986 with no ill effects on rice production.

Fossil fuel-derived energy used in pumping irrigation water amounts to annual emissions of 81–305 g CO_2 m^{-2} (22–83 g C m^{-2}) (C = 3.67 CO_2) for irrigated agricultural lands in the United States (Schlesinger 1999). Moreover, because of high Ca and CO_2 content of groundwater of arid regions (as much as 1% Ca and CO_2 versus 0.036% in the atmosphere), use of such water for agricultural irrigation in arid lands would release 8.4–15 g C m^{-2} yr^{-1} to the atmosphere from the formation of soil carbonate ($CaCO_3$) (Schlesinger 1999).

In the whole chain of manufacture, transport, and consumption, the fertilizer industry generates an atmospheric greenhouse gas emission of 1.436 mol of CO_2–C per mole of N produced (Schlesinger 1999). Emissions by the fertilizer industry include CO_2, NH_3, nitric and nitrous oxides (NO_x and N_2O), hydrogen sulfide (H_2S), sulfur dioxide (SO_2), sulfur trioxide (SO_3), fluorine (SiF_4 and HF), and radiation (from phosphogypsum). They cause global warming, ozone depletion, acidification of soils, depletion of oxygen in water (eutrophication), toxicity for organisms (such as cadmium), and aerosols.

More than 99% of world N fertilizer production is based on ammonia (NH_3), and about 77% of world ammonia production capacity is currently based on a chemical process using natural gas (methane) as a feed stock (IFA, UNEP, and UNIDO 1998) (Table 12.3). This method of NH_3 production emits 1.15–1.3 tons of CO_2, 1.3 kg of NO_2, and 0.01 kg of SO_2 per ton of NH_3 produced (IFA, UNEP, and UNIDO 1998) (Table 12.4). Fertilizer production is energetically expensive, with current world production consuming 4.4 billion GJ yr^{-1} (ca. 1.2% of the world's energy), of which 92.5% is used for

Table 12.3
Comparative Consumption of Different Energy Resources as Feedstock and Fuel for Ammonia (NH_3) Production

Feedstock and Fuel	Requirements per 1 t of Ammonia Production	Energy ($\times 10^6$ kcal)
Natural gas	873 m^3	7.0
Coal	1.54 t	9.8
Oil	0.87 t	8.5

Source: From IFDC-International Fertilizer Development Center, 1999, Fertilizer Use by Crop, 4th ed. Jointly with IFA-International Fertilizer Industry Association, and FAO. Rome: AFA, IADC, and FAO.FDC. 72 pp.

N, 3% for P_2O_5, and 4.5% for K_2O (Konshaug 1998), exclusive of the energy consumption caused by fertilizer transport and application. Recent discourses provide detailed information on the perspectives of energy efficiency of fertilizer production (Gellings and Parmenter 2004, Rafigul, Weber, Lehman, and Voss 2009).

Biocides: Use, Persistence, and Health Effects

Biocides (or pesticides) are chemicals to kill organisms of all kinds that we consider undesirable (**pests**) due to our interference with the natural regulation and control of the population in agricultural and urban–industrial ecosystems. Biocides include fungicides (fungus killers), herbicides and weedicides (plant killers), insecticides (insect killers), rodenticides (rat and mouse killers), etc. These inorganic and organic chemicals are used mostly in agricultural production to compensate for the function of natural biodiversity. They can persist in the environment for many decades and magnify exponentially in concentration based on the food web relations among humans, plants, and animals, thus harming the health of humans and other species.

Until the 1940s, naturally occurring elements and compounds were used to deter the consumption of and damage to crops by pests.

Table 12.4
N–P–K Fertilizer Consumption by Regions

Region	N–P–K Fertilizer Consumption ($\times 10^5$ t)[a,b]		Change (%)
	1961	1997	
Africa	7.3	36.4	399
Asia and former Soviet Union	67.6	738.3	992
Europe	141.1	216.1	53
North and Central America	87.4	256.3	193
Oceania	9.7	28.6	195
South America	6.2	85.0	1271
Global total	319.5	1361.0	326

[a] FAO (Food and Agriculture Organization of the United Nations). 1998. FAOSTAT. Rome, Italy.
[b] IFA (International Fertilizer Industry Association). 1998. *Nitrogen, Phosphate and Potash Statistics*. Paris: IFA.

These included such compounds as those of arsenic, lead, and mercury and those compounds with insecticidal properties that were extracted from other plants (e.g., cocaine from cocoa leaves, nicotine extracts from tobacco leaves, and pyrethrum from chrysanthemum flowers). These are also referred to as **first-generation pesticides.** Although some compounds break down over time (**biodegradable**), others do not (**nonbiodegradable**). The latter, as we have seen in Chapter 6, accumulate through the food chain (bioconcentration and biomagnification).

The second-generation pesticides began when the Geigy chemist Paul Müller began trials with a synthetic insecticide DDT (dichloro-diphenyl-trichloroethane) to combat the potato beetle in Switzerland. DDT was later used for eliminating the cotton boll weevil and other insects and for this Paul Müller received the Nobel Prize in chemistry in 1948. Wrote Woodwell (1984):

> It was a miracle compound: highly toxic to insects, virtually insoluble in water, and of low toxicity to mammals, it did wonderful things. It eliminated lice and typhus in fighting armies in the Second World War, Colorado potato beetle, cotton boll weevil, it eliminated malaria in the tropics, it improved yields of agricultural crops and forest trees, it was a revolution in public health.

Table 12.5
Estimated Annual Pesticide Use

Country/Region	Pesticide Use ($\times 10^6$ t)
United States	0.5
Canada	0.1
Europe	0.8
Other, developed	0.5
Asia, developing	0.3
Latin America	0.2
Africa	0.1
Total	2.5

Source: From Pimentel, D. 1995. In *The Literature of Crop Science*, ed. W. C. Olsen, 49–66. Ithaca, NY: Cornell University Press.

Thus began the era of second-generation pesticides, marked by the use of DDT and its relatives. DDT was nonbiodegradable, as illustrated in Chapter 7, so it accumulated in the environment. Currently, 3 million tons of biocides comprising about 1,600 different chemicals are applied annually throughout the world (Pimentel, Greiner, and Bashore 1998) (Table 12.5). However, the use of biocides is very inefficient because only 0.1% of the pesticides applied actually are estimated to reach the target pests, causing the remaining 99.9%

Table 12.6
Total Estimated Environmental and Social Costs for Pesticides in the United States

Negative Effects	Cost ($ Millions yr⁻¹)
Human pesticide poisonings	933
Domestic animal deaths and contaminated livestock products	31
Loss of natural enemies	520
Pesticide resistance	1,400
Losses of honeybees and pollination	320
Losses of crops	959
Fishery losses	56
Bird losses	2,100
Surface water monitoring	27
Groundwater contamination	1,800
Government regulations and monitoring	180
Total	8,326

Source: From Pimentel, D. et al. 1998. In *Environmental Toxicology: Current Developments,* ed. J. Rose, 121–150. Amsterdam: Gordon and Breach Science Publishers.

to damage the environment (Pimentel et al. 1998; Table 12.6). Biocides are ineffective in that roughly 40% of food grown throughout the world is lost to pests and disease. Despite the 10-fold increase in insecticide use in the United States from 1945 to 1989, total crop losses from insect damage nearly doubled from 7 to 13% (Pimentel et al. 1998). About 100 of at least 1 million known insects cause about 90% of the damage to food crops (some estimates suggest 10 quintillion insects live on Earth at any one time). The World Health Organization (WHO) estimates that pesticides poison about 1 million people every year, causing up to 20,000 deaths.

During the last decade, agricultural pesticide use composed 75–80% of the total use in the United States; the rest was used on urban pest control, roadways, landscaping, and forests (EPA). One-quarter of all pesticides used in the United States is applied in California, although the cultivated land in California represents only 2–3% of the U.S. total. According to Pesticide Action Network (Kegley et al. 2000), about one-third of the total pesticides used in California in any given year are toxic

to humans as immediate (acute) poisons and/or chronic toxins, including carcinogens, reproductive or developmental toxicants, neurotoxins, or groundwater contaminants.

Intensity of reported pesticide use (active ingredients applied per unit time and space) in California increased 60% from 16.1 kg/ha in 1991 to 25.7 kg/ha (23 lb/acre) in 1998, with a total 51% increase in the amount of pesticides from 58.5×10^3 to 88.4×10^3 metric tons during the same period. The most toxic pesticides used in California include acute poisons (e.g., the fumigants, methyl bromide, chloropicrin, and sulfuryl fluoride), carcinogens (e.g., the soil fumigants, metam sodium, and Telone), reproductive or developmental toxicants (e.g., the soil fumigants, methyl bromide, and metam sodium), neurotoxins (e.g., organophosphates and carbamates), and groundwater contaminants (e.g., the herbicides of diuron and norflurazon and the insecticide of aldicarb). A 127% increase in use of carcinogenic pesticides has resulted in a concurrent increase in age-adjusted incidence of cancers in California during this 8-year period (including childhood leukemia, brain tumors, non-Hodgkin's

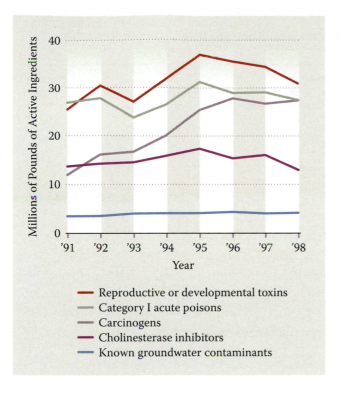

Figure 12.5 Trends in the use of different categories of California bad-actor pesticides, 1991–1998. (Kegley, S., S. Orme, and L. Neumeister. 2000. Hooked on Poison: Pesticide use in California, 1991–1998. Pesticide Action Network available online at http://www.panna.org/files/hookedAvail.dv.html)

lymphoma, testicular cancer, and some forms of breast cancer).

Total use of the most toxic pesticides increased 44% from 22.9 thousand tons in 1991 to 32.9 thousand tons in 1995, but declined 12% to 29 thousand tons in 1998 due to reduction in the use of soil fumigants, methyl bromide, and metam sodium and regulatory and public pressures from health concerns (Figure 12.5) (Kegley, Orme, and Neumeister 2000). The remaining pesticides used in California include fungicide sulfur (responsible for most farm worker poisonings), petroleum oils as insecticides, major air pollutants, and endocrine-disrupting pesticides.

Endocrine-disrupting chemicals (EDCs; also known as hormone disruptors or xenoestrogens) are a group of chemicals that are able to imitate or modify the action of natural hormones of organisms. EDCs include such industrial chemicals as alkylphenols, bisphenol A, polychlorinated biphenyls (PCBs), and phthalates and pesticides such as DDT and dioxin. People are becoming increasingly concerned about the potential adverse reproductive and developmental effects of EDCs on humans and wildlife.

Adverse reproductive and developmental human health outcomes that have been linked to these agents include breast, ovarian, and endometrial cancers; endometriosis; infertility; prolonged time to pregnancy; increased spontaneous abortion rates; decreased ratios of male-to-female birth; increased testicular cancer in young men; increased prostate cancer; decreased semen quality; increased frequency of testicular maldescent; increased prevalence of hypospadias; and precocious puberty. The variety of wildlife responses attributed to EDCs is most obvious in species inhabiting areas contaminated extensively with chemicals (Table 12.7).

There is now a global treaty—the Stockholm Convention on Persistent Organic Pollutants (POPs). This convention says that "POPs are chemicals that remain intact in the environment for long periods, become widely distributed geographically, accumulate in the fatty tissue of living organisms and are toxic to humans and wildlife." The convention targets the "dirty dozen," which include

- certain insecticides, such as DDT and chlordane, which were once commonly used to control pests in agriculture and in building materials, as well as to protect public health and are now considered POPs;
- PCBs, which were used in hundreds of commercial applications, such as in electrical, heat transfer, and hydraulic equipment, and as plasticizers in paints, plastics, and rubber products; and
- certain chemical by-products, such as dioxins and furans, which are produced unintentionally from most forms of combustion, including municipal and medical waste incinerators, open burning of trash, and industrial processes.

This treaty was ratified by the United States on May 23, 2001. We shall discuss this in more detail in Chapter 21 on environmental policy.

Table 12.7
Selected Examples of Wildlife Reproductive and Developmental Abnormalities Attributed to EDCs (Endocrine Disrupting Chemicals)

Species	Site	Observation	Contaminant
Invertebrates: gastropods	Marine	Pseudohermaphroditism, imposex, intersex sterility, population declines	tributyltin
Fish			
Trout, roach	English rivers	Hermaphroditism, vitellogenin in males altered testis development	Sewage effluent
Lake trout	Great Lakes	Early life stage mortality, deformities, blue sac disease	Dioxin and related AhR agonists
White sucker	Jackfish Bay, Lake Superior, MI	Reduced sex steroid levels, delayed sexual maturity, reduced gonad size	Bleached kraft pulp, mill effluent
Flatfish	Puget Sound, WA	Decreased hormone levels, reduced ovarian development	PAHs
Reptiles			
Alligator	Lake Apopka, FL	Decreased viability, abnormal gonadal development, decreased phallus size	DDE
Birds			
Waterbirds	Global	Egg shell thinning, mortality, developmental abnormalities	DDE
Raptors: waterbirds	Great Lakes	In ovo and chick mortality, growth retardation, deformities	PCBs, AhR agonists
Raptors: western gulls	California	Abnormal mating behavior, supranormal clutch size, skewed sex ratios	DDT and its metabolites
Mammals			
Mink	Great Lakes	Population decline, developmental toxicity, hormonal alterations	PCBs and dioxins

Source: From Council for Agricultural Science and Technology. 2000. *Estrogenicity and endocrine disruption.* Issue paper number 16, July 2000.

Disruption of Biogeochemical Cycles

Salinization

Human-induced accumulation of dissolved (soluble) salts in the soil (known as **secondary salinization**) is a major concern, especially in arid and semiarid regions of the world, where clearance of vegetation together with inadequate leaching, drainage, and dilution of salts disrupts the hydrological regime. The high content of dissolved salts (such as chlorides and sulfates of sodium, calcium, magnesium, and potassium) in irrigation water coupled with high evapotranspiration rates causes salinization/alkalinization over time that eventually reaches levels deleterious to plant growth (Figure 12.6).

Salt accumulation near the soil surface results in two main types of salt-affected soils in that sodium carbonates and bicarbonates lead to **alkaline (sodic) soils,** while sodium chloride and sodium sulfates lead to **saline soils** (Table 12.8). As irrigated agriculture increases, human-induced salinization and alkalization increase, adversely affecting about 20% of irrigated lands and 77 million ha worldwide, and this is spreading at a rate of up to 2 million ha yr^{-1} (Ghassemi, Jakeman, and Nix 1995; Umali 1993; Oldeman, Hakkeling, and Sombroek 1991) (Figure 12.7). Subsurface

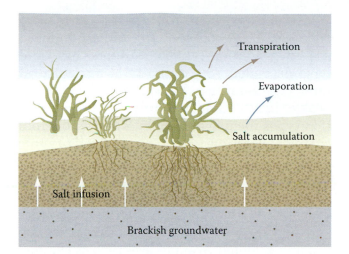

Figure 12.6 The process of waterlogging and salinization. (Hillel, D. 1998. *Environmental Soil Physics*. New York: Elsevier Science Publishers.)

runoff and leaching from saline soils can cause increased salinization levels in surface and groundwater resources. In rain-fed croplands, where precipitation is high enough to grow crops with little or no additional irrigation, the absence of adequate drainage systems results in the rise of water tables and waterlogging (sometimes known as wet deserts).

The Aral Sea, which was once the world's fourth-largest inland body of water, is fed by annual flows of ca. 50 km³ from the rivers Amu Dar'ya and Syr Dar'ya, yet it illustrates the environmental consequences of overpumping and inefficient irrigation. Since 1960, irrigation for cotton and rice production in Kazakhstan and Uzbekistan has reduced the surface area of the Aral Sea by nearly 50% and its volume by 75%, more than tripling its salinity in only 33 years (Postel 1997). This has resulted in waterlogging, salinization, and losses of fishing industry and productive lands due to salts blown from the former sea bed (WMO 1997).

Soil Erosion

Soil erosion can occur either naturally or anthropogenically by the principal agents of water and wind. Natural (geological) soil erosion is the long-term process that wears down mountains and builds up plains and deltas during millions of years. Human activities, such as removal of vegetation, overgrazing, construction, and conventional tillage, accelerate the rate at which soil components, especially surface litter and topsoil, are removed from their

Table 12.8
Extent of Natural and Human-Induced Salt-Affected Soils[a] in Regions of the World

Regions	Saline	Sodic	Total
North America	6.2	9.6	15.8
Mexico and Central America	2.0	—	2.0
South America	69.4	59.8	129.2
Africa	122.9	86.7	209.6
South Asia	82.2	1.8	84.0
North and central Asia	91.4	120.1	211.4
Southeast Asia	20.0	—	20.0
Australasia	17.6	340.0	357.6
Europe			50.8
Global total	411.7	617.9	1080.3

Sources: From Szabolcs, I. 1989. *Salt-Affected Soils.* Boca Raton, FL: CRC Press; Abrol, I. P., J. S. P. Yadav, and F. I. Massoud. 1988. *Salt-Affected Soils and Their Management.* FAO Soils Bulletin 39, Rome, Italy.

[a] × 10⁶ ha.

Table 12.9
Generalized Relationships in the Tropics between Mean Annual Soil Erosion and Type of Ground Cover

Type of Ground Cover	Scale of Magnitude of Soil Erosion
Dense forest	1
Cultivation with copious mulch	1
Savanna or grassland in good condition	10
Rapidly developing cover plants	100
Early stage of plantation	100
Bare ground	1000

Source: After Roose, E. in UNESCO/UNEP/FAO, 1978, p. 264; UNESCO/UNEP/FAO (1978). *Tropical Forest Exosystems*. UNESCO, Paris, France.

original sites, transported, and eventually deposited at a new location (Table 12.9).

There are three types of water erosion: sheet erosion, rill erosion, and gully erosion. The loss of relatively uniform layers of topsoil by a wide flow of surface water down a slope or across a field is known as **sheet erosion. Rill erosion** occurs when the surface water forms fast-flowing little streams that cause shallow, narrow channels in the soil. **Gully erosion** is the more severe form of rill erosion in which huge rivulets of fast-flowing water carve deeper, wider channels in the soils until they become ditches and gullies.

Recent agricultural practices are responsible for 29% of the total human-induced soil degradation (1,965 million ha) and have degraded 38% of the total agricultural land (1,475 million ha) on the global scale (Table 12.10) (Oldeman 1994). Soil degradation types include erosion by water and wind, chemical degradation (e.g., loss of soil organic matter and nutrients, salinization, pollution, and acidification), and physical degradation (e.g., compaction and waterlogging) (Table 12.11). Erosion is by far the most widespread type of soil degradation worldwide, causing a global average loss rate of the fertile topsoil of 30 tons ha^{-1} yr^{-1} (13 tons ha^{-1} yr^{-1} in the United States and 40 tons ha^{-1} yr^{-1} in China) (Pimentel 1993). In the U.S. upper Midwest, precious topsoil is lost by wind erosion in both summer and winter (Figure 12.8). A single blizzard storm in late winter–early spring may blow away 30 million or more tons of topsoil (Wali 1975).

The renewal rate of topsoil is approximately 2.5 cm per 500 years under agricultural conditions and about 1.5 cm per 3,000 years under natural conditions (Pimentel 1999b). Globally, more than 10 million ha of productive croplands are severely degraded and abandoned annually due to soil erosion in excess of soil formation (Pimentel 1999b). Factors contributing to soil erosion include erosivity (the capacity of erosive agents to cause erosion in

Table 12.10
Global and Continental Extent[a] of Human-Induced Soil Degradation for Various Ecosystems

	Agricultural Land			Permanent Pasture			Forest and Woodland		
	Total	Degraded	%	Total	Degraded	%	Total	Degraded	%
Africa	187	121	65	793	243	31	683	130	19
Asia	536	206	38	978	197	20	1273	344	27
S. America	142	64	45	478	68	14	896	112	13
C. America	38	28	74	94	10	11	66	25	38
N. America	236	63	26	274	29	11	621	4	1
Europe	287	72	25	156	54	35	353	92	26
Oceania	49	8	16	439	84	19	156	12	8
World	1475	562	38	3212	685	21	4048	719	18

Source: From Oldeman, L. R. 1994. In *Soil Resilience and Sustainable Land Use,* ed. D. J. Greenland and I. Szabolcs, 99–118. Wallingford, Oxford, England: CAB International Publishers.

[a] Million hectares.

Table 12.11
Estimates of Human-Induced Soil Degradation Worldwide

Type and Degree of Soil Degradation ($\times 10^6$ ha)	Global Estimates	Tropics[a]	Dryland Zone[b]	Humid Zone[c]
Water erosion	1094	920	478	615
Wind erosion	548	472	513	36
Chemical degradation	240	213	111	130
Physical degradation	83	46	35	48
World total	1965	1651	1137	829
Light	749	671	488	261
Moderate	910	689	509	401
Severe and extreme	305	290	139	166

Sources: From Oldeman, L. R. 1994. In *Soil Resilience and Sustainable Land Use,* ed. D. J. Greenland and I. Szabolcs, 99–118. Wallingford, Oxford, England: CAB International Publishers; Oldeman, L. R., and G. W. J. van Lynden. 1998. In *Methods for Assessment of Soil Degradation,* ed. R. Lal, W. H. Blum, C. Valentine, and B. A. Stewart, 423–441. New York: CRC Press.

[a] Includes Africa, Asia, South and Central America, and Australia.
[b] Dryland zone is defined as a climatic region with an annual precipitation/evaporation ratio of 0.65 or less (UNEP 1992).
[c] Includes North and Central America.

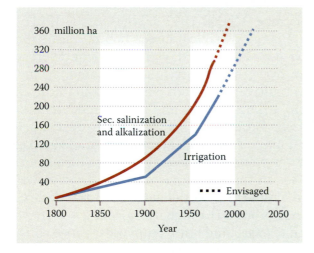

Figure 12.7 Global development of irrigation and secondary salinization of soils. (From Szabolcs, I. 1997. In *Methods for Assessment of Soil Degradation,* R. Lal et al. eds., 253–264. Boca Raton: CRC Press.)

given circumstances), erodibility (the vulnerability of soil to erosion), land form (slope and aspect), land cover, and land management (Figure 12.9).

In **conventional-tillage farming** using a moldboard plow followed by a disk, the upper 15 cm of the soil is inverted and broken up with previous crop residues and any cover vegetation is plowed under. This tillage method results in the removal and/or burial of crop residues in the field and not only leaves the soil surface vulnerable to soil erosion but also increases the decomposition and mineralization of soil organic matter (SOM), which in turn contributes to the loss of SOM, and the emission of greenhouse gases (e.g., CO_2, N_2O, and CH_4). Within the first 50 years of tilling, an average of 40–50% of the original level of SOM is lost. The contribution of agriculture to global climate change and sustainability-oriented agricultural management practices will be discussed in Chapter 19 and Section C, respectively.

Losing the topsoil rich in SOM reduces the formation of soil aggregates, water- and nutrient-holding capacities, and, as a consequence, soil fertility. The resulting sediments have significant effects on water quality and quantity, which will be discussed in Chapter 16. Loss and reduced quality of agricultural lands due to soil erosion make the challenge of meeting food and fiber needs of rapidly growing human populations even worse. Mismanagement and overuse of agricultural ecosystems transcend their human-made physical boundaries by impoverishing surrounding ecosystems with which agricultural land use exchanges energy

Figure 12.8 Topsoil being blown away from plowed agricultural fields in the upper Midwest in summer (left) and ending winter (right). The mixture of snow and rich organic matter is referred to by the native people as "snirt." (snow + dirt) (Photos by M. K. Wali.)

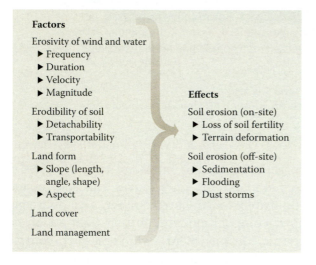

Figure 12.9 Factors that create susceptibility for soil erosion and their multiple effects.

and materials. Agricultural production is the primary source of income, employment, and the alleviation of poverty, especially in developing countries; thus, degradation and loss of productive agricultural lands reduce both food security and economic stability.

Desertification

Desertification, a term introduced by Aubreville in 1949 (UNEP 1997; Barrow 1995), can be described as a human-induced process that permanently diminishes biological productivity, diversity, and self-regenerative capacity of ecosystems, especially in arid, semiarid, and dry, subhumid regions (drylands) of the globe. With a limited supply of water, these areas (which cover about 47.2% of the Earth's ice-free land surface) are more susceptible to desertification than humid regions because of their low resistance to and slow recovery rates from disturbance (Figure 12.10). About 20% (some $1,035 \times 10^6$ ha) of the world's vulnerable drylands and the livelihoods of more than a billion people are directly impoverished by desertification in this zone (UNEP 1997; UNEP/ISRIC 1991).

Primary driving forces behind desertification (degradation of drylands) include deforestation, overgrazing, agricultural mismanagement (overcultivation and salinization/alkalinization), urban sprawl, and urban–industrial pollution. Desertification results in dust storms that may carry large amounts of eroded top soil for very long distances. The dust storms of the 1930s in the area known as the Dust Bowl in the United States demonstrated the desertification process through overgrazing and conversion of grasslands into croplands, which caused droughts, wind erosion, and, consequently, loss of ecosystem productivity.

Animal Agriculture

In the foregoing discussion we emphasized crop growth in relation to soils and the energy subsidies crops receive in their multiple manifestations. A large component of agriculture is the animal industry—that is, animals such as beef cattle, chickens, turkeys, lambs, goats,

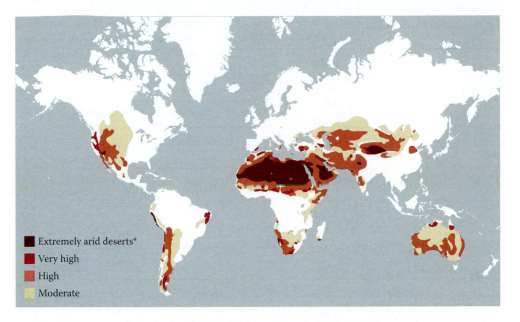

■ Extremely arid deserts*
■ Very high
■ High
■ Moderate

Figure 12.10   Worldwide distribution of lands at risk for desertification. Extremely arid deserts (*) have little or no plant cover or organic soil and infrequent human activity; thus, in effect, there is no risk of further desertification. (*UNEP—World Atlas of Desertification.* 1997. N. Middleton and D. Thomas, eds. London: Arnold; New York: Co-published with John Wiley.)

and others. Worldwide, notes a recent comprehensive report (FAO 2006), livestock constitutes 40% of agricultural gross domestic product and employs 1.3 billion people. "The global production of meat is projected to more than double from 229 million metric tons in 1999/01 to 450 million metric tons in 2050 and that of milk to grow from 580 to 1,043 million metric tons." Yet, as the report emphasizes, "livestock's contribution to environmental problems is on a massive scale." The report suggests that livestock production should receive major attention from policy-makers when they are "dealing with problems with land degradation, climate change, water shortage and water pollution, and loss of biodiversity."

Challenges of Food Security

Since 1970, global food production has doubled, and livestock production has tripled. Despite the increase, we still fail to feed 800 million people of the current human population even as the world's human population grows at a rate of 1.3% annually. **Food security** refers to production of adequate and nutritious food

that is accessible to everyone, especially the poor, by securing the natural resource base of agricultural productivity in the long term (e.g., Pinstrup-Andersen and Pandya-Lorch 1996; Swaminathan 1996). The State of Food Security in the World report (FAO 1999) reveals that the number of undernourished people is 800 million in developing countries, 8 million in the industrialized countries, and 26 million in countries in transition (Table 12.12).

The resulting starvation, undernutrition, micronutrient deficiency, and nutrient-depleting sickness pose a severe threat to the health of at least one in every six people and about one of every three children (Table 12.13). Sub-Saharan Africa is home to almost a quarter of the developing world's hungry people. The problem varies in severity across the continent. Although West Africa has the largest total population of any of the African regions, it has the fewest undernourished. By contrast, East Africa, with a slightly smaller total population, has more than twice as many undernourished people. The numbers in central and southern Africa are also proportionately larger, although both have much smaller total populations.

A majority of undernourished people in developing countries live in Asia and the Pacific. This region is home to 70% of the

Table 12.12
Prevalence of Undernourishment by Regions

Region	Hunger (Millions of People in 1995)
Sub-Saharan Africa	179.6
Central Africa	35.6
West Africa	31.1
Southern Africa	35
East Africa	77.9
Near East and North Africa	32.9
Near East	27.5
North Africa	5.4
Latin America and Caribbean	53.4
Caribbean	9.3
Central America	5.6
South America	33.3
North America	5.1
Asia and Pacific	525.5
East Asia	176.8
Oceania	1.1
Southeast Asia	63.7
South Asia	283.9
Total	791.4

Source: From FAO. 1999. State of Food Insecurity in the World 1999. Rome, Italy. Retrieved from http://ftp.fao.org/docrep/fao/007/x3114e/x3114e00.pdf. Accessed February 13, 2009.

total population of the developing world and accounts for almost two-thirds (526 million) of the undernourished people as well. India alone has more undernourished people (204 million) than all of sub-Saharan Africa combined. With India's neighbors added in, the South Asian region accounts for more than one-third of the world's undernourished total (284 million). Another 30% (240 million people) live in Southeast and East Asia and more than 164 million in China.

The Green Revolution, with the introduction of new wheat and rice cultivars, resulted in large increases in crop production. It undoubtedly gave more than the "breathing space" in food production envisioned by its proponent.[3] Starvation was avoided for many in South Asia and Latin America, where the new varieties and new methods were enthusiastically adopted. The Green Revolution was a landmark in agricultural development. So long as water for irrigation, fertilizers and pesticides could be assured, so would be the bountiful harvests. One problem-one solution became a success.

After less than a couple of decades, concerns were being voiced. On hunger in Africa: "Despite all our achievements, I think it is fair to say that we failed in Africa along with everybody else. ... We have not fully understood the problems. We have not fully designed our projects to fit the agroclimatic conditions of Africa, and the social, cultural and political framework of African countries."[4]

On the effects in India, the *Indian Express* reported:

The Green Revolution has transformed the nature of agriculture in this country [India]. It is no longer a means of subsistence, a livelihood. It means seeds to be purchased, fertilizers to be purchased, pesticides, weedicides, micronutrients, bacteria promoters, etc. all to be purchased. ... All the forests are gone. How can one expect rain? This is the response of most villagers: "The future is bleak. On one hand, drought and what is worse, drinking water scarcity. On the other, in the irrigated areas, water tables are rising rapidly, threatening the land ... Clearly the Green Revolution has failed, modern technology has slipped somewhere, development has backfired."[5]

These words, now more than 20 years old, reflect the articulation by a working farmer of the fundamental ecological connections among forests, the hydrologic cycle, water availability, crop growth, drought and salinization of soils, and, of course, monetary affordability. Therein lie some of the challenges of agriculture in the twenty-first century. To the shrinking arable land in the world and its decreasing productivity, the prices of energy inputs will loom large. Two of these challenges—one old and the other new—are provided briefly here as examples.

Water Availability

The availability of water for agriculture will emerge as *the* prime limiting factor for crop production in many parts of the world. Global

Table 12.13
Recent Estimates of World Hunger

Dimension of Hunger	Population Affected		Year	Source
	(Millions)	%		
Starvation				
Famine (population at risk)	15–35	0.3–0.7	1992	WHP
Related deaths per year	0.15–0.25	≤0.01	1990s	RPG
Undernutrition (Chronic and Seasonal)				
Household	786	20	1988–90	FAO
Children	184	34	1990	ACC/SCN
Micronutrient Deficiencies				
Iron deficiency (women aged 15–49)	370	42	1980s	ACC/SCN
Iodine deficiency (goiter)	655	12	1980s	ACC/SCN
Vitamin A deficiency (children under 5)	14	3	1980s	ACC/SCN
Nutrient-Depleting Illness				
Diarrhea, measles, malaria (deaths of children under 5)	4.7	0.7	1990s	UNICEF
Parasites (infected population)[a]				WB
Roundworm	785–1300	15–25	1980s	
Hookworm	700–900	13–17	1980s	
Whipworm	500–750	10–14	1980s	

Source: From Kates, R. W. 1996. *Consequences* 2 (2): 1–11.
Notes: ACC/SCN = Advisory Committee on Coordination–Subcommittee on Nutrition of the United Nations; FAO = Food and Agriculture Organization of the United Nations; RPG = Refugee Policy Group; WHP = World Hunger Program, Brown University; WB = World Bank.
[a] Includes people expected to have multiple infections.

climate change will exacerbate the problem, largely in areas where climate will get even drier than it is now. To a lesser extent, it may pose problems where the precipitation amounts will increase but pour down as intense rain showers.

Hoekstra and Chapagain (2007) provide a useful study in which "the water footprint of a country is defined as the volume of water needed for the production of goods and services consumed by the inhabitants of a country." Their calculations included domestic water availability and water that is imported from other countries. For the period 1997–2001, average per-capita water use was 2,480 m^3 yr^{-1} for the United States, 700 m^3 yr^{-1} for China, and 1,240 m^3 yr^{-1} as the global average. These authors consider four facts important: (a) local climatic conditions, (b) volume and (c) pattern of consumption, and (d) water use efficiency. The footprints of growing crops (Figure 12.11) are instructive: for example, rice, which is the staple for nearly 2 billion people, has the largest water footprint.

Biofuel Dilemma

Converting crops to biofuels has emerged recently as a major national and global problem. Of particular emphasis has been the conversion of corn (maize) to ethanol. This conversion received a tremendous impetus when the U.S. Congress, mandating production of 15 billion (U.S.) gallons of biofuels by 2015, provided lucrative subsidies. To achieve that level of biofuel production, estimates are that

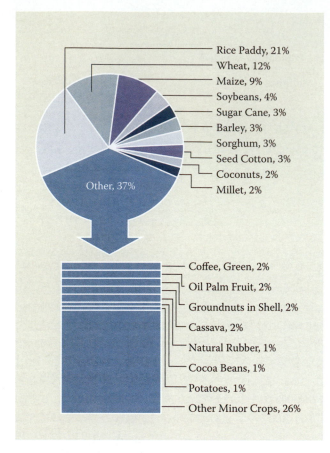

Rice Paddy, 21%
Wheat, 12%
Maize, 9%
Soybeans, 4%
Sugar Cane, 3%
Barley, 3%
Sorghum, 3%
Seed Cotton, 3%
Coconuts, 2%
Millet, 2%

Other, 37%

Coffee, Green, 2%
Oil Palm Fruit, 2%
Groundnuts in Shell, 2%
Cassava, 2%
Natural Rubber, 1%
Cocoa Beans, 1%
Potatoes, 1%
Other Minor Crops, 26%

Figure 12.11 Use of water for growing different crops, also referred to as the water footprint of crops. (Hoekstra, A. Y., and A. K. Chapagain. 2007. Water footprints of nations: Water use by people as a function of their consumption pattern. *Water Resources Management* 21: 35–48.)

40% of the U.S. corn crop would be needed. An overwhelming number of commentaries, editorials, and scientific papers have questioned the wisdom of converting an energy-intensive crop like corn into biofuel.[6]

The second concern is what we might call the "food void" that will be created by such conversion. The concerns have been voiced all over the world. In the United States, for example, corn not only is food and feed, but its derivatives are also the mainstay of many other products. Thus, food price increases are inevitable and, indeed, they are occurring (Elliott 2008; Mitchell 2008) (Figure 12.12). On July 23, 2008, the governor of Texas, citing escalating feed prices for cattle (an important industry in Texas), requested the U.S. Environmental Protection Agency to lower

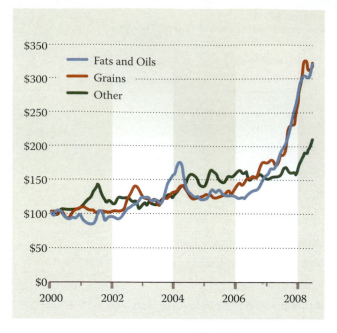

Figure 12.12 Rise in prices of fats and oil, grains, and other commodities from January 2000 to January 2008 (Mitchell, D. 2008. A note on rising food prices. Policy Research Working Paper 4682. Washington, D.C.: The World Bank.)

the production target.[7] The rise in prices was acknowledged but the request was denied.

How do we ensure sustained agricultural productivity and food security for all living beings? The importance of food production is not debatable; therefore, what must be debated seriously is disconnection among increasing human population growth, industrial agriculture, energy inputs, and conflicts in land use. Jason Clay sums it best:

> There are two basic truths that will shape the future of farming—there is a steady increase in the consumption of food and fiber produced by agriculture, while at the same time there is a steady decline in the quality and productivity of soil around the world. The two trends are on a collision course. This collision will not be avoided by a single solution. (Clay 2004, p. vii)

Along with the preceding considerations, this collision will be exacerbated if crop production decreases occur with climate change (Evrendilek and Wali 2001, 2004; Cline 2007). The necessary strategies for sustainability of crops will be discussed in Chapter 26 on sustainable development.

Notes

1. The following table summarizes the structural changes in U.S. agriculture from 1900 to 2000 (Dimitri, C., and A. Effland, 2005. Milestones in U.S. farming and farm policy. Amber Waves. Available at www.ers.usda.gov/Amber Waves/June05/DataFeature accessed February 16, 2009, modified by Miller 2007).

Structural changes in U.S. agriculture, 1900 to 2000.[a]

Category	1900	1930	1945	1970	2000
No. of farms ($\times 10^6$)	5.7	6.3	5.9	2.9	2.1
Farm size, ha	59	61	79	152	179
Commodities per farm	5.1	4.5	4.6	2.7	1.3
Farm share of population, %	39	25	17	5	<1
Rural share of population, %	60	44	36	26	21
Work force in agriculture, %	41	22	16	4	1.9
Agriculture's share of GDP, %	na[b]	7.7	6.8	2.3	0.7
Farmers working off-farm, %	na	na	27	54	93

[a] *Source:* Modified from Dimitri and Effland, 2005.
[b] na. not available.

2. In his Nobel Peace Prize Lecture, "The Green Revolution, Peace, and Humanity," on December 11, 1970, Norman Borlaug noted: "The green revolution has won a temporary success in man's war against hunger and deprivation; it has given man a breathing space. If fully implemented, the revolution can provide sufficient food for sustenance during the next three decades. But the frightening power of human reproduction must also be curbed; otherwise, the success of the green revolution will be ephemeral only. Most people still fail to comprehend the magnitude and menace of the "Population Monster."
3. See also Evenson and Gollin (2003).
4. Stern, quoted by J. Walsh (see references).
5. *Indian Express,* New Delhi, India, December 27, 1987.
6. Papers have appeared in such journals as *Nature, Proceedings of the National Academy of Sciences,* and *Science.*
7. "Gov. Rick Perry of Texas is asking the Environmental Protection Agency to temporarily waive regulations requiring the oil industry to blend ever increasing amounts of ethanol into gasoline. A decision is expected in the next few weeks.
   "Mr. Perry says the billions of bushels of corn being used to produce all that mandated ethanol would be better suited as livestock feed than as fuel.
   "Feed prices have soared in the last two years as fuel has begun competing with food for cropland.
   "When you find yourself in a hole, you have to quit digging," Mr. Perry said in an interview. "And we are in a hole" (Streitfeld 2008).

Questions

1. Discuss the importance of productive agricultural lands to human well-being.
2. What human activities can lead to the loss and degradation of prime farmland?

3. Discuss the implications of the Green Revolution for food quantity and quality.
4. What natural processes in croplands are altered through the use of fertilizers and pesticides? How does this influence food quantity and quality?
5. Discuss the environmental impacts that have resulted from an increased use of fossil fuels and biocides in agricultural production.

Land Use— Forest Resources

13

Forest Ecosystems in Perspective

Forests, as we have seen in several preceding chapters, not only represent some of the most fascinating ecosystems in world, but also have a tremendous role in the functioning of many other ecosystems. These unique life forms, with relatively long life and the capability of producing wood over many years, support a whole mosaic of smaller ecosystems dependent on several tree parts and underlying vegetation. Our descriptions of the typical forest ecological processes that result in the development of various types of soils (Chapter 4), their role in biogeochemical cycles (Chapter 6), their community structure (Chapter 9), and global distribution in several biomes (Chapter 10) should provide the necessary background to evaluate the contribution of forest ecosystem services to mankind and other biota (Figure 13.1). Here, we examine the extent of global and national forest resources, the historical and contemporary utilization of forests by humans, and the current state of the forest resource base, and describe changes that have been and are brought about by natural pests and pathogens.

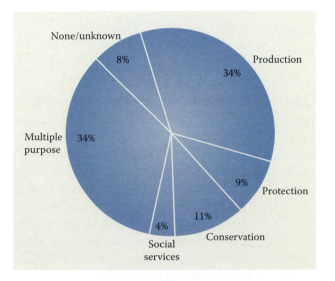

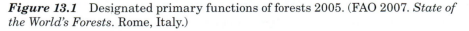

Figure 13.1 Designated primary functions of forests 2005. (FAO 2007. *State of the World's Forests.* Rome, Italy.)

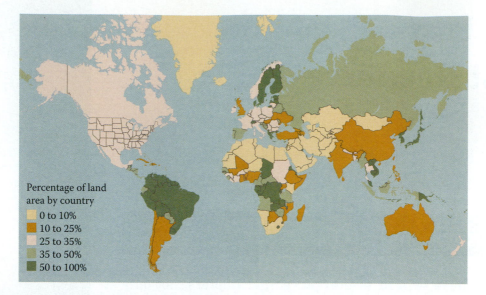

Figure 13.2   The distribution of global forest resources. (FAO 2007. *State of the World's Forests*. Rome, Italy.)

Global and National Extent of Forests

Global Distribution

A forest, according to Kimmins (2004), may refer to a stand-level ecosystem (e.g., 1–100 ha) or to a local (e.g., 100–1,000 ha) or regional landscape (1,000–100,000 ha or more), all dominated by trees. The AFPA (2001) defines "forestland" as an area ≥ 0.4 ha (= 1 acre) and 10% forest cover. Forests occupy approximately 3.5 billion ha or about 30% of the world's land area (Figure 13.2). Nearly two-thirds of the forests are in 10 countries (Table 13.1), whereas less than 10% of the area is forested in 53 countries (FAO 2005).

Worldwide, forests are a direct source of livelihood for more than 350 million people, and nearly three times that number are indirectly dependent on forests. As we note later, there has been a progressive loss of forested areas to grazing, agricultural, and other developmental uses.

National

In the United States, forests occupy over 300 million ha or approximately 33% of the land area (Table 13.2). The predominant vegetation of trees and other woody plants with at least a

Table 13.1
World's Top 10 Forested Countries, by Rank

Country	Total Forest Land (× 10³ ha)
Russia	808,790
Brazil	477,698
Canada	310,134
United States	303,089
China	197,890
Australia	163,678
Democratic Republic of Congo	133,610
Indonesia	88,495
Peru	68,742
Angola	59,104

Source: From FAO. 2007. *State of the World's Forests 2007*. Rome, Italy.

10% crown cover characterizes these forest ecosystems, ranging from closed-canopy forests to open woodlands. The Forest Service, part of the United States Department of the Agriculture, is the major agency that reports on all aspects of forest resources, grouping the 50 states by regions and subregions (Figure 13.3). The agency also has within it an excellent cadre of professionals and researchers dealing with a multitude of research and applied-management forest-related issues.

Historical data for the United States of a few representative years (Table 13.3;

Table 13.2
Land Area in the United States by Major Class and Region, 2002

Region	Total Land Area	Forest Land — Total Forest Land	Timberland	Reserved[a]	Other[b]	Other Land
North	167,334	68,669	64,230	3,204	1,236	98,666
South	216,301	86,648	82,019	1,822	3,005	129,453
Rocky Mountains	300,365	58,414	28,582	7,670	22,163	241,952
Pacific Coast	231,897	89,149	28,945	18,487	41,717	142,749
United States, total	915,897	303,079	203,776	31,183	68,122	612,819

(× 10^3 ha — Land Class)

Source: From Smith, W. B. et al. 2004. Forest resources of the United States, 2002. General technical report NC-241, North-Central Research Station, USDA, Forest Service, St. Paul, MN.

[a] For 2002, reserved forest includes lands previously classified as unproductive reserved and tabulated under the other forest category.

[b] For 2002, other forest no longer includes lands classified as unproductive reserved. This area, amounting to about 4.9 million ha (12 million acres) in 1987, is now included in the reserved forest category (see Smith et al. 2004).

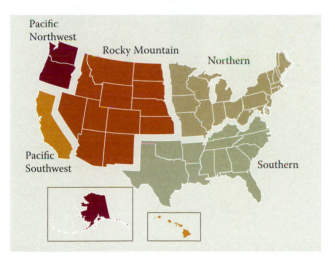

Figure 13.3 Forest resources reporting regions and subregions of the United States. (Modified from http://www.fs.fed.us/contactus/regions.shtml.)

Figure 13.4) reveal some interesting features. Of the four regions shown, maximum change expectedly has been in the north followed by the south. Both are related to increasing human numbers and associated development such as use of forest resources and clearing of forests for other uses. Forested land ownership (Figure 13.5) shows that most of the forested area in the northern and southern parts of the United States is privately owned, whereas in the Rocky Mountain and on the Pacific coast, regions are public lands.

Types and Functions of Forests

The kind and character of forest present at any particular location are the result of multiple, interacting factors, most important of which are the macro- and microclimate, geomorphology, hydrology, and soil; associated plants and animals; and historical natural and human-induced disturbance regimes of the area such as fire, flood, weather extremes, timber cutting, and deforestation. Depending on combinations of these factors, forests differing in such characteristics as appearance, species composition and diversity, density, age and size structure, and growth habit exist throughout the world.

Based on their structure and age, forest ecosystems are generally classified as old growth, second growth, or plantation. Mature forests that have reached their relatively steady-state stages of primary and secondary succession and have been left intact for several hundreds of years are referred to as

Table 13.3
Forest Area in the United States by Region—A Historical Perspective[a]

Region	× 10³ ha				
	2002	1997[b]	1987[b]	1907[b]	1630[b]
North	68,697	68,929	66,977	56,130	120,433
South	86,848	86,643	85,447	95,396	143,267
Rocky Mountains	58,414	57,969	56,508	60,141	62,520
Pacific Coast	89,149	88,743	89,625	95,546	96,854
United States, total	303,079	302,283	298,557	307,213	423,073

Source: From Smith, W. B. et al. 2004. Forest resources of the United States, 2002. General technical report NC-241, North-Central Research Station, USDA, Forest Service, St. Paul, MN.

[a] Data in original source are in acres; these have been converted to hectares (1 acre = 0.404686 ha).

[b] For historical data, see footnotes to Table 3 on page 34 in Smith et al. 2004.

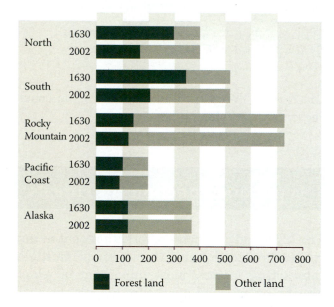

Figure 13.4 Land and forest area distribution in the United States in 1630 and 2002. (Redrawn from Smith, W. B. et al. 2004. Forest resources of the United States, 2002. General technical report NC-241, North-Central Research Station, USDA, Forest Service, St. Paul, MN.)

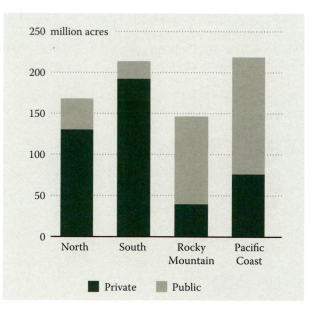

Figure 13.5 Distribution of forest land by major area and ownership group. (Redrawn from Smith, W. B. et al. 2004. Forest resources of the United States, 2002. General technical report NC-241, North-Central Research Station, USDA, Forest Service, St. Paul, MN.)

old-growth forests, late-successional forests, or **frontier forests.** The World Resources Institute defines frontier forests as "large, ecologically intact, and relatively undisturbed natural forests which are likely to survive indefinitely without human assistance" (WRI Web site http://forests.wri.org). As characterized by WRI, these forests are usually multilayered (stratified) and contain relatively large (often decadent), long-lived trees; large snags

(standing dead trees); large fallen logs; high amounts of coarse, woody debris; and a relatively high diversity of habitats.

Second-growth forests include the areas where forests have grown back relatively recently through secondary succession after timber cutting or major natural disturbances. **Plantations** are forests composed primarily of trees established by planting or artificial seeding following a human-induced or natural disturbance (**reforestation**) or on land where the previous ecosystem was some other non-forest ecosystem (**afforestation**). Plantations are usually established with the objective of producing a specific mix of products, benefits, or amenities, and their structure and species composition reflect that objective.

Forests are **renewable resources** as long as the rate of forest cutting and degradation does not exceed the rate of forest regeneration and growth (also called **sustainable harvest**) (discussed further in Section C). Forest ecosystems provide a vast array of **nonmarketed ecosystem goods and services** and **marketed products.** Examples of marketed products include wood for fuelwood, construction, and other lumber; pulp and paper products; engineered wood products (e.g., glulam timbers, glued laminated beams, particleboard, waferboard, and fiberboard); utility products (e.g., poles, posts, and pilings); and food, fiber, pharmaceuticals, and employment.

Examples of nonmarketed goods and services include the production of shelter and food for wildlife and recreational opportunities; assimilation and retention of nutrients, wastes, and greenhouse gases; regulation of the hydrological cycle (e.g., the control of floods and drought) and extreme climatic events; soil stabilization; and protection of biodiversity, seed banks, and bioindicators. The relative importance placed on specific forest products, benefits, and amenities within a particular place of the world depends, in large part, on economic and social development and cultural values in that place.

Forest Products

Although the world's forests as a whole have been decreasing in size and quality, demand for forest products has been increasing and is projected to continue to increase as a result of increases in both world human population and per-capita consumption. In the mid-1990s, total global wood consumption was about 3.4 billion m^3—8% higher than the estimated 1984–1985 consumption (WRI 1998). Approximately 54% of the wood consumed in the world in the mid-1990s was used for fuelwood, 26% for industrial round wood products (logs and sawn wood for construction), 11% for paper making, and 9% for processed wood products (veneers, chipboard, and plywood) (WRI 1998).

The growing stock and removals (Figure 13.6) reveal a healthy surplus of net growth. The percentage of different forest products harvested in the United States is very different from the preceding world figures. For example, in 1991, 49% of the wood harvested was sawlogs or veneer, 28% was pulpwood, 18% was fuelwood, and the remaining 5% was other products (Smith, Faulkner, and Powell 1994; Figure 13.7). Softwoods (gymnosperms) coming primarily from the southeast and western United States made up 67% of the volume

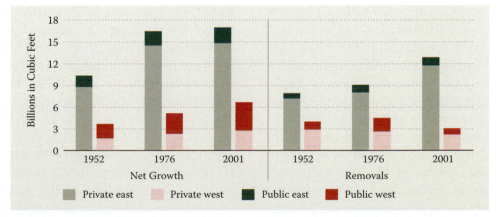

Figure 13.6 Growing stock growth and removals by major region, 1952–2001. (Redrawn from Smith, W. B. et al. 2004. Forest resources of the United States, 2002. General technical report NC-241, North-Central Research Station, USDA, Forest Service, St. Paul, MN.)

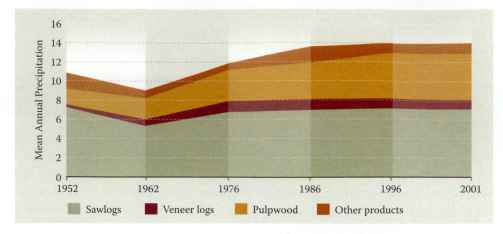

Figure 13.7 Trends in growing stock harvested for timber products output in relation to the amount of precipitation received, 1952–2001. (Redrawn from Smith, W. B. et al. 2004. Forest resources of the United States, 2002. General technical report NC-241, North-Central Research Station, USDA, Forest Service, St. Paul, MN.)

harvested; hardwoods (angiosperms) coming primarily from the eastern United States comprised 33% (Smith et al. 1994). A large proportion of the harvested softwoods are used for construction lumber, pulp, and paper, and a large proportion of the harvested hardwoods are used for furniture, cabinets, flooring, and similar products.

The production of wood products contributed about US$400 billion per year to the global economy (1.8% of the global economy and 4.1% of developing countries' economies) (Myers 1996b). By the year 2010, global wood consumption is predicted to range from 3.6 to 6.3 billion m³ (Mathews and Hammond 1999). In the United States, more than 400 million m³ of wood are harvested annually and used by more than 43,000 forest product companies to produce about $US300 billion in forest products each year (Smith et al. 1994; NHLA 1998). The forest products industry is the fourth largest in the nation and consumes 15% of the nation's raw materials (metals, fuels, and timber), employs 5% of the workforce, and produces 5% of the gross national product.

Because some forest land has been withdrawn from timber production (e.g., for a state or national park), of the approximately 298 million ha of forest land in the United States, 66% is considered commercially productive (producing 1.4 m³ of wood ha⁻¹ yr⁻¹). The majority of commercially productive forest land (58–60%) is owned by nonindustrial private individuals; only 14% of commercially productive land belongs to forest product companies.

In the early 1900s, timber growth on lumbered forest lands was only about half the rate of harvest, but by the 1940s, growth and harvest came into approximate (sustainable) balance. The timber growth nationwide exceeded harvest by 17% in 1952, 54% in 1976, 38% in 1986, and 33% in 1991. For the entire United States, hardwood growth exceeded removals by 80% in 1991, and softwoods exceeded removals by 9%. However, for the first time since 1952, softwood removals in the southeastern United States once again exceeded growth by 14% in 1991 (Smith et al. 1994).

The production and utilization of many nonmarketed goods and services are also essential to well-being and health of humans and their economies. According to one estimate, the worldwide value of nonmarketed ecosystem goods and services provided by forests is as much as US$90 billion per year (Pimentel et al. 1996). In the Pacific Northwest of North America, for example, old-growth forests serve to protect habitats for about 112 fish stocks, and the salmon industry alone is worth US$1 billion per year (U.S. Forest Service 1993). In situations such as this, the timber value and production potential of the forest might well be far less important in determining land management than the needs of other economically important resources.

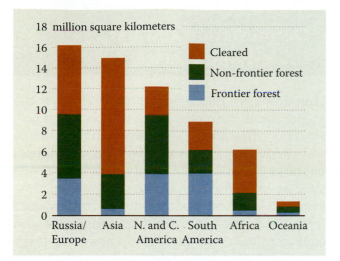

18 million square kilometers

Cleared
Non-frontier forest
Frontier forest

Russia/Europe Asia N. and C. America South America Africa Oceania

Figure 13.8 Old-growth (frontier) forests by world regions and what happened to those that once covered the Earth. (Bryant, D., D. Nielson, L. Tangley. 1997. *The Last Frontier Forests: Ecosystems and Economies on the Edge.* Washington, D.C.: Forest Frontier Initiative, World Resources Institute.)

Harvesting

Historically, timber harvesting of frontier forests and nonfrontier forests has occurred throughout the world (Figure 13.8). In many forests, the harvesting is part of an overall management plan that provides for the artificial or natural regeneration of the forest following the cut. However, in recent years, land conversion from forested to other land uses (such as crop and pasture growth or human habitation) has been unprecedented. Even today, "deforestation continues at the alarming rate of 13 million hectares a year" (FAO 2007).

There are a number of methods of harvesting timber, depending on the objective of the harvest and the character of the stand (e.g., species composition, age and age structure, soil characteristics, slope, etc.), including clear-cutting, shelter wood, seed tree, and selection harvesting. Clear-cutting, shelter wood, and seed tree harvests result in forests that have an **even-aged** structure: All of the trees are essentially the same age. Selection harvesting results in forests that are **uneven aged,** with trees of several age classes. When a timber harvest is performed as part of a forest management plan, it usually has multiple

objectives, including the removal of "mature" trees and the regeneration of the stand.

Clear-cutting, as the name suggests, is the removal of all the trees from a forested area at one time. Obviously, this is the most common harvesting method used in exploitative logging and land conversion. When properly applied, however, clear-cutting is an important forest management tool used by foresters and other land managers to

regenerate forests composed primarily of trees requiring full or near-full sunlight to survive and thrive;

create openings and early successional vegetation for wildlife habitat;

create vistas, campsites, roads, etc.; and

clear the area for tree planting or seeding.

Clear-cutting is the harvesting method quite commonly used in the management of forest stands devoted primarily to timber production because it is usually the most cost-effective and efficient method of harvesting. When clear-cutting is employed as a management tool to produce forest products, stands are usually harvested at regular intervals termed rotations. Rotation lengths are determined by a number of factors, including the product being produced, tree species, and growing conditions (including site quality and climate); they commonly range from 100 or more years for sawtimber to 30 or fewer years for fiber. Stands harvested at rotation age are regenerated utilizing young trees that were already established prior to the harvest or that became established following the harvest, or by seeding or planting.

The **seed tree** method of harvesting can be thought of as a modified clear-cut in which a few widely spaced residual trees are left throughout the cut area to provide seed for the purpose of establishing the next forest. This method is often used with trees having lightweight seed (e.g., many conifers) and that regenerate successfully in an open and unprotected environment. Commonly, 5–20 trees per acre are left, depending on the species. The seed trees may or may not be harvested.

The **shelter wood** method of harvesting consists of a series of two or more partial cuts spaced over several years (commonly, 5–15 years, depending on species and site). The major objectives of a shelter wood are to provide an adequate seed source for the future stand and create a partially shaded and protected environment in the understory of the forest where young trees can become established and grow. Once the desired reproduction is well established, the remaining larger trees are harvested. One advantage of the shelter wood method is that it produces far less negative visual impact than a clear-cut because the harvested area always is dominated by trees. However, shelter wood harvests are not common—except in areas where the visual impact of a timber harvest is undesirable—because they are not as economically efficient as alternative methods, and they require a long-term commitment to the management of the forest, which is not a characteristic of many forest landowners.

The **selection** method is characterized by the periodic removal of individual trees or small groups of trees from a stand based on their size, species, quality, condition, and spacing. Selection harvesting is commonly applied in forests containing species that survive and thrive in partial or complete shade. In such a forest, trees of desired species are present in the understory and lower crown to replace the harvested trees, resulting in a forest that varies little in structure and density over time. When species composition and stand structure permit, selection is often the harvesting method of choice when aesthetics is an ownership objective.

One should not leave any discussion of timber harvesting without at least a brief discussion of a particularly devastating form of exploitative logging: **high grading.** In high grading, all or almost all of the trees of economic value are removed from the stand. Such a harvest ignores the sound ecological principles on which established harvesting methods are based and generally leaves a forest composed of poor-quality, low-vigor trees with very little value or ability to meet most landowners' objectives in the future. Depending on the quality of the stand prior to harvest, high-graded stands may range in appearance from a poorly performed clear-cut to a poorly performed selection harvest.

Deforestation and Its Underlying Causes

Humans have had a substantial effect on the amount and character of the world's forest ecosystems throughout history. Early humans were primarily gatherers and had little impact on the forest as they obtained subsistence needs such as wood for fuel, timber for construction, foods, medicines, and often, spiritual values from the forest. As world population size and consumption have increased, however, large areas of forest have been and currently are being deforested, cleared by exploitative commercial timber harvesting and cleared for agriculture and ranching, urban–industrial development (housing, mining, transportation networks, dams), and other permanent nonforest land uses.

Deforestation refers to the removal of a forest where the land is put to other land uses and covers. Today, the world's forests and woodlands occupy approximately 30% of the land area compared to about 47% in the period before agriculture (Figures 13.8 and 13.9). Only one-fifth of the remaining forests are old-growth forest ecosystems in an unmanaged state (Bryant, Nielsen, and Tangley 1997) (Figure 13.8 and 13.9). Per-capita forest area has fallen by more than 50% from a global average of 1.2 ha in 1960 to 0.6 ha in 1995. In the United States, forest land area has fallen almost 30%, from about 421 million ha around 1,600 to 298 million ha in 1992 (Smith et al. 1994), although it should be pointed out that the United States has increased forested area from a low of about 235 million ha in 1920.

About 2 billion people, especially in Africa, parts of Andean South America, the Caribbean islands, and most of the Indian subcontinent, rely on wood as a fuel for warmth and cooking. In 1990, just more than half of all wood extracted from the world's forests was used for fuelwood. Overharvesting of forested areas in these regions has created an acute scarcity of fuelwood (Figure 13.10). More than 10 million ha of forests are cleared worldwide annually for agricultural production to support the increase in the world population size.

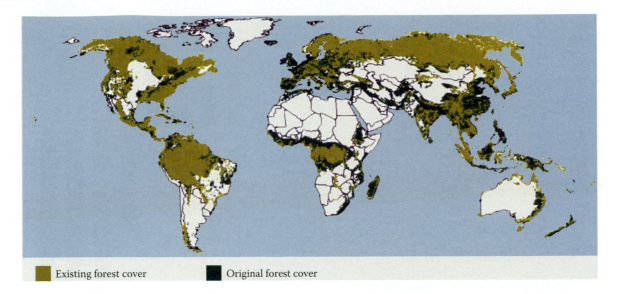

Existing forest cover   Original forest cover

Figure 13.9 Historical reduction in global forest cover from the original forest cover of 8,000 years ago. Obviously, more remains than has been lost, but the rate of loss is accelerating in many regions, and localized losses have devastating effects on those living in the area of the loss. (WRI. 1987. *Frontier Forests of the World. World Resources 1994–95.* New York: Oxford University Press.)

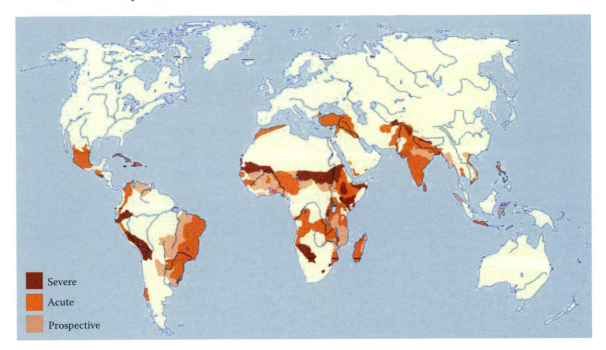

Severe

Acute

Prospective

Figure 13.10 Areas of fuelwood shortages in the world. (Redrawn from Williams, M., ed. 1993. *Planet Management.* New York: Oxford University Press.)

The area of the world's forests (including plantations) has been decreasing at an annual rate of 11.3 million ha, from 1980 to 1995 (Figure 13.11). Regional deforestation between 1980 and 1995 caused a net loss of forest resources of 200 million ha in developing countries, but there was a net increase of 20 million ha in developed countries. The rates of global deforestation were 12, 15, and 13 million ha yr^{-1} in the 1970s, 1980s, and 1990s, respectively (Watson et al. 1998) (Table 13.4).

Natural forests were lost at an annual rate of 14.6 million ha between 1990 and 2000; 1.5 million of the 14.6 million ha were converted to forest plantations (FAO 2005). Generally, the forest plantations grew by

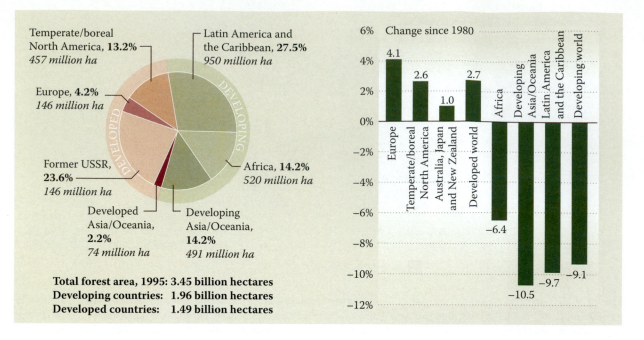

Figure 13.11   Forest areas by regions in 1995 and change as percentage relative to 1980 forest areas. (FAO. 1999. State of the Worlds Forests 1999. Retrieved from http://www.fao.org/docrep/W9950E/w9950e02.htm#TopOfPage.)

Table 13.4
Estimates of Deforestation Rates for the 10 Countries with Largest Tropical Forest Areas

Country	Forest Area (10^6 ha)	% of World Total tropics	Deforestation Rates (10^6 ha yr^{-1})		
			[a]Late 1970s	[b]Late 1980s	[c]Late 1990s
Brazil	356	30.7	1.36	5.0	8.0
Indonesia	113	9.8	0.55	1.2	0.9
Zaire	106	9.1	0.17	0.4	0.18
Peru	69	6.0	0.24	0.35	0.27
Colombia	46	4.0	0.8	0.65	0.82
India	46	4.0	0.13	0.4	1.5
Bolivia	44	3.8	0.06	0.15	0.09
Papua, NG	34	2.9	0.02	0.35	0.02
Venezuela	32	2.7	0.12	0.15	0.12
Burma	31	2.7	0.09	0.8	0.68
Total	877	75.7	3.54	9.45	12.58

Source: From Melillo, J. M., R. A. Houghton, and A. D. McGuire. 1996. *Annual Reviews of Energy and the Environment* 21:293–310.

[a] FAO/UNEP.

[b] Myers, N. 1996b. *Environmental Conservation* 23 (2): 156–168.

[c] WRI (World Resources Institute). 1998. *World Resources: A Guide to the Global Environment.* New York: Oxford University Press.

about 3.1 million ha per year. Some caution is needed in recording these numbers because planting an area with forest tree species does not instantly make an ecologically functional system or a commercially viable plantation. Even under the best climatic and soil conditions, it takes several decades to restore a forest to a viable and profitable community.

Deforestation in the tropical zone occurred most rapidly and caused a net loss of 450 million ha (an area half the size of the United States) between 1960 and 1990; about two-thirds of this amount was due to slash-and-burn agriculture (FAO 1999). Temperate forests are now more or less in steady state after losing half their original areas over a period of centuries. Boreal forests, which form the Earth's largest terrestrial biome with 13 million km², are also being extensively deforested, primarily by exploitative clear-cut logging. Exploitative logging and fires destroy 40,000 km² of Siberia's boreal forests annually, and as much as 70,000 km² of Canada's boreal forests were burned annually in the 1990s.

In addition, certain boreal forests, especially in Siberia, northeastern North America, and northern and central Europe, are experiencing acid deposition (Chapter 20). Industrial pollution (along with logging and fires) depletes another 65,000 km² of Siberia's forests, an amount twice as much as recent annual deforestation in Brazilian Amazonia and four times as much as the area logged each year in Canadian boreal forests (Myers 1996b). Furthermore, acid deposition is expected to affect as many as 1 million km² of tropical forests adversely (Rodhe 1994) (discussed in Chapter 20).

Timber overharvesting, overgrazing, urban–industrial development and transportation, and fragmentation of forests for agriculture have been and continue to be major causes of deforestation and forest ecosystem degradation worldwide (Figure 13.12). Behind these direct actions is an array of underlying socioeconomic forces or causes that must be recognized to understand the full scope of the problem of deforestation, including population and consumption growth, poverty, land shortage and maldistribution of farmlands, market and policy failures such as short-term profit maximization, greed, and lack of valuation of nonmarket

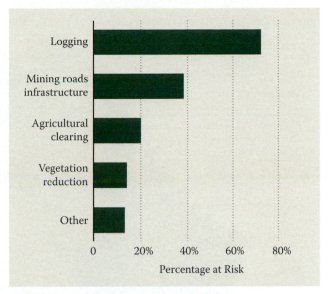

Figure 13.12 Types of threats to old-growth (frontier) forests and their percentage at risk. (Bryant, D., D. Nielson, and L. Tangley. 1997. *The Last Frontier Forests: Ecosystems and Economies on the Edge*. Washington, D.C.: Forest Frontier Initiative, World Resources Institute.)

ecosystem goods and services (externalities). Misplaced incentives also encourage overharvesting, as do unequal economic growth and limited access of the poor to the processes of policy- and decision-making (also called distributive injustice) and the lack of institutional regimes to promote long-term sustainability (see Section C).

Deforestation and degradation have substantially reduced the capacity of the world's forests to provide needed commercial products and nonmarketed goods and services to an ever-increasing population with ever-increasing consumption. Existing and new forest management strategies should, therefore, emphasize the increased production of both market and nonmarket goods and services.

Impacts on the Hydrological Cycle

Forests play a vital role in regulating the hydrological cycle by buffering the rate at which water that falls as precipitation is discharged through the soil into ground and

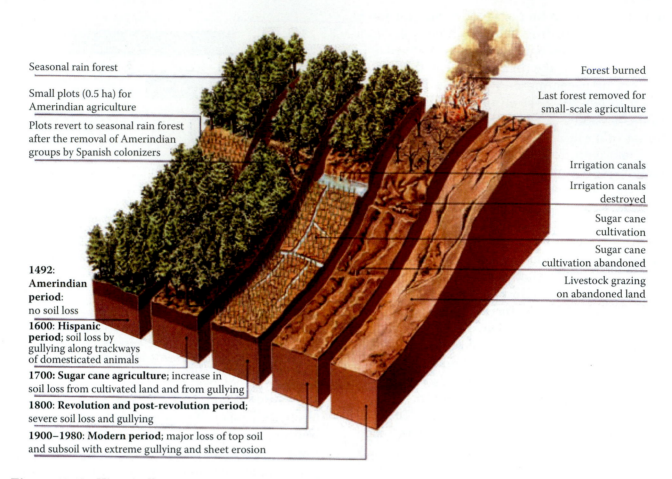

Seasonal rain forest

Forest burned

Small plots (0.5 ha) for
Amerindian agriculture

Last forest removed for
small-scale agriculture

Plots revert to seasonal rain forest
after the removal of Amerindian
groups by Spanish colonizers

Irrigation canals

Irrigation canals
destroyed

Sugar cane
cultivation

Sugar cane
cultivation abandoned

1492:
Amerindian
period:
no soil loss

Livestock grazing
on abandoned land

1600: Hispanic
period; soil loss by
gullying along trackways
of domesticated animals

1700: Sugar cane agriculture; increase in
soil loss from cultivated land and from gullying

1800: Revolution and post-revolution period;
severe soil loss and gullying

1900–1980: Modern period; major loss of top soil
and subsoil with extreme gullying and sheet erosion

Figure 13.13 Historically progressive deterioration of health of a part of watershed in a seasonal rain forest. (From Williams, M., ed. 1993. *Planet management.* New York: Oxford University Press.)

surface water, by preventing the soils from drying out by evaporation, and by returning water by transpiration to the atmosphere from where it will ultimately return as precipitation. When forest cover substantially decreases due to deforestation, so does the precipitation provided by those forests, creating local and regional water scarcity for drinking and other domestic uses, irrigation, livestock, industry, and wildlife. A historical perspective of the conversion of a seasonal rain forest to agricultural production as presented by Williams (1993) shows the degradation of the watershed over time (Figure 13.13).

The loss of forest cover also adversely affects the amount and distribution of solar radiation absorbed (albedo effect) and can accelerate soil erosion by wind and water, runoff, and the leaching of nutrients in the soil. These in turn result in droughts; floods; landslides; loss of soil fertility and ecosystem productivity;

siltation of rivers, harbors, irrigation systems, and reservoirs; and deterioration of watershed. All combine to degrade the resources and quality of life (Figures 13.13 and 13.14).

Deforestation and its attendant impacts on the hydrological cycle have important socioeconomic consequences on human quality of life and in, some instances, on life itself (Figure 13.14). For example, logging on step slopes in Bacuit Bay in the Philippines increased soil erosion by 235 times; the resulting sediments destroyed the bay's coral reef and its fisheries, causing a loss of half of the commercial revenues in the mid-1980s (Hodgson and Dixon 1988). Deforestation, particularly in Latin America and China, is worsening floods, resulting in the loss of thousands of lives and creating millions of homeless people and environmental refugees. In Venezuela, the damage from flooding and mudslides in December 1999 killed 50,000 people.

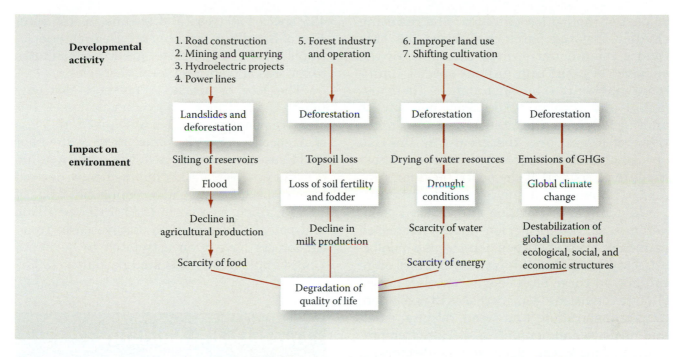

Figure 13.14 Social–economic–ecological consequences of deforestation for the quality of life. (Modified from Shiva, V., and V. Bandopadhyay. 1979. *Neglect of Ecological Factors in Development: The Case of Tehri Garwhal.* Unpublished.)

Impacts on Gaseous and Sedimentary Cycles

Forest ecosystems constitute about 60% of 2,000 billion tons (Gt) of carbon (C) sequestered in all terrestrial plants and soils. About half of the total carbon in forest ecosystems is stored in boreal forests, one-third in tropical forests, and about one-seventh in temperate forests, with more than two-thirds of the total residing in the forests' soils and peat deposits (Dixon et al. 1994). Boreal forests alone contain more carbon than all the Earth's proven fossil fuel reserves (oil, coal, and natural gas). Therefore, local and regional forest ecosystems cumulatively play a major role in stabilizing the cyclic exchanges of gases (CO_2, N, O_2, and H_2O) and sediments (e.g., phosphorus and iron) among the atmosphere, biosphere, lithosphere, and hydrosphere at the global scale (see Chapter 6 for discussion).

Deforestation causes more CO_2 emissions to the atmosphere than disturbance of any other ecosystem. For example, the amount of C sequestered in forest vegetation and soil is generally 20–50 times more per unit area than agricultural lands. Destruction and degradation of local and regional forests therefore disrupt gas and sediment cycles, releasing gases to the atmosphere that change the circulation, chemical composition, and heat balance of local, regional, and global climates. Average annual deforestation rates of the tropics have increased significantly, accounting for nearly 1.9 ± 0.6 Pg yr^{-1} of global CO_2 emissions. The resulting changes at the global scale are called global climate change, and the gases emitted that disrupt the heat balance of the Earth are called greenhouse gases (discussed further in Chapter 19).

Habitat Fragmentation and Loss of Biodiversity

Deforestation and forest fragmentation both contribute to the loss of biodiversity. **Fragmentation** is the breaking of forests into smaller patches through human activities (e.g., transportation or utility corridors, urbanization, and large clear-cuts), thereby reducing area, integrity, connectivity, and population sizes of habitats and altering all or part of their microclimates. The resulting smaller, isolated patches of forest not only reduce the amount of area available for the survival of plant and animal species (minimum viable area) but also impede their dispersal and recolonization from one site to another. The smaller the size and connectivity of forests are, the greater is the risk of loss of species requiring extensive forest areas (e.g., several of the neotropical migrant birds) (Robbins, Dawson, and Dowell 1989).

In addition to isolation and reduced population size, fragmentation causes increased boundary areas between forests and nonforest ecosystems that exhibit very distinct **edge effects.** In these forest edges, harsher microclimate conditions (e.g., increases in insolation, evapotranspiration, and wind velocity) exist, and invasive and non-native species often prevail. In addition, the edges can have deleterious effects on forest-interior and endemic species, changing the forest structure (species diversity, distribution, and abundance) and altering biogeochemical cycles and ecosystem stability.

According to a 20-year study in the central Amazon of Brazil in a rain forest ecosystem containing more than 64,000 trees composed of about 1,300 species, fragmentation caused tree mortality rates of 1.50% yr^{-1} within 300 m of forest edges and of 0.43% yr^{-1} in forest interiors (Laurance et al. 2000). In addition, forest fragmentation resulted in a disproportionately severe effect on large trees (stem diameters larger than 60 cm), killing them nearly three times faster in forest edges than in forest interiors because larger trees in forest interiors experienced increased wind turbulence, desiccation, intense sunlight and evapotranspiration, and parasites when forest edges were brought near them through fragmentation. Because a host of animals and microorganisms are dependent on trees for their required microclimate, shelter, and food, destruction of large trees has detrimental effects on their composition, abundance, and distribution in forest ecosystems.

Tropical forests are uniquely rich in species diversity, covering 6% of the Earth's land surface and containing two-thirds of all plant species on the Earth (about 170,000) and at least 50% of the Earth's plant, animal, and microorganism species. There are, for example,

425 species of trees and shrubs in less than 1 ha of Brazil's Atlantic forest. Deforestation of tropical forests, therefore, greatly contributes to the loss of plant and animal species.

The loss of species represents not only important ecological and economic losses, but also tremendous social costs, given the loss of a species with pharmaceutical value that could save lives or improve the quality of life of countless people. Existing medicines derived from tropical forests had a commercial value of US$43 billion per year in 1985 (Principe 1996), and potential drugs to be discovered in tropical forests have been estimated to be worth US$147 billion (Mendelsohn and Balick 1995).

Forest Death: Mortality, Insects, and Pathogens

Forest tree mortality increased between 1976 and 2001 in all regions of the United States, on all ownerships, and for hardwoods and softwoods (Smith et al. 2004). Insect outbreaks in the southern and western United States and hurricane impacts were cited as "significant factors in sharply rising mortality rates." Most plants and animals have, over time, developed measures to combat naturally occurring indigenous pathogens and pests. Also, over time, many organisms have developed resistance to many diseases through natural selection. But over the last several centuries, as modes of human transportation have become more diverse, the proverbial and literal germs for many diseases have "hitched a ride" to distant lands. In their new environments, these diseases become epidemic and have caused unprecedented devastation. We discuss the role of invasive species in Chapter 20; here, we concentrate on examples of species responsible for the devastation of huge tracts of native forests and many individual tree populations occupying large geographic areas.

Insects

The **gypsy moth** (*Lymantria dispar*), originally brought into the United States in 1868 to breed with silkworms to improve disease resistance, has taken a heavy toll on eastern U.S. forests. The moth began its spread near Boston, Massachusetts; areas most affected include the southern Appalachian Mountains, the Ozark Mountains, and northern lakes states. The rate of advance is estimated to be about 21 km per year. Although it feeds on many trees, it particularly feeds on oaks and aspen, defoliating the trees completely.

The **emerald ash borer** (EAB) (*Agrilus planipennis*) is a non-native insect (native to portions of Asia) that is currently attacking ash trees throughout portions of the northeastern United States and threatens to spread to the remainder of states within this region and beyond.[1] EAB is a selective pest; it attacks only ash trees, members of the genus *Fraxinus*. The larvae are responsible for tree mortality by feeding on the trees' inner bark (phloem and outer xylem), resulting in a reduction of their ability to transport food and water.

Because of their ability to survive in a wide variety of site conditions, ash trees have been among the most common species planted in the urban environment and for reforestation. To date (autumn 2008), EAB has killed more than 40 million trees in Michigan (where it was first discovered in 2002) and tens of millions of trees in other infested states. It threatens to reduce or eliminate ash dramatically throughout much of the United States unless effective, economical methods for controlling and managing it are developed.

The **spruce budworm** and **pine beetles** devastate the forests in western Canada, and the U.S. western spruce budworm (*Choristoneura* spp.) was reported from British Columbia, Oregon, and Idaho at the beginning of the twentieth century. Spruce and fir trees are defoliated and treetops die. Mountain pine beetle (*Dendroctonus ponderosae*), a native of the same area, bores the crevices of bark of all western pines. After it gets established, the foliage changes from green to yellow to red. In the last 12 years, the beetles have destroyed trees on 8.9 million ha—1.58 million ha in 2007 alone—in Colorado, Idaho, Montana, Oregon, Washington, and Wyoming. With increasing air temperatures, the infestations have become widespread and intense.

Diseases

Chestnut blight (*Cryphonectria parasitica*) is a fungal disease that has virtually eliminated American chestnut trees (*Castanea dentata*) from the upper slopes and ridges of forest ecosystems in the eastern United States, where it was once the dominant species. These once magnificent trees were among the fastest growing, attaining heights in excess of 30 m and crown spreads approaching 30 m. They were the raw material for a thriving forest industry, now gone. The disease infects the inner bark (phloem) and cambium, ultimately girdling the tree. Because the root collar and roots of chestnut are relatively resistant, a large number of small American chestnut trees still persist as sprouts from existing root systems. Today, only two small surviving populations of the original stands are known, one in northern Georgia and another in northeastern Ohio (Figure 13.15).[2] It is interesting to note that, although this fungus originated in China, its virulence is absent in the Chinese chestnut (*Castanea molissima*).

Dutch elm disease (*Ophiostoma* sp.) is a fungal disease of elm trees that is spread by elm bark beetles (*Scolytus multistratus,* the European elm bark beetle, and *Hylurgopinus rufipes,* the native American elm bark beetle). It was first introduced into the United States around 1930 and spread rapidly thereafter, resulting in the mortality of the infected trees. Once infected, elm trees produce "plugs" (tyloses) in their xylem cells as a defensive response to block the spread of the fungus. These plugs ultimately destroy the ability of the xylem to transport water and nutrients from the roots to the crown of the tree, and the tree dies. Dutch elm disease has had a devastating impact on both the native and urban ecosystems of the United States. Elms are an important component of many natural ecosystems, particularly in the eastern United States and in riparian zones and areas with wet soils. Because of its majestic, vase-like form, the American elm was among the most commonly planted trees in the urban forest and most of them are gone today.

Sudden oak death is a disease of native oaks and tanoaks and a variety of shrubs, including some important ornamental species such as viburnums (*Viburnum* spp.) and

Figure 13.15 Left: one of the last few surviving American chestnut (*Castanea dentata*) trees from the stock of once-extensive forests, somewhere in Trunbull County, Ohio (first reported in June 2007). Right: growing less than 20 m away is the Chinese chestnut (*Castanea mollissima*). Note (a) the upright and spreading nature of the two trees; (b) leaves of the American chestnut are already infected with the chestnut blight, but not so the Chinese chestnut. (Photos by M. K. Wali.)

rhododendrons (*Rhododendron* spp.). It is caused by a fungal-like organism (*Phytophthora rmaorum*) (actually an oomycete or water mold) whose symptoms include leaf infection, which creates black or brown lesions (dead areas) along the edges or tips of the leaves and, ultimately, bleeding cankers (infected areas on the stems or trunk that seep red to black gum, depending on the species infected). Ultimately, the infected tree or shrub dies. Sudden oak death was first reported in 1995 on native oaks and tanoaks in California. Since that time, it has been reported to infect additionally almost 30 species that move in the nursery trade, as well as in several other states, perhaps as a result of the movement of infected nursery stock.

Figure 13.16 A 2007 wildfire in the western United States. (Photo: CBS News.)

Forest Fires

Periodic forest and grassland fires, or wildfires, are a significant ecological factor in many parts of the world and especially so in the western United States (Figure 13.16). Fire has been such a pervasive factor that ecologists coined the term "fire climax," referring to those communities that are maintained by fire. Major (1974) deemed it important enough to include it in the state factor equation of soil and vegetation development: s and $v = f(cl, o, r, p, t, py)$, where py refers to *pyric* or fire (remember Chapter 4 on soils). For many northern tree species, fire is important for regeneration, scarification of seeds, reducing otherwise long-term litter degradation, making available the ash elements, and opening canopies to let the light in. Thus, fire in such cases has been an integral part of the ecological complex.

Figure 13.17 Anatomy of a wildfire: 1. Fire in the foliage on the forest floor can climb to lower tree branches. This is called a ladder fire. 2. Spreading into the crown of the tree, the fire nourishes itself with oxygen drawn from below (like a chimney). 3. Hurled by the convective drafts, floating embers can ignite dry fuel far from the fire. 4. Hot gases rising in a column from the inferno prime foliage above for ignition. (Graphic: CBS News.)

What has changed in recent years is that fires have become more frequent, have lasted longer, have burnt trees that were considered fire hardy, and have caused the destruction of property and death of humans and wildlife. Fires have become more intense (Figure 13.17). Recent forest fires have been fewer in number but much larger in areal extent. Wildfires burned more than 5 million acres in Alaska during July and August of 2004. More than 190,000 acres burned in Texas in a 24-hour period, March 6–7 in 2006. In 2007, more than 500,000 acres burned in October in Southern California. The losses have been staggering, amounting to billions of dollars in lost timber resources, lost built property, and human death and displacement. Health hazards from pollution-related effects have been acute, causing asthma, bronchitis, and skin and eye irritations.

How forest fires have become a major concern is illustrated by the attention the issue has received from scientists, policymakers, and media. CBS's *60 Minutes* devoted a program (which was repeated) to the "age of mega fires" (2007).[3] The anatomy of a fire illustrated in Figure 13.17 is from CBS Graphics and should be self-explanatory.

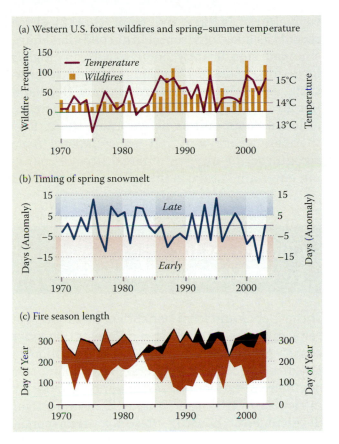

Figure 13.18 (a) Annual frequency of large (>400 ha) western U.S. wildfires (bars) and mean March through August temperature for the western United States (line). (b) First principal component of center timing of stream flow in snowmelt-dominated streams. Low (pink shading), middle (no shading), and high (light blue shading) tercile values indicate early, middle, and late timing of snowmelt, respectively. (c) Annual time between first and last large-fire ignition and last large-fire control. (From Westerling, A. L., H. G. Hidalgo, D. R. Cayan, and T. W. Swetnam. 2006. Warming and Earlier Spring Increase Western U.S. Forest Wildfire Activity. *Science* 313: 940–943.)

Several explanations have been offered for both the intensity and size of fires, especially in the United States. One is the suggestion that forest policies in the past have discouraged forest fires, which has resulted in the buildup of litter on the forest floor and allowed forest densities to increase, thus providing the fuel for the fire. Also, human habitation has gone perilously close to forest boundaries, which have shrunk and made a hop for a forest fire to human dwellings easy. Both suggestions have merit. Equally important is the recent analysis of forest fires in the western United States by Westerling et al. (2006) (Figure 13.18). Their analysis compellingly makes the point that in three recent decades, environmental conditions have changed significantly enough with (a) higher large-fire frequency, (b) longer wildfire durations, and (c) longer wildfire seasons. Repeated like-fire conditions threaten the very existence of forest communities as we know them. It is therefore gratifying that policymakers have recognized the urgency of forest fires in changing environmental conditions.[4]

Challenges for Forest Management

If the global and local benefits of forest ecosystems are to be retained, net deforestation must cease. As already discussed, deforestation is the direct result of such actions as exploitative commercial timber harvesting and clearing for agriculture and ranching, urban–industrial development (housing, mining, transportation networks, dams), and other permanent nonforest land uses. However, behind these direct actions are not only technical forest management problems but also the staggering national and international socioeconomic challenges that must be addressed by national and international policy and action.

How, then, will the forest ecosystems of the world provide the predicted increase in demand for forest products while at the same time meet the increasing demand for nontimber products, amenities, and benefits? Certainly the world's existing old- and second-growth forests will be important contributors. Depending on the specific forest area, their utilization for wood products will undoubtedly range from none (except, perhaps, for fuelwood gathering) to those fairly intensively managed to produce multiple resources, including timber.

The emergence of a social sensitivity to the destruction and degradation of forest ecosystems will undoubtedly lead to the use of less intense and less obtrusive management practices in old- and second-growth forests. As just one example, in many public and private nonindustrially owned forests in the northern and central hardwood regions of the eastern United States, timber harvesting by individual tree selection, group selection, and patch cutting has replaced more extensive clear-cutting. One obvious extremely important management question for old-growth and second-growth forests is who will make the decisions concerning how and for what resources each forest area will be managed and on what criteria those decisions will be made.

Plantations are most likely to play an increasing role in providing timber products because they would be managed primarily for the production of timber products while providing other benefits and amenities, thus reducing demand for such products from old- and second-growth forests. Plantations established specifically for timber product production can be managed on shorter, more economical rotation lengths using more intensive silvicultural techniques such as site preparation, fertilization, integrated pest management, intermediate cutting, and harvesting and regeneration techniques that would be less appropriate for natural forests managed for a broader mix of products and nonmarketed benefits and amenities (Figure 13.19).

Currently, plantations worldwide occupy an area equivalent to 3.4% of the total forest area but produce nearly 25% of the total industrial wood supply (Mathews and Hammond 1999). Sedjo and Botkin (1997) estimated that the world's current demand for industrial roundwood could potentially be met from 0.15 billion ha of forest plantation (equivalent to 4% of current global forest) growing at 10 m^3 ha^{-1} yr^{-1} (see Table 13.5). Although this may not be attained in the short term, it does put into perspective the potential contribution that plantations can make.

Better utilization of harvested trees and wood manufacturing by-products can contribute to meeting future timber demand effectively by utilizing harvesting residue currently left in the forest as branches and low-quality logs (estimated as high as 38% of volume in some instances) as well as the manufacturing residue that is the by-product of sawing and milling consumer products from wood (e.g., sawdust and wood scraps). Much of this residue could be used to produce engineered wood products, increasing the total yield from the forest and

Naturally regenerated forests				Planted forests		Trees outside forests
Primary	Modified Natural	Semi-Natural		Plantations		
		Assisted natural regeneration	Planted component	Productive	Protective	
Forest of native species, where there are no clearly visible indications of human activities and the ecological processes are not significantly disturbed	Forest of naturally regenerated native species where there are clearly visible indications of human activities	Silvicultural practices by intensive management: ▶ Weeding ▶ Fertilizing ▶ Thinning ▶ Selective logging	Forest of native species, established through planting or seeding, intensively managed	Forest of introduced and/or native species established through planting or seeding mainly for production of wood or non-wood goods	Forest of introduced and/or native species, established through planting or seeding mainly for provision of services	Stands smaller than 0.5 ha; tree cover in agricultural land (agroforestry systems, home gardens, orchards); trees in urban environments; and scattered along roads and landscapes

Figure 13.19 Planted forests in the continuum of forest characteristics. (FAO. 2007. *The State of the World's Forests*. Rome, Italy.)

Table 13.5
Global Industrial Wood Harvests by Forest Type

Forest Type	Global Industrial Wood Harvest (% Total)
Old growth[a]	30
Second growth, minimal management[b]	14
Indigenous second growth, managed[c]	22
Industrial plantations, indigenous[d]	24
Industrial plantations, exotic[e]	10
Total	100

Source: From Sedjo, R. A., and D. Botkin. 1997. *Environment* 39 (10): 15–20, 30.
Note: Values in this table are for illustrative purposes only.
[a] Includes forests in Canada, Russia, the Amazon, Indonesia, and Malaysia.
[b] Includes forests in parts of the United States, Canada, and Russia.
[c] Includes forests in North America, Europe, and Russia.
[d] Includes plantations in the Nordic regions, much of the rest of Europe, a significant portion of the United States (particularly in the South), Japan, and parts of China and India.
[e] Includes plantations in Brazil, Chile, Venezuela, Uruguay, Argentina, New Zealand, Australia, South Africa, Indonesia, Thailand, and the Iberian countries of Europe.

providing a suitable substitute for solid wood for many applications. Expanded recycling and product substitution will also reduce future demand for virgin fiber and timber products. We have to make transitions seamless.

Recent developments in forest research have been most encouraging in several ways.

The field of **agroforestry** recognizes the interplay of croplands with forests thus:

Agroforestry is a collective name for land use systems in which woody perennials are deliberately integrated with crops and/or animals on the same land management unit. The integration can either be in a spatial mixture or in a temporal sequence. There are normally both ecological and economic interactions between woody and nonwoody components in agroforestry. (World Agroforestry Center, ICRAF 1993)

Similarly, an initiative was launched in social forestry that recognizes the intimate connection of rural people to fuelwood, timber, medicinal plants, and fodder. That these services come from the native forests has been well known to all indigenous people. The field of social forestry and community forestry has gained wide recognition and utility in some regions—for example, India.

The importance of forest watersheds in the existence, life span, and survival of freshwater fishes has recently been explored in an elegant way (Northcote and Hartman 2004). Such fishes "include over 10,000 freshwater species," note these authors. Thus, the interconnections of forest watersheds to streams, rivers, estuaries, fish, and humans are significant. We shall elaborate on challenges of forestry and their linkages to other ecosystems in our discourse in Chapter 26 on sustainable development.

Notes

1. As of autumn 2008, EAB infestations had been identified in Michigan, Ohio, Indiana, Illinois, Wisconsin, Missouri, Pennsylvania, West Virginia, Virginia, and Maryland.
2. In summer 2007, several Ohio newspapers—Dayton *Daily News,* Cincinnati *Enquirer,* and the Columbus *Dispatch*—carried stories on finding some American chestnut trees still growing in Trumbull County in northeastern Ohio. One of us (MKW) visited and walked through the area. It seems that a stand of American chestnuts was perhaps growing there until recently (1970s–1980s) but now only isolated and scattered trees remain. It seems that extensive stone quarrying in the area (in some cases exposing the roots) may have considerably reduced the vigor of these trees, making them susceptible to the blight; all were found infected.
3. Scott Pelley presented the "age of mega fires" on CBS's *60 Minutes* on October 27, 2007; this program was presented again on December 30, 2007. (Seen and accessed December 30, 2007.)
4. The U.S. House of Representatives Committee on Energy and Nature Resources held a hearing on **scientific assessment of effects of climate change on wildfire** on September 24, 2007, in which the committee received the expert testimony of several scientists.

Questions

1. Discuss the ecosystem goods and services provided by forests in your region.
2. Explore the environmental and economic implications of forest losses in your region.
3. What factors are responsible for the destruction, degradation, and fragmentation of forests?
4. Discuss the cumulative effects of local deforestation events.
5. How can harvesting practices be designed to protect the ecosystem goods and services that forests provide?

Land Use—Mining for Mineral and Coal Resources

14

Introduction

Humans have been extracting minerals and metals from the Earth for millennia. Copper ingots of 99.5% purity were produced in the Negev Desert area more than 13,000 years ago. The art of glass-making and glass-blowing has been dated as long ago as 4000 B.C. What has changed in this century, however, is the phenomenal scale and diversity of our modern-day machines, from jumbo jets and earth-moving cranes several stories tall to silicon chips smaller than a dime. These, like all other machines, contain varying quantities of processed metals and minerals. For example, automobiles come in myriad shapes, sizes, and colors, and each contains hundreds of kilo-grams of metals and plastics. Similarly, appliances such as washers and refrigerators used to be white until the 1960s, and then as fash-ions of the day changed, these appliances became avocado or harvest gold in the 1970s, burnt pumpkin in the early 1980s, and stainless steel or copper by the 1990s. Now we are back to eggshell!

We do not know why appliance colors change every few years or who sets the color change in motion. The important point is that all these machines, big and small, entail the use of materials manufactured by complex processes. All of these materials must be extracted from the Earth, processed in numerous ways, reshaped and reformulated into a desired product, and then disposed of or recycled after use (Figure 14.1). Each step entails the transforma-tion of the land surface, use of energy, and many local and regional impacts on air, water, and biota.

Types of Minerals

Minerals are mostly **nonrenewable** natural resources; that is, their total stock availability to humans is considered fixed on long-time scales. Minerals are naturally occurring inorganic substances (ele-ments, compounds, and an aggregate of elements and compounds)

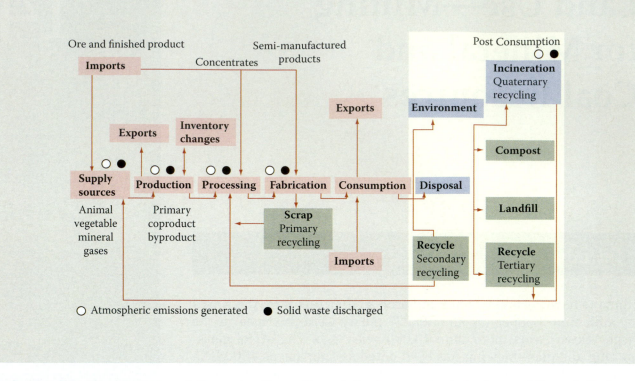

Figure 14.1 The life cycle of minerals from extraction and importation to processing to final use and disposal. (Redrawn from Brown, W. M., III, G. R. Matos, and D. E. Sullivan. 1998. Materials and energy flows in the earth science century. A summary of a workshop held by the USGS in November 1998. U.S. Geological Survey Circular 1194.)

with an orderly atomic arrangement; they are both inorganic and organic in origin. **Mineral resources** include materials, metals, fossil and nuclear fuels, and nutrients that are vital to industrial, agricultural and technological processes, and the production of consumer goods. "A mineral **resource** is a concentration of naturally occurring solid, liquid, or gaseous material, in or on the Earth's crust, in such form and amount that economic extraction of a commodity from the concentration is currently or potentially feasible" (U.S. Geological Survey).

For a resource to be considered a **mineral reserve,** it must be economically, environmentally, and legally extractable now (Figure 14.2). **Proven mineral reserves** are an estimate of those that can be extracted profitably for human use with currently available technology, while **recoverable reserves** are estimates of mineral resources that are likely to be available for future human use. Some resources are called **economic minerals** when they are central to the economy of a nation; others are **strategic minerals** when they are of extraordinary importance to a nation's military

defense (Figure 14.3). Metals can be mainly classified into two groups: those that are geochemically abundant and those that are geochemically scarce. **Geochemically abundant metals** individually constitute 0.1% or more of the Earth's crust by weight and include iron, aluminum, silicon, manganese, magnesium, and titanium. **Geochemically scarce metals** individually constitute less than 0.1% of the Earth's crust by weight and include copper, lead, zinc, molybdenum, mercury, silver, and gold.

Fossil and nuclear fuels are nonrenewable mineral resources and include petroleum, coal, and natural gas as fossil fuels and uranium and thorium as nuclear fuels (these are discussed in Chapter 19). The remaining mineral resources include all of those **materials** (e.g., granite, basalt, and sand) and **nutrients** (e.g., N, P, K, and Ca) that are used for purposes other than their metallic and energy-containing properties. The mineral deposit is known as an **ore** when the amount of metal-bearing rock per given volume (**grade**) is high enough to make mining economically feasible.

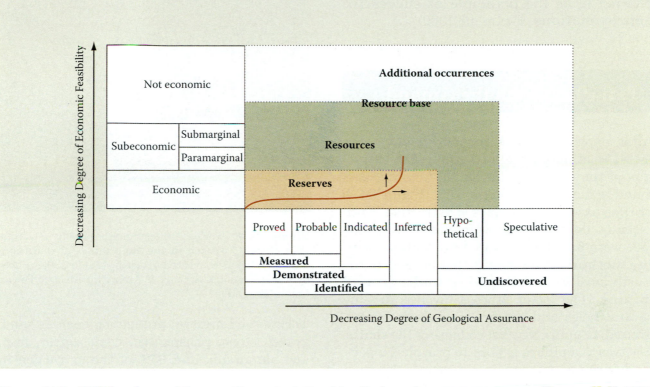

Figure 14.2 McKelvey box and the quantity–cost relationship of hydrocarbon resource recovery. (Rogner, H. H. 1997. An assessment of world hydrocarbon resources. *Annual Review of Energy and the Environment* 22: 217–262.)

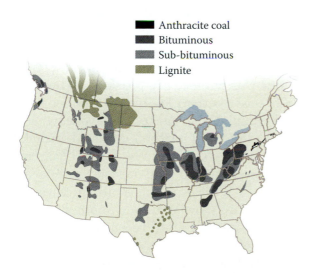

Figure 14.3 Coal map of the United States. (Courtesy of USDI-Geological Survey.)

Mineral Exploration

An ore must be found in large enough quantities to make it economically profitable to mine and process, so the establishment of mining begins with **mineral prospecting.** Trained professionals and amateur prospectors familiar with rock formations with which specific mineral ores are associated do the initial step of locating ores. One clue might come from vegetation. Some plants grow in habitats with high concentrations of elements such as nickel, copper, or selenium, and others show preferential uptake by virtue of specialized biochemical pathways that they have developed through evolutionary history. These are also referred to as **indicator plants** because their occurrence and abundance may signal the quantity, quality, and identity of mineral ores. Examples are the California poppy (*Eschscholtzia mexicana*), an indicator of copper deposits in Arizona; locoweed (*Astragalus* spp.), an indicator of selenium; *Papaver macrostomum,* an indicator of zinc; and *Salsola nitraria,* an indicator of boron.

The initial exploration is followed by detailed geological, geophysical, and geochemical field and laboratory analyses that focus on the most concentrated deposits in an area. The last stage is ascertaining the economic dimensions—that is, the profitability of mining the ore. This three-step process of identification, feasibility, and development has been referred to as the **principle of successive approximations** (Fortescue 1980).

Extraction from the Ground

After exploring the economic viability of a metal or mineral, the next step is the extraction of the ore from the Earth. Coal deposits found in many states in the U.S. (Figure 14.3) provide excellent examples of extraction. Until recently, deposits that were close to the surface were usually excavated by **surface mining methods,** but deposits that lie deep in the Earth have been extracted by **underground mining.** The latter process involves digging several holes and constructing shafts underground; to guard against collapse of the shafts, a network of pillars is built to provide support. Underground mining is expensive, does not allow for the extraction of more than 40–60% of the ore, and is inherently dangerous for the miners because of the possibility of collapse. In the case of coal, underground mining has taken a big toll on the health of miners, both from black lung disease caused by breathing coal dust and deaths of miners by the collapse of mines and explosions.

Coal deposits are found in many States in the U.S. (Figure 14.3). Modern long-wall mining techniques remove essentially all support pillars and allow overlying materials to collapse as the mining progresses. Resulting massive subsidence of the land surface is a serious environmental concern.

In recent years, **surface mining** has become widespread all over the world for several reasons. It is less expensive than underground mining, the recovery of the desired material is better than 90%, and it is less of a threat to the health and life of miners. Surface mining entails the removal of **overburden**—the material overlying the mineral or coal deposit. Surface mining operations generally consist of **open pit mining** (large amounts and high rates of mineral extractions) and **strip mining** (extractions of flat-lying layers of minerals just beneath the surface). Together, the two processes account for two-thirds of the

Figure 14.4 World's largest earth-moving machine, "Big Muskie," worked in the strip mines of southeastern Ohio. It was converted to scrap steel in 1999. (Photo by Floyd Hivnor. Courtesy of American Electric Power, Columbus, Ohio.)

world's solid mineral production such as sand, gravel, stone, phosphates, coal, copper, iron, and aluminum, and 95% of the extraction in the United States.

The removal of overburden is carried out by three main methods: use of **draglines, bucket-wheel excavators,** and **power shovels.** A dragline is a massive, self-propelled machine with a boom that operates a huge bucket that removes the overburden. The world's largest dragline was named "Big Muskie" and weighed 27,000 tons, with a 94-m (310-ft) boom and 168-m^3 (220-yard^3) bucket. The bucket could remove 325 tons of earthen material, including rocks, in one scoop (Figure 14.4). The boom could swing from 90 to 180° and deposit the excavated materials in a series of ridges and furrows, creating trenches parallel to one another, referred to as **spoil banks.** Until recent legislative controls took effect, the topmost layers of the Earth were buried deep and the deepest layers formed the new surface. This requirement made Big Muskie obsolete.

Power shovels work in the mine pits and dig materials from the face of a deposit using stripping shovels. These shovels have smaller buckets, usually not exceeding 30 m^3 (40 yards^3). This method allows for material segregation, whereas the dragline cannot do so. Bucket-wheel excavators are crawler-mounted machines that continuously dig, transport, and deposit overburden. They have a wheel

Figure 14.5 Mountaintop removal coal mining in Boone County, West Virginia, near the town of Bob White, December 10, 2005. (Photo courtesy of Vivian Stockman.)

Figure 14.6 A massive coal sludge impoundment permitted to hold 9 billion gallons of coal sludge near Whitesville, West Virginia, October 19, 2003. (Photo courtesy of Vivian Stockman.)

Figure 14.7 This coal seam from the Powder River Basin of Wyoming is tens of meters thick. (Photo by M. K. Wali.)

and several buckets to excavate overburden and drop it onto a continuous belt conveyor that transports the material to a spoil bank (Brown et al. 1986). It is generally agreed that draglines create the most land disturbance. New laws in several countries now require the segregation of topsoil materials for replacement later (see Chapter 25), though enforcement is often lax.

Although surface mining methods of the past have changed only in their extent and magnitude in this century, we have now added a new dimension—**mountaintop removal.** In the United States, the practice of physically decapitating mountains to expose coal seams has increased considerably in extent. The removed material is never replaced where it originated but, rather, is pushed into adjacent valleys. In West Virginia alone, a November 2005 report notes, more than 120,000 ha of mountains were "decapitated … and [miners] filled in about 1,000 miles of streams with rubble" (Lydersen 2005). Though not so widespread, this form of mining is also found in Kentucky, Ohio, and Pennsylvania (Figure 14.5). The shearing away of the tops of mountains is done in 30-m (100-ft) segments, "like layers of a birthday cake, until most of the mountain is gone" (Warrick 1998). In 1997, the state of West Virginia granted permits for 20 new projects covering 52 km^2 (20 mi^2). The justification for this type of mining is the lower cost of obtaining coal. Environmental constraints on this practice were removed by the federal government in 2002 so that this type of mining could proceed.

Sometimes the rubble from the mountain top is used to build a dam in the adjacent valley and the reservoir produced is used to collect water for use in coal washing. Removal of impurities from coal increases its value as a commodity, but leaves huge volumes of toxic waste (coal sludge) behind (Figure 14.6).

Coal mining in the western United States has expanded considerably since the 1973 energy crisis. A number of reasons have contributed to both the feasibility and profitability of such expansion. First, the coal seams in the West are reported to be 10–30 m thick (Figure 14.7). Second, the earthen material on top of coal seams is relatively shallow (30–50 m) (as seen in Figure 14.7). Third, surface mining mines 95% or more of the coal to be excavated from the ground. Fourth, the population density relative to the eastern United States is considerably lower in the coal-bearing regions of the West. This point has evoked resentment in the native and naturalized people in the West. Thus, in several essays, the West has been characterized as a colony of the East (Dix 1975; Josephy 1975). These feelings, however, abated considerably for two reasons:

- Economic development generated taxes for the government and profits for the individuals.
- Revised and new laws provided the assurance that mined land would be brought back to biological productivity and not abandoned, as was the case until the early 1970s (Figure 14.8). Indeed, the federal law created the **abandoned mine land** (AML) program to reclaim and restore such lands to some use; these aspects are discussed in Chapter 20 on environment policies and laws.

Another form of mining, **quarrying,** refers to an open pit mine from which ore, coal, stone, or gravel is extracted. This method can leave an immense hole in the ground and huge piles of waste rock (tailings). Other extraction methods include the removal of unconsolidated material from rivers, streams, lakes, and seas

Figure 14.8 An abandoned coal mining area in the early 1970s from the upper Midwest of the United States. After the mid-1970s, the laws do not permit such abandonment. (Photo by M. K. Wali.)

(termed **dredging**), the extraction of liquid and gaseous mineral resources (**drilling**), and hydraulic and solution mining.

Although mining activities do not disturb a significant area per se on the basis of overall land use, the impacts on the environment—land, water, and air—are enormous. The earth-moving activity is indeed so large that it has earned humans the characterization of "a mighty geological agent." What has now become a classic example is the story of Bingham Canyon Mine near Salt Lake City, Utah. Ever since the 1848 discovery of gold there, the mine has been in continuous production of gold, silver, and copper. Its current annual production of metals stands at more than 282,000 tons of copper, 98,000 kg of silver, and 9,800 kg of gold. The mine has gone from a mountain to a 0.8-km deep hole in the ground (equivalent to five times the height of the Eiffel Tower), 4 km wide and visible from space. During its history, 5 billion tons of rock have been removed to get the ore—10 times the amount of material removed in building the Panama Canal. Each ton of ore produces 5.4 kg (12 lb) of copper; to expose each ton of ore, 10 tons of rock overburden must be removed (Gnidovec 2000).

In the last 20 or so years, crude oil squeezed from **tar** or **bitumen sands** has been proceeding at an accelerated pace in the province of Alberta in Canada.[1] Huge deposits of these tar sands are found in the Athabasca region of Alberta. These sands are found more than 400 m belowground; the aboveground vegetation is boreal accompanied by the characteristic elements of flora and fauna of this biome (covered in Chapter 10). These forests and overburden material must be torn and excavated by surface mining (Figure 14.9). Crude oil from these sands is separated by mechanical means and by the use of steam and solvents. It is a very water-intensive process, using three barrels of water to produce one barrel of crude oil. Thus, it generates vast quantities of wet sludge and toxic wastewater (Figure 14.10), characterized as "a giant slow-motion oil spill" and a colossal environmental disaster for indigenous people of Canada (Hatch and Price 2008).[1] Also, the mining and processing of tar sands "would significantly increase global risks of dangerous climate change" (WWF-UK 2008).[2] The

Figure 14.9 Surface mining operations for tar sands in northern Alberta, Canada. (Photo courtesy of Environmental Defense, Canada, 2008.)

Figure 14.10 Waste water that contains toxic materials is a by-product of tar sand mining. (Photo courtesy of Environmental Defense, Canada, 2008.)

resultant vast quantities of mine tailings contain some bitumen and thus far revegetation of these residual materials has been elusive.

Mineral Processing

Although the removal of overburden is similar for coal and other mineral and metal mining, there are differences in handling and processing. In the case of coal, the material is loaded onto trucks or railway cars and transported to the site, where it is burned to produce electricity. When the water content of coal is high, as in the case of lignite, whose water content may be as high as 35%, it is uneconomical to transport water large distances. In such cases, the electricity-generating plants are located close to the mines, referred to as **mine-mouth operations.** However, all the metal-bearing minerals and most materials require **processing** after their extraction because they are not in concentrated, pure forms and are interwoven with undesirable minerals such as carbonates or silicates, known as **gangue.**

Processing involves breaking down the ore to liberate the ore mineral particles (**comminution**), separating the ore and gangue minerals (**beneficiation**), using heat to recover the metal (**smelting**), and recovering the pure metal from impurities using chemicals (**heap leaching, hydrometallurgy**). Gangue minerals after processing, which make up about 80% or more of mined ores, become wastes, called **tailings.** Laws govern the handling of tailings at mine operations; a spill can have most deleterious consequences, as was the case in the cyanide-tailings spill in Romania (see Chapter 18).

Consumption Patterns and Environmental Impacts

Metals provide indispensable benefits to human society in many forms, ranging from heat exchangers, to catalysts for chemical reactions, to electricity wires, to rails for trains, and to chassis for automobiles. Global use of mineral materials rose 2.4-fold to 9.8 billion tons in 1995, relative to the early 1960s (Tables 14.1 and 14.2). Resource use in the United States increased from 2 billion to 2.8 billion tons between 1970 and 1995 and 18-fold since 1900 (Table 14.3) (Figure 14.11). With 5% of the global population, the United States consumed nearly 30% of global mineral resources. Both

Table 14.1
Primary Metal Resource Consumption and Reserve Base on a Global Scale in 1996

Resource	Global Mine Production in 1996 ($\times 10^6$ t yr^{-1})	Reserve Base ($\times 10^6$ t)	Ratio of Reserve Base to Annual Mine Production (Years)
Ores			
Iron ore	1000	232,000	232
Bauxite	111	28,000	252
Metals			
Copper	10	610	61
Lead	2.8	120	43
Zinc	7.2	330	46
Magnesium	0.347	Recovered from natural	Brines and dolomite
Nickel	1.1	110	100
Tin	0.190	10	53
Tungsten	0.030	3.3	110
Cobalt	0.024	9	375
Mercury	0.003	0.24	80

Source: From Wernick, I. K., and N. J. Themelis. 1998. *Annual Reviews of Energy and the Environment* 23:465–497.

Table 14.2
Growth in World Materials Production, 1960–1995

Material	Production in 1995[a] (Million t)	Increase over Early 1960s (Factor of Change)[b]
Minerals[c]	7641	2.5-fold
Metals	1196	2.1-fold
Wood products[c]	724	2.3-fold
Synthetics[d]	252	5.6-fold
All materials	9813	2.4-fold

Source: From Gardner, G. T., and P. Sampat. 1998. Worldwatch paper 144. Washington, D.C.: World Watch Institute.
[a] Marketable production only; does not include hidden flows.
[b] Minerals and total materials data are for 1963; wood products data are for 1961.
[c] Nonfuel.
[d] Fossil fuel-based.

magnitude and composition of resource use have important implications on environmental degradation and depletion. For example, the U.S. economy increased its use of nonrenewable resources from 59% in 1900 to 92% in 1995 (Matos and Wagner 1998). Approximately 7% of all copper, 13% of all nickel, 14% of all steel, 19% of all aluminum, 23% of all zinc, 35% of all iron, 41% of all platinum, and 50% of all lead used by the United States in 1992 went into the automotive industry.

Growing populations and economies demand more goods and services, land, and physical infrastructure from stock-limited

Table 14.3
Growth in U.S. Materials Consumption, 1900–1995

Material	Consumption in 1995 (Million t)[a]	Increase Over 1900 (Factor of Change)
Minerals[a]	2401	29-fold
Metals	132	14-fold
Wood products[a]	170	Threefold
Synthetics[b]	131	82-fold
All materials	2843	18-fold

Source: From Gardner, G. T., and P. Sampat. 1998. Worldwatch
        paper 144. Washington, D.C.: World Watch Institute.
[a] Nonfuel.
[b] Fossil fuel-based.

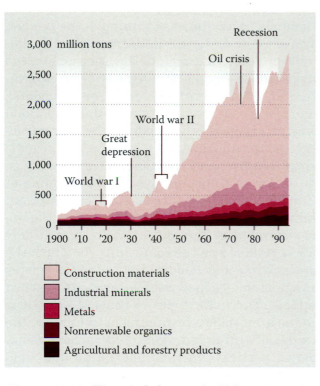

Figure 14.11 Historical changes in U.S. consumption of materials, 1900–1995. (Matos, G., and L. Wagner. 1998. Consumption of materials in the United States, 1900–1995. *Annual Review of Energy and the Environment* 23: 107–122.)

and rate-limited natural resources (see Chapter 13) (Figure 14.11). Between 1950 and 1990, the rate of increase in human consumption of six major base metals (aluminum, copper, lead, nickel, tin, and zinc) was more than eightfold, while world population only doubled. The total quantities of waste generated due to mineral and metals extraction are enormous in the form of low-grade mineral-bearing

Table 14.4
World Ore and Waste Production for Selected Metals, 1995

Metal	Ore Mined (Million t)	Share of Ore That Becomes Waste (%)[a]
Iron	25,503	60
Copper	11,026	99
Gold[b]	7,235	99.99
Zinc	1,267	99.95
Lead	1,077	97.5
Aluminum	856	70
Manganese	745	70
Nickel	387	97.5
Tin	195	99
Tungsten	125	99.75

Source: From Gardner, G. T., and P. Sampat. 1998. Worldwatch paper 144. Washington, D.C.: World Watch Institute.
[a] Does not include overburden.
[b] 1997 data.

rock that becomes waste and the overburden removed to reach it in the mine (Table 14.4). The quantity of materials consumed in the United States grew from 2 tons per person per year in 1900 to 10 tons per person per year in 1995. Heavy reliance on technological innovation and substitution cannot sustain current per-capita use and intensity of use of nonrenewable minerals (volume relative to gross national product). Social costs of depletion and degradation of natural resources should

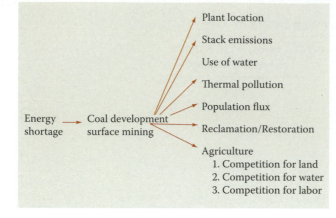

Figure 14.12   Some problems attendant on coal development as a response to energy shortage. (From Wali 1975.)

be reflected in the market prices of goods and services (see Chapter 23).

The extraction, processing, transportation, and disposal of minerals are accompanied by multiple environmental impacts, such as the pollution of air, water, and soil; losses of productive farmlands, soil fertility, and biodiversity; and fragmentation of habitats. Especially, the use of coal as a dominant energy source has important adverse effects on environmental quality (Figure 14.12) and public health. As we deplete most accessible mineral reserves, the increasing consumptive demand will, without question, make it necessary to exploit less accessible and lower-grade deposits. This will mean an increase in the rate of environmental degradation and increased needs, as well, for the disposal of wastes.

Mine lands, through a relatively short-term land use activity, cause long-term deleterious effects on air quality and the quantity and quality of soils, water, and vegetation, unless rehabilitation plans for making them productive are in place after their abandonment (see Chapter 24 on ecosystem restoration). Mining activities can be a cause of land use conflicts if mineral resources are found in ecologically sensitive and important protected natural areas such as national parks and wildlife refuges. Destruction of landforms and vegetation and consequent modification of microclimates are the most obvious impacts of mining on the structure of terrestrial and aquatic ecosystems at local and regional levels. Such areas are called **derelict** lands. On the other hand,

at the global level, mining activities contribute to the increasing atmospheric emissions of greenhouse gases, consequent changes in biogeochemical cycles and thermal properties of the atmosphere due to increasing consumption of fossil fuels, and disturbance of vegetation and soils (see Chapter 21).

Waste Production and Disposal—Eastern United States and Canada

Common to all stages of mining is the production of waste materials in the form of solids, liquids, or gases that pollute the atmosphere, the hydrosphere (fresh- or saltwater or groundwater), the lithosphere (soils, rocks, or sediments), and the biosphere (plants and animals). Solid wastes are generally produced by the extraction phase and include waste rocks, tailings, spoil banks, and particulate matter (e.g., dust). For example, an average metal mine in Canada immediately rejects 42% of the total mined material as waste rock, 52% of the ore is separated at the mill as tailings, and only 2% is retained as the commodity for which the ore is extracted (Godin 1991). Liquid wastes range from leached soluble acids to leached metals. The main types of gaseous emissions to the atmosphere by mining activities are the oxides of sulfur, nitrogen, and carbon; radioactive particles; metals; and liquid and solid particulate matter (aerosols).

Upon their exposure to air and water, chemically reactive minerals in ores and coal deposits generate acids and soluble metal compounds. Especially, iron sulfide (FeS_2) minerals (such as pyrite, marcasite, and pyrrhotite) in the presence of air and water form sulfuric acid (H_2SO_4), other sulfate (SO_4) compounds, and iron oxides that are, in turn, washed into surface waters by runoff and into soils and groundwater by leaching and seepage. The resulting acidification, which degrades and destroys the quality of surface water, groundwater, and aquatic life is known as **acid mine drainage** (AMD). The principal chemical reactions that result in AMD are summarized as follows (Singer and Stumm 1970; Nordstrom and Alpers 1999a, 1999b):

$$FeS_{2(s)} + 7/2\ O_2 + H_2O = Fe^{++} + 2SO_4 + 2H^+$$

$$Fe^{++} + 1/4O_2 + H^+ = Fe^{+++} + 1/2H_2O$$

$$Fe^{+++} + 3H_2O = Fe(OH)_{3(S)} + 3H^+$$

$$FeS_{2(S)} + 14Fe^{+++} + 8H_2O = 15Fe^{++} + 2SO_4 + 16H^+$$

Iron is generally accompanied by aluminum released through acid attack of gangue minerals. Aluminum complicates the acid mine drainage picture by its tendency to form aluminum hydroxides with the release of additional hydrogen ions, thereby exacerbating the acidity problem:

$$Al^{+++} + HOH = Al^{++}(OH) + H^+$$

$$Al^{++}(OH) + HOH = Al^{+}(OH)_2 + H^+$$

$$Al^{+}(OH)_2 + HOH = Al(OH)_3 + H^+$$

Although the pH of AMD is generally between 2 and 4, a recent analysis of drainage from an abandoned copper mine at Iron Mountain in California recorded such low pH values (−3.6 to 1) that new tests had to be devised to measure them (Nordstrom et al. 2000).

Acid mine drainage is fairly common in abandoned coal mining areas in temperate climates (Western Europe, eastern United States). Until recently, it was generally believed that the bacterium *Acidothiobacillus ferrooxidans* was responsible for generating and accelerating acid production from pyrites. Recent studies show that the acid mine drainage picture is a little more complex than believed earlier: *A. ferrooxidans* seems to have a lesser role, whereas *Leptospirillum ferrooxidans* plays a greater role at very low pH (Shrenk et al. 1998). Further studies have shown that a specialized group of prokaryotes, the Archaea, plays a fundamental role in the generation of AMD at elevated temperatures (40°C) and low pH (Edwards et al. 2000). Mining activities release about 150 million tons of sulfur (the primary cause of acid mine drainage) annually, constituting about 50% of the sulfate

input into the ocean (Brimblecombe and Lein 1989). These studies underscore the role of these Archaea in the global cycling of iron and sulfur.

Waste Production and Disposal—Western United States and Canada

In semiarid and arid areas where evapotranspiration most often exceeds precipitation and the newly excavated parent materials are high in easily weathered fine clays, the problems are different. Here, the predominance of clays, a preponderance of sodium and calcium salts, sodium–clay complexes, and alkaline materials hinder vegetation growth. For such overburden materials, the electrical conductivity, saturation percentages, and the role of exchangeable sodium require evaluation (Table 14.5). Mining for coal involves large areas, and a large body of research work has accumulated in the last 30 years. The rehabilitation of overburden materials—both of acid and alkaline–calcareous–sodic parent materials—in temperate, semiarid, and arid climates is described in Chapter 24.

Metal mining entails a few more processing steps than coal, and the following example may be illustrative. Mining operations that separate gold from the ore and extract gold from the tailings left over from mining other minerals use the deadly poison cyanide. After extraction, the cyanide-laced waste is stored in diked-off lagoons. In January 2000, an estimated 100,000 m³ of such cyanide waste escaped from lagoons in the Romanian mining town of Baria Mare. It contaminated the Tisa and Danube rivers (Europe's largest waterways), killing virtually all aquatic life, including microorganisms (Koenig 2000). This accidental spill is considered the biggest environmental catastrophe since the explosion at the Chernobyl nuclear power plant in Ukraine in 1986 (see Chapter 17).

Such poisonings of the food chain of plants, animals, and humans through bioaccumulation

Table 14.5
General Characteristics of Overburden Materials in the
Western United States Sampled before Mining

Layer	Depth (m)	Saturation (%)	Electrical Conductivity (S m⁻¹)	SAR[a]	Plant Environment
1	0–0.6	70	0.1	1	Good
2	0.6-3	60	0.2	2	Fair
3	3–6	90	0.5–1.0	5–10	Fair
4	6–11	100	0.2–0.5	1–20	Poor
5	11	150	0.2	20+	Very poor

Source: From Wali, M. K., and F. M. Sandoval. 1975. In *Practices and Problems of Land Reclamation in Western North America,* ed. M. K. Wali, 133–153. Grand Forks: University of North Dakota Press.

[a] Sodium adsorption ratio ($SAR = Na/[(Ca + Mg)/2]^{1/2}$).

of metals and other toxic materials (including copper, zinc, and lead) cause long-term health problems. Sedimentation and turbidity also reduce the biological productivity and diversity of water bodies and destroy habitats of many benthic and freshwater organisms. Other major spills include the flow of 3 million m³ of wastewater with cyanide and copper from a mine in Guyana, South America, into the Essequibo River in 1995, and cyanide spills occurring in Latvia and Kyrgyzstan in the 1990s and in Spain in 1998.

In areas around mines, suspended particulate materials in the air (aerosols) are absorbed by or deposited on vegetation and animals in dry or wet form. These deposits may impede the growth and productivity of plants and animals, and they can produce serious effects on human health (discussed further in Chapter 20). Some mining towns in Montana are surrounded by blackened, treeless landscapes. Particularly, the release of gaseous sulfur dioxide during smelting processes accounts for much of the acidification-related loss and degradation of forests, soil fertility, aquatic habitats, biodiversity, and human health. Smelters emit arsenic, mercury, and other toxic materials, as well as fluoride gas that causes a disease called fluorosis in animals (pain in an animal's joints). The results of smelting near Copperhill and Ducktown in Tennessee, Aspen in Colorado, and Sudbury in Ontario are some examples of how well-functioning terrestrial and aquatic ecosystems can be turned into barren and toxic wastelands.

Radioactive waste materials, waste heat, noise, and aesthetic distortions constitute other important environmental effects of mining. Radioactive waste materials, especially from uranium mill tailings, can be in the form of gases, liquids, or solid particulate matter that remain radioactive for a few hours to thousands of years. When managed improperly, they pose a substantial threat to human and environmental health. Five principal pathways include the diffusion of radon gas directly into indoor air in the case of tailings misused as construction materials; diffusion of radon gas from the piles into the atmosphere; production of gamma radiation in the immediate vicinity of tailings; the dispersal of radioactive materials from tailings by wind or water, or by leaching into surface or groundwater; and biomagnification in the food chain (see Chapter 7).

Improper disposal of such hazardous wastes and accidents in their transportation endanger human and ecosystem health because of their toxicity, ignitability, corrosivity, and reactivity, especially when institutional regulatory regimes are unable to respond quickly to public health concerns (Section C). Waste heat emissions from mining can also raise the temperatures of water bodies sufficiently to kill some organisms, although electric power generation is responsible for much of the waste heat released to aquatic environments. Such waste heat contributes to modification of the local climate, creating islands of higher temperatures relative to surrounding areas, termed heat islands (discussed in Chapter 15). Residents living in close proximity to a mining operation are also adversely affected by noise and visual ugliness.

Industrial Ecology

Realizing the importance of the need for economic and strategic minerals and the impact that their extraction and use has on the environment, a new field of *industrial ecology* has been ushered in.[2] It "involves designing industrial infrastructure as if they were a series of interlocking man-made ecosystems interfacing with the natural global ecosystem....[It] takes the pattern of the natural environment as a model for solving environmental problems" (Tibbs 1992). A recent illustration of "industrial ecology of the automobile" clearly mimics the ecological food webs (Keoleian, Kar, Manion, and Bulkley 1997).

It is now clearly recognized that for "sound industrial ecology, technological fixes" alone will not be enough (Patel 1992):

(1) During the production of manufactured goods, U.S. industry generates 300 million tons or more of hazardous wastes that need to be treated, stored, or disposed. (2) During production, U.S. industry generates more than 600 million tons of non-hazardous wastes that have to be disposed. (3) To meet existing regulations, U.S. manufacturing spends more than $40 billion annually on pollution control. (4) The disposal of wastes generated in 2 and 3 above is an expensive proposition, and these costs are rising faster than the rate of industrial production. This raises the specter of disposal costs becoming the dominant cost of production.

Whether the introduction of new terminology or a new branch of ecology will solve the environmental problems related to mining remains to be seen. What is indisputable is that there is great need for experts in different fields to come together to resolve problems that involve many disciplines. Our efforts should be to make models complete so that they reflect all environmental impacts and attendant costs of manufacturing processes—from resource exploitation to final disposal of the product and ecosystem restoration.

The United States possesses vast deposits of coal (see also Chapter 17 on fuel energy) and over 1 billion short tons are mined each year. Comprehensive legislation to regulate coal mining (only) was enacted at the federal level in 1977 (see Chapters 21 and 24). But several laws pertain to other types of mining: for example, the Mining Law of 1872, the Mining Leasing Act of 1920, and the Mining in the Parks Act of 1976. Even state laws such as the Alaska Native Lands Conservation Act of 1980 guarantee reasonable access to mining claims.

Overall, the mining laws are a "mixed bag." Particular mention must be made of the 1872 mining law that, even today, regulates the mining of hard rock minerals such as silver, gold, lead, uranium, and copper. Under this law, a person or company can lay claim to a 20-acre (8-ha) land parcel (or multiple parcels) at a nominal fee of $5 an acre (0.4 ha). The mining award also includes such surface resources as trees and water needed for mining. Approximately 270 million acres (109 million ha) of land have been mined under this law.

The devastating results from this 128-year-old law have been reported widely. In 1995, a Danish firm obtained rights to minerals worth an estimated $1 billion, but paid the U.S. government a mere $275 in fees. Likewise, in 1994 American Barrick Corporation paid $5,140 for 1,000 acres (405 ha) of public land in Nevada that contained minerals worth $10 billion. Many times, the lands are not mined, but once patents are established, they may be used for other purposes, such as developing resorts (Wuerthner 1998).

An estimated 557,000 abandoned hard rock mine sites have been left unreclaimed and more than 19,000 km of rivers have been poisoned by mining wastes—all of which are exempt from federal regulation and rules to govern and regulate rehabilitation of disturbed ecosystems. The estimated minimum of $32–72 billion in cleanup costs will be borne by the taxpayers. Even within the national parks, there are more than 4,000 abandoned mining sites. Although Congress in 1994 passed a temporary moratorium on the Mining Law of 1872, it has been ineffective.

As was noted earlier, this is not the case for the mining of coal, oil, and gas on federal lands. Other federal laws, including the Mineral Leasing Act of 1920, govern their extraction. Under this act, lands are not sold but leased, and companies pay a 12.5% royalty for these resources to be removed. Government agencies can deny leasing if environmental values are likely to be diminished by mineral extraction. What the preceding examples clearly illustrate is the need for laws that are comprehensive and contemporary in valuation for the overall public good.

Legislative Measures

A major law, the Surface Control and Reclamation Act, was passed in the United States in 1977. This legislation applies only to coal. We discuss the policy aspects of this law in Chapter 20. The act also governs ecosystem rehabilitation restoration in its manifest dimensions (Imes and Wali 1977, 1978), and we discuss this in Chapter 25 on restoration.

Notes

1. Several authoritative reports are now available documenting the enormous environmental costs of extracting crude oil from tar sands and oil shales. These include the 2008 report from Environmental Defense Canada and WWF-UK's 2008 report, "Unconventional Oil: Scraping the Bottom of the Barrel," a study conducted in conjunction with Cooperative Investments.
2. Jelinski et al. (1992) note that the industrial ecology concept seeks to optimize all steps from production to waste disposal and emphasize that the characteristics of the concept of industrial ecology should be "1. proactive, not reactive; 2. designed in, not added on; (3) flexible, not rigid, encompassing, not insular."

Questions

1. What happens to soil and water conditions near mining operations? Do the effects depend on which method is used to extract minerals? How does this affect plants, animals, and people who live in an area where mining occurs?
2. Name two minerals or metals that are mined in your region. What are some possible environmental impacts that result from these operations? Discuss what can be done to curb these impacts, both during and after an area is mined.
3. Discuss the environmental impacts of coal mine waste, its production, and its disposal techniques.
4. What are some of the "heavy metals" contained in mine runoff? What are some of their potential effects on human and environmental health?
5. Discuss the importance of reclamation and rehabilitation practices on mined lands.

Land Use—Urbanization and Transportation

15

Forces Driving Urbanization

Throughout this book, we have emphasized the interconnectedness of ecosystem structure and function. The topics discussed in this chapter are no exception. The connections between land use change and ecosystem response were succinctly captured by Grimm et al. (2008), who stated, "Land changes associated with urbanization drive climate change and pollution, which alter properties of ecosystems at local, regional, and continental levels. … urbanization alters the connectivity of ecosystems, energy, and information among social, physical, and biological systems." We discuss some of these aspects here.

As the centers of trade and commerce; of political governance; and of consumption, production, and the distribution of economic goods and services, cultural–political enrichment, and technological innovation, **urban–industrial ecosystems** play a vital function in the progress of human civilization. Such dense population centers cannot be sustained within their own geographic limits. The sustenance and well-being of urban ecosystems depend on the ecological goods and services provided from regional life-supporting ecosystems such as agriculture, grasslands, forests, and wetlands and what can be imported from distant areas. It was the realization that urban dwellers appropriate a large share of the Earth's resources that led to the development of the **ecological footprint** concept (Rees 1992). This concept demonstrates that the land area needed to support a city is many times larger than land area of the city itself. This concept is explained in detail in Chapters 22 and 26.

The transformation from a rural–agrarian system to an urban–industrial system in which a city's population increases along with related economic, social, and political changes is referred to as **urbanization.** The growth of urban populations is a function of the growth of the urban population over time and net migration from the countryside. Current rates of urban population growth now number more than 1 million people per week globally. Migration into cities results from the interaction of **push factors** (conditions perceived by people to be detrimental to their well-being that induce them to

BOX 15.1   The Second Wave of Urbanization (from United Nations Population Fund 2007)

The huge increase in urban population in poorer countries is part of a "second wave" of demographic, economic, and urban transitions that is much bigger and much faster than the first. The first wave of modern transitions began in Europe and North America in the early eighteenth century. In the course of two centuries (1750–1950), these regions experienced the first demographic transition, the first industrialization, and the first wave of urbanization. This produced the new urban industrial societies that now dominate the world. The process was comparatively gradual and involved a few hundred million people.

In the past half-century, the less developed regions have begun the same transition. Mortality has fallen rapidly and dramatically in most regions, achieving in one or two decades what developed countries accomplished in one or two centuries, and the demographic impacts of these mortality changes have been drastically greater. Declines in fertility are following—quite rapidly in East and Southeast Asia and Latin America and more slowly in Africa.

In both waves, population growth has combined with economic changes to fuel urban transition. Again, however, the speed and scale of urbanization today are far greater than in the past. This implies a variety of problems for cities in poorer countries. They will need to build new urban infrastructures—houses, power, water, sanitation, roads, commercial and productive facilities—more rapidly than cities anywhere during the first wave of urbanization

Two further conditions accentuate the second wave. In the past, overseas migrations relieved pressure on European cities. Many of these migrants, especially to the Americas, settled in new agricultural lands that fed the new cities. Restrictions on international migration make it a minor factor in world urbanization today.

Finally, the speed and the size of the second wave are enhanced by improvements in medical and public health technology, which quickly reduce mortality and enable people to manage their own fertility. Developing and adapting forms of political, social, and economic organizations to meet the needs of the new urban world is a much greater challenge.

migrate) and **pull factors** (perceived conditions that attract people to move to a new location) (Box 15.1). Examples of push factors generally include environmental degradation, inadequate agricultural land, maldistribution of agricultural land, and economic insecurity. Examples of pull factors generally include better opportunities for jobs and a higher quality of life, with access to social services such as health care and education.

Trends of Urbanization

Rapidly growing urban populations coupled with the often uncontrolled growth of urban land lead to the expansion of residential, commercial, industrial, and transportation land uses into what were once life-supporting ecosystems such as productive farmlands, forests, wetlands, and nature conservation areas; this process is called **urban sprawl.** Such urban–industrial land uses have the potential to affect public health and environmental quality drastically by increasing demands on limited energy, materials, land, and water as well as the need to handle the wastes generated and the resulting pollutants released into air, water, and soil (Figure 15.1).

The city represented in Figure 15.1 was originally built on a river floodplain close to agriculture, water supply, and transportation (a typical landscape position for urban areas). As the city grew, the air increasingly warmed

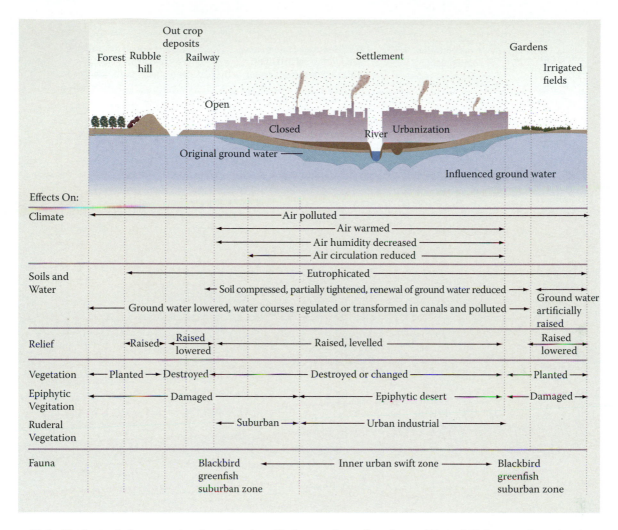

Figure 15.1 Ecological characterization of a city. (Redrawn from Starfinger, U., and Sukopp. 1994. Assessment of urban biotopes for nature conservation. In *Landscape Planning and Ecological Networks*, ed. E. A. Cook and H. N. van Lier. New York: Elsevier.)

and became polluted. Buildings interrupted the flow of air to the soil, effectively producing a closed, warm, polluted canopy of air overhead. Human wastes and fertilizer discharged to the river cause eutrophication, while wells drew out groundwater faster than it was replenished, lowering the water table under the city (Figure 15.1). In Mexico City, Mexico, withdrawal of groundwater has actually caused the ground to sink 10 m in the past century. In contrast, irrigation of agricultural fields outside the city artificially raises the water table there. The city also tends to degrade nearby ecosystems, leading to declines in biodiversity (Alberti 2005). Increased rates of urbanization also have important global implications for altering local climatic patterns (Figure 15.2).

Growing Global Urban Population

The global urban population was 750 million in 1950, 2.6 billion in 1995, and projected to rise to nearly 5 billion (60% of the world's population) by 2030 (Figure 15.3). Almost all of this increase (>80%) will occur in developing countries (Figure 15.4). Global urban population increased from 37% of the world population in 1975 to 47% of the world population in 2000. The urban share is projected to rise to 57% in developing countries by 2025. The urban population in developing countries is growing on average at 3.5% per year, compared to less than 1% in developed countries (WRI 1994).

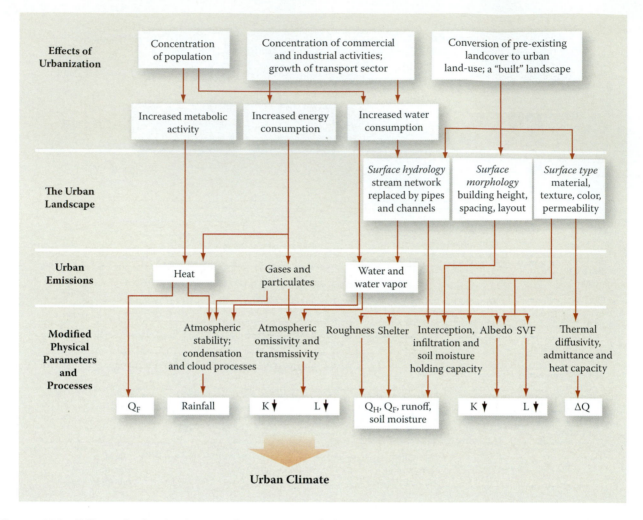

Figure 15.2 Effects of urbanization on radiation, energy balance, and local climate. (Redrawn from Cleugh, H. 1995. In *Future Climates of the World; a Modeling Perspective,* ed. A. Henderson-Sellers, 477–509. Amsterdam: Elsevier Science B.V.) $K\downarrow$, incoming shortwave radiation; $L\downarrow$, downwelling longwave radiation; SVF, sky view factor (the percentage of a point's field of view that is occupied by sky as opposed to buildings, trees, or any other object in the landscape); Q_F, anthropogenically generated heat flux; and ΔQ_S, heat storage flux.

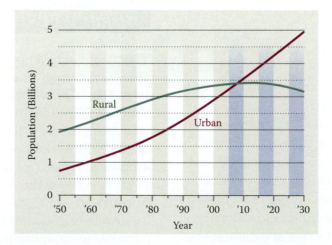

Figure 15.3 Urban and rural population of the world, 1950–2030. (Redrawn from Figure 1, p. 9, *UN World Urbanization Prospects: the 2005 Revision.*)

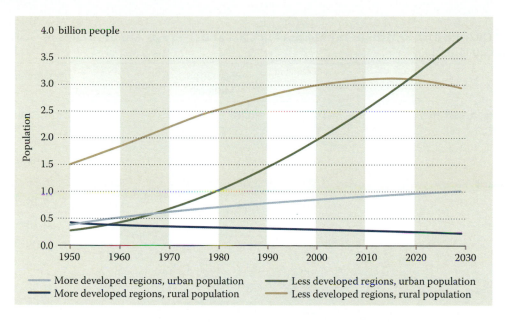

Figure 15.4 Urban and rural population of more developed regions and less developed regions, 1950–2030. (Redrawn from fact sheet 5, Figure 6, in *UN World Urbanization Prospects: The 2005 Revision.*)

Of people in developed countries, 60–80% already live in cities. The affluent urban population (20% of the world's population) consumes more than 70% of the world's primary energy (oil, coal, and natural gas) and almost 80% of the raw materials.

Migration-driven population growth can outstrip job availability; the supply of housing; the capacity of physical infrastructure such as waste disposal and sanitation facilities, distribution systems of water, electricity, and roads; and social services such as health, education, and social security. This imbalance between the demand and supply gives rise to increases in poverty, homelessness, crime, environmental degradation, and deterioration of infrastructure and public health conditions and to scarcities in drinking water and electricity.

Cities Expand

The expansion of cities, as noted earlier, will continue to dominate in developing countries. Population growth between 1975 and 2005 ranged between 2 and nearly 6% in developing countries, while it was 1% or less in the developed countries (Table 15.1). Most of the cities with high urban growth rates are in Asia. In the 40-year period between 1975 and 2015, Lagos and Dhaka are expected to have grown eight times; the population of Delhi will

have more than quadrupled, and it will have tripled for Mumbai and Jakarta and doubled in a number of cities. The megacities, as they are called now, will need resources in unprecedented numbers and quantities (see Figure 15.5 for regional trends in urban populations).

The number of **urban poor** is predicted to reach 1 billion over the next 25 years, more than double today's value (Watson et al. 1998). Between 1950 and 1990, urban population in developing countries expanded by more than 1 billion inhabitants and will have increased by another 2 billion by 2015, exceeding the urban population in developed countries (Piel 1997). About 30–70% of the 1 billion people who have migrated from villages into urban areas in developing countries live in shelters that they have built illegally (on land they do not own) called **shantytowns** (or "citas miserias," "bustees," "bidonvilles," or "gecekondu") (Piel 1997).

Currently, 220 million urban residents in developing countries lack access to potable drinking water, 350 million have no access to basic sanitation, and 1 billion have no solid waste collection service (Watson et al. 1998). Piel (1997) states that in Latin America, where the population is fully 70% urban, 30% of households lack piped water and sanitation; in the cities of Asia, 50% lack these amenities,

Table 15.1
Population of Cities with ≥10 Million Inhabitants in 2005 and Average Annual Rates of Growth, 1975–2005 and 2005–2015

Urban Area/City[a]	Population (Millions)				Avg. Annual Rate of Change (%)	
	1975	2000	2005	2015	1975–2005	2005–2015
Tokyo	26.6	34.4	35.2	35.5	0.93	0.08
Mexico City	10.7	18.1	19.4	21.6	1.99	1.05
New York–Newark	15.9	17.8	18.7	19.9	0.55	0.60
Sao Paulo	9.6	17.1	18.3	20.5	2.15	1.13
Mumbai (Bombay)	7.1	16.1	18.2	21.9	3.15	1.84
Delhi	4.4	12.4	15.0	18.6	4.08	2.12
Shanghai	7.3	13.2	14.5	17.2	2.28	1.72
Kolkatta (Calcutta)	7.9	13.1	14.3	17.0	1.98	1.73
Jakarta	4.8	11.1	13.2	16.8	3.37	2.41
Buenos Aires	8.7	11.8	12.6	13.4	1.20	0.65
Dhaka	2.2	10.2	12.4	16.8	5.81	3.04
Los Angeles–Long Beach–Santa Ana	8.9	11.8	12.3	13.1	1.07	0.63
Karachi	4.0	10.0	11.6	15.2	3.56	2.67
Rio de Janeiro	7.6	10.8	11.5	12.8	1.39	1.0
Osaka–Kobe	9.8	11.2	11.3	11.3	0.45	0.04

Source: http://www.peopleandplanet.net/graphs/Megacities (accessed February 8, 2008).

[a] In the 40-year period between 1975 and 2015, the population of Lagos (Nigeria) is expected to have grown eight times, and more than two times for Beijing (China), Cairo (Egypt), and Manila (Philippines).

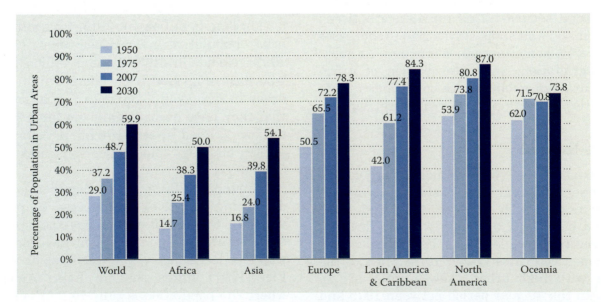

Figure 15.5 Percentage of population residing in urban areas by major area, 1950, 1975, 2007, and 2030. (Redrawn from Figure 3, p. 12, in 2006 *UN World Urbanization Prospects: The 2005 Revision.*)

and in Africa, where the cities are smaller but the fastest growing, 70% lack them.

Land Use

These local and regional activities have global consequences for the well-being of humans and their environment, such as loss of productive ecosystems and greenhouse gas (GHG) emissions (discussed later in Chapter 21). Human population growth is not the only primary factor that stimulates urban sprawl. For example, urban growth in the United States encroaches on surrounding lands as residents abandon inner cities due to the decline in their environmental quality and move to the suburbs. Between 1960 and 1990, urban sprawl in Ohio increased more rapidly than population growth (Thompson 2000).

In total, the world's total urban area grew from 21 million ha in 1982 to 26 million ha in 1992. This resulted in the irreversible conversion of slightly more than 2 million ha of forestland, 1.5 million ha of cropland, 943,500 ha of pastureland, and 774,000 ha of rangeland to urban land uses in one decade (WRI 1994). In developing countries, 476,000 ha of arable land are being transformed annually to urban–industrial uses, threatening their food security.

Generally, the loss of productive farmlands leads to the intensification of agricultural production on the remaining arable lands with higher inputs of fossil fuel energy, fertilizers, pesticides, irrigation water, and mechanization (see Chapter 12). In addition, to compensate for this loss of cropland, marginal (agriculturally less suitable) but ecologically significant and vulnerable ecosystems are increasingly being converted for food production, which in turn results in salinization, soil erosion, loss of biodiversity, and the depletion of aquifers.

Coastal ecosystems, such as wetlands, tidal flats, salt-water marshes, mangrove swamps, beaches, lagoons, and dunes, are one of the ecologically important and fragile ecosystem types in which an urban population of about 1 billion already lives worldwide. As an example, in the past 150 years, filling for urban-related development along San Francisco Bay, the most highly urbanized estuary in the United States, has reduced the spatial extent of the bay by one-third and caused an 80% loss of the 81,000 ha of coastal marshes. Similarly, 4,000 ha of lagoons and swamps in India were filled to provide housing for 100,000 middle-class families, causing local flood problems and an annual loss of 25,000 tons of fish production.

Draining of wetlands; construction of houses, resorts, and roads; and filling along the shoreline are the main urban activities that threaten roughly half of the world's coasts (WRI 1995). Coastal areas also suffer from heavy human impacts due to pollutants in roadway and urban runoff. Though the composition, fate, transport, and toxicity of these pollutants change temporally and spatially, polluted runoff results in the decreased quality of receiving water bodies as well as an increased potential for bioaccumulation in the food chain (Table 15.2).

Urban Needs

Currently, urban–industrial ecosystems use about 75% of the world's natural resources and generate wastes of similar magnitude, although they are spatially restricted to about 2% of the Earth's land surface (Girardet 1999). For example, the Sears Tower in Chicago uses more energy in 1 day than an average American city of 150,000 inhabitants or an Indian city of more than 1 million inhabitants (Hahn and Simonis 1991). The increasing energy and material consumption in urban–industrial areas is coupled with the increasing extraction and processing of energy and mineral resources with the accompanying deleterious effects on the environment.

Urban Transportation and Impacts

Increasing numbers of roads and vehicles are linked in a positive feedback loop. The factors that control the number of vehicles and roads

Table 15.2
Total Heavy Metal Concentrations and Other Pollutants in Roadway and Urban Runoff

Pollutants	Roadway Runoff	Urban Runoff	Drinking Water Standard (MCL)[a] (mg/L)
Heavy metals (µg/L)			
Cadmium	0.17–12	0.02–13,730	0.005
Chromium	1.5–110	1–2300	0.1
Cobalt	0.05–13.7	1.3–5.4	
Copper	3–1200	0.06–1410	1.3
Iron	130–45,000	0.07–440,000	0.3
Lead	3–2100	0.053–26,000	0
Nickel	1–57	0.06–1410	
Zinc	10–1200	7–4600	5.0
Mercury	0.076–5.6	0.05–67	0.002
Aluminum	30–4,000	1–49,000	0.05–0.2
Organic chemicals			
Total PAH (mg/L)	1.86–18.2	2.4E-4–1.3E-2	0.0002[b]
Benzo(pyrene)		2.5E-6–1E-2	0
Fluoranthene		3E-5–5.6E-2	3.96
Naphthalene		3.6E-5–2.3E-3	0.62
Phenanthrene		4.5E-5–1E-2	
Tetrachloroethylene		0.0045–0.043	0
Heptachlorepoxide		<0.0002	0
Oil and grease	<1–480	0.001–110	

Source: From Grant et al. 2003.
[a] Maximum contaminant levels for drinking water quality, as of the *Federal Register* dated March 21, 2000.
[b] U.S. aquatic regulation (freshwater chronic).

include growing urban population, affluence (increasing income levels), consideration of the automobile industry as an important generator of economic growth, lack of public transportation, and the dispersed forms of cities. Globally, the number of motor vehicles was 580 million in 1990 and is expected to be 816 million by 2010, a 16-fold increase relative to 1950.[1] Wealthier regions of the world have higher car ownership rates than low-income countries. For example, in the United States, average car ownership in 1993 was 561 cars per 1,000 residents, 58% of households own two or more cars, and 20% own three or more cars (WRI 1994). By 2006, the number of cars exceeded 250 million (U.S. Bureau of Transit Statistics 2006) In 1993, average car ownership in developing countries ranged from about 68 cars

per 1,000 residents in Latin America and the Caribbean to 29 cars per 1,000 residents in East Asia and the Pacific area to about 14 cars per 1,000 residents in Africa. Worldwide, the number of cars and light trucks is estimated at 622 million (Figure 15.6) (Renner 2008).[2]

Reliance on privately owned cars is increasing, especially in developing countries where consumer (market) potential toward motorization increases with increasing affluence. The numbers are rising at a phenomenal rate. China produced 8.1 million vehicles in 2007 and expected to produce 9.3 million in 2008 (equivalent to the number in the United States) (Renner 2008). India is manufacturing vehicles at a fast pace (see Box 15.2); the cheapest car introduced there in 2008 was priced at about $2,800. High structural (involuntary)

BOX 15.2   Car Production in India

"Indian market analysts," notes automobileindia.com,[3] "are jubilant over the fact that India has the fastest growing automobile industry (passenger cars segment) in the world today. This is seen as the result of the rapid increase of the middle-class population in the country combined with the government's pro-development policies."

The findings of the International Organization of Motor Vehicle Manufacturers reveal that in India, car production had grown to 30% in 2004; Brazil came in a distant second with an increase of 17%.

The following table attempts to capture in figures the rapid growth of production of the automobile sector in India. Values refer to the number of vehicles produced.

Type of Vehicle	2002–03	2003–04	2004–05	2005–06
Passenger	557,400	782,550	960,480	1,045,880
Utility	114,470	146,330	182,020	196,380
Multipurpose	51,450	60,670	67,370	66,665
Total	723,320	989,550	129,870	1,308,925

What will be the impact of increasing vehicles on Indian cities? As an example, the veteran environmental commentator Sunita Narain (2007) writes thus of the "Silicon Valley" of India: "Bangalore adds more vehicles than Delhi each day—over 1,000. It has less road space than Delhi and more green space. The future the city has is ugly and hopefully not inevitable." She characterizes the hope of orderly city development as an "infantile illusion."

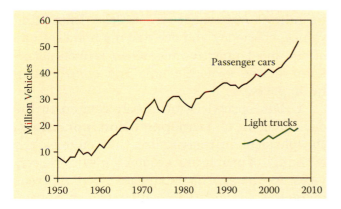

Figure 15.6 World light-vehicle production, 1950–2007. (Renner, 2008. World light-weight vehicle production, 1950–2007. http://www.worldwatch.org/vsonline)

unemployment and abundant and cheap labor constitute major incentives for domestic and transnational automobile industries to intensify automobile manufacturing in developing countries. The increase in vehicle numbers projected just for the next two decades is phenomenal (Table 15.3). Economic policies that view the increasing per-capita car ownership as a means for national economic growth can also discourage the design and spread of urban transportation systems that facilitate walking, bicycling, and public transportation infrastructure such as minibuses/buses, railroads, and subways.

Urban sprawl also promotes reliance on car and gasoline consumption, replacing trips formerly made by foot, cycling, or public transportation. The dispersed form of cities results from a combination of factors, including changing modes of economic production, low transportation costs, abundant resources, issues of land planning and policy, and land markets.

North America is the best example of the dispersed patterns of urban land uses: Land is abundant, fuel costs have been low until recently, the availability of lower-priced housing at the urban periphery (suburbs) in effect subsidizes automobile use, and service sectors dominate the economy. The transition from mass manufacturing industrial production to service- and high-technology-based sectors in response to the globally keen competition among nations has reduced the dependence

Table 15.3
Urban Transportation Characteristics in Global Cities, 1990

Global Cities	Mean Transportation Energy Consumption (GJ per capita)	Mean Public Transportation (%)	Mean Transportation Fuel Efficiency (MJ km^{-1})	Metropolitan Density (people ha^{-1})
American[a]	64.3	1	5.03	14.2
Australian[b]	39.4	2	5.11	12.2
Canadian[c]	39.1	3	4.85	28.5
European[d]	25.6	5	3.79	49.9
Asian[e]	12.8	11	3.53	161.9

Source: From Newman, P., and J. Kenworthy. 1999. *Sustainability and Cities: Overcoming Automobile Dependence.* Washington, D.C.: Island Press.

[a] Includes Portland, Sacramento, San Diego, Phoenix, Denver, Boston, Houston, Chicago, Washington, D.C., San Francisco, Detroit, Los Angeles, and New York.

[b] Includes Canberra, Perth, Adelaide, Brisbane, Melbourne, and Sydney.

[c] Includes Calgary, Winnipeg, Edmonton, Vancouver, Toronto, Montreal, and Ottawa.

[d] Includes Frankfurt, Amsterdam, Zurich, Brussels, Munich, Stockholm, Vienna, Hamburg, Copenhagen, London, and Paris.

[e] Includes Singapore, Tokyo, Hong Kong, Kuala Lumpur, Surabaya, Jakarta, Bangkok, Seoul, Beijing, and Manila.

on centralized workplaces and transportation schemes. This has in turn shifted job opportunities to the suburbs, where service sectors (capital) have moved, promoting the decentralization of urban life and leaving city centers (downtown) as "ghost cities." This has been followed by the recent attempts to revitalize them.

The current energy-intensive and ecologically incompatible trends of urban and transportation growth are not sustainable in the long term. Worldwide, at least a third of urban land is allocated to roads, parking lots, and other automobile-related land uses. In the United States, more land is devoted to transportation than to housing. About 20% of all energy produced globally is used for transportation. From 1971 to 1992, worldwide energy use in the transportation sector alone grew an average 2.7% per year. Increasing motor vehicle use is the primary cause of the increasing dependence on foreign petroleum by countries with scarce energy supplies (Table 15.4). This is correlated with the commuting distances using automobiles and the availability of mass transit (Figure 15.7).

The urban demand for fuels (e.g., gasoline, diesel, electricity, and natural gas) continues to grow as supplies diminish. Rising prices, supply disruptions, and local and cumulative environmental impacts of fossil fuels will eventually make us change our habits of transportation and rely on a more diversified range of transportation solutions—from higher efficiency vehicles to use of alternative fuels, including renewable resources. Improving vehicle efficiency and our driving habits is one of the most effective means to reduce petroleum-related socioeconomic and environmental impacts.

Urban sprawl is intimately related to the development of **transportation corridors,** the physical infrastructure that enables the exchange of people and economic goods and services. As the spatial expansion of urban areas increases, so does the need for transportation and the space to accommodate it. Growing urban areas and motorization adversely affect quality and quantity of natural resources in three main ways: (1) increased loss and degradation of forests, prime farmlands, and coastal ecosystems; (2) increased depletion of nonrenewable fossil fuels and other mineral resources; and (3) increased wastes and emissions released into air, water, and soil.

The impact of transportation can be visualized in a wider context of the interconnectedness of economic growth, transport services, and environmental impacts—linkages that

Table 15.4
Passenger Car Ownership per 1,000 Persons and Total Absolute Numbers in Major World Regions in 2006 and Numbers Predicted for 2010, 2020, and 2030

Region	Cars per 1,000 Persons				Millions of Cars				Growth (%)/year
	2006	2010	2020	2030	2006	2010	2020	2030	2006–2030
North America	474	476	490	503	210	219	246	269	1.0
Western Europe	433	454	490	511	233	248	275	290	1.0
OECD Pacific	416	439	470	481	84	89	95	95	0.7
OECD	446	460	487	503	527	557	616	654	1.0
Latin America	113	128	151	174	47	56	73	91	2.8
Middle East and Africa	25	29	39	49	20	25	42	64	4.9
South Asia	10	15	32	67	15	24	59	135	9.4
Southeast Asia	92	108	138	167	37	45	65	85	3.6
China	18	30	53	86	23	41	76	126	7.5
OPEC	37	45	66	94	22	28	47	75	5.2
DCs	33	41	60	87	164	219	362	576	5.3
FSU	130	143	167	192	37	40	46	52	1.4
Other Europe	221	232	252	270	12	12	13	13	0.5
Transition economies	144	157	180	204	49	53	60	65	1.2
World	113	121	136	157	740	828	1037	1296	2.3

Source: From OPEC World Outlook. 2008.

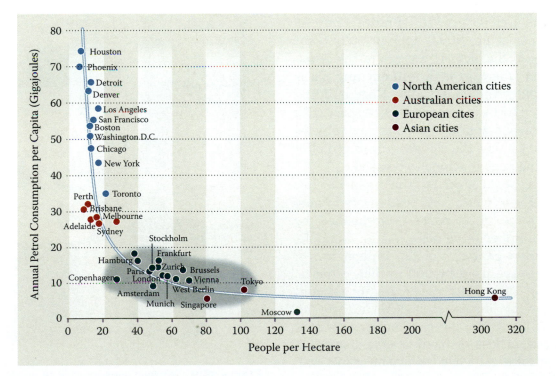

Figure 15.7 Gasoline consumption versus urban density: Low gasoline use is associated with residential crowding. (Redrawn from Figure 4.1, p. 87, Newman and Jennings, 2008. Gasoline consumption versus urban density: low gasoline use is associated with residential crowding.)

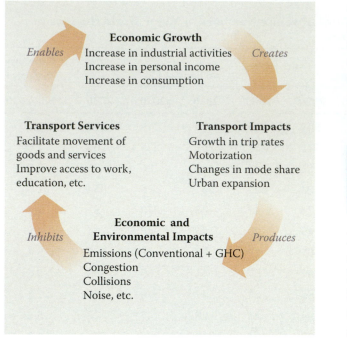

Figure 15.8 The vicious cycle of growth. (Ganken-heimer, R., L. T. Molina, et al. 2002. The MCMA transportation systems: Mobility of air pollution. In *Air quality in the Mexico Megacity*. Ed. L T. Molina and M. J. Molina, 213–284. Dordrecht, the Netherlands: Kluwer Academic Publishers.)

have been appropriately characterized as "the vicious cycle of growth" (Figure 15.8). As Gakenheimer et al. (2002) appropriately point out, this cycle characterizes cities in the developing world where growth without any plans occurs "rapidly and chaotically." Road-building and orderly parking areas cannot keep pace with such rapid urbanization. Although new vehicles are added each day, the old, gas-guzzling, pollution-emitting ones continue to be used.

In developed countries, the problems are different. Even when the road system may seem adequate, it is often overwhelmed by the sheer number of vehicles. As an example, in the United States, the 6.2 million km of public roads are used by 200 million vehicles, and 10% of the public roads are in national forests, having direct ecological impacts on 15–20% of the U.S. land area such as habitat fragmentation; loss of biodiversity; disruption of the hydrological cycle; pollution of air, soil, and water; and soil erosion and sedimentation (Forman and Alexander 1998). As the number of privately owned cars increases, the provision

and facilitation of access to people, goods, and services decrease due to clogged urban roads and limited parking spaces, also called **traffic congestion.** For example, if current trends continue, U.S. motorists will spend an average of 2 years of their lifetimes sitting in traffic jams.

Urban–Industrial Wastes and Biogeochemical Cycles

Increased levels of consumption of fossil fuels and materials by urban population naturally lead to increased levels of **waste generation** in the form of gases, solids, and liquids (Figure 15.9). In addition, improper waste disposal into water bodies, open dumps, and poorly designed landfills is another principal cause of disruption of biogeochemical cycles. Cycles of gases, sediments, and water and flows of energy form the link not only among the atmosphere, hydrosphere, lithosphere, and biosphere but also among the local, regional,

Figure 15.9 Average consumption of goods and generation of wastes in an urban ecosystem. (Redrawn from Williams, M., ed. 1993. *Planet Management.* New York: Oxford University Press.)

and global levels. Therefore, the generation of one form of waste adversely affects not only the particular ecological component in which its form prevails but also the other ecological components. Waste generation that exceeds assimilative capacities at the local and regional levels is capable of altering biogeochemical cycles and energy budgets at the global level (discussed in Chapter 21).

The collective production of all solid, liquid, and gaseous waste is known as the **waste stream** and includes agricultural waste (see Chapter 12), mineral waste (previously discussed in Chapter 14), industrial and military waste (e.g., radioactive and hazardous or toxic materials, ashes, soot, and dust), and municipal wastes including paper products, yard wastes (e.g., grass clippings, leaves, brush, and branches), food wastes, plastics, metals, glass, textiles, and rubber. Waste streams can be classified based on the materials' **toxicity, persistence,** and potential for **bioaccumulation** and **biomagnification** as nonhazardous, hazardous, and radioactive wastes.

Chemicals with high persistence (slowly biodegrading or nonbiodegradable), high potentials of bioaccumulation and biomagnification in the food chain, and high toxicity are called **persistent organic and inorganic pollutants** and include a diversity of organic compounds and metals. **Hazardous waste** comprises wastes that exhibit one of the following features: ignitability (spontaneously combustible), corrosivity, reactivity (explosiveness and toxicity when mixed with water), and toxicity (harmful or fatal when ingested or absorbed). In the United States, more than 13 billion tons of waste stream are generated collectively each year in response to an increase in human population and consumption and the use of nonrecyclable materials, and 2% (273 million tons) of the total are legally regulated hazardous waste. **Nonhazardous wastes** include urban–industrial and agricultural wastes that do not meet the definitions of the other types of wastes. An increasing awareness about methods to manage hazardous and nonhazardous waste cost effectively, with minimal risk to human health and the environment, would help to tap the vast potential of world urban centers to reduce, reuse, and recycle the amount of energy and materials consumed.

Liquid Wastes

Two main types of liquid waste are generated by urban–industrial activities: municipal sewage and domestic wastewater and industry-originated wastewater. **Municipal sewage** is 98% water and a variety of organic and toxic compounds and is usually discharged, whether treated or untreated, into bodies of water near urban–industrial ecosystems with the assumption that the solution to pollution is dilution. For example, the city and county of Los Angeles, California, discharge 2.6 million m^3 of wastewater per day into estuaries and coastal waters. However, the dilution assumption no longer works with increasing volumes of wastes discharged because of the limited assimilative capacities of receiving waters and their incomplete mixing with coastal waters. Aquatic microorganisms feeding on the organic pollutants discharged into water bodies use oxygen in the water, leading to low oxygen levels in the water, and the nitrogen and phosphorus in sewage often lead to **eutrophication,** in which extensive algal growth further increases the biological oxygen demand. Proper mechanical, chemical, and biological treatments are used to clean up sewage water to yield outgoing water (**effluent**) that will not harm the environment and can be of high (even drinkable) quality.

Although the designs of sewage treatment plants vary, the most common municipal type provides secondary treatment (Figure 15.10). In Figure 15.10, sewage enters at the left through a series of screens, followed by primary treatment where oil and grease are skimmed from the top and solids are settled. Secondary treatment in this design involves suspended bacteria provided with oxygen injected from the bottom. A clarifier allows the bacteria and particles to settle out as sludge, part of which is recycled to the digester to maintain a large inoculum of bacteria there. The remainder is further digested and dewatered prior to disposal on land as fertilizer. The clarified effluent is treated (usually with chlorine) to kill disease bacteria and viruses that can cause

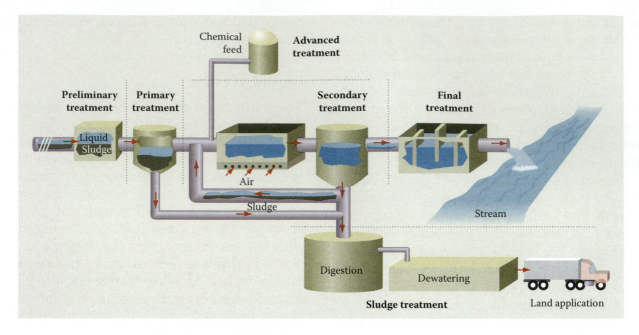

Figure 15.10 Diagrammatic view of a secondary sewage treatment plant. (Redrawn from K. Mancl, *Wastewater Treatment Principles and Regulations.* The Ohio State University Extension Service Fact Sheet AEX-0768-96. http:// ohioline.osu.edu/aex-fact/0768.html)

serious illnesses such as typhoid, dysentery, diarrhea, and cholera (see Chapter 16) before discharge to surface waters. Sometimes, additional (tertiary) treatment is done to remove specific compounds, especially the nutrients phosphorus and nitrogen, prior to chlorination and discharge. Starting in the 1970s, the removal of phosphate by precipitation with aluminum sulfate at the Detroit, Michigan, sewage treatment plant significantly reduced the phosphorus input to Lake Erie, helping decrease the blooms of algae and increasing the amount of oxygen in the lake.

Most municipal and industrial pollutants (or wastes) are relatively easy to identify or detect, typically discharging from a pipe, and hence are called **point sources of pollution;** pollutants from agriculture, uncollected sewage, and urban stormwater are **nonpoint sources of pollution.** These are much more difficult to regulate or reduce. There are two basic ways to deal with the wastes and pollutants that we produce: mitigation and prevention. **Prevention** is ecologically long term and provides the most economically cost-effective solutions to environmental problems if environmental consequences of human activities (externalities) are valued (internalized into) properly in the process of decision-making (discussed

in Section C). In the case of our ignorance of how to prevent environmental degradation, **mitigation** becomes unavoidable in order to minimize adverse effects and buy us some time until we develop preventive measures.

Solid Wastes

Solid wastes (also called **trash** or **refuse**) are any mixture of solid materials discarded, and more than 5 billion tons of solid waste are generated every year in the United States. **Sources of solid waste** generation are generally classified under two broad groups: nonmunicipal waste (agricultural waste, mining waste, and industrial waste) and municipal waste. In the United States, **nonmunicipal wastes** constitute about 97% of all the solid waste generated; 51% is from agriculture, 36% from mining, and 10% from industry, and the rest (3%) is made up of municipal solid wastes (Figure 15.11). **Municipal solid waste** (**MSW**) refers to solid materials disposed of by residential, commercial, and institutional sectors, and it increased from 88 million tons in 1960 to 222 million tons in 2000 in the United States (U.S. EPA 1996b).

As a mitigative measure, the disposal of solid wastes is carried out in the world in six basic ways: dumping, burying, burning, exporting,

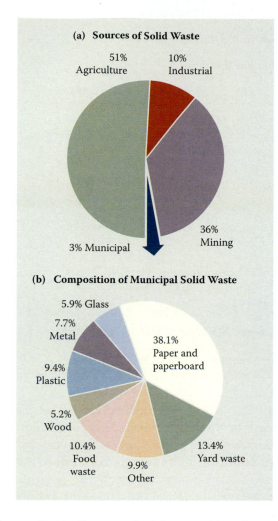

(a) Sources of Solid Waste

51% Agriculture

10% Industrial

36% Mining

3% Municipal

(b) Composition of Municipal Solid Waste

5.9% Glass

7.7% Metal

9.4% Plastic

5.2% Wood

10.4% Food waste

9.9% Other

13.4% Yard waste

38.1% Paper and paperboard

Figure 15.11 Sources of solid waste (a) and composition of municipal solid waste (b) in the United States. (Redrawn from Kaufmann, D. G., and C. M. Franz. 2000. *Biosphere 2000: Protecting Our Global Environment.* 3rd ed. Dubuque, IO: Kendall-Hunt.)

detoxifying, and composting. The disposal of waste by **open dumping** into aquatic and terrestrial ecosystems is humankind's most primitive and unsanitary means; this once accounted for more than half of the waste generated in the United States and the world. The land and the ocean seemed so vast that little thought was given to the environmental effects of open dumping.

Every method of waste disposal necessitates the selection of a biogeochemically suitable landfill site for the technological storage of wastes, through an ecological analysis of abiotic (e.g., prevailing wind; geological stability and impermeability; proximity of water table, surface water, and organisms; and environmental vulnerability) and biotic (e.g.,

protection of human health, biodiversity, and food chain) factors of the site. However, limited suitable sites for landfills; limited carrying capacity of landfills; leaching, runoff, and emissions from landfills; and people's opposition to the location of landfills nearby prove the mitigation of waste problems by waste disposal to be a short-term remedy.

One issue associated with the establishment of landfills is the production of methane gas within them. Methane gas is released from the anaerobic decomposition of organic matter as the solid waste accumulates, and unless proper vents are created to release the gas to the atmosphere, explosions, fires, and offensive odors can result. Disease-carrying vermin such as rats and flies proliferate in such open dumps, facilitating the spread of diseases among people and animals. **Leachate** (liquid that seeps through the solid waste) eventually ends up in the soil and groundwater by leaching and in surface water by runoff. Open dumping into the oceans and seas causes toxicity to aquatic life and food poisonings in the food web.

Burial of hazardous and nonhazardous solid waste is carried out in **secure landfills** and **sanitary landfills,** respectively, which are lined with impermeable layers of clay and plastic to minimize the pollution of surface water and groundwater. Currently, about 60% of the solid waste in the United States is disposed of in sanitary landfills. Sanitary landfills have an average lifetime of 10 years because the oxygen-depriving conditions created in these landfills prevent decomposition of organic matter, thus more rapidly filling up their limited volumetric capacity. Profit maximization, corruption, and the lack of sufficient landfill capacity and regulations are causing both legal and illegal exports of wastes from regions. In an example of legal exports, Canada and Mexico receive almost 98% of U.S. hazardous waste exports.

Land-based disposal of hazardous liquid wastes can also take the form of **deep-well disposal** (injection into dry, porous geologic formations isolated from overlying groundwater), which accounts for 90% of disposal, or **soil impoundment** (pits or ponds), which,

Table 15.5
Major Air Pollutant Emissions from Transportation in Major Cities

Cities	Mean Urban Transportation Emissions[a]					
	CO_2	NO_x	SO_2	CO	Hydrocarbon	Particulates
American[b]	4541.2	22	1.6	204.5	22.3	1.0
Australian[c]	2788.9	22	0.6	185.8	23.0	1.4
Canadian[d]	2434.3	27	2.3	160.6	21.7	3.9
European[e]	1887.9	13	2.0	72.6	11.6	0.8
Asian[f]	997.5	8	1.3	40.8	7.9	2.3

Source: From Newman, P., and J. Kenworthy. 1999. *Sustainability and Cities: Overcoming Automobile Dependence.* Washington, D.C.: Island Press.

[a] Kilograms per capita.
[b] Includes Phoenix, Denver, Boston, Houston, Washington, D.C., San Francisco, Detroit, Chicago, Los Angeles, and New York.
[c] Includes Perth, Adelaide, Brisbane, Melbourne, and Sydney.
[d] Includes Toronto (metro).
[e] Includes Frankfurt, Amsterdam, Zurich, Brussels, Munich, Stockholm, Vienna, Hamburg, Copenhagen, London, and Paris.
[f] Includes Singapore, Tokyo, Hong Kong, Kuala Lumpur, Surabaya, Jakarta, Bangkok, Seoul, Beijing, and Manila.

together with secure landfills, accounts for 9% of disposal in the United States.

The burning of municipal solid waste (called **incineration**) has increased as a way to reduce landfill needs and waste mass by up to 90% by volume and 75% by weight, leaving much of the solid waste as residual **bottom ash.** At the same time, incineration causes intensified air pollution by releasing poisonous gases, GHGs, **fly ash,** and fine particulates (aerosols) that concentrate many of the toxic materials and trace metals from the waste. The smoke stacks require air pollution control devices (e.g., limestone scrubbers to remove sulfur dioxide and electrostatic precipitators to remove fly ash) and contribute to the reduction in visibility (**haze**) and public health in urban areas (discussed later in this chapter). Incineration comprises 17% of all solid waste disposal in the United States.

Biodegradable solid wastes such as agricultural and yard waste can be transformed by microbial action into soil humus, known as **composting;** this can serve the purposes of enhancing soil productivity if problems with toxic materials and odor are overcome. Conversion of wastes into less hazardous or nonhazardous materials by biotechnological means is called **detoxification.** The use of natural or bioengineered microorganisms to break down wastes in order to clean up soil and water pollution is known as **bioremediation,** and plants are used to absorb and accumulate toxic materials from the soil and later harvested and disposed of in hazardous waste landfills, called **phytoremediation.** Unless care is taken, however, these methods pose significant threats to animal, plant, and human health if they are allowed to introduce toxins or toxic by-products into the food chain through bioaccumulation and biomagnification.

Waste disposal causes societal conflicts and environmental injustices when people do not want disposal sites near where they live—known as **not in my back yard** (**NIMBY**) and **locally unwanted land uses** (**LULUs**) responses or syndromes—due to health concerns for the present and future generations and declining property values. For these reasons, the number of landfills in operation in the United States decreased from 20,000 in 1973 to 2,400 in 1996, but their capacity remained the same due to the larger size of current landfills. The air pollutant load from the transportation sector American cities with the most and Asian with least emissions (Table 15.5).

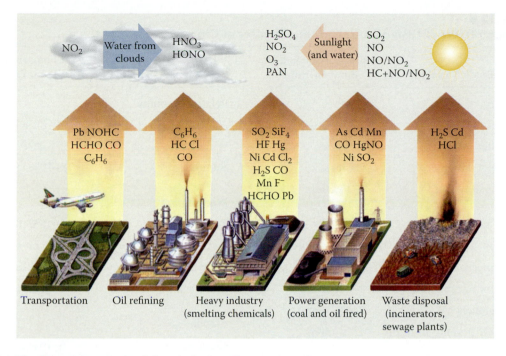

Figure 15.12   The five main causes of chemical air pollutants in urban areas. (Redrawn from Williams, M., ed. 1993. *Planet Management.* New York: Oxford University Press.)

Urban Climate and Air Pollutants

Transportation, industrial activities, waste disposal, and power generation are all major sources of air pollution, producing substances such as oxides of carbon, nitrogen, and sulfur, CO_2, and CH_4 (Figure 15.12). Landfills are the largest (36%) methane (CH_4) source, one of the most important of GHGs, and the majority (90–95%) of methane emissions results from municipal solid waste (MSW) landfills, with the remaining from the disposal of industrial wastes in the United States.

The most common primary air pollutants in urban areas include sulfur oxides (SO_2, SO_3), nitrogen oxides (NO, NO_2, and N_2O), carbon oxides (CO_2 and CO), aerosols (particulates), and hydrocarbons (CH_4 and C_6H_6). At the local scale, emissions by vehicles contribute 12% of the global anthropogenic GHG emissions (Faiz 1993). The transportation sector is, at present, responsible for about 20 and 40% of the global CO_2 and NO_x emissions, respectively, during the construction, use, and maintenance of roads (Hahn and Simonis 1991). In city centers, road traffic accounts for 90–95% of carbon monoxide (CO) and lead (Pb), 60–70% of nitrogen oxides (NO_x) and hydrocarbons (HCs) and a major share of particulates (Johnstone and Karousakis 1999).

Human populations, especially the children and the elderly, exposed to high concentrations of aerosols are adversely affected as a result of breathing and respiratory symptoms, aggravation of existing respiratory and cardiovascular disease, alterations in the body's immune system against foreign materials, damage to lung tissue, carcinogenesis, and premature death.

Urban–industrial ecosystems are distinctively different from those of surrounding agricultural and seminatural/natural ecosystems. The urban climate is characterized by the higher average air temperatures and air pollution levels, rapid runoff of precipitation due to impermeable surfaces, and high evaporation in built-up areas (Table 15.6). Air pollution, reduced vegetation cover, and increased heat emissions are the major factors that cause thermal anomalies in urban climates such as urban heat island effects and thermal inversions.

Table 15.6
Average Effects of Urbanization on the Climates of Cities

Climatic Variable	Comparison with Rural Environs	Climatic Variable	Comparison with Rural Environs
Radiation		**Relative humidity**	
Global	2–10% less	Annual mean	20–30% less
Ultraviolet, winter	30% less	Winter	2°C less
Ultraviolet, summer	5% less	Summer	8–10% less
Sunshine duration	5–15% less		
		Precipitation	
Temperature		Total	5–15% more
Annual mean	0.5–3°C more	Days with less than 5 mm	10% more
Mean winter minimums	1–2°C more	Snowfall, inner city	5–10% less
Mean summer maximums	1–3°C more	Snowfall, lee of city	10% more
Sunshine days	2–6°C more		
Greatest difference at night	11°C more	**Cloudiness**	
		Cloud cover	5–10% more
Wind speed		Fog, winter	100% more
Annual mean	20–30% less	Fog, summer	30% more
Calms	5–20% more		
Extreme gusts	10–20% less	**Contaminants**	
Thunderstorms	10–15% more	Condensation nuclei	10 times more
		Gaseous admixtures	5–25 times more
		Particulates (aerosols)	50 times more

Sources: From Landsberg, H. E. 1981. *The Urban Climate.* New York: Academic Press; Sukopp and Werner. 1983.

The term **urban heat islands,** first observed by Luke Howard in 1833, refers to elevated air temperatures in cities relative to surrounding rural areas (Figure 15.13). The dome of heat over cities traps suspended and gaseous pollutants, increasing urban air pollution discussed previously by up to 1,000 times. Wind can elongate the dome downwind to rural areas, thus creating a **regional heat island** (Figure 15.14). **Thermal inversions** occur when cool air is trapped below a layer of warm air and thus prevents the polluted air from rising, unlike the normal situation in which the hot air rises into the atmosphere, carrying the pollutants with it. Topographic features that prevent normal air circulation (such as valleys, coasts, and leeward sides of mountains) and urban heat islands increase the likelihood of thermal inversions in urban areas.

Urban Vegetation

Urban parks and open spaces such as backyards, riverfronts, and neighborhood parks play a significant role not only in betterment of urban environmental conditions but also in making rapidly growing urban populations feel connected to the environment. Naturally growing and planted trees, shrubs, and grasses are found in open spaces of urban areas and form **urban vegetation** that ranges from remnants of natural habitats, municipal recreation areas, greenbelts, greenways, and gardens to cemeteries and abandoned vacant lots. In general, urban vegetation can be classified, based on types of urban land use, as residential, institutional, and commercial areas; industrial sites; inner city vacant land; recreational areas;

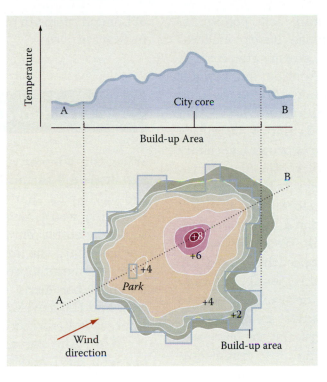

Figure 15.13 Air temperature structure in an urban heat island. (Redrawn from Cleugh, H. 1995. In *Future Climates of the World: A Modeling Perspective,* ed. A. Henderson-Sellers, 477–509. Amsterdam: Elsevier Science B.V.)

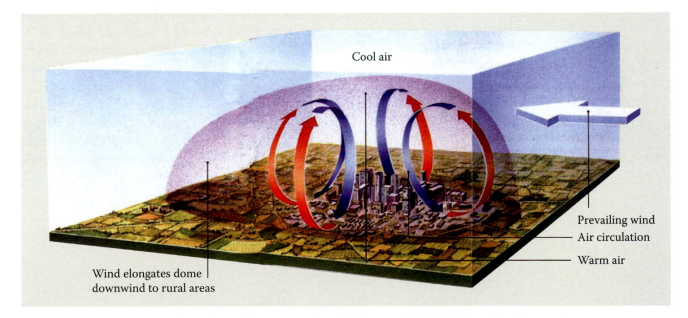

Figure 15.14 Formation of a regional heat island. (Redrawn from Williams, M., ed. 1993. *Planet Management.* New York: Oxford University Press.)

cemeteries; traffic corridors (e.g., roads, railways, and waterways); waste disposal areas; water bodies; and forests, fields, and orchards. Urban vegetation has many ecological and psychological functions that operate within cities and also transcend their political boundaries.

The **ecological services** of urban vegetation include production of oxygen, shelter and food for animals, biodiversity, urban–rural connectivity by migration and dispersal corridors, buffer zones and control of urban sprawl, absorption of air pollutants and noise, assimilation of water and soil pollutants, sequestration of greenhouse gases, moderation of climatic extremes, stormwater control, soil stabilization, water retention, and bioindication of signs of environmental change. The **psychological functions** include prevention of alienation of people from nature, cognitive development and education of people (especially children) about nature, recreational opportunities for all age groups, symbolic representation of cities, aesthetic pleasures, and spiritual enrichment. The degradation and loss of urban vegetation generally result from the lack of ecologically integrative approaches and their implementation in city and regional planning.

Market forces tend to determine where to locate land uses in cities unless guided by ecological principles and analyses in the process of physical planning, thus leading to fragmentation of urban vegetation along the urban–rural gradient and its conversion to built-up areas. On the other hand, introduction of exotic species that are not suitable to climate conditions of urban areas requires higher inputs of fertilizer, pesticides, and water. Such mismanagement practices increase not only the pollution of air, water, and soil but also the costs of maintaining urban vegetation.

A few cities in the world are now concentrating on accounting quantitatively on their resource use and ecological footprint. The City of London seems to have made a major advance in that respect.[4]

An Afterword

When this chapter was written several months ago, little did the authors imagine the magnitude of impact that the gargantuan economic downturn would have on every sector of human society, notably on the housing market. Not unexpectedly, the impact on car industry, from parts to manufacture to sales and collectively on the labor force, has been equally significant and the downward numbers are charting unprecedented records.

While the car sales have fallen steadily between 2007 and the first three-quarters of 2008, the end of 2008 and the beginning of 2009 showed precipitous drops (Figure 15.15). Between February of 2008 to February 2009, the cars sales dropped by 38.4 percent and light-duty trucks by 44.1 percent. Total SUV sales for the same period decreased by 52.6 percent. For the first time in six decades, Toyota Motor Company reported a revenue loss.

At this time few can predict what the future holds for the automobile industry. Several trends appear to have a considerable degree of certainty: the new automobiles will

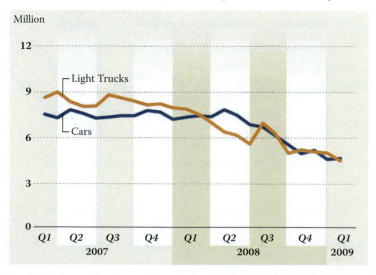

Figure 15.15   Sales of cars and light trucks in the U.S.; in millions of units in seasonally adjusted rate. Based on data courtesy of www.motorintelligence.com

be downsized, will be more fuel-efficient, and powered by multiple energy (fuel) resources. How fast will this industry be able to turn around the downward spiral and reach its hey-day sales numbers is unclear. But the lessons to be learned are obvious. "Restructuring a car business doesn't mean just taking a red pen to the budget or designing new line of green cars although they'll promise to do that", reported Yuki Nagochi.[5] She added: "For GM and Chrysler, revamping their business means revamping their relationship with the entire auto ecosystem." For the whole industry, that will inevitably mean much more than negotiating labor costs and wages and overcoming the debt burden.

Notes

1. Worldwide, the number of cars that are added each year is reaching staggering proportions. The introduction of low-cost cars in both India and China is going to have many impacts in such sectors as fossil fuel consumption, urban infrastructure, and loss of more habitats to roads and parking areas. We will discuss the sustainability of such developments in Chapter 26.
2. For the latest data and trends, Worldwatch Institute's updated *Vital Signs* is highly recommended (http://www.worldwatch.org/vsonline).
3. AutomobileIndia.com provides information on international statistics and other aspects of automobiles in India.
4. The city of London seems embarked on a futuristic trend. See Best Foot Forward Ltd, 2002, City Limits: A Resource Flow and Ecological Footprint Analysis of Greater London. http://www.citylimitslondon.com/downloads/Complete%20report.pdf
5. Yuki Nagochi, Auto Executives Prepare To Outline Plans, National Public Radio's *All Things Considered*, February 16, 2009. http://www.npr.org/templates/story/story.php?storyId=100754644

Questions

1. What forces drive urbanization and sprawl?
2. Discuss the relationship between urban–industrial ecosystems and surrounding rural ecosystems.
3. Research how land use decisions are made by institutions in the city where you live.
4. How do urban areas differ from seminatural or natural areas?
5. What causes the urban heat island effect, and how does it affect air quality and atmospheric circulation around a city?

Table 16.2
Estimated Global Water Use and Consumption by Sector, ca. 1990

Sector	Use (km³ yr⁻¹)	Consumption (km³ yr⁻¹)	Ratio of Use to Consumption (%)
Agriculture	2880	1870	65
Industry	975	90	9
Municipalities	300	50	17
Reservoir evaporative losses	275	275	100
Subtotal	4430	2285	52
Instream flow needs	2350	0	0
Total	6780	2285	
Percent of the total accessible freshwater (12,500 km³ yr⁻¹)	54	18	

Source: From Postel, S. L. et al. 1996. *Science* 271:785–788.

The relative amounts of water withdrawn, consumed, and used vary widely for the different uses and the sources of that water. In total, 77.6% of the available water comes from surface flow, and the remaining 22.4% is pumped from groundwater sources (aquifers). Of the 9.5% used for public water supply, almost twice as much comes from surface water as does that from groundwater, and 87.9% of the public supply goes to domestic–commercial use. Most (80.8%) of the domestic–commercial water is eventually released in return flow as sewage. Thermoelectric generation plants return 97.5% of the water they extract for cooling, consuming only 2.5%. Agricultural uses (irrigation, livestock), however, consume 60.7% of the water they remove because little of the irrigation water returns to the irrigation source; the water that does return may be degraded by addition of salt, nutrients, or pesticides and herbicides.

These differences are represented in Figure 16.3, where the relative amounts of water withdrawn by agricultural, municipal, and industrial users are compared at the top and the relative amounts of net consumption by the same three sectors are compared at the bottom. Not only does agriculture withdraw the largest amount of water, but it also returns the smallest fraction of water it uses to runoff.

Postel et al. (1996) performed a series of mass balance analyses for the world's fresh water, estimating the amount of the runoff

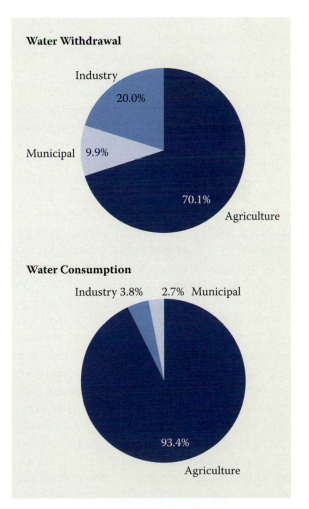

Figure 16.3 Water withdrawal and consumption by sectors. (Shiklomanov, I. A. ed. 1996. *Assessment of water resources and water availability in the world.* Background report to the comprehensive freshwater assessment. St. Petersburg, Russia: State Hydrological Institute.)

Table 16.3
Estimated Appropriation of Total Terrestrial Evapotranspiration by Human-Dominated Land Uses

Land Use	Net Co-Opted[a] Primary Production (× 10⁹ tons)	Evapotranspiration Co-Opted[b] (km³ yr⁻¹)
Croplands	15.0	5,500[c]
Grasslands	11.6	5,800
Forests	13.6	6,800
Human-occupied areas (lawns, parks, golf courses, etc.)	0.4	100
Appropriated and % total terrestrial evapotranspiration of 40,700 km³ yr⁻¹	40.6	18,200 (45%)

Source: From Postel, S. L. et al. 1996. *Science* 271:785–788.

[a] From Vitousek, P. M. 1986. In *Ecology of Biological Invasions of North America and Hawaii,* ed. H. A. Mooney and J. A. Drake, 163–175. New York: Springer–Verlag.

[b] Assumes 1 L of water evapotranspired for each 2 g of biomass produced.

[c] Adjusted for the amount supplied by irrigation (2,000 km³ yr⁻¹).

and groundwater extractions (use) by various human activities and how much of that was not returned and hence could be counted as consumption (Table 16.2). Much of the evapotranspiration processes that return water to the atmosphere are controlled (co-opted) by humans because they harvest the plants that are transferring the water, thus interfering with the evapotranspiration process. To make this calculation, Postel and colleagues used estimates for the amount of plant production multiplied by 1 L of water transpired for every 2 g of plant biomass; they determined that fully 45% of evapotranspiration was "co-opted" by humans for various uses (Table 16.3).

Summarizing Postel and colleagues' analysis (Figure 16.4), the total terrestrial renewable freshwater supply (110,300 km³ yr⁻¹) can be allocated into evapotranspiration and total runoff. Some of the runoff is in remote locations like far northern Canada or Russia or in the Amazon jungle, where plentiful water flows in areas of sparse population, so it cannot be exploited. Similarly, floodwater is water that is too abundant, often at the wrong time of the year; therefore, much of it (20,426 km³ yr⁻¹) cannot be used unless dams and reservoirs are constructed.

Reservoirs represent high consumption use due to their losses to evaporation (Table 16.2). Of the total precipitation captured by vegetation on land (i.e., total evapotranspiration of 69,600 km³

yr⁻¹), 26% is appropriated (co-opted) by man, and of the total *readily accessible* precipitation, 30% is appropriated by humans (Figure 16.4).

As mentioned earlier, runoff is not distributed on the Earth proportionally to local populations. For example, with 60.5% of the global population, Asia has only 35.8% of global river runoff (Table 16.4). In contrast, South America has 25.6% of the global river runoff and only 5.5% of the world's population. In general, a nation is adversely affected by **water scarcity** and **water stress** when its accessible runoff levels drop below 1,700 and 1,000 m³ per person per year, respectively (Figure 16.5). The competing demands of industries, cities, households, agriculture, wildlife, and other ecosystem functions for the limited water supply make food self-sufficiency very difficult, if not impossible (Table 16.5). Human population growth in semiarid and arid regions puts pressure on the total amount of available water in aquifers and watercourses, leading to what is called demographic water scarcity. Overexploitation of freshwater resources is referred to as technical water scarcity (Shiklomanov 1996; Falkenmark 1997; Penning de Vries, Rabbinge, and Groot 1997) (Figure 16.6).

A United Nations report states that nearly 2.3 billion people face water shortages. Only at levels exceeding 2000 m³ per person per year are regions considered to have abundant water

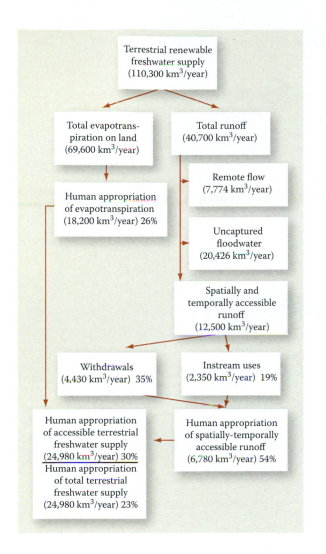

Figure 16.4 A mass-balance model of human appropriation of the terrestrial renewable freshwater supply. (Postel, S. L., G. C. Daily, and P. R. Ehrlich. 1996. Human appropriation of renewable freshwater. *Science* 271: 785–788.)

supplies (Fischer and Heilig 1997; Postel 1998) (Table 16.6). About 100 L of water per person per day (equivalent to 36 m³ per person per year) is assumed to be the long-term household water required to maintain a realistic quality of life. Developing a self-sufficient food production system in a semiarid climate corresponds to 720–800 m³ per person per year (Higgins et al. 1988; Falkenmark 1997). About 1 and 5 m³ water per 1,000 kcal are generally needed for crop and animal production, respectively. Based on this, obtaining a sound nutritional level of 2,700 kcal per person per day—2,300 kcal based on plants and 400 kcal based on animals—requires 4.3 m³ of water per person per day (or 1,570 m³ per person per year) (Klohn 1996). Depending on the climate, all of this may have to be provided by irrigation water in an arid climate, by soil moisture in a humid climate, and by 50% irrigation water and 50% soil moisture in a semiarid climate. The water needed for industry may vary from 36 to 360 m³ per person per year, depending on the water-use efficiency of industries and the water quality and quantity in countries (Falkenmark 1997).

Irrigated agricultural area expanded by 107% from 129.4 million ha in 1961 to 267.7 million ha in 1997, comprising about 18 and 33% of the world's cropland and annual food production, respectively. It is expected to grow to 330 million ha by 2025. The overall efficiency of agricultural irrigation worldwide is only 40%, which means that fully 60% of the water is wasted in agricultural production due

Table 16.4
Share of Global River Runoff and Population by Region

Region	Total River Runoff (km³ yr⁻¹)	Share of Global River Runoff (%)	Share of Global Population (%)
Europe	3,240	8.0	13.0
Asia	14,550	35.8	60.5
Africa	4,320	10.6	12.5
North and Central America	6,200	15.2	8.0
South America	10,420	25.6	5.5
Australia and Oceania	1,970	4.8	0.5
Total	40,700	100.0	100.0

Source: From Postel, S. L. et al. 1996. *Science* 271:785–788.

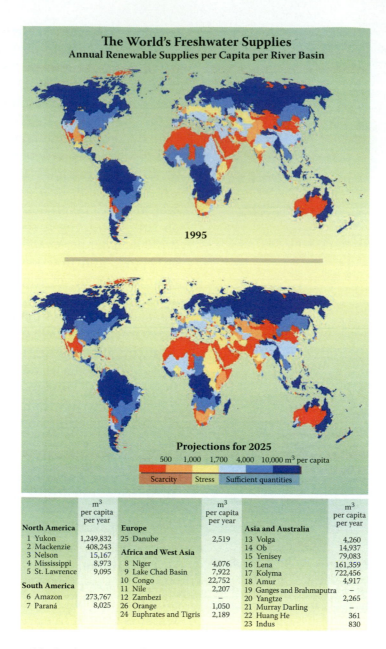

Figure 16.5 Annual renewable freshwater supplies per capita per river basin in 1995 and a projection of freshwater supplies for the same areas in 2025. (Copyright 2002, United Nations Environment Program.)

to seepage from the distribution systems and evaporation from bare soils (Postel 1997).

Moreover, government subsidies of agricultural production contribute to waste of water supplies. For example, in Tunisia, water is priced at one-seventh of what it costs to pump it, thus discouraging the conservation and efficient use of water resources. Inefficient and increasing water use have caused overexploitation of the world's freshwater supply through aquifer depletion, falling water tables, and reduced discharge rates into rivers, wetlands, and the seas.

Groundwater Use

Withdrawal or pumping in excess of the recharge rate of an aquifer causes a progressive drop of the water table (called water mining), creating a waterless zone (known as a cone of depression) in the short term and depletion of aquifers in the long term (Figure 16.7). Aquifer depletion threatens the world's most important food-producing regions, including the northern plain of China, the Punjab in India, and large areas of North Africa, the Middle East, Southeast Asia, and the western United

Table 16.5
Grain Import Dependence of African, Asian, and Middle Eastern Countries[a]

Country	Internal Runoff per Capita $(m^3 \ yr^{-1})$[b]	Net Grain Imports as Share of Consumption (%)
Kuwait	0	100
United Arab Emirates	158	100
Singapore	200	100
Djibouti	500	100
Oman	909	100
Lebanon	1,297	95
Jordan	249	91
Israel	309	87
Libya	115	85
South Korea	1,473	77
Algeria	489	70
Yemen	189	66
Armenia	1,673	60
Mauritania	174	58
Cape Verde	750	55
Tunisia	393	55
Saudi Arabia	119	50
Uzbekistan	418	42
Egypt	29	40
Azerbaijan	1,066	34
Turkmenistan	251	27
Morocco	1,027	26
Somalia	645	26
Rwanda	808	20
Iraq	1,650	19
Kenya	714	15
Sudan	1,246	4
Burkina Faso	1,683	2
Burundi	563	2
Zimbabwe	1,248	2
Niger	380	1
South Africa	1,030	–3
Syria	517	–4
Eritrea	800	Not available

Source: From Postel, S. L. 1998. *BioScience* 48 (8): 629–637.

[a] With per-capita runoff of less than 1,700 $m^3 \ yr^{-1}$ in 1995.

[b] Runoff Figures do not include river inflow from other countries, in part to avoid double-counting.

States, thus intensifying global food insecurity. Excessive extraction of freshwater in coastal areas causes the movement of salt water into aquifers, a process called **saltwater intrusion.** In addition, aquifer depletion from porous rocks can cause the surface level to sink over time, called **subsidence.** For example, aquifer depletion in the San Joaquin Valley of California has led to a 10-m subsidence of some areas in the past 50 years. Major rivers that discharge little or no water to the sea during the dry seasons due to diversion for irrigation include the Ganges and most rivers in India, China's Huang He (Yellow River), Thailand's Cha Phraya, the Amu Dar'ya and Syr Dar'ya in central Asia's Aral Sea basin, the Nile River in northeast Africa, the Murrumbidgee in Australia, and the Colorado River in southwestern North America.

The water levels in inland water bodies such as Lake Chad have been reduced substantially due to the intense use of their waters for agricultural irrigation. Information gathered on the Aral Sea (Kazakstan and Uzbekistan),[1] the Dead Sea (Israel and Jordan), and the Salton Sea (Southern California) is particularly instructive. For example, the waters of two main rivers, the Syr Dar'ya and Amu Dar'ya, that feed the Aral Sea 400 km east of the Caspian Sea were diverted to provide irrigation to cotton fields. The result was a drastic decline in the surface area of the Aral Sea—from 66,900 km^2 in 1960 to 28,000 km^2 in 1994—and a drop in water level from 53 m in 1960 to 34 m in 1994. These, in turn, created a domino effect on all limnological characteristics of the Aral Sea and a total loss of its economic value (Table 16.7) (Mainguet and Létolle 1998; Stone 1999).

The Dead Sea has no outlet, so water entering it (primarily from the Jordan River entering at its north end) evaporates, leaving salty water, especially in the shallow southern basin. In fact, shallow evaporation ponds dug there in the 1930s have been used for salt production. Diversion of the waters of the Jordan River for agricultural irrigation has decreased the flow to the Dead Sea, so its level has dropped 20 m since 1930. Water currently reaches the evaporation ponds only because it is pumped there from the deeper north basin. California's Salton Sea, a 984-km^2 lake created by the intentional diversion of the Colorado River in 1901, exists in

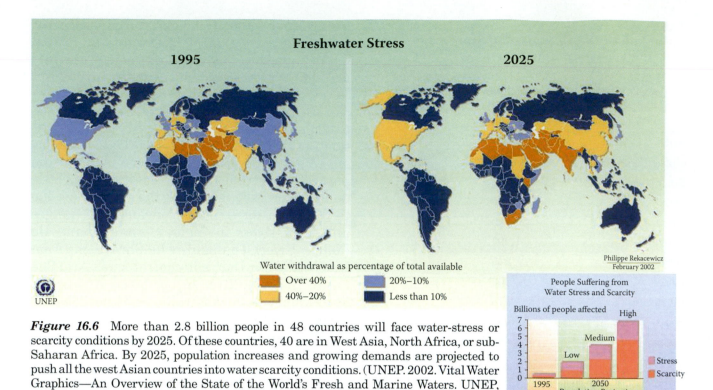

Figure 16.6 More than 2.8 billion people in 48 countries will face water-stress or scarcity conditions by 2025. Of these countries, 40 are in West Asia, North Africa, or sub-Saharan Africa. By 2025, population increases and growing demands are projected to push all the west Asian countries into water scarcity conditions. (UNEP. 2002. Vital Water Graphics—An Overview of the State of the World's Fresh and Marine Waters. UNEP, Nairobi, Kenya. http://maps.grida.no/go/graphic/freshwater-stress-1995-and-2025)

a large basin more than 200 ft below sea level, which had, in fact, been flooded many times by the Colorado River before humans became involved. It, too, is salty due to evaporation; however, beginning in the 1970s, several major storms washed effluents from agricultural fields into the lake, adding to its salt content and increasing its nutrient richness, which led to eutrophication, fish kills from low oxygen, and bird die-offs from avian botulism (Kaiser 1999a, 1999b, 2000). Obviously, deprivation of water from rivers leads to lowered water levels in rivers and lower groundwater tables, increased salinity, desiccation of wetlands, diversion of water from wildlife, and degradation and loss of aquatic life.

Impounding Water

Building dams serves to create water storage for irrigation, livestock, recreational activities, and electric power generation, and as a temporary water barrier to hold back high flow of rivers to prevent floods. Impounding of water

has perhaps been going on for millennia, but nothing like the enormous extent and magnitude that has been witnessed since 1900. Hoover Dam on the Colorado River between Arizona and Nevada, a spectacular example, was the highest (221 m) and longest (379 m) dam in the world when it was completed in 1936. Its impounded water forms Lake Mead, which is 1,885 km long and 1.6–12.9 km wide, with a shoreline length of 885 km. Its completion seems to have ushered in a new era in dam building worldwide and may have created the illusory impression that humans can control both the time and space dimensions of water supply. Other dam projects have followed in the United States and in the rest of the world. The Aswan Dam on the Nile River in Egypt—114 m high and 3,600 m long—formed Lake Nasser, which has enough water to irrigate more than 2.8 million ha of farmland.

Dam construction has become so widespread that in the United States only 2% of rivers are free flowing (Abramovitz 1996). Globally, the average construction rates of large dams at least 15 m high were about 885 yr^{-1} between 1950 and the mid-1980s and about 500 yr^{-1} in the 1990s. Construction rates are expected

Table 16.6
Annual Renewable Freshwater Resources[a]

Region	1995	2030	2050
United States and Canada	18,141	14,386	14,012
Eastern Europe	2,403	2,586	2,749
Northern Europe	10,858	10,619	10,771
Southern Europe	3,580	3,823	4,287
Western Europe	2,178	2,170	2,355
Russian Federation	28,769	33,400	37,361
Japan	4,374	4,611	4,993
Australia and New Zealand	31,269	22,826	21,926
Caribbean	2,806	1,989	1,779
Central America	7,919	4,908	4,243
South America	30,019	20,267	18,199
Eastern Africa	2,351	982	745
Middle Africa	23,563	9,371	6,885
Northern Africa	546	318	272
Southern Africa	1,304	702	578
Western Africa	4,966	2,121	1,628
Western Asia	1,850	940	769
Southeastern Asia	10,883	7,256	6,460
South-central Asia	3,032	1,879	1,641
Eastern Asia	2,282	1,854	1,834
Central Asia	4,881	3,353	2,966

Source: From Fischer, G., and G. K. Heilig. 1997. *Philosophical Transactions of the Royal Society of London Series B* 352:869–889.

[a] Cubic meters per person per year.

to drop to 350 yr^{-1} for the next 30 years. The decline in construction is occurring for several reasons. First (and perhaps foremost), few rivers remain on which dams can be built. Second, the evaporative loss from the surface area of the 500,000 km^2 of the world's reservoirs is staggering, especially in semiarid and arid areas where evaporation can exceed precipitation by more than 1 m yr^{-1}. More importantly, spectacular as many of these engineering structures are, studies from many regions of the world show that the environmental, societal, economic, and cultural costs associated with dams are enormous (Figure 16.8).

Costs such as the displacement of local people, degradation and depletion of aquatic habitats, heavy pollution loads, loss of biological diversity, and such other economic losses as the decimation of fish populations are now entering the cost/benefit calculations performed before dam construction. Furthermore, dams are often designed to have a useful life of only 50 years due to their rapid rates of filling with sediment carried by the rivers entering the reservoir. In the case of the Three Gorges Dam in China, there is little agreement about how well the management plan will prevent the rapid filling of the reservoir with sediment.

The damming of the upper stretches of the Columbia River in the American Northwest—initially by Grand Coulee Dam and now with a total of 14 dams—destroyed 40% of the most productive salmon-spawning habitat in the world. About 10–15 million adult salmon used

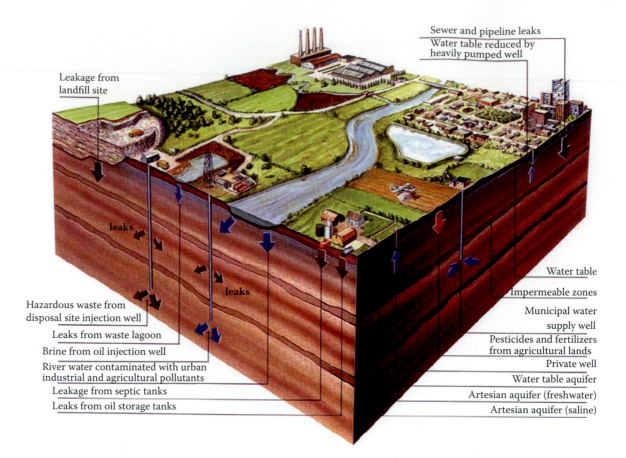

Figure 16.7 Groundwater extraction and pollution. (Williams, M., ed. 1993. *Planet Management*. New York: Oxford University Press.)

to swim upriver from the ocean to spawn in a year. Even ferrying hatchery-bred juveniles back to the ocean by barge or truck has failed to restore the system's productivity. The construction of the Mississippi's shipping channel in the late 1920s, together with dozens of upriver dams mainly erected on the Missouri River, managed to reverse the 7,000-year geomorphologic expansion of coastal Louisiana in New Orleans by blocking sediments behind the dams. This in turn caused encroachment of the Gulf of Mexico on coastal Louisiana (i.e., land loss) at a rate of 100 km^2 yr^{-1} by the 1980s, which continues now at a rate of 65 km^2 yr^{-1} (US-EPA 1999; Reisner 2000). Recent research suggests the storm surge damage from hurricanes Dennis and Katrina in 2005 was greatly exacerbated by the loss of coastal wetlands in the Gulf of Mexico and by canals through the wetlands. These were dredged by oil companies wanting a straighter, shorter route through the wetlands between their offshore oil platforms and their land-based facilities.

There are other documented examples. The first phase of the James Bay project in Quebec, which cost $6 billion (of which only $20,000 was spent on assessing environmental impacts), brought serious hardship to the Cree Indians. The Three Gorges Dam on the Yangtze in China, expected to be completed in 2009 at a cost of $24.5 billion (at 1997 dollar value), will be 2 km across and 200 m high. The reservoir that will submerge 600 km^2 and will result in the displacement of 1.2 million people. As of this writing, similar environmental impacts of Sardar Sarovar Wonder Canal on Narmada River in central India are being hotly debated. A detailed study of the 54 dams in that country "conservatively" estimates that, during the past 50 years, 33 million people have been displaced by dam construction alone (Roy 2000). The Sardar Sarovar, a 460-km-long concrete-lined dam with a 75,000-km network of branch and sub-branch canals, is designed to irrigate 2 million ha of land.

Table 16.7
Changes in the Aral Sea between 1960 and 1994

Parameter	1960	1994	2006[a]
Size, world ranking	Fourth	Seventh	
Area, km^2	66,900	28,000	17,382
Volume, km^3	1,064	310	108
Water level, m	53	34	
Shoreline length, km	4,430	3,950	
Lake length, km	428	240	
Lake width, km	234–292	100	
Depth max, m	68–69	68–69	
Depth average, m	16.1	10.2	
Salinity, g/L	14	34	>100
No. of islands in the sea	12	2	
Total area of islands, km^2	2,230	4,000	
Fishery, t yr^{-1}	30–40	0	
Flow of Amu Dar'ya and Syr Dar'ya to Aral Sea, km^3 yr^{-1}	53–56	20	

Source: From Zonn 1995; see also Mainguet, M., and R. Létolle. 1998. In *The Arid Frontier: Interactive Management of Environment and Development,* ed. H. J. Bruins and H. Lithwick, 129–142. Dordrecht, the Netherlands: Kluwer Academic Publishers.

[a] Data for 2006 are from Micklin, P. 2007. *Annual Review of Earth and Planetary Sciences* 35:47–72.

One-fifth of the world's irrigated land is salt affected and this is exemplified by projects on the Indus River in Pakistan. The Mangla and Tarbela dams, built in Pakistan with the aid of the World Bank in 1967 and 1977, respectively, are salt affected, waterlogged, or both. These areas have now received a loan of $785 million from the same agency for a drainage project; the same is true for the states of Haryana and Punjab in India (Roy 2000).

Just like air, water (especially in rivers) is a transboundary natural resource that is shared by both upstream and downstream regions. Coordination and cooperation among regions and nation states that utilize such resources are essential to the conservation and protection of the resources, particularly in water-limited countries and regions. Examples of the international conflicts over water resources include three shared river basins in the Middle East: the Jordan, which flows between Jordan, Syria, Israel, and Palestine; the Tigris-Euphrates, which flows from Turkey through Syria and into Iraq; and the Nile, which is shared by Egypt, Sudan, and Ethiopia. In southern Asia, international conflicts simmer in the Indus River basin between India and Pakistan and the Ganges basin between India and Bangladesh. Of these, the Southeast Anatolia Project in Turkey (called GAP after the Turkish *Guneydogu Anadolu Projesi*) has received considerable attention. When completed, the $20-billion GAP will have 88 dams and enough water for irrigation of 2 million ha. The regional conflicts from impounding this water and the hydroelectric power generation are discussed in Chapter 17. In addition to the international conflicts throughout the world, scarce water is a source of land use conflicts and competition because of inadequate understanding and valuation of social, environmental, and economic consequences, as well as of profit-maximization motives that eventually

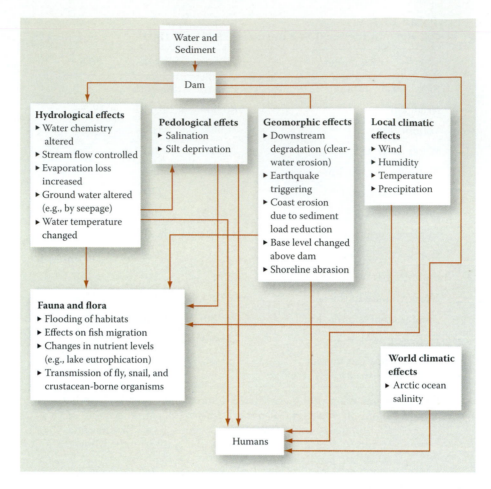

Figure 16.8 Ecological effects of dams. (Pickering, K. T., and L. A. Owen. 1994. *An Introduction to Environmental Issues.* New York: Routledge.)

result in the degradation of water resources at local and regional scales.

In 1995, about 40% of the 613 cities in the world with more than 500,000 residents were located in coastal areas, and 75% of the **megacities** (those with more than 10 million residents) were coastal (WRI 1996). When coupled with lack of physical infrastructures and mismanagement, increasing pressures from coastal urban–industrial activities (such as housing, tourism, marine transportation, and waste disposal) result in losses of healthy coastal ecosystems (wetlands, mangroves, and coral reefs) and ecosystem productivity, biodiversity, stability, and aesthetic values. In 1997, half of the nearly 269 million people in the United States lived in coastal counties with an average population density of 132 persons per square kilometer—more than four times greater than in noncoastal counties (EHC 1998). One consequence has been

that the estimated expanse of wetlands in the United States decreased from 894,355 km² in the 1700s to 408,328 km² in 1995, of which 95% were inland, freshwater wetlands and 5% were coastal or estuarine wetlands (Table 16.8; Mitsch and Gosselink 2007). A documented example is that of water diverted for urban and agricultural consumption in Florida from the Everglades—the "river of grass" that shrank from 11,700 to 5,960 km², thus endangering 68 of the native species, including the manatee and the Florida panther.

Competition for Water

Water use may pit constituencies against each other (e.g., natural areas vs. human land uses or agriculture vs. cities), and this can bring

Table 16.8
Largest Changes in Area of Wetlands by State in the United States

State	Original Wetlands ca. 1780 ($\times 10^3$ ha)	National Wetland Inventory, Mid-1980s ($\times 10^3$ ha)	Change (%) From Original to Mid-1980s
California	2024	184	−91
Ohio	2024	195	−90
Iowa	1620	171	−89
Indiana	2266	304	−87
Missouri	1960	260	−87
Illinois	3323	508	−85
Kentucky	634	121	−81
Connecticut	271	70	−74
Maryland	668	178	−73
Arkansas	3986	1119	−72

Source: From Mitsch, W. J., and J. G. Gosselink. 2007. *Wetlands,* 4th ed. New York: Van Nostrand Reinhold.

to the surface tensions that have been growing for some time. In California, for example, it has been cogently argued that although the agricultural industry accounts for 85% of water withdrawals in the state, agriculture represents only a small portion of the gross state product ($14 billion of the $550 billion in 1988) (Reisner 1989). Irrigated pasture alone consumes as much water as is needed for the entire population of the state. Thus, the subsidies given to agricultural water users have been seriously challenged because water there is such a scarce commodity. Similarly, in New Mexico, agriculture accounts for only 18% of the state's income but uses 92% of the water. Worldwide, such concerns are going to become widespread because the rate of decline of freshwater is increasing at 6% a year (Cambridge University, "The Living Planet Report," 1998).

In addition to water scarcity that threatens the well-being of society and ecosystems, too much water can put areas bordering rivers and streams (called **floodplains**) at extreme risk of flooding. **Floods** (the spread of the water onto the floodplain and land areas beyond) occur when increasing surface runoff during times of high precipitation exceeds the discharge capacity of stream channels. The removal of water-absorbing vegetative cover due to land uses such as clear-cutting, urbanization, and agriculture is one of the most important anthropogenic factors that cause floods by decreasing the rate of evapotranspiration and increasing the rate of surface runoff (see Chapter 15). Stream management also has an impact; streams are deepened and straightened to increase the rate at which water is carried out of an area so that they do not rise to flood levels, a procedure known as **channelization.** However, in the United States, the adverse effects of channelization have become well known since the enactment of the National Environmental Policy Act of 1970. As a consequence, much less channelization is now done in the nation.

Water Quality—Freshwater and Marine Ecosystems

Initially enacted in 1948, the U.S. Federal Water Pollution Control Act has been amended numerous times since then; however, nothing could have awakened the nation more to the ineffectiveness of regulatory controls over the pollution of surface waters than a historic and spectacular event that took place on June 22, 1969. The Cuyahoga River caught fire! Located in northeastern Ohio and flowing through the city of Cleveland, Ohio, and into Lake Erie, the river had become a dumping place for assorted industrial wastes that included oils, grease, wooden logs, and debris. Actually, it had caught fire many times before, especially in the 1950s, but the nation was galvanized by the cumulative effects of air and water pollution and this time people paid attention. Lake Erie was faring no better. It had been declared "dead" because it, too, was receiving appreciable quantities of pollutants, particularly excess nutrients that were causing eutrophication (see later discussion) and led to fish kills and declines in other products of economic value to the region.

Human uses of land and water and myriad management practices result in the release of materials, ranging from solid waste to toxic compounds, into water resources. Collectively, the materials that alter the biological and chemical composition of water so as to harm the health, survival, or activities of aquatic life and human populations are referred to as water pollutants. Sources of water pollutants include improper disposal of solid and liquid wastes such as untreated sewage, industrial and municipal discharges, mine and nuclear

wastes, and leakage from fuel storage tanks; agricultural fertilizers and pesticides; acid deposition; and soil erosion and sedimentation. **Point sources of water pollution** are relatively easy to identify; they include discharges from municipal sewage treatment plants and most industrial factories and are regulated by federal laws designed to control the amount of pollutants entering a given stream—specifically, the National Pollution Discharge Elimination System (NPDES) laws. Agricultural (and some mining) activities, roads, septic tanks, and urban stormwater are **nonpoint sources of pollution** (also called diffuse sources) that not only pollute surface water but also contaminate groundwater through seepage. Nonpoint sources are largely unregulated by federal laws in the United States.

Degradation of fresh and marine waters may come from human stresses of different types (Figure 16.9). Water pollutants include particulate matter, inorganic and organic chemicals, and biological organisms, grouped respectively as physical, chemical, and biological pollutants (Table 16.9). **Sedimentation** (or siltation) is a process by which soil particles are carried away by erosion from uplands into a body of water

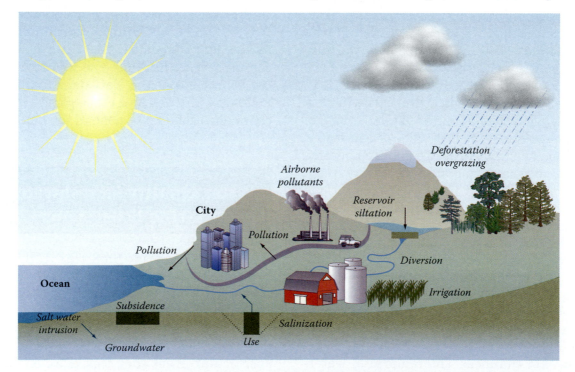

Figure 16.9 Human stresses on water resources. (Modified from IPCC. 1996. *Climate Change 1995: The Science of Climate Change,* ed. J. T. Houghton, L. G. M. Filho, et al. Cambridge, England: Cambridge University Press.)

Table 16.9
Water Pollutant Categories and Some of Their Ecological Effects

Pollutant Category	Composition	Example
Chemical pollutants		
Organic chemicals	Organic	Biocides, PCBs, oil, acids
Inorganic chemicals	Inorganic	Salt, nutrients, metals, acids, radioactive substances
Organometallic chemicals	Organic and inorganic	Methylmercury
Biological pollutants	Organic	Pathogens, invasive species
Physical pollutants	Organic and inorganic	Sediments, solid trash, thermal wastewater

where the particles may move downstream and/or accumulate on the bottom. The more mismanaged the land use practices are (deforestation, overgrazed rangelands, cultivated agricultural lands, strip mines, urbanization, and deforested stream banks), the more accelerated are the erosion and sedimentation rates (Figure 16.9). Sediment-laden water carries suspended particles; has increased turbidity that prevents deeper penetration of light, thus reducing primary productivity; has elevated inorganic nutrient levels, thus increasing the rate of eutrophication; coats stream and river bottoms, effectively suffocating stream organisms; and can carry loads of toxic materials. Each of these adversely affects aquatic life. In addition, sedimentation clogs waterways, reservoirs, lakes, and irrigation ditches. Nearly all suspended materials come from nonpoint sources—predominantly agriculture and construction activities.

Chemical water pollutants or contaminants include a variety of inorganic and organic elements and compounds. Inorganic pollutants include elements such as P and N (and their compounds); water-soluble chemicals such as salts; a great variety of metals such as Mn, Zn, Pb, Cr, As, Cu, Ni, Cd, Hg, and Se; in some areas, even water-soluble radionuclides such as radium and thorium; and the decay products of uranium. Nearly 82% of nitrogen and 84% of phosphorus come from nonpoint sources, again primarily from agricultural activities.

In the United States, about 8.7 km³ (2.3 trillion gallons) of effluents are discharged from municipal and industrial sewage treatment facilities into surface waters annually. Industries use about 20% of the water they typically withdraw in their processes that treat and discharge their wastewater back to coastal and inland aquatic ecosystems (EHC 1998). Runoff of fertilizers from agricultural and urban (lawn) areas, nutrient-rich effluents from sewage treatment plants, and accelerated erosion of nutrient-rich topsoil are the major sources of **eutrophication,** a process that results in excessive growth of algae and aquatic plants. The microbial decay of these plants depletes biologically required oxygen that is dissolved in surface water bodies and can cause **hypoxia**. In the marine environment, deoxygenated water with dead aquatic organisms can be circulated back to shore with offshore winds. For example, fertilizer runoff into the Mississippi River from Iowa, Ohio, and Illinois has caused the eutrophication of a 17,500-km² area in the Gulf of Mexico, destroying commercial fisheries (Rabalais, Turner, and Wiseman 2001). This area is now known as the "dead zone."

The most recent U.S. EPA's toxics release inventory (1998) shows that U.S. companies reported the release of 544,300 tons of chemicals into the air and water, of which 33,500 tons were released into surface waters. Of these chemicals, 53% are known or suspected developmental or neurological toxins. The report states further that because chemical companies report only an estimated 5% of all chemical releases, the total estimated release might be as high as 10.8 million tons (NET 2000).

There is a great diversity of organic pollutants as well. These include gasoline, plastics, detergents, oils, fats, proteins, and carbohydrates from industrial plants; agricultural pesticides; organic chemicals; and organometals. It has been estimated that more than 150,000 industrial and commercial entities discharge such pollutants into water bodies. Some of these may be very hazardous to the health of living organisms and are designated as toxic wastes. Industrial discharges of organic compounds into surface water and the leaching of pesticides and fertilizers into groundwater threaten not only human health when people use contaminated surface water and groundwater but also domestic animals, wildlife, and aquatic life due to their toxicity.

Combined Sewer Overflows and Biological Issues

Contamination of drinking water supplies by the discharge of inadequately treated sewage poses serious threats to public health in the United States and elsewhere. A **combined sewer system** (CSO) is one in which sewage, industrial wastes, and urban storm water runoff are mixed in the underground system of storm sewer pipes. During storms, the pipes must be big enough to handle huge volumes of runoff in a relatively short time. If the capacity of the sewer system is exceeded by the high flows, then the overflow is routed directly to nearby water bodies, including any raw sewage and industrial waste that is present. It is estimated that 30% of the U.S population is served by combined sewers. Many cities (e.g., Chicago) are now spending billions of dollars to separate their sewage and storm water draining systems. They are doing this by providing huge underground tanks to store the excess storm water so that it can be intercepted during storms and held until a later time when it can be pumped into the sewage treatment plant at a rate that will not disrupt normal treatment plant function (Laws 1997).

Biological water pollutants include invasive species and disease-causing microorganisms (pathogens) such as viruses, bacteria, protozoa, and parasitic worms. Zebra mussels (*Dreissena polymorpha*) are an example of the ability of an invasive species to change the structure and function of an ecosystem dramatically. These are small, fingernail-sized mussels native to Asia. They were originally transported to the Great Lakes via ballast water (vast tanks of water held on the boat to balance its weight in the water) from a transoceanic vessel. Ballast water taken on in a freshwater port was subsequently discharged into Lake St. Clair, near Detroit, where the mussel was first discovered in 1988. Since then, the mussels have spread rapidly to all of the Great Lakes and waterways in many states, including much of the Mississippi River drainage basin. They have few natural predators in North America, so their populations have exploded dramatically. Zebra mussels are filter feeders, and each individual can filter up to a liter of water per day. Based on their population densities in Lake Erie, it is estimated that the entire volume of the lake is filtered every day! This has dramatically increased the water clarity of the lake, something that years of water pollution control efforts had not accomplished. Algal biomass has decreased significantly, with potential impacts on the entire food web. It has also been shown that zebra mussels are accumulating the toxic compounds PCBs (DePinto and Narayanan 1997).

Pathogens are a constant concern in drinking-water supplies. Common examples of pathogens that contaminate water include *Guardia labia*, *Cryptosporidium* spp., *Shield* spp., *Campylobacter* spp., *Salmonella* spp., and *Escherichia coli*. In tests for public-water safety, the predominant fecal coliform bacterium of *Escherichia coli* is used as an indicator microorganism. Safe drinking water should contain no more than one coliform bacterium per 100 mL of water and safe swimming water no more than 200 coliform bacteria per 100 mL of water. Currently, the ocean and lake waters at many U.S. public beaches are closed as much as 50% of the time due to coliform counts above the swimming criterion. In 1993, a microorganism known as *Cryptosporidium* caused a major outbreak of a waterborne disease (cryptosporidiosis) that made 370,000 people ill when it entered the surface water used to supply drinking water in Milwaukee, Wisconsin. The World Health Organization (WHO) estimates that more than 5 million people die each year from diseases caused by unsafe drinking water and lack of water for sanitation and hygiene (Table 16.10).

Table 16.10
Estimates of Global Morbidity and Mortality of Water-Related Diseases in the Early 1990s

Disease	Morbidity (Episodes yr^{-1} or People Infected)	Mortality (Deaths yr^{-1})
Diarrheal diseases	1,000,000,000	3,300,000
Intestinal helminths	1,500,000,000 (people infected)	100,000
Schistosomiasis	200,000,000 (people infected)	200,000
Dracunculiasis	150,000 (in 1996)	—
Trachoma	150,000,000 (active cases)	—
Malaria	400,000,000	1,500,000
Dengue fever	1,750,000	20,000
Poliomyelitis	114,000	—
Trypanosomiasis	275,000	130,000
Bancroftian filariasis	72,800,000 (people infected)	—
Onchocerciasis	17,700,000 (people infected; 270,000 blind)	40,000 (mortality caused by blindness)

Source: From WHO. 1995. 48th World Health Assembly, A48/INEDOC./2,28 April, Geneva, Switzerland.

Oil Spills in Marine Systems

Marine and coastal ecosystems accommodate commercial and private transportation activities and facilities such as shipping, ports, and harbors as well as the utilization of mineral resources and reserves such as oil and gas, gold, cobalt, phosphorites, sand, and gravel that are transported over surface waters. Offshore energy sources account for 11.8 and 25% of oil and natural gas production worldwide and for 18.6 and 26% of U.S. oil and natural gas production, respectively (EHC 1998). These human activities are also sources of oil spills into marine ecosystems through oil-drilling blowouts and oil-tanker accidents. Such spills have occurred worldwide at a rate of three to five per year since 1967 (U.S. OTA 1990). There have been 30 oil spills larger than 100,000 barrels worldwide (Figure 16.10). Although the *Exxon Valdez* oil spill of 240,500 barrels received great attention, it accounts for only 18.8% of the total of 1,278,843 barrels spilled in 1989. The largest oil spill in history was during the Persian Gulf War (1990–1991), when an estimated 6 million barrels of oil—23 times the

amount from the *Exxon Valdez*—were spilled (EHC 1998). Oil pollution can have long-term and short-term adverse effects on marine life, depending on the types and magnitude of oil products spilled, climatic factors, geographical factors, and the vulnerability of marine organisms in a given area.

The spread of a heavy, opaque **oil slick** (a smooth and slippery film on the surface of the water) inhibits photosynthesis of marine algae below the oil spills, destroying the sources of food that support the growth and reproduction of all organisms that depend on marine algae. Oil slicks can also cause the direct death or abnormal behavior of marine animals and birds that come in physical contact with the spilled oil. Oil products are complex mixtures of mostly hydrocarbon-based substances (composed of the elements of hydrogen and carbon) and widely differ in their toxicity in the food chain. Lighter oil products such as kerosene and gasoline have greater toxic effects, but they evaporate more quickly than heavy oil products. Oil products evaporated into the atmosphere eventually come down as precipitation, thus polluting land and water bodies (see Chapter 20). The heavier components of oil left behind by evaporation can sink to the bottom of water bodies through

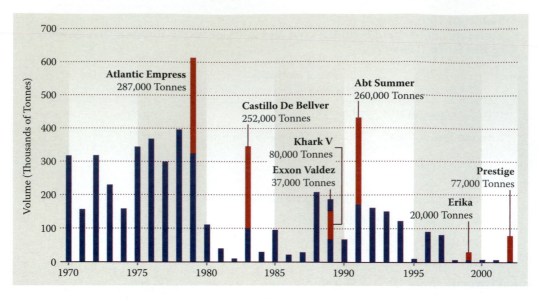

Figure 16.10 Volume of oil spills recorded between 1970 and 2002, highlighting major spills. Note that the *Exxon Valdez* is the small part of the bar for 1989. (From http://www.informaction.org/cgi-bin/gPage.pl?menu=menua.txt& main=oil_history.txt&s=Oil.)

weathering and dispersion (breakdown of a slick into droplets by wave action). Their incorporation into sediments increases their residence time in the environment. Mixtures of oil and water droplets (**emulsification**) can cause "chocolate mousse" emulsions that linger in the environment for years.

Decline of Fisheries

Essential spawning, nursery, and feeding grounds of marine species can be destroyed by the overharvesting of fish stocks, chemical pollution (e.g., land-based pollution, accidental oil spills, and deliberate dumping of litter), ecologically incompatible fishing methods (e.g., bottom trawling and dredging for scallops and clams), destruction of coastal ecosystems, and introduction of invasive non-native species. The total catch of all types of marine species accounts for 16% of the global animal protein intake in human diets. In 2005, the total world fish harvest reached 141.6 million metric tons, of which 60% was from marine capture, 13% from marine aquaculture production, and 20% from inland aquaculture production, leaving about 7% from freshwater capture. (Table 16.11). This amounts to 16.6 kg of fish

for every person on the globe. The UN Food and Agriculture Organization (FAO) estimated that the unintended catch of nontargeted species (called **bycatch**), including protected and immature species, amounts to 27 million tons each year (about 30% of the total world fish production). Much of this bycatch is used to produce fish meal, oil, and other industrial products or is simply dumped overboard.

Commercial fishing removes approximately 8% of the global net primary production in the sea and 24–35% of the upwelling and continental shelf production (Pauly and Christensen 1995). A recent analysis of world marine fish production indicates that about 44% of the major fishery resources are intensively to fully exploited, 16% are overfished, 6% are depleted, and 3% are slowly recovering (Botsford, Castilla, and Peterson 1997) (Figure 16.11). The National Marine Fisheries Service (1996) estimates that 36% of U.S. fisheries are overexploited and 44% are fully exploited. **Overfishing** occurs when the harvest rate of fishes exceeds their reproductive capacities to replace themselves. The symptoms of overfishing include a diminution in the average size of fish caught, an increasing proportion of catch composed of smaller and younger fish, and a decline in yield (Figure 16.12). The size of the fleets of fishing vessels that obtain most of the world's marine catch is 30% larger than that required to take the present world catch, thus intensifying both overfishing and international conflicts among fishing fleets over the harvesting of dwindling fish populations in the oceans and seas. One of the policy tools some nations have adopted to prevent overfishing is to establish an **ocean enclosure,** putting the management of marine organisms within 320 km (200 miles) of the shore under the jurisdiction of the country bordering the ocean.

Worldwide, the catch is declining for about one-third of major commercial fish, directly hurting 1 billion people, particularly in Southeast Asia. Examples of harvesting fish populations in excess of their growth, survival, and reproductive rates include the sardine stocks off California and Japan in the late 1940s, the anchovy off Peru and Chile in 1972, and the Canadian cod fishery and several New England groundfish stocks in recent years (Gulland 1977; Hutchings 1996; Meyers, Hutchings, and Barrowman 1997). The collapse of the North Atlantic cod fishery put 30,000 Canadians out of work and ruined the economies of an estimated 700 communities. Fisheries in the Atlantic Ocean, the eastern central and northeast Pacific Ocean, and the Mediterranean and Black Seas are now showing declining and stabilizing trends in terms of annual catches. The eastern and western Indian Ocean and the western Central and northwest Pacific Ocean have the largest incidence of fish stocks whose state of exploitation is unknown but are expected to have relatively more underexploited or moderately exploited fish stocks.

About 80% of marine degradation is estimated to be caused by human activities on land. Coastal ecosystems play an important role in marine productivity because between 60 and 80% of all commercially important ocean fishes spend part of their lives in coastal

Table 16.11
World Fisheries Production and Utilization

	1994	1995	1996	1997	1998	1999[a]
	(million tons)					
Production						
Inland						
Capture	6.7	7.2	7.4	7.5	8.0	8.2
Aquaculture	12.1	14.1	16.0	17.6	18.7	19.8
Total inland	18.8	21.4	23.4	25.1	26.7	28.0
Marine						
Capture	84.7	84.3	86.0	86.1	78.3	84.1
Aquaculture	8.7	10.5	10.9	11.2	12.1	13.1
Total marine	93.4	94.8	96.9	97.3	90.4	97.2
Total capture	91.4	91.6	93.5	93.6	86.3	92.3
Total acuaculture	20.8	24.6	26.8	28.8	30.9	32.9
Total world fisheries	112.3	116.1	120.3	122.4	117.2	125.2
Utilization						
Human consumption	79.8	86.5	90.7	93.9	93.3	92.6
Reduction to fishmeal and oil	32.5	29.6	29.6	28.5	23.9	30.4
Population (billions)	5.6	5.7	5.7	5.8	5.9	6.0
Per capita food fish supply (kg)	14.3	15.3	15.8	16.1	15.8	15.4

[a] Preliminary estimate.

continued

Table 16.11 (continued)
World Fisheries Production and Utilization

	2000	2001	2002	2003	2004	2005[a]
	(million tons)					
Production						
Inland						
Capture	8.8	8.9	8.8	9.0	9.2	9.6
Aquaculture	21.2	22.5	23.9	25.4	27.2	28.9
Total inland	30.0	31.4	32.7	34.4	36.4	38.5
Marine						
Capture	86.8	84.2	84.5	81.5	85.8	84.2
Aquaculture	14.3	15.4	16.5	17.3	13.3	18.9
Total marine	101.1	99.6	101.0	98.8	104.1	103.1
Total capture	95.6	93.1	93.3	90.5	95.0	93.8
Total acuaculture	35.5	37.9	40.4	42.7	45.5	47.8
Total world fisheries	131.1	131.0	133.7	133.2	140.5	141.6
Utilization						
Human consumption	96.9	99.7	100.2	102.7	105.6	107.2
Non-food uses	34.2	31.3	33.5	30.5	34.8	34.4
Population (billions)	6.1	6.1	6.2	6.3	6.4	6.5
Per capita food fish supply (kg)	16.0	16.2	16.1	16.3	16.6	16.6

Source: From FAO. 2007. *The State of World Fisheries and Aquaculture 2007.* FAO, Rome, Italy.
Note: Excluding aquatic plants.
[a] Preliminary estimate.

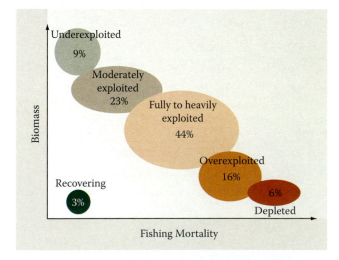

Figure 16.11 Distribution of degree of exploitation of world fisheries with their relative biomass and fishing mortality indicated. (Botsford, L. W., J. C. Castilla, and C. H. Peterson. 1997. The management of fisheries and marine ecosystems. *Science* 277: 509–515.)

areas such as coastal wetlands, tidal marshes, mangrove marshes, estuaries, seagrasses, and coral reefs. However, human activities such as urbanization and the construction of resorts in coastal ecosystems have destroyed about 50% of all coastal wetlands and 58% of all coral reefs worldwide in the past century (see Chapter 15). As a result of rearing marine and freshwater inland fishes by humans, fish production increased by about 80% between 1990 and 1997. **Aquaculture** (fish farming) is practiced in inland freshwaters, in estuaries, and near shore in oceans, and it is growing currently at an annual rate of 10–15% worldwide. Aquaculture can result in a progressive loss of genetic diversity in aquatic ecosystems, in their vulnerability to diseases, and in the spread of diseases when farmed individuals are mixed with wild populations (see Chapter 20).

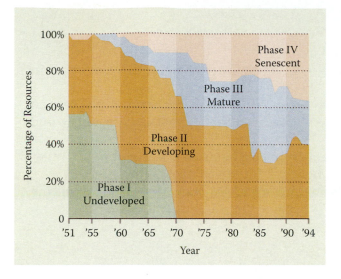

Figure 16.12 Percentage of major world marine fish resources in different phases of development, 1951–1994. (Garcia, S. M., and R. Grainger. 1996. Fisheries. FAO fisheries technical paper 359.) Undeveloped = a low and relatively constant level of catches; developing = rapidly increasing catches; mature = a high and plateauing level of catches; senescent = catches declining from higher levels.

As noted, phytoplankton (algae) are the base of the food web in aquatic ecosystems. They are the main energy producers, without which aquatic life would not exist. Human-induced disturbances such as elevated nutrient levels (especially phosphorus and nitrogen), reduced salinity, and warmer surface temperatures can cause the explosive population growth of algae (**algal bloom**), thus depleting dissolved oxygen in aquatic ecosystems. In addition, the bloom of large numbers of certain algal species (called dinoflagellates) with reddish pigments makes water appear red—hence the term **red tide.** Red tide organisms produce toxins that can poison the food web, thereby adversely affecting and even killing zooplankton, shellfish, fish, birds, marine mammals, and humans. These are called harmful algal blooms (HABs).

Major HAB-related events in the coastal United States include fish kills and neurotoxic

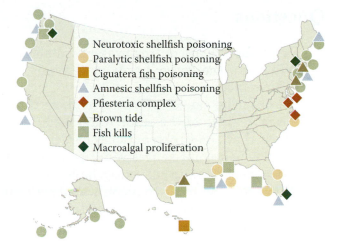

Figure 16.13 Spatial distribution of major HAB-related events of fish kills and poisonings in the coastal United States. (HARRNESS, 2005. Harmful Algal Research and Response: A national environmental strategy 2005–2015.)

shellfish poisoning (NSP) from red tides in Florida and other Gulf Coast states; paralytic shellfish poisoning (PSP) and amnesic shellfish poisoning (ASP) on the West Coast, including Alaska; and ciguatera fish poisoning (CFP) in the Florida Keys, Hawaii, and Puerto Rico (Figure 16.13). Toxins produced by harmful algae (such as ASP, CFP, diarrheic shellfish poisoning [DSP], NSP, and PSP) adversely affect human health through the consumption of contaminated seafood products. Each of these natural algal toxins, except DSP, has caused illnesses in the United States regularly since 1978 (Anderson 2000).

Legislation on Water Issues

The United States has been leading the world in enacting strong environmental legislation. We discuss these aspects in Chapter 21 on environmental policy and law.

Note

1. Several papers on the Aral Sea have been published recently; these include Micklin (2007) and Micklin and Aladin (2008).

Questions

1. What factors affect freshwater availability and quality?
2. What are the water needs of different sectors (e.g., agriculture, industry, residential) in your region?
3. Discuss the causes of water conflicts that have occurred between residents of upper and lower regions of the same watershed. Examine watersheds that lie within a country and those that cross international borders.
4. How is non-point-source water pollution different from point-source water pollution? What effect does the type of source have on a government's ability to control pollutants?
5. What are the most common water pollutants in your area? Do they affect your drinking water supply or any local industries?

Fuel Energy

17

Energy as Fuel

The energetics of natural ecosystems, governed by the laws of thermodynamics and driven by the capture of solar energy by primary producers, and the transfer of that energy to consumers were discussed in Chapter 7. Here we turn our attention to how different sources of energy drive the myriad machines—from electric toothbrushes to jumbo jets—that form the linchpin of modern-day life and the economic and military might of nations.

Energy resources can be classified as **nonrenewable (fossil) fuels** (including coal, oil, natural gas, and nuclear energy) and **renewable** (or **alternative**) **energy** sources, such as biomass (fuelwood and other plant and animal materials), solar (photovoltaic cells or thermal), hydropower, tidal energy, wind, and ocean thermal and geothermal energy. Primary energy (the energy contained in raw materials) is extracted from sources such as crude oil and then is transformed by a chain of processes (with an overall efficiency of about 30%) into useful energy forms that reach residential, commercial, industrial, and transportation end-users. The world's total consumption of primary energy—petroleum, natural gas, coal, and electric power (hydro, nuclear, biomass, geothermal, solar, and wind)—increased by about 945%, from 38.1×10^{18} J in 1900 to 400.3×10^{18} J (or exajoules, EJ) in 2007 (EIA 2007).

Energy from Biomass

As humans discovered fire and its many uses, they used plant materials as energy sources: wood, charcoal derived from wood, and straw. Today, these sources, together with energy obtained from animal wastes, are collectively referred to as **biomass fuels.** Biomass is still the predominant form of energy used by the people in developing countries for cooking and heating, cumulatively accounting for 14% of world energy use. In the United States before 1850, two-thirds of all energy used came from animal and human labor, supplemented by the mechanical work done by windmills and water. Over 90% of the total U.S. energy consumption in 1850 was biomass in the form of wood. The progression from the mid-1800s to

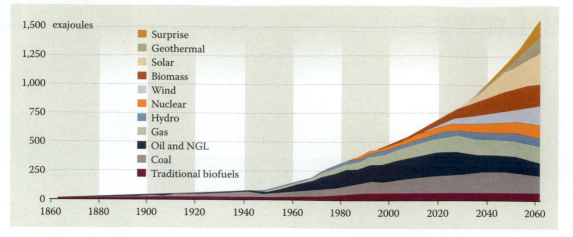

Figure 17.1 The form and magnitude of renewable and nonrenewable energy use in the human systems beginning in 1860 and projected to 2060. (Watson, R. T., J. A. Dixon, et al. 1998. Protecting our planet—securing our future: linkages among global environmental issues and human needs. United Nations Environment Program, US National Aeronautics and Space Administration, The World Bank. Washington, D.C.)

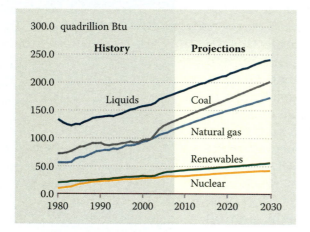

Figure 17.2 World marketed energy use by fuel type, 2005–2030. (U.S. DOE, Energy Information Administration. 2008. International Energy Outlook, 2008. DOE/EIA-0484(2008). Washington, D.C.)

the present, from the use of wood to coal to oil and, finally, to the harnessing of other sources of energy is quite a story of human ingenuity (Figures 17.1 and 17.2).

The use of biomass fuels results in the emissions of greenhouse gases, aerosols, acidic gases, and dust. Their use also leads to the removal of organic matter from agricultural and forest areas and the conversion of natural ecosystems to managed energy monoculture plantations with their associated loss of biodiversity. Simple wood stoves and fireplaces in many developing and developed countries are inefficient, converting only about 5–10% of the potential energy of wood into useful heat and

thus contributing a great deal to indoor and outdoor air pollution. Today, there are basically three distinct sources of biomass energy: the organic leftovers in municipal and industrial wastes, agricultural residues and wastes, and energy (tree) plantations. Most recently, energy from biomass has acquired new and larger dimensions in developed countries. This aspect is considered under the subject of biofuels later in this chapter.

Energy from Nonrenewable Resources

Coal

Coal, oil, natural gas, tar sands, and oil shales were formed from plant and animal materials over hundreds of millions of years of geological history. They are collectively referred to as **fossil fuels.** Given such long periods of time to be transformed, they are, from a human perspective, **nonrenewable resources.** The use of coal as a fuel source dates back many millennia. Historical mention of its use for heat and producing pottery by the North American Hopi Indians dates back 10,000 years. In China, coal was used for copper smelting 2,000–3,000 years ago and in the manufacture of porcelain 600–700 years ago. Coal use in Europe, particularly England, increased steadily from the thirteenth century. With the development of canals and railroads, coal mining expanded rapidly in Great Britain; production is estimated to have been 2.25 million tons in 1660, 2.5 million tons in 1700, and more than 10 million tons in 1800.

In the United States, coal use rose dramatically in the late 1850s as extensive deforestation led to wood shortages. It got a further boost during the Civil War. Coal deposits were being discovered at an accelerated pace in many states, including Illinois, Ohio, Pennsylvania, Virginia, and West Virginia. The transport of coal by barges in major waterways and by railroads accelerated industrial growth. From about 49,000 tons in 1822, coal production increased to about 342.3 million tons in 1929. In that year, "each of the four states—Kentucky, Illinois, Pennsylvania, and West Virginia—produced a greater [monetary] value of bituminous coal than the total [global] gold production" (Eavenson 1935). Coal was dubbed "king."

Even though coal production was more than 342.3 million tons in 1929, coal's share of the annual energy consumption in the United States was only 60%, having fallen from 90% in 1899. Even with that, Hubbert (1973) notes:

> During the eight centuries prior to 1860, it is estimated that the world's cumulative production of coal amounted to 7 billion tons. By 1970, cumulative production reached 140 billion metric tons. Hence, the coal mined during the 110-year period between 1860 and 1970 was approximately 19 times that of the preceding eight centuries. The coal produced during the 30-year period from 1940 to 1970 was approximately equal to that produced during all preceding history.

The oldest commercial coal deposits in the United States are from the Carboniferous Period and are more than 300 million years old. In terms of energy content, **anthracite** (**hard coal**), with the highest carbon content and the lowest water content, releases the largest quantity of heat per unit weight when burned (hence, it is used for steel-making), followed by **bituminous** (**soft coal**), **sub-bituminous,** and **lignite coals.** Coal with the least energy content (from the Paleocene Epoch nearly 60 million years ago) is lignite (also called **brown coal**). The **sulfur** in coal occurs in combination with the organic compounds it contains or is attached to the inorganic portion. Simple washing techniques and other coal preparation processes before combustion can remove most of the inorganic sulfur but cannot remove the organic sulfur, which requires expensive postcombustion removal methods such as flue-gas desulfurization devices. In the United States today, all four kinds of coal are mined for electric power generation.

Most coal throughout the world is burned directly to produce steam that drives electric generators. Coal may also be burned in order to convert it to gas and liquid forms. **Gasification** by the "Lurgi process" is the conversion of coal to a gaseous form in a reaction with steam, oxygen, and carbon dioxide. **Liquefaction** occurs by the destructive distillation of coal to make liquid hydrocarbons yielding such products as gasoline, diesel oil, alcohols, waxes, and other products. These two processes have been in widespread use; for example, they sustained the fuel supply of the Germans during World Wars I and II. Similarly, South Africa used coal gasification and liquefaction products during the 1980s when the world imposed an economic embargo for South Africa's apartheid policies.

The possibilities for the conversion of coal to gas or liquid were discussed widely in the western United States after the 1973 energy crisis. A quarter of a century later, new and "clean" coal conversion technology again came to the forefront. This took the form of the "FutureGen concept," which began in 2003. After nearly 30 months of discussion, a site for an experimental coal plant design was selected in Illinois. While the jubilation was just beginning by people at the plant site (Mattoon, Illinois), the FutureGen Industrial Alliance (a public–private partnership of 12 U.S. and international companies with the U.S. Energy Department), and the renewable energy enthusiasts, the U.S. Department of Energy withdrew the funding and cancelled the project on January 30, 2008, because of rising costs. Thus, although the use of oil and natural gas prevails, coal continues to be a major energy source in the world today.

Table 17.1
Proved Reserves of Fossil Fuels in Major Regions of the World

Region/Country	Proved Reserves		
	Crude Oil ($\times 10^9$ t oil equivalent)	Natural Gas ($\times 10^{15}$ m³)[a]	Coal[a] ($\times 10^6$ t)[b,c]
North America	9.5	7.98	250,510
Venezuela	12.5	5.15	479
Kazakhstan	5.3	1.90	31,300
Russian Federation	10.9	44.65	157,010
Middle East	102.9	73.21	1,386
Africa—Libya	5.4	1.50	—
Africa—Nigeria	4.9	5.30	—
China	2.1	1.88	114,500
India	0.7	1.06	56,498
World	1237.9	177.36	847,488

Source: Data selected from BP (British Petroleum). 2008. BP Statistical Review of Energy. June 2008 (http://www.bp.com/productlanding.do?categoryId=6929 &contentId=7044622).

[a] Norway's natural gas reserves are 2.96, Iran 27.8, and Qatar 25.6 (all $\times 10^{15}$ m³).

[b] Includes all anthracite, bituminous, sub-bituminous, and lignite coal.

[c] Brazil's coal deposits are at 7,068, Colombia 6,959, Germany 6,708, and Ukraine 33,783 (all $\times 10^6$ t).

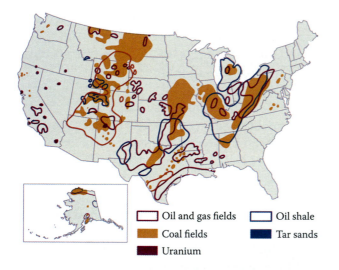

Oil and gas fields   Oil shale
Coal fields   Tar sands
Uranium

Figure 17.3 Major fossil fuel deposits in the United States. (USDI. Geological Survey. 1967. *Surface mining and our environment—a special report to the nation.* Washington, D.C.: US Government Printing Office.)

The world has substantial proven reserves of coal deposits relative to the proven reserves of oil (Table 17.1; Figure 17.3). Of the major sources of energy in the past 150 years, coal has been dominant globally, especially for electricity generation. Globally, coal accounted for 26% of total world energy consumption in 2004, second behind oil as a source of fossil fuel and as a source of carbon dioxide emissions. In 2006, global coal production stood at 5.37 billion tons, with China, the United States, and India as the leading producers. In 2007, coal produced in the United States was more than 1 billion tons (hard coal about 990 million tons and brown coal about 76 million tons).

These quantities may be put in the context of the annual national energy budget of the United States for a better understanding of their relative contributions. In 2005, the United States produced 68% (2,154 TWh) of its electricity from coal. More recent coal extraction methods, particularly surface mining, have significantly increased the recovery of coal from the ground and substantially decreased the health hazards involved in underground mining (as was noted in Chapter 14). Internationally, coal is the single largest source of electricity generation worldwide, and its use is increasing (Figure 17.4). Recent news reports estimate that a coal-fired power plant sufficient to meet the needs of a city the size of

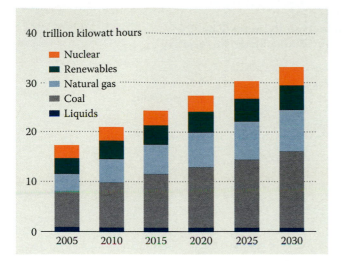

Figure 17.4 World electricity generation by fuel, 2005–2030. (U.S. DOE, Energy Information Administration. 2008. International Energy Outlook, 2008. DOE/EIA-0484(2008). Washington, D.C.)

San Diego is built every 10 days in China and every 14 days in India. Such expansion has enormous consequences that we shall discuss in forthcoming chapters.

The basic chemical composition of **coal** is the same as that for all fossil fuels. It is predominantly made up of carbon (55–90% by weight) and hydrogen with small amounts of nitrogen and sulfur compounds, plus other trace elements. Upon combustion, carbon, nitrogen, and sulfur combine with oxygen to form carbon dioxide, nitrogen and sulfur oxides, and carbon monoxide. These are emitted to the atmosphere, where they are converted into various forms, causing acid precipitation; air, water, and soil pollution; and increases in the atmospheric concentration of greenhouse gases (discussed in Chapters 18 and 19). Despite its current importance, the future of coal depends on how well carbon dioxide is captured from the power plants (Figure 17.5).

Petroleum

What stole 30% of the U.S. energy share from coal in the 30 years between 1899 and 1929, and why does coal account for only 28% of the national energy budget now? This is the story of how the **"hydrocarbon age"** and the era of international commerce were ushered in and how both have transformed our lives. It is also the story of how petroleum (oil) came to dominate the world's energy supply that is

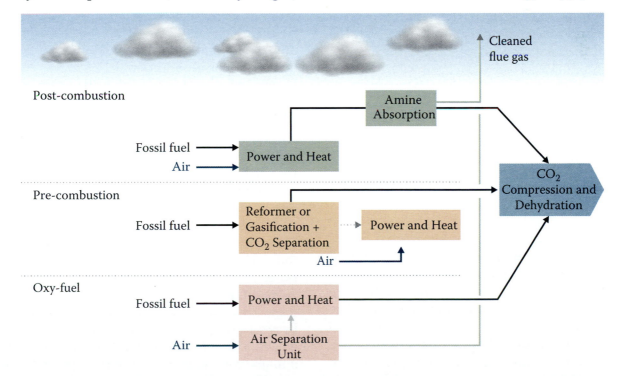

Figure 17.5 Overview of main technology options for carbon dioxide capture from power plants. (From International Energy Agency. 2008. Clean Coal Technologies—Accelerating Commercial and Policy Drivers for Deployment. Paris, Switzerland.)

Table 17.2
Energy Used in the 1997–2007 Decade from Primary Energy Sources by Top 10 Countries of the World and Their Respective 2007 World Share

	Millions of Tons Oil Equivalent		World Share (%)
	1997	2007	
United States	2204.8	311.54	21.3
China	961.4	1863.4	16.8
Russian Federation	611.7	692.0	6.2
Japan	506.6	517.5	4.7
India	260.5	404.4	3.6
Germany	337.8	311.0	2.8
France	241.0	255.1	2.3
Brazil	167.5	216.8	2.0
United Kingdom	240.4	215.9	1.9
Italy	163.9	179.9	1.6

Source: Data from U.S. Department of Energy, Energy Information Administration, May 2008 (http://www.eia.doe.gov/fuelrenewable.html) and British Petroleum 2008 statistics.

Note: Includes only commercially traded fuels; data do not include energy from biomass fuels or generated from geothermal, solar, and wind.

so elegantly documented by Daniel Yergin in his book, *The Prize: The Epic Quest for Oil, Money and Power* (1991). The intensity with which exploration for oil began throughout the United States is legend. Here is a vignette of the landmark discovery as captured by George F. Will (2001):

"One hundred years ago, around 10:30 a.m. on Jan. 10, 1901, near Beaumont, Texas, on a hillock called Spindletop, the first great gusher of the East Texas oil fields roared in. Before long, the population of Beaumont was such that water was selling for $6 a barrel, while oil was selling for 3 cents a barrel."

In the 1930s and 1940s, the United States had a major share of global oil reserves, but used its reserves so fast that by 1970 this share had fallen to 6%. The words of Hubbert (1973) are as instructive for oil as they were for coal: "The world's cumulative production of crude oil up to 1970 amounted to 233 billion barrels. Of this, the first half required the 103-year period from 1857 to 1960 to produce, the second half only the 10-year period from 1960 to 1970." The total amount of crude oil produced in the 113-year period doubled only in the next 12-year period and reached about 594 billion barrels in 1998. In 2006, Saudi Arabia, Russia, and the United States produced 32.9% of the world's crude oil, and production from Iran, China, and Mexico accounted for an additional 14.8%.

As more and more oil wells were discovered in the Middle East, Venezuela, Mexico, Indonesia, and other countries (Table 17.1), enormous quantities were being consumed by industrialized nations, making oil trade the lifeblood of world economies (Table 17.2). Although the United States uses energy from many sources (Table 17.3), its share of worldwide oil consumption in 2007 was a little more than 24%. Realizing this dependence of industrialized countries and their own importance in it, the oil-rich countries of the world formed **OPEC**—the Organization of Petroleum Exporting Countries[1]—to control the price of oil by restricting the amount sold by the cartel. Oil will continue to be a major energy source, with a major portion going to the transportation sector (Figure 17.6). As with overall energy use, consumption will continue to intensify in developing countries, particularly China and India.

Table 17.3
U.S. Energy Consumption by Energy Source, 2003–2007

Energy Source	2003	2004	2005	2006	2007
			10^{15} Quadrillion Btu		
Total	98.209	100.351	100.503	99.861	101.605
Fossil fuels	84.078	85.830	85.816	84.662	86.253
Coal	22.321	22.466	22.795	22.452	22.786
Coal coke net Imports	0.051	0.138	0.044	0.061	0.025
Natural gas[a]	22.897	22.931	22.583	22.191	23.625
Petroleum[b]	38.809	40.294	40.393	39.958	39.818
Electricity net imports	0.022	0.039	0.084	0.063	0.106
Nuclear	7.959	8.222	8.160	8.214	8.415
Renewable	6.150	6.261	6.444	6.922	6.830
Biomass[c]	2.817	3.023	3.154	3.374	3.615
Biofuels	0.414	0.513	0.595	0.795	1.018
Waste	0.401	0.389	0.403	0.407	0.431
Wood-derived fuels	2.002	2.121	2.156	2.172	2.165
Geothermal	0.331	0.341	0.343	0.343	0.353
Hydroelectric conventional	2.825	2.690	2.703	2.869	2.463
Solar/PV	0.064	0.065	0.066	0.072	0.080
Wind	0.115	0.142	0.178	0.264	0.319

Sources: From U.S. DOE, Energy Information Administration. Non-renewable energy. 2008. Monthly energy review. DOE/EIA-0035 (2008/03). Washington, D.C. March 2008, Tables 1.3, 1.4a, and 1.4b; renewable energy: Table 2 of this report.

Notes: Ethanol is included only in biofuels. In earlier issues of this report, ethanol was included in both petroleum and biofuels, but counted only once in total energy consumption. Totals may not equal sum of components due to independent rounding. Data for 2007 are preliminary.

[a] Includes supplemental gaseous fuels.

[b] Petroleum products supplied, including natural gas plant liquids and crude oil burned as fuel.

[c] Biomass includes: biofuels, waste (landfill gas, municipal solid waste biogenic, and other biomass), wood and wood-derived fuels.

Crude oil can be refined to a diversity of products—from gasoline, to middle distillates (kerosene, heating oils, diesel oil, and jet fuels), to wide-cut gas oil (waxes, lubricating oils), and to residual oils (asphalt); however, all these production processes have environmental impacts. Refining produces particulate matter, hydrocarbons, and oxides of nitrogen and sulfur, and it releases elements like chromium, copper, lead, and zinc. Oil is transported by pipelines (e.g., Alaska oil pipeline) or by tankers (e.g., from the Middle East to the United States), and tanker accidents and the recent leaks in the Alaskan pipeline have all polluted the environment with crude oil.

Although the number of tankers that carry oil within and among continents may not have increased significantly (from 3,500 in 1954 to 4,024 in 2007), the maximum carrying capacity of these ships has increased phenomenally (46,000 tons per ship in the early 1950s to the current capacities exceeding 555,000 tons per ship). The larger the carrying capacity of a ship is, the greater will be the consequences of oil spills (as was discussed in Chapter 16).

Natural Gas

Both globally and in the United States, natural gas ranks third as a source of energy. Natural

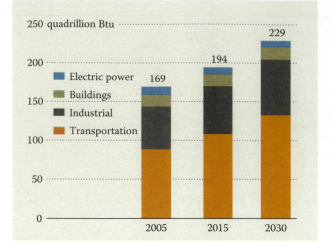

Figure 17.6 World's liquids consumption by end-use sector, 2005–2030. (U.S. DOE, Energy Information Administration. 2008. International Energy Outlook, 2008. DOE/EIA-0484(2008). Washington, D.C.)

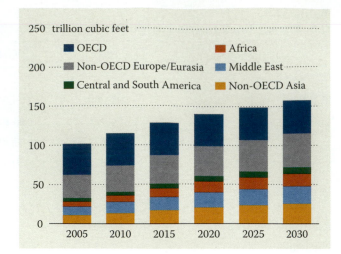

Figure 17.7 World natural gas production, 2005–2030. (Redrawn from U.S. DOE, Energy Information Administration. 2008. International Energy Outlook, 2008. DOE/EIA-0484(2008). Washington, D.C.)

gas is typically a mixture of methane (80–85%), ethane (10%), and the nitrogen (5%), propane (4%), and butane (1%). It is colorless, odorless, and highly flammable. Because it is odorless, leaks are difficult to detect, so gas companies add tracer compounds (mercaptans) that are easy to smell. Natural gas is a relatively clean-burning fuel and thus is widely used, including for electricity generation. Stocks occur naturally, deep beneath the surface of the Earth, most often in association with petroleum deposits. In 2007, the known proven reserves of natural gas worldwide were about 177.36×10^{15} m³ (Table 17.1). Producing 52.4% of the world total, Russia is the leading producer, followed by the United States, Canada, Iran, and Norway (Figure 17.7). As with oil, most of the increase in natural gas consumption is predicted to be in non-OECD countries.[2]

Natural gas is most efficiently transported after being pressurized to **liquefied natural gas** (**LNG**), but compounds such as sulfur and heavier hydrocarbons must be removed first. Liquefied natural gas is transported by pipelines using compressors or by specialized tankers, which are very complex and expensive to build. Like the oil tankers, the number of LNG tankers has also increased significantly (1 in 1960 to 100 in 1995), with a carrying capacity averaging 135,000 m³ for the newer tankers.

Tankers connect to pipelines to discharge their loads, and a great deal of heat must be added to the liquid to reconvert it to a gaseous form.

Nuclear

Unlike chemical reactions where only the electrons of an atom take part, in nuclear reactions, the nucleus is either **split** or **fused;** the former is termed **fission** and the latter **fusion.** In both of these processes, a small amount of mass is converted into energy following the relationship in Einstein's theory of relativity: $E = mc^2$, where E is energy, m mass, and c the speed of light. Hence, conversion of a small amount of mass generates a huge amount of energy. The most commonly used isotopes are uranium-235 (^{235}U) and processed plutonium-239 (^{239}Pu). Pellets of ceramic uranium dioxide sealed in metal rods are the fuel for most nuclear reactors and are bundled together to form a **fuel assembly.** A fuel assembly can weigh up to 680 kg, depending on the nature of reactors, and fuel assemblies of about 100 tons of uranium bundled together in a heavy steel vessel constitute the **nuclear reactor core.** After the fissionable uranium content in a fuel assembly becomes too low to support the reaction, used fuel rods are sent to a reprocessing plant where some of the ^{235}U is recovered and recycled into new fuel rods (Figure 17.8).

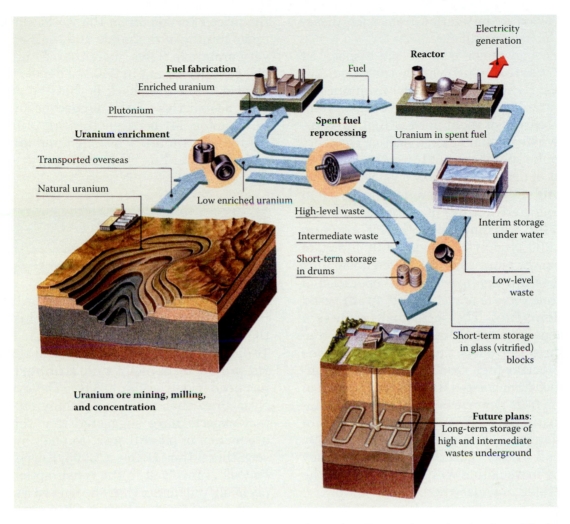

Figure 17.8 The nuclear cycle. Uranium, a radioactive fuel for nuclear power plants, is mined, enriched, and made into fuel rods. Once used, the fuel rods from the reactor are sent to a reprocessing plant where some of the uranium is recovered. (Redrawn from Williams, M., ed. 1993. *Planet Management*. New York: Oxford University Press.)

Nuclear power plants provide 80% of the electricity in France, 30% in Japan, and 20% in the United States, where 104 plants are in operation. Except for the minor leakage at the Three Mile Island plant in Pennsylvania in 1979 and the major meltdown of a nuclear reactor in Chernobyl (115 km north of Kiev, capital of Ukraine) in 1986, the world's 430 power plants have been running safely. Because of concerns about global climate change, nuclear energy has entered discussions for supplementing the increasing energy demands of developed countries, as well as for China and India (for example, building as many as 45 nuclear power plants in the United States has been suggested recently). The advantages touted by proponents of nuclear power include the fact that, unlike coal and oil, nuclear power does not add carbon to the atmosphere. These power plants and their emissions are regulated strictly. Additionally, advances in design (e.g., Advanced Generation III) are now in place, with many advances in safety and efficiency. These include a simpler and more rugged design, easy operation, the reduced possibility of core melt accidents, minimal effects on the environment, longer life, and less fuel waste (Hore-Lacy 2008).

But concerns remain. Of the ravages of radioactive materials, Feshbach and Friendly (1992) wrote that the explosions at Chernobyl "poured more radioactive material into the atmosphere than had been released in the atomic bombings of Hiroshima and Nagasaki" (p. 12). The suffering in human health, the contamination of water and agricultural lands and associated crop losses, and the estimated costs involved in mitigation from such disasters can be astronomical.

The second major issue is the disposal of **nuclear wastes,** which can be classified into four types: (1) spent nuclear fuel from nuclear reactors and high-level radioactive waste from the reprocessing of spent nuclear fuel, (2) transuranic waste (defined as radioactive elements with atomic numbers and weights greater than that of uranium), (3) low-level radioactive waste, and (4) uranium mill tailings from the mining and milling of uranium ore (see Chapter 16). Current global use of uranium is about 200 tons yr^{-1} for every 1,000 MW_e of nuclear power plant capacity. When about 75% of ^{235}U goes through fission in the operation of nuclear reactors, it is considered **spent** (used up) **nuclear fuel.** The spent fuel produces a tremendous amount of heat that must be dissipated in water pools: currently, 65,333 tons of spent nuclear fuel, 4,667 tons of liquid high-level radioactive waste (HLW),[3] and 50 tons of plutonium from defense liquid high-level radioactive waste (HLW). Wastes are disposed of in Yucca Mountain, Nevada (U.S. DOE 1999). Despite all these concerns, serious as they are, many more plants will be built here in the United States and elsewhere in the world.

Tar Sands

Another source of nonrenewable fuel, tar sands, has gained attention in North America. Huge deposits of these sands contain a thick crude oil, bitumen. More than 80% of the world's supply of bitumen is in the province of Alberta in Canada. The known reserves of bitumen are estimated to be 280×10^9 m^3 (equivalent to 1.75 trillion barrels).

The bitumen or tar sands region in the taiga biome, or boreal zone, is mined by surface mining methods. The sands are treated with sodium hydroxide and hot water to free up the slurry of bitumen. The slurry is agitated and upgraded to crude oil; about 16 tons yield a barrel of oil. Several other methods, such as steam stimulation, vapor extraction, and cold flow, are also used to separate bitumen from the sands. The 1,980-mile Keystone Pipeline carries the crude oil from Hardisty, Alberta, to Port Arthur, Texas. The unprecedented rise in the price of crude oil on the world market has made mining of tar sands and their extraction very profitable. Thus, more than $7 billion will be spent beginning in 2008 to double crude oil recovery. One drawback is that the process is very destructive of the landscape (discussed in Chapter 14) because the tar tailings left behind contain 5–7% bitumen and land restoration thus far has been elusive.

Fossil Fuel Energy— Retrospect and Prospect

Nearly 60 years ago, U.S. geologist M. King Hubbert declared on the basis of his studies that U.S. oil production would peak in the early 1970s and show a downward trend thereafter. He proved to be right and **Hubbert's peak** has been widely acclaimed (see Deffeyes 2001). Ever since that prediction, many oil exploration experts have suggested that the world's production of oil will start to peak by 2015 or soon thereafter (Deffeyes 2005; Korpela 2006). Few of us involved in the consumption or supply of oil, particularly in the United States and in the rapidly developing China and India, have heeded those warnings. Most predictions are that fossil fuel use will continue to increase until 2030 (Figure 17.9).

Two aspects of worldwide energy use have tremendous consequences on world stability and geopolitics. First is the movement of oil and gas from resource-rich to consuming countries. The international trade movements in oil and natural gas and their major routes among the different regions of the world (Figure 17.10) illustrate the obligations that trading partners have with each other that may not have been there historically. Second, there is the likelihood of price instability, as was starkly demonstrated in the first 6 months of 2008: The price for a U.S. barrel of oil escalated to more than $145 (Figure 17.11), with predictions that it would be $200 by the end of 2008. This has had a tremendous impact on the cost of living in all countries of the world. Although lower supply and greater demand are important ingredients of the price fluctuation scenario,

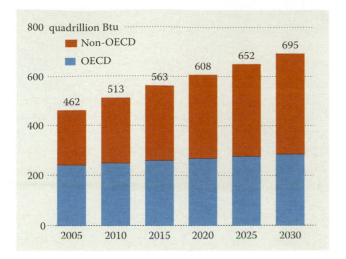

Figure 17.9 Total world marketed energy consumption in OECD (the Organization of Economic Cooperation and Development) and non-OECD countries, 2005–2030. (U.S. DOE, Energy Information Administration. 2008. International Energy Outlook, 2008. DOE/EIA-0484(2008). Washington, D.C.)

the opportunity to reap hefty and quick profits by speculators and futures markets in the capitalistic economic system is no less a force. In developed and developing countries alike, these aspects have provided a tremendous impetus to serious pursuit of the development of renewable energy sources.

Energy from Renewable Resources

Water

For millennia, humans have known that the mechanical force of falling water can be converted into energy. In recent human history, this principle has been used on a massive scale by creating dams to impound large quantities of water to run the hydraulic turbines. With an overall energy conversion efficiency of 80–90%, the world's theoretical hydropower potential is about 40,000 TWh yr^{-1} (1 tera watt hour [Twh] = 10^{12} Wh), of which about 14,000 TWh yr^{-1} is technically feasible for development and about 2,600 TWh yr^{-1} is feasible using current technology (WEC 1998). About 700 GW of hydropower capacity are actually in

operation worldwide, amounting to 28% of the economically available energy and generating about 19% of the world's electricity production. Engineering marvels such as hydroelectric dams have an immense impact on the ecology of an area, including on its human inhabitants.

The generation of hydroelectric power increased from 2,088 TWh in 1989 to 2,994 TWh in 2005; China (397 TWh), Canada (364 TWh), Brazil (337 TWh), the United States (290 TWh), and Russia (175 TWh) were the five largest producers. Collectively, these countries accounted for 52% of the world total (IEA 2007).

One of the world's major hydroelectric projects is the Aswan Dam on the Nile River in Egypt, which has a power generation capacity of more than 10 MW. The impounded water forms Lake Nasser, 114 m deep, 550 km long, and 35 km wide at its widest point. The lake has enough water to irrigate more than 2,809,400 ha of farmland. In the United States, the Hoover Dam on the Colorado River between Arizona and Nevada is 221 m high and 379 m long, with a power-generation capacity of 1.3 MW. The resulting impoundment of water, called Lake Mead, has a staggering shoreline length of 885 km, is 185 km long, and is between 1.6 and 12.9 km wide.

A series of dams constructed in the Tennessee River basin, under the Tennessee Valley Authority, was created to provide flood control; a navigable waterway from Knoxville, Tennessee, to the Mississippi River; and to a cheap electricity supply. These were supplemented later by three nuclear power plants and 11 coal-fired power plants. The environmental costs of these dams, now with 400,000 km^2 of impounded water, have been enormous. Changing the flow of rivers alters the hydrological cycle of a given area. Within the river channel, as well as in the surrounding watershed, the biological diversity, soil productivity, erosional processes, and sediment movement all are affected (see Chapter 16).

Power-generation projects also have regional impacts. In hydropower, the impounded waters have a major role in the generation of electricity for a given nation, but deprive the downstream populations of their water supply. Examples illustrative of this have been the Colorado River in the United States and

Major trade movements 2007
Trade flows worldwide (million tons)

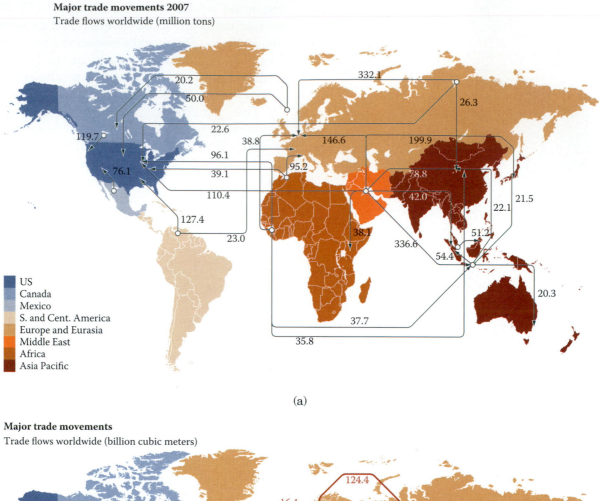

(a)

Major trade movements
Trade flows worldwide (billion cubic meters)

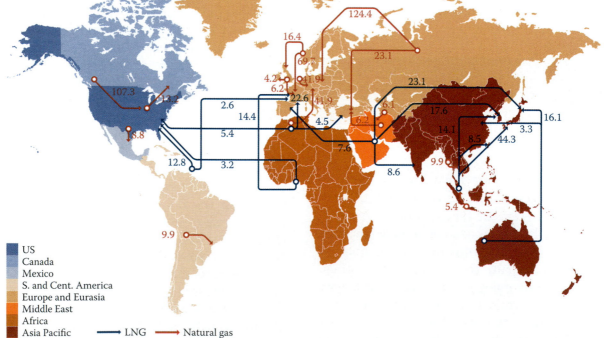

(b)

Figure 17.10 Worldwide major movements of (a) oil (× 10⁶ t) and (b) natural gas (10¹² m³). (Redrawn from BP. 2008. http://www.bp.com/productlanding.do?categoryId=6929&contentId=7044622.)

Crude oil prices 1861-2009
US dollars per barrel
World events

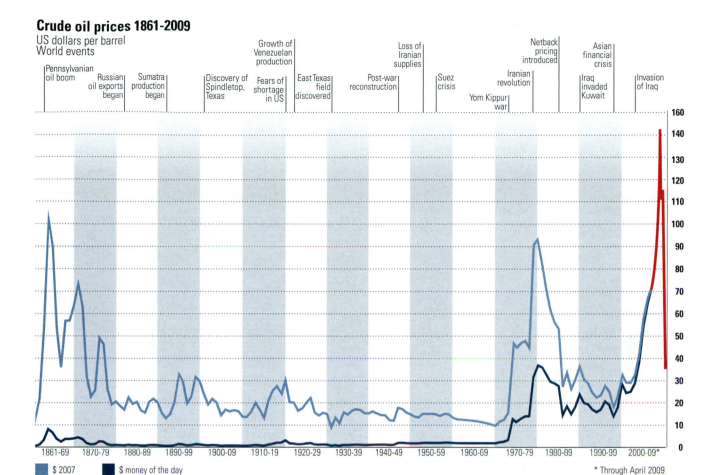

Figure 17.11   The historical trends of oil prices, their escalation in the first 6 months of 2008, and their subsequent fall. (Modified and redrawn from BP. 2008. http://www.bp.com/productlanding.do?categoryId=6929&contentId= 7044622.)

Mexico and the rivers that flow through India and Pakistan. Turkey's development of the GAP project on the Euphrates River, while a major marvel for that country, is fraught with serious water-supply questions for downstream Iraq and Syria. Even within a single country's borders, the inundation of huge tracts of land and the displacement of people and wildlife have been shown to have significant impacts. The development of Quebec Hydro's James Bay project in Canada, which diverted water from several rivers northward to James Bay to produce hydroelectric power, has meant serious hardship for Cree Indians because the increased flow has altered ice coverage patterns on Hudson Bay, affecting hunting. This $6 billion construction enterprise had $20,000 allocated for environmental impact assessment.

The Three Gorges Dam on the Yangtze River in China, at an estimated cost of $24.5 billion (in 1997 dollars), is 2 km long and 200 m high.[4] Scheduled to be completed in 2009, the reservoir will be about 640 km long, submerging more than 600 km². The project will result in the displacement and hence the need for relocation of 1.2 million people. It will also benefit air quality in China, a country known for its pollution problems. Cleveland (2008) writes:

> The government [of China] also notes that the dam's power generation potential of 84.7 billion kWh/yr is the energy equivalent of burning 50 million tons of coal or 25 million tons of crude oil. Thus, the switch to cleaner hydroelectric power would have the effect of cutting 100 million tons of carbon dioxide, up to two million tons of sulfur dioxide, ten thousand tons of carbon monoxide, 370,000 tons of nitrogen oxide, and 150,000 tons of particulates annually from the atmosphere.

As of this writing, similar environmental impacts of the Sardar Sarovar Dam on the

Narmada River in southern India are being hotly debated. A detailed study of the 54 dams by the Indian Institute of Public Administration "conservatively" estimates that during the past 50 years, 33 million people have been displaced just by dam construction alone. The battle on the cost-benefit analyses, in which the economic benefits of these projects are supposed to exceed their costs, rages on.

Energy from Other Major Renewable Resources

About 51% of the world's biomass, geothermal, solar, wind, and ocean (tidal) electric power of 116 GW in 2005 was accounted for by Germany (21.4 GW), the United States (20 GW), Spain (10.3 GW), India (5.3 GW), and Japan (2.5 GW). In the United States, renewable energy accounted for only 7% of the nation's total energy budget (Figure 17.12).

Renewable energy (such as solar and wind) provides immediate environmental benefits by avoiding the environmental impacts of fossil fuels, such as global climate change; acid precipitation; pollution of air, water, and soil; loss and degradation of ecosystems; and their associated adverse impacts on public health. Increasing use of renewable energy can have other positive effects, such as increased energy diversity and security through increased use of indigenous energy supplies, income generation from less dependence on the import of fossil fuels, and increased employment in renewable

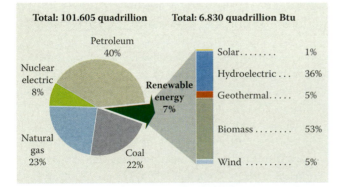

Figure 17.12 Contribution of renewables in the overall energy budget of the United States (U.S. DOE, Energy Information Administration 2008; http://www.eia.doe.gov/cneaf/alternate/page/renew_energy_consump/reec_080514.pdf).

energy technologies. However, the low cost of fossil fuels and the lack of evaluation of the environmental and social benefits of renewable energy, as well as the environmental costs of fossil fuels in their full cycle (from extraction to disposal) and the lack of information by consumers, create **market barriers** (or obstacles) to the rapid and widespread development of renewable energy technologies.

Solar Photovoltaic Systems

Solar energy can be harnessed by active or passive thermal systems or by photovoltaic systems. Two intrinsic difficulties in utilizing solar energy are the low energy density (low concentrations of solar radiation) at high latitudes and the variability of the source due to clouds. Direct solar radiation can be concentrated by mirrors or lenses, but diffuse solar radiation cannot be concentrated. The term **passive solar thermal systems** refers to the exploitation of solar energy for heating and cooling buildings by design, not by using solar collectors.

Active solar thermal systems use solar collectors of the flat plate type, generally installed on the roof for heating, cooling, and the production of hot water for residential, commercial, and industrial uses. Photovoltaic (PV) systems are semiconductor solar cells in which incident solar radiation is converted directly into electrical energy for lighting and other purposes. Current annual world production has reached 150 MW and is growing at about 20% per year. PV systems are easy to install and maintain and are particularly suitable to areas remote from a conventional source of electricity.

The power output capacity of large-scale PV systems has increased worldwide from 15.3 MW in 1995 to 951.2 MW in 2007 (Figure 17.13; Lenardic 2008). With 47% of world's solar power share, Germany leads the world, followed by Spain (28%) and the United States (15%). Worldwide, the generation of solar power increased in 2007 (489.2 MW) by more than 2.6 times over that of 2006 (185.6 MW). Europe collectively has 81% of the current solar power share. Several major utilities in the United States are considering the German model. Currently, PV technologies have an average

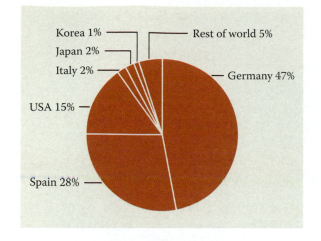

Figure 17.13 Large-scale photovoltaic power plants and their output as of the end of 2007. (Lenardic, D. 2008. http://www.pvresources.com/en/top50pv.php)

energy conversion efficiency of 12–15% with no built-in storage capacity.

Solar Thermal Systems

In solar thermal power (abbreviated **CSP**— concentrated solar thermal power), unlike the photovoltaics, the sun's energy heats water or other fluids to generate steam, which in turn is used to drive a turbine. The cumulative power worldwide generated by CSP in 2007 was 457 MW, up from a mere 1 MW in 1980. By 2012, the capacity is expected to rise to 6,400 MW (Figure 17.14) (Dorn 2008). The United States and Spain are the leaders in CSP development (Dorn 2008).

Wind Energy

At many locations throughout the world, average wind speeds are above the 7 m s^{-1} threshold needed to effectively turn wind turbines. Wind turbines to provide electricity are the world's fastest growing energy technology. Higher outputs are achieved with larger turbines mounted on taller towers because wind speeds generally increase with height. Total world wind capacity in 2007 was 93.8 GW (Figure 17.15), generating 200 TWh yr^{-1} (1.3% of the global electricity consumption) and distributed as follows: Europe: 61%; North America (including Canada and Mexico): 20%; Asia: 17%; South and Central America: 0.6%; Australia Pacific: 1.2; and Middle East and Africa: 0.4%. Wind energy is being used in more than 70 countries,

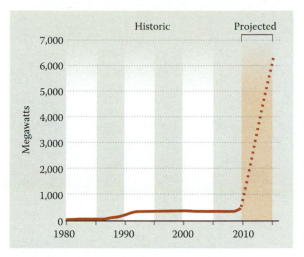

Figure 17.14 World cumulative installed solar thermal power capacity, 1980–2007, with projections for 2012. (Dorn, J. G. 2008. Solar Thermal Power Coming to a Boil. Washington, D.C.: Earth Policy Institute.)

with the United States, Spain, and China in the lead; 19.7 GW capacity was added in 2007. Wind energy can accomplish rural electrification of homes, villages, and small industries cheaply and quickly without extending the utility grid.

The wind energy industry employs 350,000 people worldwide (WWEA 2008). The World Wind Energy Association (WWEA) has increased its prediction for 2010 and now expects that 170,000 MW will be installed by the end of 2010 (WWEA 2008). The growth rate for wind energy increased on an annual average by more than 27% between 2000 and 2007. Turkey's wind energy grew by 220% in 2007, China's by 127.5%, Czech Republic's by 105%, and Poland's and New Zealand's by more than 80% (WWEA 2008). In the United States, wind energy has shown an unprecedented and welcome increase that produces in excess of 18 GW of energy (Table 17.4).

Energy from many renewable resources cannot be stored, so it must be used as it is captured, for example, from photovoltaic cells or wind turbines. This means that energy collected at no or low consumption areas (such as open lands, high wind areas, and deserts) must be transferred to areas of high consumption, such as industrial or urban areas. This requires a power transmission grid so that power can be transmitted from one area to another. Such a national grid will be expensive

Figure 17.15 World wind energy: total installed capacity (MW), 1997–2007, and predicted increases for each year, 2008–2010. (WWEA 2008; http://www.wwindea.org/home/images/stories/pr_statistics2007_210208_red.pdf)

Table 17.4
U.S. Wind Energy Projects

| State | National Total Power Capacities (MW) | | Rank |
	Existing (18,302.68)	Under Construction (6,736.3)	
Texas	5,316.65	1.997.10	1
California	2,483.83	290.00	2
Minnesota	1,299.75	46.40	3
Iowa	1,294.78	549.10	4
Washington	1,195.38	94.00	5
Colorado	1,066.75	0.00	6
Oregon	887.79	201.60	7
Illinois	735.66	171.00	8
Oklahoma	689.00	0.00	9
New Mexico	495.98	0.00	10

Source: From American Wind Energy Association. 2008. http://www.wwindea.org/home/images/stories/pr_statistics2007_210208_red.pdf

and will take time to build. Texas took such a path-breaking step on July 17, 2008, when the Texas Public Utility Commission approved a $4.9 billion plan to build transmission lines that will carry West Texas wind-generated electricity to urban areas in East Texas such as Dallas (Vertuno 2008).

The main environmental concerns surrounding the use of wind energy include local impacts such as the deleterious effects on bird and bat populations, increased noise levels, the visual aesthetics of wind farms, and disruption of radio transmissions.

Geothermal Energy

Geothermal energy is obtained by extracting thermal energy from hot water found underground at depths ranging from a few hundred meters to a few kilometers beneath the Earth's surface. The managed exploitation of heat from the Earth began about 100 years ago with the first piped heating systems in Europe and the United States. More than 8,000 MW of geothermal power are produced today. Unlike wind and solar resources, geothermal resources provide firm, uninterrupted power. But most large geothermal installations are short-lived relative to other alternative energy sources because geothermal power production involves depleting quantities of stored thermal energy beyond its renewal rate.

The average geothermal plant occupies only 400 m² for the production of a gigawatt hour (GWh) over 30 years. Using geothermal energy has higher environmental costs than any other renewable energy source but considerably lower environmental costs than any fossil fuel. High costs are associated with drilling, exploitation, and the potential for release of gaseous pollutants and toxic elements during electricity production.

Tidal, Wave, and Ocean Thermal Energy Resources

Tides originate from the motions of the Earth, Moon, and Sun. Energy is obtainable by damming the entrance to a bay or estuary in a region of large-amplitude tides, which drive turbines as the tidal basin fills and empties. Of about 22,000 TWh yr^{-1} dissipated by the tides, 200 TWh is now considered economically recoverable, and less than 0.6 TWh is actually produced by existing plants (WEC 1998). Damming a bay or estuary without careful planning can adversely affect water quality, sedimentation, and aquatic life of neighboring coastal and aquatic ecosystems, as well as navigation and recreation.

Wave Energy

Wave energy is generated when winds pass over open bodies of water and transfer some of their energy to form waves. The amount of energy stored in waves is both potential energy (in the mass of water displaced from the mean sea level) and kinetic energy (in the motion of the water particles), and it depends on the wind speed, the length of time for which the wind blows, and the distance over which it blows (fetch). Even after the winds die down, storm waves continue to travel great distances from the point of origin with minimal loss of energy in deep water. Current wave energy converters extract energy from the sea and convert it with an energy efficiency of 20% to a more useful form as fluid pressure or mechanical motion.

Biofuels

Organic plant residues as well as animal wastes (excrement) can be used for energy production through direct combustion or biochemical and thermochemical (heating) conversion such as pyrolysis, anaerobic digestion, gasification, alcohol fermentation, and landfill gas. **Biochemical conversion** of biomass is completed through **alcoholic fermentation** to produce **liquid fuels** (called **biofuels**) such as ethanol and methanol and through **anaerobic digestion** or fermentation to produce **biogas** fuels such as methane (plus carbon dioxide as a by-product) (Figure 17.16). Wood and many similar types of biomass (e.g., agricultural wastes, cotton gin waste, wood wastes, and peanut hulls) can be converted through **thermochemical processes** into solid, liquid, or gaseous fuels. **Pyrolysis** has been used to produce **charcoal** from wood since the dawn of civilization, and it is still the most common thermochemical conversion of biomass to commercial fuel. The distillation of charcoal produces hydrogen and carbon monoxide, which can be converted (with applied heat, pressure, and appropriate catalysts) to **methanol,** a liquid with properties similar to gasoline.

Crops that are used for energy production are called **energy crops** and include sugarcane, corn, sugar beets, grains, elephant grass, kelp (seaweed), and many others. The two main factors that determine the suitability of a crop for energy use are (1) a high yield of biomass (dry weight) per unit space and time (tons per hectare per year), which reduces land requirements and lowers the cost of producing energy from biomass; and (2) the amount of energy actually produced from the energy crop, which must be greater than the amount of energy required to grow the crop.

As oil prices have skyrocketed, alternative sources in biofuels have received widespread attention. The conversion of corn to ethanol has received tremendous stimulus by recent subsidies from the U.S. government. In 2006,

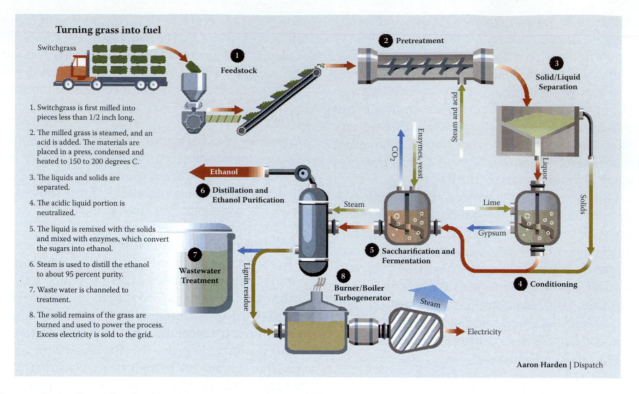

Turning grass into fuel

Switchgrass

① Feedstock

② Pretreatment

③ Solid/Liquid Separation

Steam and acid

CO₂

Enzymes, yeast

Steam

Liquor

Lime

Gypsum

Solids

⑥ Distillation and Ethanol Purification

Ethanol

⑤ Saccharification and Fermentation

④ Conditioning

Lignin residue

⑦ Wastewater Treatment

⑧ Burner/Boiler Turbogenerator

Steam

Electricity

1. Switchgrass is first milled into pieces less than 1/2 inch long.

2. The milled grass is steamed, and an acid is added. The materials are placed in a press, condensed and heated to 150 to 200 degrees C.

3. The liquids and solids are separated.

4. The acidic liquid portion is neutralized.

5. The liquid is remixed with the solids and mixed with enzymes, which convert the sugars into ethanol.

6. Steam is used to distill the ethanol to about 95 percent purity.

7. Waste water is channeled to treatment.

8. The solid remains of the grass are burned and used to power the process. Excess electricity is sold to the grid.

Aaron Harden | Dispatch

Figure 17.16 Steps involved in the processing of plant materials for producing ethanol and other useful by-products. (Courtesy Fred Michel and the *Columbus Dispatch,* 2008.)

U.S. corn production had risen to more than 260 million tons; Iowa led the states with more than 52 million tons. Recent studies show that using corn for ethanol production has serious implications, both economic and environmental. The drawbacks are several:

- The energy inputs into the production and processing of ethanol are essentially a zero-sum undertaking (i.e., it takes as much energy to make it as it contains as a fuel).
- Because corn is used for food products, its diversion into ethanol production has already resulted in higher food prices.
- Croplands used for the production of soybeans are now used for growing corn, resulting in shortages of soybeans.
- Laurance (2007) notes that "Amazon deforestation and fires are being aggravated by U.S. farm subsidies" because more land is being cleared for soybean production.

We discuss the impacts of such land conversions in Chapter 26 on sustainable development.

For several years, research has been under way to investigate the feasibility of other plants that can be processed into ethanol or biofuels. The premises are that some plants (1) grow naturally without any energy inputs; (2) grow from root stocks every year, thus sparing the cost and labor of seeding them; and (3) are capable of producing significant biomass per unit area per unit time, thus providing excellent sources. One such plant is switchgrass (*Panicum virgatum*), which grows in the wild and does well on degraded substrates (Choi and Wali 1995). This plant came into prominence when it was mentioned as an energy source in a U.S. presidential speech.[5] Switchgrass, it was reported, "could get almost 5 1/2 units worth of ethanol."[6] *Miscanthus,* also a C4 perennial grass, has been found even more productive (Heaton, Dohleman, and Long 2008). Research on the production of biodiesels from other plants and other waste products (such as chicken fat, used cooking oils, and municipal wastes) is proceeding vigorously. Given the diversity of plant and other materials, biofuel production presents a set of research challenges (Figure 17.17).

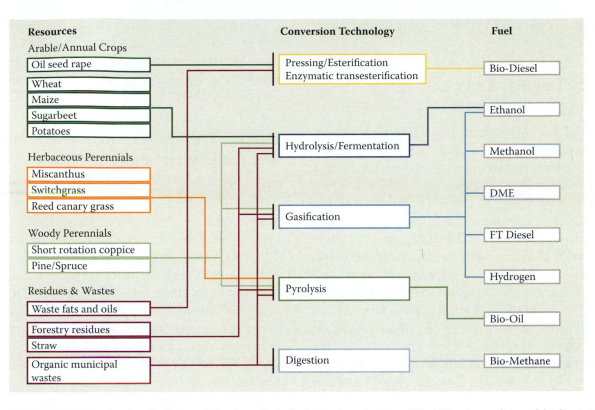

Figure 17.17 Multiple plant materials and biomass waste for biofuel production. (World Business Council for Sustainable Development, 2006. *Mobility 2030: Meeting the Challenges to Sustainability*, Overview 2004. Geneva, Switzerland.)

Current Global and National Energy Profile

World energy systems are dependent on oil as never before in human history. It is the number-one energy source with, a 35% share of the global fuel energy budget in 2005 (Figure 17.2). The three largest producers in 2006 were Saudi Arabia, the United States, and Russia. Given that people all over the world are heavily dependent on fuel energy for food production, transportation, and the continued operation and growth of industries, it is important that we examine the multitude of dimensions of energy use. These include (1) energy and politics, (2) energy imports and national economies, (3) energy and food production, (4) energy and transportation, (5) energy and the environment, and (6) energy and recreation.

Energy and Politics

Politics, rather than environmental issues, have had a telling effect on energy prices and movement. At least six post-Second World War oil crises have been experienced by the United States. These included nationalization by Iran of the Anglo-Iranian (British-owned) oil industry in 1951 and nationalization of the French-built and British-owned Suez Canal in 1956, resulting in the loss of control of that shipping trade route by Western nations. After the 1967 Six-Day War between Egypt and Israel, the Suez Canal was closed to all shipping between 1967 and 1975. In 1973, the Yom Kippur War brought on a serious Arab oil embargo. This was the fourth Arab–Israeli war, the most severe of these crises, and the most consequential. The U.S. hostage crisis in Iran during 1979–1981 shut down the oil supplies from that country and oil prices jumped from $13 a barrel to what was then a staggering $34. In 1990, Iraq invaded Kuwait, the United Nations imposed a trade embargo, and the flow of oil from that country stopped. Disruption of supplies was also affected in the first years of the twenty-first century by the Iraq war in 2003 and tensions with Iran, which continue. These examples illustrate the changed face of trade relations of oil and its close relationship with political policies.

Energy Imports and National Economies

With the exception of a few oil-rich countries, most countries must import oil to meet their needs. Both basic living and modern amenities depend upon the availability of fuel energy. Equally important is the capacity for nations to sustain and expand their industrial infrastructure and the overall enhancement of their standards of living. The relationship between the growth in gross national products and energy use in all countries is intimate. Recent economic growth in China and India, at an annual growth rate of 7–12%, has produced a staggering demand for oil and natural gas worldwide, sending the prices up to unexpected levels (Figure 17.11). The historical behavior of **energy intensity**—the ratio of energy use to gross national productivity (GNP)—has strongly influenced the links between energy efficiency and conservation, economic growth, and the environment. As more nations embrace free trade, the issues of energy supplies and the environmental impacts of energy use are likely to increase the tension between least-developed countries and most-developed countries. To illustrate, energy consumption fell by 2.2% for the United States and increased by 7.7% for China in 2007 (BP 2008).

Energy and Food Production

We have already discussed the vital need of energy inputs necessary for ensuring modern agricultural productivity. From the preparation of the soil to fertilization, irrigation, biocide use, harvesting, processing, packaging, and shipping, food systems are dependent on sustained and reliable supplies of energy. In the United States, as noted, 20% of the national energy budget goes to the agricultural sector from farm to marketable product. The Green Revolution of the developing countries depends heavily on energy inputs. The recently enacted policies of converting grains (especially corn) to

fuel, as discussed earlier, have increased food prices. This concern has been raised in several studies and commentaries, including by governors and legislative leaders (as an example, see Blaney 2008).[7]

Energy, Transportation, Recreation, and the Environment

The automobile has transformed the mode of living in Western societies, particularly in the United States. Transportation issues and the growth of suburbs are discussed in Chapter 15. Now, more than ever, tourism is a global phenomenon. Accessibility to places, near and far, has been tremendously enhanced by multiple and cheap modes of transportation. How large the "mass" movement of people actually is may be seen by one or two examples. During the Memorial Day 2000 weekend in the United States, 34 million people traveled—2 million by air, and 32 million by automobile. For the Labor Day 2000 weekend, the numbers traveling by air and car were estimated at 33 million. In the state of Ohio alone, tourism generates $15 billion in expenditures and revenues.

The extraction, processing, transport, and, finally, use of energy—from running industrial plants, automobiles, and flying in airplanes to domestic use—all have an impact on the environment. Pollution of air, water, and soil that results from extraction, processing, transportation, and consumption of fossil fuels has been discussed in Chapters 14, 15, and 17. Two of the major environmental impacts to which human consumption of fossil fuels contributes considerably—atmospheric deposition of pollutants and global climate change—are discussed in Chapters 18 and 19.

In the next 50 years, energy consumption is expected to increase by 361% in Asia, 340% in Latin America, and 326% in Africa, where economic growth is projected to be highest (Brown et al. 1999). Although annual emissions from developed countries are currently twice as high as those from developing ones, emissions from developing countries will nearly quadruple, while those from developed countries are predicted to increase by 30% over the next 50 years (Brown et al. 1999). The environmental effects, particularly of global climate change, are being felt universally and are discussed in Chapter 19.

Energy Policy

Given the immense importance of energy in human affairs all over the world, one would expect that all nations would have energy policies. Unfortunately, this is not the case. It may seem incredible that the United States, whose citizens use the most energy per capita, does not have a national energy plan or policy. To make matters worse, we are wasteful of both energy and other resources.

We believe that the agenda of environmental issues in the twenty-first century will be topped by the problems directly related to energy use and the growing human population. The problem of energy is a nondenominational, nonsectarian, non-political-party problem, the solutions to which depend on our ability to meet incredible challenges and, for many of us, moderate our lifestyles. We know the problems; now we need the will to find solutions.

Notes

1. OPEC includes Algeria, Indonesia, Iran, Iraq, Kuwait, Libya, Nigeria, Qatar, Saudi Arabia, the United Arab Emirates, and Venezuela.
2. OECD (Organization for Economic Cooperation and Development) includes 30 countries—Australia, Austria, Belgium, Canada, Czech Republic, Denmark, Finland, France, Germany, Greece, Hungary, Iceland, Ireland, Italy, Japan, Korea, Luxembourg, Mexico, Netherlands, New Zealand, Norway, Poland, Portugal, Slovak Republic, Spain, Sweden, Switzerland, Turkey, United Kingdom, and the United States (http://www.oecd.org/home/).
3. High-level waste is 1 billion times more radioactive than low-level waste (LLW), which has a radioactive activity level of less than 100 nCi g^{-1}. About 56% by volume of LLW comes from nuclear power plants; if it is not solid already, it must be solidified before storage.
4. Three Gorges Dam, China (http://www.eoearth.org/article/Three_Gorges_Dam,_China). Accessed July 7, 2008.
5. In his January 2006 State of the Union address, President George W. Bush noted a number of energy sources that his administration would fund to reduce America's oil dependence. He noted, "We'll also fund additional research in cutting-edge methods of producing ethanol not just from corn but from wood chips and stalks and switchgrass" (Vertuno 2008).
6. Shortly after the presidential speech, it was reported: "For every unit of energy used to grow the switchgrass, Ken Vogel of USDA says, he could get almost 5 1/2 units worth of ethanol. That's a lot more efficient than making ethanol from corn, he says" (Joyce 2008).
7. A story by Betsy Blaney of the Associated Press (2008) reported the denial by the U.S. Environmental Protection Agency of "a request from Texas Gov. Rick Perry to cut the federal ethanol mandate in half for a year. ... Between January and June, cattle feeders nationwide lost $1.5 billion, officials said. ... [The] spokesman for the Amarillo-based Texas Cattle Feeders Association said the EPA's decision will mean cutbacks by cattle producers and higher priced beef in meat cases in about a year.

    "[EPA Administrator] Johnson said the agency's assessment looked at the livestock issue and found feed prices have increased because of biofuel production. ... More than four dozen House Republicans and two dozen GOP senators, including presidential candidate John McCain, wrote the EPA in support of a waiver. The state of Connecticut also supported Texas' request."

Questions

1. How are fossil fuels different from renewable energy resources?
2. What are the consequences of relying solely on nonrenewable energy resources in terms of environmental health and national security?
3. How are humans altering biogeochemical cycles by burning fossil fuels such as coal, oil, and natural gas?
4. Discuss the significance of renewable energy sources and technologies in spurring economic development and reducing environmental degradation.
5. What are the main obstacles to widespread use of renewable energy technologies and energy efficiency practices?

Air Quality and Stratospheric Ozone

18

Introduction

The haze and smoke in the atmosphere and their attendant ill health effects, particularly on the breathing of royal personages, have been recorded in the British Isles since the thirteenth century. The earliest record is of Eleanor, queen of Henry III, who, reportedly in 1257, could not bear the smoke and left Nottingham. Clear instructions against coal burning came in 1307 from King Edward I in England, where on first offense there would be "great fines and ransoms," a second offense would result in the demolishing of coal furnaces (Galloway 1882), and the third offense would be punishable with death. But even with the royal powers behind these proclamations, there was not much success. Historically, records of air pollution episodes have been many in number with significant loss of life (Table 18.1).

A **pollutant** is any extraneous waste material that is put into the air, water, or soil that is likely to produce a harmful effect. Pollutants degrade the quality of air that we breathe, and many health effects are very serious (see Table 15.5). Thus, whereas smoke in the examples cited earlier is a pollutant in itself, it also carries several other pollutants, particularly chemicals, with it. If such problems have existed for centuries, why is there increased concern now? The answer lies simply in the much accelerated rate at which human activities have added these materials to the atmosphere, the multiple sources that generate them, and the resultant diversity in the types of pollutants. Current estimates suggest that the problem continues worldwide and that air pollution results in the deaths of 53,000 people in America alone each year.

Sources of Air Pollutants

Emissions into the atmosphere originate from a number of sources, including such natural sources as volcanic eruptions. But the

Table 18.1
Major Air Pollution Episodes[a]

Year/Date	Location	Excess Deaths
1873	London	500
1880, Jan. 26–29	London	1,000
1930, December	Meuse Valley, Belgium	63
1948, October	Donora, PA	20
1948, Nov. 26–Dec. 1	London	700–800
1950, Nov. 21	Poza Rica, Mexico	22
1952, Dec. 5–9	London	4,000
1953, November	New York	250
1956, Jan. 3–6	London	1,000
1957, Dec. 2–5	London	700–800
1959, Jan. 26–31	London	200–250
1962, Dec. 5–10	London	700
1963, Jan. 7–22	London	700
1963, Jan. 9–Feb. 12	New York	200–400
1966, Nov. 24–30	New York	168
1984, Dec. 2–3	Bhopal, India	2,000 dead[b]; 300,000 injured; 1,000 animals killed; 6,000 affected

[a] Based on Bach, W. 1972. *Atmospheric Pollution.* New York: McGraw–Hill Book Company; Anderson, H. R. 1999. Health effects of air pollution episodes. In *Air pollution and health*, S. T. Brown, Holgate, J. M. Samet, H. S. Koren and R. L. Maynard, ed. 461–482. San Diego, CA: Academic Press; Choukiker, S. 2005. *Major air pollution episodes: Environmental distortions that kill.* http://www.visionriviewpoint.com/article.asp?articleid=26.

[b] Choukiker notes that this number may only be one-quarter of actual deaths.

following human activities have surpassed all natural sources with serious attendant consequences (Rodhe et al. 1988) (Table 18.2):

- generation of electricity;
- smelting and refinement of metals;
- production and use of concrete, glass, ceramic, and plastics materials;
- refinement and use of petroleum and petrochemicals;
- use of transportation vehicles;
- space and water heating;
- decomposition and incineration of sanitary and solid wastes;
- production and application of fertilizers, pesticides, and other agricultural and silvicultural chemicals;
- disposal of excreta from humans and domestic animals;
- burning of biomass and farm and forest residues; and
- use of explosive devices in peace and war.

Among the major sources of air pollutants, the burning of fossil fuels—primarily coal, oil, and natural gas—tops the list (Figure 18.1). Of the users, the transportation sector (such as automobiles and trucks) is the largest contributor of pollutants. In the 1970s, for every 1,000 gal (3,785 L) of gasoline consumed, the automobile emissions into the atmosphere were enormous: 1,434 kg of carbon monoxide, 90–180 kg of organic vapors, 9–34 kg of nitrogen oxides, 8 kg of aldehydes, 7.5 kg of sulfur compounds, 0.9 kg of organic acids, 0.9 kg of ammonia, and 0.13 kg of such solids as zinc and other metal oxides. Local vehicular emissions contribute 12% of the global anthropogenic

Table 18.2
Sources of U.S. Emissions of Criteria Air Pollutants, 1985–1996[a]

Air Pollutant and Year	Fuel Combustion[b]	Transportation[c]	Industrial Processes[d]	Miscellaneous[e]
Carbon Monoxide				
1985	8.5	92.0	7.2	7.8
1996	5.9	69.9	5.8	7.0
Nitrogen Oxides				
1985	10.0	12.2	0.8	0.3
1996	10.4	11.7	0.8	0.2
VOC[f]				
1985	1.5	11.6	10.4	0.5
1996	1.0	7.9	9.4	0.5
Sulfur Dioxide				
1985	20.0	0.7	2.4	0.01
1996	16.7	0.6	1.6	0.009
PM-10[g]				
1985	1.5	0.9	1.3	41.7[h]
1996	1.1	0.8	1.2	28.0[h]
Lead[i]				
1985	0.5	18.9	3.4	N/A
1996	0.4	0.5	2.8	N/A

Source: From U.S. EPA. 1996a. Hazardous waste characteristics scoping study. Washington, D.C.: U.S. Environmental Protection Agency, Office of Solid Waste.
[a] Million tons.
[b] Includes electrical utilities, industry, and other.
[c] Includes on-road vehicles and nonroad engines and vehicles.
[d] Includes chemical industries, metals-processing, petroleum industries, solvent utilization, storage and transportation, waste disposal and recycling, and other industries.
[e] Forest wildfires.
[f] Volatile organic compounds
[g] Particulate matter with a diameter ≤ 10 μm.
[h] Includes wind erosion, fugitive dust from agriculture, forestry, (un)paved roads, construction, and wildfires and other combustion.
[i] Expressed as thousand tons.

gas emissions (Faiz 1993). During construction, use, and maintenance of roads, the transportation sector, at present, is responsible for about 20 and 40% of the global carbon dioxide (CO_2) and oxides of nitrogen (NO_x) emissions, respectively (Hahn and Simonis 1991). In city centers, road traffic accounts for 90–95% of carbon monoxide (CO) and lead (Pb), 60–70% of nitrogen oxides (NO_x) and hydrocarbons (HC),

and a major share of particulates (Johnstone and Karousakis 1999) (Table 18.2).

Although the emissions are substantial from a fleet of vehicles exceeding 100 million, several major developments have resulted in significant reductions in vehicle emissions in the last 30 years, including (1) improved internal combustion engine technology, (2) major changes in the refinement of petroleum products, and

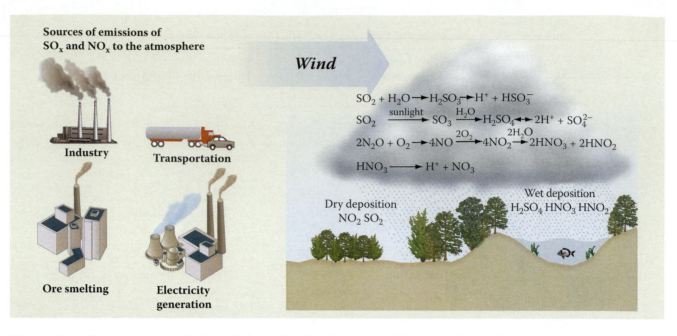

Figure 18.1   Sources of atmospheric emissions of oxides of nitrogen (NO_x) and sulfur (SO_x) for acid wet/dry deposition.

(3) enforcement of major air pollution regulations in many countries. These changes are reflected in air quality in many cities (Table 18.3). However, given the major increases in recent decades in the numbers of vehicles—not only in the United States but also all over the world—these major gains have been partially offset and the transportation sector still contributes more than about 55% of the total gases emitted.

The second major source of air pollutants is power-generating utilities, which contribute over 20%. Plants fueled by coal are particularly large polluters. For example, the Muskingum River Power Plant in Ohio emits 2,090,305 tons of carbon per year (Pechan 1999)—approximately four times more carbon than is sequestered in aboveground woody biomass in the forests of all of southeastern Ohio each year (Watts 2001). Other industries contribute about 15% of total emissions, with the rest coming from several sources, including waste disposal.

Types of Air Pollutants

Air pollutant emissions from human activities can be divided into primary and secondary categories. **Primary air pollutants** are those that enter directly into the atmosphere, whereas the **secondary air pollutants** are formed by chemical reactions with other substances in the atmosphere. Air pollution can be classified under five types (Table 18.4):

- particulate air pollution: primary and secondary aerosols;
- gaseous air pollution: primary and secondary gases involved in photochemical smog, acid deposition, ozone depletion, and global climate change;
- tropospheric ozone and other photochemical oxidants: secondary air pollutants such as peroxyacyl nitrates (PANs), nitric acid (HNO_3), formaldehyde, and other aldehydes;
- radioactive and toxic air pollution: carbon monoxide (CO), radionuclides (e.g., radon), fibers (e.g., asbestos), trace organics (e.g., polychlorodibenzo-dioxins and -furans), and selected heavy metals (e.g., beryllium, cadmium, and mercury); and
- thermal and noise pollution: urban heat islands and loud or high-pitched sounds.

Aerosols (particulate matter) are a complex chemical mixture of solid and liquid particles suspended in the atmosphere that can be produced by two distinct mechanisms: (1) by the

Table 18.3
Improving Air Quality Trends as Indicated by Number of Days When PSI Exceeded 100 (Unhealthful) in Selected U.S. Urban Areas (1987 vs. 1996)

Primary Metropolitan Area	No. Monitoring Sites	No. Days with PSI > 100[a]	
		1987	1996
Los Angeles	36	201	88
Houston	28	67	28
San Diego	20	61	4
New York	26	44	4
Phoenix	25	42	5
Denver	21	37	1
Philadelphia	37	35	5
El Paso	17	32	9
Baltimore	15	28	3
Atlanta	8	27	6
Washington, D.C.	34	26	2
St. Louis	53	17	4
Chicago	44	17	3
Minneapolis–St. Paul	23	14	1
Seattle	14	14	0

Source: From U.S. EPA. 1996a. Hazardous waste characteristics scoping study. Washington, D.C.: U.S. Environmental Protection Agency, Office of Solid Waste.

[a] PSI (pollutant standard index) integrates information from many pollutants across an entire monitoring network into a single number that represents the worst daily air quality experienced in an urban area. PSI index ranges are 0–50 (good); 51–100 (moderate); 101–199 (unhealthful); 200–299 (very unhealthful); and >300 (hazardous).

direct injection of particles into the atmosphere (**primary aerosols**), and (2) by the conversion of gaseous precursors into liquid or solid particles (**secondary aerosols**). Natural and anthropogenic aerosol sources include dust, sea salt, volcanic eruption, fossil fuel combustion, biomass burning, nitrates, and sulfates (Table 18.5). Aerosols have a cooling effect on the global climatic system directly by the backscattering of incoming solar (short wavelength) radiation and indirectly by acting as nuclei for the formation of cloud droplets and the enhancement of cloud albedo (reflectivity). Aerosols such as sea salt and desert dust also have a relatively minor warming effect by the trapping of outgoing terrestrial (long-wavelength) radiation. Aerosols impair visibility by scattering light and cause haze. Evidence from satellite data indicates that aerosol emissions from urban and industrial activities inhibit rain and snowfall due to the suppression of droplet coalescence and ice precipitation in clouds polluted by aerosols (Rosenfeld 2000; Baker and Peter 2008) (Figure 18.2). The polluted clouds adversely affect the hydrological cycle and consequently gaseous and sedimentary cycles and biological productivity of organisms at the local, regional, and global scales.

Atmospheric deposition includes not only acids (converted from sulfur and nitrogen oxides), but also organometals, metals, and organic pollutants that, as we shall read, cause

Table 18.4
Types and Characteristics of Air Pollutants and Their Ecological Effects

Types of Air Pollution	Characteristics	Ecological Effects
Particulate Air Pollution (Aerosols)		
Sea salt, biological debris and dust, soot, charcoal, lead, and asbestos	Solid and liquid particles	
Primary aerosols	Reduction in visibility and incoming solar radiation, and cooling effect	
Sulfuric acid (H_2SO_4), sulfates (SO_4^{2-}), nitrates(NO_3^-), and nonmethane hydrocarbons	Secondary aerosols	Global climate change
Gaseous Air Pollution		
Carbon dioxide (CO_2), nitrogen oxides (NO_x), methane (CH_4), water vapor (H_2O), ozone (O_3), and chlorofluorocarbons (CFCs)	Primary and secondary pollutants	Stratospheric ozone depletion
CFCs, hydrochlorofluorocarbons (HCFCs), carbon tetrachloride, methyl chloroform, halons, methyl bromide (CH_3Br), and hydrofluorocarbons (HFCs)		
Oxides of nitrogen and sulfur		Acid deposition
Nitrogen oxides, and nonmethane hydrocarbons		Photochemical smog
Tropospheric Ozone and Other Photochemical Oxidants		
Peroxyacyl nitrates (PANs), nitric acids (HNO_3), formaldehyde, and other aldehydes	Secondary air pollutants	Photochemical smog
Radioactive and Toxic Air Pollution		
Radionuclides (e.g., radon), fibers (e.g., asbestos), carbon monoxide (CO), trace organics (e.g., polychlorodibenzo-dioxins and -furans), and selected heavy metals (e.g., beryllium, cadmium, and mercury)	Hazardous air pollutants	Carcinogenic and poisonous chemicals and biomagnification
Thermal and Noise Pollution		
Heavy traffic, air compressors, and jet airliners	Temperature anomalies and high sound intensity	Urban heat islands, hearing impairments, and physiological effects

acidification of water and soils, tree death, and losses of biodiversity and ecosystem stability.

Together with the pollutants discussed earlier, we must also consider the role of ozone. Ozone (O_3) can occur in the upper atmosphere (the stratosphere), where it protects life on the Earth from the sun's harmful ultraviolet rays (UV-B) (this is discussed in the next chapter), and in the lower atmosphere (the troposphere) by a chemical reaction between oxides of nitrogen (NO_x) and volatile organic compounds (VOCs) in the presence of sunlight. Motor vehicle exhaust, industrial emissions, gasoline vapors, and chemical solvents are some of the major sources of NO_x and VOCs, known as ozone precursors. The formation of harmful concentrations of

tropospheric ozone is facilitated by strong sunlight and hot weather and degrades air quality, human health, vegetation, and many common materials. The larger its concentration in the tropospheric region is, the greater is its detrimental impact on many physiological processes of animals, plants, and humans. These latter aspects are discussed here.

Air Pollutant Transformations

Globally, current human activities mobilize about 150 Tg (teragrams) of nitrogen and 150 Tg

Table 18.5
Global Emission Estimates for Major Aerosol Sources in 1995

Source of Emissions (Tg yr⁻¹)		Low	High	Lifetime (Days)	Direct Forcing (W m⁻²)[a]
Natural		2,144	23,475	4	−2.3
Primary	Soil dust (mineral aerosol)	1,000	3,000	1	−0.75
	Sea salt	1,000	10,000	4	−0.09
	Volcanic dust	4	10,000	4	−0.05
	Biological debris	26	80	5	−0.07
Secondary	Sulfates from biogenic gases	60	110	5	−0.68
	Sulfates from volcanic SO_2	4	45	5	−0.09
	Organics from biogenic NMHC[b]	40	200	7	−0.55
	Nitrates from NO_x	10	40	4	−0.03
Anthropogenic		245	555	4	−2.5
Primary	Industrial dust (w/o black carbon)	40	130	6	−0.14
	Soot and charcoal (black carbon)	10	30	5	−0.21
Secondary	Sulfates from SO_2	120	180	8	−1.06
	Biomass burning (w/o black carbon)	50	140	4	−0.91
	Nitrates from NO_x	20	50	7	−0.05
	Organics, anthropogenic NMHC	5	25		−0.10
Total		2389	24,030		−4.8
Natural (%)		90	98		48

Source: Modified from Andreae, M. O. 1995. In *Future Climates of the World: A Modeling Perspective,* ed. A. Henderson-Sellers, 347–389. Amsterdam: Elsevier Science B.V.
[a] Refers to global cooling or an imposed change in energy input that modifies the Earth's heat balance.
[b] Nonmethane hydrocarbons.

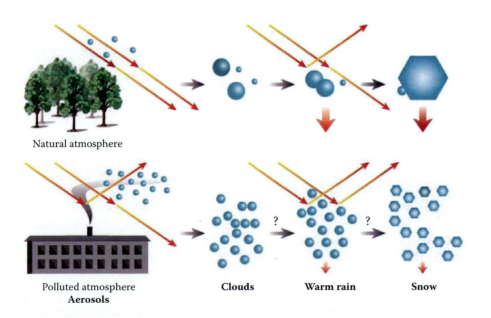

Figure 18.2 Pollution of clouds by aerosol emissions. (Baker, M. B., and T. Peter. 2008. Small-scale cloud processes and climate. *Nature* 451: 229–300.)

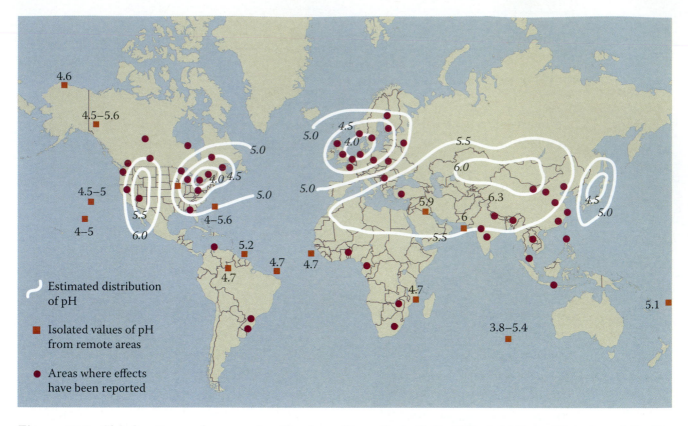

Figure 18.3 Global patterns of mean rain pH values. (From Park, C. C. 1987. *Acid Rain: Rhetoric and Reality.* Methuen: London, McCormick, J. 1985. *Acid Rain.* Toronto: Gloucester Press.)

of sulfur annually (Galloway 1996). As an illustrative example, in 1997 human-induced nitrogen additions disrupted the nitrogen cycle through N fertilizer (which added 80 Tg N), fossil fuel combustion (30 Tg N), and cultivation of legumes and rice (40 Tg N). About 80 and 20% of total anthropogenic nitrogen emissions are deposited in oceans and on lands, respectively. About half of human-mobilized sulfur comes from fossil fuel combustion and mining activities, and the rest comes from sulfur fertilizer, animal husbandry, and wetlands. Thus, the oxidation of SO_x, NO_x, and NH_3 emissions to nitric and sulfuric acids increases the acidity of natural precipitation. Precipitation naturally has acidic pH of about 5.6 because it is in equilibrium with atmospheric CO_2, which forms H_2CO_3 (carbonic acid) upon combination with water. Therefore, any pH lower than 5.6 is due to other acids (Figure 18.3). In the absence of human activities, biological N fixation and lightning create reactive N from atmospheric N_2, while mineral weathering and volcanic eruptions create reactive S. When acid

deposition exceeds the capacity of biotic (plant and microbes) uptake and retention, N saturation takes place in terrestrial and aquatic ecosystems. The effects of acid deposition are discussed in more detail later in this chapter.

Modes of Air Pollutant Action

As these pollutants are generated, they rise to the atmosphere and are simultaneously transported globally as well as transformed chemically. The height at which a pollutant is discharged into the atmosphere will determine how far it will be transported. Raising the height of power plant and other industry gas stacks has brought about the discharge into the atmosphere of air pollutants at greater heights from the ground, sparing local effects but ensuring long-range transport. In turn, the long-range transport is significantly dependent on patterns of airflow (wind speed) and its prevailing direction.

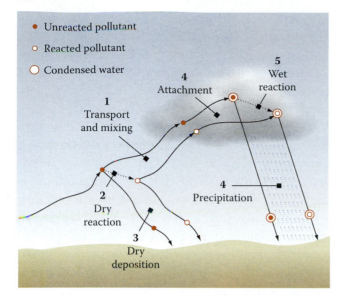

Figure 18.4 Atmospheric pathways leading to acid deposition. (NRC. 1983. *Acid deposition: Atmospheric processes in eastern North America*. Washington, D.C.: National Academy Press.)

The nature and concentration of a pollutant will determine its chemical reactivity with other components of the atmosphere such as its attachment to water vapor in the atmosphere (Figure 18.4), the speed with which such reactions take place, and varieties of the resulting compounds. Meteorological conditions such as relative humidity, cloud cover, and intensity of solar radiation are also very important. The attachment of a pollutant to water in the atmosphere has been called "**precipitation scavenging,**" which is a composite process involving four steps. First, the pollutant and the water in the atmosphere (cloud, rain, or snow) must intermix within the same airspace. Second, the pollutant must attach to the condensed-water molecule. Third, the pollutant may react physically and/or chemically within the aqueous phase. The final step is that pollutant-laden water elements must be delivered to the Earth's surface via precipitation.

For example, a number of pathways have been identified by which sulfuric (H_2SO_4) and nitric (HNO_3) acids are formed. The reaction rates of SO_2 and NO_x oxidation vary considerably in the homogenous gas and liquid phases, from a few to 20–30% per hour in the gas phase to 100% in water. Species associated with acid–base formation include SO_2, SO_3, HNO_3, NH_3, Cl, and metallic ions. Nitrous oxide (N_2O) is relatively inert in the troposphere, so it is not readily deposited on the Earth from the atmosphere, but the opposite is true of NO_x (NO and NO_2). The short atmospheric lifetime of NO_x and the absence of evidence that it accumulates in the atmosphere dictate that surface NO_x sources and sinks must balance. Although soil-emitted NO is not readily deposited in precipitation, the first product of its photochemical oxidation in the troposphere, NO_2, is strongly absorbed by both soils and plants. Following further oxidation, the resulting HNO_3 is extremely efficiently deposited to almost all surfaces (Hutchinson 1995).

Criteria Air Pollutants

The U.S. EPA has set national air quality standards for six principal pollutants (referred to as "criteria pollutants"): carbon monoxide (CO), lead (Pb), particulate matter (PM), nitrogen dioxide (NO_2), sulfur dioxide (SO_2), and ozone (O_3). The EPA's most recent toxics release inventory (1998) shows that U.S. companies reported the release of 544.3 thousand tons of chemicals into the air and water, of which 514.2 thousand tons were released directly into the air. Of these chemicals, 53% are known or suspected developmental or neurological toxins. The report states further that because chemical companies report only an estimated 5% of all chemical releases, the total estimated release might be as high as 10.8 million tons (National Environmental Trust [NET] 2000).

The criteria air pollutants affect physiology and biochemistry of cells, growth, anatomy, and morphology of organisms and thus the structure, function, and dynamics of communities and ecosystems (Figure 18.5). For example, the ozone in the troposphere, particularly in the land–atmosphere contact region, is not emitted directly into the air but, rather, is formed as a strong photochemical oxidant by the reaction of VOCs and NO_x in the presence of heat and sunlight. The larger its concentration in

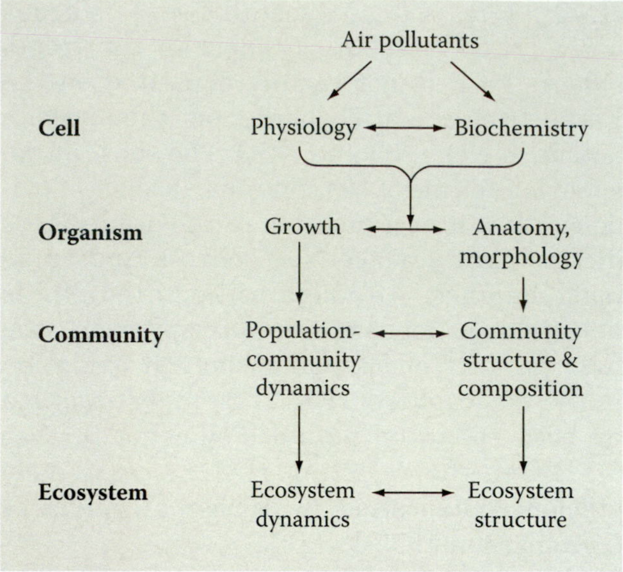

Air pollutants

Cell   Physiology ←——→ Biochemistry

Organism   Growth ←————→ Anatomy,
                                  morphology

Community   Population- ←——→ Community
                community   structure &
                dynamics   composition

Ecosystem   Ecosystem ←——→ Ecosystem
                dynamics   structure

Figure 18.5 Effects of air pollutants at different ecological levels. (Westman, W. E. 1990. Detecting early signs of regional air pollution injury to coastal sage scrub. In G. M. Woodwell, ed. *The Earth in Transition: Patterns and Processes of Biotic Impoverishment*, 323–346. New York: Cambridge University Press.)

this region is, the greater its impact is on many physiological processes of animals, plants, and humans. Short-term (1–3 hours) and prolonged (6–8 hours) exposures to ambient ozone have been linked to a number of health effects of concern, such as respiratory infection, lung inflammation, asthma, chest pain, and cough. Elevated surface-level ozone has been shown experimentally to decrease photosynthesis, crop maturation, and dry matter production and yield and to increase plant susceptibility to disease, pests, and other environmental stresses (Table 18.6). Plant species sensitive to ozone showed even greater impacts: decreased leaf conductance, leaf area, and water use efficiency (Krupa and Jager 1996). Other impacts are discussed later in this chapter.

The remaining criteria pollutants have a diversity of impacts on humans. Carbon monoxide enters the bloodstream through the lungs and reduces oxygen delivery to the body's organs and tissues and can be poisonous at

Table 18.6
Effects of Elevated Surface-Level UV-B Radiation or O_3 on Crops[a]

Plant Characteristic	Effect of UV-B	Effect of O_3
Photosynthesis	Reduced in many C_3 and C_4 species (at low light densities)	Decreased in most species
Leaf conductance	Reduced (at low light intensities)	Decreased in sensitive species and cultivars
Water-use efficiency	Reduced in most species	Decreased in sensitive species
Leaf area	Reduced in many species	Decreased in sensitive species
Specific leaf weight	Increased in many species	Increased in sensitive species
Crop maturation rate	Not affected	Decreased
Flowering	Inhibited or stimulated	Decreased floral yield, fruit set and yield; delayed fruit set
Dry matter production and yield	Reduced in many species	Decreased in most species
Sensitivity between cultivars (within species)	Response differs between cultivars	Frequently large variability
Drought stress sensitivity	Plants become less sensitive to UV-B, but sensitive to lack of water	Plants become less sensitive to O_3 but sensitive to drought
Mineral stress sensitivity	Some species become less sensitive, while others become more sensitive, to UV-B	Plants become more susceptible to O_3 injury

Sources: From Krupa, S. V., and H-J Jager. 1996. Adverse effects of elevated levels of ultraviolet (UV)-B radiation and ozone on crop growth and productivity. In *Global climate change and agricultural production*, ed. F. Bazzaz and W. Sombroek, 141–169. New York: John Wiley and Sons. Runeckles, V. C., and S. V. Krupa. 1994. The impact of UV-B radiation and ozone on terrestrial vegetation. *Environmental Pollution* 83: 191–213.

[a] These are summary conclusions from artificial exposure studies; however, there can be exceptions.

higher levels of exposure. Visual impairment, reduced work capacity, reduced manual dexterity, poor learning ability, and difficulty in performing complex tasks are all associated with exposure to elevated CO levels. Exposure to Pb occurs mainly through inhalation of air and ingestion of Pb in food, water, soil, or dust. It accumulates in the blood, bones, and soft tissues, adversely affecting the kidneys, liver, nervous system, and other organs. Lead can also be deposited on the leaves of plants, presenting a hazard to grazing animals. Exposures to NO_2 and SO_2 may increase respiratory illnesses, alterations in the lungs' defenses, and susceptibility to respiratory infection. Atmospheric transformation of NO_x can lead to the formation of ozone and nitrogen-bearing particles, exacerbating adverse health effects. Inhalable PM includes both fine and coarse particles; exposure to coarse particles is primarily associated with the aggravation of respiratory conditions such as asthma. Exposure to fine particles is most closely associated with increased hospital admissions and emergency room visits for heart and lung disease, increased respiratory symptoms and disease, decreased lung function, and even premature death. In addition to health problems, PM is the major cause of reduced visibility in urban areas and can also cause damage to paints and building materials.

Nature of Atmospheric Deposition

Some of the pollutants discussed earlier have been shown conclusively to increase the acidity of rain (that is, reduce its pH value). Although the importance of acid deposition has been realized only in the last 40 or so years, the recognition of such problems is much older. Scientific knowledge of acidification is reported to have been known since the middle of the eighteenth century and specifically so since 1845. That these pollutants may mix with the rain, alter its chemistry, and come down to Earth was recognized in 1852 by an English chemist,

Robert Angus Smith, who was the first to use the term "acid rain" (Gorham 1998).

When it was thought that atmospheric pollutants reached the ground only when they mixed with water, the term widely used was **acid rain.** However, snow and hail also act as conveyors of pollutants to the ground, so a better term is **acid precipitation.** Furthermore, research has shown that the pollutants also settle down in dry forms on the surfaces of buildings, animals, and plants; the dry deposition portion may range from 20 to 60% of total atmospheric deposition. Thus, the current preferred term, **atmospheric deposition,** was introduced to reflect the combination of wet deposition (rain, snow, and mist droplets—**wet fall**) and dry deposition (dust, aerosols and gases—**dry fall**). Emissions by burning of fossil fuels, electric utility plants, and transportation of oxides of nitrogen and sulfur in the atmosphere react with water, oxygen, and oxidants to form acidic (secondary) compounds that acidify precipitation.

Atmospheric deposition affects biota and man-made objects in multiple ways (Figure 18.6). To assess effects of acidic deposition and air quality on public and ecosystem health, emission inventories of changes over time and space in the chemical composition of the precipitation and air must be carried out. The National Atmospheric Deposition Program/National Trends Network (NADP/NTN) (1977) and the Clean Air Status and Trends Network (CASTNET) (1987) were established in the United States to monitor wet and dry atmospheric depositions. Comparison of emissions of six principal pollutants shows an appreciable and significant decrease in their levels from 1980 to 2006 (Figures 18.7 and 18.8). The atmospheric nitrogen pathways illustrate the several steps involved from source to effects (Figure 18.9).

Factors Affecting Atmospheric Deposition

Several factors affect atmospheric deposition; these may be grouped into meteorological and nonmeteorological factors (Figure 18.10). To the first group belong such factors as wind

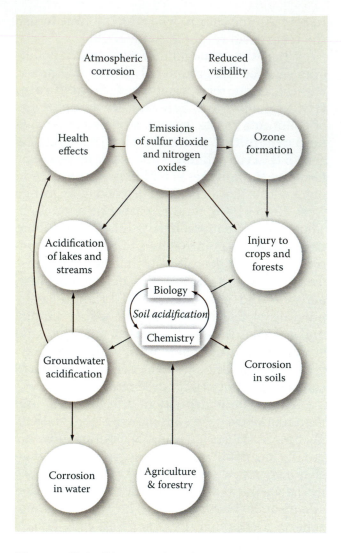

Figure 18.6 Direct and indirect effects of emissions of oxides of sulfur and nitrogen. (From Rodhe, H. et al. 1988. In *Acidification in Tropical Countries,* ed. H. Rodhe and R. Herrera, 3–39. Chichester, England: John Wiley & Sons.)

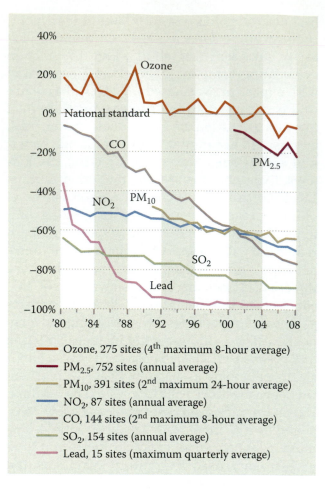

- Ozone, 275 sites (4th maximum 8-hour average)
- PM$_{2.5}$, 752 sites (annual average)
- PM$_{10}$, 391 sites (2nd maximum 24-hour average)
- NO$_2$, 87 sites (annual average)
- CO, 144 sites (2nd maximum 8-hour average)
- SO$_2$, 154 sites (annual average)
- Lead, 15 sites (maximum quarterly average)

Figure 18.7 Comparison of national levels of the six principal pollutants to national ambient air quality standards, 1980–2006. National levels are averages across all sites with complete data for the time period (air quality data for PM$_{10}$ and PM$_{2.5}$ start in 1990 and 1999, respectively). (From U.S. EPA. The Latest Findings on National Air Quality—Status and Trends Through 2006. EPA-454/R-07-007, Research Triangle Park, NC. January 2008.)

speed and direction, insolation (amount of sunlight), lapse rate (temperature variation with height), mixing depth, and precipitation. The height at which the emissions are discharged into the atmosphere and the prevailing wind direction and speed determine how far the pollutants will be carried. For example, in the eastern parts of the United States and Canada, the combined influences of the airstream from the Canadian Arctic, the westerlies from the Pacific Ocean, and the tropical airstream from the Gulf of Mexico (Figure 18.10) move the emissions from the U.S. Midwest to the northeastern United States and to the southern

parts of Ontario and Quebec, Canada. The wind patterns in Asia, with two of the most populous countries in the world—China and India—whose economic developments guarantee major increases in emissions, provide another illustration of the movement of pollutants (Figure 18.11). The summer wind patterns that include the monsoon clouds ensure a flow of emissions from India to China and a movement in the opposite direction in the winter.

Factors in the second, nonmeteorological group include the types and quantity of pollutants, physical features of aquatic bodies, topography of land, vegetation, and susceptibility

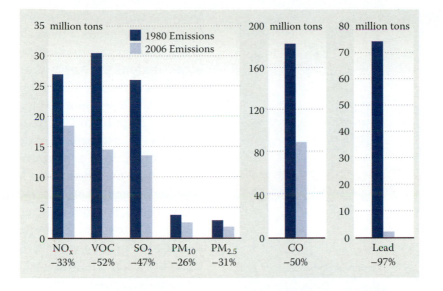

Figure 18.8 Comparison of U.S. emissions, 1980 versus 2006 (U.S. EPA 2008). $PM_{2.5}$ estimates are for 1990 and 2006; PM_{10} estimates are for 1985 and 2006. (U.S. National Science and Technology Council Committee on Environment and Natural Resources, Air Quality Research Subcommittee. 1999. The role of Monitoring Networks in the Management of the Nation's Air Quality. EPA-454/R-07-007.)

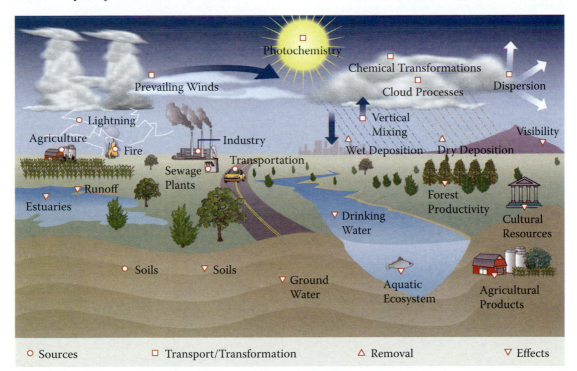

Figure 18.9 Atmospheric nitrogen pathways. (U.S. National Science and Technology Council Committee on Environment and Natural Resources, Air Quality Research Subcommittee. 1999. The Role of Monitoring Networks in the Management of the Nation's Air Quality. EPA-454/R-07-007.)

of organisms—plants, animals, and humans. Depending upon the region of the world, each of these factors will vary in its predominance of action by pollutants. Of paramount importance is the ability of water in lakes, streams, and soils to neutralize the effects of acidity. As is discussed later, when carbonates and bicarbonates are present in soils or lakes due to underlying marine limestone rocks (central United States), the incoming acidity is neutralized and

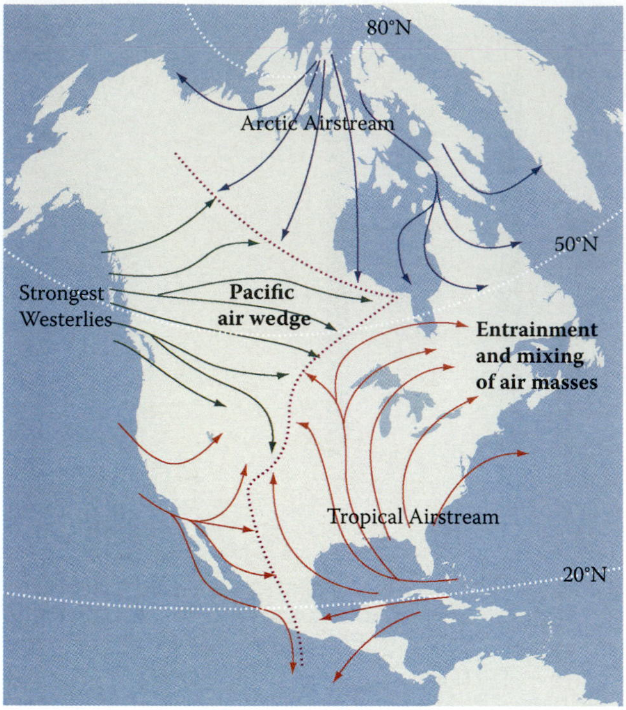

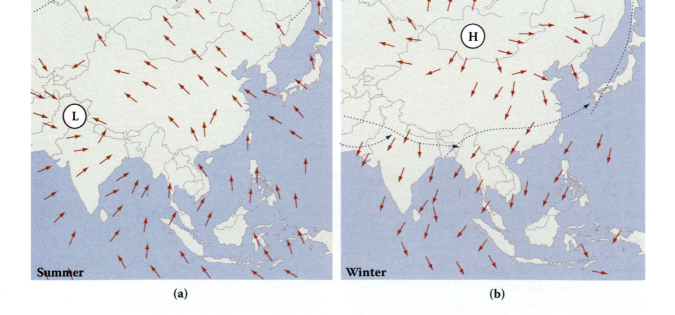

Figure 18.10 Surface flows across North America, illustrating the area of complex entrainment and mixing of air masses in the eastern portion of the continent. (From Bryson, R. A., and F. K. Hare. 1974. *Climates of North America. World Survey of Climatology,* vol. 11. New York: Elsevier Scientific Publishing.)

thus the impact of atmospheric deposition is low. But in other areas where the geochemical buffering capacity to neutralize acidity is low (e.g., some regions in North America with volcanic [west] or granitic [Canada] rock substrates—Figure 18.12—and also old metamorphic rocks in U.S. Appalachia and in Scandinavia), the impacts of atmospheric deposition are more pronounced. Not surprisingly, because of the widespread acidification of lakes in Canada, the problem of acid atmospheric deposition was, for many years, a major political issue between the United States and Canada. Most recent news items also bring focus on U.S. interstate relations between the Midwest and eastern states. Obviously, proper assessment of effects requires that the climatic features be coupled with geochemical capacity of acidified materials.

Acidification of Water and Soils

In discussing the changes that occur during many soil-forming processes, we learned in

Figure 18.11 (a) Summer wind flow patterns during the monsoon in Asia; (b) winter wind flow patterns in Asia. (From MacKinnon, J., and K. MacKinnon. 1986. *Review of the Protected Areas System in the Indo-Malayan Realm.* Gland, Switzerland and Cambridge, UK: IUCN/UNEP publication.)

Figure 18.12 Regions of North America with low geochemical capacity for neutralizing acid deposition. (From Galloway, J. N., and E. B. Cowling. 1978. *Journal of Air Pollution Control Association* 28:229–235.)

Chapter 4 that when leaching of soluble salts occurs in the soil surface layers, soils become progressively more acidic (Figure 18.13). This process of acidification is slow and subtle. In natural undisturbed systems, precipitation passes through the soils, and the resulting acid leachates reach the streams and lakes, thereby increasing the acidity of water. However, the resulting acidification is very slow and slight. On the other hand, unprecedented and diversely large human activities create emissions of high loads of acid-forming gases into the atmosphere. Thus, the acidification process that simultaneously affects all forms of precipitation, lakes, streams and rivers, soils, and groundwater is much more rapid. This acidification results directly from oxides of sulfur (mainly SO_2 and SO_4^{2-}, referred to collectively as SO_x), nitrogen (mainly NO_x, NH_3, NH_4, and NO_3, referred to collectively as NO_x), other acid and alkaline substances,

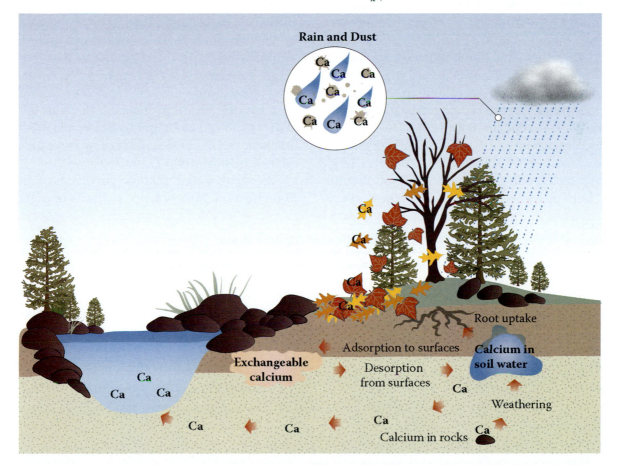

Figure 18.13 Calcium cycle in forest ecosystems showing leaching of calcium and increasing soil acidification. (Lawrence, G. B., and T. G. Huntington. 1999. Soil-carbon depletion linked to acid rain and forest growth in the eastern United States. WRIR 98-4267. Denver, CO: US Geological Survey.)

and the resultant products of these gases. Human-induced acidification has resulted in lowering the pH of rain to as low as 4. In the discussion that follows, we illustrate the intensity and the extent of acidification, as well as their impacts on aquatic, forest, and agricultural systems and monuments.

Acidification of Lakes and Streams

The serious concerns about acid deposition during the last 30 years were first concentrated on its effects on lakes. Indeed, the effects on aquatic fresh water systems were elegantly demonstrated through quantification of a number of parameters. The physical and chemical nature of the aqueous medium is such that it ensures quick diffusion and mixing of extraneous material reaching water bodies. In contrast, in soils, as solid media, the diffusion and mixing of pollutants is a much slower process, as are the chemical reactions (e.g., cation exchange between hydrogen ions and soil calcium) that may take place in soils among organic matter, clay complexes, and the pollutants.

The Scandinavian region and lakes in Canada are two of the world's most acidified regions. In Sweden alone, 20,000 of 90,000 lakes were acidified and 4,000 of these were totally devoid of fish life. In Norway, 80% of the lakes and streams in the southern half of the country were classed as technically "dead" or were placed on the critical list. More than 13,000 km^2 of Norway's lakes were left without any fish life (Hinrichsen 1988).

In Canada, scientists began to conduct experiments in acidification and eutrophication of whole lakes in their experimental lake area, Kenora, Ontario, in the late 1960s. The work was important and the lake systems also represented ideal sites for long-term experimentation. In one of the experimental lakes (lake 223), the pH was artificially decreased from >6.5 to 5.1 between 1976 and 1980, held at 5.0–5.1 between 1981 and 1983, and then raised to 5.8 by 1988. In another experimental lake, the pH was decreased from 6.5 to 4.6 from 1981 to 1988. Eutrophication experimental lakes received additions of phosphorus and nitrogen. Both eutrophication and acidification reduced species diversity of lakes (Schindler 1990a). These and other studies have shown that each successive reduction in pH showed impacts on individual species. For example, trout species were adversely affected at pH 6 and below, and rainbow trout were completely lost by pH 5.5. The lowest tolerable level for fish was found to be pH 5. Below pH 4.5, many insects, snails, and crustaceans were also completely eliminated (Table 18.7).

Several major factors govern how quickly the effects of atmospheric pollutants will occur:

Table 18.7
Summary of Known Effects of Water Acidification on Aquatic Biota

Taxonomic Group	Distribution Density	Reproduction	Species Richness	Trends with Recovery
Zooplankton	Reduced	Impaired	Reduced	Invasion/improvement
Macroinvertebrates	Reduced	Impaired	Reduced	Invasion/improvement
Fish	Reduced	Impaired	Reduced	Return to successful reproduction
Amphibians	Reduced	Impaired	Unknown	Unknown
Mammals	Usually unaffected	Minimal	Unaffected	Unknown
Birds	Usually reduced	Usually impaired	Minimal; possibly reduced	Breeding densities increase

Source: From Environment Canada. 1997. 1997 *Canadian Acid Rain Assessment.* Volume one: *Summary of Results.* Volume two: *Atmospheric Science Assessment Report.* Volume three: *The Effects on Canada's Lakes, Rivers and Wetlands,* ed. D. S. Jeffries. Volume four: *The Effects on Canada's Forests,* P. Hall, W. Bowers, H. Hirvonen, G. Hogan, N. Forster, I. Morrison, K. Percy, and R. Cox. Volume five: *The Effects on Human Health,* ed. L. Liu. Ottawa, Ontario, Canada.

- the quantity and nature of pollutants that a body of water receives; among gases, sulfur dioxide has been shown to have the largest effect on acidification;
- the nature of watersheds (such as terrestrial vegetation type and degree of vegetation cover) through which precipitation traverses before reaching the surface water; the nature of vegetation will also determine the rate at which the precipitation water will reach the ground;
- the nature of soil will determine the rate and infiltration and percolation of water and the rate at which it will be discharged into the lakes and streams; thus the chemical nature of soil is important because, with the appropriate chemical composition, some of the acidity may be neutralized; and
- length of time for which water stays in the soil system and geochemical reactivity of soils, parent materials, and bedrock (NRC 1983).

Acidification primarily reduces the variety of life inhabiting a lake and alters the balance among the surviving populations (Environment Canada 1997).

It has been shown that aquatic ecosystems associated with soils that have a high sulfate adsorption capacity will acidify much more slowly than those that have a low sulfate adsorption capacity. But once acidification of aquatic ecosystems occurs, biological changes are rapid and extensive (Galloway 1988).

Acidification of Forests

Serious concern on the effect of atmospheric deposition began in 1970 with the death of fir trees (*Tannensterben, Tannen* fir, *sterben* death) in Europe, and this was followed by a major decline in several other tree species. When more than 5 million ha of forest (38,000 km^2) in Germany, representing 52% of the country's forests (Hinrichsen 1988) showed a pronounced decline across Europe—termed *Waldsterben* (forest death) by the Germans—a great deal of attention was paid to finding the causes for such a major decline. Later, in 1987, conifers in the British Isles were also reported to have severe damage. Generally, it was the conifers rather than broadleaved species, older trees rather than younger trees, and trees at high altitudes that showed the greatest decline. Similar declines in red spruce growth were later reported from the Adirondack Mountains of New York. Since then, several hypotheses have been posed to explain the possible mechanisms by which such damage to tree growth occurs. Rather than being a simple process, tree decline is the result of the cumulative and synergistic effects of a combination of factors, as Manion (1990) suggested. These include a spiral of predisposing factors, inciting factors, and contributing factors (Figure 18.14), all leading to tree death. This "spiral of decline," we believe, applies not only to forest trees but also to other biota and ecosystems overall.

Several major factors likely determine the effects of atmospheric deposition on forests: (1) the chemical nature and loading of elements, (2) the ion exchange characteristics of soils, (3) the residence times and hydrological pathways of water through the watershed and the nature and extent of existing vegetation, and (4) the geochemical activity of bedrock and soils. Not all forest ecosystems are expected to respond to acid deposition in the same way. Effects are likely to be site specific and dependent on the relative contributions of external and internal sources of acidity (NRC 1983).

Evaluation of acid precipitation effects on forest ecosystems is much more difficult than it is for lakes, where (1) boundaries are more or less defined, (2) healthy pH is between 6 and 7, and (3) diffusion of acidity is relatively rapid in the liquid medium. The complex species compositions of forest ecosystems cause them to be stratified and even patchy; low soil pH is characteristic of many forest ecosystems and the adaptation of organisms to such conditions already exists. Evaluations of atmospheric pollution effects thus involve long-term studies. Nevertheless, some conclusions have already been made.

Increased acidity damages trees directly and indirectly. Acidification of soils that have

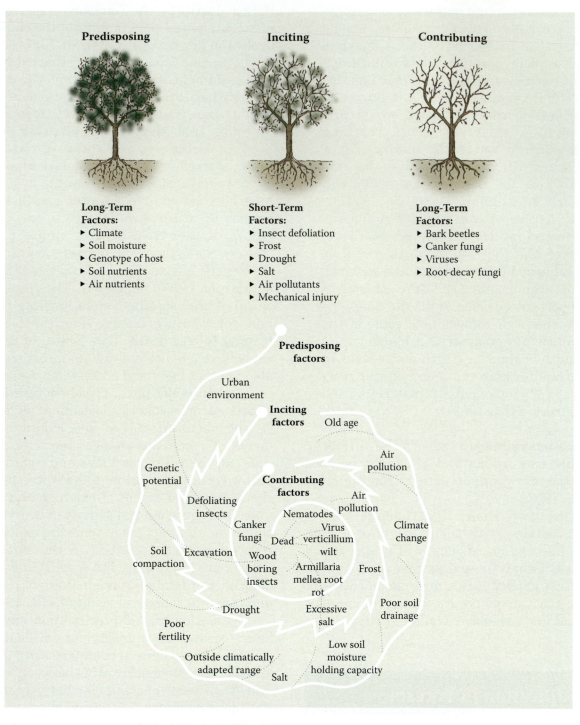

Figure 18.14 Cumulative and synergistic influence of multiple factors on the decline of trees. (Adapted from Manion, P. D. 1990. *Tree Disease Concepts,* 2nd ed. Upper Saddle River, NJ: Prentice Hall.)

low buffering capacity results in the lowering of pH and all its attendant effects even more quickly. Lowering of pH mobilizes aluminum. The predominant availability of Al ions at low pH has direct effects on the availability and uptake of essential nutrients, especially phosphorus and calcium, by plants. In most plants, roots become stubby and are unable to translocate phosphorus and other elements to the aboveground parts. Low pH and high aluminum concentration have been implicated experimentally in the export of calcium from forest soils. Excess nitrogen compounds have the opposite effect; that is, they can accelerate

Table 18.8
Recovery Time from Effects of Atmospheric Acid Deposition

Ecosystem	Time Period for Recovery
Acute human health effects	Hours to week
Episodic effects on aquatic resources	Days to months
Chronic effects on aquatic resources	Years to decades
Soil nutrient reserves	Decades to centuries

Source: From NAPAP (National Acid Precipitation Assessment Program). 1998. *National Acid Precipitation Assessment Program Report to Congress—An Integrated Assessment.* Washington, D.C.: Executive Office of the President.

growth, but that is only possible if the uptake of other nutrients and water can keep pace—but they cannot. Also, excess nitrogen accumulates in the form of ammonium ion, which, upon oxidation, releases even more hydrogen ions and organic acids, consumes plant carbohydrates, lowers leaf pH, and further acidifies the soils (Nihlgard 1985). Overall, the result is a stressed tree growth environment.

Markewitz et al. (1998, 1999) documented a dramatic increase in acidity and a steady depletion of nutrients over a 30-year period in an experimental forest in South Carolina. They showed that acid rain hampers the ability to buffer trees from toxic substances such as aluminum and other heavy metals. They calculated the hydrogen ion budget—net gains and losses in hydrogen—by measuring accumulated nutrients, tree root respiration, and the organic acids produced by the breakdown of pinestraw, roots, and litter. During a 28-year period between 1962 and 1990, soil pH dropped by one pH unit in the top 35 cm of soil and by half a pH unit in the lower 35 cm. These authors' recommendation was that, to ensure sustainable growth from managed forests, major nutrient fertilization would also have to include calcium and potassium in addition to nitrogen and phosphorus. Calcium loss has now been confirmed from 10 other states in the eastern United States; both red spruce and sugar maple are showing stress in growth and reduced resistance to insect attacks (Lawrence and Huntington 1999) (Figure 18.13). Thus, although the acidification in soils occurs slowly, the ecosystems will be even slower to recover once the atmospheric loads of acid are curtailed (Table 18.8).

In addition to the effects of acid atmospheric pollutants, plant responses to photochemical oxidants, particularly ozone, play a major role in forest dynamics. Ozone, as was noted previously, is a very important oxidant accumulating in the lower atmosphere from brief stratospheric intrusions and its more chronic photochemical synthesis from nitrogen dioxide and unsaturated hydrocarbons. With the exception of peroxyacyl nitrate (PAN), other oxidants, such as formic acid, peroxypropionyl nitrate (PPN), and hydrogen peroxide, formed by photochemical reactions in the troposphere have not been shown to damage forest vegetation (McBride and Miller 1987). Data recorded since 1974 suggest that the Pacific Southwest and the Atlantic Northeast have the greatest potential for tree damage from ozone. Other areas documented to show ozone injury include southern Ontario, the greater Vancouver area, and the Valley of Mexico.

Plumes containing high ozone concentrations may be carried a long distance, such as the Los Angeles urban plume that moved 350 km east to the Colorado River Valley near Needles, California. Both topographic features and meteorological factors have important influences on ozone concentrations. The land–water interface where temperature inversions persist well into the day (e.g., the Pacific coast, northeast coast, and Great Lakes regions) are more prone to develop high ozone concentrations (McBride and Miller 1987), provided nitrous oxide and hydrocarbon pollution are present. Ozone has caused injury to eastern white pine trees in the Blue Ridge Mountains and southern Appalachia, sugar maples in the upper Ohio River Valley, white ash and eastern white pines in southern Ontario, and pine and fir forests 20–30 km southwest of Mexico City.

Ozone concentrations damage cell membranes (a phenomenon similar to aging) and decrease photosynthesis and stomatal function (McBride and Miller 1987). Experimental evidence shows that ozone, formed from high concentrations of oxides of nitrogen, is directly injurious to leaves of broadleaved species (such as oaks) and needles of conifers (such as pine, spruce, and fir) and results in the leaching of nutrient ions, especially calcium and magnesium. Loss of magnesium results in the yellowing of leaves because magnesium is an important component of the chlorophyll molecule. Nitrogen oxides contribute to acidity, but transformations in the soil may, in part, enhance nitrogen availability for plant growth. Increased acidity, on the other hand, appears to reduce the rate of soil nitrification and also favors the ammonium form of N over time. This has a strong influence on root growth.

Acidification Impacts on Agriculture

As the preceding discussion on lakes, streams, and forests has illustrated, soils hold the key to acidification processes in relation to agriculture as well (Table 18.9). Generally, agricultural soils are maintained at pH greater than 5 in ranges that are good for a given crop growth. As long as soils have a buffering capacity (that is, enough acid-neutralizing capacity), acidification is far slower and less inimical to plant growth. Coarse-textured, low-organic-matter, and shallow soils will be acidified more quickly than soils that are deeper and have more organic matter and clay content. Thus, for example, Mollisols (the dominant agricultural soils in the United States), with a high content of bases, have a high buffering capacity and show fewer deleterious effects from acid deposition. Alfisols and Ultisols, too, have a high base saturation and hence will show lower impacts. In contrast, forest soils like Spodosols, tropical soils like the Oxisols, and developmentally

Table 18.9
Changes That Occur as a Soil Becomes Acidified

Process	pH 6 to pH 4
Dominant soil microbial groups	Change from bacteria to fungi
Decomposition	Marked decrease in activity
Respiration	Marked decrease in activity
Enzyme activity	Marked decrease in activity
Nitrogen mineralization	Marked decrease in activity
Nitrification	Marked decrease in activity
Denitrification	Slow in aerated soils
Nitrogen fixation	
Legume symbionts	Inhibited
Nolegume symbionts	Some reduced
Nonsymbionts	Some groups sensitive, some resistant
Mycorrhizae	Changes in type but not abundance
Earthworm activity	Marked decrease in activity
Soil microfauna	Diversity reduced and changes in dominant groups

Source: From Hutchinson, T. C. 1980. In *Effects of Acid Precipitation on Terrestrial Ecosystems,* ed. T. C. Hutchinson and M. Havas, 617–627. New York: Plenum Press.

recent soils like the Entisols have low organic matter content and cation exchange capacities, as well as a predominance of aluminum and iron oxides; hence, they will be affected greatly. As with forests, the process of acidification is not only the lowering of pH in soils but also a cumulative effect of pH, buffering ability, the solubilization of aluminum, and, as importantly, the export of calcium and magnesium from the system.

Tropical soils are generally quite acid due to their age and humid environment. However, because of their acidity, the high concentrations of iron and aluminum oxyhydroxides give them a high sulfate adsorption capacity. In these types of systems, acidification of aquatic ecosystems may be extensively delayed after the onset of the mobile ion (Galloway 1988). It has been suggested that some of the leafy salad crops, such as lettuce and spinach, may serve as sensitive indicator crops for detecting damage (Hutchinson 1980).

Atmospheric Pollution and Human Health

From the preceding discussion, it is evident that human support systems are greatly affected by atmospheric deposition (i.e., through the decreased growth of forest trees, yields of food crops, aquatic species, and increased global warming from enhanced greenhouse gases). However, pollutants have a direct effect on human health and, like the forest tree decline, can be caused by the synergistic result of a number of factors. Such factors include air pollutants as well as those ingested from drinking water and food. High doses of each element can have differential effects on different parts of the human body. Air pollutants specifically include carcinogenic effects of organic pollutants such as benzene and metals like nickel, respiratory ailments resulting from oxides of nitrogen and sulfur, irritation of eyes, and brain and nervous disorders. High ozone concentrations alone have been shown to cause painful

irritation to the respiratory system, reduced lung function, asthma, and emphysema. Ozone concentration alerts are now a part of the daily weather reports in the United States and in major cities of the world (see Chapter 20 on environmental policy).

Atmospheric Deposition Impacts on Monuments

An excellent example of the deterioration process by air pollutants and weathering over time is illustrated by an angel statue in Cologne, Germany (Figure 18.15). Several processes are involved: these include the deposition of dust, smoke and soot, and continuous atmospheric deposition that accelerates considerably the deterioration of the statue material both by physical and chemical weathering. In the case of the Taj Mahal, seventh wonder of the world, acid deposition has resulted in the degradation of the all-marble (calcite) structure. On more than one occasion, the Indian courts have decried and, in some cases, shut down

Figure 18.15 Deterioration of the angel statue at the "Peters" portal on the Cathedral of Cologne, Germany as seen by comparing: (a) a photograph taken in 1880 by Anselm Schmitz of Cologne, and (b) a 1993 photograph by Professor Dr. A. Wolff of Cologne. (Warscheid, T., and J. Braams. 2000. Biodeterioration of a stone: A review. *International Biodeterioration & Biodegradation* 46: 343–368.)

coal-burning power generation plants whose sulfate and nitrogen oxides have reached this famed monument. All rocky materials that contain calcium carbonate are prone to dis-solution from acid deposition, and buildings made of marble show discoloration and deteri-oration over time. The reaction of sulfuric acid with marble and limestone results in the for-mation of water-soluble gypsum (calcium sul-fate), which is washed off after a rain shower. Thus, the surface layers bearing ornate chis-eled work are lost and their loss exposes deeper layers to degradation as well.

In contrast, granite and sandstone rocks that are primarily composed of silica and sili-cate materials are not damaged by acid. Some sandstones, however, do have carbonate and are damaged. Studies have shown that dry deposition (the deposition of particles between rain events) is more damaging than wet depo-sition (the deposition of particles during rain events) (U.S. National Science and Technology Council 1998).

Not only are monuments affected by atmo-spheric deposition, but its chemical constitu-ents also affect all like materials, and the relatively immediate effect of pollutants is most pronounced in areas that are closest to where these originate. The automobile industry uses a composite term called "environmental fall-out" to denote "damage [to cars] caused by air pollution (e.g., acid rain), decaying insects, bird droppings, pollen, and tree sap." Sulfur diox-ide has the greatest corrosive effect on exposed materials made of steel and zinc. Although NO_x compounds have been shown to have limited effects, their combination with SO_2 promotes corrosion of copper and electrical contact mate-rials like gold and copper coatings.

Through a complex process, soil acidifica-tion has important corrosivity implications for building materials that include cast iron, lead, metal water pipes, and concrete (Kucera 1988). An annual loss of more than $2 billion resulting from the degradation by atmospheric deposition of paints, plastics, and other poly-mer materials is expected in the United States alone (U.S. EPA).

No good estimates exist on the total dam-age to human infrastructure from atmospheric pollution. However, one estimate suggests that damage to metals, building exteriors, and painted surfaces costs the 24-member OECD countries about $20 billion a year. Hinrichsen (1988) notes that if the costs of dead and dying forests, acidified lakes and streams, and crop losses were factored into the equation, the price of atmospheric pollution might very well prove to be astronomical.

The Ozone "Hole" and the Montreal Protocol

All of the foregoing discussion on air pollutants is related to the troposphere, which extends up to about 8 km (5 miles) above the Earth's sur-face, but there is more to consider even higher in the upper atmosphere. The stratosphere, up to about 59 km, harbors a layer of O_3 that is densest between 11 and 20 km. It filters out much of the harmful ultraviolet radiation com-ing from the sun before it reaches the Earth.

In the discussion of spectral attributes of incoming solar radiation in Chapter 3, we noted that ultraviolet (UV) radiation may be characterized as UVA (315–400 nm) and UVB (280–315 nm), both of which are progressively detrimental to the growth of organisms, and UVC (<280 nm), which kills both animals and plants. Thus, the deleterious effects of UV radi-ation increase considerably with decreasing wavelength. Radiation below 10 nm is high-frequency short wave radiation and includes x-rays and gamma rays, which are capable of penetrating matter.

A specialized group of compounds used by humans for the past 75 or so years rises into the upper atmosphere (the stratosphere) and causes a unique change that has an immense impact on life on the Earth. This group of halo-genated compounds, known collectively as the chlorofluorocarbons (CFCs), brings about a deg-radation of ozone in the stratosphere, thinning it over a large area to a point of destruction of the layer that has been referred as the "ozone hole." In areas where the O_3 is lost, UVB and UVC reach the Earth.

Where did stratospheric ozone come from in the first place? The formation of the O_3 layer is

Table 18.10
Characteristics of Chlorine-Containing Compounds, Their Relative Contribution to Total Atmospheric Chlorine in 1985,[a] and Ozone Depletion and Global Warming Potentials[b,c]

Sources	Chemical Formula	Atmospheric Lifetime (yr)	Relative Contribution (%)	Ozone Depletion Potential	Global Warming Potential
Industrial Sources					
CFC-11	$CFCl_3$	60	22	1.0	1.0
CFC-12	CF_2Cl_2	120	25	0.92	3.4
CFC-113	$C_2F_3Cl_3$	90	3	0.83	1.4
CFC-114	$C_2F_4Cl_2$	200	< 1	0.63	4.1
CFC-115	C_2F_5Cl	400	< 1	0.36	7.5
HCFC-22	CHF_2Cl	15	3	0.057	0.37
Carbon tetrachloride	CCl_4	50	13	1.16	0.34
Methyl chloroform	CH_3CCl_3	6	13	0.14	0.022
Natural Sources					
Methyl chloride	CH_3Cl	1.5	20		

[a] Prather, M. J., and R. T. Watson. 1990. *Nature* 344:729–734.

[b] Ozone depletion and global warming potentials are the relative potentials of these trace gases to destroy stratospheric ozone, based on the reference gas of CFC-11 = 1.0, and to enhance global warming based on the reference gas of CO_2 = 1.0, respectively; they are based on an atmospheric model of Atmospheric and Environmental Research Inc.

[c] Fisher, D. A. et al. 1990a. *Nature* 344:508–512; Fisher, D. A. et al. 1990b. *Nature* 344:513–516.

directly related to the evolution of photosynthesis. As photosynthetic organisms in the oceans became abundant, more oxygen was produced. The abundant oxygen diffused out of the water and into the atmosphere, eventually reaching the stratosphere, where it was converted into O_3. This stratospheric O_3 layer absorbs the incoming UV radiation, thereby reducing its harmful effects on the organisms and allowing photosynthetic organisms to survive on the continents as well.

The CFCs accelerate the breakdown rate of O_3 into molecular (O_2) and atomic (O) oxygen under the influence of energetic UV radiation. Called **ozone depletion substances** (**ODS**), these include chlorine-containing compounds such as CFCs, hydrochlorofluorocarbons (HCFCs), bromine-containing compounds such as halons and methyl bromide (CH_3Br), hydrofluorocarbons (HFCs), and nitrogen oxide (N_2O) (Table 18.10). Their use has multiplied enormously since 1930, when Thomas Midgley, Jr., a General Motors chemist, discovered the first CFCs and showed their diverse industrial utility.

The CFCs (e.g., Freon) are stable, odorless, nonflammable, nontoxic, noncorrosive compounds. They are used as refrigerants (home and automobile air conditioners), propellants (in aerosol spray cans), foam-blowing agents, solvents, cleaning agents for computer chips, hospital sterilants, and fumigants for granaries; to create bubbles in styrofoam; and in insulation and packing. Given so many uses for these compounds, the unprecedented rise in atmospheric CFCs and thus the chlorine levels was inevitable.

Because they are nonreactive, CFCs may stay in the atmosphere for 60–400 years. It has been estimated that a single Cl atom can convert as many as 100,000 molecules of O_3 to O_2. To get an idea of the enormity of these numbers, a single small polystyrene coffee cup contains more than 1 billion (i.e., 10^9) molecules of CFCs that, if released, could convert 10^{14} molecules of ozone in the stratosphere. The total stratospheric chlorine concentrations increased from 0.6 ppbv (parts per 10^9 molecules of air by volume) in the 1800s to 2 ppbv in the late 1970s. Chlorine and bromine concentrations in the stratosphere are now more than 3 and 0.02 ppbv, respectively (Prather and Watson 1990).

The amount of ozone above a location on the Earth varies naturally with latitude and season and from day to day. The role of CFCs as O_3-destroying agents was first proposed in 1974 by Mario Molina and F. Sherwood Rowland, who were the first to show how chlorine from CFCs degraded O_3 to O_2 and O, whereas Paul Crutzen elaborated the role of nitrogen compounds. All three shared the 1995 Nobel Prize in chemistry—the first time the prize was shared for work involving a problem in the environment. It was only in the late 1970s that computer models were developed that showed that severe human-induced ozone depletion over Antarctica had been occurring since 1979. Global ozone levels declined an average of about 3% between 1979 and 1991.

During September and October of 1987, 1989, and 1990, there was an O_3-depleted area over the Antarctic whose size was larger than the area of the continental United States; this has been referred to as the Antarctic ozone hole. The Antarctic ozone hole expanded to a record size of approximately 28.3 million km² (11 million mi²) on September 3, 2000, after the second largest hole of about 27.2 million km² (10.5 million mi²) was recorded on September 19, 1998. A combination of cold temperatures, water droplets in clouds, and the CFCs creates the largest ozone depletion at high latitudes in both hemispheres. In both the polar regions, the depletion is in winter and spring—in the Antarctic, beginning in late August and intensifying from September to November (see Figure 18.16).

In October 2006, Geir Braathen, ozone specialist at the UN's Geneva-based World Meteorological Organization, reported that the "ozone hole" over Antarctica in 2006 had matched the record size of 29.5 million km² (11.4 million mi²). The area of the hole is slightly larger than the record year of 2000, according to measurements by NASA. It was of greater concern to Braathen that the amount of ozone gas particles remaining in the hole was even lower than in 2000, a measurement called "the mass deficit." According to the European Space

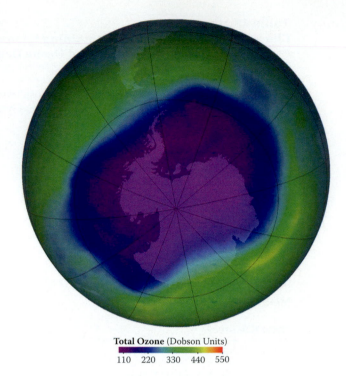

Total Ozone (Dobson Units)

110   220   330   440   550

Figure 18.16 Low ozone levels in the stratosphere over the South Pole in September 2000 shown in blue in a false-color picture produced by NASA. (SVS-Scientific Visualization Studio, TOMS-Total Ozone Mapping Spectrometer.)

Agency, the loss has been 39.8 megatons. "In a way this mass deficit is a better measure of how much ozone is depleted … because it counts how many tons of ozone are lacking," Braathen said (Associated Press 2006).

Braathen noted that the hole had been forming in the extremely low temperatures that mark the end of Antarctic winter every year since the mid-1980s. Generally, the hole is biggest around late September. This year's Antarctic winter has been very cold, the weather agency said earlier this month, which has led to greater ozone depletion. Although there has been a decrease in ozone-depleting substances over the last few years, the atmosphere is still saturated with them. According to the agency, it will take until 2065 for the ozone layer to recover and the hole over the Antarctic to close. That estimate is 15 years longer than previous predictions by the agency.

In the Arctic polar regions, depletion begins in January and peaks in April and May. It appears with the return of the warmer temperatures of spring and summer. Cl and

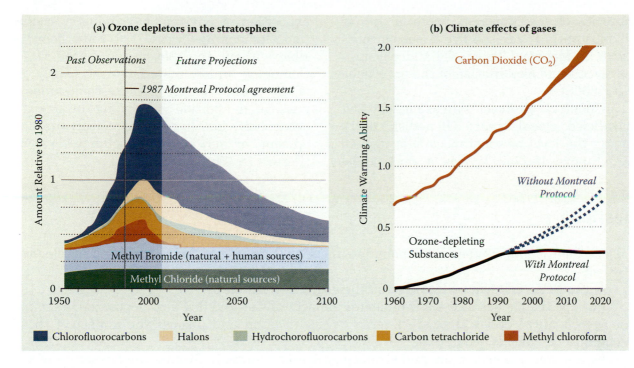

Figure 18.17 The Montreal Protocol that reduced worldwide use of ozone-depleting substances in the stratosphere and its impact in reducing the use of these substances (left panel). Because ozone-depleting substances are also greenhouse gases, the Montreal Protocol and its amendments gave an early start to slowing climate warming (right panel). The left axis (climate warming ability) gives the climate forcing in units of watts per square meter (Wm^{-2}). (From USGCRP. 2008. *Our Changing Planet*. Climate Change Science Program for fiscal 2008. The White House, Washington, D.C. Based on multiple sources.)

Br become more active and degrade O$_3$ some more. The effects of loss of ozone and incoming UV radiation have been most noticeable in large parts of Australia, New Zealand, and the tips of South America and Africa.

To a lesser extent, O$_3$ depletion also occurs from the emissions of reactive N$_2$O gases. These gases interact with natural and anthropogenic hydrocarbons and carbon monoxide, thus contributing to the photochemical production of ozone, air pollution (aerosols), the greenhouse effect in the lower atmosphere (troposphere), and ozone depletion in the stratosphere.

The U.S. Environmental Protection Agency estimates that 5% stratospheric ozone depletion would cause the following adverse effects in the United States:

- an extra 170 million cases of skin cancer;
- a sharp increase in eye cataracts;
- suppression of the human immune system;
- increase in U.S. health care—$3 billion (in 1996 dollars);

- increase in eye-burning photochemical smog;
- decreased yields of important food crops (e.g., corn, rice, soybeans, wheat);
- reduction in growth of ocean phytoplankton;
- an additional loss of $2 billion from degradation of paints, plastics, and other polymer materials; and
- increased global warming from an enhanced greenhouse effect.

The Montreal Protocol

In 1987, concern about the enormity of the depletion of the stratospheric ozone layer led to action by the world community. The result was the acceptance, on September 16, 1987, of an agreement to phase out (over time) the production and use of all CFCs and halons, carbon tetrachloride, and methyl chloroform. By 1989, 29 countries, accounting for 82% of the world's consumption of such compounds, had ratified the treaty; as of 2006, 191 nations are now signatories of the Montreal Protocol. It has been substantially revised five times since its passage. A Montreal Protocol handbook provides details and explanation of the provisions (UNEP 2005).

The Montreal Protocol is an exemplary and unparalleled global treaty (Benedick 1991), reducing the level of substances that deplete ozone (Figure 18.17a). It was a remarkable exposition of environment–industry cooperation. Its passage would not have been possible without the cooperation of the DuPont Company, which manufactured nearly 80% of the world's supply of these compounds. Its passage, in addition to reducing the size of the ozone hole, has been a most welcome development in slowing down climate warming (Figure 18.17b).

Questions

1. Describe the major types of air pollution and give examples of their sources and ecological effects.
2. List the EPA's criteria pollutants and their effects on humans.
3. What are the causes and environmental impacts of wet and dry acid deposition? List the biogeochemical cycles that are altered by acidification and discuss the impact of these changes on various ecosystems.
4. How are tropospheric ozone and stratospheric ozone formed? Discuss the formation of the ozone hole and the implications of the Montreal Protocol.
5. Research the indoor and outdoor air quality indices for where you live. What are the health effects of common indoor and outdoor air pollutants in your area?

The Greenhouse Effect and Climate Change

19

Introduction

During the 1970s and early 1980s, debates on environmental issues were largely concentrated on air and water pollution and the effects of acid rain on lakes and terrestrial ecosystems. The concerns, as we studied in the last chapter, were serious and the effects involved local, regional, and transboundary pollution into neighboring nation-states. Lurking not too far from this environmental scenario was yet another effect that some of these gases might have on the global climate as a whole. This involved less infrared radiation leaving the troposphere and thus the retention of greater amounts of heat in the Earth's atmosphere. A perusal of the scientific literature shows that the effects of such a phenomenon—that is, heat retention by gases—were mentioned as early as 1827 by the Frenchman Jean-Baptiste Fourier. It was Fourier who introduced the concept of **greenhouse effect** to identify the natural processes whereby the Earth sustains a heat balance. Later, the Swedish chemist Svante Arrhenius in the 1890s wrote specifically on the role of carbon dioxide, as did Roger Revelle at the Scripps Institution of Oceanography in 1957. (For an excellent narrative, read *The Discovery of Global Warming* by Spencer Weart, 2003.)

Greenhouse Gases and the Greenhouse Effect

The term "greenhouse effect" came into vogue because the atmosphere bears a similarity in trapping heat to a glass greenhouse. In the absence of any greenhouse effect, the Earth's average surface temperature would be 33°C colder. It was global warming after the last ice age that made present-day agriculture and other activities possible. The Earth has experienced many climatic changes throughout its history.

In the past couple of million years, the Earth has gone through several episodic cycles caused primarily by its rotation around the sun. The three cycles most commonly recognized are

(1) **eccentricity**—the shape of Earth's orbit around the sun, with a periodicity of about 100,000 years; (2) **obliquity** or **axial tilt**—the inclination of the Earth's axis in relation to its plane of orbit around the sun, with a periodicity of 40,000 years; and (3) **precession**—the "slow wobble" as Earth spins on its axis, with a periodicity of about 23,000 years. These three cycles are referred to as Milankovich cycles.[1] Together, the "three Milankovitch cycles impact the *seasonality* **and** *location* of solar energy around the Earth, thus impacting contrasts between the seasons."[2] When the cycles coincide, cold periods set in. It is important to note that **albedo,** the reflectance of the sun's radiation by the surface of the Earth back to the atmosphere, is an important variable in the cooling cycles.

Geological History and Temperature

In the last 1.5 million years or so, known in geological terms as Pleistocene, the Earth has gone through several cold periods, called collectively **glaciations** or **ice ages.** The common feature of all glaciations was that large portions of the Earth were covered by ice sheets. Each glacial period lasted for about 70,000–100,000 years, and at least four such glaciations alternated with interglacial periods of 10,000–12,000 years.

The time from the last glaciation, the Holocene, is the period in which we live. The Earth's mean surface temperatures during this period seem to have fluctuated only moderately, up or down by 0.5–1°C (0.9–1.8°F) over 100- to 200-year periods. Except for drastic local or regional weather events, there have been no major phenomena. Thus, climatic stability during the Holocene has created a great degree of confidence in the human mind regarding weather patterns. For example, in temperate climates, a winter will always be followed by spring, then by summer, and finally by autumn, with more or less expected variations in temperature for each season.

In Chapter 7, in discussing energy flows, we illustrated the quantity of energy coming from the sun and the portion of it that is radiated back into the atmosphere (Figure 7.2). The amount of heat in the atmosphere depends mostly on the concentrations of various heat-trapping gases known as the **greenhouse gases** (GHGs). These include carbon dioxide (CO_2), water vapor (H_2O, the dominant greenhouse gas), tropospheric ozone (O_3), methane (CH_4), nitrous oxide (N_2O), and human-made chlorofluorocarbons (CFCs) (Table 19.1). GHGs are transparent to incoming short-wave solar radiation but absorb outgoing infrared (long-wave) radiation emitted by the Earth's surface and atmosphere in the range of 7–19 μm (Figure 7.2). This infrared radiation is part of the **atmospheric "window"** through which more than 70% of the radiation emitted from the Earth's surface escapes into space.

Radiative Forcing and Global Warming Potential

Global climate represents the overall expression of a multitude of processes driven by the sun, oceans, and seas that cover 71% of the Earth's surface and 21% of its land area (Figure 19.1). If the chemical composition of the atmosphere changes, the natural intensity of incoming solar radiation affects the average radiative (cooling and heating) balance at the top of the atmosphere (342 W m^{-2}) and is termed **radiative forcing.** Increases in the atmospheric concentrations of GHGs enhance the radiative heating of the Earth (a positive radiative forcing). The radiative forcing effect of a greenhouse gas depends on its concentration, **global warming potential** (**GWP**), and **atmospheric lifetime** (residence time).

Carbon Dioxide and Temperature

Since the Industrial Revolution (in particular, the past 70 or so years), the rate of release of GHGs into the atmosphere (through the

Table 19.1

Global Atmospheric Concentration, Rate of Concentration Change, and Atmospheric Lifetime of Selected Greenhouse Gases

Atmospheric Variable	CO_2	CH_4	N_2O	SF_6	CF_4
Preindustrial atmospheric concentration	278 ppm	0.715 ppm	0.270 ppm	0 ppt	40 ppt
Atmospheric concentration[a]	379 ppm	1.774 ppm	0.319 ppm	5.6 ppt	74 ppt
Rate of concentration change	1.4 ppm yr^{-1}	0.005 ppm yr^{-1a}	0.26% yr^{-1}	Linear[b]	Linear[b]
Atmospheric lifetime (years)[c]	50–200[d]	12[e]	114[e]	3200	>50,000

Source: U.S. EPA 2008. Inventory of U.S. Greenhouse Gas Emissions and Sinks. EPA 430-R-08-005. Washington, D.C.

Notes: Preindustrial atmospheric concentrations, current atmospheric concentrations, and rate of concentration changes for all gases are from IPCC, 2007. *The Physical Science Basis,* 93–127. Working Group I Report. Frequently asked questions. New York: Cambridge University Press.

[a] The growth rate for atmospheric CH_4 has been decreasing from 1.4 ppb yr^{-1} in 1984 to less than 0 ppb yr^{-1} in 2001, 2004, and 2005.

[b] IPCC (2007) identifies the rate of concentration change for SF_6 and CF_4 as linear.

[c] Source: IPCC (1996).

[d] No single lifetime can be defined for CO_2 because of the different rates by which removal occurs.

[e] This lifetime has been defined as an "adjustment time" that takes into account the indirect effect of the gas on its own residence time.

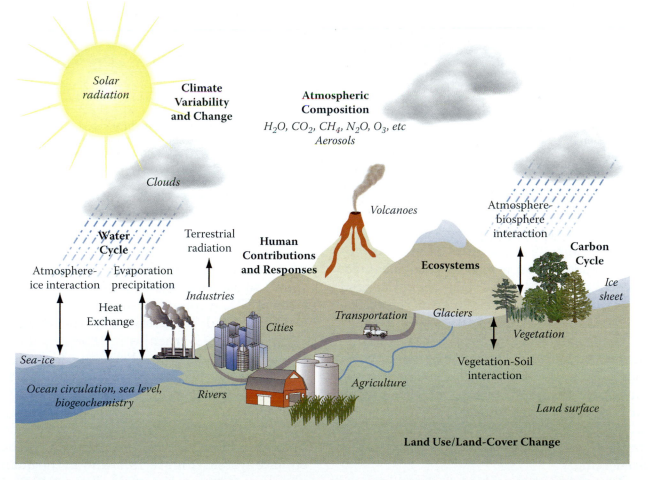

Figure 19.1   Major components to understand the climate system and climate change. (From IPCC, 2007. The Physical Science Basis, 93–127. Working Group I Report. Frequently asked questions. New York: Cambridge University Press.)

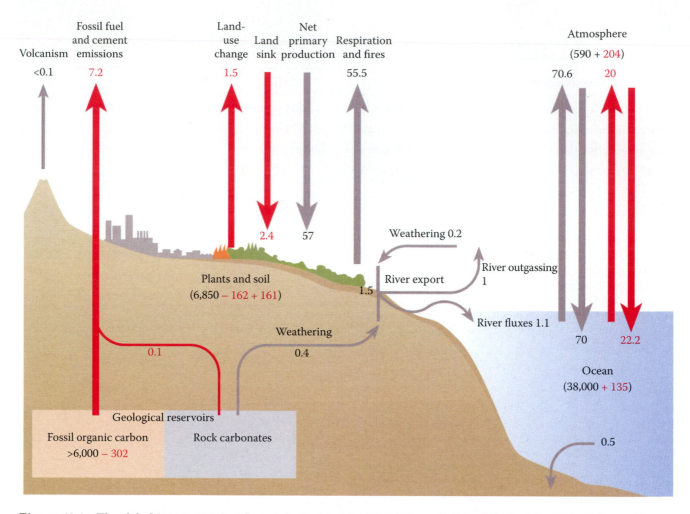

Figure 19.2 The global (2000–2005) carbon cycle: Pools of C are in billion metric tons and are bracketed. Annual fluxes are in billion tones C per year. Background or preanthropogenic pools and fluxes are shown in black. Human perturbation to the pools and fluxes is shown in red. (Redrawn from Figure 2.19 in GEO₄, UNEP 2007; *Global Environmental Outlook: Environment for Development*. Valletta, Malta: Progress Press Ltd.)

consumption of fossil fuels and the degradation and loss of vegetation and soils) has been faster than known at any other time in human history (Table 19.1). What is of great consequence and concern is whether the global biophysical capacities of the soils, vegetation, and aquatic ecosystems can sequester them. The rate at which warming will proceed depends on GHG emissions, the residence times in various sinks, and the rapidity with which the sources and sinks (sequestration) grow. Thus, over the past century, the global mean air temperature has increased by about 0.6°C (±0.2), despite increases in sulfate particles that produce cooling effects (a negative radiative forcing).

An enhanced greenhouse effect increases the Earth's average surface temperature,

creating a domino effect on other abiotic and biotic processes by virtue of their ecological interdependence and interaction. This linkage was recognized more than 100 years ago when an increase in surface temperature was shown to correlate with increased concentrations of CO_2 in the atmosphere—an increase linked to coal combustion by the Swedish chemist Svante Arrhenius in 1896 (see Rodhe et al. 1997). Thus was born the first clear link between human activities and global warming: specifically, the link between CO_2 and temperature. Our scientific knowledge has advanced much in the last few decades to shed light on factors that affect the global climate system (Figure 19.1) and in the quantification of the carbon cycle (Figure 19.2). It increasingly appears that our actions are not limited by scientific knowledge so much as by economic, political, and social considerations.

Climate changes cannot be absorbed without risk. The greatest threat to human food production, vegetation, wildlife, and economic systems is a rapid climate change that may involve the rise of just a few degrees in the Earth's mean air temperature over a few short decades. The more quickly such a rate of change occurs, the less prepared organisms will be to adapt to these changed environmental conditions. Thus, temperature, climate, and the chemical composition of the troposphere and stratosphere become very important factors in determining the mean temperature of the Earth's atmosphere.

To establish clear relationships between the atmospheric concentration of CO_2 and the Earth's surface temperature, one would need instrumental records of both parameters. In their absence, a historical relationship has been established based on studies of ice cores obtained from glaciers from many parts of the world. These studies have been complemented by deciphering climate and weather conditions using cores obtained from the oldest trees, a subject area called **dendrochronology.** A synthesis of these data reveals that, over the past 150,000 years, a close relationship has existed between CO_2 concentration and temperature (Figure 19.3).

Based on thermometer readings over the past 100 years, the direct relationship between CO_2 and temperature seems to be borne out (Figure 19.4). Note in Figure 19.4 that temperatures have dipped with volcanic eruptions: Santa Maria in the Sierra Madre Range in Guatemala in 1902; Agung in Bali, Indonesia, in 1963; and Pinatubo in the Zambales Range of the Philippines in 1991. This is because volcanic eruptions spew out dust and gases (mainly sulfur dioxide). These gases mix with water vapor in the atmosphere and the resulting aerosols and haze reflect back some of the sun's radiation, causing a cooling effect and lowering the temperature. Data from the past 25 years or so clearly show that as concentrations of CO_2 and other greenhouse gases have been rising, so has the temperature, particularly in the Northern Hemisphere (Figure 19.5).

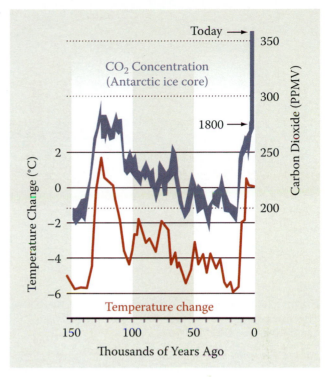

Figure 19.3 Historical relationship between global air temperature and carbon dioxide concentration in the atmosphere. (http://clinton4.nara.gov/media/gif/Figure5.gif accessed February 25. 2009.)

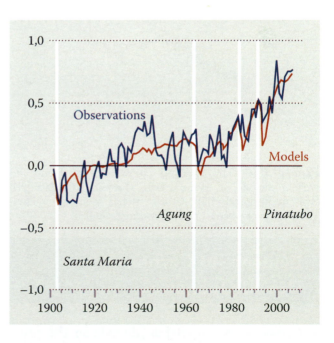

Figure 19.4 Rise in Earth's atmospheric temperature during 1990–2005. Temperature increase has been 0.7°C with the volcanic eruptions Santa Maria, Agung, and Pinatubo producing aerosols that cooled down the Earth temporarily. (IPCC, ARW4, vol. 1, p. 62.)

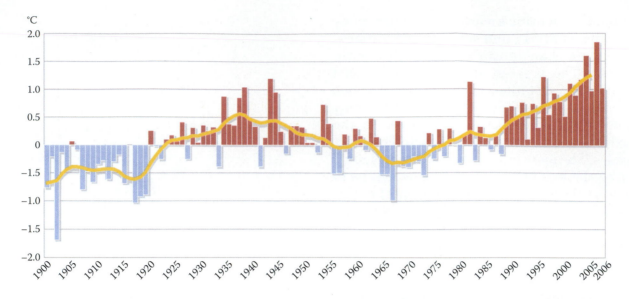

Figure 19.5 Annual average change in arctic (between 60 and 90° N) temperatures between 1990 and 2006; the zero line represents average temperatures between 1961 and 1990. (Redrawn from Figure 6.54 in GEO₄–UNEP 2007; original source CRU 2007.)

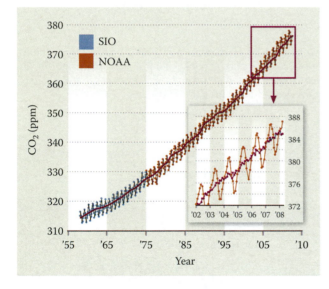

Figure 19.6 The Keeling curve. Recent trends in carbon dioxide at Mauna Loa, Hawaii. NOAA–Earth System Research Laboratory, Global Monitoring Division. (http://www.esri.noaa.gov/ccgg/trends/)

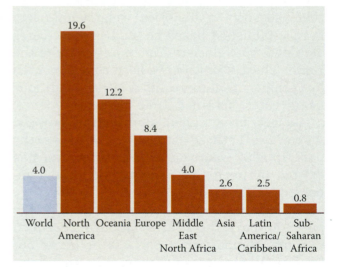

Figure 19.7 World carbon emissions per capita for 2002 showing the gap between developed and developing countries. (Based on data from pages 11–14 of Population Reference Bureau. 2007 World population data sheet. Washington, D.C.)

Evidence of Carbon Dioxide Increase

Any discussion on carbon dioxide in the present environmental context must begin with the meticulous work of the Scripps Institution scientist, the late David Keeling. He recorded both seasonal and yearly fluctuations in the concentration of atmospheric CO_2 at Mauna Loa, Hawaii. His illustration, known as the **Keeling curve** (Figure 19.6) is perhaps the most cited and recognizable example of atmospheric CO_2 change. The curve illustrates that (a) CO_2 concentrations have, without exception, been rising since Keeling started his measurements in 1958, and (b) when photosynthesis is at a maximum in summer north of the equator (with most land mass), the atmospheric CO_2 is low and increases

Table 19.2
CO_2 Emissions from Fossil Fuel Combustion by Fuel Consuming End-Use Sector[a]

End-Use Sector	1990	1995	2000	2005	2006
Transportation	1,488.1	1.602.5	1,801.6	1,874.5	1,861.0
Combustion	1,485.1	1,599.4	1,798.2	1,869.8	1,856.0
Electricity	3.0	3.0	3.4	4.7	4.9
Industrial	1,527.5	1,589.5	1,645.1	1,579.6	1,567.1
Combustion	844.9	876.5	860.3	847.3	862.2
Electricity	682.5	713.1	784.7	732.3	704.9
Residential	929.5	995.5	1,129.7	1,206.4	1,151.9
Combustion	340.1	356.5	372.1	358.5	326.5
Electricity	589.4	639.0	757.6	847.9	825.4
Commercial	750.8	810.0	964.6	1,017.3	1,003.0
Combustion	216.1	225.8	228.0	221.9	210.1
Electricity	534.7	584.2	736.6	795.4	792.9
U.S. territories	28.3	35.0	36.2	53.2	54.9
Total[b]	4,724.1	5,032.4	5,577.1	5,731.0	5,637.9
Electricity generation	1,809.6	1,939.3	2,282.3	2,380.2	2,328.2

Source: From U.S. EPA 2008. Inventory of U.S. Greenhouse Gas Emissions and Sinks. EPA 430-R-08-005. Washington, D.C.

Note: Carbon dioxide equivalent is a simple calculation to evaluate the impact of GHGs in relation to CO_2; thus, $1 \times CO_2$ for CO_2, $21 \times CO_2$ for CH_4, and $310 \times CO_2$ for N_2O.

[a] Teragram CO_2 equivalent.

[b] Totals may not sum due to independent rounding. Combustion-related emissions from electricity generation are allocated based on aggregate national electricity consumption by each end-use sector.

in winter. Measurements in recent years have shown CO_2 increases coming mostly from developed and industrialized countries (Figure 19.7). Although the overall global carbon emissions per capita are modest, North America leads the world in per-capita emissions.

Fossil fuel consumption shows that in the United States, as in nearly all other countries, carbon emissions come primarily from transportation and electricity generation (Table 19.2). In addition to the industrial and transportation sectors, agriculture, land use, forestry, and the generation of waste contribute significantly to CO_2 as well as to CH_4 and N_2O emissions (Tables 19.3 and 19.4). In the United States, the overall levels of NO_x, CO, nonmethane volatile organic compounds (NMVOCs), and SO_2 have decreased significantly (Table 19.5).

Overall, the United States, China, Russia, Japan, India, Germany, the United Kingdom, Canada, Italy, and South Korea were the world's largest sources of fossil fuel-related CO_2 emissions, producing 64% of the world total in 1998. Petroleum (43%), coal (36%), and natural gas (21%) consumption was the world's primary source of CO_2 emissions. Coal contains the highest amount of C per unit of energy, exceeding that of petroleum and natural gas by 25 and 45%, respectively. Globally, the total net flux of C to the atmosphere from changes in land use was 124 Gt C from 1850 to 1990, and the flux rate increased from about 0.4 Gt C in 1850 to 2.0 Gt C in 1990 (Houghton 1999) (Tables 19.3 and 19.4). Contributions of land use changes to the total net flux of atmospheric CO_2 emissions were 68% from agricultural expansion and 13% by biomass burning

Table 19.3
Recent Trends in U.S. Greenhouse Gas Emissions and Sinks by Chapter/IPCC Sector[a]

Chapter/IPCC Sector	1990	1995	2000	2005	2006
Energy	5,203.9	5,529.6	6,067.8	6,174.4	6,076.9
Industrial processes	299.9	315.7	326.5	315.5	320.9
Solvent and other product use	4.4	4.6	4.9	4.4	4.4
Agriculture	447.5	453.8	447.9	453.6	454.1
Land use, land use change, and forestry (emissions)	13.1	13.6	30.0	23.2	36.9
Waste	179.6	176.8	155.6	158.7	161.0
Total emissions	6,148.3	6,494.0	7,032.6	7,129.9	7,054.2
Net CO_2 flux from land use, land use change, and forestry (sinks)[b]	(737.7)	(775.3)	(673.6)	(878.6)	(883.7)
Net emissions (sources and sinks)	5,410.6	5,718.7	6,359.0	6,251.3	6,170.5

Source: From U.S. EPA 2008. Inventory of U.S. Greenhouse Gas Emissions and Sinks. EPA 430-R-08-005. Washington, D.C.
Note: Totals may not sum due to independent rounding. Parentheses indicate negative values or sequestration.
[a] Teragram CO_2 equivalent.
[b] The net CO_2 flux total includes both emissions and sequestration and constitutes a sink in the United States. Sinks are only included in the net emissions total.

Table 19.4
Emissions from Land Use, Land Use Change, and Forestry[a]

Source Category	1990	1995	2000	2005	2006
CO_2	7.1	7.0	7.5	7.9	8.0
Cropland/remaining cropland: liming of agricultural soils and urea fertilization	7.1	7.0	7.5	7.9	8.0
CH_4	4.5	4.7	19.0	12.3	24.6
Forest land/remaining forest land: forest fires	4.5	4.7	19.0	12.3	24.6
N_2O	1.5	1.8	3.5	3.1	4.3
Forest land/remaining forest land: forest fires	0.5	0.5	1.9	1.2	2.5
Forest land/remaining forest land: forest soils	0.1	0.2	0.3	0.3	0.3
Settlements/remaining settlements: settlement soils	1.0	1.2	1.2	1.5	1.5
Total	13.1	13.6	30.0	23.2	36.9

Source: From U.S. EPA 2008. Inventory of U.S. Greenhouse Gas Emissions and Sinks. EPA 430-R-08-005. Washington, D.C.
Note: Totals may not sum due to independent rounding.
[a] Teragram CO_2 equivalent.

(from the conversion of forest to pastures and croplands; Table 19.3).

The atmospheric concentration of CO_2 has increased from about 285 parts per million by volume (ppmv) in 1850 to more than 380 ppmv in 2007 (33%). In 2005, carbon emissions into the atmosphere were more than 28 billion metric tons and came equally from developed and developing countries (Figure 19.8). By the year 2030, worldwide emissions of more than 42 billion tons are mostly going to come from developing countries (non-OECD), with China and India leading in emissions.

CO_2 emissions in 1998 accounted for 1,494 Mt C (81.5%) of the U.S. total GHG emissions, equivalent to 1,834 Mt C [C equivalents = GHG $\times$ GWP $\times$ C(12)/CO_2(44)]. Burning of fossil fuels comprised 98% (1,468 Mt C) of the CO_2 emissions with the remainder coming from such industrial activities as cement production,

Table 19.5
Emissions of NO_x, CO, NMVOCs, and SO_2 (Gg)

Gas/Activity	1990	1995	2000	2005	2006
NO_x	21,645	21,272	19,203	15,569	14,869
Mobile fossil fuel combustion	10,920	10,622	10,310	8,739	8,287
Stationary fossil fuel combustion	9,883	9,821	8,002	5,853	5,610
Industrial processes	591	607	626	519	515
Oil and gas activities	139	100	111	316	315
Municipal solid waste combustion	82	88	114	97	97
Agricultural burning	28	29	35	39	38
Solvent use	1	3	3	5	5
Waste	0	1	2	2	2
CO	130,461	109,032	92,777	72,365	68,372
Mobile fossil fuel combustion	119,360	97,630	83,559	63,154	59,213
Stationary fossil fuel combustion	5,000	5,383	4,340	4,860	4,844
Industrial processes	4,125	3,959	2,217	1,724	1,724
Municipal solid waste combustion	978	1,073	1,670	1,437	1,437
Agricultural burning	691	663	792	860	825
Oil and gas activities	302	316	146	321	322
Waste	1	2	8	7	7
Solvent use	5	5	46	1	1
NMVOCs	20,930	19,520	15,228	14,444	14,082
Mobile fossil fuel combustion	10,932	8,745	7,230	6,289	5,991
Solvent use	5,216	5,609	4,384	3,846	3,839
Industrial processes	2,422	2,642	1,773	1,890	1,849
Stationary fossil fuel combustion	912	973	1,077	1,545	1,538
Oil and gas activities	554	582	389	528	523
Municipal solid waste combustion	222	237	257	235	232
Waste	673	731	119	111	110
Agricultural burning	N/A	N/A	N/A	N/A	N/A
SO_2	20,935	16,891	14,829	13,114	12,258
Stationary fossil fuel combustion	18,407	14,724	12,848	11,573	10,784
Industrial processes	1,307	1,117	1,031	797	793
Oil and gas activities	390	335	286	213	207
Municipal solid-waste combustion	38	42	29	22	22
Waste	0	1	1	1	1
Solvent use	0	1	1	0	0
Agricultural burning	N/A	N/A	N/A	N/A	N/A

Source: From EPA 2008 (disaggregated based on EPA 2003) except for estimates from field burning of agricultural residues. Inventory of U.S. Greenhouse Gas Emissions and Sinks. EPA 430-R-08-005. Washington, D.C.

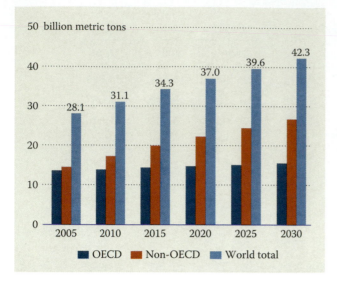

Figure 19.8 World carbon dioxide emissions 2005–2030 (in metric tons) comparing developed (OECD) and developing (non-OECD) countries. (From U.S. DOE EIA. 2008. Monthly energy review. DOE/EIA-0035 (2008/03). Washington, D.C. March 2008.)

natural gas flaring, lime, limestone, iron, steel, soda ash, and landfills. Forest regrowth may provide a good sink, but it should be borne in mind that sequestration by trees declines with forest maturation.

Sources and Sinks of Carbon

An important question is where CO_2 (and other gases too) goes after it enters the atmosphere. Equally important to ask is if this gas and others have a long residence time, what will the cumulative response of chronic accumulations over time be? Lastly, how serious will be the positive feedbacks beyond human control (referred to recently as the "tipping point")?

Under normal natural conditions—that is, with minimal emissions from human activities—the carbon cycle is more or less balanced with one-half exchanged with oceans and seas and the other half exchanged with vegetation and soils. Thus, to appreciate the effect of human-induced emissions, the role of carbon has been visualized as a source-and-sink problem and can be quantified as such (Figure 19.2). Dauncey (2001) describes three

sources and sinks. The first sink is the oceans. In general, half of photosynthesis in nature is from phytoplankton and the other half from land. Every year, the oceans absorb 92.4 Gt of carbon and release 90 Gt, storing 2.4 Gt as dissolved inorganic carbon in the deep ocean.

The second sink is the soil. Every year the world's soils absorb 50 Gt of C from dying vegetation and release 50 Gt through decomposition. Over millennia, the world's soils have accumulated 1,500 Gt of C, of which 500–800 Gt is locked up in the world's peatlands, including 500 Gt in the Arctic tundra. The third sink is vegetation. The world's forests and vegetation store about 550 Gt of C, 40% in tropical forests. Every year forests lose 50 Gt of C to the soil and 50 Gt to the atmosphere through respiration; however, they absorb 101.5 Gt from the atmosphere, reducing its load by 1.5 Gt. Not all the carbon emissions generated from fossil fuel burning and other land disturbances can be sequestered, leaving about 3 Gt annually to the atmosphere (Figure 19.9).

Other Greenhouse Gases

Water Vapor

Water vapor is the most abundant and dominant greenhouse gas. In the opinion of some (R. Essenhigh, personal communication), it is the gas responsible for most warming. This view is not shared by IPCC-Intergovernmental Panel on Climate Change (2007).[6] Most of the water vapor in the atmosphere is due to the hydrologic cycle. Although human activity is not directly linked to the concentration of water vapor in the atmosphere, its state, mixing, abundance, and radiative properties are affected by other greenhouse gases.

Methane (CH_4)

Almost 20% of the total direct radiative forcing of GHGs (245 Wm^{-2}) is attributable to CH_4, the second biggest contributor to anthropogenic GHG emissions (Table 19.1). Since the mid-1700s, the atmospheric concentration of CH_4 has increased

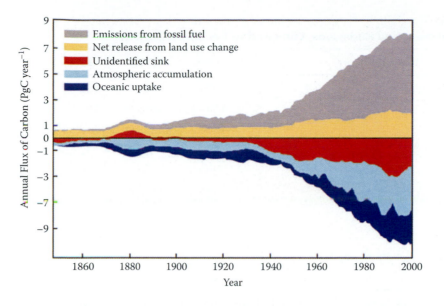

Figure 19.9 Annual sources and sinks of carbon from 1850 to 2000 for a balanced carbon budget (total sources balanced by total sinks). The unidentified sink is residual terrestrial sink. (From Houghton, R. A. 2007. Balancing the global carbon budget. *Annual Review of Earth and Planetary Sciences.* 35: 313–347.)

by about 143% from 722 parts per billion (ppb) to 1,774 ppb in 2005. Bacteria that decompose organic matter in oxygen-poor (anaerobic) environments produce methane that remains in the atmosphere for 12–17 years. Each molecule of CH_4 is about 25 times more effective at trapping heat in the atmosphere than a molecule of CO_2 over 100-year time horizon (thus, the GWP value of CH_4 is 25). Methane emissions (about 181 million tons of C-equivalent) constituted 10% of U.S. total GHG emissions.

Anthropogenic sources in the United States are agricultural activities such as rice paddies, enteric fermentation, manure management, and agricultural residue burning (33%); landfills (32%); fossil fuel-related activities such as natural gas and petroleum systems, coal mining, and petrochemical production (32%); stationary and mobile sources (2%); and wastewater treatment (1%) (Table 19.6). Fossil fuel use, agriculture, and waste disposal account for more than 50% of the CH_4 in the atmosphere (IPCC 2007). Livestock (primarily enteric fermentation in cattle) overtook rice farming in the early 1980s as the leading agricultural source.

Nitrous Oxide (N₂O)

Atmospheric N_2O concentration has increased by 18%, from 270 ppb in 1750 to 319 ppb in

2005 (IPCC 2007). Released by such human activities as the combustion of fossil fuels and solid waste, breakdown of nitrogen fertilizers, livestock wastes, contaminated groundwater, and biomass burning, nitrous oxide contributes about 6% of GHGs. Its average residence time in the troposphere is about 120 years. The GWP for each molecule of N_2O is about 298 times that of CO_2 over 100-year horizon.

All gaseous NO_x are considered important radiatively, chemically, and ecologically and their exchange across the soil–atmosphere boundary is directly or indirectly related to such events as acid deposition, global warming, stratospheric O_3 depletion, groundwater contamination, deforestation, and biomass burning (Hutchinson 1995). Whereas the high reactivity of NO_x ($NO + NO_2$) is a factor, the important global atmospheric consequences of N_2O exchange result from its long lifetime and spectral properties.

Hutchinson (1995) notes that

"bacterial processes of nitrification and denitrification are of paramount importance and represent a major source of atmospheric N oxides. The available data suggest that net production of both gases, as well as the $NO:N_2O$ emission ratio, are strongly dependent on soil temperature, soil N availability, and soil water content."

Table 19.6
U.S. Greenhouse Gas Emissions Allocated to Economic Sectors[a]

Implied Sectors	1990	1995	2000	2006	2006
Electric power industry	1859.1	1989.7	2328.9	2430.0	2377.8
Transportation	1544.1	1685.8	1917.5	1987.2	1969.5
Industry	1460.3	1478.0	1432.9	1354.3	1371.5
Agriculture	506.8	524.1	528.0	521.3	533.6
Commercial	396.9	404.5	390.3	400.4	394.6
Residential	346.9	370.9	387.7	376.0	344.8
U.S. territories	34.1	41.1	47.3	60.5	62.4
Total emissions	6148.3	6494.0	7032.6	7129.0	7054.2
Land use, land use change, and forestry (sinks)	(737.7)	(775.3)	(673.6)	(878.6)	(883.7)
Net	5410.6	5718.7	6359.0	6251.3	6170.5

Source: From U.S. EPA 2008. Inventory of U.S. Greenhouse Gas Emissions and Sinks. EPA 430-R-08-005. Washington, D.C.
Note: Totals may not sum due to independent rounding. Emissions include CO_2, CH_4, N_3O, HFCs, and SF_6.
[a] Teragrams CO_2 equivalent.

N_2O emissions in 1998 made up 6.4% (up from 5.7% in 1992) of the U.S. total GHG emissions; agricultural activities, such as soil and manure management, and residue burning accounted for 74% and energy-related activities for about 18% (Table 19.6).

CFCs and Their Substitutes

Greenhouse gases that are not naturally occurring include by-products of CFCs, as well as hydrofluorocarbons (HFCs), perfluorocarbons (PFCs), and sulfur hexafluorane (SF_6) generated by industrial processes. HFCs and PFCs are categories of synthetic chemicals that have been introduced as alternatives to ozone-depleting substances (ODSs) that are being phased out under new international and national mandates (see the section at the end of this chapter). In addition to their use as substitutes for ODSs, these gases are emitted by aluminum production, hydrochlorofluorocarbon (HCFC-22) production, semiconductor manufacturing, electrical transmission and distribution, and magnesium production and processing.

PFCs have an extremely long atmospheric lifetime (up to 1,000 years). CFCs are responsible for about 12% of anthropogenic GHG emissions and are likely to rise to about 25% by 2020. Depending on the type, CFCs remain

in the atmosphere for 60–400 years and generally have 5,000 to 10,000 times the impact per molecule on global warming compared to CO_2. HFCs, PFCs, and SF_6 in 1998 constituted 2% of total U.S. greenhouse gas emissions (U.S. EPA 2000).

Indirect GHGs

Carbon monoxide (CO), nitrogen oxide (NO_x), nonmethane volatile organic compounds (NMVOCs), and sulfur dioxide (SO_2) do not have a direct global warming effect. However, indirectly, these gases affect the absorption of terrestrial radiation by influencing the formation and destruction of tropospheric and stratospheric ozone. In the U.S. Clean Air Act, these gases are commonly referred to as ozone precursors or criteria pollutants.

Aerosols can also affect the absorptive characteristics of the atmosphere (see Chapter 18). Carbon monoxide is produced when carbon-containing fossil fuels are combusted incompletely. Nitrogen oxides (i.e., NO and NO_2) are created by lightning, fires, and fossil fuel combustion, and in the stratosphere from N_2O. NMVOCs—which include such compounds as propane, butane, and ethane—are emitted primarily from transportation and industrial processes such as the manufacture of chemical

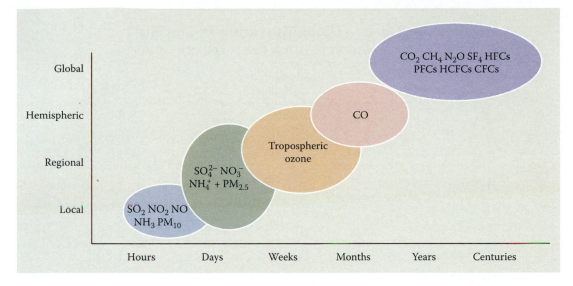

Figure 19.10 Selected air pollutants, their average residence times in the atmosphere, and maximum extent of their impact. (Redrawn from Figure 2.1, p. 43 in GEO_4–UNEP 2007; *Global Environmental Outlook: Environment for Development*. Valletta, Malta: Progress Press Ltd.)

and allied products, metals processing, and the use of solvents.

In the United States, SO_2 is primarily emitted from the combustion of fossil fuels and by the metals industry. Sulfur dioxide affects the Earth's radiative budget through its photochemical transformation into sulfate aerosols in the atmosphere. This, in turn, can (1) scatter sunlight back to space, thereby reducing the radiation reaching the Earth's surface; (2) affect cloud formation; and (3) affect atmospheric chemical composition (e.g., stratospheric ozone, by providing surfaces for heterogeneous chemical reactions). As discussed in the last chapter, sulfur dioxide is also a major contributor to the formation of urban smog, and electric utilities are the largest source of SO_2 emissions.

In sum, for major greenhouse gases, evaluation of impact must be based on their residence time in the atmosphere (Figure 19.10), the rate at which they increase each year, and their relative **global warming potential** (**GWP**) (Table 19.7). The GWP is the ratio of both direct and indirect radiative forcing from one unit mass of a greenhouse gas to that of one unit mass of the reference gas (CO_2) over a period of time. The concept of GWP has been developed to compare the ability of each GHG to trap heat in the atmosphere relative to another gas. GWPs are not provided for other pollutants, such as CO, NO_x, NMVOCs, and

SO_2, because there is at present no agreed-upon method to estimate the indirect radiative forcing contribution to climate change of these gases. In the most recent IPCC report (2007), their values are compared with those reported in three earlier IPCC reports.

Evidence of Global Warming

The scientific community, as will be discussed in the following pages, is convinced that an increase in GHGs resulting from human activities is already changing the climate and its weather attributes. Although it seems to be the notion in some segments of the general citizenry that global climate change is mostly a product of theoretical modeling, evidence from field observations proves otherwise. From discussions on ecosystem structure and function and the attendant human impacts on them, we have noted throughout this book that ecosystem processes are complex and that scientific knowledge based on long-term experiments for each ecological process or compartment is not available to make definitive conclusions. To make up for these limitations, scientists have been using both direct observations and innovative computer models in which atmospheric

Table 19.7
Greenhouse Gases—Global Warming Potentials (GWP) And Atmospheric Lifetimes (Years)

Gas	Atmospheric Lifetime	GWP[a]
CO_2	50-200	1
CH_4	12±3	21
N_2O	114	298
HFC-23	270	11,700
HFC-32	4.9	650
HFC-125	29	2,800
HFC-134	14	1,300
HFC-143	52	3,800
HFC-152	1.4	140
HFC-227	34.2	2,900
HFC-236	240	6,300
HFC-4310	15.9	1,300
CF_4	50,000	6,500
C_2F_6	10,000	9,200
C_4F_{10}	2,600	7,000
C_6F_{14}	3,200	7,400
SF_6	3,200	23,900

Source: From EPA. 2008. Inventory of U.S. Greenhouse Gas Emissions and Sinks. EPA 430-R-08-005 April. Washington, D.C. (see http://ipcc-wg1.ucar.edu/wg1/Report/AR4WG!_Print_Ch02.pdf, accessed December 12, 2008.)

[a] 100-year time horizon.
[b] Each molecule of CH_4 is about 25 times more effective in trapping heat in the atmosphere than a molecule of CO2 over a 100-year time horizon. Besides direct effects and those indirect effects due to the production of tropospheric ozone and stratospheric water vapor. The indirect effect due to the production of CO_2 is not included.

and biological data are simulated to project the changes that are likely to be brought about by one or multiple human impacts.

However, modeling provides just one line of evidence. Based on direct measurements of a number of ecosystem parameters, a large majority of scientists are convinced that global warming is indeed taking place; the uncertainty lies only in its extent and magnitude. That is, how severe will the impact of warming in various biomes across the globe be? Evidence from physical and biological systems (direct and indirect) on climatic warming has been

mounting in the last decade. In the following, some of the direct evidence is presented first, and then some projected effects from general circulation models are summarized.

Direct Evidence

1. Global air temperature. Data collected from a number of stations around the world show that surface air temperatures have been warming (Figures 19.3–19.5). On a global basis, 22 of the hottest years in recent history were in the period between 1983 and 2007, and

23 consecutive years have been above the average. The Northern Hemisphere in particular is warming faster than the Southern Hemisphere. The White House report, "Our Changing Planet" (USGCRP 2001), notes: "Regardless of which approach was adopted, all reconstructions (plus two others from mid-latitude summers) agree that the late twentieth century temperatures are the highest in at least the last 1000 years" (Figure 19.3). Global temperatures have warmed at the rate of 1°C per century.

2. U.S. air temperature. Temperatures for the calendar years 1998 and 1999 for the United States were the warmest since 1900 (average in 1998 of 56.4°F and in 1999 of 56°F). These temperatures even exceeded those of the warm decade of the 1930s. These trends were similar to those found for global temperatures—that is, a trend toward long-term warming in the United States (0.5°C per century), with much of the warming occurring during two periods: 1910–1955 and 1976–present (NOAA National Climatic Data Center 2000).

3. Arctic air temperature. The northern ecosystems, particularly the Arctic, are reported to have warmed, with an average of 2°C increase in temperature over the last 100 years of recorded data. In the Arctic, this warming has had a marked impact on the cryosphere (a collective term for glaciers, ice caps, sea ice, and permafrost), which is in retreat. Indeed, the Arctic Council, comprising eight arctic nations (Canada, Denmark/Greenland/Faroe Islands, Finland, Iceland, Norway, Russia, Sweden, and the United States), brought together several hundred scientists to assess the effects of global climate change in the region. In a clear and well-documented report (ACIA 2004), they were unanimous that the ongoing changes in melting permafrost, sea ice, and biota will be devastating.

4. Warming of ocean waters. A number of studies have shown that surface and intermediate depth ocean waters have been warmer in the last two decades. Fukasawa et al. (2004) completed a trans-Pacific survey in the deepest waters of the North Pacific Ocean in 1985 and 1999. They found that "the deepest waters of the North Pacific Ocean have warmed significantly over the entire width of the ocean basin in a 10-year period. Our observations imply that changes in water properties are now detectable in water masses that have long been insulated from heat exchange with the atmosphere." Southern Pacific ocean water level rise has also been attributed to warming temperatures (Byrnes 2008).

5. Borehole temperatures. From 600 deep boreholes drilled into the Earth, temperature readings showed a marked increase in the latter half of the twentieth century. In comparison with trends of the last 400–1,000 years, the temperature increases were thought to be without precedent. Scientists reporting on these increases note that no combination of natural mechanisms explains this warming phenomenon (Pollack 2000; Overpeck et al. 2000).

6. Melting glaciers. On all continents, glaciers are on the retreat. In North America, Alaska's Columbia Glacier has retreated 13 km since 1982, and all 14 other glaciers in the area are retreating. Over 100 of the 150 glaciers in Glacier National Park have completely melted since 1850, and the rest are projected to be gone in 30 years. In South America, Argentina's Upsala Glacier has retreated 60 m yr^{-1} for the past 60 years, and the rate is accelerating. The South Patagonian Ice Field in Argentina and Chile has decreased by >500 km^2 in the past 50 years. Quelccaya Ice Cap in the Andes of Peru increased its rate of retreat from 3 m yr^{-1} in the 1970s to 30 m yr^{-1} in the 1990s.

In Europe, glaciers in the Alps have lost 50% of their volume since 1850, 14 of Spain's 27 glaciers have disappeared since 1980, and glaciers of the Caucasus have lost 50% of their volume in the past century. In Africa, Mt. Kenya's largest glacier has lost 92% of its mass since the late 1800s. Glaciers of Mt. Kilimanjaro in Tanzania have shrunk by more than 70% since the late 1800s.

In Asia, Duosuogang Peak glaciers in the Ulan Ula Mountains of China have shrunk by 60% since the early 1970s. Tien Shan Mountain glaciers of Central Asia have lost 22% of their volume in the last 40 years. In the eastern Himalayas, about 2,000 of the glaciers have disappeared in the past century. Scientists have recently reported that four glaciers in North India are "facing 'terminal retreat,' while 15 others are undergoing a substantial reduction in area" (Sharma 2004).

Their models suggest "that these four glaciers will disappear by 2040 ... [or] much faster if atmospheric temperature increases over the next 40 years" (Sharma 2004).[3]

The evidence from glaciers for climate change is summed up by the pioneer glaciologist, Lonnie Thompson, thus:

"Glaciers are among the first responders to global warming, serving both as indicators and drivers of climate change. Over the last 35 years, ice core records have been recovered systematically from both polar regions as well as twelve high-elevation ice fields, eleven of which are located in the middle and tropical latitudes. Analyses of these ice cores and of the glaciers from which they have been drilled have yielded three lines of evidence for abrupt climate change both past and present. They are: (1) the temperature and precipitation histories recorded in the glaciers as revealed by the climate records extracted from the ice cores; (2) the accelerating loss of the glaciers themselves, specifically Quelccaya ice cap, Peru, Kilimanjaro, Africa, and Naimona'nyi, Himalayas will be updated with 2009 results and; (3) the uncovering of ancient plants and animals from the margins of the glaciers as a result of their recent melting, thus illustrating the significance of the current ice loss. The current melting of high-altitude, low latitude ice fields is consistent with model predictions for a vertical amplification of temperature in the tropics." (personal communication)

7. Melting ice. A number of studies from several locations in the world have methodically documented changes in the cryosphere. Arctic sea ice is reported to have shrunk by almost 58,800 km² (23,000 mi²) per year between 1978 and 1998—a 6% reduction overall and an area equal to that of Maryland and Delaware combined (see Kerr 1999; Vinikov et al. 1999; Johannessen, Shalina, and Miles 1999). Recession of sea ice is now a consistent phenomenon (Figures 19.11 and 19.12).

In Greenland, sea ice has also thinned by about 40% in recent decades, from an average of 3 m in the period between 1958 and 1976 to about 2 m between 1993 and 1997. On average, current thinning is estimated at about 10 cm per year. Since 1993, the Greenland Ice Sheet—the Arctic's largest—has thinned by more than a meter a year on its southern and eastern edges. The Greenland Ice Sheet has thinned in some areas by more than 6 m since 1992. The margin

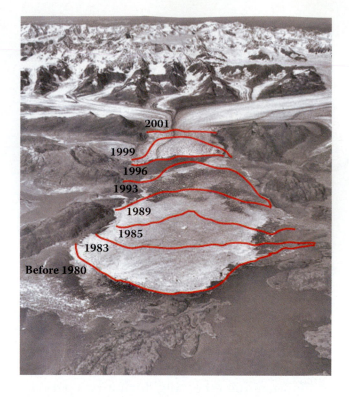

Figure 19.11 Receding sea ice 1983–2001. (Meier, M. F., and M. B. Dyurgerov. 2002. How Alaska affects the world. *Science* 297: 350–351.) Alaskan glaciers produce more water than previously thought and "may be important in characterizing ... climate change."

of the fast-flowing outlet glacier Jacobshavn Isbrae in western Greenland and a large number of icebergs from the fast-flowing terminus of Kangerdlugssuaq Glacier in eastern Greenland are fast melting away (Dowedswell 2006).

Alaskan permafrost is reported to be thawing across the state. Near Barrow, permafrost was 20–25 cm (8–10 in.) thinner in 1991–1997 than it was between 1964 and 1968. Several reports are now available showing large chunks of broken glaciers floating on the ocean (include US-NSIDC; on file).

8. Precipitation. The United States showed decreased precipitation during 1999. The year was the 22nd driest year out of the last 100 years, with nationally averaged precipitation of 76 cm (29.93 in.) (NOAA National Climatic Data Center 2000). Satellite data revealed plumes of reduced cloud particle size and suppressed precipitation originating from major urban areas and from industrial facilities such as power plants (Rosenfeld 2000). Urban and industrial pollution can completely shut off precipitation over large areas from

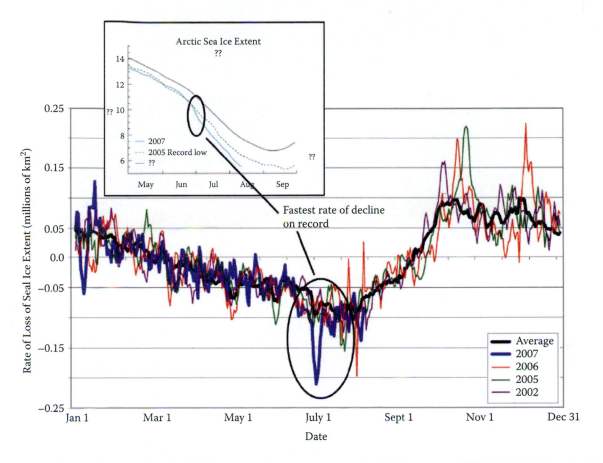

Figure 19.12 Rate of sea ice loss in 2007 compared to recent record-breaking years. (NSIDC 2007; http://nsidc.org/news/press/2007_seaiceminimum/images/20070814_rate.jpg accessed February 26, 2009.)

clouds that have temperatures at their tops of about −10°C.

Indirect Evidence

Climate change has also been deciphered by changes that have been documented in the morphological and physiological responses over time displayed by plants and animals. These changes in **phenology** include timing of various phases of plant and animal growth, change in seasonal responses, and modifications of other responses to temperature change (Figure 19.13). Several of these are illustrated next:

1. For the decade 1981–1991, an increase in plant growth was reported for vegetation growing in northern high latitudes (45–70° N). There was a significant increase in photosynthetic activity of 10–12%, attributed to a warm spring (Myneni et al. 1997).

2. Grassland vegetation has been shown to react differently to increases in minimum temperatures (T_{Min}) compared to maximum temperatures (T_{Max}). In the shortgrass prairie, a dominant C_4 blue grama grass (*Bouteloua gracilis*) was shown to have slower growth than the native and invasive C_3 forbs. This may make the grass more vulnerable to invasion by other species and less tolerant of drought and grazing. Thus, such ecosystems may be sensitive to increases in T_{MIN} (Alward, Detling, and Milchunas 1999).

3. In the desert ecosystems of Chihuahua, Mexico, woody shrubs have been observed to show a threefold increase in growth. A shift in animal species in these ecosystems could be correlated with a shift in regional climate since 1977, rather than changes generally associated with historical desertification brought about by livestock grazing or drought. The changes apparently were caused by a shift in regional climate: Since 1977, winter precipitation throughout the region has been substantially higher than the average of this century (Brown et al. 1997; *PNAS*, Reorganization of

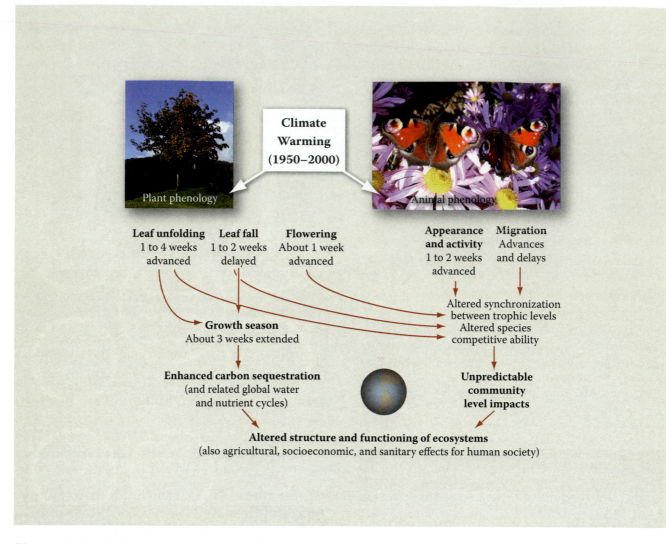

Figure 19.13 Ecological consequences of climate warming on plant and animal phenology. (From Peñuelas, J., and L. Filella. 2001. Responses to a warming world. *Science* 294: 793–795.)

an arid ecosystem in response to recent climate change).

4. Several studies have recently documented the seasonal activity of both animals and plants. These include the changes in migratory patterns of birds in England, early egg-laying by tree swallows in North America, early reproduction by birds and amphibians, changes in moth phenology, early breaking of leaf buds, and so on. In each case, these activities have been related to changes in temperatures at low altitudes (Inouye et al. 2000).

5. In two separate studies, a meta-analysis[4] technique was used to reveal if common patterns or responses of species to global climate change could be revealed. In one study, 1,700 species were analyzed and it was found

that 279 species showed a distinct "diagnostic fingerprint" of global climate (Parmesan and Yohe 2003). Species showed a 6.1-km shift in their ranges toward the poles. In another meta-analysis investigation, 143 studies that included animal and plant species "ranging from mollusks to mammals and from grasses to trees" revealed that the effects of global climate change are "already discernible" (Root et al. 2003). Similar findings have been reported from British animals and plants (Thomas et al. 2004).

6. Coral reef studies reveal that they are more sensitive to warmer temperatures than they are to other environmental changes, including pollution. Increased temperature causes "bleaching" of coral reefs (loss of their symbiotic algae's ability to carry out photosynthesis), affecting as many as 27% of corals in the year 2000. Most damaged are 59% of coral reefs of the Indian Ocean, 35% in the Middle East, and 34% in Southeast and East Asia. The corals are given a "reasonable chance" of recovery in the Indian Ocean and a lower chance for reefs of the other regions. Degradation is apparent through the bleaching of the coral reefs, which is correlated with El Niño events and much more so with chronic warmer temperatures (Normile 2000; Pockley 2000).

7. To date, no studies have demonstrated a link between the frequency of hurricanes and increasing temperature. However, because they have a positive correlation with water temperature, a link to hurricane intensity has been suggested (see Webster et al. 2005; Emanuel 2005). Who can forget the massive damage from hurricanes Katrina, Rita, and Wilma in 2005?

8. Lastly, global climate change has been attributed to activities of the sun. This does not seem to be feasible for the following reasons: (1) the stratosphere gets cooler as the surface gets warmer, (2) night temperatures (also referred to as T_{min}) are rising faster than day temperatures (T_{max}), (3) northern latitudes are warmer than lower latitudes, (4) higher altitudes are warmer than lower altitudes, and (5) temperatures rise more quickly in winter than in summer. Harte[5] refers to these as "fingerprints of human-induced global warming."

General Circulation Models

The science of global warming is most complicated because it involves many interactions of land–water–atmosphere that are individually and collectively very complex. Thus, to assess the overall impact of GHGs on temperature and other relationships, scientists over the last decade have employed what are known as the **general (or global) circulation models (GCMs)**. Over time, as is illustrated in Figure 19.14, more and more variables of the processes of the Earth have been included in these simulation models.

These models utilize current knowledge of the variables of atmosphere, generalized (crude) land use patterns, oceans, sea ice, and atmospheric

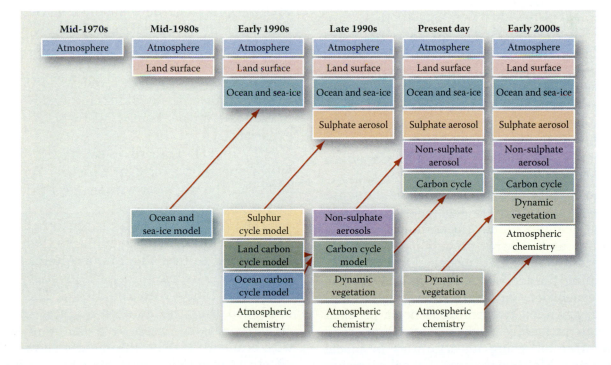

Figure 19.14 The inclusion of more and more important variables in general circulation models ensures the comprehensive coverage of the Earth systems. (From IPCC. 2001. *Climate Change 2001: The Scientific Basis,* ed. J. T. Houghton, Y. Ding, et al. Cambridge, England: Cambridge University Press.)

and oceanic circulation. Equilibrium GCMs compare the results with current atmospheric CO_2 concentrations ($1 \times O_2$) and doubled ($2 \times CO_2$) scenarios to predict what a doubling of atmospheric CO_2 level over 100 years would do and estimate that a global mean surface temperature increase in the range of 1.5–4.5°C would occur. In approximate proportionality to the global mean temperature change, global mean amounts of water vapor are expected to increase in the lower troposphere (6% for each 1°C of warming), also contributing to global warming. A sea level rise of 50 ± 25 cm by the year 2100 is projected to occur due to the thermal expansion of seawater and ice cap melting.

Given the lack of actual quantitative data on temperature and greenhouse gases from the past, scientists in the previous decade or so have intensified research and devised ways in which they are able to simulate changes based on known and even fragmented information. Using known numbers from the immediate past for which there are records and fragmented evidence from the distant past, scientists simulate interactive changes among four compartments: *atmosphere, land, ocean,* and *sea ice.* Thus, when the models use equilibrium states at $1 \times CO_2$ or $2 \times CO_2$ (and are hence primarily land based), these are called *equilibrium* GCMs. In contrast, *transient* models use changes in oceanic parameters and therefore are more sophisticated and very complex. Among several, these major modeling centers are cited in the literature: Geophysical Fluid Dynamics Laboratory in the United States, Hadley Center of the U.K. Meteorological Office, and the Canadian Climate Center.

The climate modelers have made immense progress in a relatively short period of time. Several evaluations have been made as to the certainty and relative probability of outputs generated by the climate models. After carefully examining the models, Mahlman (1997) came to the conclusion that "it is obvious that human-caused greenhouse warming is not a problem that can rationally be dismissed or ignored." Even after assessing uncertainties in model projections, Mahlman's overall conclusion was that "it is virtually certain that human-caused greenhouse warming is going to continue to

unfold, slowly but inexorably, for a long time into the future." Another study examined 928 scientific papers and concluded: "Remarkably, none of the papers disagreed with the consensus position" that global climate change is the result of human activities (Oreskes 2004). When several modeling approaches yield similar results, the conclusions drawn based on the models' evidence are strengthened.

Climate–Weather Uncertainties

In their established tradition, scientists will proceed to seek not only refinement but also the validation of their models; one trend is apparent already: the unpredictability of weather patterns in several places around the world. Here are a few examples. Between 1988 and 1997, nearly 80 million people were severely affected by weather-related disasters. In 1998, Hurricane Mitch dropped 91–183 cm (3–6 ft) of rain within 48 hours in Central America, killing more than 10,000 people in landslides and floods, triggering a cholera epidemic, and virtually wiping out the economies of Honduras and Nicaragua (Sarewitz and Pielke 2000). In the United States, several abnormalities have been reported. In 1998 and 1999, (1) Mount Baker in Washington received 28 m (93.2 ft) of snow (a record), (2) Oklahoma had a record 512 km (318 mi) per hour windstorm, (3) Alaska experienced a record low temperature of −53°C (−64°F), and (4) a record 46 cm (18 in.) of snow fell in one snow storm in Chicago. However, 1998 was the warmest winter in recorded history in the lower 48 states!

The insurance industry is concerned about all these changes and the damages that it will be required to pay. The Munich Reinsurance Company of Germany, in its 1999 study, compared the 1960s with the 1990s. The study concluded that the number of great natural catastrophes increased by a factor of three, with economic losses—taking into account the effects of inflation—increasing by a factor of more than 8 and insured losses by a factor of no less than 16 (Sarewitz and Pielke 2000).

Response of Biota to Climate Change

Global warming will manifest itself in diverse ways, depending upon the present climatic regimes and the types of biota that have adapted to them during the course of evolution. Overall, the predictability of weather conditions will override interactions with other factors in determining the impact on ecosystem productivity. For agriculture, two positive results could ensue from climate warming in the northern latitudes. First, there would be a higher CO_2 level, thereby increasing photosynthesis. Second, general warming would result in a longer growing period and more frost-free days. In warmer climates, higher evapotranspiration would result in lower water availability. This would be coupled with the alteration of phenology due to warmer temperatures and, for some species, the lack of a cold period necessary for the vernalization of seeds and hence decreased productivity. Increasing CO_2 concentration would also have a differential effect upon the production of vegetative and reproductive parts of plants in general and crops in particular (Bazzaz and Sombroek 1996).

For the forests, GCMs clearly depict a change in the survival and migration of forest species brought about by global warming. It is very important to remember that trees have a long generation time, and their migration and subsequent establishment in new areas depend upon a number of factors. These factors include the distance to which the seeds can be carried (by wind, water, and animals) in the face of accelerated fragmentation of habitats, their arrival at a hospitable environment, and their ability to compete with the resident species. The phrase "northward march of species" is used sometimes in the literature, as if all trees could pack their bags and avail themselves of public transportation, but this may breed undue complacency. Fossil pollen studies have shown that only a small number of species were able to migrate after the last glaciation (Davis et al. 2000).

Scientists have made a number of projections on the response of biota in agricultural, forest, and other ecosystems to the warming of climate. One GCM-based study is cited here as an example because we believe it to be a comprehensive study. Using an extensive database from the eastern United States, Prasad et al. (2007) have carefully analyzed forest inventory data that included more than 100,000 forested plots from more than 2,100 counties east of the 100th meridian. Included in the analysis were climate, soils, land use, elevation, and species composition of major communities found in this region. Overall, a loss of biodiversity and homogenization of community types was expected (Figure 19.15). From their studies, we

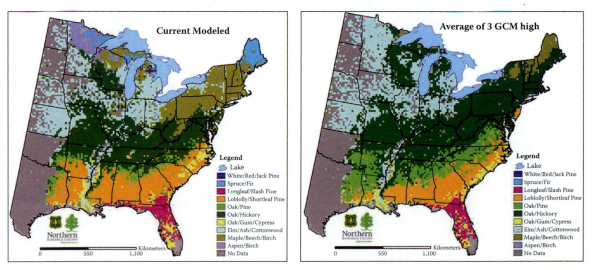

Figure 19.15 Vegetation types in the eastern United States as of 2007. Left: current model; right: modeled average of three GCMs high. (From Prasad, A. et al. 2007. *Climate Change Tree Atlas*, a spatial database of 134 tree species of the eastern United States. Courtesy of the authors; see http://www.nrs.fs.fed.us/atlas/tree/tree_atlas.html)

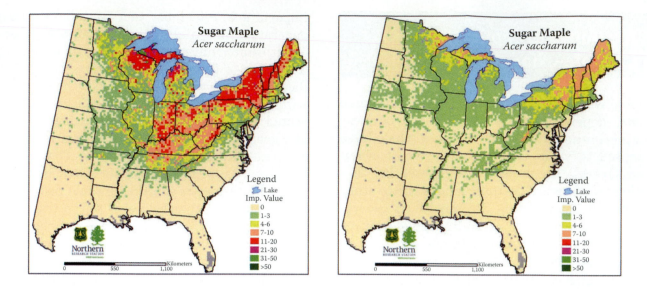

Figure 19.16 Sugar maple. Left: current; right: modeled average of three GCMs high. (From Prasad, A. et al. 2007. *Climate Change Tree Atlas,* a spatial database of 134 tree species of the Eastern United States. Courtesy of the authors; see http://www.nrs.fs.fed.us/atlas/tree/tree_atlas.html.)

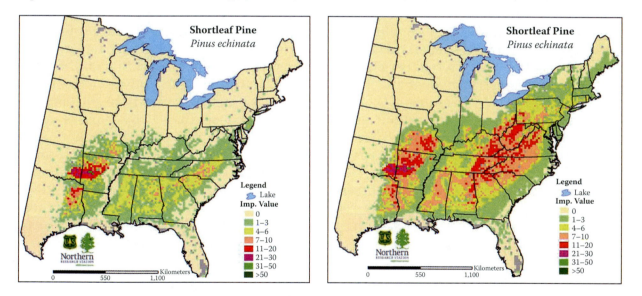

Figure 19.17 Shortleaf pine: left, current; right, modeled average of three GCMs high. (From Prasad, A. et al. 2007. *Climate Change Tree Atlas,* a spatial database of 134 tree species of the Eastern United States. Courtesy of the authors; see http://www.nrs.fs.fed.us/atlas/tree/tree_atlas.html)

selected two plants (sugar maple and shortleaf pine) (Figures 19.16 and 19.17) and two birds (Blackburnian warbler and summer tanager) (Figures 19.18 and 19.19) to illustrate the contrasting responses to a climate change. Sugar maple and Blackburnian warbler will show a great shrinkage in current areas of distribution, whereas both shortleaf pine and summer tanager will expand their areas of distribution considerably. Advances in refining GCMs and their use to predict potential biotic responses

to possible climatic changes are going on at a feverish pace. It is significant that output from models is becoming vastly closer to ground reconnaissance and field observations.

For aquatic ecosystems, warming of the surface layers of water has been shown to result in the migration of phytoplankton to deeper (= cooler) depths; the distribution may be to 23 m (75 ft) in contrast to the normal 15 m (50 ft). But this also means that the phytoplankton will grow more slowly because less light penetrates

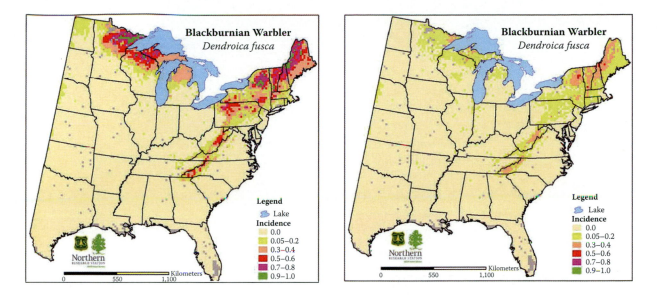

Figure 19.18 Blackburnian warbler. Left: current; right: modeled. (From Matthews, S. et al. *Climate Change Tree Atlas,* a spatial database of 147 bird species of the Eastern United States. Courtesy of the authors; see http://www.nrs.fs.fed.us/atlas/bird/.)

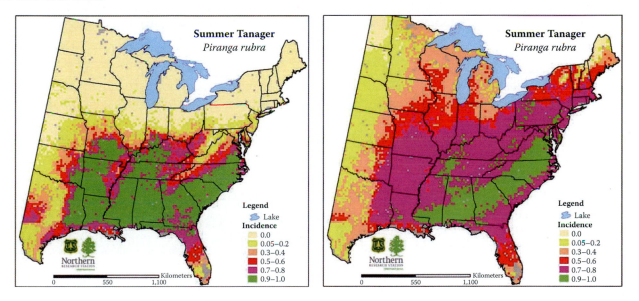

Figure 19.19 Summer tanager. Left: current; right: modeled. (From Matthews, S. et al. *Climate Change Tree Atlas,* a spatial database of 147 bird species of the Eastern United States. Courtesy of the authors; see http://www.nrs.fs.fed.us/atlas/bird/)

as the depth increases. The reduced phytoplankton growth means less food for the higher trophic levels. "If something suddenly made the plankton population crash, the whole food chain could start to collapse" (Ralph 2000). For northern and western North America, warm air temperature coupled with an increase in CO_2 could impact lake and stream ecology in several ways. Two of these ways appear to be major: greater evaporative and transpirational loss relative to incoming precipitation and

faster water flows from higher fire frequency in terrestrial ecosystems. Altered hydrological cycles and stream flows will directly affect the physical, chemical, and biological features of lake ecosystems, including changes in thermocline, and hence lake turnover, and eutrophication (Schindler 1997). Collectively, all these factors will have a great impact on both indigenous and non-native species.

From the preceding discussion it is clear that all human support systems, or what have

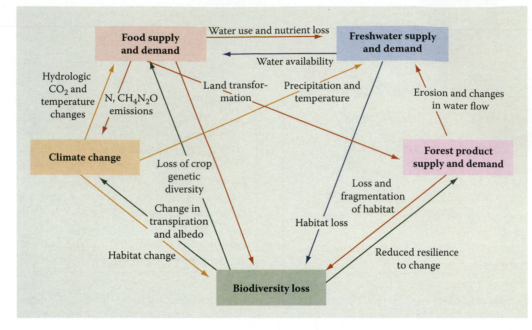

Figure 19.20 Linkages among ecosystem goods and services and climate change. (Ayensu, E., D. V. R. Claasen, M. Collins, A. Dearing, et al. 1999. International ecosystem assessment. *Science* 286: 685–686.)

been termed "ecosystem services/environmental services," depend vastly upon the proper functioning of ecosystems (Figure 19.20). Natural resources of food production, of forests and grazing lands, and the habitats for life forms other than humans are all related to each other and to atmospheric factors that will be discussed later (Chapter 26).

There is widespread agreement among the scientific community that, given all the changes that human activities have brought about, the impacts superimposed by climate change are likely to be varied, intense, and highly complex. Several examples may illustrate the point. First, the Arctic Climate Impact Assessment (2004) shows that a change in global temperature may considerably alter thermohaline conditions (Figure 19.21) and thereby accelerate climatic conditions that will have a tremendous impact on Europe. It notes further, "Changes in global ocean circulation can lead to abrupt climate change. Such change can be initiated by increase in arctic precipitation and river runoff, and the melting of arctic snow and ice because these lead to reduced salinity of ocean waters in the North Atlantic, evidence of which is shown above" (pp. 32, 36–37). Shellnhuber and Held

(2002) note that there may indeed be a multiplicity of changes that will occur through the world, from changes in the hydrological cycle in northern Asia, to suppression of monsoons in southern Asia, to biome changes in South America, as is shown in Figure 19.22.

Climate Change and Human Health

A factor of great importance is human health. Scientific reports have suggested, for example, that as ozone depletion occurs and ultraviolet radiation exceeds a certain threshold, skin diseases, the formation of eye cataracts, and even blindness increase in humans and wildlife, especially in Australia and South America. It also appears that the spread of pathogens and the weakening of immune systems coupled with air and water quality will have adverse effects on human health. "Climate change could have far-reaching impact on health patterns in the United States," note Patz et al. (2000), and this will take many forms and follow many routes (Figure 19.23).

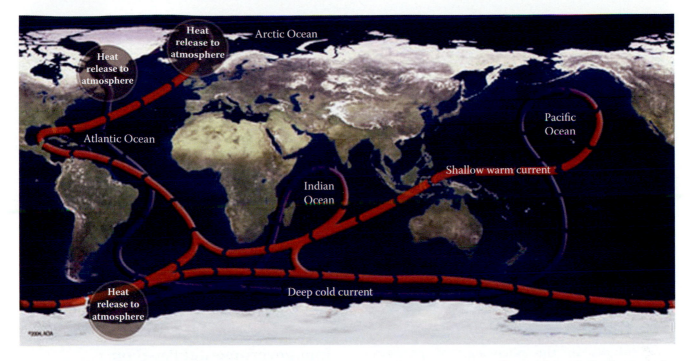

Figure 19.21 The North Atlantic heat pump (i.e., circulation of warm water to the northern latitudes may be disrupted by changes in thermohaline circulation). (From Arctic Climate Impact Assessment. 2004. Impacts of a Warming Arctic: Arctic Climate Impact Assessment. ACIA Overview report. http://amap.no/workdocs/index.cfm?dirsub=%2FACIA%2Foverview)

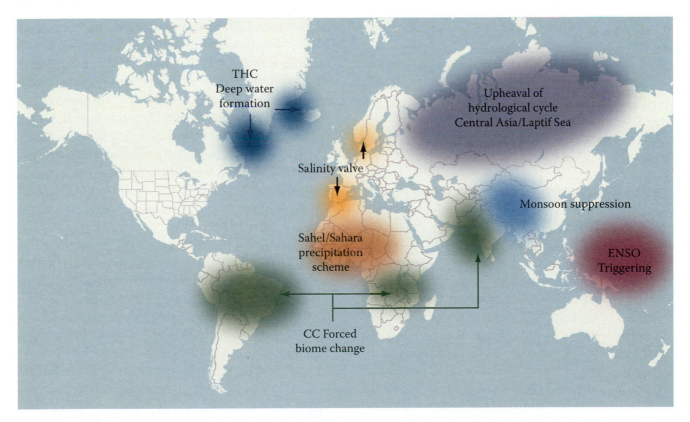

Figure 19.22 Critical switch and choke points within the Earth system. (Redrawn from Shellnhuber, H.-J. and H. Held, *GIAM* 2002.)

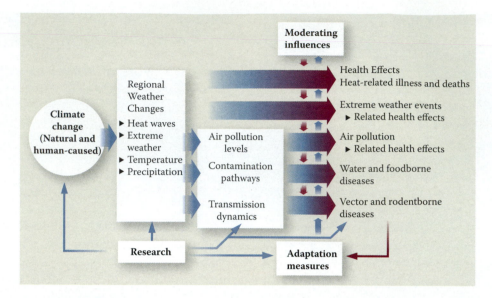

Figure 19.23 Potential health effects of climate variability and change. (From Patz, J. A. et al. 2000. *Environmental Health Perspectives* 108:367–376.)

Changes in the phenology of parasites brought about by warmer temperatures and longer growing seasons, as well as the spread of diseases and disease vectors, loom large on both the global and national scenes. As will be discussed in the next chapter, invasive plant species seem to produce more pollen (e.g., ragweed) and have a great potency of toxins (e.g., those in poison ivy that cause dermatitis). Disease epidemics such as mountain pine beetle (e.g., Ootsa Lake, British Columbia) and spruce beetle (e.g., southern Kenai Peninsula, Alaska) can cause significant damage to forests.

Recent National and International Responses

The United Nations has proposed a Kyoto Protocol to curb greenhouse emissions, and this has been accepted by a number of nations. In addition, several countries have taken their own initiatives, as have many states in the United States, to achieve similar goals. Although these aspects will be discussed in Chapter 21 on environmental policy and law, here the rigorous scientific work on global climate change by an internationally constituted organization, the IPCC, is discussed.

Intergovernmental Panel on Climate Change (IPCC)

In 1989, the United Nations Framework Convention on Climate Change (UNFCC) recognized "a change of climate which is attributed directly or indirectly to human activity that alters the composition of the global atmosphere and which is in addition to natural climate variability observed over comparable time periods." But even before the passage of the UNFCC, the World Meteorological Organization and the United Nations Environment Program recognized the profound impact such a climate change was likely to have on humans and other biota alike; they jointly established IPCC—the Intergovernmental Panel on Climate Change—in 1988.

The task assigned to IPCC was to study and synthesize the likely changes that would occur as a result of greenhouse gas accumulation and the ensuing global climate change. The panel has a threefold task: to assess available scientific information on climate change, to assess the environmental and socioeconomic impacts of climate change, and to formulate strategies by which mitigating such changes could be addressed. In 2007, the IPCC issued its fourth rigorously researched and extensively peer-reviewed report, which has been cited frequently in this chapter.[6]

Although the IPCC defines climate change as any sort of change in the Earth's climate regardless of the cause, it has found overwhelming evidence of the significant increase in GHGs from human activities, particularly fossil fuel use, land use change, and agriculture. The IPCC has provided a great impetus to the further development and refinement of the GCMs that predict a temperature increase of 1.0–4.5°C by the end of this century.[7] Few scientific project reports can lay claim to hundreds of authors and hundreds of peer reviewers. For its exemplary work, the IPCC shared the Nobel Peace Prize in 2007.

Notes

1. The cycles are named after Milutin Milankovich (1879–1958), a Serbian astrophysicist and mathematician who devoted his life to developing a mathematical theory of climate.
2. http://www.homepage.montana.edu/~geol445/hyperglac/time1/milankov.htm
3. Reading climatic shifts through ice cores taken from glaciers in different parts of the world has come from the pioneering work of Lonnie Thompson and Ellen Mosely-Thompson at the Ohio State University. Beginning in 1974 with the first exploration of the Quelccaya ice cap in Peru, Lonnie Thompson visited the same glacier again in 1983 to obtain the first deep core of a tropical glacier. Since then, he extended his work to many other glaciers obtaining cores that bear yearly records of dust and stable oxygen isotopes. Collectively from these studies, he has reconstructed a climatic history from such glacial records. He is convinced that human activities have considerably altered the climate; the rapidity with which the glaciers are melting bear testimony to his conclusions (see Krajick, K. 2002).
4. "1. Meta-analysis is a statistical technique for amalgamating, summarizing, and reviewing previous quantitative research. By using meta-analysis, a wide variety of questions can be investigated, as long as a reasonable body of primary research studies exists. Selected parts of the reported results of primary studies are entered into a database, and this 'meta-data' is 'meta-analyzed,' in similar ways to working with other data—descriptively and then inferentially to test certain hypotheses. 2. Meta analysis can be used as a guide to answer the question 'Does what we are doing make a difference to X?' even if 'X' has been measured using different instruments across a range of different people. Meta-analysis provides a systematic overview of quantitative research that has examined a particular question. 3. The appeal of meta-analysis is that it in effect combines all the research on one topic into one large study with many participants. The danger is that in amalgamating a large set of different studies, the construct definitions can become imprecise and the results difficult to interpret meaningfully." See http://wilderdom.com/research/meta-analysis.html.
5. John Harte, public lecture at The Ohio State University, June 27, 2006.
6. United Nations Framework Convention on Climate Change (full text) (March 23, 2007) http://www.eoearth.org/article/United_Nations_Framework_Convention_on_Climate_Change_(full_text)
7. (www.ipcc.ch). This Web site is an excellent source for the study of climate change.

Questions

1. Compare carbon sequestration capacities for different biomes and different stocks within ecosystems.
2. What are the most important greenhouse gases and how are their levels influenced by human activity? Rank them in terms of their residence times in the atmosphere and their climate forcing values.

3. What are the likely consequences of climate change for agricultural, forested, and coastal regions? Discuss possible responses to climate change by plants, animals, microbes, and human societies.

4. What information do we need to better understand natural climate variability and extended weather forecasting so that we can better reconstruct past climates and predict future climates?

5. Does global climate change have the same impact across the globe? Do nations have the same level of responsibility in contributing to climate change? What implications do these considerations have for reaching a solution?

Biological Diversity Loss and the March of Invasive Species

20

Biological Diversity

Ever since the dawn of history, the nature of life and the variety of life forms found in the tropical jungles of the Amazon, in the high-altitude forests and alpine systems of the Himalayas, in the coniferous boreal forests of Canada and northern Europe, in the hot and cold world deserts, on the coral reefs and other marine life—indeed life all over the Earth—have fascinated the nonprofessional and professional alike. To both groups, the spectacular variety of plants and animals; their forms, sizes, and functions; the places where they grow; their intricate relationships to their own kind and to other organisms; and their overall exchange with the physical components of the Earth have been intriguing. Humans at large have marveled at the beauty and aesthetics, but professional ecologists have always posed the question: Why do populations and species grow where they do? The quest to understand the why, where, and how of growth in particular environments, we believe, shall remain unquenched in the future as it has been throughout the history of ecology.

The variety of all of the Earth's life forms is collectively referred to as **biological diversity,** or **biodiversity.** For biologists, a species—plant or animal—is a group of individuals that possess characteristic common morphological, anatomical, and physiological features, and they interbreed or have the potential to do so. Worldwide thus far, biologists have described a total of between 1.5 and 1.8 million species; hence, we use the term "species diversity." Experts believe that many more species have yet to be discovered and described, with estimates of the potential number of species ranging from 3.6 million to more than 100 million (Table 20.1).

Although the recognition of **species diversity** is useful from a scientific and practical standpoint, in nature, many species have subspecies and many diversified populations that occupy different (isolated) habitats. Hence, populations are the groups of focus for ecologists (see Chapter 8), rather than species. We mention this to emphasize that the diversity of organisms is much larger than may

Table 20.1
Estimated and Recorded Numbers of All Species and Species That Have Become Extinct since 1600

Selected Groups of Organisms	Potential Numbers of Total Species	Recorded Numbers of Total Species	Recorded Numbers of Extinctions since 1600	Extinct (%)
Insects	8,00,000	100,000	73	0.007
Bacteria	1,000,000	4,000		
Arachnids	750,000	75,000		
Viruses	400,000	5,000		
Chordates	50,000			
Mollusks	200,000	100,000	286	0.29
Crustaceans	150,000	40,000	4	0.10
Protozoans	200,000	40,000		
Nematodes	400,000	15,000		
Algae	200,000	40,000		
Vertebrates	50,000	47,000	257	0.55
Fishes		24,000	36	0.15
Birds		9,500	114	1.20
Reptiles		6,000	20	0.33
Mammals		4,500	83	1.84
Amphibians		3,000	4	0.13
Total animals	13,000,000	1,151,000	620	0.05
Gymnosperms		758	2	0.26
Dicotyledons		190,000	120	0.06
Monocotyledons	400,000	52,000	462	0.89
Palms		2,820	4	0.14
Total plants		245,578	588	0.24
Grand total	13,400,000	1,396,578	1208	0.09

Source: Combined from Stork, N. E. 1999. In *The Living Planet in Crisis: Biodiversity Science and Policy*, ed. J. Cracraft and F. T. Grifo, 3–33. New York: Columbia University Press; Groombridge, B., ed. 1992. *Global Biodiversity: Status of the Earth's Living Resources*. London: Chapman & Hall.

be reflected by the term "species diversity." Given that the total species complement of the world numbers as high as 3.6 to 100 million, if their subspecies and populations were to be counted, they may number a thousand times more (Myers 1996a). Thus, when species, subspecies, and populations are all considered, the importance of biological diversity at all spatial scales that range from local to regional to global becomes apparent.

The expression of biological diversity is fundamentally based on the genetic composition of animal, plant, and microbial populations; hence, **genetic diversity** should logically be the starting point in understanding the dimensions of the issues of biological diversity. For reasons of convenience and scientific investigation, considerations of diversity can be analyzed by recognizing three of its components. **Composition** diversity includes the commonly recognized populations and species and the ecosystems of which they are a part. **Structural diversity** pertains to the spatial arrangement of physical units. For example, structural diversity of trees at the stand level can be characterized by the number of vertical strata within the forest (e.g., overstory canopy, subcanopy, shrubs, herbaceous plants). At the landscape level, the distribution of age classes in a forest or the spatial arrangement of different ecosystems provides a measurement of structural diversity. The third component, **functional diversity,** represents variation in ecological processes such as nutrient cycling or energy flow. The processes driving the latter component are the most difficult to measure and understand and require long-term ecological studies. Other aspects of biodiversity are discussed in Chapter 25.

Values of Safeguarding Biological Diversity

In the preceding chapters, we have learned of the major changes that human activities have brought about in nearly all ecosystems of the world. The conversion of land for food production, grazing lands, and pastures; the overharvesting of lakes, seas, and oceans; the release of heretofore unprecedented loads of chemicals into the atmosphere, and urbanization and road-building have resulted in the degradation of habitats. Degradation and, often, the complete destruction of habitats leads to the exclusion of organisms from habitats to which they have adapted over the course of their evolution. Given the unprecedented growth of the human population and the resultant resource conversions, such disturbance events will continue and further modification, deterioration, and destruction of habitats and a loss of biological diversity will result.

Overall, the loss of biodiversity of an ecosystem may diminish its resilience to disturbance, increase its vulnerability to diseases, and decrease its productivity. Noting that the Earth is going through an abrupt transition under the impact of human degradation of many habitats, including the tropics, Woodwell (1990) noted:

The current tragedy of the tropics is not simply the loss of species, but the transformation of a highly productive, self-maintaining landscape of great versatility and considerable resilience into a barren landscape of limited potential for support of life, including people. The tragedy is compounded by the widespread assumption that human interests are advanced by the transformation.

Therefore, in assessing the importance of any natural population, species, or community, one's attention is invariably drawn to its economic valuation. That is, what is the economic justification for saving and safeguarding biological diversity? As should be clear from the previous discussion, the protection of biodiversity has two important economic features: the direct usefulness of species of economic importance and the loss of economic value due to species that are able to invade habitats, causing the loss of economically important species.

The economic worth of individual species and populations that supply food, fodder, fiber, timber, and medicinal and industrial materials is well known and may be illustrated by two examples. First, global food production alone was worth US$1.3 trillion in 1997. The improvement of food crops is largely dependent on incorporating genetic material from wild varieties, and ongoing scientific investigations of wild varieties from all over the world (e.g., corn, tomatoes) bear testimony to the value of nature's genetic diversity.

The second important example comes from the pharmaceutical industry, which is largely dependent on plants and animals for the extraction of new drugs. In 1984, U.S. consumers spent more than $8 billion for prescriptions whose active ingredients came from plants. Drugs extracted from plants are contained in 25% of all prescriptions, and about 120 "pure chemical substances extracted from higher plants are used in medicine throughout the world" (Farnsworth 1988). The most recent and well-publicized illustration is provided by the production of Taxol® from extractions from Pacific yew (*Taxus brevifolia*). This compound was discovered in the early 1990s and has been characterized "as one of the most promising anticancer drugs to be discovered in 20 years" for treating breast cancer (Hansen 1999). It is no surprise that the pharmaceutical industry all over the world is investing billions of dollars in the research and development of new products from plants, bacteria, and animals.

The pollination of flowers by honeybees illustrates the double role of some species in maintaining and ensuring ecosystem function while they contribute extensively to the human economy. Beginning in the 1990s, honeybees in the United States were first infected by a mite (*Varroa jacobsoni*) that feeds on the developing bee larvae and usually causes death of untreated colonies within a year or two. In fact, most feral bee populations have likely died without the protection of a beekeeper. The resulting decline in the bee population in North America caused alarm by scientists and farmers alike, but it also caused them more formally to assess the overall contributions of insect pollination for crops. Vegetables such as asparagus, carrots, celery, onions, radishes, and turnips; vine crops such as melons, squash, pumpkins, and cucumbers; and fruits such as sweet cherries, apples, plums, prunes, and nuts (especially almonds) all require cross-pollination, preferably by bees. Worldwide, more than 390 species and varieties of economically important plants are pollinated by bees (Dana Wrensch, the Ohio State University, personal communication). For the United States alone, a loss of

$9–19 billion was estimated from lack of pollination by the bee populations in the 1990s.

In summary, there is widespread agreement within the scientific community that extensive habitat loss and fragmentation decrease the average species' lifetime and, as a consequence, drastically increase the rate of extinction (by a factor of several hundred times the rate of the geological extinctions). Many species have become so rare that they may be lost forever (Pimm and Brooks 1997). Biological diversity has numerous values, ranging from its ecological and evolutionary functions to economic values and ethical and aesthetic considerations (Table 20.2). To safeguard biodiversity, humans will have to choose actively to do so.

Biological Invasions

In addition to the loss of habitats and direct effects on organisms of air, water, and soil pollution, yet another subtle and insidious agent brings about and/or accelerates the loss of species: biological invasions of non-native species. Non-native, nonindigenous, or exotic species that invade an area and persist to the detriment of native species are referred to as **invasive species.** A U.S. presidential executive order

Table 20.2
Some Values and Benefits of Preserving Biodiversity

Value Category	Explanation
Economic	As yet unidentified species may provide valuable food, fiber, drugs, and other products for human use
Evolutionary potential	The genetic diversity contained in all species constitutes the basis for future natural evolution and artificial breeding/selection programs
Natural laboratory	Natural ecosystems with their full complement of species represent museums and laboratories for the study of the Earth's natural history
Aesthetic	Natural landscapes and wild species provide many amenities and recreational values to the public
Ethical	Humans have a moral responsibility to be stewards of the natural environment and protect all species
Ecosystem integrity/function	Diversity must be maintained in order to preserve critical ecosystem services and the integrity of the Earth's life-support system

Source: From Pitelka, L. F. 1993. In *Biodiversity and Ecosystem Function,* ed. E-D. Schulze and H. A. Mooney, 481–493. New York: Springer–Verlag.

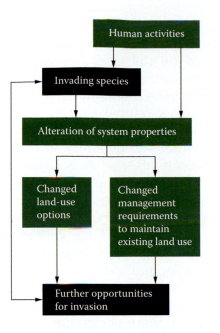

Figure 20.1 Alteration of land use by human activities provides a foothold for invasive species. (Hobbs, R. J. 2000. Land-use changes and invasions. In *Invasive Species in a Changing World*, ed. H. A. Mooney and R. J. Hobbs, 55–64. Washington, D.C.: Island Press.)

(1999) defines invasive species as "plant, animal, or microbial species that are not native to the United States or to the affected area (i.e., an ecosystem—a community of organisms and their environment) and whose introduction causes harm to the economy, the environment, or human health." The human-driven introduction of invasive plants, animals, and microorganisms into new regions where they adversely affect the structure and function of native species and ecosystems is referred to as **biological invasion.** Invasive species are also facilitated by changing land use (Figure 20.1). It is fairly clear from an impressive body of ecological literature that the loss of native species and the spread of exotic species are problems so closely related that one cannot be considered without considering the other; they are indeed two sides of the same coin.

Major Factors That Facilitate Invasions

Major factors that contribute to biological invasions include arrival of non-native species in new locales without their natural associates, low species richness in recipient ecosystems, disturbance, and human assistance in their dispersal. Escape from native biotic constraints such as competitors, predators, grazers, and parasites can enable introduced species to proliferate and spread easily and quickly in their new habitats. Habitats with relatively low biodiversity are more vulnerable to the successful establishment of non-native species due to the availability of vacant niches. Low species richness can result from biogeographical barriers such as oceans, mountain ranges, rivers, and inhospitable climate zones (as in the case of oceanic islands) and from human-induced disturbances. Humans can incidentally or deliberately facilitate biological invasion by disrupting the structure and function of ecosystems, altering natural disturbance regimes of ecosystems, dispersing organisms across biogeographical barriers, or simply favoring some species over others through technological means, such as in the cases of aquaculture and agriculture.

Invasive species alone are estimated to have contributed to the decline of 42% of species on the U.S. threatened and endangered species list, lowering biodiversity and the function of ecosystems. Currently, the introduction of invasive species is second only to the destruction of habitats in its contribution to an unprecedented loss of biodiversity at a global scale. Non-native species can be introduced accidentally or deliberately by humans from their native ranges into new regions through facilitated migration, direct transport, and commerce (e.g., Table 20.3). Humans help the invaders cross numerous barriers (Elton 1958; Ehrlich 1986). For example, the brown tree snake (*Boiga irregularis*), a native Australian species, reached Guam, an island with no snakes, likely in a ship's or airplane's cargo. It has been accused of reducing the biological diversity by eating the lizards and birds on the island (Fritts and Rodda 1998), and specimens have fallen out of the wheel wells of airplanes landing on Hawaii, another island without snakes.

Humans have thus been a most significant agent in the spread of invading species, either introducing them deliberately for use as forage or horticultural purposes or involuntarily by acting as carriers in their modes of transport

Table 20.3
Modes of Weeds' Entries into North America

By ballast in ships

By impure seed crop

By adhesion to domesticated animals

In surrounding soils and roots of nursery stock

Deliberate introduction as:

  Forage plants

  Fiber plants

  Medicinal plants

  Ornamentals

  Erosion controls (e.g., kudzu [*Pueraria*] from Japan); timber plantations (e.g., introduction of Australia's *Melaleuca* into Florida and *Eucalyptus* into California)

Source: From Baker, H. G. 1988. In *Ecology of Biological Invasions of North America and Hawaii*, ed. H. A. Mooney and J. A. Drake. New York: Springer–Verlag.

(Table 20.3). Travel around the world has increased by several orders of magnitude in the last four or five decades and will continue to grow, so the transfer of species from one region to the other is also going to grow.

Temporal Attributes of Invasive Species

Non-native species can go through a time lag after introduction before becoming invasive. This temporal lag may be brief (weeks) or quite long (decades). Examples of biological invasions with a brief lag phase include Africanized bees in the Americas and Eurasian zebra mussels in the Great Lakes. Africanized bees (also known as "killer" bees due to their aggressive nature and irritability) were accidentally released in the mid-1950s when a swarm escaped from a hybridization research project in Brazil, and they have since been moving northward at about 320–480 km (200–300 miles) per year. They reached Los Angeles in 1998, colonizing more than 88,000 km² of Southern California, including Orange, Imperial, Riverside, and San Bernardino counties and parts of San

Diego County. They produce much less honey and wax than do European honeybees and are not calmed by the smoke canisters used by beekeepers to control bees in their hives.

The Eurasian zebra mussel (*Dreissena polymorpha*) and a related species, the "quagga" mussel (*D. bugensis*), from the Caspian Sea were introduced accidentally in ballast water from ships into all the Great Lakes as well as several river systems and lakes in the eastern United States about 1986 and thereafter. They attach to hard surfaces with sticky "byssal" threads and have caused problems such as clogging major pipes of treatment facilities for human water consumption as well as cooling systems for power plants. They have also smothered populations of native unionid mussels, and their shells have caused serious cuts in humans swimming and wading in habitats where they are abundant. Because few native animals feed on these dreissenid mussels, their populations of extremely efficient grazers on plankton are thriving, depriving native mussels of their food supply.

Zebra mussels have been reported to have spread to 19 states in less than 10 years from their original locations in Lake St. Clare and Lake Erie. The mussels were distributed thereafter in large part by transport on commercial barges traveling on navigable rivers, reaching New Orleans via the Mississippi River and New York state via the New York State Canal System and the Hudson River (Pimentel 2005). Others have been intentionally introduced into quarries by recreational scuba divers in the ungranted wish that they would "clear up" the water. Dreissenids illustrate many characteristics of invasive species, including high reproductive potential (as many as 1 million eggs produced per year by a 20-mm female), high genetic variability, and the ability to survive out of water for long periods and to thrive in a range of water salinities from fresh to brackish (20% as salty as sea water).

On the other hand, Brazilian pepper trees (*Schinus terebinthifolius*) illustrate a much slower invasion. Introduced for roadside planting in Florida in the nineteenth century, in part because their green leaves and red berries look somewhat like American holly, they did not become widely noticeable until the early

1960s. They are now established on more than 28,000 ha in South Florida in dense stands, excluding all other vegetation, increasing water loss through increased transpiration, and intensifying fire risk through their production of flammable plant and litter biomass.

Effects of Invasions on Ecosystem Structure and Function

From the preceding examples and some more that follow, it should be clear that species invasions are numerous and worldwide. Within vascular plants alone, invasive species currently represent a large fraction of plant communities (Table 20.4). A few additional invasion examples point to the mechanisms by which these species come to dominate an area over time, to the partial or complete exclusion of indigenous and endemic species.

Deforestation of the tropical rain forests has received much attention. Among the reasons for such deforestation—for example, in the Brazilian states of Rondonia and Eastern Acre in the Amazonia area of Brazil—have been the establishment of small farms and large cattle ranches and land speculation within the rain forest. These activities are always accompanied by human population increases, building of infrastructure such as roads, and an emphasis on short-term agricultural profits. These, in turn, have led to more speculative land development and increased deforestation and high inflation in Brazil. As a result, the losses of biological diversity have been very significant.

Naveh and Kutiel (1990) studied the Mediterranean region, where there is a long history of human land use, and the resulting biotic impoverishment. The preagricultural period during the Pleistocene era included a long period of coevolution for the people, flora, and fauna of the Mediterranean region, including the domestication of plants and animals. This period was followed by "a very long prehistoric and historic agricultural period, during which agropastoral cultural landscape was shaped,

reached its peak, then gradually declined, and seminatural vegetation formations were maintained in a metastable state" (Naveh and Kutiel 1990). But the expansion of the present period of agricultural intensification, urban–industrial development, and population growth has been unprecedented in causing the losses of biodiversity as well as the historical sites in the Mediterranean landscape.

For example, the original oak and conifer forests were cut down, and erosion from agriculture, grazing by feral goats, and the impacts of frequent fires decreased the plant and animal diversity. To compound this problem, biodiversity losses are predicted to be among the highest in Mediterranean regions of the world as a result of global change (Sala et al. 2000). "Nowhere else has it been more convincingly demonstrated [but in the denuded Mediterranean uplands that] humanity has not only the power to destroy its habitat and deplete its flora and fauna, but also, with sufficient motivation and skill, to reclaim it" (Naveh and Kutiel 1990, pp. 259–299).

A few examples from the United Sates are also very instructive. The first is the well-documented invasion of European cheatgrass (*Bromus tectorum*) into the plant communities of the Great Basin in Utah and Nevada that were originally dominated by sagebrush (*Artemisia tridentata*), a microphyllous shrub, bunchgrass, and other herbs. A thorough historical and ecological documentation reveals that cheatgrass, an annual-grass invader, first made its appearance in the late 1880s along railroads and eventually along highways; it was then spread by sheep, cattle, and horses as they were driven to rangelands (it was even spread by automobile tires). Once established, upon its death in late May or early June of each year, cheatgrass dries and becomes an excellent fuel for fire. Wildfires had been rare in the original open bunchgrass–sagebrush vegetation in the nineteenth century, but more cheatgrass meant more fires. Even without fire, cheatgrass has been able to invade overgrazed sagebrush, dominating and replacing the original diverse communities (Billings 1990, pp. 301–322).

California, perhaps the richest North American state in biologically diverse species

Table 20.4
Relative Number of Native and Invasive Non-Native Species of Vascular Plants in Various Parts of the World, Including Islands[a]

Location	Number of Native Species	Number of Non-Native Species	Non-Native Species (%)	Location	Number of Native Species	Number of Non-Native Species	Non-Native Species (%)
Alaska	1,229	144	11	Hawaii	956	861	47
Australia	15,000–20,000	1500–2000	9	Italy	5,599	294	5
Bahamas	1,104	246	18	Mediterranean	23,000	250	1
Bermuda	165	303	65	Mexico	14,140	639	4
Britain (+ Ireland)	1,623	442	21	Namibia	3,159	60	2
California	4,844	1025	18	New Caledonia	3,250	500	13
Canada	9,028	2840	24	New England	1,995	877	31
Cook Islands	284	273	49	Newfoundland	906	292	24
Cuba	5,790	376	6	New Zealand	1,790	1570	47
Czech Republic	2,288	194	8	Poland	1,958	293	13
Denmark	1,250	239	16	Puerto Rico	2,741	356	12
Egypt	2,815	86	3	Rwanda	2,500	93	4
Europe	11,000	1568	13	Senegal	1,980	120	6
Florida	4,994	1210	20	Solomon Islands	3,172	200	6
Fiji	1,628	1000	38	South Africa	20,263	824	4
Finland	1,006	221	18	South Africa	8,500	379	4
France	4,200	438	9	Switzerland	3,030	280	9
French Polynesia	959	560	37	Tanzania	1,940	19	1
Germany	1,718	429	20	Texas	4,498	492	10
Germany	2,572	339	12	Turkey	8,575	79	0.9
Great Plains	2,495	394	14	Turkey	1,245	86	6.5
Greenland	427	86	17	Uganda	4,848	152	3.0
				United States	—	2000	—

Source: From Vitousek, P. M. et al. 1996. *American Scientist* 84:468–478.
[a] Synthesized from multiple sources.
[b] Islands are particularly prone to invasion.

(5,720 vascular plants, more than 200 mammals, 28,000 insect species, 525 birds, 129 amphibians and reptiles, and 132 inland fishes), has been characterized as an area of unusual biotic diversity. Its wide range of natural ecological communities of forest, woodland, scrub, grassland, and aquatic ecosystems has been seriously affected by both habitat destruction and invasive species. The replacement of native grasslands such as perennial bunchgrasses (e.g., *Stipa pulchra*) by invading species (particularly those imported and grown as forage, such as wild oats, wild ryegrass, brome grass, and clovers) has been called "most dramatic" (Mooney, Hamburg, and Drake 1986).

A more spectacular example involves the impact of an introduced species on a single tree species, the American chestnut (*Castanea dentata*) in the eastern United States (Ronderos 2000). Its decline changed not only its own growth form but also the relationships among all plants and animals growing in the chestnut ecosystem. Once a dominant tree of the eastern forest communities, the chestnut has been documented to have comprised 25% of the forest from Massachusetts to Alabama and westward to Ohio and central Tennessee. These trees could exceed a height of 33 m (100 ft), attain an age of 600 years, and reach an average diameter of 1.5 m (5 ft), although individual trees reached 2.6–3.3 m (8–10 ft) in diameter. The tree was an important industrial species as fuelwood; for furniture, poles, pilings, posts, and railroad ties; for tannin extracted from its bark; and, of course, for its "copious production of sweet nuts." It contributed more than $10 million annually to the local economies during 1907–1910, and that reached $16 million in 1920.

Then the catastrophe struck. A fungus (*Cryphonectria parasitica*), believed to be on the nursery stock imported from China in about 1905, infected the chestnut. The fungus enters through holes and cracks in the bark and spreads through the living portions of the bark, ultimately girdling the tree and thus cutting off the nutrient flow from the roots to branches and leaves until the tree dies. It is reported to have spread quickly by 1950, and the tree had disappeared from 36,420 km^2 (9 million acres) of the eastern deciduous forest. The tree is tree no more because infected trees only produce small sprouts from their roots. Before they can reach maturity, the fungus again kills the branches and leaves. Here, then, is a remarkable example showing how one invading species, a pathogen, reduced a tree to nothing more than a shrub; altered the entire community structure of the American chestnut forest, including its wildlife relations; and created an economic nightmare for those whose livelihood depended upon it. In addition, replacement of the chestnuts by red oaks increased the occurrence of oak wilt disease in many native oak communities. Similarly, Dutch elm disease entered the United States from Europe in about 1930 in imported elm logs and led to the destruction of more than 4 million elm trees between 1933 and 1940. In an attempt to control the elm bark beetles that spread the disease, the insecticide DDT was used to spray elm trees in the 1950s; this not only was ineffective, but also spread the disease and resulted in widespread poisoning of birds, especially robins.

Other documented examples of invasive species and the toll they take on particular ecosystems are now abundant. They range from algae and higher aquatic plants choking streams, lakes, and seas (Table 20.5) to agricultural fields with resistant weeds and invaders of the forest. The construction of the Welland Canal in 1883 allowed ships to circumvent Niagara Falls between Lakes Erie and Ontario, but it also allowed the sea lamprey, a native parasitic fish in the Atlantic Ocean, to enter the Great Lakes. Its subsequent spread, destroyed 97% of the trout population in the Great Lakes between 1921 and 1940. The intentional introduction in the 1950s of a very large (to 200 kg) predatory fish, Nile perch (*Lates niloticus*), into Africa's Lake Victoria, the largest freshwater lake in the world and the source of the Nile River, caused the extinction of 300 species of indigenous fish in the 1980s. This exotic species has eliminated a major food source for 30 million people, fundamentally altering the fishery and fish industries.

On many oceanic islands, feral and domestic cats have become devastating predators of small native mammals and ground-nesting birds, causing the extinction of at least six species of endemic birds in New Zealand and of at least six species of native, rodent-like Australian marsupials. Goats and donkeys introduced to St. Helena Island (west of Angola) and the Galapagos Islands, Ecuador, not only caused the extinction of endemic plant and animal species but also altered the islands' hydrological cycle due to their grazing and trampling. Similarly, an invasion of shrub communities in South Africa's Cape Province by eucalyptus, pine, and acacia trees has dried up and reduced the flow of entire river systems, threatening the agricultural productivity of the region and leading to the extinction of many endemic plant genera such as *Protea*. An invasion of Hawaii Volcanoes National Park by a tall shrub, *Myrica faya* (native to the Canary Islands), is changing the entire structure and function of the ecosystem because the invader fixes nitrogen, increasing

Table 20.5
Examples of Invasive Marine and Freshwater Species Introduced from Their Native Regions to Multiple Regions in the World

Taxon	Source Region	Recipient Region
Annelida		
Ficopomatus enigmaticus	SWP	MED, HAW, NEA, NEP, NWA, SWA
Coelenterata		
Haliplannella luciae	NWP	MED, NEA, NEP, NWA
Cordylophora caspia	CAS	AUST, BALT, MED, NEA, NEP, NWA
Chordata		
Molgula manhattensis	NWA	AUST, NEP, NWP
Styella clava	NWP	AUST, NEA, NEP, NWA
Acanthogobius flavimanus	NWP	AUST, NEP
Tridentiger trigonocephalus	NWP	AUST, NEP
Crustacea		
Balanus improvisus	NWA	AUST, BALT, NEP, NWP
Mytilicola orientalis	NWP	MED, NEA, NEP
Oithona davisae	NWP	NEP, SEP
Pseudodiaptomus marinus	NWP	HAW, NEP, IO
Carcinus maenas	NEA	AUST, NEP, NWA, SAFR
Eirorchier sinensis	NWP	BALT, MED, NEP
Paaemon macrodactylus	NWP	AUST, NEP
Rhithropanopeus harrisii	NWA	BALT, CAS, MED, NEA, NEP
Sphaeroma walkeri	IO	AUST, HAW, MED, NEP, NWP, SAFR
Mollusca		
Dreissenia polymorpha	CAS	BALT
Musculista senhousia	SWP	AUST, MED, NEP, NWA
Mya arenaria	NWA	BALT, NEP
Crepidula fornicata	NWA	MED, NEA, NEP

Source: Modified from Grosholz, E. D., and G. M. Ruiz. 1996. *Biological Conservation* 78:59–66.

Notes: AUST: Australia; BALT: Baltic Sea; CAS: Caspian Sea; HAW: Hawaii; IO: Indian Ocean; MED: Mediterranean; NEA: northeast Atlantic; NEP: northeast Pacific; NWA: northwest Atlantic; NWP: northwest Pacific; SAFR: South Africa; SWP: southwest Pacific.

its supply in the nitrogen-poor volcanic soils at a rate that is 90-fold greater than that available to native plants. The availability of *Myrica* seeds paves the way for invasions of other non-native species in Hawaii such as the introduced Japanese white-eye, the most destructive invasive bird species on the island and a competitor of several native bird species in native forests. This bird in turn facilitates the dispersal of *Myrica* seeds.

Zebra mussels (discussed earlier) also illustrate what Simberloff and von Holle (1999) call **invasional meltdown,** in which the successful introduction and establishment of one nonindigenous species (e.g., the zebra mussel) facilitate the establishment of other species from their home communities, including those that attach to them (the quagga mussel), feed on them (the round goby fish), hide under them (an amphipod crustacean), or perhaps parasitize them. (Could the parasite of the European pike perch that uses zebra mussels as an intermediate host come to infect the North American walleye fish species?)

Fire-prone invasive species such as palatable African grasses, Asian cogongrass, Australian *Melaleuca*, and Australian pine can increase the frequency of forest and grassland burning and contribute to global climate change by releasing more CO_2 into the atmosphere and sequestering less carbon. Through crossbreeding (hybridization), invasive species can also drive native species to extinction and reduce the number of new offspring added to the species' own population. For example, North American mallard ducks introduced for hunting threaten the existence of both the Hawaiian duck and the New Zealand gray duck due to hybridization. North American cordgrass accidentally introduced to southern England hybridized with the native cordgrass there, producing a fertile, highly invasive cordgrass species. Hybridization of the rainbow trout introduced into western watersheds of the United States endangered two endemic species, the Gila trout and the Apache trout.

How Do Species Invade So Successfully?

Invasive species expand their ranges through a number of means. Of the total number of invading species, most are plants and not animals. This is in part due to their powers of dispersion by wind and the fact that animals tend to require more deliberate introductions to become established than do plants (Vitousek et al. 1996). Once established, plant invaders are often capable of (1) obtaining nutrients and water under a wide range of conditions, (2) growing rapidly, and (3) producing a large number of seeds that are able to germinate and become established quickly (Table 20.6).

Table 20.6
Characteristics of Successful Invasive Plant Species

Rapid growth and early flowering

Able to cope with a range of nutritional conditions

High phenotypic plasticity

Different growth forms depending on the ecosystem type

Large numbers of seeds produced that are easily dispersed

Seed set in a wide range of temperatures and photoperiods

Germination "polymorphisms" (i.e., some germinate early, some late)

Source: Compiled from Pimentel, D. 1986. In *Ecology of Biological Invasions of North America and Hawaii*, ed. H. A. Mooney and J. A. Drake, 149–162. New York: Springer–Verlag.

Discussing the attributes of the invasive species garlic mustard (*Alliara petiolata*), Rodgers, Stinson, and Finzi (2008) provide a comprehensive view of the mechanisms that contribute to the success of this invasive species; these range from phenotypic plasticity and high reproductive output to synthesis of allelochemicals and high competitive ability (Figure 20.2). Equally, the species' impacts occur directly on indigenous species themselves and indirectly through the soil, diminishing their and overall ecosystem productivity. Such analyses could apply to a large number of invasive species.

Successful animal invaders tend to have similar characteristics, including (1) short generation times; (2) a great deal of genetic variability and, like plants (3) the ability to function in a wide range of environmental conditions (Table 20.7). One species can displace another in various ways without direct interaction. One of the most obvious ways is niche invasion: A new bird can use all the food or all the nesting sites of a previously established bird.

Among plants, competitive exclusion can take the form of shading out the seedlings of a less vigorous predecessor. A dramatic example is kudzu (*Pueraria lobata*), a fast-growing leguminous vine. Introduced from Japan to control erosion in the southeastern United States, kudzu has literally blanketed thousands of hectares of successional forest, shading even mature trees. These kinds of dominance can

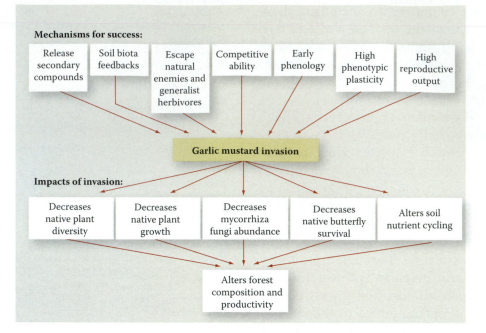

Figure 20.2 Mechanisms by which the invasive species garlic mustard achieves success and the impacts of its invasion. (Redrawn from Figure 1 in Rodgers, V. L., K. A. Stinson and A. C. Finzi. 2008. *BioScience* 58:426–436.)

Table 20.7
Characteristics of Successful and Unsuccessful Invasive Animal Species

Successful Invaders	Unsuccessful Invaders
Abundant in original range	Rare in original range
Polyphagous (eat a diversity of foods)	Monophagous or oligophagous
Short generation times	Long generation times
Great genetic variability	Little genetic variability
Fertilized female able to colonize alone	Fertilized female not able to colonize alone
Larger than most relatives	Smaller than most relatives
Associated with *Homo sapiens*	Not associated with *Homo sapiens*
Function well in a wide range of physical conditions	Function well only in a narrow range of physical conditions

Source: From Ehrlich, P. 1986. In *Ecology of Biological Invasions of North America and Hawaii,* ed. H. A. Mooney and J. A. Drake, 79–95. New York: Springer–Verlag.

be viewed as a plant version of "territoriality" (Lande 1987). Plants can compete for more than just space, water, or nutrients. For example, the introduced purple loosestrife (*Lythrum salicaria*), a popular garden perennial, has spread to the wetlands across eastern North America (Figure 20.3). It has now been established that this exotic species attracts native pollinators more efficiently than does the native winged loosestrife, *L. alatum* (Brown et al. 2002). Plants can also "compete" for agents of dispersal: The bright red fruits of introduced honeysuckle (*Lonicera*) shrubs can attract

birds that would otherwise prefer the fruits of native shrubs, vines, or trees.

Does disturbance affect the establishment of invasive plant species? Westman (1990) performed a fumigation experiment that showed that air pollutants reduced the abundance of native species of a coastal sage community and increased the relative abundance of exotics. When species in the sage community were exposed to ozone, sulfur dioxide, or a combination of the two, premature senescence of leaves and pollution-induced water stress, followed by leaf drop, occurred, eventually causing

Figure 20.3 The non-native species purple loosestrife has showy, purple flowers and the ability to displace native wetland plant species. (Photo courtesy of the National Park Service.)

plant death from water loss. The space left open by the death of sage was occupied by an exotic annual grass (*Bromus rubens*) that had evolved a pollution-tolerant strain after living for 25 years downwind of a source of industrial pollution. "In the Riverside–San Bernardino Basin, which collects pollutants blown inland from Los Angeles, the richness of native species has declined, and exotic annuals have increased" (Westman 1990). Grazing, fire, and chronic air pollution exacerbate the situation.

Invasions as Agents of Environmental Change

Current human activities facilitate biological invasions that are occurring at an unprecedented pace and constitute a threat to global biodiversity. The invasions exert a domino effect not only on abundance, composition, and kinds of species present, but also on the hydrological cycle, gaseous cycles, and sedimentary cycles and on the ecosystem goods and services on which well-being of humans depends

(Figure 20.4). Species invasions negatively impact ecosystem functions such as those noted by Vitousek (1986); and Vitousek et al. (1996):

- extinction of native species due to breakdown of biogeographical barriers (e.g., oceans and mountain ranges), competitive elimination, grazing, predation, and hybridization;
- homogenization of ecosystems with an increased vulnerability to pathogens, parasites, and diseases;
- emergence and spread of new pathogens and diseases;
- irreversible changes in species composition and abundance because many invasions are difficult and expensive to eradicate, if not impossible;
- changes in vital ecosystem processes such as primary productivity, decomposition, movement of nutrients through the ecosystem, geomorphology (e.g., dune formation, soil erosion, and drainage patterns), successional development, and decomposition and mineralization of soil organic matter (e.g., litter quality);

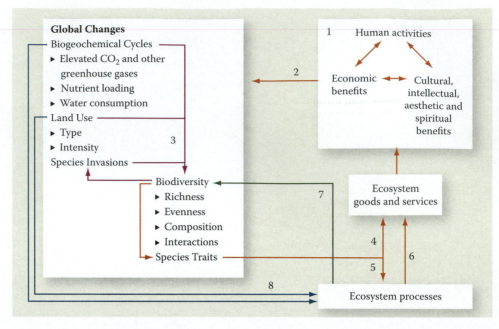

Figure 20.4 The links between biodiversity and human activities that lead to global change. Numbered pathways indicate that: 1) Human actions are influenced by economic, cultural, intellectual, aesthetic, and spiritual interests, 2) Human activities, through a variety of mechanisms, contribute to global changes including changing biodiversity and altered land use, which in turn can 3) lead to species invasions. 4) Changing biodiversity leads to altered species traits and therefore 4) the ecosystem goods and services that provide direct economic benefits and contribute to human well-being. 5) Ecosystem processes are also affected by changes in biodiversity, and hence 6) ecosystem services. 7) Changes in ecosystem processes 'feedback' to cause further changes in biodiversity. Finally, (8) the effects of global changes in biogeochemical cycles and patterns of land use may directly alter ecosystem processes. In sum, this system demonstrates that biodiversity losses can directly diminish human welfare. (Redrawn from Chapin, F. S. III, E. S. Zavaleta, V. T. Eviver, et al. 2000. Consequences of changing biodiversity. *Nature* 405: 234–242, Fig. 1, p. 235.)

- changes to local microclimate (e.g., soil temperature and moisture);
- changes in trophic structure and energy budgets (e.g., keystone species and competition);
- changes in natural disturbance regimes (e.g., fire frequency and intensity); and
- the introduction of diseases.

Economic Losses from Invasions

Given the diversity of ways in which invasive species affect ecosystem function, it is not surprising that the losses from invasive species all over the world amount to billions of dollars annually. A 1993 report of the Congressional Office of Technology Assessment (OTA) estimated that the total cost of damage from 79 invasive species in the United States (plants, animals, and microbes) totaled about $97 billion from 1906 through 1991. Cornell University scientists estimated total economic costs at $137 billion per year (U.S. GAO 2000; Mack et al. 2000). Florida federal departments spent the greatest amount of funds ($94.5 million in 1999 and $127.6 million in 2000), followed by California ($82.6 million in 1999 and $87.2 million in 2000) (U.S. GAO 2000). The total cost of controlling the boll weevil in U.S. cotton croplands, which arrived from Mexico in the 1890s, now exceeds $50 billion. Leafy spurge that has invaded western rangelands caused losses of $110 million in 1990 alone.

The costs of eradication campaigns in the United States include $764 million for the invasion of eastern forests by European gypsy moths in 1981, $100 million annually for Sri Lankan *Hydrilla* and Central American water hyacinth, $200 million for fire ants in 1977, and $200,000 for Asian citrus blackfly in 1979 (Simberloff and von Holle 1999). McNeely et al. (2001) provide

Table 20.8
Economic Costs of Some Invasive Alien Species

Species	Economic Variable	Economic Impact[a]	Ref.
Introduction disease organisms	Annual cost of human, plant, and animal health in U.S.	$41 billion per year	Daszak et al. 2000
A sample of alien species of plants and animals	Economic costs of damage in U.S.	$137 billion per year	Pimentel et al. 2000
Salt cedar	Value of ecosystem service lost in western U.S.	$7–16 billion over 55 years	Zavaleta 2000
Knapweed and leafy spurge	Impact on economy in three U.S. states	$40.5 million per year direct costs; $89 million indirect costs	Bangsund 1999; Hirsch and Leitch 1996
Zebra mussel	Damages to U.S. and European industrial plants	Cumulative costs 1988–2000 = $750 million to 1 billion	National Aquatic Nuisances Species Clearinghouse 2000
Most serious invasive alien plant species	Cost of 1983–1992 of herbicide control in Britain	$344 million per year for 12 species	Williamson 1998
Six weed species	Cost in Australia agroecosystems	$105 million per year	CSIRO 1997, cited in Watkinson, Freckleton, and Dowling 2000
Pinus, Hakeas, and *Acacia*	Costs on South Africa floral kingdom to restore to pristine state	$2 billion	Turpie and Heydenrych 2000
Water hyacinth	Costs in seven African countries	$20–50 million per year	Joffe-Cooke, cited in Kasulo 2000
Rabbits	Costs in Australia	$373 million per year (agricultural losses)	Wilson 1995, cited in White and Newton-Cross 2000
Varroa mite	Economic cost to beekeeping in New Zealand	$267–602 million	Wittenberg et al. 2001

Source: From McNeeley. 2005.
[a] U.S. dollars.

further estimates on the economic toll of the invasive species (Table 20.8).

Getting Rid of Invasive Species

How easy is it to control the growth of invasive species or to eradicate them? The simple answer is that, once they are established, few measures will successfully eradicate invaders. Thus, drastic measures have been suggested to control the spread, and the following two examples (one animal and one plant) might illustrate one such approach.

More than 150 years ago, a dozen or so rabbits were imported into Australia. Presently, the estimates are that the progeny of those original parents now exceeds 300 million. Previous attempts to control rabbits involved introduction (from Uruguay) of the myxomatosis virus, which caused death of massive numbers of rabbits in the 1950s, followed by the increase of disease-resistant strains of the rabbit and less virulent forms of the virus through natural evolutionary processes. Myxomatosis was intentionally introduced into France in 1952 and quickly spread throughout Europe and Great Britain. It exists in both Australia and Europe today. To further combat the problems created by such a huge population of rabbits, the government of Australia launched an experiment on Wardang Island off Australia's south coast to test whether a different potent virus, which causes rabbit hemorrhagic disease (RHD), would successfully infect and kill the rabbit population. In 1995, the virus escaped and became established in Australia (Drollette 1996) and eventually New Zealand. When the feared infection of other species did not occur, the government approved release of the virus elsewhere on the Australian continent (Drollette 1997), where it has decreased rabbit populations profoundly, particularly at more arid sites. At the same time, a vaccine must now be used to protect pet and show rabbits in both countries.

A plant species targeted for biological control is the salt cedar or tamarisk tree introduced into the western United States from Asia to control soil erosion. The plant spread and displaced the local native and indigenous plants. Now a leaf-eating beetle imported from China is being used experimentally to control the salt cedar and assess its effectiveness is being assessed. According to Malakoff (1999):

> However, as so often happens, there are potential drawbacks to any tactic that alters the vegetation of an area. Some people are concerned that controlling the salt cedar will eliminate nesting sites for the willow flycatcher. The flycatcher was classified as an endangered species in 1995. Ironically, the spread of salt cedar has been cited as one reason for the partial recovery in the flycatcher population. Due to these concerns, the plans for initial release of the beetle have been scaled back until the risks are better understood.

Of 67,000 estimated pests, only 300 have been targeted for biological control; of these, only 120 have been "success stories" (Myers 1996a). However, the preceding discussion illustrates the fact that introducing new organisms, whether parasitic, disease-causing (pathogenic), or predatory, is not without risk and no one can tell for sure what the consequences of such introductions will be.

Policies to Lessen Biodiversity Loss

The dangers of species extinction were brought to public attention by the noted U.S. conservationist Aldo Leopold, whose efforts resulted in the passage of the 1966 Endangered Species Preservation Act. In 1969, this law was revised to form the Endangered Species Conservation Act, affording greater protection of animals and plants that were in imminent danger of "worldwide extinction," including such charismatic species as the endangered panda, tiger, and condor. The 1969 legislative measures in the United States were further updated and resulted in the landmark Endangered Species Act (ESA) of 1973. We present discussion of the ESA in Chapter 21 on environmental policy.

Questions

1. What are the consequences of invasive species introductions? What are the ecological characteristics of certain species that can make them successful invaders?
2. How do humans contribute to biodiversity loss and habitat fragmentation?
3. Provide examples of biodiversity loss in your region and discuss how they could affect your daily life.
4. Discuss the implications of habitat fragmentation, especially relating to the problems that result when trying to protect environmentally significant areas and biodiversity hotspots.
5. What ecological principles from Section A can be applied today to improve species conservation success and protect biodiversity?

SECTION C

Seeking Solutions

- *How has governmental policy lessened the impacts and improved the environment?*
- *How can the ecosystem services be made sustainable?*

Technological advances to grow and develop resources to meet increasing human demands are universal components of human civilization. Unfortunately, the rate of human population growth and the ensuing number (as discussed in Chapter 11) exceed the availability and growth in natural resources and ecosystem services. In Section B, we detailed the inputs and outputs in our ecosystems necessary to meet the demands of food, wood products, minerals and metals, and energy resources. In this section (Chapters 21–26), we examine strategies for solutions to environmental problems.

We begin this section with environmental policy and law. With the passage of the National Environmental Policy Act of 1969 came the establishment of the Council on Environmental Quality in the executive office of the president, whose task is to coordinate the environmental responses to issues that arise in individual cabinet-level departments. Shortly thereafter, by an executive order, President Richard Nixon created a new cabinet-level department, the Environmental Protection Agency (EPA). All 50 states followed suit in establishing EPA-type cabinet departments (or departments under elected officials) who advise the governors and state legislatures and oversee the application of federal and state laws and regulation for solving environmental problems. Now, most county governments and municipalities also have a dedicated environmental professional staff. Equally, one would be hard pressed to find a corporation or industry without environmental advisers and professionals. These policies, laws, and regulations are the backbone of strategies for safeguarding the environment.

All countries seek rapid industrial development. What may be unprecedented today is that two of the all-time most populated countries of the world, China and India, has been at 6–12% annual rates of economic growth. Such rates of growth use enormous resources and produce equally enormous wastes. Economics as applied to human systems is both the problem and the solution. Despite the common roots of ecology and economics, economics somehow got divorced from natural resources and environment more than a century ago. Many ecologists and some economists have decried this separation, and the crux of their criticism is succinctly summed up by a distinguished economist:

> Twentieth century economics has in large measure been detached from the environmental sciences. Judging by the profession's writings, we economists see nature at best as a backdrop from which resources can be considered in isolation; we also imagine most natural processes to be linear. Macroeconomic forecasts routinely exclude environmental resources. Accounting for nature, if it comes into the calculus at all, is an afterthought to the real business of "doing economics." (Dasgupta 2004)

But economics must also be the solution if human survival on the Earth is to remain. Thus, it is vital that we find new economic metrics and economic indicators that cement the connection with all things environmental.

We also discuss the principles of environmental management, the restoration of ecosystems, and conservation of biological diversity. Ecosystem restoration and reducing loads of pollutants into water bodies are emerging as first-order tasks to ensure the availability of food, fiber, and other ecosystem services. Conservationists are convinced that the two major factors responsible for the loss of biodiversity are loss of habitats and the spread of invasive species. Lastly, the sum of all these issues is codified in the concept of sustainable development. This concept brings all human dimensions—scientific, economic, sociocultural, and ethical—together.

Environmental Policy and Law

21

Introduction

Many environmental problems, discussed in the last 10 chapters, continue to grow at an alarming pace. As the demand for adequate food, shelter, transportation, water, and fuel to support growing human and domestic animal population increases, the need to exploit environmental resources also increases. All countries are seeking rapid industrialization and other developments. For example, two of the most populous countries, China and India, have been developing at 6–12% annual rates of economic growth in the last 5 or more years.

The environmental problems that are unleashed in the process of rampant development are not always amenable to easy solutions because of the interconnectedness and manifold linkages that exist within and among the three basic components of biosphere: aquatic, terrestrial, and atmospheric (as discussed in Section A). Furthermore, the spatial–temporal scales of problems such as atmospheric pollution, global warming, toxic waste disposal, and desertification, transcend the national boundaries within which environmental laws and rules generally govern. Thus, it is important to recognize that environmental policies and laws (a) have both national and international components, and (b) most often involve direct conflicts between public and private ownership of land and resources and the ways in which they are used or abused. This chapter examines how policies and laws that seek to address environmental problems are developed, how such policies and laws are applied and enforced, and how effective they have been, as illustrated by the history of the last 60 years or so in the United States.

Policy-Making and Science

The ultimate goal of environmental policy and law-making is to protect the environment and the public from immediate as well as long-term harm. **Environmental policy,** like all **public policy,** may be defined as broadly agreed-upon guidelines and rules that govern decisions and actions for the resolution of present and anticipated problems. Scientific knowledge and technology have a central role in formulating such policies. **Decision-making** also takes into account economic viability and social acceptance of the proposed solutions. The following considerations that apply to developing science policy also apply to environmental policy.

Pielke (2007) noted that "if 'science' refers to the scientific pursuit of knowledge, and 'policy' refers to a particular type of decision-making, then the 'science policy' involves all decision-making related to the systematic pursuit of knowledge." In this context, Pielke turns his attention to the role of scientists and their four "idealized" choices: (1) sticking strictly to scientific data and remaining aloof from social–emotional values ("pure scientist"), (2) adjudicating issues ("science arbiter"), (3) acting as an "issue advocate", or (4) being an "honest broker of policy alternatives." At their face value, these assignations may seem clear-cut, but as Jasanoff (1990) notes, "The notion that scientific advisors can or do limit themselves to addressing purely scientific issues, in particular, seems fundamentally misconceived." There also are basic differences between the emphases in the expectations in science and in policy (Table 21.1). Regardless of such considerations, the basis of policy-forming and decision-making must rest firmly on extant scientific knowledge.

Table 21.1
Science–Policy Mismatches in Research Design

Science Emphasis	Policy Need
Scientific pursuit	Ground truth
Global patterns	Regional effects
Mean conditions	Variability and extremes
Most probable outcome	Risks and options
Trends	Thresholds and timeliness
Cycles/equilibria	Rates of change
Physical impacts	Societal responses

Source: Modified from Bernabo, J. C. 1989. Testimony in the U.S. House of Representatives, 27 July 1989 (Document #104-74, 203-212).

Process of Making Policy

The more complex human societies become, the greater is the need for policies and laws at local, regional, and international levels. Thus, it is useful to review how policies are made. **Policy-making** provides a method and framework to enact laws, develop rules, and implement and enforce both. The development of policies is based on an iterative process linking the public, the scientific community, the affected industrial sector, and decision-makers. The expression of policies can be made through the articulation of goals and viewpoints by citizen groups, political leaders, and government representatives. The United States has an enviable tradition of bringing about change through citizen groups, a characteristic that did not escape the notice of the French philosopher Alexis de Tocqueville more than 170 years ago.[1]

According to Anderson (1994), policy-making is a multitiered process and involves six steps. **Policy formation** represents the totality of the creation, development, and adoption of a policy. It begins with a **policy agenda,** which provides description of a problem under consideration for discussion as an issue of concern. These discussions lead to **policy formulation,** which Anderson describes as "the crafting of proposed alternatives or options for handling a problem." Once the need for formulating a policy is agreed upon, **policy adoption** becomes a reality. Appropriate means, such as existing or new administrative units, must be in place or created specifically for **policy implementation** and **policy evaluation** through time (Figure 21.1).

Policy formulations result in laws—formal statutes, rules, and regulations. Policy implementation can be carried out through the actions of a diversity of administrative agencies and even the courts. The preceding brief description betrays the complexity and multiplicity of issues and parties involved, the number of constituencies that are or claim to be affected, and the time and effort that forming policy takes. A simplified game primer created for new legislators by a veteran journalist and his artist illustrates a view of the process[2] (Figure 21.2).

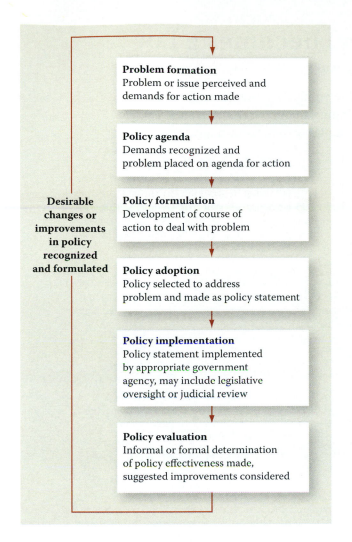

Desirable changes or improvements in policy recognized and formulated

Problem formation
Problem or issue perceived and demands for action made

Policy agenda
Demands recognized and problem placed on agenda for action

Policy formulation
Development of course of action to deal with problem

Policy adoption
Policy selected to address problem and made as policy statement

Policy implementation
Policy statement implemented by appropriate government agency, may include legislative oversight or judicial review

Policy evaluation
Informal or formal determination of policy effectiveness made, suggested improvements considered

Figure 21.1 The necessary steps in the formation, adoption, and evaluation of public policy. (Based on Anderson, J. E. et al. 1984, *Public Policy and Politics in America,* 2nd ed. Monterey, CA: Brooks/Cole Publishing Company, as adapted by Cubbage, F. W., J. O'Laughlin, and C. S. Bullock III. 1993. *Forest Resource Policy.* New York: John Wiley & Sons.)

Environmental Policy

Environmental policy determines a course of action designed to solve environmental problems by integration of activities and powers of the legislative, executive, and judicial branches of government with actions of citizens and groups. Environmental policy is both national and local, as can be illustrated by the case of regulating domestic pollution of air, water, and soil. It can also be regional and global in scope,

as in the case of the **transboundary issues** of atmospheric acid (SO_x and NO_x) deposition and management of international rivers. Its global reach is illustrated by the issues of climate change, loss of biodiversity, and thinning of the ozone layer, which are, in general, dealt with in separate bi- and/or multilateral environmental agreements.

The underpinning of a successful environmental policy is to (1) find remedies for the root causes and sources of degradation and destruction of the environment; (2) value prevention first and then mitigation, rehabilitation, and amelioration of environmental damages; and (3) bridge the gap between science and decision-makers. Linking science and policy requires recognizing that (1) uncertainty and fallibility are inevitable, and (2) policy formulation and decisions based on scientific information must be made in a context of uncertainty and be adaptive to continual correction through new information. The introduction, implementation, and enforcement of new policy instruments should have the attributes of accountability, transparency, efficiency and effectiveness, responsiveness, consistency, coherence, vision, and coordination.

Environmental Policy and Ecology

Evidence is as indispensable in the realm of ecology as in other disciplines of knowledge, including the legal profession. But ecology has to deal with multivariate models in time and space, so the proof of causality is often difficult to obtain. Indeed, the correlational evidence in ecology (the circumstantial evidence in jurisprudence) predominates over the cause–effect phenomena (or direct evidence). Further, ecologists have to deal with processes that may not show any impacts immediately as well as be able to visualize those impacts on the "potential environment" that could become operational in the near or distant future.

Therefore, in formulating public policy on the environment, the following consideration must also be addressed: Who makes the decisions that affect our environment? The goal of

How a bill (really) becomes a law

By Lee Leonard and Charlie Zimkus | © The Columbus Dispatch

Two scenarios are presented below to play a game. Here's how it is played: You be the bill. Take turns flipping a coin. Heads: Advance 1 space. Tails: Advance 2 steps. The first player to become a law wins.

In textbooks

There are two ways to play the game. In this version—for novices—you are a bill trying to become a law through the textbook legislative process. The method might sound familiar if you paid attention in school or during *Schoolhouse Rocks.*

In the ReaL WoЯLd

In this version—for advanced players—you are a bill trying to become a law through the legislative process as it *really* happens. The method might sound familiar if you pay attention to the news or watched *The West Wing.*

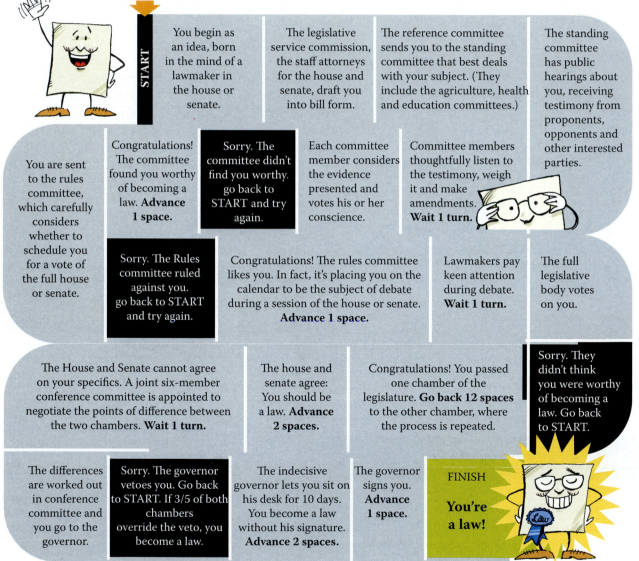

Figure 21.2 Illustration of the many iterative steps that a bill (policy formulation) must take before it can become a law. Above, next page is an expanded version of a textbook-type approach of a seemingly seamless process as it proceeds from first to the last step. At right is a simplified, real-life scenario of what actually happens. (Revised and simplified from their original 2005 illustration by Lee Leonard and Charlie Zimkus 2008.)[2]

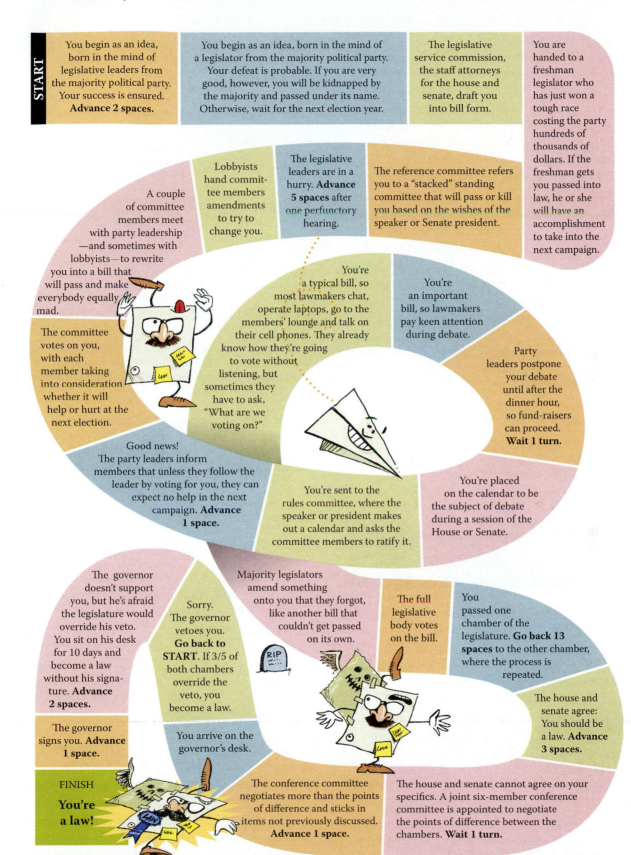

START You begin as an idea, born in the mind of legislative leaders from the majority political party. Your success is ensured. **Advance 2 spaces.**

You begin as an idea, born in the mind of a legislator from the majority political party. Your defeat is probable. If you are very good, however, you will be kidnapped by the majority and passed under its name. Otherwise, wait for the next election year.

The legislative service commission, the staff attorneys for the house and senate, draft you into bill form.

You are handed to a freshman legislator who has just won a tough race costing the party hundreds of thousands of dollars. If the freshman gets you passed into law, he or she will have an accomplishment to take into the next campaign.

A couple of committee members meet with party leadership —and sometimes with lobbyists—to rewrite you into a bill that will pass and make everybody equally mad.

Lobbyists hand committee members amendments to try to change you.

The legislative leaders are in a hurry. **Advance 5 spaces** after one perfunctory hearing.

The reference committee refers you to a "stacked" standing committee that will pass or kill you based on the wishes of the speaker or Senate president.

You're a typical bill, so most lawmakers chat, operate laptops, go to the members' lounge and talk on their cell phones. They already know how they're going to vote without listening, but sometimes they have to ask, "What are we voting on?"

You're an important bill, so lawmakers pay keen attention during debate.

The committee votes on you, with each member taking into consideration whether it will help or hurt at the next election.

Party leaders postpone your debate until after the dinner hour, so fund-raisers can proceed. **Wait 1 turn.**

Good news! The party leaders inform members that unless they follow the leader by voting for you, they can expect no help in the next campaign. **Advance 1 space.**

You're sent to the rules committee, where the speaker or president makes out a calendar and asks the committee members to ratify it.

You're placed on the calendar to be the subject of debate during a session of the House or Senate.

The governor doesn't support you, but he's afraid the legislature would override his veto. You sit on his desk for 10 days and become a law without his signature. **Advance 2 spaces.**

Sorry. The governor vetoes you. **Go back to START.** If 3/5 of both chambers override the veto, you become a law.

Majority legislators amend something onto you that they forgot, like another bill that couldn't get passed on its own.

The full legislative body votes on the bill.

You passed one chamber of the legislature. **Go back 13 spaces** to the other chamber, where the process is repeated.

The governor signs you. **Advance 1 space.**

You arrive on the governor's desk.

FINISH **You're a law!**

The conference committee negotiates more than the points of difference and sticks in items not previously discussed. **Advance 1 space.**

The house and senate cannot agree on your specifics. A joint six-member conference committee is appointed to negotiate the points of difference between the chambers. **Wait 1 turn.**

The house and senate agree: You should be a law. **Advance 3 spaces.**

Figure 21.2 (continued)

industry is to make a profit in relatively short time cycles; the future effects of pollution or the impacts of resource depletion are not always accounted for and seldom considered.

- Are our leaders planning for long-term environmental health, or are policies the sole result of economic decisions and political reactions without clear or consistent goals?
- How do policy-makers use information provided by scientists? Once environmental policies are actually legislated, how are they implemented and their success or failure evaluated?
- Who is responsible for cleaning up pollutants from the environment and restoring the ravaged lands that have resulted from human activities? Is this a function of industry or of governments (taxpayers)?
- Finally, how best can one act when some governments and industries respond positively to environmental concerns, but others deny even the existence or significance of any problem?

Environment and Energy: An Example

Environmental issues are both politically and scientifically complex. For example, environmental and energy policies are derived from higher-level policy objectives and cannot be designed or developed in isolation from one another. That is because energy policies have important impacts on the environment, and environmental policies in turn affect the pattern of energy consumption and production (Figure 21.3).

Development of environmental and energy policies must be coordinated for the following reasons:

- Energy is the fuel that drives the economy.
- Energy availability, price, conservation and efficiency, and market distortions in energy infrastructure services strongly

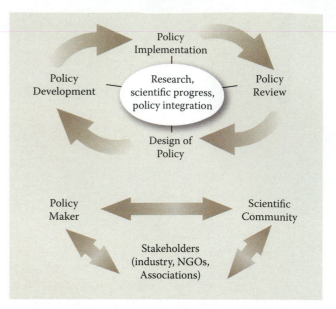

Figure 21.3 Interactions between energy and environmental policies. (Modified and redrawn from Quevauviller, Ph. 2005. *Journal of Environmental Monitoring* 7:89–102, Figure 1, p. 90.)

affect economic performance, environmental quality, and social welfare.
- Production and consumption of fossil fuels result in significant environmental impacts at local to global scales.
- Domestic energy resources usually lie above or below private and publicly owned land, so extraction results in land use conflicts.
- Dependence on imported energy resources has important implications for national and global security and the environment.
- An economical and reliable supply of energy is the key input to the production and distribution of economic goods and services, so it necessarily affects the competitive power of nations.

We already studied (Chapters 18 and 19) how the combustion of organic (fossil) energy sources releases CO_2, a major component of the gaseous causes of the greenhouse effect, as well as SO_x and NO_x. Therefore, it would seem relatively easy to pinpoint the producers and consumers of organic fuels and then legislate some kind of fee or tax to be used for carbon sequestration, better fuel efficiency, or development of cleaner, sustainable, and renewable energy resources. But even at the federal level, at least 16 agencies deal with the greenhouse effect, and sometimes these agencies themselves have conflicting goals and agendas. Furthermore, the federal government continues to subsidize fossil fuel exploration and extraction on public lands, with the profits going to private corporations and the cleanup primarily left to the taxpayers. Finally, even when laws are passed and enforced, the financial penalties for breaking environmental laws are often set so low that it is more effective for industry to pay fines than to prevent environmental degradation.

Public versus Private Rights

One of the cardinal principles of the capitalistic economies is the freedom of the marketplace— that is, open commerce with prices based on supply and demand (see Chapter 22 for discussion). However, there has been a frequently recognized need for governmental regulation to protect public welfare from private exploitation. This recognition was implicit in the establishment of national parks and other protected lands. It was recognized explicitly about the same time as the passage of the Sherman Antitrust Act (passed in 1890 to limit monopolies) and creation of the Interstate Commerce Commission (passed in 1897 to limit railroad power and later trucking power).[3]

Since then the U.S. population has grown significantly, and more and more land area has been converted to human uses. But no significant increases in national land holdings have occurred for five decades, except during the Clinton administration during the 1990s. However, designating land as "protected" is no guarantee of its preservation. Without commitment and funding for stewardship and management, public lands are vulnerable to many forms of exploitation, including some that are actually subsidized by the government itself. For example, as was noted in Chapter 14 on mining, for $25 one can obtain a permit for exploratory mining on federal lands. If the search is unsuccessful, the environmental

damage incurred is left to either nature or the taxpayers to repair.

Environmental Policy and Law: The United States as an Example

In the United States, we have three branches of government: the **legislative** to make statutes; the **judiciary** to clarify statutes written by the legislatures and examine their conformance to the Constitution, as well as to make common law; and the **executive** to carry out (or enforce) laws. The boundaries among these branches are not always clear, and the three branches occur at federal, state, and local levels as well. Both executive and legislative officials have some power over funding and spending. Overlaps of jurisdiction and authority (including funding and spending) make it difficult to create and implement clear environmental laws, especially because the rights of the individual citizen may be at odds with the needs of larger groups. The enactment of new laws often occurs in reaction to public or political pressure to address a single problem or issue. Thereafter, the enactment and application of the law are subject to various influences and changes. "Checks and balances" among the three branches of government were established to ensure that no single branch would assume too much power. But it is this very safeguard of our political freedom that makes environmental legislation so complex to formulate and difficult to enforce.

Ideally, **environmental legislation** is based on good scientific data and is free of personal or political interests. However, this ideal is difficult to achieve, especially when communication and dialogue are influenced by individuals with contradictory motivations. Neither politicians nor the general public is necessarily qualified to evaluate scientific data and draw objective conclusions. On the other hand, scientists are not always free of subjective bias in their research and findings. Therefore, the path of any environmental legislation may wander from the original intent. Nevertheless, the United States has historically enacted and adopted some comprehensive environmental protection laws. For example, publicly controlled fresh water, which is rapidly becoming recognized as the limiting factor for continued development in many parts of the world, has been under federal legislative control for more than a century in the United States. More significantly, public awareness of environmental concerns is growing. In many European nations, "green" parties are specifically organized around environmental issues and societal equity.

In a democracy, enacting environmental laws that would be considered ideal by all is not possible. However, the activities of a few enlightened and enthusiastic persons can motivate a large numbers of legislators and voters to support the right kind of environmental legislation. For example, in her book *Silent Spring* (1962), Rachel Carson suggested that DDT was responsible for reproductive failure in raptorial birds such as the bald eagle due to eggshell thinning. Her book was influential in accelerating the rate of environmental awareness in the United States, and the nation outlawed the use of DDT in 1972. In the last generation, new laws and procedures for environmental regulation and oversight have set the stage for further refinement in this area, especially through emergence of a much more environmentally conscious citizenry.

Water quality: a legislative example. The United States has been leading the world in enacting strong environmental legislation. Some of the earliest environmental laws focused on surface water, to protect both navigation and water quality. For example:

- The Rivers and Harbors Act of 1899 was designed to protect navigable waters.
- The Public Health Service Act of 1912 provided the United States with powers of investigating water pollution affecting human health.
- The Oil Pollution Act of 1924 prohibited discharge of oil into coastal waters (Kubasek and Silverman 1994).

- The Federal Water Pollution Control Act (FWPCA) of 1972 was renewed as the Clean Water Act (CWA) of 1977.
- The revised Water Quality Act (WQA) of 1987 was a renewal of the CWA that made storm water discharge subject to NPDES (discharge) permit requirements.
- The Oil Pollution Act of 1990 provided an important strengthening of the original act of 1924.

All these acts and amendments have collectively resulted in a comprehensive, though complex, system of water pollution control with the following national goal: "to restore and maintain the chemical, physical, and biological integrity of the Nation's waters" (CWA Section 101; Kubasek and Silverman 1994). The U.S. Environmental Protection Agency is responsible for the enforcement of regulations pertaining to surface water pollution. Although much has been achieved under these regulations, 40% of the nation's rivers were still considered unfit for fishing in 1999.

Despite legislative mandates in most countries to ensure surface water quality, recent reports show that, internationally, many problems remain. Here are a few illustrative examples:

- Nonpoint pollution from nitrogen and phosphorus loadings into surface waters continues to be a serious problem worldwide. Most of these pollutants come from agriculture industries and include fertilizers and wastes from high-density livestock-raising, which cause eutrophication (ESA 1998).
- As a result of bacterial contamination, Ontario is imposing even tougher rules to ensure drinking water quality. The water supply in Canada's largest province became contaminated with a virulent strain of *Escherichia coli*; 2,700 people became ill and seven individuals died from exposure to the bacterium (Hrudey et al. 2003).
- The consequences of toxic algal blooms in contaminated water seem to be a worldwide problem. A report documents the causative organisms and health issues they cause (Morris 1999).

- Severe degradation of Lake Tahoe, which borders Nevada and California and once contained what was described as crystal clear, cobalt-blue waters, has resulted from human activities such as automobile pollution, wood-burning fires, lawn fertilizers, construction site runoff, and misguided forest fire management. Unless immediate steps are taken to address the decline in water quality, the lake faces a grim future (Reuter and Miller 2000).

- Cyanide spills are still common in several parts of the world. The cyanide spill from the tailings dam of Arul gold mine in Romania on January 31, 2000, killed thousands of fish in Hungary and Yugoslavia. Water intake was banned from the Danube River. Cyanide poisoning of hundreds of birds was also reported from a gold mine in Montana (Ware, Nigg, and Doerge 2004).

The examples cited here underscore the importance of water quality: The well-being of human beings and of the world's diverse biota depends upon it. As for water availability, its importance in maintaining global crop and forest production and human existence is already becoming clear to world leaders and the general public. The protracted negotiations of the world's disputed territories (for example, the Arab–Israeli conflict) now have experts on water withdrawals and use as a foremost item on the agenda. These issues are likely to continue until we recognize the finite nature water resources and the prevalence of agents that degrade water quality.

With the passage of comprehensive landmark legislation such as the Clean Air Act (CAA) of 1970 and the Clean Water Act (CWA) of 1972, the United States began to move toward cleaning up the most egregious pollution problems. It must be noted that each environmental statute begins with the ideal (and laudatory) intentions of the Congress in passing the legislation. However, for enforcing the enacted laws, the proposed implementing regulations are swiftly and persistently challenged in the courts, and thus the intent is often weakened. Some of the other significant environmental legislative acts are briefly discussed next.

The National Environmental Policy Act (NEPA) of 1970 requires formal considerations of the environmental consequences of proposed major federal projects. NEPA championed the identification of ways for obtaining beneficial uses of the environment without causing degradation. It also established the Council on Environmental Quality, which coordinates federal environmental efforts and whose chair serves as the primary environmental advisor to the U.S. president.

The Endangered Species Act (ESA) of 1973 seeks to encourage recovery of threatened and endangered species by identifying those at risk and protecting the habitats in which they reside. The intent of ESA is clearly expressed in the "findings" of this act (see later discussion).

The Resource Conservation and Recovery Act (RCRA) of 1976 addressed and regulated pollution of the air, water, and land, respectively, by specifically controlling the disposal methods for solid and hazardous wastes. The Surface Mining Control and Reclamation Act (SMCRA) of 1977 addresses the extreme damage to land, forests, fish, and wildlife that coal surface mining has caused. It specifies that mined land must be reclaimed, but that the primary responsibility for the management of this program will fall on the state in which the mining is being done.

The National Environmental Education Act (NEEA) of 1990 seeks to improve the effectiveness of federal programs to inform and educate the public on environmental issues and train more environmental professionals. The Pollution Prevention Act (PPA) of 1990 suggests that the huge amounts of money being spent on cleaning up after discharge of pollutants could best be used by redesigning manufacturing processes so that they do not generate pollutants in the first place. For example, automobile paints and furniture varnishes are now made with water-based formulas rather than petroleum solvent-based types. This has helped decrease the solvents released to the soil, air, and water, as well as the health risks for those involved in handling the materials.

The Energy Policy Acts (NPA) of 1992 and of 2005 sought to reduce U.S. dependence on foreign sources of energy and to reduce pollution from coal; they called for many other related actions in an attempt to provide a comprehensive policy on the nation's energy resources. None of the goals were achieved. In December 2007, the Energy Independence and Security Act (EISA) of 2007 was passed. It provides goals for and regulation of major areas in energy policy. These include vehicle fuel economy (also known as CAFE—corporate average fuel economy) standards, renewable fuel standards (biofuels, building efficiency for government, commercial, and industrial sectors); and research and development in renewable sources such as solar and geothermal, as well as effective ways of carbon sequestration for cleaner use of coal. All provisions of the new law have a direct impact on the environment.

The Food, Conservation, and Energy Act of 2008 (also called the 2007 Farm Bill) includes five major areas: ensuring food security, promoting homegrown renewable energy, reforming farm programs, protecting the environment, and strengthening international food aid. It increases funding for the conservation program by $7.9 billion and provides funding for the growth of energy crops. All agricultural programs have an intimate relationship with multiple aspects of the environment.

Many would argue whether these laws go far enough or are enforced in ways that Congress originally intended. Nevertheless, progress has been made in the protection of air, land, water, wetlands, and biota. Trends in air quality since the passage of the Clean Air Act indicate that the concentrations and deposition of many airborne pollutants have decreased significantly (Figure 21.4). However, much work remains to be done. For example, under the Clean Water Act, despite the fact that most point sources of pollution have been brought under control, many of the nation's streams and rivers show chronic degradation and a reduction in in-stream diversity (Figure 21.5) (U.S. EPA 2007). In some cases, protection for special ecosystem types has increased dramatically. For example, draining the swamps was once welcomed by many in creating drier land areas for other uses. Today, development that destroys a wetland must be approved with a permit, and a replacement wetland 1.5 times the area of the

original, natural site (so-called mitigation wetlands) must be constructed. Each case, however, is site specific—and whether approvals are given or denied is based on site-specific arguments, as well as local politics.

The preceding narrative and what follows are equally relevant to developing countries as they seek better living conditions for their citizens, conservation of their resources, and protection of their biological diversity (Safaya and Wali 1992). As these countries become more industrialized, Safaya and Wali note that "their environmental problems are becoming more complex. Hence, there is much to learn from the experiences of the industrialized world. Public initiatives through nongovernmental organizations (NGOs) are gaining ground in some developed nations, and many recent NGO efforts are laudable."

Legislative Goals versus Implementation Actions

A crucial aspect of environmental policy-making must be borne in mind. The laws (statutes and rules) begin with an exhortatory expression of the general principles or general purpose intended by the legislature. It is important to distinguish between these expressions of principles or ideal goals and the specific actions that are required of many parties to fulfill those goals. Let us consider some specific examples.

The Clean Water Act, created by Congress as the modern form of the Federal Water Pollution Control Act (FWPCA) states that its objective is "to restore and maintain the chemical, physical, and biological integrity of the Nation's waters," [FWPCA, Sec. 101(a)]. It also states that the national goal is to eliminate the discharge of pollutants be eliminated by 1985 [FWPCA, Sec 101(a)(1)].

Thus, to ensure the implementation of the FWPCA, terms such as integrity, discharge, pollution, and others must be defined in detail. Until that is done and the specifics are elaborated, the conditions do not exist that establish the appropriate, enforceable mechanisms that move conditions toward the objective. These are

Average wet sulfate deposition
(1990–1994)

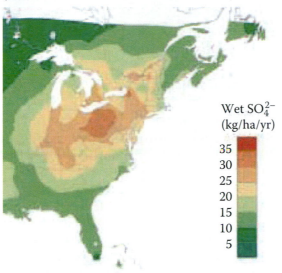

Wet SO_4^{2-}
(kg/ha/yr)

35
30
25
20
15
10
5

Annual wet sulfate deposition
(2002)

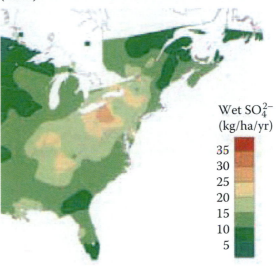

Wet SO_4^{2-}
(kg/ha/yr)

35
30
25
20
15
10
5

Average wet nitrate deposition
(1990–1994)

Wet NO_3^-
(kg/ha/yr)

35
30
25
20
15
10
5

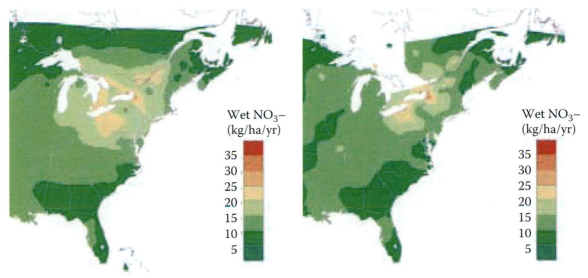

Annual wet nitrate deposition
(2002)

Wet NO_3^-
(kg/ha/yr)

35
30
25
20
15
10
5

Figure 21.4 Trends in wet sulfate and nitrate deposition over the period of 1989–2004. These two compounds are the primary agents responsible for "acid rain." Deposition dropped substantially between 1990 and 2004 as a function of clean air laws. (U.S. EPA 2007; http://www.epa.gov/castnet/mapconc.html)

the areas where the input from environmental scientists and other professionals is required.

The Clean Air Act of 1970 declares that a purpose of the statute is "to protect and enhance the quality of the Nation's air resources" [CAA, Section 101(b)(1)]. Details must be developed that define such terms as protect, quality, and air resources. For example, the *quality* of air must be prescribed precisely as a chemical concentration in milligrams per liter. The term

protect must be translated into details that prescribe precisely who will (or will not) do what with which facilities.

Section C of the Federal Endangered Species Act (ESA) of 1973 is entitled "Congressional Findings and Declaration of Policy," and Section 2(c) declares that it is "the policy of Congress that all federal agencies shall seek to conserve endangered species and shall utilize their authorities in furtherance of purposes of

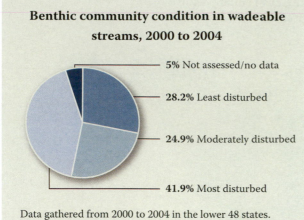

Benthic community condition in wadeable streams, 2000 to 2004

5% Not assessed/no data

28.2% Least disturbed

24.9% Moderately disturbed

41.9% Most disturbed

Data gathered from 2000 to 2004 in the lower 48 states. Categories based on the number and diversity of benthic species present, with "least disturbed" being the most diverse. Graphic shows the percentage of stream miles in each category.

Figure 21.5 Assessment programs to determine the ecological condition of streams based on the diversity of benthic invertebrates indicate that the majority of stream miles in the conterminous United States are degraded by human activity. In this case, invertebrates are used as indicators of "biological integrity" as required by the Clean Water Act. (U.S. EPA 2006. *Wadeable Streams Assessment—A Collaborative Survey of the Nation's Streams.* EPA 841-B-06-002, Washington, D.C.: Office of Water.)

this chapter." Are there binding, specific obligations established here? When it appears in a statute, what does a phrase such as "seek to conserve an endangered species" mean? These definitions highlight the generality that declarations of policy in a statute are statements of general principles or purpose. Therefore, it is necessary to look further into the statute for specific statements about who is to do what and under what conditions. The ESA Section 4(a)(1) directs the secretary of the interior to "determine whether any species is an endangered species." These are the instructions that affected parties, legislators, and officials look to for guidance.

The last example is the Surface Mining Control and Reclamation Act (SMCRA) of 1977. The statute states the purpose to be to:

- "establish a nationwide program to protect society and the environment from the adverse effects of surface coal mining operations" [Section 102(a)];

- "assure that surface mining operations are not conducted where reclamation as required by this chapter is not feasible" [Section 102(c)]; and

- "assure that surface coal mining operations are so conducted as to protect the environment" [Section 102(d)].

The law also stipulates performance standards for reclaiming and restoring coal-mined lands. Thus, the coal mining operator shall

… establish on the [regarded] areas, and all other lands affected, a diverse, effective, and permanent vegetative cover of the same seasonal variety native to the area of land to be affected and capable of self-regeneration and *plant succession* at least equal in extent of cover to the natural vegetation of the area; except, that introduced species may be used in the revegetation process where desirable and necessary to achieve the approved postmining land use plan." [Section 515(b)(19)]

The ground **cover** and **productivity** of the revegetated area shall be considered equal if they are 90% of ground cover and productivity … of the area in comparison." [Section 816.116(b)(3)]

Species diversity, seasonal variety, and **regenerative capacity** of the vegetation of the revegetated area shall be evaluated … [Section 816.117(c)(3)(ii)] (emphasis added).

In meeting the mandates of this law, one must have the scientific knowledge to evaluate terms indicated in italics in the preceding passages—terms that we learned in the previous two sections of this book.

These examples are given to make sure that readers know that, first, the laws identify all parties (industry, government, and research agencies) relevant to an environmental issue and assign responsibilities where they appropriately belong. Second, each legislative mandate must provide legally acceptable definitions of what needs to be done, where appropriate; what physical, chemical, and biological tests must be conducted; how new knowledge is incorporated into regulations; and what should be the continuing and future course of action. Thus, understanding what a statute or rule means depends on details and upon the interpretations and practices that are adopted by involved parties in the consideration of these details.

International Issues, Laws, and Treaties

Over the last two decades, international awareness of environmental problems and forecasts has increased even more rapidly than in the United States. Both governments and non-governmental agencies (NGOs) have worked together to compile and analyze data of regional and global concerns and to draft legislation and international agreements that can address these issues on a worldwide scale. However, the interests of individual governments and economic forces have prevented international accords from being effectively achieved. The United States has declined to support several global initiatives, notably the climate change treaty developed by the United Nations Conference on Environment and Development (UNCED) in Rio de Janeiro in 1992 and the Kyoto protocol in 1997, which sets emission targets for carbon dioxide and other greenhouse gases.

Sometimes, conservation efforts have succeeded on one front but caused new problems on another. Elephant populations are still endangered and declining in other parts of Africa; in Kenya, however, antipoaching legislation, the creation of reserve areas, and increased international regulation of the ivory trade have allowed elephant populations to increase dramatically. Now, as the elephants forage and migrate, they have become a much greater problem for farmers than before. But attempts to cull herds and to market ivory that has been legally harvested have met with opposition from international agencies that monitor the trade in animal products. Thus, the country that has done the best job of protecting elephants is being economically penalized for doing so.

Economic organizations such as the World Bank have perhaps the greatest impact on whether environmental concerns will be heeded. By the World Bank's promotion of sustainable practices instead of financing unregulated growth, the future can improve for many of Earth's inhabitants. Control of human population growth is a prerequisite to any sustainable policy, however. As long as so many people are hungry, homeless, and desperate to survive one day at a time, there is no hope for long-term environmental health. Although some economic organizations pretend to be uninvolved in political maneuverings, all governmental power is ultimately reduced to control of resources.

Environmental Policy Instruments

Multiple environmental **policy instruments** exist as the magnitude, severity, extent, and type of environmental issues change. In general, environmental policy instruments are distinguished under two broad categories: **command and control** instruments and **market-based** instruments. The command and control approach (also known as **standards** or **regulation**) requires government intervention in stating that polluters must not exceed a certain level of environmental degradation (command) and in monitoring and enforcing the standards (control). This approach includes two types of standards:

- **Ambient standards** stipulate the level of air, water, and soil pollutants that polluters can safely emit.
- **Performance-based emission standards** stipulate a limit to emissions to which each polluter must conform.

Both ambient and performance standards are based on the stipulation that the "best" available technology be used, in addition to placing a limit on emissions.

Disadvantages of the command and control approach include lack of incentives for polluters to provide abatement beyond the standards, inadequate enforcements, low penalties for violations of standards, and inflexible standards notwithstanding dynamics of ecosystems and evolving knowledge.

Another approach used is to seek liabilities through the judicial systems—that is, to sue the offending parties in the courts of law. A few well-known examples may make the point here. Take the case of asbestos, a magnesium silicate mineral that was used commonly for thermal insulation in several building products. It was discovered that when asbestos fibers are in the air, they can easily get into the lungs and cause asbestosis and cancer. The lawsuits that followed this discovery not only were protracted but also involved huge sums of money. This denigrates the value of common law litigation.

Two more examples that have received a lot of publicity are cited here. The Bhopal gas tragedy, or the Bhopal disaster, and the case made famous by the Hollywood production *Erin Brockovich* may be illustrative here. The first deals with emissions into the air and the second is related to discharges into drinking water supplies via groundwater.

Bhopal is a city of 800,000 in central India where Union Carbide, a multinational company, operated a pesticide plant. On December 3, 1984, in the early hours of the morning, a poisonous vapor (methyl isocyanate, a precursor of the insecticide carbaryl) escaped from the plant, instantly killing 2,000 people and rendering as many as 300,000–500,000 injured. It is claimed that eventually 20,000 died and 120,000 continue to suffer health effects. More than 1,000 animals were killed and another 7,000 were injured. A court battle ensued that lasted from 1984 until 1996, when the case was settled by Union Carbide for $470 million.

The second example received immense publicity from a Hollywood movie production, *Erin Brockovich,* a story based on a leading litigant by that name. The case involved discharges of hexavalent chromium, or chromium-6, into the water supply of the town of Hinkley, California, by Pacific Gas and Electric Company (PP&G). The suit alleged that PP&G knowingly let chromium-6 seep into the groundwater even though the company knew of the chemical's role in causing cancers. Settling what was likely to be a protracted litigation, the company paid the plaintiffs and their counsel $333 million, characterized as "the largest settlement ever awarded in a direct-action lawsuit in the United States."

These two cases are not isolated examples. In a serious indictment of the chemical industry, a former chairman of the International Joint Commission on the Great Lakes issued a scathing review of chemical spills in the Great Lakes region of Lakes Superior, Huron, Michigan, Erie, and Ontario. Gordon Durnil compared chemical companies to "child molesters" and wondered why government would allow such law-breakers to keep operating their businesses.[4] These cases raise the inescapable question: How better can we incorporate externalities into our economic models? (For a discussion on externalities, see Chapter 22.) We surely want to avoid individual or class-action lawsuits to bring about the needed changes. It is hard not to argue for a replacement of traditional economic metrics by new and innovative ones that reflect the problems of the twenty-first century.

Like other systems, laws regulating environmental systems come into being after lengthy negotiations that build in the process enforcement incentives and sanctions—the proverbial carrot-and-stick approach. Nevertheless, the environmental laws have some important limitations in their application as tools for achieving environmental goals. These include:

- lack of effectiveness beyond national boundaries;
- lack of clearly and specifically identified provisions;
- high cost of implementation and enforcement;
- incompatibility of boundaries of legal ownership rights with ecosystem management scales;
- conflicts between lengthy observation periods required by environmental management and prompt and tangible financial returns required by existing legal order;
- inflexibility of laws in the face of the dynamics and unpredictable nature of environmental processes; and
- time delays in legal solutions, new scientific knowledge, and technological developments.

Despite these shortcomings, the significance of having environmental laws cannot be discounted or underestimated.

Market-based instruments provide economic incentives for abating environmental degradation that include taxes, charges, specification of property rights, subsidies, marketable permits, deposit-refund schemes, ecolabeling, and performance rating. A charge or tax that is levied on emissions, products, users, and services in proportion to the amount of pollution is based on the **"polluter pays" principle,** which asserts that the polluter should bear the cost of any abatement taken to maintain an acceptable level of environmental quality. This type of tax is also referred to as a **Pigouvian tax** after the British economist Arthur Cecil Pigou, who first proposed the idea in 1920. The disadvantages of imposing such a tax include the difficulty of identifying non-point-source polluters; high cost of monitoring environmental degradation; tendency of polluters to raise product prices, particularly when the product in question is a necessity for low-income households; and job losses created by industry trying to minimize its costs in response to the taxation.

Environmental externalities are not the only significant economic barrier to environmental problems, argues Coase (1960). The ability of interested parties to negotiate and bargain with one another in an effort to overcome these problems is impeded when one or more of the parties are made of many individuals. Often, there is considerable expense in time and resources for these individuals to get together and act collectively. These expenses are called **transaction costs.**

Coase brought attention to these costs. He also made some suggestions to those who make and administer laws and rules. One suggestion was that property rights and liability for imposed damages be made clear and specific so that time and resources are not spent in clarifying unclear laws. He showed that once laws are clear, parties are able to negotiate with one another in an effort to find the best solution to a problem, whatever the starting property right or liability law. He then suggested that legal and administrative systems be designed to create conditions that minimize transaction costs and that efforts through the legal process be made to put parties in a position to bargain and negotiate effectively and fairly.

The Coasian solution falls short of remedying environmental problems (such as global climate change) created by the existence of the natural commons at the global, regional, and local levels whose property rights are impossible to assign. High transaction costs (such as legal fees and the cost of bringing a large population of affected people to the bargaining table) incurred in the process of negotiating, the free-rider problem, and the lack of competitive markets can hinder the prevention and mitigation of environmental damages.

Marketable (tradable) permits, first introduced in the United States in 1977 as part of the Clean Air Act regulations but not taken up until placed in the 1990 statutory amendments, recognize actions for abating below the standard or pollution allowances for emitting a certain amount of pollution. We began this text in Chapter 1 discussing how private citizens had purchased some of the sulfur dioxide emissions allowances as gifts for others ("Dear Santa: Please bring me sulfur dioxide for Christmas") to force sulfur dioxide emissions down. Marketable permits can generate incentives for innovation and investment in efficient abatement capabilities of pollution and can be applied to the regulation of certain open-access resources. This idea was first used under the name of **individual transferable quota** for ocean fisheries in New Zealand in 1986. This scheme may involve high administrative, monitoring, and enforcement costs and is difficult to operate when there are several pollutants and many industries in a given region.

A **deposit-refund scheme** involves a front-end payment (deposit) for a potential polluting activity and a guarantee of a return of the payment (refund) when it is shown that the polluting activity did not occur. The system encourages recycling and more efficient use of raw materials, proper disposal of waste products, and **environment friendly behavior of consumption and production.** For example, mining companies extracting coal by strip mining are required to purchase a bond (which can be for millions of dollars) that guarantees they will restore the mined area to its original contour and reclaim it to its premining land use following coal extraction. If they do not, their bond is paid over to the state to provide funds to reclaim the affected area. Other policy instruments include ecolabeling and performance rating to provide information on the end-use products for consumer choices.

Environmental Justice

Concerns about the disproportionate distribution of environmental costs and benefits, responsibilities, and entitlements to poor or politically weak people (intra- and intergeneration) and their residential environments were transformed into the environmental justice movement. Expressions (slogans) such as "not in my backyard!" (**NIMBY**) and "not in anybody's backyard" (**NIABY**) were coined to describe popular antipathy for residential proximity to "locally unwanted land uses" (**LULUs**). These syndromes reflect the lack of public participation in decision-making processes regarding siting of facilities and land use choices. When applied to global climate change, the issue of environmental justice between the developed and developing countries springs from (1) the disproportionate per-capita or per-country use of fossil energy resources and the assimilative capacity of the atmosphere for greenhouse gas (GHG) emissions, (2) the disproportionate spatial distribution of source and sinks for GHG emissions, (3) the disproportionate spatial and temporal distribution of climate change impacts, and (4) the share of climate protection responsibilities inconsistent with historical GHG emissions.

Both trade liberalization and economic globalization have allowed firms to shift environmentally risky investments in the form of hazardous industries and imported wastes across national boundaries—in particular, to developing nations that often welcome these transfers as "investments of foreign capital" in their economic growth. The cheap supplies of labor power and raw materials, poverty, and the absence of environmental quality and assurance standards in developing countries have facilitated multinational capital and transnational corporations to decide to locate their environmentally degrading activities there.

The Precautionary Principle

Issues of scientific uncertainty, human values, and social justice that cannot be separated from environmental problems affect relations between scientific assessments and decision-making. Demands for proof of causation and verifiable results beyond present-day scientific capability lead to lack of actions or delayed actions that need to be taken to prevent adverse effects on human health and ecosystem sustainability. Asbestos, benzene, and tetra-ethyl-lead (for addition of lead to gasoline) are some examples of preventable health hazards for which credible evidence of harm existed before it was generally recognized. In the face of uncertain knowledge about potential adverse impacts of environmental degradation on health, the justification for acting is a function of the societal perception, consensus and cost of the issue, and the precautionary principle. The introduction of the **precautionary principle** into national and international law shifts the burden of proof from demonstrating the presence of risk to demonstrating the absence of risk, thus helping to avoid possible future harm associated with suspected, rather than conclusive, environmental risks. The greater the societal consensus on an issue is, the less scientific certainty is required to force action. The greater the societal costs of large-scale or irreversible environmental impacts are, the greater the need for precaution is.

Similarly, the **polluter pays principle** is found in the earliest examples of environmental policies and represents a sound basis for the enactment and enforcement of any environmental legislation. This principle was originated in the 1970s when members of OECD countries sought a means by which pollution control costs would be financed by the polluters rather than the public in general. However, this principle calls for answers to two questions: Who is the polluter of air, water, or soil? How are all the social costs of pollution to be accounted for? Unlike point source pollution, the diffuse nature of nonpoint source pollution makes it difficult to apply the polluter pays principle. On the other hand, incorporation of social costs to arrive at the real cost of a product requires that the environmental impact (cost) of that product (incurred during its production, transportation, usage, and disposal stages) be accurately reflected in its price.

Miles Gained, Miles Remain

Granted the intricacies—economic, legal, social, and cultural—that must be overcome in the implementation of policy and regulation of laws, the achievements in the last four decades have been phenomenal. We are far ahead of our grandparents in the awareness and recognition of environmental problems and in seeking appropriate mitigation measures for them. Many of these laws have worked and are working.

With the passage of NEPA of 1969 came the establishment of the U.S. Environmental Protection Agency (EPA) at the federal government cabinet level. In addition, the Executive Office of the President has the Council on Environmental Quality, whose task is to coordinate the environmental responses to issues that arise in cabinet-level departments. This EPA action has been followed in all 50 states by the establishment of similar cabinet-level or elected positions advising the governors and state legislatures in overseeing the application of laws and regulations for safeguarding environmental problems. Additionally, most county governments and municipalities have dedicated environmental professional staff.

Equally, one would be hard pressed to fund a corporation or industry without environmental divisions staffed by environmental professionals and advisers. There are also many for-profit and nonprofit organizations that provide employment for professionals trained in basic sciences, environmental sciences, and economic and social sciences. All this is most welcome. Lazarus (2004) presents a fine narrative on how the road to environmental policy was built in the 1970s, how the road was expanded in the 1980s, and how it is being maintained beyond the 1990s.

Given the nature of ecological systems, their integration and coordination with the legal system will not be easy; indeed, it can be daunting, according to Keiter (1998). Such complexity, Keiter notes further, arises because legal standards "represent a fragmented amalgam of federal, state, and local laws, often addressing single resources rather than the ecological complex itself." The complexity of global ecology and that of globalization demand new policies and laws that are effective; even the full implementation of existing policies will need our firm commitment and unwavering resolve.

Notes

1. In 1835, Alexis de Tocqueville, in his *Democracy in America* (1969), had this to say: "The power of associations has reached its highest proportions in America. Associations are made for purposes of trade, and for political, literary and religious interest. It is never by recourse to a high authority that one seeks success, but by an appeal to individual powers working in concert."
2. The original expanded version of this illustration, by the veteran journalist Lee Leonard and artist Charlie Zimkus, first appeared in the Columbus *Dispatch* (January 9, 2005, page C1) during the 126th session of the Ohio Legislature. Although it was intended as a "refresher course" for the new House and Senate members of Ohio, steps manifest in this illustration have wide applicability for all laws (including environmental), in all the states (with bicameral or two-chambered legislature), and at the federal level in the United States

and in other countries as well. At the request of one of the authors (MKW) of this book, this figure was revised and simplified by Leonard and Zimkus for which our grateful thanks are due.

3. In 1995, the functions of the Interstate Commerce Commission were transferred to the Surface Transportation Board.

4. Gordon Durnil was appointed to the Joint Canada–United States Commission on the Great Lakes by the 41st president of the United States, George H. W. Bush. Durnil's book, *The Making of a Conservative Environmentalist* (1995), is instructive on the responsibilities of multiple sectors of society appropriate to environmental management, regulation, and overall responsibility.

Questions

1. Discuss the interplay between science and policy during the process of decision- and policy-making.
2. How can the cost effectiveness (cost ratio of observed effects to policy measures) of environmental policies be assessed?
3. How can environmental policies be designed to deal with issues of uncertainty and risk, environmental justice, and the need to move to larger scales of implementation?
4. Name the types of environmental policy instruments used in your region and discuss their advantages and disadvantages.
5. What are the primary causes behind nonimplementation of, nonenforcement of, and noncompliance with environmental laws?

Economic Systems, Growth, and Development

22

Economics and Its Beginning

Human societies function materially through the exchange of goods and services, and specific methods of exchange are the basis of various economic systems. Hunting and gathering societies live at the subsistence level, literally a "hand-to-mouth" existence. In such societies, there is often no surplus accumulation of food or goods, and any exchange between persons or groups occurs on a goods-to-goods basis, also known as barter. With the advent of agriculture, groups of humans became more stationary and were capable of producing surplus. This surplus production and population growth was the genesis of monetary economic systems that lowered transaction costs and facilitated trade.

The French social philosophers of the eighteenth century based their economic thought on agricultural production, which they believed was the only true wealth. Then, "all of France was a major source of information and instruction. ... In 1765, one saw the rich land, the intelligent, patient and good-humored men who work it and the wonderfully varied products of French soil" (Galbraith 1977, pp. 16, 18).

Across the English Channel, their contemporary, the Scottish economic philosopher Adam Smith (now considered the father of economics) was contemplating other facets of trade and production. He laid major emphasis not on the bounty of the land but rather on the individual. "The wealth of a nation results from the diligent pursuit by each of its citizens of his own interests—when he reaps the resulting reward or suffers any resulting penalties. In serving his own interests, the individual serves the public interest" (Galbraith 1977, p. 22). Smith referred to this as the "invisible hand" that guides people to pursue their own self-interests, thereby benefiting society as a whole.

He further emphasized the concept of "division of labor" by which workers could produce much more. By specializing in one aspect of a trade, a company would, in turn, become more productive. The third element of Smith's thinking was freedom of trade at national and international levels. Taken together, one finds the three pillars of economics that rule the day today: the individual self-interest, the assembly-line concept based on the division of labor, and the freedom of trade.

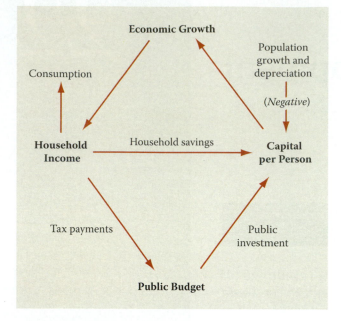

Economic Growth

Consumption

Population growth and depreciation

(*Negative*)

Household Income   Household savings   **Capital per Person**

Tax payments   Public investment

Public Budget

Figure 22.1 The basic mechanics of capital accumulation. (From Sachs, J. D. 2005. *The End of Poverty—Economic Possibilities of our Time.* The Penguin Press, New York, NY.)

With much increased production in sectors beyond agriculture came lowered costs, increased sales, and higher profits. This generated wealth and capital, and trade on a larger scale became possible (Figure 22.1). As villages grew to become cities, more and more people were occupied in trades or activities not directly concerned with subsistence food production. In time, the use of money to represent goods and services became the standard practice for all economic systems. This means that anything can be given a "value" in monetary terms—but nothing has any real value beyond what someone is willing to pay for it at a given moment. Thus, a global system of banking and investment has developed since the beginning of the Industrial Revolution.

Economic Systems

Although they cannot always be precisely defined, economic systems can be loosely placed along a continuum ranging from pure **market** (also called **capitalism,** where all production and distribution of goods and services are privately owned and controlled) to pure **command** economics (centrally planned and controlled by government). In capitalism, the law of supply and demand is fundamental; however, no existing system is "pure," and contradictory practices appear and overlap in each system. Many developing countries (e.g., India) use a combination of private enterprise and government-owned industries.

In well-functioning markets, exchange of goods or services between two persons leaves each better off than before the transaction. Markets together with government activities are the sum of a society's economic activities. Market activities are generally focused on the very short-term profits of the participants, rather than on the long-term consequences of the transactions. It is this time frame between the short- and long term—or the delay between the cause (such as atmospheric pollution by profitable heavy industries) and the effect (such as acid deposition)—that prevents markets from fully accounting for environmental impacts. This short-term view is further abetted by short-sighted political interventions.

Economic Practice and Its Major Principles

In the early nineteenth century, most regions of the world had a relatively low economic productivity of about $1,000 (in 1990 dollar value) per capita. Exceptions were found in Western Europe, the United States, Canada, and Oceania, where productivity was about $2,000 per capita (Figure 22.2). Increases in economic growth led to increasing affluence (defined as the average standard of living in a nation). Although populations in the United States increased over this time period, the gross domestic product (GDP; see later discussion) increased more rapidly, thus increasing levels of overall affluence. During this time, although land, labor, and capital have remained the mainstays of economic systems, major theories and models in economics have concentrated on eight aspects: economic growth, full employment, economic efficiency, price stability, economic freedom, equitable distribution of income, economic security, and

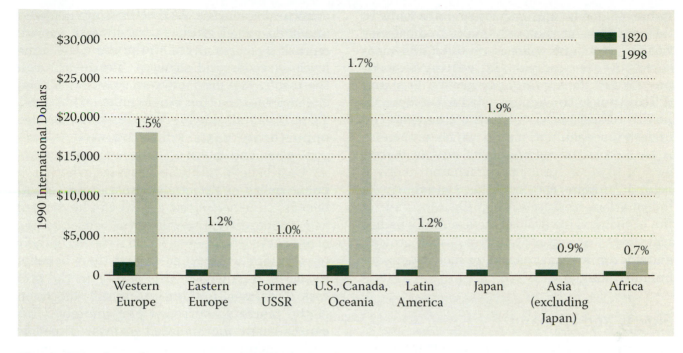

Figure 22.2 Gross domestic product (GDP) per capita by region in 1820 and 1998. (Figure 3, p. 29 in Sachs, J. D. 2005a. *The End of Poverty: Economic Possibilities for Our Time*. New York: Penguin Press.)

BOX 22.1   The Goals of Economics
(Based on McConnell and Brue, 2002, p. 9)

1. **Economic growth.** Produce more and better goods and services, or, more simply, develop a higher standard of living.
2. **Full employment.** Provide suitable jobs for all citizens who are willing and able to work.
3. **Economic efficiency.** Achieve the maximum fulfillment of wants using available productive resources.
4. **Price-level stability.** Avoid large upswings and downswings in general price level, that is, avoid inflation and deflation.
5. **Economic freedom.** Guarantee that businesses, workers, and consumers have a high degree of freedom in their economic activities.
6. **Equitable distribution of income.** Ensure that no group of citizens faces poverty while most others enjoy abundance.
7. **Economic security.** Provide for those who are chronically ill, disabled, laid off, aged, or otherwise unable to earn minimal levels of income.
8. **Balance of trade.** Seek a reasonable overall balance with the rest of the world in international trade and financial transactions.

balance of trade (see Box 22.1). With affluence comes increased demand for constructing shopping malls and housing developments, building roads, converting forests to agricultural land, and such.

Economics deals with allocations (monetary trade-offs) of scarce resources to produce **marketable (commercially priced) goods and services** available for consumption based on people's willingness to pay on individual business (micro) or societal (macro) scales. Goods are things people find useful (food, cars, baseballs), while services represent the flow of benefits over time. The **economic**

value of goods and services relates only to the contribution that they make to monetary wealth. Economic value is revealed and measured by the individual's **willingness to pay** (**WTP**) for a gain (or to avoid a loss) and **willingness to accept** (**WTA**) compensation to tolerate a loss or forego a benefit. The underlying assumptions on which economics is based are the rationality of human behavior (utility- or profit-maximization motives behind human choices), the availability of information, and (near) equilibrium state of the economy. However, these assumptions are not always representative of political, ethical, and environmental reasons or motivations for human behaviors.

Supply and Demand

A resource or an economic good or service is considered **scarce** whenever its demand exceeds its supply. **Demand** is the amount of an economic good or service that people are willing and able to buy at various possible prices, and **supply** is the amount of an economic good or service available to be purchased at different prices. The prices of marketed goods and services are controlled by the **law of supply and demand.** This states that if demand exceeds supply, then the price rises; the reverse is true if supply exceeds demand.

Markets, where the exchange of economic goods and services among people occurs freely, are economically efficient when demand and supply are in equilibrium. But market prices and shares cannot increase forever and are stabilized by negative feedbacks, thus reaching an equilibrium that becomes the most efficient allocation of scarce resources under the circumstances. The economic law of diminishing returns and the ecological concept of carrying capacity or limits to growth have the same underlying assumption, namely that positive feedbacks for consumption and production growth in a resource-limited environment cannot continue indefinitely (see Chapter 13).

Cost-Benefit Analysis (CBA)

An economy's ability to produce goods and services depends on the overall supply and quality of natural resources. As resources are depleted and/or degraded, it takes more effort to acquire more of those resources and produce the same levels of goods and services. **Trade-offs** are the result: The production of one kind of good or service means that others must be foregone. Thus, trade-offs due to scarce resources involve **opportunity costs,** which are what is foregone when one choice is made over another.

Cost-benefit analysis assesses whether a public policy or project generates more social benefits than social costs. CBA can be defined as a comparative analysis of the consequences of alternative courses of action in terms of their present (opportunity) costs and their benefits to society as a whole. This assists in the process of decision-making about the allocation of the scarce resources. Unfortunately, CBA can be easily manipulated in favor of people with monetary wealth, knowledge, and political power in terms of calculation and distribution of social costs and benefits because it can fail to account for the underlying causes of problems, such as population and consumption growth, distributive injustice, and ecologically incompatible technologies and institutional and political regimes.

Resources

Natural Resources and Marketable Resources

Resources that contribute to human well-being can be loosely divided into three categories. **Natural resources** (or natural capital) include the atmosphere, water, soil, the Earth's crust (such as minerals, ores, and fossil fuels: petroleum, natural gas, and coal), and biota (living components of the Earth's systems, such as plants that can be exploited for fuel, food, fiber, or timber and fish and land animals that can be used for human consumption). Natural resources are the totality of these biophysical assets. **Manufactured** (or **human-made capital**) **resources** are those that result from human activities, such as refined metals, dammed rivers, and all agricultural outputs.

Economics generally include the human capital (which some authors refer to as "**human resources**")—that is, the labor that fuels and controls these systems, as well as the mental energy that goes into business and information systems. In other words, resources used to produce economic goods and services include **human capital** (e.g., labor, human ingenuity, and social values), **man-made capital** (e.g., technology and other means of production), and **natural capital** (e.g., ecosystem goods and services).

Nonrenewable and Renewable Resources

All resources can be defined as either **renewable** or **nonrenewable.** Nonrenewable resources are those that, once used, cannot be replenished or are replenished very slowly by natural processes. Oil, natural gas, and coal are examples of nonrenewable resources; renewal rates of these organic fossil fuels are so slow as to be beyond comprehension on the human timescale. Renewable resources are those that can be replenished quickly—at least, theoretically—such as biological resources, water, or stratospheric ozone. Fertile soils, for example, are continually developing, but the time required to replace even a few inches of topsoil lost to poor farming practices can take centuries, making soil a nonrenewable resource for practical purposes. Timber, on the other hand, is widely cited as a renewable resource, but forest extractions require very careful management for yields to be truly sustainable. In general, renewability is a valid concept only when the rate of resource use remains below the rate or capacity for restoration. Thus, although clean water is theoretically an infinitely renewable resource, in many regions of the world, including the United States, people are using both ground and surface water at a rate much higher than it can be replenished. Similarly, overharvesting of some animal species has led to their extinction.

Property Values in Retrospect

In economic transactions, property rights play a crucial role. If someone has exclusive property rights on a good, he can exclude others from using or consuming that specific good. Exclusion refers to the ability of a producer to restrict potential consumers in using goods or services. Based on these facts, goods are broadly classified as:

- private goods: high degree of excludability and where there is rivalry in consumption;
- toll goods: high degree of excludability, but where joint consumption is possible;
- common-pool goods: no excludability, but with rivalry in consumption; and
- public goods: no excludability and no rivalry in consumption.

Unlike other species, humans "own" natural and manufactured resources, and thus they are unique in their concept of "ownership." Nomadic, preliterate (hunting and gathering) societies generally have limited concepts of personal or private ownership, viewing Earth systems as free to all members of the community. With the development of agriculture, permanent settlements, and commerce, however, personal ownership has become the rule in human society: All manufactured goods and services "belong" to someone, and this is the basis of "wealth." Individuals can acquire basic goods (food, clothing, shelter, fuel) by extracting them from natural sources. They can trade surplus of one material good for another and, in so doing profitably, may be able to pay for the result of other persons' labor—that is, to buy manufactured goods.

The United States—a society that, in the late nineteenth century, was more than 90% farmers producing the bulk of their own food—has become a society in which fewer than 2% of the population actually farms. The rest of us trade labor and brain power for the commodities we consume. Even though our concept of private ownership includes land as well as mineral and water rights, many natural resources are still considered collective property. This includes vast areas of publicly owned land, as well as some marine resources. In contrast, some created resources (for example, mechanical inventions, medical formulae, and music) have become "common" property over

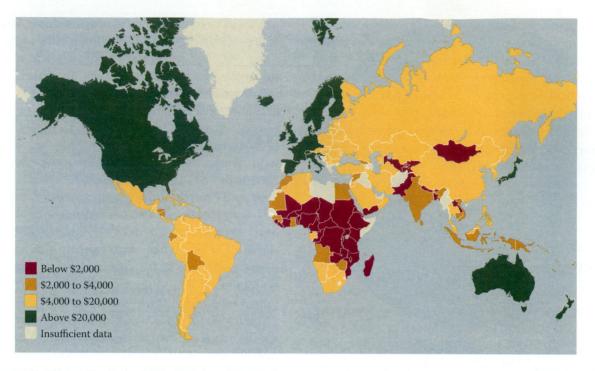

Figure 22.3 Countries in moderate poverty and extreme poverty. (From Sachs 2005a, based on data from the World Bank 2004.)

time; that is, no one has a current patent or copyright to control their use. Unfortunately, human wealth is not evenly distributed, and a large proportion of humans lives in poverty and need (Figure 22.3).

Under the current U.S. system, private interests may exploit public resources and, in some cases, federal revenues (our tax money) are used to subsidize these extractions. One example is the previously discussed 1812 Mining Law (Chapter 18) whereby the federal government issues licenses at a nominal fee to individual companies for the mining rights for metals. Another example is the use of National Forest Service funding to survey and construct roads into federally held old-growth forest so that privately owned timber companies can profit from the timber they harvest. Similarly, the American government subsidizes privately owned coal mining companies who sell the coal to privately owned utilities and industries. The problem of public resources being overused for the benefit of only a few is well illustrated in Garrett Hardin's "The Tragedy of the Commons" (1968), which was discussed in the preface to Section B.

The nonsustainable use of renewable resources lies at the heart of environmental **carrying capacity.** In the past, when the carrying capacity of the environment (such as available food supply or the ability to absorb wastes) was exceeded, famine, war, or disease would temporarily reduce the population to sustainable levels, and the cycle would begin again. This model is based on the widely observed phenomenon that humans, like all other organisms, have the ability to produce more offspring than can successfully reach reproductive age under normal (natural) environmental conditions. Human ingenuity and technology, however, have enabled far greater survival rates. Decreases in early mortality, better sanitation, control of many infectious diseases, and more productive food production have resulted in much longer human life spans than in the conditions under which humans evolved.

The Externalities

How do the current economic concepts cope with variables that are not accounted for or are external to the economic system? These are referred to as **externalities**, and they include the use of water or air for the discharge of industrial wastes; pollution such as this imposes costs on the community (costs associated with cleanup or health effects) that are not borne by the polluters. In this case, the use of the environment to get rid of unwanted waste is essentially free. These are costs not covered by the actual buying of resources or processing and selling them; instead, the costs accrue to a third party. For example, when utilities process coal for the generation of electric power, air pollution becomes an externality whose costs are borne by the society at large. Economists refer to the preceding example as an externality, or a spillover cost. Baumol and Oates (1988) define externalities as an unpriced, unintentional, uncompensated extra effect of an economic activity.

Incorporating the Externalities

With the present economic policies, externalities have been corrected with direct regulatory controls, also referred to as command-and-control regulation (explained in Chapter 21). In this case, governments simply mandate a level of acceptable environmental damage (for example, types and quantities of air pollutants emitted to the atmosphere or pollutants discharged into the water bodies). Scientific evidence of the harm from such pollutants and the best available technologies to combat them play a major role in the formulation of such controls.

Externalities may also be corrected by what are known as market approaches by which externalities are simply internalized; that is, they are incorporated as the cost of manufacturing a product. An excellent example of this is the trading of sulfur dioxide emissions mentioned in Chapter 1. Under new amendments to the Clean Air Act in 1990, companies were allotted a certain quota of sulfur dioxide emissions into the atmosphere (i.e., each company was allowed to emit a certain amount by weight of sulfur dioxide to the atmosphere) and were required to reduce these emissions to 50% of the 1990 levels by the year

2000. The incentive provided was that, if a company reduced its emission levels, it could sell or "trade" its savings (in terms of "units" of sulfur dioxide) to other utilities that might be lagging behind in retrofitting new technological devices. In this way, the total amount of SO_2 allowed into the atmosphere is capped, but companies can buy or sell excess units of sulfur credits. By all accounts, the program has been a great success and has surpassed the original expected reductions. This system of trading pollutant credits became the model for carbon trading under the Kyoto protocol.

Other Considerations

When markets fail to address ecological, political, or ethical considerations, there are **market failures.** The ecological considerations of market failures include scarcities of natural capital, common property resources, externalities, the long-term security and sustainability of life-supporting ecosystems, and conservation and use efficiency of energy, water, and materials. The **political** considerations include, for example, public participation, institutional coordination, and instabilities and inconsistencies; **ethical** considerations include those of public sovereignty, citizen responsibility, community solidarity, and sociocultural equity, integrity, and pluralism. Economists themselves suggest that "in the environmental domain, perfectly functioning markets are the exception rather than the rule" and that market failures are more instructive than market successes (Fullerton and Stavins 1998).

What is rarely considered in national economic bookkeeping is the **nonconsumptive use** value of ecosystem productivity, protecting water resources and soils, and regulation of the climate. Until ecosystem health can be quantified and integrated into our economic value system, it is inevitable that we will continue to exploit natural resources. International trade agreements rarely acknowledge the need for **environmental accounting**—recognition and inclusion of the true costs of resource exploitation and the value of intangible ecosystem benefits: clean air and water, peace, and stability.

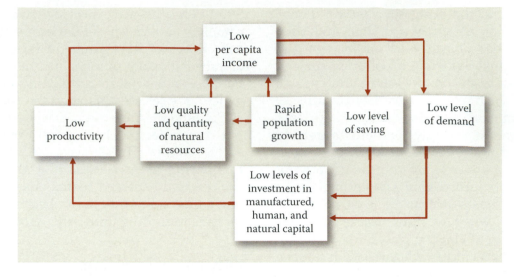

Figure 22.4 The multiple cause–effect pathways of poverty. Positive feedback exacerbates this human condition as population size increases. (Modified from Figure 39.3 in McConnell, C. R., and S. L. Brue. 2002. *Economics—Principles, Problems, and Policies,* 15th ed. New York: McGraw–Hill, Inc.)

Global Poverty and Economic Sector Subsidies

Nearly a billion people in the world today (more than 16% of the total human population) go hungry every day (Figure 22.3). Although the reasons for hunger and poverty are many, the connection with lack of access to natural resources in multiple cause–effect pathways is inescapable (Figure 22.4). Infertile soils, recurring drought conditions (as in some parts of Africa), high flood conditions followed by droughts (as in some parts of Asia), and lack of land ownership all play important roles. Several decades ago, the reason for famines was that there was not enough food to go around. Today, several countries are surplus-grain producers, but many people do not have the purchasing power to buy food. As the American expression goes, same difference: hungry still go hungry.

Although few countries have ever been truly isolated from regional economic influences, it is only in the last half-century that globalization has become a dominant force that often overrides individual national concerns. Industrially and technologically developed countries such as the United States and Western Europe increasingly export manufacturing jobs to countries where labor costs are relatively cheap and environmental regulations are lacking and then exploit the natural and human resources of less affluent populations. Such jobs do produce income for many; however, there is the ever present danger that landless wage-earners have no way of ensuring their continued employment—or providing for food and shelter—if markets for their products should shift or decline. In most developing countries, there has been increasing migration of rural people to cities in search of jobs. This movement cuts off the possibility of subsistence agricultural production and further skews the unequal distribution of wealth and human well-being.

Subsidies

In nearly all countries, governments provide direct and indirect monetary support to various segments of the society, including major economic sectors. These include individuals who cannot afford a daily living, people who are handicapped, or those who cannot afford medical care. This is part of the social responsibility.

But large subsidies also go to major economic sectors in direct infusion of money from the government and/or indirect subsidies in relaxed or write-off taxation. Several examples of the complexity of these subsidies may be illustrative. For example, the highway-related sector in the United States receives subsidies of many billions of dollars each year. At the end of July 2005, Congress passed a $286.5 billion appropriation for the 6-year spending plan on highways and mass transit and their infrastructure. Only $24 billion were allocated to causes only faintly related to the original purpose. An Associated Press story noted that there was something for everyone in this bill and quoted critics who said that this was "pork barrel spending at its worst." This is an example of "perverse subsidies" (see later discussion).

This level of subsidy has many **multiplier effects.** These effects ensure (a) a widely traveled automobile population, (b) an enhanced volume of truck traffic for moving goods and supplies, (c) service islands of gas pumps and restaurants to serve the traveling public, and (d) enhancement of tourism. The multiplier effects create constituencies who have vested interests in the perpetuation of such subsidies. It is no wonder that such large appropriations in the United States are referred to as "pork"— the expectation that elected members of the Congress will "bring home the bacon."

Other sectors receive subsidies as well. For instance, major oil companies receive subsidies for oil and gas exploration, large-scale agricultural producers receive price-support subsidies, and so on. As we noted earlier, no economic system can be considered "pure"; despite claims to the contrary, the capitalistic or free market economies with subsidies are pure no more. Also, many times subsidies invariably bring with them controls that have their impacts on the freedom to trade.

The extent, nature, and largesse of these subsidies were investigated by the Oxford environmental scientist Norman Myers (Myers 1998; Myers and Kent 1998). He examined the conventional subsidies together with the economic externalities of natural resource use and was dismayed to find that the policies were detrimental not only to the environment but also to economic systems. He coined the term "perverse subsidies" in reference to the infusion of money in five prominent leading sectors of the economy: agriculture, fossil fuels and nuclear energy, road transport, water, and fisheries (Table 22.1). Myers concluded that "a U.S. citizen pays taxes of at least $2,000 a year to fund perverse subsidies and pays almost another $2,000 through increased costs for consumer goods and through environmental degradation" (1998, p. 328).

These environmental subsidies were a major topic at the 2005 G-8 countries meeting (see note 6), drawing attention to one of the major causes of global poverty and the politics of nations as a deterrent to solving deeply entrenched problems. These subsidies have also been the major topic of discussions at the World Trade Organization meetings.

Table 22.1
Perverse Subsidies[a]

Sector	Conventional Subsidy	Environmental Externality	Total (range)	Perverse Subsidies (range)
Agriculture	325	250	575	460 (390–520)
Fossil fuels and nuclear energy	145	—	145	110
Road transport	558	359	917 (798–1041)	639
Water	60	175	235	220
Fisheries	22	—	22	22
Total	1110	785	1895	1450

Source: From Myers, N. 1998. *Nature* 392:327–328.
Notes: Conventional subsidies include established and readily recognized subsidies, including both direct financial transfers and indirect supports such as tax credits. Some of the estimates are supported by ranges (see text). In some instances, ranges are not included owing to insufficient agreement. Data are too patchy and disparate to allow even a reasonably agreed-upon value of the environmental externalities for fossil fuels and nuclear energy.
[a] Billions of dollars per year.

Environmental and Resource Economics

The preceding discussion has been on how economic systems presently work around the world. These systems collectively are referred to as neoclassical economics. The word *neoclassical* refers to the revival or continuance in a slightly modified form of a classical system. Neoclassical economics "is characterized by its marginal utility theory of value, its devotion to the general equilibrium model stated mathematically, its individualism and reliance on free markets and the invisible hand as the best means of allocating resources, with a consequent downplaying of the role of government" (Daly and Farley 2004, p. 437).

Concerned about pollution effects and the overall degradation of the environment, in 1975, several mainstream economists formally established the subdiscipline of environmental and resource economics as the application of the principles of (neoclassical) economics to the study of management of environment and natural resources incorporating some ecological principles and methodology. The focus of environmental economics is to assign appropriate values for environment and natural resources to assess the full costs and benefits of economic activity on the environment. The conceptual base of environmental economics, strongly grounded in neoclassical economics, is designed to capture the full social costs and benefits of economic activities in the utilization of natural resources. This academic field has the acceptance of mainstream economists, and several treatises are available (Gilpin 2000; Tietenberg 2003).

Economics and Ecology

Although the general impression created by some is that ecology and economics are on cross-paths, the idea that what is good for ecology will be hurtful to the economy is false. Nothing could be further from the truth.

Both disciplines originate from the same root: *oikos,* meaning house; that ecology is the study (*logos*) of the house and economics (*nomie*) its management is more than a truism and beyond semantics. More than 40 years ago, the economist Kenneth E. Boulding wrote two seminal papers (1966a, 1996b). In one, he likened the Earth to a closed system, what he called "spaceship Earth" (Boulding 1996a; see Chapter 6). He made a strong distinction between what some viewed as the limitless resources of an "open" Earth and a "closed" Earth that has finite limits. In a closed system, the outputs of all parts of the system are linked to the inputs of other parts. There are no inputs from the outside and no outputs to the outside; indeed, there is no outside at all (Boulding 1996a, p. 4).

Even more incredible to Boulding (1966b) was the fact that although economists held on tenaciously to gross national product, ecologists never conceived of thinking of gross primary or secondary productivity as a genuine measure of anything. Rather, it was the net primary productivity (gross minus maintenance costs) that was the only relevant measure. In summary, the statement from Postel (1991) becomes a truism: "While the environment and the economy are tightly interwoven in reality, they are almost completely divorced from one another in economic structures and institutions" (Postel, State of the World 1991).

Steady-State Economics

Is growth essential to economic health? Under our current economic system, the answer is an unequivocal "yes." But continued exploitation of resources depends on sustained yield, which may not be possible for most resources. In general, the cost to extract a resource increases as the supply diminishes. The concept of **optimal yield,** in which resources are never overdrawn, is an ideal for sustained production. However, given the reality of a speculative economy whose success is based on only the shortest term profits or stock market fluctuation, it is not likely that our patterns of use, consumption, production, and distribution will change substantially.

The ultimate approach to achieving a balance between natural resource use and avoidance of environmental problems (the most difficult) is the concept of steady-state growth. The concept, proposed by Herman Daly (1991), advocates a steady-state economy rather than unrestrained economic growth by distinguishing between "growth" and "development." Daly emphasizes that there are clear biophysical limits to economic growth, but not to economic development. Ecological limits on human population and consumption require that the distinction between economic growth and development be made. The term **growth** refers to increase in size and cannot be sustained indefinitely; **development** refers to betterment or progression and can be sustained definitely in our rate- and stock-limited world.

Three basic criteria for maintenance of natural capital and ecological sustainability (Daly 1990) include:

- For renewable resources, the rate of harvest should not exceed the rate of regeneration (sustainable yield).
- The rates of waste generation should not exceed the assimilative capacity of the environment (sustainable waste disposal).
- For nonrenewable resources, their depletion should require that comparable developments of renewable substitutes for that resource take place.

In sum,

"the most distinctive trait of steady-state economics is stable size ... [It] undergoes neither growth nor recession...has constant populations of people (and therefore 'stocks of labor') and constant stocks of capital ... a constant rate of 'throughput'—i.e., the energy and materials used to produce good and services" (Czech and Daly 2004).

As noted earlier, this concept is anathema to current economic thinking. Unless we drastically overhaul and abandon adherence to the contemporary economic precepts and constrain human preferences, achieving steady-state economics is unattainable (see notes).

Ecological Economics

These considerations have given rise to an emerging field called **ecological economics,** which aims at integrating economics and ecology under one umbrella. Ecological economics recognizes the economy as a subsystem that must operate within the limits of the biosphere (Figures 22.5a and 22.5b). Here, energy and raw materials flow through the economy from natural ecosystems and return to those ecosystems as "high entropy wastes." It views environmental issues holistically, acknowledging that growth has limits. Economic progress should not be measured solely as growth in GDP, but also as improvement in human welfare overall, a measure of how sustainably we live (Costanza and Martinez-Alier 2006). Because ecosystem services are not quantified in monetary terms, their contributions to human well-being have been easy to ignore when policy decisions are made. Ecological economics allows one to quantify the value of ecosystem services. Several market-based recommendations toward solving environmental problems that emerge from this approach include:

- valuation methods to place economic values on environmental goods and services that have traditionally been externalities;
- better aggregate indicators of nations' wealth (i.e., indicators that are more sensitive to changes in environmental quality than the GDP);
- economic incentives and disincentives;
- tradable or marketable permits; and
- removal of market barriers to environmentally responsible business and of environmentally destructive subsidies.

According to environmental valuation methods, the **total economic value** of an ecosystem good or service is made up of its use and nonuse values (Figure 22.6). Ultimately,

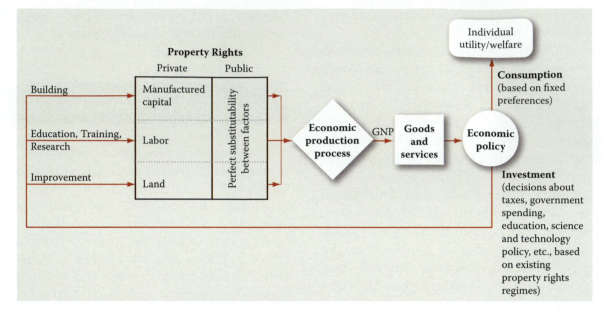

Figure 22.5a Conventional paradigm of the economy: A model depicting primary production factors of land, labor, and capital that accounts for economic production processes of marketed goods and services. (From Figure 1a, p. 5, in Costanza, R. 2000. *Ecosystems* 3:4–10.)

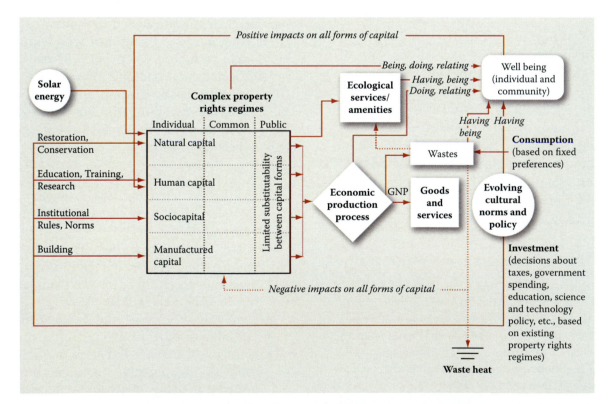

Figure 22.5b The ecological economic systems as an emerging paradigm: a model depicting primary production factors of natural, human, social, and manufactured capitals that have limited substitutability and conform to the basic laws of thermodynamics that account for economic and ecological production processes of marketed and nonmarketed goods and services, as well as solid, gaseous, and liquid wastes. (From Figure 1a, p. 5, in Costanza, R. 2000. *Ecosystems* 3:4–10.)

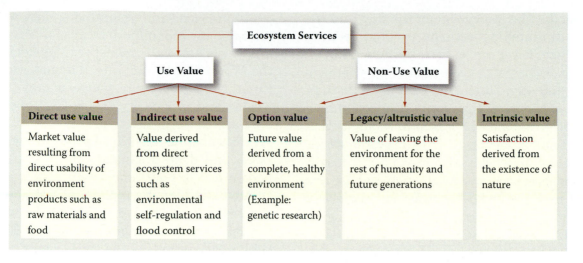

Figure 22.6 Total economic value of an environmental resource as made up of use value and nonuse value and definition of valuation tools for ecosystem goods and services. (From *Science on Sustainability* 2006.)

this allows estimates of the value (in monetary terms) of the services that ecosystems provide, including food, water, climate regulation, water purification, and aesthetic enjoyment. A recent study of the global economic value of 17 ecosystem goods and services in 16 biomes estimated their total average value of US$33 trillion per year—equivalent to twice the global GNP (Costanza et al. 1997). The average global value of the ecosystem services of wetland ecosystems alone was estimated to be nearly $15,000 per hectare per year—the highest estimated value for any biome. Most of the ecosystem services quantified in this study are not included in the market system in any way.

Nonuse values refer to individuals' WTP and WTA for reasons unrelated to their direct utility of nonmarketed ecosystem goods and services such as biodiversity, conserved habitats, and endangered species. These are estimated by the contingent valuation method. Contingent valuation is a questionnaire-based method in which nonuse economic values of nonmarketed goods and services are estimated contingent upon the nature of the constructed (hypothetical) market and of their description in the survey scenario. Travel cost and random utility methods are, in general, employed to estimate recreational values based on travel expenditures, behaviors, and choices (WTP). For example, if people are willing to drive 50 miles to a lake in order to fish, but are not willing to go if they have to drive 100 miles,

this provides an indication of the value of that lake for recreation.

Safe minimum standards (SMSs), a concept introduced by Ciriacy-Wantrup (1952), refer to the incorporation of experts' advice on ecological use capacity of natural resources into public decision- and policy-making in order to avoid irreversible environmental damage (the precautionary principle). As long as it is socially acceptable and economically not too costly, SMSs may serve to facilitate the communication between policy-makers and scientists in recognizing ecological limits beyond which further economic utilization becomes irrational, as well as uncertainties associated with loss of ecosystem goods and services and their impacts on present and future public health.

Growth and Development

Not too long after Adam Smith's death came *An Essay on the Principle of Population,* the essay from the English vicar the Reverend Thomas R. Malthus, in 1798. We have already reviewed in Chapter 13 the relevance of his postulations. It may be worthwhile to repeat them here: Human population will have a tendency to grow in a geometric progression (1, 2, 4, 8, 16, …), whereas food production will increase only arithmetically (1, 2, 3, 4, 5, …). He arrived at his conclusions by considering the knowledge of his day and the law of diminishing returns in land fertility and, hence, lower agricultural production.

That ecologists, economists, environmentalists, and social scientists should be discussing his paper more than two centuries after it was written should be a lesson in thoughts that endure to all students. He is still revered for his foresight by those who see natural limits to current models of economic production and use and reviled by others who lay scorn for his lack of foresight on what was to come in technological advances. Even today, however, his detractors commend him for "the germs of truth" in his doctrines, and he is viewed as "still important for understanding the population behavior" (Samuelson and Nordhaus 2001, p. 365).

Current economic systems and insatiable human wants promote an ever-increasing consumption of goods—consumerism—that invariably results in ever-increasing rates of natural resource exploitation, manufactured resource production, and waste generation. Thus, economic growth at such annual rates as 8–10% is not sustainable because we are eroding the natural capital upon which we depend. Sustained exponential growth is simply not realistic in any living system or in any known economic system.

Market Economics: GDP and GNP

Economists measure the wealth of a nation or growth and development of a nation by two major indices: **Gross domestic product (GDP)** is the production by companies, domestic and foreign, within a country's borders, and **gross national product (GNP)** measures the current market value of all output (goods and services) by a country's companies at home and abroad. Thus,

Consumer spending + business investment

+ government purchases + net exports

= gross domestic product (GDP)

Impressive and voluminous data sets are released by economic bureaus of all nations and international monetary agencies (like the World Bank and International Monetary Fund) as a surrogate gauge of any nation's wealth and prosperity (Table 22.2). So venerated are these indices that major government services such as military spending, social services, and debts incurred are presented as a percentage of GDP and GNP. Values are calculated similarly by individual states (termed the gross state product, GSP) or the world as a whole (gross world product, GWP: the total GNP of all nations of the world combined).

Despite the sophistication and complexity of models used, both indices are calculated on easily tractable variables. When real economic costs are calculated in their entirety (as explained later), both indices miss the degradation of natural assets and the welfare puzzle. As is now recognized by thoughtful economists, as an aggregate economic measure of a nation's wealth, GNP or GDP does not address nonmarketed goods and services, environmental degradation and destruction, income distribution between the rich and the poor, scarcities of natural resources, and intergenerational needs of clean air, water, land, and fertile soils. For example, the resulting deaths and illnesses of people due to environmental pollution of air, water, and soil (such as oil spills, poisoning of the food chain by pesticides and heavy metals, and runoff of excess nutrients into the water supply) increase the annual GNP of a nation due to vast sums of money spent on producing goods and services, such as emergency medical help, hospital bills, and cemetery bills.

In a country with a high GNP, some people starve chronically, while others overconsume chronically. Scandinavian countries have a high GDP per capita, but they redistribute relatively more of that income and wealth; thus, their economic policies help to ensure the well-being of their people by securing **long-term social** and **natural productive capacities.** Examples include distributive policies (e.g., agricultural land reform, minimum wage legislation, safety net, and progressive income tax) and using ecologically compatible

Table 22.2
**The Top Ten—Ranked Countries by (a) Gross Domestic Product (GDP) and
(b) Gross National Income (GNI)**

	(a) GDP Top Ten, 2007 (in millions of US dollars)			(b) GNI Top Ten, 2007 (in millions of US dollars)	
Rank	Country	GDP	Rank	Country	GNI
1	United States	13,811,200	1	United States	13,888,472
2	China	7,055,079	2	Japan	4,813,341
3	Japan	4,283,529	3	Germany	3,197,029
4	India	3,092,126	4	China	3,120,891
5	Germany	2,727,514	5	United Kingdom	2,618,513
6	Russian Federation	2,088,207	6	France	2,447,090
7	France	2,061,884	7	Italy	1,991,284
8	United Kingdom	2,046,780	8	Spain	1,321,756
9	Brazil	1,833,601	9	Canada	1,300,025
10	Italy	1,777,353	10	Brazil	1,333,030

Source: Adapted from Gross Domestic Product 2007. World Development Indicators Database, World Bank. Accessed from http://www.worldbank.org January 16, 2009; Gross National Income 2007, Atlas Method. World Development Indicators Database, World Bank. Accessed from http://www.worldbank.org January 16, 2009.

land use policies, energy- and water-efficient technologies, and renewable energy.

Perhaps in no way has the picture of nation-states and the world been so distorted as by the measurement of productivity and welfare of humans by the much widely tracked economic metric of the GNP. This measure is fundamentally—indeed, seriously—flawed because it measures only market value of goods and services and not the environmental and human consequences.

GNP has been described as an "irrelevant metric" (Hall 1992) for several reasons. First, it is only a partial measure of those conditions that contribute to human happiness and well-being—the supposed goal of economic activity. Second, GNP does not account for the distribution of wealth among the rich and the poor. Third, it is not an accurate measure of production for developing countries. Fourth, GNP does not measure nonmarket transactions and thus undervalues environmental services. Therefore, it takes into account neither a properly functioning ecosystem nor a degraded one. Lastly, although the efficiency of production

has risen, the flow of goods encourages production of more, but inferior, goods. Thus, economic analysis based narrowly on market prices can lead to increasing destruction of nature.

Genuine Progress Indicator

In 1995 an article by Cobb, Halstead, and Cobb appeared in the *Atlantic Monthly* with a provocative title: "If the GDP is up, why is America down?" Reliable accounts and measures of natural capital are at the heart of a sustainable course of economic development that would allow policy-makers, the public, and researchers to monitor the state of life-supporting ecological services, set goals, and make choices consistent with biophysical capacity at a given time and space scale. Lack of biophysical indicators paves the way for depletion and degradation of natural resources more rapidly than they are replenished or restored because time lags exist between human activities of environmental degradation and destruction and their immediate consequences.

The real economic measures must include degradation of the environment and the costs needed to mitigate it, as well as the social costs of both unemployment and unpaid work. The labor of women who contribute substantially to the sustenance of their families in developing countries is not included in any labor or wealth statistics of their respective countries or international agencies. Inclusion of these results in what has been called a genuine progress indicator (GPI) is calculated as:

GDP – (unpaid work + crime and

social breakdown + ecological damage)

= genuine progress indicator (GPI).

The true economic costs can be calculated in other ways. One example is the $I = P.\ A.\ T$ formula, where I is the environmental impact, P is the human population size, A is affluence, and T is the impact of technology. This formula estimates environmental impacts as a function of both population size and affluence and the technology needed to use resources and generate waste to produce the goods and services that we consume.

GPI and GDP show quite a variance (Figure 22.7). For example, it has been shown that, despite a growing economy of the United States with an annual GNP increase of 1.8% in the 1980s, the sustainable economic welfare actually declined by 0.8% per year for the same period.

Economic disincentives are taxes or fines on the depletion and degradation of natural resources. The purpose of the economic disincentives is to discourage environmentally destructive activities (e.g., underpriced ecosystem goods and services derived from limited resources, ozone layer-depleting chemicals, deforestation, overgrazing, and fossil fuel consumption). At the same time, they promote more efficient use of natural resources (e.g., conservation and efficient use of energy, materials, and water) and mitigate the environmental and social impacts of economic activities through green tax-based revenues (e.g., restoration and rehabilitation of damaged ecosystems and the development of infrastructure required by the additional population and consumption growth that accompanies the economic activities).

Economic incentives serve to encourage the production and consumption of ecologically compatible goods and services and range from government-provided research and development grants and tax breaks to donations for the development of environmentally beneficial technologies. Economic disincentives and incentives directly and indirectly reorient both

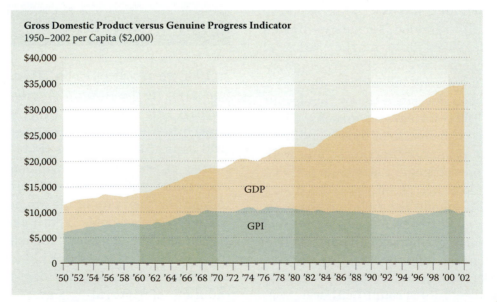

Figure 22.7 The genuine progress indicator, 1950–2003. (From *Science on Sustainability 2006. Summary Report.* A view from Japan. RSBS-Research on the Scientific Basis for Sustainability. http://www.sos2006.jp. Accessed January 15, 2009.)

supply and demand sides toward taking the externalities into account.

Tradable or **marketable permits** refer to the permits or licenses issued by governments to businesses to release certain amounts based on safe minimum standards of pollution into the air and water; these can be bought and sold in the marketplace (see Chapter 1). Sulfur is the only pollutant that is covered by the marketable permit in the United States. This incentive system allows companies to meet and reduce their required pollution limits flexibly and creatively. The downside of the permit approach is that it does not take into account that vulnerability (sensitivity) of ecosystems to pollutants differs considerably and that it must be regionally constrained to avoid geographic concentrations of polluting producers.

Human Development Index

This index has been developed by the United Nations Development Program and has been in currency for several years. Its development is based on the premise that the overall well-being of a society is not really represented by a measure that considers only the production of goods and services in a society. As dimensions of human development will go beyond mere economic accumulation, the aspects affecting human development such as a long and healthy life, knowledge, and the standard of living of people need to be addressed to arrive at a more comprehensive measure of human development. The human development index (HDI) emerged as a measure that considers the broader aspects of human development. The HDI is a composite index that measures the average achievements of a society as a provider of basic capability enhancement opportunity for the member in that society:

- A long and healthy life can be viewed as a single variable that can explain the citizens' opportunity for accessing and utilizing a healthy life in a society.
- Knowledge transforms people to carriers of economic development rather

than passive receivers of a development process. This is measured by the adult literacy rate (with two-thirds weight) and the combined primary, secondary, and tertiary gross enrollment ratio (with one-third weight).

- A decent standard of living, as measured by GDP per capita in US$ and expressed in purchasing power parity, is a measure of relative purchasing power across the countries. (The calculations for the HDI are given in the notes.)

The Concept of the Ecological Footprint

The **ecological footprint** (**EF**) is an accounting of how much ecologically productive land and water are needed to sustain one person's lifestyle. This includes the land and water bodies needed to produce all the goods and to assimilate all wastes involved (Wackernagel and Rees 1996) (Box 22.2). By simple division, with the current population of more than 6 billion people on the Earth, approximately 1.9 ha of land area is available to support each person. Current estimates are that the average U.S. citizen has an ecological footprint of 9.5 ha; thus, five Earths would be needed for the rest of the world's population to consume at this rate. The goal of a footprint analysis is to present, in clear terms, the ecological overshoot in resource use (i.e., how much higher the demands for resources are compared to that which nature can continually supply). This makes tangible what it means for each of us to live sustainably.

An analysis of the average footprint of different countries reveals that some are ecological creditors (i.e., each citizen uses, on average, less than 1.9 ha per person, so there is a net surplus) and some are ecological debtors (where the average citizen has a footprint larger than 1.9 ha). Patterns of world consumption illustrate differences in resource use that contribute to footprint size (Figure 22.8) and remind us of a quote from Malthus (1798): "As population doubles and redoubles, it is as if the globe

BOX 22.2   Definition, Significance, and Assumptions of Ecological Footprint Analysis (Based on Wackernagel et al. 2002)

Sustainability requires living within the regenerative, assimilative, productive, protective, and regulative capacities of the ecosphere. In an attempt to account for the extent to which humanity uses the ecosphere, Wackernagel and Rees (1996) developed an accounting framework to measure "the area of biologically productive land and water required to produce the resources consumed and to assimilate the wastes generated by humanity, under the predominant management and production practices in any given year." According to a Wackernagel et al. assessment (2002), humanity's load grew from 70% of the capacity of the global biosphere in 1961 to 120% in 1999. This 20% overshoot means that it would require 1.2 Earths, or 1 Earth for 1.2 years, to regenerate what humanity used in 1999.

Economic significance of the ecological footprint analyses can be summarized as follows:

1. Ecological footprint analysis is an easy-to-determine biophysical indicator of sustainability that can assist in informing humans of production choices.
2. Ecological footprint analysis informs economic actors of the total availability of a resource to be used over time and signals whether or not the resource will run out at the time the price of the resource rises to that of a substitute resource and technology.
3. Ecological footprint analysis provides biophysical data that can enable adjustments in market prices to reflect social costs, including costs to future generations (adjustments referred to as "shadow prices").
4. Ecological footprint analysis provides biophysical indications of the consequences of the current distribution of resource access within and between generations from which, along with ethical criteria, new distributions of rights might be made.

Assessments of the ecological footprint of humanity are based on six assumptions:

were halving and halving again in size—until finally it has shrunk so much that food production has shrunk below the level necessary to support the population." Ultimately, the ecological footprint is an index of sustainability.

The central public policy implication of this concept is that human population and consumption grow exponentially, but the ecological goods and services that support them are rate- and stock limited. The ecological footprint represents human-appropriated biologically productive stock—the inverse of carrying capacity (defined as the population that can be supported indefinitely by a given habitat).

1. It is possible to keep track of most of the resources humanity consumes and the wastes humanity generates.

2. Most of these resource and waste flows can be measured in terms of the biologically productive area necessary to maintain these flows (those resource and waste flows that cannot are excluded from the assessment).

3. By weighting each area in proportion to its usable biomass productivity (that is, its potential production of biomass that is of economic interest to people), the different areas can be expressed in standardized hectares. These standardized hectares, which are called "global hectares," represent hectares with biomass productivity equal to the world average productivity that year.

4. Because these areas stand for mutually exclusive uses and each global hectare represents the same amount of usable biomass production for a given year, all mutually exclusive resource-providing and waste-assimilating areas can be added up to a total representing the aggregate human demand.

5. Nature's supply of ecological services can also be expressed in global hectares of biologically productive space.

6. Area demand can exceed area supply. This phenomenon is called "ecological overshoot."

Human demands on nature's services and nature's supply of services (the economy of nature) change over time in response to innovations in technology, changes in resource management practices, land use and land cover, patterns of human consumption and production, increase in environmental awareness, and past, present, and future damages of environmental issues. However, ecological footprint analyses need to be equipped with information concerning (1) how rapidly natural resources are becoming depleted and degraded, (2) how quickly limiting factors are coming into play for the welfare of humans and their life-supporting ecosystems, and (3) how long such environmental degradation and depletion can continue.

By definition, the difference between the ecologically productive size of a region and the footprint of that region's population must be offset by trades of ecological surpluses from other regions or the depletion of natural capital stocks at the expense of future generations (Table 22.3). The indicator also points out how the ecological footprint can be reduced through a combination of (a) lower population growth, (b) lower consumption, (c) more efficient technologies, (d) higher ecological productivity, (e) rehabilitation and restoration of damaged ecosystems, and (f) conservation of natural resources (see Box 22.2).

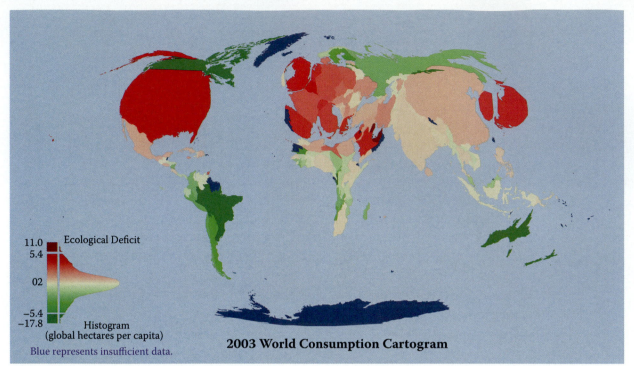

2003 World Consumption Cartogram

Countries have been stretched to indicate their effective consumption based upon 2006 Global Footprint Network and corresponding 2003 CIA World Fact Book data. The basemap is an edited ESRI ArcIMS world shapefile in the Mollweide equal area projection. Created with ArcMap 9, MAPresso 1.3, OpenOffice 2.2 and Perl 5.8. http://pthbb.org/natural/footprint/ Accessed February 24, 2009.

Figure 22.8 World map showing the national ecological footprints (indicated by relative size) and the degree to which each country is an ecological debtor (red) or creditor (green).

Table 22.3
Ecological Footprint of Nations[a]

Country	Footprint (global hectares per capita)	Country	Footprint (global hectares per capita)	Country	Footprint (global hectares per capita)
United States	9.57	Croatia	2.76	Cameroon	1.24
United Arab Emirates	8.97	Botswana	2.70	Senegal	1.23
Canada	8.56	Macedonia	2.69	Ghana	1.23
Norway	8.17	Bulgaria	2.65	Guinea	1.22
New Zealand	8.13	Turkmenistan	2.60	Sudan	1.20
Kuwait	8.01	Mexico	2.59	Burkina Faso	1.19
Sweden	7.95	Namibia	2.52	Egypt	1.16
Australia	7.09	Romania	2.46	Mali	1.16
Finland	7.00	Korea Republic	2.43	Moldova Republic	1.13
France	5.74	Venezuela	2.42	Philippines	1.11
Mongolia	5.68	Brazil	2.39	Nigeria	1.10
Estonia	5.37	Lebanon	2.37	Kyrgyzstan	1.10
Portugal	5.34	Mauritania	2.36	Laos	1.09

Table 22.3 (continued)
Ecological Footprint of Nations[a]

Country	Footprint (global hectares per capita)	Country	Footprint (global hectares per capita)	Country	Footprint (global hectares per capita)
Denmark	5.32	Paraguay	2.29	Kenya	1.08
Switzerland	5.26	Turkey	2.20	Zimbabwe	1.05
Belgium and Luxembourg	5.11	World	2.18	Guinea-Bissau	1.05
Ireland	4.97	Jamaica	2.15	Cambodia	1.03
Spain	4.90	Costa Rica	1.91	Zambia	1.02
Austria	4.87	Azerbaijan	1.91	Gambia	1.01
Greece	4.78	Panama	1.89	Indonesia	0.98
United Kingdom	4.72	Gabon	1.87	Madagascar	0.97
Latvia	4.40	Iran	1.85	Benin	0.92
Russia	4.28	Ecuador	1.77	Morocco	0.92
Germany	4.26	Syria	1.74	Tanzania	0.89
Czech Republic	4.24	Trinidad and Tobago	1.73	Sri Lanka	0.88
Korea DPRP	4.07	El Salvador	1.72	Sierra Leone	0.88
Saudi Arabia	4.05	Dominican Republic	1.69	Georgia	0.85
Israel	3.97	Algeria	1.67	Liberia	0.85
Japan	3.91	Bolivia	1.67	Eritrea	0.81
Lithuania	3.87	Cote d'Ivoire	1.60	Congo	0.80
Netherlands	3.81	Nicaragua	1.57	Rwanda	0.78
Kazakhstan	3.75	Honduras	1.54	Vietnam	0.76
Ukraine	3.53	Cuba	1.53	Myanmar	0.76
Slovenia	3.52	Tunisia	1.51	India	0.76
South Africa	3.52	Colombia	1.51	Angola	0.76
Poland	3.40	Bosnia Herzegovina	1.49	Armenia	0.75
Uruguay	3.32	Central African Republic	1.48	Pakistan	0.67
Slovakia	3.27	Thailand	1.41	Ethiopia	0.67
Italy	3.26	Jordan	1.39	Tajikistan	0.65
Hungary	3.26	China	1.36	Malawi	0.64
Mauritius	3.25	Chad	1.31	Burundi	0.63
Libya	3.21	Guatemala	1.30	Congo Dem. Rep.	0.62
Argentina	3.18	Uganda	1.29	Haiti	0.62
Belarus	3.17	Peru	1.26	Nepal	0.57
Chile	3.04	Albania	1.25	Mozambique	0.56
Malaysia	2.99	Papua New Guinea	1.25	Bangladesh	0.50

Source: From Venetoulis, J., D. Chazan, and C. Gaudet. 2004. Available at www.RedefiningProgress.org; Table 2 on p. 12.
[a] Sustainability Indicators Program. Redefining Progress for People, Nature, and the Economy.

An Afterword

The economic and financial events of the last six or so months have been riveting. The warnings of several economists and financial experts of looming crises in several financial sectors have proved to be correct. These crises, unprecedented in both their extent and magnitude, have resulted from the mismanagement of financial institutions. The causes are many, but stem in a large extent part from loosening of government regulation of the financial system, low interest rates and pressure to make loans, particularly housing loans, from unpredictable economic engineering, greed, and the application of models that have proven to be poor predictors of contemporary conditions. With the disbanding of most regulation and the laxity of other government controls, economics has gone into a "death spiral" causing worldwide financial confusion and misery. Invariably, among the first casualties of such serious economic downturns are not the environmental advances and regulation alone (Figure 22.9), but even the routinely-scheduled services such as cleanup and recycling.

The economic death spiral is no different than the forest/tree death spiral discussed in Chapter 18 (Figure 18.14). While the latter results from the synergy of a multitude of physical, chemical, genetic and human-induced factors, the economic death spiral is solely orchestrated by man. The economic system displays a pattern of cyclic ups-and-downs that seem to come about every few decades or so, but the recent phenomenon of economic "bubbles"— for example the savings-and-loan bubble of the 1980s, the dot.com bubble of 1990s, the current subprime- and banking bubble, result from market speculation that the value of something will continue to rise. Bubbles burst when investors recognize that the price of something is far beyond its intrinsic worth. The misery that the recent economic events have brought to millions of people not just in the U.S. but worldwide has not been seen since the Great Depression. What has been a first is that banks thought to have the most balanced credit-and-debit books,

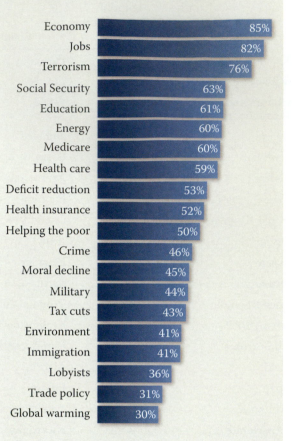

Percentage of people who rated each a 'top priority:'

Economy	85%
Jobs	82%
Terrorism	76%
Social Security	63%
Education	61%
Energy	60%
Medicare	60%
Health care	59%
Deficit reduction	53%
Health insurance	52%
Helping the poor	50%
Crime	46%
Moral decline	45%
Military	44%
Tax cuts	43%
Environment	41%
Immigration	41%
Lobyists	36%
Trade policy	31%
Global warming	30%

Figure 22.9 Of the 20 priorities for the U.S. citizens in a survey conducted by the Pew Research Center, the environment ranked 16th and global warming 20th. The top priorities were economy (85%) and jobs (82%). Source: Economy, jobs trump all other policy priorities in 2009, January 22, 2009. Washington, D.C.: The Pew Research Center for the People and the Press.

today do not know how much they are worth and how much and to whom they owe.

Thoughtful economists and other professionals now agree that things got out of control. "[T]here's no question that with hindsight, stronger regulation of [financial institutions] would have been appropriate. ... Large swaths of economics are going to have to be rethought on the basis of what has happened" (quoting Lawrence Summers quoted in "Reeducation of Larry Summers" by Michael Hirsch and Evan Thomas, *Newsweek*, March 2, 2009, p.26). Such rethought is the hope of environmental scientists as well to ensure sustainability of all resources, including financial. One means by which this might occur is through government intervention into the markets. Interestingly, the recent stimulus package in the U.S. will pump billions of dollars into environmental programs, including 7.2 billion in funds to boost programs run by the U.S. Environmental Protection Agency. These funds will help pay for programs in brown-field clean up, development of clean fuels, and habitat restoration (Environmental News Service, February 20, 2009).

Notes

1. Galbraith (1977, p. 23) quotes the following from Adam Smith's "division of labor" after Smith had carefully observed how pins are made. "One man draws out the wire, another straights it, a third cuts it, a fourth points it, a fifth grinds it at the top for receiving the head; to make the head requires two or three distinct operations; to put it on is a peculiar business, to whiten the pins is another; it is even a trade by itself to put them into the paper." Galbraith adds: "Ten men so dividing the labor, Smith calculated, could make 48,000 pins a day, 4,800 apiece. One man doing all the operations would make maybe one, may be twenty" (p.23).

2. Currently, the Kuznet's aggregates are GDP and expenditure, national income, personal income, and personal disposable income (Kuznet 1948).

3. The two distinguished economists, Nobel Prize winner Paul A. Samuelson of MIT and William D. Nordhaus of Yale, write this (2001): **"Flawed Prophecies of Malthus.** Despite Malthus's careful statistical studies, demographers think that his views were oversimplified. Malthus never fully anticipated the technological miracle of the Industrial Revolution. Nor did he foresee that population growth in most Western nations would begin to decline after 1870 just as living standards and real wages grew most rapidly.

   "In the century following Malthus, technological advance shifted out the production-possibility frontiers of countries in Europe and North America. Indeed, technological change occurred so quickly that output far outpaced population, resulting in a rapid rise in real wages. Nevertheless, the germs of truth in Malthus's doctrines are still important for understanding the population behavior of India, Ethiopia, Nigeria, and other parts of the globe where the race between the population and food supply continues today" (p. 365).

4. This history of "pork barrel spending" dates back to mid-1800s and seems to have become entrenched. The practice undermines the decision-making process.

5. "By drawing the boundaries within which their exchange and production occur, human communities label certain subsets of their surrounding ecosystems as resources, and so locate the meeting place between economics and ecology" (Cronon 1983).

6. The International Joint Commission, or IJC, is a bilateral semiautonomous commission of the United States and Canada responsible for overseeing the environment and welfare of the Great Lakes region.

7. G-8 refers to the world's major economic powers and includes Great Britain, Canada, France, Germany, Italy, Japan, Russia, and the United States. The group was formed after the 1973 oil crisis and, as G6, met in 1975; Canada was added in 1976 to form G7, and at the 1998 Birmingham summit, Russia was added to the group to form G8. At the 2005 meeting in Scotland, G8 leaders were joined by Brazil, China, and India, and discussions were held to seek common ground in addressing the problem of global climate change.

8. Neoclassical economics "is characterized by its *marginal utility* theory of value, its devotion to the *general equilibrium* model stated mathematically, its individualism and reliance on free markets and the *invisible hand* as the best means of allocating resources, with a consequent downplaying of the role of government" (Daly and Farley 2004, p. 437). In the

definition, *marginal utility* refers to "the additional pleasure or satisfaction to be gained from consuming one or more unit of a food or service" (p. 436); *general equilibrium model* refers to "the vision of the economy as a giant system of thousands of simultaneous equations balancing the supply and demand, and determining the price and quantity for each commodity in the economy" (p. 433); *invisible hand* (after Adam Smith) refers to the firm belief "that individuals seeking their own benefit will automatically serve the common good" (p. 224).

9. HDI: United Nations Human Development Report 2007/2008 http://hdr.undp.org/en/media/ HDR_20072008_EN_Complete.pdf

Questions

1. What do ecology and economics have in common?
2. How is economic growth different from economic development?
3. What are market failures? Give an example of how a market failure might affect the environment. Pick a specific market failure and discuss options for solving it.
4. How can environmental concerns be incorporated into economic decisions and policy-making?
5. Discuss the importance of the economic valuation of ecosystem services, especially in improving environmental and public health.
6. Explain the key aspects of each of the following principles in guiding environmental policies: polluter pays, user pays, precautionary, intergenerational, equity, and environmental justice.

Environmental Management

23

Principles of Adaptive Ecosystem Management

To manage an ecosystem properly, all ecological and biophysical factors must be brought together with the economic and social factors. Therein lie the complexity and the challenge of management. The illustration in Figure 23.1 provides a generalized overview of the spatial and, implicitly, the temporal scales involved in management decisions.

It is no surprise that there is no agreement on a precise definition of ecosystem management. However, several definitions have been proposed (Box 23.1) that show that there is a considerable degree of consensus about a set of essential elements that must be incorporated into all environmental management activities:

- sustainability of ecological processes that produce ecosystem goods and services;
- systems thinking—interdependence, interconnectedness, and dynamism of ecological, social, and economic systems;
- hierarchy and heterogeneity of spatial and temporal scales;
- ecologically compatible boundaries;
- adaptive management to deal with ecosystem dynamics as well as uncertainty; and
- collaborative and participative process of decision-making.

Ecosystem management is not only about managing vegetation, wild life, land, soil, water, and air but also about managing institutions and human patterns of consumption and production in an integrated way. Institutional regimes strongly shape interferences and interactions of humans with nature through formal and informal rules and norms, planning, management, policies, and education. **Ecosystem management** can be defined as the system of ecologically compatible resource governance that mediates the biophysical, social, and economic relationships or trade-offs among societal needs, economic development, and sustenance of life-supporting ecological processes through cooperation, coordination, collaboration, and consensus (the four Cs) (Figure 23.2).

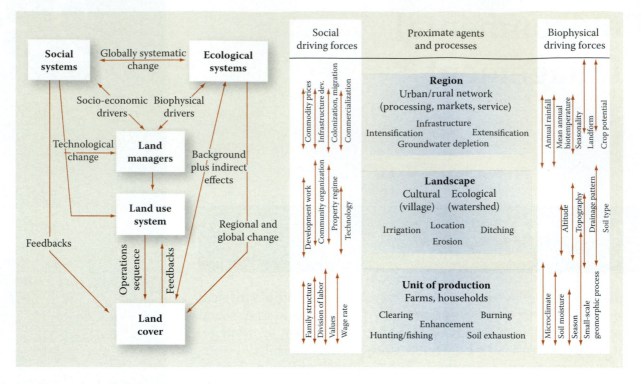

Figure 23.1   A framework, *left*, of land use/ land cover change that integrates ecological and social sciences and, *right*, multi-scale driving forces of land use/ land cover change. (Redrawn from Turner, B. L., II, D. L. Skole, et al. 1995. Land use and land cover science/ research plan. IGPP report 35, IHDP report 7. International Geosphere Biosphere Program, Stockholm, Sweden.)

We use "adaptive ecosystem management" to emphasize that management goals include not only the amelioration of externalities but also the incorporation of social, normative values and ecological principles into the process of decision- and policy-making so as to prevent environmental degradation. Adaptive ecosystem management systems are such dynamic and flexible management systems so that they can respond to ecological, social, and economic feedbacks; their different spatio-temporal scales; and the inherent uncertainties coupled with them. They are also reliant on local communities and indigenous ecosystem goods and services in a regional and global context because the severity of social and environmental problems and the potentials for their solutions are best recognized from local communities. Furthermore, community-based policies can empower and mobilize local communities and their institutions to combat the underlying causes of environmental degradation and their negative impacts voluntarily, through the decentralization of decision-making.

Land cover is what covers the surface of the Earth, such as bare soil, grassland, deciduous and/or coniferous forests, water, ice, and snow. **Land use** describes how the land is used for human purposes such as nature conservation area, recreational area, agricultural land, rangeland, and urban–industrial area. Classification systems of land uses/land covers provide significant information for environmental decision-making and will be discussed later in this Chapter.

Humans change patterns of land uses/land covers through time to meet their needs. In so doing, humans resort to resource-extractive or harvesting activities (such as mining, forestry, agriculture, grazing, and fishery), urban–industrial activities (such as construction of infrastructures and building of residential, industrial, and commercial areas), transportation activities, recreational activities, and nature conservation activities. Changes in patterns of land uses/land covers through time and space are called **land use/land cover dynamics.**

In addition, humans manage land use/land cover in a variety of ways to meet their economic

BOX 23.1   Selected Definitions of Ecosystem Management

... the careful and skillful use of ecological, economic, social, and managerial principles in managing ecosystems to produce, restore, or sustain ecosystem integrity and desired conditions, uses, products, values, and services over the long term (Overbay 1992).

... a resource management system designed to maintain or enhance ecosystem health and productivity while producing essential commodities and other values to meet human needs and desires within the limits of socially, biologically, and economically acceptable risk (American Forest and Paper Association 1993).

... integrating scientific knowledge of ecological relationships within a complex sociopolitical and values framework toward the general goal of protecting native ecosystem integrity over the long term (Grumbine 1994).

... integration of ecological, economic, and social principles to manage biological and physical systems in a manner that safeguards the ecological sustainability, natural diversity, and productivity of the landscape (USDI-BLM 1994; Wood 1994).

... management driven by explicit goals, executed by policies, protocols, and practices, and made adaptable by monitoring and research based on our best understanding of the ecological interactions and processes necessary to sustain ecosystem structure and function. Ecosystem management does not focus primarily on the "deliverables" but rather on sustainability of ecosystem structures and processes necessary to deliver goods and services (Christensen et al. 1996).

... the process of land-use decision making and land-management practice that takes into account the full suite of organisms and processes that characterize and comprise the ecosystem and is based on the best understanding currently available as to how the ecosystem works. Ecosystem management includes a primary goal of sustainability of ecosystem structure and function, recognition that ecosystems are spatially and temporally dynamic, and acceptance of the dictum that ecosystem function depends on ecosystem structure and diversity. Coordination of land-use decisions is implied by the whole system focus of ecosystem management (Dale et al. 2000).

interests and in response to local environmental conditions. These are referred to as **land management practices** (e.g., clear-cut and selective-cut harvesting, conventional and conservational till agriculture, and irrigated and rain-fed agriculture, which will be discussed later in this Chapter). Incorporating ecological principles into the process of decision-making regarding land use/land cover dynamics and land management practices, as well as understanding their implications for ecological processes, is essential to sustainable and adaptive ecosystem management. Decision-making levels and their regulatory and nonregulatory powers of land use and land management in the United States are many (Table 23.1).

According to Dale et al. (2000), the broad principles of ecosystem management are summarized as follows:

- **Time principle.** Ecosystems change through time as their ecological processes function on the scale of seconds to centuries.
- **Species principle.** Particular species and networks of interacting species can have key, broad-scale, ecosystem-level effects. The relationships between species composition and abundance and ecosystem productivity and stability (resistance and resilience) may take the form of a dynamic "Jenga" game rather

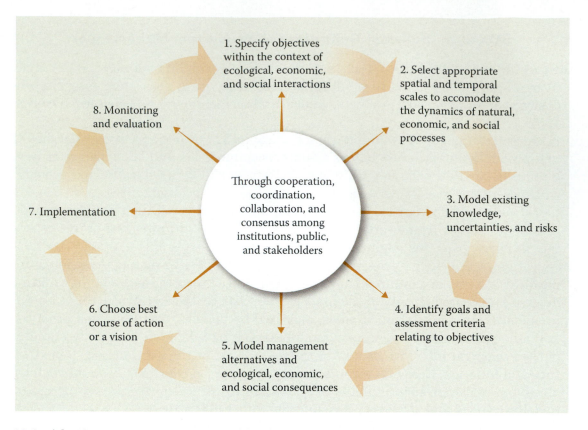

Figure 23.2 Adaptive ecosystem management process. (Amended from figure 1, page 1710 of Bearlin, A. R., E. S. G. Schreiber, S. J. Nicol, A.M. Starfield, and C.R. Todd. 2002. Identifying the weakest link: simulating adaptive management of the reintroduction of a threatened fish. *Canadian Journal of Fisheries and Aquatic Science* 59: 1709–1716.)

than a static "stone arch" (de Ruiter et al. 2005).

- **Place principle.** Local physical, chemical, and biological factors change from one place to another along gradients of longitude, latitude, elevation, and aspect. Spatial changes strongly affect environmental processes, biophysical limits, and physiographic appearances of ecosystems.
- **Disturbance principle.** Natural disturbance regimes—their type, intensity, magnitude, frequency, and duration—are integral characteristics of populations, communities, and ecosystems.
- **Landscape principle.** Spatial patterns, relationships, and arrays of land uses/land covers influence the dynamics of populations, communities, and ecosystems.

Similarly, Zipperer et al. (2000) suggested six ecological considerations key to managing

urban–industrial ecosystems and guiding land use decisions:

- **Content** refers to biological, physical, and social components and their attributes and spatial arrangements (ecosystem structure), as well as interactions of the components (ecosystem function) for ecosystems.
- **Context** refers to spatial and temporal relationships of ecosystems with their surrounding landscapes.
- **Connectivity** refers to spatial and temporal integrity of ecosystems.
- **Dynamics** refers to spatial and temporal changes in ecosystems.
- **Heterogeneity** refers to spatial and temporal distribution or mosaics of ecosystems.
- **Hierarchy** refers to a system contained in another system—that is, nestedness of ecosystems.

Table 23.1
Levels of Decision-Making in the United States and Examples of Their Regulatory and Nonregulatory Powers of Land Use/Land Management

Powers	Federal	State	Local
Direct regulatory	Clean Water Act Endangered Species Act National Flood Insurance Program Surface mining reclamation Wetlands/Waterways Reclamation Act	Regulation and permitting (e.g., pollutant discharge, siting power plants, landfills, reservoirs, and mines) State endangered-species acts Growth-management statutes Programs (e.g., coastal zone management)	Storm water management Land use zoning (e.g., lot size, housing density, structural dimensions, and landscaping) Agricultural land use regulations
Indirect regulatory	Tax policy (e.g., estate taxes and home mortgage deduction) Clean Air Act Transportation funding and development Agricultural programs (e.g., Conservation Reserve Program and Farmland Protection Program) Subsidies (e.g., gasohol program, land and production credit, and crop insurance)	Property-tax exemptions (e.g., for farmland or commercial property) Transportation policy Economic development programs	Property-tax rates Water-use ordinances Local service placement and development (e.g., water and sewer systems, schools and roads)
Management of publicly owned lands	Land use planning (e.g., national parks, national forests, and Bureau of Land Management [BLM] properties) National Wilderness Act Wild and Scenic Rivers Act Siting and design of roads and other facilities	State parks and forests State roads and rights of way Regulation of mining and reclamation activities	Municipal parks and recreation areas County roads and rights of way Green-space systems Greenways

Source: From Dale, V. H. et al. 2000. *Ecological Applications* 10 (3): 646 (Table 2).

Environmental Information Systems

No matter what specific ecosystem management objectives are, analysis, evaluation, synthesis, and monitoring of essential ecosystem attributes provide the basis of environmental decision-making concerning use and management of natural resources (Figure 23.3).

Technological developments in **remote sensing** and **geographic information systems** (**GIS**) now facilitate our ability to estimate past, present, and future changes in ecosystems at the local, national, regional, and global scales. Geographer Roger Tomlinson developed the world's first true operational GIS to determine the land capability for rural Canada for the Federal Department of Energy, Mines and Resources in Ottawa, Ontario, in 1967. The initial development and success of GIS and the advent of computer and remote sensing technologies spurred the growing use of such technologies to look at the spatial patterns of environmental problems. GIS combines maps with a means of entering, storing, retrieving, transforming, measuring, combining, classifying, and displaying information that has been registered to a common coordinate system. Remote sensing records unprecedented amounts of data about spatial patterns and changes—whether qualitative or quantitative—on the Earth's surface on a continuous basis via satellites.

Monitoring landscape conversion from one land use/land cover to another or tracking

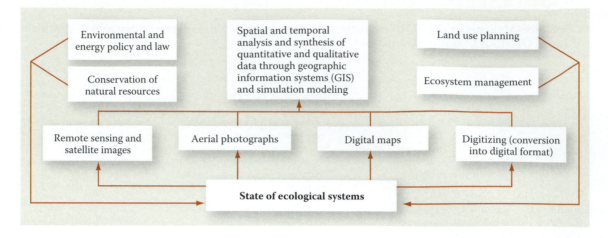

Figure 23.3 Relationship among environmental information systems, ecosystem conditions, and environmental decisions.

modifications within a given land use/land cover for a given time and place through remotely sensed data and GIS analysis is referred to as **detection of land use/land cover change. Conversions** can be of a reversible or irreversible nature, such as conversion of forest to cropland or to urban, respectively. **Modifications** within a given land use/land cover include loss of biodiversity; pollution of air, water, and soil; and impoverishment of habitats and ecosystem productivity. However, classification, monitoring, and processing systems of data through GIS and remote sensing technologies need to be standardized among countries in order to enhance their contributions to the solution of global environmental issues. Geographic information systems are an important part of management strategies that improve our ability to:

- analyze spatial relationships among human actions, their environmental impacts, and vulnerability of humans and ecosystems;
- assess the effectiveness of management strategies taken;
- monitor and detect environmental risks;
- classify ecosystems, and land uses/land covers; and
- predict future implications of environmental changes on ecosystem productivity and health.

Ecosystem Analyses

All the environmental decisions concerning environmental and energy policy and laws, conservation of natural resources, land use planning, ecosystem management, and environmental designs require collection, analyses, syntheses, and evaluation of information about past and current environmental conditions. These can be grouped into six categories used to define ecoregions: landscape, biotic and abiotic conditions, natural and human-induced disturbance regimes, and ecological processes (Figure 23.4).

In his seminal work *Design with Nature,* McHarg (1969) provided a new method for evaluating and implementing an ecologically compatible approach to land use planning, called ecological suitability analysis of lands. His land suitability analysis aims at determining compatibility (1) between actual and potential capacities of lands and proposed land uses, and (2) among proposed and actual land uses, based on the overlaying of different layers of geographically referenced (georeferenced) information concerning sociocultural, economic, and ecological characteristics of a given place. His use of overlay analysis paved the way for the development of the current overlay analysis in GIS and land evaluation methods.

The Food and Agriculture Organization (FAO) of the United Nations defines land

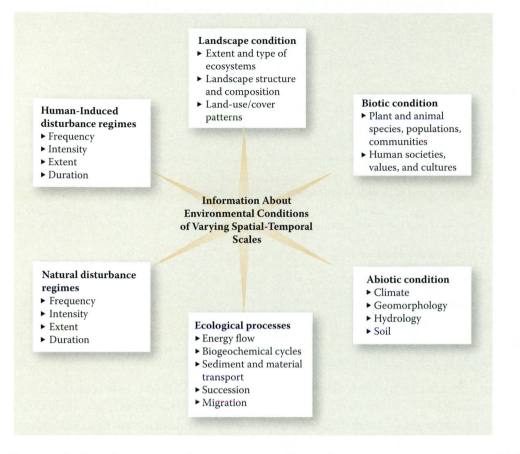

Figure 23.4 Data gathering about essential ecosystem attributes for ecosystem management. (Modified from Table ES-1, p. 3 in Young, T. F. and S. Sanzone, eds. 2002. A framework for assessing and reporting on ecological condition: An SAB Report. Prepared by the Ecological Reporting Panel, Ecological Process and Effects Committee, EPA Scince Advisory Board. Washington, D.C.: US Environmental Protection Agency, EPA-SAB-EPEC-02-009).

evaluation as "the process of assessment of land performance when [the land is] used for specified purposes" (FAO 1985). The rationale behind the necessity and utility of land evaluation is summarized as follows (Rossiter 1996):

(1) Land varies in its physical, social, economic, and geographic properties ("land is not created equal"); (2) this variation affects land uses: For each use, there are areas more or less suited to it, in physical and/or economic terms; (3) the variation is at least in part systematic, with definite and knowable causes, so that … ; (4) the variation (physical, political, economic and social) can be mapped by surveys (i.e., the total area can be divided into regions with less variability than the entire area); (5) the behavior of the land when subjected to a given use can be predicted with some degree of certainty, depending on the quality of data on the land resource and the depth of knowledge of the relation of land to land use; therefore, … ; (6) land suitability for the various actual and proposed land uses can be systematically described and mapped so that … ; (7) decision makers such as land users, land-

use planners, and agricultural support services can use these predictions to inform their decisions.

Ecosystem Classification Systems

A classification system for terrestrial, marine, and freshwater ecosystems is one of the most significant tools that facilitate the understanding of the structure, function, composition, and distribution of ecosystems and the application of ecosystem management principles. There are many classification schemes designed separately for the compartments (components) of an ecosystem such as soils, parent materials, geomorphology, vegetation, land use, land cover, and climate. However, **ecosystem classification systems** are based on the ecosystem

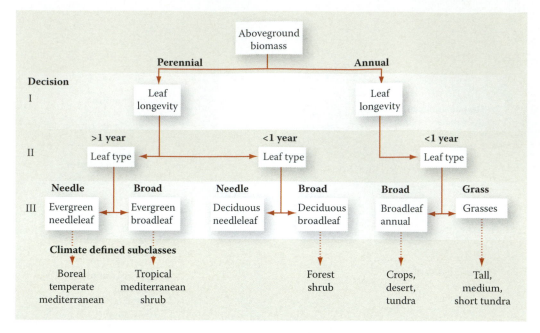

Figure 23.5 A vegetation classification logic for monitoring of global land cover. (Adapted from Running, S. W., T. R. Loveland, and L. L. Pierce. 1994. *Ambio* 23:77–81.)

approach that integrates all the interactive ecosystem components.

Because it is impossible to document the ecological features of each hectare of land, classification of ecosystems and ecoregions provides a most useful management tool. The U.S. Geological Survey delineates areas (called ecoregions) based on similar ecosystem types and environmental factors such as climate, soils and topography; it is meant to offer a framework to help to make environmental decisions (Figure 23.4). Another significant classification system is used particularly by dynamic global vegetation models (DGVMs) to simulate changes in different vegetation communities at the global scale (Figure 23.5). This classification system required to simplify the complex assembly of global vegetation is the aggregation of individual species with similar characteristics into the category of plant functional types (DeFries and Townshend 1994).

Remote sensing allows both for mapping of the density and productivity of green vegetation and for the differentiation among global vegetation types. Based on remotely sensed data of spectral reflectance response characteristics for green plants, the most common measurement of vegetation density is **normalized difference vegetation index**

(**NDVI**) (often referred to as the greenness index). The greenness index ranges from –1 (corresponding to the lack of vegetation such as barren areas of rock, sand, water bodies, or snow) to +1 (corresponding to highest density of green vegetation), with moderate values (0.2–0.3) representing shrub and grassland. Similarly, remotely sensed data of permanence of vegetation (woody vs. nonwoody and annual vs. perennial), leaf longevity (evergreen vs. deciduous), and leaf types (needleleaf vs. broadleaf) assist in derivation of a vegetation classification logic to monitor spatial and temporal changes in land covers of local to global scales, and for use in biogeochemical models (Figure 23.5).

Ecosystem Integrity and Stability

Scientific research and development, institutional regimes, environmental awareness, and societal values strongly shape environmental decisions—namely, how individuals and societies separately and collectively use and manage their life-supporting ecosystems for their needs.

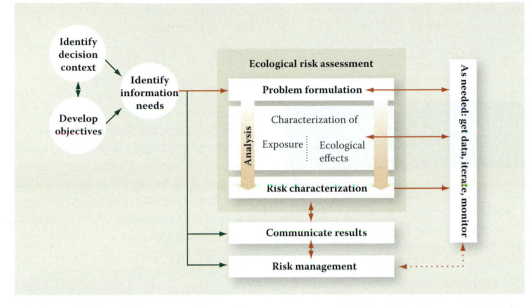

Figure 23.6 Planning for ecological risk assessment. (Figure 2, p. 14, in U.S. EPA. 2001. Planning for ecological risk assessment: Developing management objectives [external review draft]. U.S. Environmental Protection Agency, Risk Assessment Forum, Washington, D.C., EPA/630/R-01/001A.)

Therefore, environmental decisions about use and management of natural resources require not only long-term sustainability of ecosystem services but also cooperation among communities, economic sectors, and nations. Long-term sustainability of ecosystem services cannot be secured by designation of nationally and globally protected areas without the consideration of their ecological integrity and stability.

Even though many protected areas are currently designated both nationally and globally, these areas may not represent proportionally their original ecosystems. For example, less than 1% of original temperate grassland in the United States remains in existence today. In the face of growing human-induced disturbances, ecosystems can show strong resistance (the degree of persistence to function despite disturbances) or dramatic resilience in recovery (the rate of recovery after disturbances), but they can also be destroyed without hope of reestablishment by human actions. Land use/land management strategies aware of ecosystem integrity and stability concepts are better designed to foresee the consequences of homogenization of ecosystems, loss of biodiversity, and spatial fragmentation of habitats for ecosystem health.

Assessment of Environmental Health and Risks

Assessment of **ecosystem health** necessitates characterization of human disturbances and air, water, and soil pollutants as well as their environmental impacts upon humans, fauna, and flora. Pollutants can have individual and synergistic impacts on organisms, their interactions, and their biogeochemical environment. Pollutant (contaminant) characteristics determine their fate, transport, and toxicity and include physical and chemical behavior, degree of degradability (persistence), degradation (transformation) products, biological accumulation (bioaccumulation), and biological magnification (biomagnification) in the food chain (discussed in Chapter 6).

Ecological risk assessment evaluates potential cause–effect relationships between multiple stressors and their impacts on ecosystem components. According to the U.S. EPA (1998), **ecological risk assessment** is (Figure 23.6):

a process that evaluates the likelihood that adverse ecological effects may occur or are occurring as a result of exposure to one or more stressors (U.S. EPA 1992a). This process is used to systemically evaluate and organize data, information, assumptions, and uncertainties in order to help to understand and predict the relationships between stressors and ecological effects in a way that is useful for decision making.

Management of Damaged versus Intact Ecosystems

Management of land that is not actively being used for human profit presents a complex set of issues. Although public attention is more often focused on the preservation of "unspoiled" lands, vast tracts of land have been abandoned after one or more kinds of human use. These lands require even greater efforts in management (see Chapter 24 on rehabilitation and restoration). Here we focus on the issues and possible solutions for management of those areas still composed of largely intact ecosystems.

The recognition of intrinsic values in our natural landscape is almost as old as the U.S. government. At roughly the same time that Thomas Jefferson's Louisiana Purchase doubled the U.S. land area and catalyzed our national "manifest destiny" to impose an agricultural landscape across the continent, Henry Thoreau was writing persuasively about the necessity for untouched "nature" for human well-being. The dichotomy between valuing nature for its own sake and seeing nature as the source of tangible resources for human consumption is exemplified by the long debate between Gifford Pinchot and John Muir, beginning in the late 1880s. These two influential figures gave form to arguments that are still passionately voiced today.

Several books detail the controversy between these men, both of whom were passionate about the conservation of natural resources of our growing country. The fundamental difference was that Muir increasingly viewed natural landscapes as valuable for their intangible qualities: scenic beauty and a necessary contrast to the cities and farms of "civilization." Pinchot, in contrast, saw the value of the landscape in its potential to yield useful products to humans: timber, mineral

ores, fuels, and hydroelectric power. Both men were influential.

Muir was a founding member of the Sierra Club, one of the oldest and most visible of nonprofit conservation organizations. Established in 1872 while the American West was still being "tamed" by ranchers and farmers, the National Park at Yellowstone was the first in a system of national parks unequalled anywhere else in the world. In the larger sense, however, Pinchot can be said to have "won" the argument because the federal government has continued to manage the vast majority of the area under its control as a bank of resources to be exploited by private industry. In fact, taxpayers still subsidize these for-profit extractions. For example, the U.S. Forest Service surveys and builds roads for private companies to cut timber on public lands, and mineral extraction is also heavily supported by tax dollars—from initial exploration to cleanup after processing. It is very difficult to change these practices at the local or the federal level.

Design and Management of Natural Protection Areas

The World Conservation Union (International Union for Conservation of Nature; IUCN) has provided a foundation upon which the international system for categorizing protected areas by primary management objectives and varying degrees of human intervention was developed in 1978 and updated in 1994 by the IUCN (Box 23.2). A protected area is defined as "an area of land and/or sea especially dedicated to the protection and maintenance of biological diversity, and of natural and associated cultural resources, and managed through legal or other effective means" (IUCN 1994). The biological distinctiveness index (BDI) and conservation status index (CSI) are indicators of biodiversity and an ecoregion's vulnerability to human-induced disturbances; they are designed to sustain the ecological processes that produce not only genetic and species diversity but also ecosystem diversity at varying ecoregion scales (Figure 23.7).

BOX 23.2   Management Categories of Protected Areas (Based on IUCN 1994)

Category I. (a) Strict nature reserve: area of land and/or sea possessing some outstanding or representative ecosystems, geological or physiological features and/or species, available primarily for scientific research and/or environmental monitoring.

  (b) Wilderness area: large area of unmodified or slightly modified land and/or sea retaining its natural character and influence, without permanent or significant habitation, which is protected and managed so as to preserve its natural condition.

Category II. National park: natural area of land and/or sea designated to **(a)** protect the ecological integrity of one or more ecosystems for present and future generations, **(b)** exclude exploitation or occupation inimical to the purposes of designation of the area, and **(c)** provide a foundation for spiritual, scientific, educational, recreational, and visitor opportunities, all of which must be environmentally and culturally compatible.

Category III. Natural monument: area containing one (or more) specific natural or natural/cultural feature which is of outstanding or unique value because of its inherent rarity, representative or aesthetic qualities, or cultural significance.

Category IV. Habitat/species management area: area of land and/or sea subject to active intervention for management purposes so as to ensure the maintenance of habitats and/or to meet the requirements of specific species.

Category V. Protected landscape/seascape: area of land, with coast and sea as appropriate, where the interaction of people and nature over time has produced an area of distinct character with significant aesthetic, ecological, and/or cultural value, often with high biological diversity. Safeguarding the integrity of this traditional interaction is vital to the protection, maintenance, and evolution of such an area.

Category VI. Managed resource protected area: area containing predominantly unmodified natural systems, managed to ensure long-term protection and maintenance of biological diversity, while providing at the same time a sustainable flow of natural products and services to meet community needs.

Regardless of human impacts and incursions, the most critical aspect of land preservation is usually the reserve size. Most species cannot exist independently of a complex functioning ecosystem, and both scale and shape of preserved land areas may determine whether or not the system can function. Most landscapes are, at the regional scale, a mosaic of different vegetation types, or stand age within a type, based on the underlying soils, moisture, and history of natural disturbances (which can include windstorms, flooding, or erosion, for example). Human impacts, especially agriculture, have further broken native ecosystems into an array of patches in various conditions and stages or recovery.

Based on the theories of island biogeography, it has been demonstrated that there is an interactive effect between the size of a given habitat area, its distance from other habitat patches and the number of species that area can support. In some cases, there is a direct correlation between habitat size and number of species it can support: The bigger the area is, the more organisms it has. But this is largely true for sedentary organisms, such as plants and those animals that cannot fly or otherwise get to another patch. Other species, such as flying insects and vertebrates (and the seeds they may disperse), can cross over boundaries or discontinuities in habitat, meaning that the size of any given patch may be less important than its relative distance from other patches. **Edge effects** are also significant in habitat configuration: For many species, the zones near a change in landscape pattern are undesirable. Therefore,

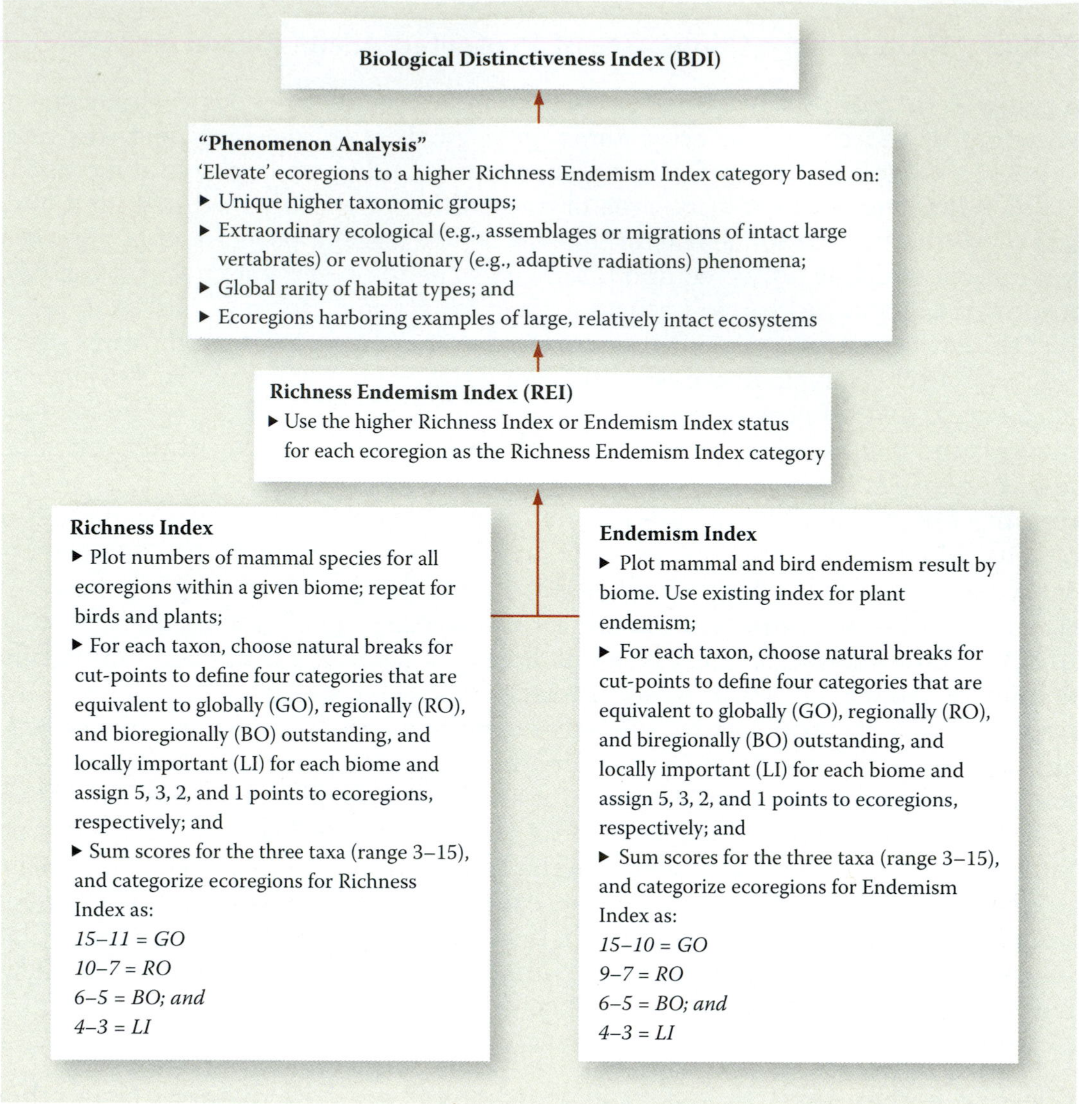

Biological Distinctiveness Index (BDI)

"Phenomenon Analysis"
'Elevate' ecoregions to a higher Richness Endemism Index category based on:
▸ Unique higher taxonomic groups;
▸ Extraordinary ecological (e.g., assemblages or migrations of intact large vertebrates) or evolutionary (e.g., adaptive radiations) phenomena;
▸ Global rarity of habitat types; and
▸ Ecoregions harboring examples of large, relatively intact ecosystems

Richness Endemism Index (REI)
▸ Use the higher Richness Index or Endemism Index status for each ecoregion as the Richness Endemism Index category

Richness Index
▸ Plot numbers of mammal species for all ecoregions within a given biome; repeat for birds and plants;
▸ For each taxon, choose natural breaks for cut-points to define four categories that are equivalent to globally (GO), regionally (RO), and bioregionally (BO) outstanding, and locally important (LI) for each biome and assign 5, 3, 2, and 1 points to ecoregions, respectively; and
▸ Sum scores for the three taxa (range 3–15), and categorize ecoregions for Richness Index as:
15–11 = GO
10–7 = RO
6–5 = BO; and
4–3 = LI

Endemism Index
▸ Plot mammal and bird endemism result by biome. Use existing index for plant endemism;
▸ For each taxon, choose natural breaks for cut-points to define four categories that are equivalent to globally (GO), regionally (RO), and biregionally (BO) outstanding, and locally important (LI) for each biome and assign 5, 3, 2, and 1 points to ecoregions, respectively; and
▸ Sum scores for the three taxa (range 3–15), and categorize ecoregions for Endemism Index as:
15–10 = GO
9–7 = RO
6–5 = BO; and
4–3 = LI

Figure 23.7 Flow diagram depicting the steps taken to estimate the biologically distinctiveness index (BDI). (Based on Figure C.1, p. 211, in Wikramanayake, E. et al. 2002. *Terrestrial Ecoregions of the Indo-Pacific: A Conservation Assessment,* 209–217. Washington, D.C.: Island Press.)

the greater the proportion of edge is, the less functional the habitat patch is.

The first principle is to minimize edge and fragmentation of the land parcels. When possible, corridors or links should be established between fragments, although in many cases, relatively narrow corridors are perceived as completely "edge" and cannot be used by many species dependent on forest interior habitat. In other cases, human systems (railways, road shoulders, irrigation networks) may provide unintended corridors for dispersal of species—both native and exotic. Park and preserve design is a dynamic area in the chang-

ing disciplines of land planning and landscape architecture (Figure 23.8).

Thus far, we have considered the practice of conservation in the context of "natural" systems or habitats, called ***in situ* conservation.** However, other valid methods for the conservation, protection, or banking of organisms and their genetic materials involve the removal of individuals or groups from their natural habitat into captivity such as aquaria, botanical gardens, zoological gardens, arboreta, and seed banks, called ***ex situ* conservation.** Botanical gardens and arboreta are living museums of plant material, and there is a global network of organizations whose aim is

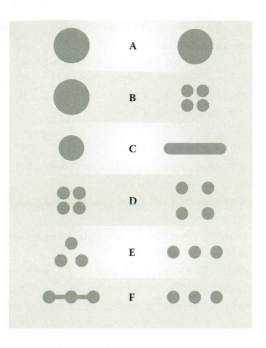

Figure 23.8 Rules of refuge design, in which each configuration on the left is seen as superior to that on the right. (Figure 22.1, p. 276, in Simberloff, D. 1997. In *The Ecological Basis of Conservation: Heterogeneity, Ecosystems, and Biodiversity,* ed. S. T. A. Pickett, R. S. Ostfeld, M. Shachak, and G. E. Likens, 274–284. New York: Chapman & Hall, International Thomson Publishing.)

to preserve all existing vegetation—in some cases, with the intention of being able to re-introduce species into localities where they have been extirpated. Zoological garden "zoos" are the counterpart for animal species, and many captive breeding and re-introduction programs are being carried out with varying degrees of success.

Sustainable Watershed Management

A watershed is a hydrologically delineated geographic area where all the water captured through precipitation flows over and under the ground into a common waterway, such as river, sea, ocean, estuary, wetland, or lake. All the biotic and abiotic components (including human communities) of a given watershed are inextricably linked through the use and discharge of water. Land use planning and environmental protection areas cannot

be a matter of land ownership only. Ideally, all land planning should take place at the **watershed** scale: those areas belonging to a single drainage system or basin. Sustainable watershed management therefore serves the following considerations:

- recognition of the interconnectedness of ecosystem components, including humans, rather than the components in isolation from one another;
- elimination of the lack of fit between human-made and biophysical boundaries; and
- integration of both land capability and the capacities and needs of local communities in the watershed.

Sustainable water management practices can be grouped into five categories (e.g., Wagner et. al. 2002):

- **Minimization of alterations to the hydrological cycle** necessitates increases in conservation and efficiency-oriented measures on both sides of water demand and supply, referred to as demand- and supply-side water management, respectively. Demand-side management involves the minimization of water use and wastewater by households, industry, and agriculture, such as use of more water-efficient technologies and consumption patterns. Supply-side management involves the maximization of water infiltration and groundwater recharge, especially in urban areas, and reduction of interbasin water transfers.
- **Reduction of point-source pollution** necessitates improvement of wastewater treatment, use of environmentally cleaner technologies, and rehabilitation of degraded lands such as waste and mining sites (see Chapter 24).
- **Reduction of non-point-source pollution** necessitates the judicious use of agricultural chemicals and fertilizers, reduction in air pollution, rehabilitation of degraded ecosystems (see Chapter 24), and the application of best

land management practices (discussed later in the chapter).

- **Enhancement of the potential to sustain/restore ecosystem processes** necessitates not only the protection of biodiversity, productivity, stability, and integrity of natural resources but also the rehabilitation and restoration of damaged terrestrial and aquatic ecosystems.
- **Improvement of community involvement and environmental education** necessitates increased access to education and decision-making; the identification of attitudes, values, demographics, and economic status of stakeholders; the strengthening of community solidarity; the eradication of poverty; and the betterment of the position of women.

Best Land Management Practices

Population and consumption growth will continue to increase the depletion and degradation of limited natural capital unless human-induced disturbances can be minimized through preventive and mitigative measures. **Mitigative measures** include the identification of ecologically suitable and compatible land uses/land covers, the protection of ecologically significant habitats, best management practices, and rehabilitation of damaged ecosystems. **Preventive measures** include decreases in population and consumption growth, distributive and participative policies, and resource conservation and efficiency.

Best management practices (BMPs) for agricultural ecosystems include conservation tillage; crop rotations; residue management; additions of mulch, manure, and compost; cover crops; wise use of fertilizers; erosion control; water management; integrated pest management; selection of ecologically compatible crops; agroforestry; intercropping; and windbreaks or conservation buffers. **Conservation tillage** methods consist of reduced till (mulch- and ridge-till) and no-till systems that disturb soils through planters, drills, row cleaners,

injection knives, row-crop cultivators, rotary hoes, or harrows as little as possible and leave more than 30% of the soil surface covered by crop residue. These methods have the following beneficial impacts (Conservation Technology Information Center, www.ctic.purdue.edu):

- reduced rates of soil erosion;
- increased water-holding capacity;
- nutrient-enriched soil;
- increase in earthworms and beneficial soil microbes;
- reduced fossil fuel consumption to operate farm machines;
- return of beneficial insects, birds, and other wildlife in and around fields;
- reduced runoff of sediments and chemicals;
- reduced potential for flooding;
- reduced aerosol pollution (e.g., dust and smoke); and
- reduced emissions of carbon dioxide into the atmosphere.

Residue management refers to keeping previous crop residues as a protective cover on the soil surface of a field to maintain soil organic matter, reduce evaporation by shading the soil surface, control both water and wind erosion, and improve infiltration of water into the soil. Mulching serves the same functions but can be in two forms: organic and inorganic materials. Improvements of both soil structure and crop productivity by these practices depend upon the quality (easily decomposable to decomposition-resistant materials) and quantity of the residues added. The practice of adding farmyard manures, green manure (undecomposed green plant tissue), and compost (a mixture of decomposed remnants of organic materials with animal and plant origins) to the soil provides organic matter and a more readily available supply of nitrogen and other nutrients. **Crop rotations** involve successive planting on the same field of different crops that do not share similar nutrient needs, diseases, pests, and root growth patterns. Crop rotations and cover crops improve soil organic matter content, pest management, regulation of deficient or excess plant nutrients, and erosion control.

Conservation buffers or windbreaks/shelterbreaks are relatively narrow strips of permanent vegetation that exist naturally or are planted strategically to intercept runoff or wind; they play a significant role in the protection of water, soil, and air quality and the enhancement of wildlife. **Integrated pest management (IPM)** is defined by the Food Quality Protection Act of 1998 as "a sustainable approach to managing pests by combining biological, cultural, physical, and chemical tools in a way that minimizes economic, health, and environmental risks."

According to the International Center for Research in Agroforestry, **agroforestry** is defined as

> a collective name for land use systems and practices where woody perennials are deliberately integrated with crops and/or animals on the same land management unit. The integration can be either in spatial mixtures or temporal sequences. There are normally both ecological and economic interactions between the woody and non-woody components in agroforestry.

In addition to agroforestry, another biodiversity-enhancing practice is planting of two or more mutually beneficial crops in proximity on the same field either simultaneously with or without a row arrangement (**mixed or row intercropping**) and in alternate strips of uniform width (**strip intercropping**) or in a temporal sequence (**relay intercropping**). Also, selection of crops that are suitable to local ecological conditions has significant implications for lowering management costs and risks.

Excessive use of fertilizers and water can impoverish environmental health, soil quality, and economic benefits. Judicious use of fertilizers and water requires analyses of soil properties, requirements of crop growth, stages of crop development, microrelief, and spatial heterogeneity of field conditions. The use of a global positioning system (GPS), GIS, and remote sensing allows individual farmers to apply the right quantities of fertilizers, chemicals, and water at the right places at the right times, depending on the recognition of spatially and temporally varying field conditions; this is called **precision farming.**

Best management practices for forest ecosystems include protection of ecosystem services and adjacent ecosystems, sustainable yield, prescribed burning, proper intensity and frequency of thinning, ecologically compatible rotation cycle of production, environmentally sensitive harvest and removal types, retention of vegetation stratification, safeguarding of biodiversity, sustenance of forest integrity and stability, harmony with forest life cycles, and integration of indigenous knowledge and forest communities into management decisions.

Now let us explain each best management practice and how it contributes to long-term productivity and quality of forest ecosystems. Forest resources provide humans with not only timber, fuelwood, and other economic goods but also vital ecosystem services such as sequestration of atmospheric carbon dioxide; protection of water, air, and soil quality; minimization of landslides and floods; and biodiversity at the local, regional, or global scales. Forestry best management practices, therefore, focus on the sustenance of both economic benefits and ecological processes from which ecological and economic goods and services are derived for a given watershed or a larger scale of terrain. Given the benefits of forest ecosystem services locally, regionally, and globally, focusing only on economic aspects threatens the stability, integrity, and health of forests. The **sustainable yield** for forests is the level of harvesting timber at a rate that will not reduce the amount and quality of forest resources in the future. Prescribed burning is critical to reducing the risk of unexpected fires that would otherwise occur due to accumulation of high fuel loads (the amount of combustible dry organic matter on the forest floor), which in turn feed dangerous fires.

Thinning intensity and frequency, rotation period, harvest methods, and removal types of trees all have significant implications for forest stability and health. **Thinning intensity** refers to percentage removal of basal area, **thinning frequency** to the number of years between two successive thinnings, and **rotation length** to the number of years required to establish and grow trees to a specified size or condition of maturity. Thinning is initiated when stands are old enough to form a degree of vegetation cover that reduces stand productivity by an increase in tree density and the associated competition for requirements of limited resources such as sunlight, nutrients, water, and space. Thinning is continued periodically to remove a certain percentage of the total basal area when timber harvested exceeds an economic criterion.

In timber harvest, **selective cutting** is ecologically more compatible than clear-cutting because clear-cutting causes loss of biodiversity and soil nutrients, soil erosion, landslide, flooding, habitat fragmentation, and increased emissions of carbon dioxide into the atmosphere. Similarly, removal of specific trees as stems only is more environmentally friendly than that of whole trees including roots. Long-term carbon storage in forest pools depends upon the selection of proper management practices and can be made possible by a longer thinning frequency, an increased rotation length, and retention of old-growth forests and understory vegetation. Apologists for nonsustainable resource extraction, especially the old-growth timber harvesters, like to claim that all forests are renewable resources. However, they often overlook the clear differences between plantation forestry—usually just a form of monoculture in even-aged stands, harvested every 30–50 years—and the complex and diverse forest ecosystems that may take centuries to recover from a single clear-cutting. Quality and quantity of ground vegetation and vegetation stratification also help to mitigate landslides, promote percolation of water, minimize soil erosion, retain nutrients, and contribute to biodiversity of autotrophic and heterotrophic organisms.

Choosing plant material, species composition, and site is the key to the success of reforestation and afforestation activities. Selection of species and provenances adapted to local site-specific abiotic and biotic conditions, enhancement of genetic variation within and between individuals, and mixture of evergreen and deciduous species increase ecosystem stability (resistance and resilience) (e.g., reducing the outbreak risks of pests and diseases in forests). Forest management practices should be in harmony with natural life cycles of forests, such as harvesting after seed fall and reducing grazing and hunting during critical reproductive periods. Last but not least, involvement of indigenous knowledge and local forest communities in forest management should be strengthened to ensure the success of forest management practices.

We conclude this discussion with the "seven pillars of ecosystem management" as given by Lackey (1998) that put in proper perspective the areas of action, their required attributes, and their limitations:

1. Ecosystem management reflects a stage in the continuing evolution of social values and priorities; it is neither a beginning nor an end.
2. Ecosystem management is place based and the boundaries of the place must be clearly and formally defined.
3. Ecosystem management should maintain ecosystems in the appropriate condition to achieve desired social benefits.
4. Ecosystem management should take advantage of the ability of ecosystems to respond to a variety of stressors, natural and man-made, but all ecosystems have limited ability to accommodate stressors and maintain a desired state.
5. Ecosystem management may or may not result in emphasis on biological diversity.
6. The term *sustainability,* if used at all in ecosystem management, should be clearly defined—specifically, the timeframe of concern, the benefits and costs of concern, and the relative priority of the benefits and costs.
7. Scientific information is important for ecosystem management; it is only one element in a decision-making process that is fundamentally one of public and private choice.

Questions

1. What institutional barriers and incentives exist in your region for the adoption of broad, cooperative, and integrated approaches to conservation, restoration, and maintenance of ecosystem processes?

2. Discuss the importance of information systems and monitoring programs to detect changes in environmental quality. How can such information be used to devise management practices that prevent environmental degradation or help mitigate damage that has already occurred?

3. Describe management practices that reduce greenhouse gas emissions from agricultural lands.

4. What is the best management approach to maintain old-growth forests? Discuss the importance of the concept of "viable population analysis" to determine how much of an ecosystem should be conserved in order to maintain the ecological integrity of an area.

5. What are the main principles of managing renewable and nonrenewable natural resources?

Restoration of Disturbed Ecosystems

24

Restoring Degraded Systems

In the preamble to Section B, we provided an overview of land use worldwide. In it, we provided a narrative of how ecosystems have been manipulated and disturbed throughout human history. We also pointed out that what sets the last century or more apart is the extent and magnitude of disturbance throughout the world. Areas of the world that are not physically disturbed by agricultural activities or by deforestation nonetheless receive pollutants from somewhere else or are changed in their character by the arrival of invasive species. However, the demand for productive land is increasing and will continue to increase as long as human population numbers increase. Thus, reconstructing or restoring disturbed systems to biological productivity attains prime importance such that food, fiber, and other natural resources continue to be available for the survival of humans and other biota. Understanding ecosystem processes becomes central to restoration, as the following narrative will show.

British ecologist Anthony Bradshaw provocatively noted in 1987 that restoration is the acid test of our understanding of ecology. He wrote:

> The successful restoration of a disturbed ecosystem is an *acid test* of our understanding of that system. In other words, there can be no more direct test of our understanding of the functioning of ecosystems than when we put back, in proper form and amount, all the components of the ecosystem we infer to be crucial, and then find that we have created an ecosystem that is indistinguishable in both structure and function from the original ecosystem, or the ecosystem that served as our model. ... The basic principles of land and ecosystem restoration are the same as the basic principles of ecological succession, although we must remember that in many situations we may be dealing with primary rather than secondary succession (pp. 27–28).

Restoration, Rehabilitation, or ...

Given its importance, restoration ecology is one of the hotly pursued areas of research and teaching today. Although terminology per se

493

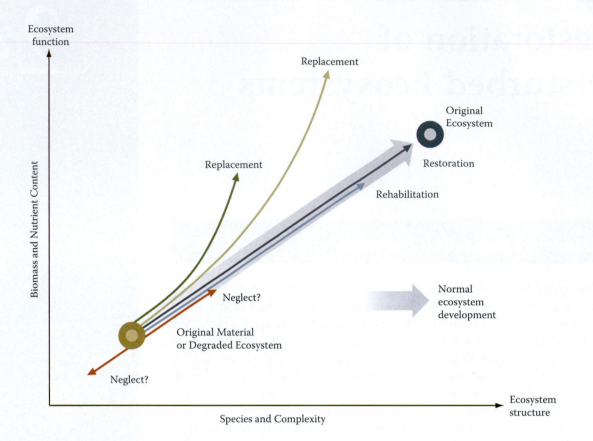

Figure 24.1 Ecosystem development quantified in the two dimensions of structure and function. Reaching the original conditions is restoration and somewhat short of it is rehabilitation. If the use is replaced completely, the replacements may be more productive than before but with less ecological diversity and complexity. (From Bradshaw, A. D. 1984. *Landscape Planning* 11: 35–48.)

should not distract us from pursuing a proper ecological course, it is important nonetheless to grasp the meaning of several terms that are commonly used. The term *restoration* came into ecological lexicon with a meaning much the same as it was used in the restoration of a painting, a building, or a statue. In that sense, restoration of an ecosystem would be more or less a replica of what it was before. Such, indeed, were the efforts in prairie restoration where, based on careful ecological studies, original species were located (even in such refugia as cemeteries) and transplanted along with the soils in which these plants were growing. A well-documented and successful example of such restoration is the Curtis Prairie, part of the University of Wisconsin—Madison Arboretum. Examples of prairie restorations are numerous and exemplary, but the scale of these restorations is miniscule compared to the extent of disturbance in prairie states in the United States and everywhere else in

the world. Nonetheless, these studies provide much useful information.[1]

At the very outset we must recognize that, given the magnitude of human disturbance globally, the total restoration of any ecosystem is not completely feasible. Aside from unimaginable costs in putting every sand grain in its original place, it would be impossible, with current technology, to do so on such a large scale. As recognition of ecosystem degradation has grown, the science of restoration ecology has grown rapidly, and the terminology used to describe restoration efforts has proliferated (Box 24.1). In all their forms, these terms refer to efforts designed to assist the recovery of an ecosystem that has been degraded or destroyed by human actions. As with many lofty ideas, the devil is in the detail!

Typically, restoration efforts focus on both the structure and function of ecosystems and are designed to mimic natural succession as much as possible. If one can recreate what was

BOX 24.1   Evolution of Terms in Use for Rehabilitating/Restoring Ecosystems

Restoration implies "the conditions of the site at the time of disturbance [that] will be replicated after a developmental activity" (U.S. National Academy of Sciences 1974, Box 1978).

Restoration is the "process of reestablishment to the extent possible a structure, function and integrity of indigenous ecosystems and the sustaining habitats that they provide" (SER 2004).

Restoration is "the process of returning ecosystems or habitats to their original structure and species composition" (Helms 1997).

"Ecological restoration is the process of assisting the process of recovery of an ecosystem that has [been] degraded, damaged, or destroyed" (SER 2004).

Restorative development is "a mode of development that increases the health or value of existing assets without (or with minimal) destruction of other assets, and without significantly increasing the restored assets' geographic or ecological footprints" (Cunningham 2002).

Reclamation implies that a site is habitable to organisms that were originally present or to others that approximate the original inhabitants (U.S. National Academy of Sciences 1974, Box 1978).

Ecodevelopment has been preferred in the developing countries, and conveys the erroneous impression that development is ecologically compatible. In most cases, that is not so. "Ecosystem redevelopment" has been preferred by some international scientific organizations.

Ecosystem reconstruction was given credibility by Bradshaw (1983) in his presidential address to the British Ecological Society.

Ecological engineering and **ecotechnology** are defined as "the design of human society with its natural environment for the benefit of both" (Odum 1962; Mitsch and Jørgensen 2004).

Rehabilitation implies that the land will be returned to a form and productivity in conformity with an approved land use plan, ensuring that the system will remain in a stable ecological state that does not contribute substantially to environmental deterioration and is consistent with the surrounding aesthetic values (U.S. National Academy of Sciences 1974, Box 1978).

Rehabilitation is bringing the land to a physically stable state to ensure (a) long-term biological productivity if the area is to be used for agriculture, rangeland, or forest resources, and (b) biological diversity if the area is for aesthetic beauty, with least impact on nearby areas.

Bioremediation refers generally to the use of microorganisms to convert harmful substances into nontoxic substances in soil and water systems.

Phytoremediation involves the use of plants to absorb contaminants from soil or water or to enhance biological activity in soil, which enhances conversion of harmful substances into nontoxic substances.

---

there before, then restoration has occurred; if it falls short, then it can be referred to as rehabilitation. Another option is replacement, in which a different type of ecosystem from the original is the result (Figure 24.1). An example of this might be a freshwater marsh replaced by a pond (Bradshaw 1987). In many cases, ecosystem reconstruction (Box 24.1) is an appropriate term that describes what is actually done.

As engineering methodologies have made inroads into applied ecology, the terms **ecological engineering** and **ecotechnology** have been preferable and are applied most often to wetland restoration. The reshaping and redesign of any disturbed ecosystem require engineering skills. The danger lies in creating the illusion that for every technological fault there is a technological fix. Nonetheless, many institutions are now hiring ecological engineers in the hopes of creating better-built environments. Other terms are used less frequently. For example, the term "ecodevelopment," which is preferred in some developing countries, is used to convey the erroneous impression that development is ecologically compatible. In most cases, that is not so.

Two terms that have had the widest usage for terrestrial systems are reclamation and rehabilitation (Box 24.1). **Reclamation** was originally used by soil scientists for the process of draining salt-laden soils of excess salts and making them suitable for growing crops. This term has also been used extensively in the repair of areas mined for coal. It became codified in the 1977 Surface Mine Control and Reclamation Act (SMCRA). As noted in Chapter 21, the federal law requires that strip- or surface-mined lands be restored to a productive land use.

However, we are trying to achieve **rehabilitation** of land areas. To do this successfully, we must first understand what is wrong (the nature of the damage) and then work to put it right (repair that damage). In so doing, two objectives should be met. One is to return the land to a form (structure) and productivity (function) to meet an approved land use plan, ensuring a stable ecological state such that the disturbance does not contribute substantially to further environmental deterioration (Figure 24.1). The second is that the proposed restoration measures are consistent with the surrounding ecological state and aesthetic values of a given area (NRC 1974, 1980; see Wali 1992).

In recent years, the term **restoration** has gained a foothold in both ecological and non-ecological literature. It has gained widespread use in referring to afforestation and reforestation, grassland rehabilitation, requirements for mitigation wetlands, the removal of dams to restore the original flows of rivers and streams, and repair of drastically disturbed ecosystems. Given their habitats, flora, and fauna and the

climatic–geographic areas in which they are found, each of these efforts will demand its own unique approach. We have adopted the term "restoration" in this chapter solely for the reason of its widespread usage. However, it should always be borne in mind that, regardless of the types of ecosystems with which we are dealing anywhere in the world, we are often rehabilitating them, rather than truly restoring them.

The centrality of ecology in restoration should be obvious: Sound ecological knowledge of an area is indispensable in any endeavor. Thus, it is important to underscore that ecology is to environmental science and resource management what anatomy, biochemistry, genetics, physiology, and such applied sciences as pharmacology are to the practice of medicine. Over the last 30 or so years, restoration plans and strategies based on ecological principles and quantitative data have been successfully demonstrated from many regions of the world under a wide range of ecogeographic conditions.

Initiating the Process of Restoration

A fundamental understanding of ecological succession (described in Chapter 9) provides the basis for the planning and management of restoration projects. This will lead restorationists from the start of a project through a period of evaluation and monitoring (see details in Wali 1999b). A long history of studies on ecological succession has established our understanding of the conceptual basis and the specific information on succession in a diversity of ecosystem types around the world. This should make the application of such basic knowledge easier.

For most types of disturbance, such as deforestation, abandonment of agriculture, and grazing, the recovery process is classified as secondary succession. Here, the soils, although disturbed, are largely intact. In some cases, when disturbance is severe, methods to promote recovery include manipulating (such as reshaping) the disturbed land to mimic natural topography. In some cases, the topsoil is replaced and subsidies such as irrigation, fertilization, and the seeds of desirable species are provided to speed up the recovery process.

If left to recover by natural processes alone, drastic disturbances, like surface mining for coal, that result in disturbing the soils completely will follow a course of primary succession. This is similar to succession that may be observed on recently deglaciated areas, on bare rocks, or after volcanic activities (pyroclastic flows and eruptions). Later in this chapter, we will discuss an example of natural succession on abandoned mined lands and succession on those mined lands, which, under the purview of legislation, must be reclaimed and managed (see SMCRA in Chapter 21).

The Restoration Protocol

The complexity of ecosystems notwithstanding, it is imperative for reasons of time, effort, and costs to establish criteria for judging the success of restoration. Students of ecology know well that no successional pathway can be replicated exactly, and hence ecologists use the word "trajectory" to describe the path of ecosystem development over time. Typically, vegetation and soil parameters are two of the most useful indicators.

Realizing this, the Society for Ecological Restoration (SER) codified essential criteria for considering a disturbed ecosystem "restored" (SER 2004). Appropriately, SER pointed out that these criteria are more the guidelines to successional trajectories than the end points of the restoration process. These criteria include the following:

1. The restored ecosystem contains a characteristic assemblage of the species that occur in the reference (undisturbed) ecosystem and that provide appropriate community structure.
2. The restored ecosystem consists of indigenous species to the greatest practicable extent. In restored cultural ecosystems, allowances can be made for exotic domesticated species and for noninvasive … species that presumably co-evolved with them …

3. All functional groups necessary for the continued development and/or stability of the restored ecosystem are represented or, if they are not, the missing groups have the potential to colonize by natural means.
4. The physical environment of the restored ecosystem is capable of sustaining reproducing populations of the species necessary for its continued stability or development along the desired trajectory.
5. The restored ecosystem apparently functions normally for its ecological stage of development, and signs of dysfunction are absent.
6. The restored ecosystem is suitably integrated into a larger ecological matrix or landscape, with which it interacts through abiotic and biotic flows and exchanges.
7. Potential threats to the health and integrity of the restored ecosystem from the surrounding landscape have been eliminated or reduced as much as possible.
8. The restored ecosystem is sufficiently resilient to endure the normal periodic stress events in the local environment that serve to maintain the integrity of the ecosystem.
9. The restored ecosystem is self-sustaining to the same degree as its reference ecosystem, and has the potential to persist indefinitely under existing environmental conditions. Nevertheless, aspects of its biodiversity, structure and functioning may change as part of normal ecosystem development, and may fluctuate in response to normal periodic stress and occasional disturbance events of greater consequence. As in any intact ecosystem, the species composition and other attributes of a restored ecosystem may evolve as environmental conditions change.

These criteria provide general guidelines and as such can be adapted for diverse ecosystems. For practitioners, the question often is how these goals can be made specific and quantifiable (i.e., what can be measured) when a restoration project is under way.

Few owners or operators of strip or surface mines in the United States prior to 1970 were required to restore mined lands. As the process of mining proceeded, the top layers of soil, which were removed first, were deeply buried by materials excavated from deeper depths. By the time the shovel reached the coal seam, the deepest parent materials ended up on top. We discussed in Chapter 14 the generalized nature of these parent materials, contrasting those in

(a)   (b)

Figure 24.2 Abandoned coal-mined areas showing (left) typical ridge-and-furrow shapes left after dragline operations and (right) burning bush growing on first-year spoil banks; note the cracks in the clay materials. (Photos by (left) M. K. Wali and (right) courtesy of R. H. Pemble.)

the eastern United States and Canada with those in western areas. These mined areas were left abandoned, often as a series of ridges and furrows (Figure 24.2). Restoration and the establishment of vegetation on these drastically disturbed sites occurred solely through natural processes. The Surface Mining Control and Reclamation Act of 1977 (discussed in Chapter 21) established an abandoned mined lands (AMLs) fund from a tax levied on each ton of coal mined. These monies are allocated to the states for the rehabilitation or restoration of mine lands abandoned prior to the passage of SMCRA.

Grassland Restoration on Mined Lands

In the absence of human intervention and aid, these abandoned surface-mined sites proceed through a slow process of primary succession. Such areas provide excellent outdoor laboratories from which important steps in

the development of both soils and vegetation can be studied. Two specific (decade-long) research studies of grassland restoration from sites selected in the Northern Great Plains of western North Dakota are presented here. The pattern and process of soil and vegetation development described in these two case studies can be useful and directly comparable to any such sites in semiarid and arid climatic regions because of general similarities in the properties of parent materials, climatic conditions, and, especially, precipitation. Such areas would include Montana, Wyoming, Colorado, Utah, New Mexico, Arizona, and the provinces of Alberta and Saskatchewan.

One study was on abandoned mines that had Paleocene shales as parent materials older than 60 million years. The second study was on sites in the same general area where the mandates of the 1977 SMCRA were strictly followed. To achieve postmining land use goals, a thorough understanding of successional processes and patterns is necessary to determine the extent of human intervention that may be necessary.

Primary Succession (Natural)

The overburden (i.e., the layer of soil and parent material found on top of coal seams, typically 30–50 m thick) excavated during the process of mining had no organic matter. Once exposed, this overburden was subject to the forces of weathering and became the substrate upon which new soil development would begin. The properties of such materials were presented in Chapter 14 and, as was noted, show high clay, sodium, and calcium contents.

To learn about primary succession, the study had to be framed along a time gradient. Using historical records of mining activities in the area, a 45-year time sequence (or chronosequence) was established. This sequence represented five age classes (or time since mining was stopped) of 45, 30, 17, 7, and 1 years. For each age site, a nearby unmined area was also studied for comparison. Field data were collected on the type and abundance of plant species growing on three slope positions, four cardinal compass directions (N, E, S, W), and their intermediate positions (NE, SE, SW, NW). At each site, the parent materials were studied and samples collected were analyzed for their physical and chemical properties.

The initial conditions of the parent material upon which plant colonization began showed high surface instability, making them prone to erosion. When plants became established, erosion lessened. Synthesis of data revealed that the number of different species increased from 24 on 1-year-old sites to 56 on 45-year-old sites, compared to 115 species on unmined sites. Mean species numbers after 45 years of succession were only half those of the unmined ("control") areas. The 1-year-old site was colonized by non-native, annual pioneer species (see Figure 24.2, right), whereas the 7-year-old site was dominated more by perennial non-natives. By year 30, non-native annuals had mostly disappeared and biennials, perennials, and native species were becoming more abundant. By 45 years, majority of the species present were natives and species numbers were higher than at any of the younger mined sites; however, species diversity was lower than at unmined sites.

Next to the time factor, topography (such as slope angle and exposure) was the major factor controlling growth, distribution, and abundance of species. Because of the way solar light is received in the Northern Hemisphere, south and west slopes are the warmest and north and east slopes the coolest. This has a significant effect on water relations of plants. Chemical properties of spoils provided clear indications of changes over time and within the spoil by depth. Electrical conductivity soluble salt concentration, Na, Cl, and SO_4 show a consistent and appreciable decrease over time, indicating that the leaching of soluble salts occurs. Carbon and N, indicators of soil fertility and productivity, increased with time.

Organic C at the mined sites was initially very low, but then showed a rate of increase of 131 kg ha^{-1} yr^{-1}; this accumulation rate was about one-half that reported for mine spoils in Saskatchewan (282 kg ha^{-1} yr^{-1}) and for eastern Montana (256 kg ha^{-1} yr^{-1}) (see details in Wali 1999b). Carbon/nitrogen ratios, a parameter of considerable value, indicate the trend toward rehabilitation of mined sites. The C–N relationship is based on the accumulation of C and N compounds during the mineralization of organic matter over time, and its importance reflects the dynamics of soil systems. Carbon/nitrogen ratios in this study were the highest reported for young mine spoils. But considerable recovery took place over a relatively short time; 70% of the 45-year-old sites showed values comparable to the soil of the unmined sites—that is, C/N ratios below 15. Carbon/nitrogen ratios can be a useful indicator to assess the relative dynamics of the recovery of mined systems over time.

To summarize the findings, the major trends of primary succession of abandoned mine spoils showed that (1) pioneer species persisted for long periods (decades) and replacement by other species was slow; (2) topographic variables, including aspect, slope angle, and slope position, were the most dominant in terms of their influence on plant species establishment; (3) there was the significant increase of N and P in the soil over time, but 45 years of natural

Figure 24.3 Areas under the new laws are required to receive topsoil and are fertilized and seeded. Left: first-year growth dominated by weeds; right: weeds are completely replaced by seeded grass cover in 4 years. Inset in the middle shows a cocklebur weevil larva; heavy insect infestation leads to synthesis of chemicals by weeds, which in turn cause their own demise and change the successional trajectory. (Photos courtesy of L. R. Iverson.)

vegetation establishment did not approximate the unmined conditions.

Secondary Succession (Human-Aided)

Prior to the onset of mining and immediately following it, SMCRA regulations require the following:

1. Prepare a mining plan and get it approved.
2. As mining commences, remove the topsoil—A-horizon first, B-horizon next; stockpile the two separately.
3. Reach the coal seam and then remove the coal.
4. As the dragline moves across the surface, withdrawing coal as it goes, start regrading mined areas, returning them to approximately the original contour (this requirement was struck down in areas such as West Virginia).
5. Replace the B-horizon first, then the A-horizon.
6. Apply appropriate fertilizers.

7. Seed with a mixture of species native to or naturalized in the area.
8. Monitor the area for a specified period (5 years' bond release time in the eastern United States and 10 years in the western United States) (Figure 24.3).

Because the reshaped postmined substrates have topsoil spread on them, they receive organic matter, nutrients, seeds, and propagules from the original vegetation, which makes vegetation and soil development more akin to secondary succession (Table 24.1; Figure 24.4). But some differences must be noted between mined areas that receive topsoil and those soils where secondary succession can be expected to proceed as a natural process (such as abandoned agricultural fields, for example, where the soil has not been "mixed"). Topsoil that is replaced on contoured spoil materials does not have the same structural integrity or the horizonation seen in naturally developed soil profiles. These topsoil-receiving areas are continually susceptible to wind and water erosion on the surface and to the upward migration of salts from the spoil materials below. Thus, changes occurring

Table 24.1
Density, Height, and Aboveground Biomass of *Kochia scoparia* **and Grasses[a] on Topsoiled Areas in the First Four Areas after Mining**

Years after mining	1	2	3	4
Density				
Kochia, individuals m^{-2}	52	10,464	5	0
Grass, shoots m^{-2}	24	267	603	934
Height, cm				
Kochia, individuals m^{-2}	54	11	<1	0
Grass, shoots m^{-2}	15	55	48	44
Biomass, g m^{-2}				
Kochia, individuals m^{-2}	222	71	0	0
Grass, shoots m^{-2}	2	36	89	96

Source: From Iverson, L. R., and M. K. Wali. 1982. *Reclamation & Revegetation Research* 1:123–160.

[a] Grass group included wheatgrasses (*Agropyron smithii, A. caninum,* and *A. elongatum,* in that order); numbers have been rounded off.

on areas receiving topsoil represent a gradient between primary and secondary succession.

In this study, field observations made shortly after the application of topsoil, fertilization, and seeding showed that the initial composition of the vegetation was determined by the diversity of viable seeds present in the topsoil or those that arrived at a site through dispersal. Of the 95 species found at these sites during the first 2 years, the dominant species (*Kochia scoparia*) found here was the same primary colonizer that was found at the abandoned mine sites. However, *Kochia* density, height, and biomass decreased sharply with time and *Kochia* was eliminated by the fourth year, whereas grasses increased concomitantly (Table 24.1; Figure 24.3).

Allelopathy (mentioned in Chapter 4) was suspected in the decline of *Kochia* from its robust growth in the first year of colonization through stunted growth to complete disappearance by the third year, but in somewhat different form. Rather than being inimical to its competitors, the plant's chemicals induced autotoxicity. Several growth chamber experiments later revealed that small quantities of decaying *Kochia* leaves and roots depressed its own growth but not that of other species such as wheatgrass (*Agropyron caninum*) or yellow

sweetclover (*Melilotus officinalis*) (Iverson and Wali 1982). Biochemical analysis revealed the presence in *Kochia* of allelopathic chemicals, including quercetin, myrcetin, and chlorogenic and caffeic acids. These substances may act as autoinhibitors and cause an imbalance in the uptake of P, Mn, and Zn in the plants. These autoinhibitors are primarily produced by the plants as a defense mechanism against herbivores. This suggestion is supported by field observations; six of the other pioneering species were 73–100% infested with cocklebur weevil larvae (*Rhodobaenus tredecimpunctatus*), whereas only 40% of the *Kochia* plants were found to be so. Allelochemical compounds have an important bearing on community development and differentiation (see Rice 1984).

Chemical parameters of the spoils/soils over the same period of time showed decreases in electrical conductivity and in the concentrations of soluble Ca, Mg, Na, Li, Sr, and SO$_4$, and an increase in OM content. To summarize, major trends noted in this study that are akin to old-field (secondary) succession included the following:

- The process of pioneering species replacement was unusually rapid.
- Dispersal in space (immigration) of species was more important than dispersal in time (establishment of species from seed banks).
- Successional trends along the pathways of Connell and Slatyer (1977) were facilitation and inhibition pathways. The latter were from autoinhibition by chemicals that inhibit perpetuation of pioneering species produced originally as defense mechanisms against herbivory. These chemicals seem to disrupt translocation of P, Mn, and Zn and hence the P/Mn and P/Zn ratios.
- Rates of leaching of ions were three to five times faster than in natural (primary) succession on abandoned areas.
- Rates of ion mineralization were five to eight times faster than in natural (primary) succession on abandoned areas.

The application of new rules, when applied properly, has been a boon to ecosystem restoration. Rehabilitation of these drastically

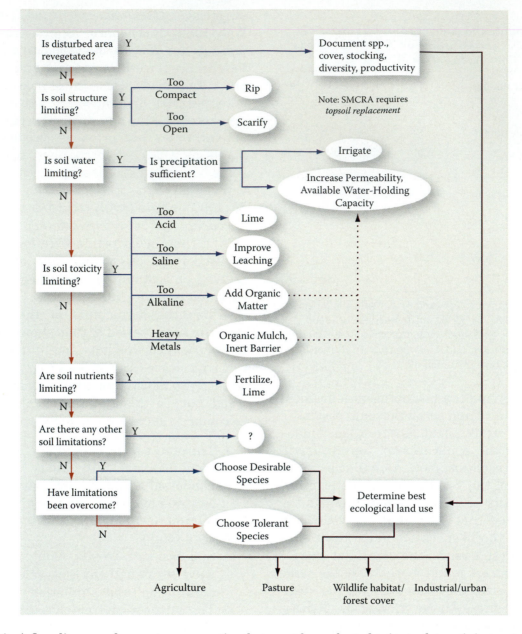

Figure 24.4 A flow diagram of ecosystem properties that must be evaluated prior to determining restoration strategy and end uses such as agriculture, pasture, forest cover/wildlife habitat, or industrial/urban use. (From Wali, M. K. 1992. In *Ecosystem Rehabilitation: Preamble to Sustainable Development*, Vol. 1. *Policy Issues*, ed. M. K. Wali, 3–23. The Hague: SPB Academic Publishing.)

disturbed-soil land areas becomes a reality in a much shorter period of time than if restoration were left to nature—with far fewer adverse effects on surrounding ecosystems. The law has provided an excellent opportunity whereby we can evaluate properties of the substrate and correct its deficiencies or excesses and select appropriate plant species and other specific treatments for multiple land use planning (agriculture, pasture, wildlife habitat/forest cover, urban/industrial) (Figure 24.4).

Forest Restoration on Mined Lands

Temperate Deciduous Forests

As in the West, there is a similar scenario in the eastern United States: namely, abandoned coal mines and current mining activities under the purview of SMCRA that together provide excellent examples of forest restoration. In the western United States, the low quantity of precipitation is the problem; however, high precipitation is the problem in eastern states such as Illinois, Kentucky, Tennessee, Ohio, Pennsylvania, and West Virginia. The materials brought to surface in this region are of the Carboniferous age, more than 250 million years old. The overburden material is made of mostly pyritic material that is high in sulfur and iron.

These overburden materials are coarse and unconsolidated, with low water-holding capacity. Upon exposure to the atmosphere and oxidative processes (as given in Chapter 14), these materials undergo weathering and produce soils and water with low pH, ranging from 1.8 to 3.5. Both the coal spoil materials and mine drainage water have a high concentration of Fe, Al, Mn, and sulfates. There is little or no N and P, as well as low Ca, thus making conditions for plant growth daunting. In most of Western Europe, the conditions for restoring mined lands are very similar. Natural colonization (primary succession) of these sites is due to herbaceous pioneering species, which include grasses; wind-dispersed seeds such as the species of ash, birch, box elder, and poplars; and animal-dispersed seeds of trees (such as species of hackberry, hawthorn, cherry, and oak). In all cases, species numbers are low. But over time, substrate conditions do improve.

As in the West, several approaches to restoration are mandated by SMCRA. As mentioned earlier, sites are now required to be reshaped and receive topsoil and seeding of desirable species. The restoration process is hastened considerably by the presence of organic matter and essential nutrients (N, P, and K). A review of restoration studies spanning several decades is given in Leopold and Wali (1992). An example that blends abandonment with some management is provided by the following example.

The Wilds, Ohio

More than 3,700 ha of abandoned surface mine land in southeastern Ohio were donated by American Electric Power company in the early 1970s for wildlife conservation, scientific studies, and education programs. The abandoned mine site had a rugged terrain with steep ravines and extensive erosion. Initially, seeding/planting was done directly into the spoil material, but later the area was recontoured and topsoil that could be salvaged was respread at the site. The reshaped site was seeded with a mixture of grasses, forbs, and legumes. Currently, the vegetation at the site consists of non-native species of European alder, black locust, and autumn olive. Although the vegetation for the most part is indigenous or cultivated varieties of the temperate regions, a number of tropical animals, such as elephants, giraffes, rhinoceros, and zebras, have been introduced in the area as a zoological park. By 1998 more than 350,000 persons had visited The Wilds. Although the efforts in the 1970s and 1980s were on ensuring surface stability and rehabilitation, The Wilds now maintains an active program of research in conservation biology and ecosystem restoration and management.

Other Examples

Restoration of Longleaf Pine Ecosystem[2]

At one time in the southeastern United States, the longleaf pine (*Pinus palustris*) ecosystem covered an estimated 36 million ha. Today, that area has been reduced to about 1 million ha, representing a 97% reduction. This is due to the exclusion of fires, natural areas converted to tree farms where commercial species are grown, and urban development. As a result, over 30 plant and animal species of these ecosystems are endangered or threatened, including the red-cockaded woodpecker (*Picoides borealis*) and gopher tortoise (*Gopherus polyphemus*).

Through partnerships of private, government, and academic agencies, the disturbance to the longleaf pine ecosystem has abated and is now showing an increase in area. These partnerships have initiated longleaf pine habitat restoration projects on 20 different sites, totaling 525 ha, across the Southeast. Several thousand additional hectares of potential restoration projects, involving more than 20 private landowner partners, have been identified. The

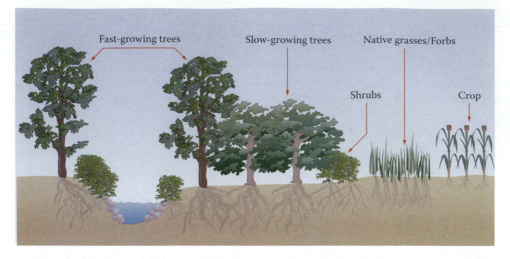

Figure 24.5 A two-zone multispecies riparian buffer with a woody and grass/forb zone. The whole woody zone is managed to keep banks stable and prevent erosion. Shrubs are added to the system for a multitude of functions. (Redrawn from Schultz, R. C., T. M. Isenhart, W. W. Simpkins, and J. P. Colletti. 2004. Riparian forest buffers in agroecosystems—lessons learned from the Bear Creek Watershed, central Iowa. *Agroforestry Systems* 61: 35–50.)

restoration plans include the reintroduction of fire and management techniques that favor longleaf pine. Widespread efforts are under way to produce large numbers of longleaf pine seedlings for plantings on sites once occupied by the pine. The Longleaf Alliance, an organization of researchers, academicians, private groups, individuals, and public agencies, is devoted solely to the restoration of longleaf pine and coordinates restoration and research activities.

Restoration of Riparian Areas

The word *riparian* refers to the long, narrow zones that lie adjacent to stream and river water bodies. When the zone, swath, or a corridor is forested, it is also referred to as a gallery forest. These areas are important for maintaining the integrity of watersheds, preventing soil erosion, retaining nutrients, and serving as wildlife habitats and recreational areas (Wikum and Wali 1974; Killingbeck and Wali 1978). The biological, economic, and social values of riparian systems are described in detail by Naiman, Decamps, and McClain (2005). Emphasizing their importance, Schultz et al. (2004) describe the major objectives of a riparian buffer as:

(1) to remove nutrients, sediment, organic matter, pesticides and other pollutants from surface runoff and groundwater by deposition, absorption, plant uptake, denitrification, and other processes, and

thereby reduce pollution and protect surface water and subsurface water quality … , (2) to create shade to lower temperature to improve habitat for aquatic organisms, and (3) to provide a source of detritus for aquatic organisms and habitat for wildlife …

These authors used a two-zone, multispecies riparian buffer with fast- and slow-growing trees, shrubs, native grasses, and forbs that intervene between cropland and river (Figure 24.5).

Restoration of Wetlands

Estimates of the global extent of wetlands range from 5.3 to 12.8 million km^2; approximately half of this has been lost due to human activities. Wetland losses include freshwater, coastal, and estuarine systems. The high rates of loss have led to the recognition of the ecosystem services that wetlands provide, such as improving water quality, mitigating floodwaters, cycling nutrients and sequestering carbon, and providing habitat for a high diversity of wetland-dependent species. In an effort to stem or reverse losses, wetlands are being restored or created in many parts of the world. In the United States, although wetlands continue to be lost to development, agriculture, or other landscape alterations, by law, many of these losses must be compensated for by the restoration (or mitigation) of wetlands elsewhere in the landscape. This is due to section 404 of the Clean Water Act, which requires that "replacement" or "mitigation" wetlands be restored in order to compensate for wetland loss. As with other ecosystem types, wetland restoration is complex and must take into account the source and amount of water entering the site, the type and condition of the soils, and the landscape setting of the project.

Wetlands are defined by the presence of three parameters: soil inundation or saturation for at least some part of the growing season (wetland hydrology), soils that are periodically deficient in oxygen (hydric soils), and vegetation adapted to life in saturated soils that are periodically anoxic or anaerobic (hydrophytic vegetation). Wetland hydrology describes the timing, duration, and frequency of inundation, which, because it varies both temporally and spatially, can be difficult to recreate (Zedler and Kercher 2005). Getting the hydrologic conditions right is a key component of aquatic restoration because water is considered the "master variable" that determines aquatic ecosystem structure and function. The hydroperiod (the pattern of water levels over time) is one of the most important predictors of overall restoration success (Mitsch and Jørgensen 2004); even small variations in hydroperiod can lead to different ecosystem types. Wetlands that are temporarily flooded, such as tropical or subtropical sites with pronounced wet–dry cycles, vernal pools in Mediterranean climates, seasonal marshes in the temperate zone, or intertidal marshes, present particular restoration challenges because of their pattern of seasonal or daily inundation followed by drydown. In these systems, even small errors in water inputs or elevation can result in sites that are too wet or too dry (Zedler and Kercher 2005).

Restoration projects located on former wetland soils (called **hydric** soils) are likely to be more successful than those on upland, nonhydric soils. As with the mined sites, wetland soils are the physical foundation of the ecosystem and have a strong influence on the development and maintenance of ecosystem processes and the composition of the biological communities. Many factors affect hydric soil development,

including the hydrologic regime, the activity of organisms (particularly vegetation), topography, salinity, climate, parent material, and time (Jenny 1958). Water is crucial to the formation of these soils by adding material through deposition of eroded sediment, leaching dissolved materials, and influencing the breakdown of plant litter into organic matter. Wetland soils are unique because they store large amounts of carbon. This is due in large part to soil saturation, which leads to anaerobic soil conditions and slow rates of decomposition and hence the buildup of organic matter. Organic matter plays an important role in plant community dynamics by supplying nutrients, retaining moisture, and reducing evaporation.

For a restoration project to be successful, suitable soils must be present. When soils are deficient in C and N, plant communities are often limited in their ability to colonize successfully or persist at a site. For example, an estuarine restoration site in San Diego Bay was constructed in order to provide habitat for the light-footed clapper rail (*Rallus longirostris levipes*), a species now listed as endangered due to the loss of marsh habitats where it resides. This project ultimately failed to support any suitable habitat for the clapper rail because the soils in the restored marsh had a higher percentage of sand compared to natural marshes in the region (Zedler and Callaway 2000). Those sandy soils did not retain or supply sufficient levels of nitrogen for plant growth. As a result, the created marsh did not develop structural (height of the desired plant species) and functional (habitat for the endangered clapper rail) properties similar to the natural marsh it replaced. Low soil nitrogen has been similarly observed for freshwater wetlands; this was associated with the slow establishment of some ecosystem processes (Fennessy et al. 2008).

Because plant communities play a central role in ecosystem dynamics, much effort has been placed on the successful establishment and development of vegetation in wetland restoration projects. Plant community development in restored wetlands is a function of two principal factors: the species pool present in the region and the environmental conditions that regulate plant establishment, growth, and reproduction over time. In any restoration project, desirable or "target" species may be planted, arise from the seed bank, or arrive through natural dispersal mechanisms. When soil alteration has not been extensive, wetland species are present in the seed bank and colonize naturally. For example, in coastal salt marshes in the United States, target species include *Spartina alterniflora* on the east coast and *Salicornia virginica* along the South California coast. Both of these species will colonize naturally if elevations are appropriate, although other marsh species may require planting to become established (Mitsch and Jørgensen 2004).

Wetland Conservation Efforts

Because it contains a major component of ecosystem restoration, we include here the North American Waterfowl Management Plan (NAWMP). NAWMP was initiated in 1986 by the United States and Canada in recognition of the need for international cooperation to restore wetlands and associated grassland habitats for declining migratory bird populations. With the addition of Mexico as a signatory in 1994, the plan became a continent-wide one. The three governments, with cooperation from nonfederal partners, are implementing a strategy to restore waterfowl populations to levels of the 1970s by protecting, restoring, and enhancing wetland and adjoining habitats.

The plan focuses on regional "joint venture" areas that are designed to conserve wetland habitat complexes identified as critical to sustaining populations of breeding, migrating, and wintering waterfowl. Private conservation organizations like The Nature Conservancy, Ducks Unlimited Inc., the Delta Waterfowl Association, and the California Waterfowl Association play a major role in wetland restoration and conservation. To date, 14 joint ventures have used nearly $5.7 billion to conserve more than 14.5 million acres of wildlife habitat in North America. Nearly 9.8 million acres have been affected in biosphere reserves by education and management plan projects in Mexico. The 2004 NAWMP implementation plan calls for protection, restoration, and enhancement of an additional 28.3 million acres of waterfowl habitat.

The North American Wetlands Conservation Act (NAWCA) was passed in part to support wetland conservation under NAWMP. More than $918 million in NAWCA grants have been used to leverage $1.8 billion in matching funds and $1 billion in nonmatching funds to conserve 24.4 million acres of wetlands and associated uplands across North America since 1990. More than 4,000 partners have been involved in 1,829 habitat conservation projects supported with NAWCA funds. The NAWMP has proven effective in large-scale habitat conservation and ecological restoration, serving as a model for similar continental bird habitat conservation programs for land birds, shorebirds, wading birds, and seabirds under the North American Bird Conservation Initiative.

Waterfowl hunters also support wetland conservation by purchasing duck stamps, contributing over $700 million to conserve 5.2 million acres of wildlife habitat since 1934. Ninety-eight cents from every Duck Stamp dollar is used to purchase, lease, or protect wetland habitat in the National Wildlife Refuge System. In addition to the ecological values that accrue from hydrological, biogeochemical, and trophic processes within wetlands, the values of North American wetlands accrue to the U.S. economy. Waterfowl hunting and viewing generated $13.4 billion that supported an estimated 135,000 jobs in 1991.

The Cache River Ecosystem

A partnership of federal, state, and private interest groups formed a joint venture to establish the Cypress Creek National Wildlife Refuge in southern Illinois. A watershed management plan was developed to improve water quality by reducing erosion and sedimentation and to preserve and restore the natural resources of more than 194,253 ha in the Cache River watershed in a manner compatible with a healthy economy and high quality of life. Situated in an abandoned channel of the Ohio River, the Cache River has been adversely impacted by channelization and sedimentation associated with forest clearing, flood control, and agricultural development over the last century.

The Cache River gained international recognition in 1996 when it and the associated Cypress Creek wetlands were added to UNESCO's list of 15 "wetlands of international importance." Located at the crossroads of mid-continental climate zones, the Cache River basin's 20 natural community sites include the most notable cypress-tupelo swamps, southern flatwoods and bottomland forests, and upland forests and limestone/sandstone glades. The area supports 104 state and 7 federally threatened or endangered species. Wetland restoration efforts in the region include revegetating forested bottomlands, establishment of riparian corridors and filter strips, and restoration of natural flow regimes to the Cache River and its tributaries.

The Tensas River Basin Area

The Tensas River is part of the Lower Mississippi Valley and, through a joint venture of NAWMP, it includes 290,570 ha in northeastern Louisiana. Once a forested watershed, 85% of the area has been cleared for row-crop production since the 1950s. About 26,305 ha of bottomland swamp remain in the Tensas River National Wildlife Refuge and Big Lake Wildlife Management Area. Loss of wetlands and riparian areas has degraded water quality, increased sedimentation and flooding, and caused loss of wildlife habitat and biodiversity. Federal, state, and nongovernmental agencies and local citizens formed collaborative partnerships to develop a watershed restoration action strategy. Best management practices, erosion control structures, and reforestation measures have been implemented through the U.S. Department of Agriculture's Environmental Quality Incentives and Wetland Reserves Programs (WRP). The USDI Fish and Wildlife Service (USDI-FWS) and state and federal partners work with private landowners to protect and restore bottomland forests in the region voluntarily. To date, an estimated 22,663 ha of farmland and 1,619 ha and more than 9 km of riparian area have been reforested, with another 19,425+ ha enrolled in WRP.

Everglades Restoration Project

This is perhaps the most ambitious restoration project under way in South Florida. The Everglades, an area of 3.6 million ha, includes rivers, lakes, and estuaries, and a human population exceeding 7 million. It is also the home of diverse plant and animal species, more than 60 of which are considered endangered and threatened. The Comprehensive Everglades Restoration Plan (CERP) is a joint (50/50) state–federal partnership to implement a $10.9 billion restoration plan covering 16 counties over a 46,000-km^2 area. To date, it will be largest restoration project in U.S. history.

Brownfields, Bioremediation, and Phytoremediation

A vast number of sites throughout the world, particularly industrialized countries, lie barren and unproductive. These are classified as **brownfields,** which are defined by the U.S. EPA (2008) as "real property, the expansion, redevelopment, or reuse of which may be complicated by the presence or potential presence of a hazardous substance, pollutant, or contaminant." This definition was adopted when Public Law 107-118, Small Business Liability Relief and Brownfields Revitalization Act, was passed on January 11, 2002. In the United States, brownfield sites are estimated to exceed 500,000 "contaminated at levels below the 'Superfund' caliber (the most contaminated sites) in the country." Brownfields result from abandonment of industrial sites and excessive contamination by chemicals. The U.S. EPA is charged with the responsibility of managing and restoring brownfield sites; the agency notes that "cleaning up and reinvesting in these properties increases local tax bases, facilitates job growth, utilizes existing infrastructure, takes development pressures off of undeveloped, open land, and both improves and protects the environment."

Bioremediation is a process by which hazardous chemicals are broken down by the addition of bacteria, fungi, or yeast that break these compounds chemically into simpler, nontoxic or less toxic compounds. The first use of introducing such microorganisms into the environment was approved when bioengineered bacteria were specially designed for breaking crude oil spills. In that case, the aerobic biodegradation broke the oils into carbon dioxide and water. However, bioremediation can also take place under anaerobic conditions if the microorganisms and the substrate are well matched. Bioremediation is also used in decontaminating soils and composting biodegradable waste materials.

The prefix *phyto* refers to plants (remember the discussion of phytoplankton in Chapter 5). Thus, **phytoremediation** refers to the use of plants in decontaminating sites that have high concentrations of elements that can become toxic to humans and animals or sites that have organic compounds. A generalized model of phytoremediation (Figure 24.6) shows that plants do so in a number of ways: by degrading pollutants (phytostimulation or

rhizodegradation), accumulation of pollutants in tissues that are then harvested (phytoextraction), and breaking and release of some compounds in a volatile form (phytovolatalization) (see Pilon-Smits 2005).

Burying of toxic materials is an expensive process that poses dangers of contaminating underground water and leaching into neighboring sites. Thus, a number of groups are studying ways in which appropriate plants capable of the processes listed are selected because "green technology is a simple concept and cost effective." Phytoremediation weaves together a number of fields, such as plant science, physiology, genetics, and soil–plant relationships. Researchers are using the technique for decontaminating areas of aluminum, uranium, zinc, and other compounds.

Evaluating Success

Given the state of knowledge today, there should be ample confidence that if we have the will, disturbed ecosystems will be restored to yield sustained ecosystem services. However, defining what makes a "successful" restoration project is challenging. In many cases, restoration success is assumed if the plant community structure at the restored site resembles that of a natural or "reference" site (i.e., sites that are in sound ecological condition). However, the establishment of food webs, the movement of carbon and energy, nutrient recycling, and other ecosystem functions may never be restored or may take many years to develop. In general, metrics or objective criteria need to be used to answer the question "Is it working?"

For example, one way to evaluate how well restored wetlands are performing is to compare them with natural wetlands of the same type using a performance curve (Figure 24.7). Typically, performance curves display data on the structure (e.g., vegetation diversity) or function (primary productivity), and it is expected that, over time, restored

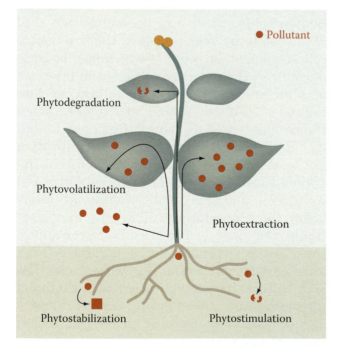

Figure 24.6 Possible fates of pollutants during phytoremediation. The pollutant (represented by red circles) can be stabilized or degraded in the root zone, sequestered or degraded inside the plant tissue, or volatilized. (Redrawn from Pilon-Smits, E. 2005. *Annual Review of Plant Biology* 56:15–39.)

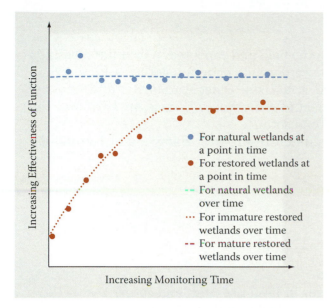

Figure 24.7 An idealized performance curve comparing natural wetlands and the performance of restored wetlands of similar types and size, and in the same land-use setting. (Courtesy U.S. Geological Survey; http://water.usgs.gov/nwsum/WSP2425/restoration.html)

sites will increasingly look like natural sites. Indicators that provide a measure of wetland condition have proved effective in their ability to monitor the outcomes of restoration projects (Fennessy et al. 2004). This will also increase our understanding of restoration ecology, ultimately improving the quality of projects. The use of ecological indicators to measure the performance of restoration projects is increasing as these indicators (e.g., the Index of Biotic Integrity) are being developed.

Global Imperative of Restoration Economy

From all the discourse in Sections B and C, it is clear that, given the extent and magnitude of disturbance, the imperative of restoration is global. Recently, this imperative has been cast in the most welcome view of *restoration economy* by Cunningham (2002). He states that "maturing civilizations stand on three legs—new development, maintenance/conservation,

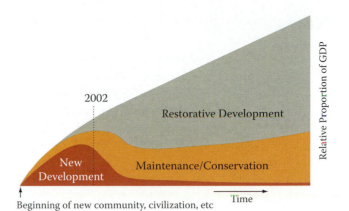

Figure 24.8 The "trimodal development perspective" of a restoration economy showing the comparative importance of restorative development in relation to new development, and maintenance/conservation in relation to time, and relative proportion of GDP. (Redrawn from Cunningham, S. 2002. *The Restoration Economy—the Greatest New Growth Factor*. San Francisco: Berrett–Koehler Publishers.)

and restorative development. Dominance periodically shifts from one leg to another, fundamentally altering technology, culture, and commerce. We are now in such a transition." He then weaves together the three modes of the development cycle (Figure 24.8):

1. New development: This pioneering mode launches most communities and civilizations, but can destroy irreplaceable assets if prolonged. New development is fast becoming less profitable, less desirable, and less possible.
2. Maintenance/conservation: This calmer mode is always present, seldom dominant.
3. Restorative development: This dynamic, high energy mode restores the existing built environment and natural environment. Restorative development is nearing dominance—in construction, ecology, government, and business.

Cunningham appropriately notes that new area development has been shrinking (Figure 24.8), making such development very difficult and divisive among communities (for example, the case of urban sprawl). Restorative development is already the basis of a staggering economy accounting for trillions of dollars. This emphasizes the view of many scientists that restoration ecology is one of the leading fronts in environmental science.

Notes

1. Restoration of ecosystems has been conducted in the field and reported in biennial meetings of the North American Prairie Conference. The conference brings together a large group of plant, animal, and microbial ecologists, soil scientists, conservationists, and (mostly) prairie enthusiasts. The first conference was held at Galesburg, Illinois, in 1970 and the 21st conference in 2008 was held in Winona, Minnesota.
2. Case studies cited in this section have been updated from Wali, Safaya, and Evrendilek (2002).

Questions

Land Use (See the Preamble to Section B)

1. What is land use planning and why is it needed?
2. What human-induced disturbances drive changes in land use and land cover?
3. How are natural disturbances different from human-induced disturbances?
4. Discuss the relationship between preventive and mitigative actions, especially to protect and revitalize the large areas of land that have been subjected to environmental degradation and destruction.

Restoration

1. Investigate ecological restoration actions that have been taken to repair major human disturbances in your region. How have they influenced the biodiversity and ecosystem services of the region?
2. What are the advantages of human-assisted recovery over natural recovery of damaged ecosystems?
3. Discuss the role of succession in the rehabilitation and restoration of damaged ecosystems.
4. How can the success of restoration practices be assessed?

Conservation of Biological Diversity and Natural Resources

25

Principles for Conservation

Changes to the biosphere have been more rapid over the past 50 years than at any other time in human history. As discussed in Section B, the consequences of such changes have been severe, ranging from the depletion of natural resources to the loss of biodiversity and the concomitant degradation of ecosystem services provided by fully intact ecosystems. Recognition of these issues leads to the question of what can be done to stem environmental degradation. The conservation of both living and nonliving resources represents complex challenges for humans all over the globe. Even defining the goals of conservation—save "nature" for its own sake? Keep reserves for future use by humans?—are heavily influenced by economic and political considerations. There are two primary arguments for the conservation of living species such as rain forest insects or old-growth trees in the Pacific Northwest of the United States. The first argument is philosophical: All species have as much right to exist as we do. Equally, as one of many species, humans do not have the right to exploit and degrade resources such that others are driven to extinction. This argument is hard to sustain when people are hungry and suffering; immediate human need almost always takes priority over other considerations.

The second argument is solely economic: It makes no sense to overuse (or extirpate) any resource that is currently or potentially useful to humans. This is a standard argument used to support "saving the rain forest": We may not care about the "right" of a tree or an herb to exist, but the fact that it may have future commercial potential in medicine or food production makes it worthwhile to conserve it for its genetic material. The same approach is taken to conserve or at least to regulate the use of natural resources. The proverbial old tale of slaughtering the goose that laid a golden egg for quick riches may apply here. It is useful to ponder upon how technological advancement has been measured—for instance, by cutting down an entire old-growth forest for timber and pulp or hunting a species to extinction. It is equally obvious

BOX 25.1   Principles of Conservation of Living Resources and Mechanisms for Their Implementation (Mangel et al. 1996)

Principle I. Maintenance of healthy populations of wild living resources in perpetuity is inconsistent with unlimited growth of human consumption of and demand for those resources.

Principle II. The goal of conservation should be to secure present and future options by maintaining biological diversity at genetic, species, population, and ecosystem levels; as a general rule, neither the resource nor other components of the ecosystem should be perturbed beyond natural boundaries of variation.

Principle III. Assessment of the possible ecological and sociological effects of resource use should precede both proposed use and proposed restriction or expansion of ongoing use of a resource.

Principle IV. Regulation of the use of living resources must be based on understanding the structure and dynamics of the ecosystem of which the resource is a part and must take into account the ecological and sociological influences that directly and indirectly affect resource use.

Principle V. The full range of knowledge and skills from the natural and social sciences must be brought to bear on conservation problems.

Principle VI. Effective conservation requires understanding and taking account of the motives, interests, and values of all users and stakeholders, but not by simply averaging their positions.

Principle VII. Effective conservation requires communication that is interactive, reciprocal, and continuous.

---

that, as human populations grow, the direct and indirect impacts on other species increase dramatically.

For those who deal in economics, models can put monetary values on ecosystems and the services they provide. That is, we can calculate the value of clean water and air in terms of health care or the dollar value of recreation time spent in a forested park, and we can predict the rapidly increasing cost of resources whose supply is diminishing. In most human societies, there is also an intrinsic or nonquantifiable value in living things, as well as in such nonliving resources as water, air, and minerals. Here, we consider the need for conservation of biological (living) diversity that represents a genetic as well as a cultural resource.

Biological diversity, or biodiversity, refers to the variety of life on the Earth at all levels, from genetic diversity to landscape diversity. Most commonly, it refers to the number of species found in a given area—plants, animals, and microorganisms (see also Chapter 20).

For conservation purposes, a broader definition is typically used that encompasses the variety of life on the Earth at all levels, from genes to ecosystems, and the ecological and evolutionary processes that sustain it (Gaston 1996). The World Resources Institute, World Conservation Union, and United Nations Environment Program agreed in their "Global Biodiversity Strategy" of 1992 that diversity can be described at three hierarchical levels: genes, species, and ecosystems. In efforts to conserve diversity, each level must be considered and planned for:

1. *Genetic diversity* refers to the variation of genes within species. This covers distinct populations of the same species (such as the thousands of traditional rice varieties in India) or genetic variation within a population (high among Indian rhinos and very low among cheetahs) ...

2. *Species diversity* refers to the variety of species within a region. Such diversity can be measured in many ways, and scientists have not settled on a single best method. The number of species in a region—its species "richness"—is

one often used measure, but a more precise measurement, "taxonomic diversity," also considers the relationship of species to each other. For example, an island with two species of birds and one species of lizard has a greater taxonomic diversity than an island with three species of birds but no lizards …

3. *Ecosystem diversity* is harder to measure than species or genetic diversity because the "boundaries" of communities—associations of species—and ecosystems are elusive. Nevertheless, as long as a consistent set of criteria is used to define communities and ecosystems, their numbers and distribution can be measured …

The question of how to quantify diversity is important; species numbers, as well as trends in those numbers over time, must be tracked, and this can be done in many ways. The most basic measure of diversity is simply to count the number of species in a given area; for example, one might say, "This forest has, on average, five tree species per acre." This is on element of diversity known as species richness. A second factor that is typically accounted for in measures of diversity is species evenness, which refers to the relative abundance of species within a community. For example, if there are 50 trees in a 1-ha plot of forest, of which 45 are sugar maples, two oaks, one ash, one hickory, and one beech, then this would not be considered the same level of species diversity as 1 ha with 10 individuals of each species.

In ecological terms, however, just the numbers of species are not representative of overall diversity. For example, if one prairie supported only closely related grass species from the same plant family, it would be considered to be of lower diversity than a second prairie with the same number of species from several plant families. Therefore, ecologists often evaluate a community in terms of functional diversity, using higher taxonomic levels than species. They might count genera, families, or even orders to get a broader view of community complexity.

It has long been recognized that the distribution of species is uneven, making the patterns of diversity around the globe "patchy." One consistent pattern that has been observed is the latitudinal gradient in the distribution of species and the complexity of communities.

In general, there are more terrestrial plant and animal species per unit area as one moves from the poles toward the equator (Mittelbach et al. 2007). This "latitudinal gradient in species diversity" describes the observation that areas at low latitudes typically have more species than do areas at high altitudes. The causes of this pattern are not well understood, although this effect is also mirrored on a smaller scale with changes of altitude: the higher the elevation is, the fewer the species found. This is one explanation why efforts to protect diversity are often focused on tropical regions, which contain an estimated 80% of all species on the Earth.

This brings us to one of the central questions in ecology: Just how many species are there in the world? As discussed in Chapter 20, over the past 250 years between 1.5 and 1.8 million species have been described and assigned a formal scientific name based on their taxonomic relationships (Figure 25.1). (Note that the lack of a comprehensive and central database on all species thus far described keeps us from knowing even how many have been identified.) The exact number may evade us; however, biological systematists are in agreement that millions of species have not been

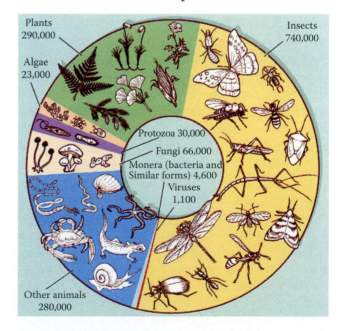

Figure 25.1 Distribution of known species across different taxonomic groups. (From *Encyclopedia Britannica* 1996.) http://media-2.web.britannica.com/eb-media/35/6535-004-8E8AB956.gif

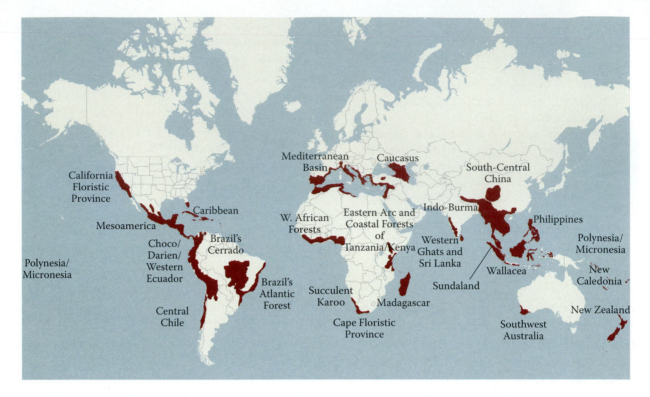

Figure 25.2 The 25 biodiversity hotspots of the world. (Myers, Norman, R. A. Mittmeier, C. G. Mittmeier, G. A. da Fonesca, et al. 2000. Biodiversity hotspots for conservation priorities. *Nature* 403: 854–858.)

described. Although more are continually being added to the list, it is difficult to protect those that are not known. We know the least about the microbial world. For example, an estimated 69,000 species of fungi have been identified, but up to 1.6 million more species are thought to exist (Wilson 2002b). As for soil fauna, there is a ubiquitous group of soil microbial species, known as the nematodes or roundworms, that contains 80,000 species, meaning that by current estimates, four out of every five animals on the Earth are nematodes. However, potentially millions more nematode species await discovery.

If the sheer number of a given group of species (or a taxonomic group) were used as a criterion for success, then beetles (Coleoptera) win the day, with about 350,000 described species. The dominance of beetles on the Earth has long been recognized. When, early in the twentieth century, the famous British biologist J. B. S. Haldane was asked what he could infer about the work of the Creator from a study of His creations, he is said to have replied, "An inordinate fondness for beetles" (Hutchinson 1959).

Some regions of the world are particularly species rich, with high numbers of endemic species (species of limited range that do not occur elsewhere) in what are known as "biodiversity hotspots" (Myers et al. 2000). To qualify as a hotspot, a region must meet two criteria:

- It must have at least 1,500 species of endemic vascular plants (which amounts to >0.5% of the world's total).
- It must have lost more than 70% of its original habitat area.

Myers et al. (2000) identified 25 such places around the globe (Figure 25.2). Understandably, these areas have become high-priority targets for conservation. Conservation experts now give top priority not only to hot spots, but also to the protection of ecologically significant habitats such as the remaining fragments of the late successional, or old-growth, forests (see later discussion). Terrestrial hotspots worldwide occupy only 1.44% (2.1 million km^2) of the world's land surface, but they contain at least 66% of all vascular plants, or one-third of the plant and vertebrate species (Table 25.1)

Table 25.1
Habitat Reduction in 25 of the Earth's "Biodiversity Hotspots" with Their Thousands of Endemic Species of Plants and Animals

Hotspot Location	Original Extent, km² (%)[a]	Area Remaining Intact, km² (%)[a]	Portion of Original Extent Still Remaining Intact, %	Area Protected, km² (%)[b]
1. Mediterranean basin	2,362,000 (1.6)	110,000 (0.074)	4.7	42,123 (1.8)
2. Indo-Burma in Southeast Asia	2,060,000 (1.4)	100,000 (0.067)	4.9	160,000 (7.8)
3. Brazilian cerrado in Central Brazil	1,783,169 (1.2)	356,634 (0.24)	20.0	22,000 (1.2)
4. Sundaland in Malaysia and Indonesia	1,600,000 (1.1)	125,000 (0.084)	7.8	90,000 (5.6)
5. Guinean forests of West Africa	1,265,000 (0.9)	126,500 (0.085)	10.0	20,224 (1.6)
6. Tropical Andes in Venezuela to Argentina in west coastal South America	1,258,000 (0.8)	314,500 (0.212)	25.0	79,687 (6.3)
7. Atlantic forest on the Brazilian Coast	1,227,600 (0.8)	91,930 (0.062)	7.5	33,084 (2.7)
8. Mesoamerica, from southern Mexico to Panama	1,154,912 (0.8)	230,982 (0.156)	20.0	138,437 (12.0)
9. Mountains of south-central China	800,000 (0.5)	64,000 (0.043)	8.0	16,562 (2.1)
10. Madagascar and Indian Ocean islands	594,221 (0.4)	59,038 (0.04)	8.0	11,546 (1.9)
11. Caucasus, in Europe between the Black and Caspian Seas	500,000 (0.3)	50,000 (0.034)	10.0	14,050 (2.8)
12. Wallacea in Indonesia	346,782 (0.2)	52,017 (0.035)	15.0	20,415 (5.9)
13. California floristic province	324,000 (0.2)	80,000 (0.054)	24.7	31,443 (9.7)
14. Southwest Australia	309,840 (0.2)	33,336 (0.022)	10.8	33,336 (10.8)
15. Philippines	300,780 (0.2)	24,062 (0.016)	8.0	3,910 (1.3)
16. Central Chile	300,000 (0.2)	90,000 (0.061)	30.0	9,167 (3.1)
17. New Zealand	270,534 (0.2)	59,400 (0.04)	22.0	52,068 (19.2)
18. Caribbean Islands	262,535 (0.2)	29,840 (0.02)	11.3	41,000 (15.6)
19. Chocó-Darién in coastal Colombia and Ecuador, and the Galapagos Islands	260,595 (0.2)	63,000 (0.042)	24.2	16,471 (6.3)
20. Western Ghats (western India) and Sri Lanka	182,500 (0.1)	12,445 (0.008)	6.8	18,962 (10.4)
21. Succulent Karoo in western Namibia and South Africa	112,000 (0.1)	30,000 (0.02)	27.0	2,352 (2.1)
22. Cape floristic province in coastal South Africa	74,000 (0.05)	18,000 (0.012)	24.3	14,060 (19.0)
23. Polynesia/Micronesia islands	46,012 (0.03)	10,024 (0.007)	21.8	4,913 (10.7)
24. Eastern arc mountains in Kenya and Tanzania, Africa	30,000 (0.02)	2,000 (0.001)	6.7	5,083 (16.9)
25. New Caledonia Island, Australasia	18,576 (0.01)	5,200 (0.004)	28.0	527 (2.8)
Mini hotspots				
Galápagos Islands in Ecuador	7,882 (0.005)	4,931 (0.003)	62.6	7,278 (92.3)
Juan Fernández Islands in Ecuador	100 (0.00006)	—	—	91 (91.0)
Total[a]	17,452,038 (11.76)	2,142,839 (1.44)	12.28	888,789 (0.6)

Source: From Mittermeier, R. A., N. Myers, and C. G. Mittermeier. 1999. Hotspots: Earth's biologically richest and most endangered terrestrial ecoregions. CEMEX Conservation International.
[a] Based on a global total land area of 148,429,000 km² for planet Earth.
[b] Percentage of original extent of each spot.

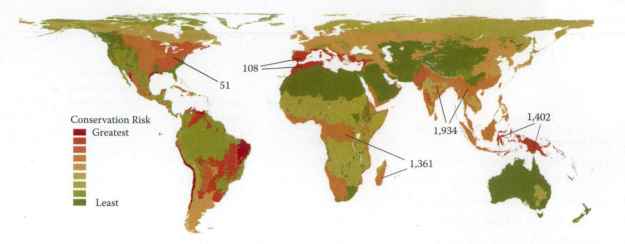

Figure 25.3 In the face of impending global change, some regions are more in need of protected lands than others. (Courtesy of W. Jetz; see also Lee and Jetz. 2008.)

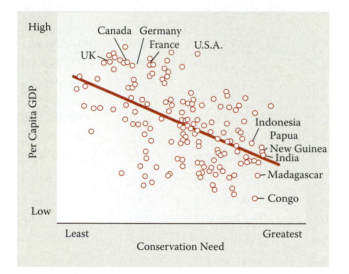

Figure 25.4 Countries with the greatest projected conservation need are those with the least resources available for conservation. Note that wealth (per-capita GDP) increases in 10-fold increments vertically, while conservation need increases linearly. (Courtesy of W. Jetz; see also Lee and Jetz. 2008.)

The need for conservation varies by country as a function of the country's inherent level of biodiversity and the resources it has to apply to issues of conservation; some nations are more in need of conservation land than others (Figure 25.3). High-conservation-need nations are those with little protected area in relation to projected land use change and with a lot of threatened biodiversity (Figure 25.4). Countries on the bottom right of Figure 25.4 have the greatest projected conservation need, but they are also the least wealthy, potentially hampering urgently needed conservation efforts. Many countries on the top left are wealthy, but from a global perspective focus on minimizing total species loss, they have relatively minor conservation needs in their own backyards. This study offers an attempt to quantify and set into context future conservation needs at the global scale (Jetz 2008). But the results offer only broad trends, and the potential effects of, for example, geographical shifts in distributions and the special fate of high-altitude species were not considered in the analysis.

Biological Diversity and Ecosystem Services

The benefits of biodiversity are many. Diverse systems provide food, fiber, medicines, and other "goods" that are indispensable for human and environmental health. Studies on terrestrial ecosystems strongly suggest that higher diversity is correlated with ecosystem function and, by extension, to the ecosystem services they provide (Figure 25.5). The ecosystem processes (also known as "supporting services") that underlie all other ecosystem services include the production of plant material (primary productivity), nutrient cycling (the movement of materials through ecosystem components), and the rates at which dead organic matter decomposes. These have all been shown to respond to different levels of diversity, although the

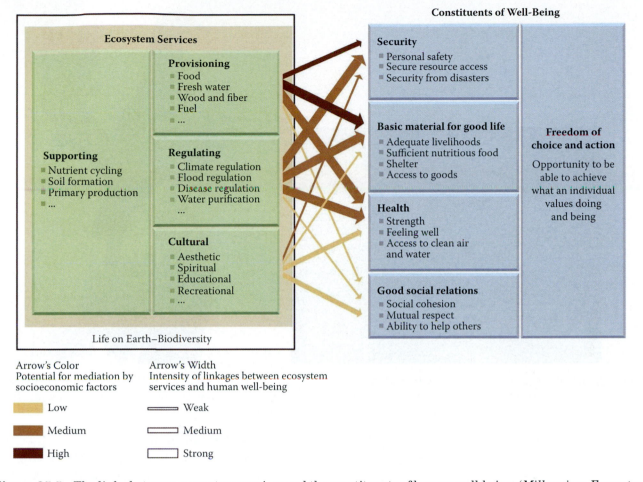

Arrow's Color
Potential for mediation by
socioeconomic factors

Arrow's Width
Intensity of linkages between ecosystem
services and human well-being

Low — Weak

Medium — Medium

High — Strong

Figure 25.5   The links between ecosystem services and the constituents of human well-being. (*Millennium Ecosystem Assessment, Volume 1*. Ecosystems and Human Well-being: Current State and Trends, edited by R. Hasan, R. Scholes and N. Ash. 2005. Island Press, Washington, D.C.)

direction and magnitude of this response varies. Changes in any one of these processes can create "ripple effects" throughout the ecosystem. For example, in wetland systems that have been altered by human disturbance, diversity declines as species composition shifts toward invasive, monoclonal species such as cattails (*Typha* sp.). This in turn results in increased productivity and altered nutrient cycles. Ultimately, biodiversity contributes to the resilience and functioning of ecosystems and may enhance their ability to retain their functions in a changing environment.

When these supporting services (or ecosystem processes) are altered, the capacity of ecosystems to contribute provisioning, regulating, and cultural services is compromised as well (Figure 25.5). Provisioning services include an ecosystem's ability to supply food (through the conversion of land to agriculture for crop

growth or the exploitation of natural systems such as freshwater and marine fisheries), fresh water and air, and fuel (wood). Regulating services are those that help regulate climate (as do the equatorial rain forests), water supply, and the spread of disease; cultural services include the chance for recreation and education, even spiritual and aesthetic fulfillment. Different ecosystem types provide a different combination of services and thus are valued for different reasons.

There is also an innate value to biodiversity for its own sake. Wilson (1984) defined the underlying value that humans place on the diversity of their environment as **biophilia,** or "love of living things." He suggested that, as a species, humans are bound to other living things through the process of evolution and that our love of diversity is a legacy of the environment that we share with other organisms.

Figure 25.6 The Carolina parakeet (left) and the passenger pigeon of North America, both now extinct due to hunting and habitat destruction. (Photo of Carolina parakeet by Fritz Geller-Grimm, mounted specimen in the Museum Wiesbaden, Wikipedia Commons; passenger pigeon lithograph not attributed.)

He added that the "biodiversity of a country is part of its national inheritance—the product of the deep history of the territory extending long back before the coming of [humans]."

Extinction Is a Natural Process

As we know from principles of population ecology, species extinction is a natural consequence of long-term environmental change, short-term catastrophic disturbance, or competitive exclusion, among other contributing factors. The Earth's fossil record can be read as a long list of plant and animal species that were once numerous and now no longer exist. In fact, of all the species that have ever existed, more than 99.9% of them have become extinct. However, it is unarguable that the rate of species extinctions has increased almost exponentially in only the last few centuries—largely due to human population growth and consequent resource depletion and habitat conversion.

In the United States, species extinctions began almost immediately after *Homo sapiens* colonized the continent (Diamond 1997); the rate of extinctions increased dramatically after European colonists arrived and began to hunt and farm, converting the landscape as they moved westward. Species from all taxonomic groups and with a variety of life-history strategies have been lost. It is perhaps not surprising that a species such as the Carolina parakeet (*Conuropsis carolinensis*), with relatively small, localized distributions and low population numbers, would be eliminated as its habitat of old-growth, riparian forests was converted to agricultural land. It is shocking that a bird species as numerous and widespread as the passenger pigeon (*Ectopistes migratorius*), whose population numbered in the billions, was completely wiped out in a few decades due to overhunting and habitat destruction (Figure 25.6).

The Sixth Mass Extinction

During the Earth's history, the rate of species extinction has not been a constant; several

Geologic Timescale of Earth
and rise and fall of plant and animal groups[1]

Era	Period	Epoch	Millions of Years Ago*	Plant Life	Animal Life	
Cenozoic	Quaternary	Holocene	0.01	Increase in number of herbs	First true humans	6
		Pleistocene	1.8			
	"Tertiary"	Pliocene	5.3	Dominance of land by angiosperms	Dominance of land by mammals, birds, and insects	
		Miocene	23.0			
		Oligocene	33.9			
		Eocene	55.8			
		Paleocene	65.5			
Mesozoic	Cretaceous		145.5	— Building of Ancestral Rocky Mtns — Angiosperms expand as gymnosperms decline	Last of the dinosaurs; first mammals and birds	5
	Jurassic		199.6	Gymnosperms (especially cyads and conifers) still dominant; last of the seed ferns, a few angiosperm-like plants appear.	Dinosaurs abundant; second great radiation of insects	
	Triassic		251.0	Dominance of land by gymnosperms; decline of club mosses, horsetails	First dinosaurs	4
Paleozoic	Permian		299.0	Forests of club mosses, horsetails, seed ferns, and conifers	Great expansion of reptiles; decline of amphibians; last of the trilobites	3
	Carboniferous**		359.2	Great coal forests, dominated at first by club mosses and horsetails and later also by ferns and seed ferns; first conifers	Age of the amphibians; first reptiles; first great radiation of insects	
	Devonian		416.0	Expansion of primitive vascular plants; first liverworts	Age of fishes; first amphibians and insects	2
	Silurian		443.7	Invasion of land by primitive vascular plants	Invasion of land by a few arthropods	
	Ordovician		488.3	Marine algae abundant	First vertebrates (jawless fish)	1
	Cambrian		542.0	Primitive marine algae (especially cyanobacteria and probably green algae)	Marine invertebrates abundant (representatives of most groups)	?
Precambrian	Proterozoic Eon		2,500	— Interval of Great Erosion —		
	Archean Eon		4,000	3,000: Extensive development of cyanobacterial reefs 3,900: First one-celled organisms		
	Hadean Eon		~4,600	No life forms		

*Millions of years ago (approximate) from start of period to present

[1] compiled from various sources, ages are from Gradstein et al. (2004)

** In North America, the Lower Carboniferous is often called the Mississippian period and the Upper Carboniferous is called the Pennsylvanian period.

Extinction events

Figure 25.7 Geologic timescale of Earth showing major evolution of major taxonomic groups and timing of mass extinction events.

events have resulted in the loss of species, referred to as **species extinction.** Note that all major extinction events before the present era have been a result of geological changes.

The fossil record shows that in the past 500 million years, there were at least five mass extinction events (Figure 25.7) in which most major taxonomic groups showed a sharp decrease in the number of species present in what is a geological blink of an eye. These events range from the Ordovician–Silurian extinctions of approximately 440 million years ago to the most recent Cretaceous–Tertiary (or "K–T") extinction about 65 million years ago that led to the demise of the dinosaurs. The first five mass extinctions were from natural causes (Eldredge 2001):

- The first mass extinction occurred at the end of the Ordovician Period about 440 million years ago (mya), due to climate change in the form of global cooling. There was essentially no life on land at this time; about 25% of all marine families became extinct.
- The second mass extinction, at the end of the Devonian Period, took place about 370 mya. The suspected cause is climate change; 19% of families were lost.
- The third mass extinction happened about 250 mya, at the end of the Permian Period. This was the greatest mass extinction of all, with an estimated 54% of families becoming extinct; evidence points to climate change and perhaps asteroid impacts as the cause.
- The fourth mass extinction, at the close of the Triassic Period, was 210 mya. The cause is not yet understood, but about 23% of families became extinct.
- The fifth mass extinction occurred about 65 mya at the K–T boundary. Seventeen percent of families were lost, including the terrestrial dinosaurs. As described previously, this event was caused by one or more asteroid collisions with Earth.

What is increasingly recognized as the sixth mass extinction today is caused solely by humans (Leakey and Lewin 1995). We have earned the title of a "mighty geological agent."

When geologic conditions are less volatile—that is, when a mass extinction event is not under way—the normal, background rate of extinction is accepted to be one to two species per year (Pimm et al. 1995). It is difficult to estimate current extinction rates in the absence of a clear idea of the total number of species on Earth. Wilson (2000b) states that the rate of species extinction is now 100 to 1,000 times the background level (about one species becomes extinct per million people per year); similarly, the birthrate of new species (speciation) has declined as the natural environment is destroyed. The United Nations report estimates that 20% of freshwater species have become extinct or been driven toward extinction in recent decades.

Extensive loss of habitats and reduction in the size of the gene pools of species living in them can first put species in a state of what biologists call "living dead" before their extinction. For example, many North American unionid mussels can no longer successfully reproduce in their native streams because the fish species on which their larvae are temporary parasites are extinct, so the remaining populations of some mussel species consist solely of 30- to 40-year-old specimens in a few rivers and streams. At this rate, it is likely that up to 50% of the Earth's species may disappear by 2100.

The estimates of human-caused species loss in the immediate past have been so staggering that the following dire predictions have been made:

> During the next 20–30 years, the world will lose more than a million species of animals and plants—primarily because of environmental changes due to humans. At 100 species per day, this extinction rate will be more than 1,000 times the estimated "normal" rate of extinction. The list of lost, endangered, and threatened species includes both plants and animals. About 10 percent of temperate regional plant species and 11 percent of the world's 9,000 bird species are at some risk of extinction. In the tropics, the destruction of forest threatens 130,000 species which live nowhere else. (UNEP 2005)

What are the causes of such biological devastation? The primary agents of biodiversity loss must be addressed if the extinction rates are to be reduced:

- overexploitation of species (overhunting and fishing);
- habitat destruction through land use changes and fragmentation of the remaining habitat;
- pollution of existing habitats by toxic compounds or by nutrient enrichment; and

- introduction of non-native, invasive species that displace native ones; the diseases carried by non-native species into their new environments are also responsible for native species dieback.

At present, these human activities worldwide threaten 11% of the remaining birds, 18% of the mammals, 5% of the fish, and 8% of the plant species on the Earth with extinction (Barbault and Sastrapradja 1995). Forests, which harbor about two-thirds of the known terrestrial species, have been reduced by more than 20% worldwide. The biodiversity of freshwater ecosystems is much more threatened than that of terrestrial ecosystems, such that 20% (more than 10,000 species) of the world's freshwater fish have become extinct, threatened, or endangered in recent decades (WRI 2000).

Note that all of the anthropogenic forces that lead to biotic impoverishment result from increasing human populations as well as ecologically incompatible technological advances. The cumulative effects of these patterns are perhaps most clearly seen in the reduction of numbers for large animals, or the "charismatic megafauna": Whale species in the oceans, red deer in Great Britain, and many ungulate species in Asia have been hunted to extinction. For island habitats, the picture is even grimmer; often lacking native predators and confined to relatively small areas, island species are far more vulnerable to extirpation, such as the well-known case of dodo of Mauritius in the Indian Ocean.

Human conversion of habitat and shrinking range or territory often have a less direct impact than hunting on species numbers, but the results can be more catastrophic. For instance, relatively few plant species are threatened by overcollecting; the biggest impact on plant species in general is caused by habitat loss due to (1) conversion to agriculture or the replacement of complex and diverse plant communities with monocultures (whether crop cultivars or plantation forestry) and (2) the incursion of non-native plants into the niches formerly occupied by local species. In North America, the combination of habitat conversion and deliberate introduction of some plant and animal species—and accidental introduction of others—along with new

diseases has completely altered vast areas of the landscape. One tragic example is the loss of 99% of the once seemingly limitless tallgrass prairies and the species they supported in the Midwest.

Introduction of diseases into habitats where they were previously unknown has had a great impact on the numbers of other animals and plants as well as human populations (Diamond 1997). Over half of many Amerindian Indian tribes died of smallpox and other diseases introduced from Europe in the first generation of colonization; the native population of Tasmania, Australia's largest island, was completely wiped out. Although disease introductions may rarely have been deliberate, the consequences have almost always been unimaginably severe, both to species and the communities they inhabit.

The Control of Invasive Species

Integration of Invasives

What are the limits to species richness? That is, why can the new species not be integrated into existing assemblages? In several cases, this has happened, and we simply accept the result. Tumbleweeds in the western United States, along with dandelions in the eastern part, are considered part of the American landscape, but both are imports from Eurasia. The common honeybee throughout the United States is a European introduction; therefore, the interbreeding of the "Africanized" strain from South America is not really a further threat to native diversity because native bee species had already been pushed aside by the honeybees (although they are a threat in other ways). As we have seen in earlier discussions of community structure, ecosystem diversity can be examined at several scales or levels. **Alpha diversity,** or local habitat diversity, is the most likely to be affected by extinction or invasion. **Beta diversity,** or diversity between ecosystems or along gradients is affected next; **gamma diversity,** measure of diversity over large regions/landscapes may be affected least by any is rarely threatened by any single species. However, composition of plant communities worldwide has been altered through human actions, and the consequences remain unpredictable.

The control of invasive species entails eradicating or reducing growth and preventing spread. Control measures also include restoring native species and habitats to prevent further invasions. The most effective control for an invasive species is to prevent its introduction to new ranges entirely; generally, this is an impossible task. Keeping invasive species from spreading requires legislation and international cooperation on its enforcement; such laws are not in practice worldwide. The mechanical control of nuisance plants is done by harvesting or mechanically removing them from the soil.

In one interesting example during the lead-up to the 2008 Olympic Games, there was an internationally embarrassing pollution event in Qingdao, China, as the site for the World Olympics sailing competition became covered with noxious algae (Figure 25.8). In this case, the work of several thousand people was used to control the outbreak:

> Huge patches of algae mar the coastline of Qingdao, the host city for sailing events at the 2008 Olympic Games, in eastern China's Shandong province. The Qingdao government has organized 400 boats and 3,000 people to help remove the algae after Olympic organizers ordered a cleanup. Experts say the algae [are] partially a result of climate change and heavy rains.

Figure 25.8 A huge algae bloom clogs the coastline of Qingdao, the host city for sailing events at the 2008 Olympic Games, in eastern China's Shandong province. (Photo by EyePress via APImages.)

The alga was *Enteromorpha prolifica* and the reason for the algal bloom appears to be due to sewage and agricultural affluents.

Chemical controls such as pesticides and herbicides are used throughout the world to control nuisance species. Currently, about 200 herbicides are registered in the United States, although fewer than a dozen are labeled for use in aquatic sites. The widespread use of what were considered relatively safe organic herbicides began in the United States in the 1940s, with a newly developed herbicide, 2,4-D, used as a control agent for control of water hyacinth (*Eichhornia crassipes*). As early as the late 1800s, the growth of *Eichhornia crassipes* was reported to be so thick that it impeded river traffic in the southeastern United States. In response, the U.S. Congress initiated the "Removal of Aquatic Growths Project" as part of the Rivers and Harbors Act of 1899. Unfortunately, water hyacinth remains a problem wherever it has spread (Cronk and Fennessy 2001).

The ballast water of ships is one common way that aquatic organisms move around the globe. Ballast water is pumped into the holds of a ship to stabilize it in the water; however, that water in the hold also contains marine species from the area. After traveling across the ocean, the water is discharged, and any organisms that have survived the journey can colonize the new environment. An open ocean exchange of ballast water, where organisms are much less common, is one way to slow the spread of species. The U.S. Congress recently passed legislation placing limits on the discharge of ballast waters in waters under U.S. territorial jurisdiction.

Conservation Biology

Concern about the loss of biological conservation began in earnest in the 1960s when it became apparent that human activities, especially in neotropical areas (Central and South America), were leading to high rates of species loss. Tropical scientists began to realize that their field research sites might disappear

Table 25.2
Contrasting Conservation Principles Derived from a Shift in Ecological Thinking from the Classical Viewpoint or "Equilibrium Paradigm" to the Contemporary Viewpoint or "Nonequilibrium Paradigm"

Components	Classical Viewpoint ("Equilibrium Paradigm")	Contemporary Viewpoint ("Nonequilibrium Paradigm")
Goal	Preserving valued objects of ecological heritage; unique individuals, communities, habitats, and vistas	Preserving the ecosystem processes and context (ecological integrity and biological diversity) by representation in natural and seminatural landscapes
Focus	Fixed natural areas surrounding closed and static communities (i.e., "dioramas")	Heterogeneous landscape mosaic sustaining open and dynamic communities (i.e., "functional landscape units")
Theme	The consequences of life; a fixed and final outcome	The circumstances of life; multiple causation
Emphasis	Stability and persistence of objects, structural completeness	Structural context, and dynamic processes; historical contingency
Humans as	Cultural landscapes; humans are excluded components	Seminatural landscapes; humans are incorporated
Scale	Generally small; set by size of object (fine grain; small extent)	Generally large; set by range of processes (variable grain; large extent)
Metaphor	The "balance of nature"; nature is constant or self-sustaining (i.e., "equilibrium")	The "flux of nature"—nature is manifold and dynamic (i.e., nonequilibrium)
Knowledge	Ecological understanding not essential	Knowledge of ecological systems is critical
Partnership	Competitive or isolated "part lines"; cooperation not emphasized	Transdisciplinary communication and cooperation vital
Management	From nonintervention ("benign neglect") to passive or limited management	Active management of processes and implications context (structure and linkages)
Application (example of overview of reserve plan)	Preserve the biological diversity of the Connecticut River tidelands by establishing several reserves	Preserve and protect ecosystem processes and context that maintain ecological integrity and perpetuate the biological diversity of the Connecticut River tidelands ecosystem by establishing and managing a regionalized, watershed-based network of integrated core-dominant reserves with cocentric buffers (i.e., a bioreserve)

Source: From Barrett N. E., and J. P. Barrett. 1997. In *The Ecological Basis of Conservation: Heterogeneity, Ecosystems, and Biodiversity,* ed. S. T. A. Pickett, R. S. Ostfeld, M. Shachak, and G. E. Likens, 236–251. New York: Chapman & Hall.

(Sarkar 2004). In response, the nascent field of conservation biology was born and work to establish nature reserves was begun to counter habitat destruction.

Design of Nature Reserves

The work of conservation biologists has focused on using ecological principles to design reserves that will effectively preserve the greatest diversity possible (Table 25.2). Initially, many reserves in the United States were established using the theory of island biogeography (McArthur and Wilson 1976). This theory was initially devised to explain the number of species that an oceanic island could hold, where species numbers (richness) are a function of the size of the island (larger islands hold more) and its distance from the mainland (more distant islands hold fewer). These two factors define the equilibrium number of species as the balance between dispersal of species from the mainland to the island (immigration) and the extinction rates of species on it. Despite its lack of testing, the theory of island biogeography was applied to natural areas that were seen as an island of natural habitat in an "ocean" of a human-dominated landscape. Drawing on its lessons, small, isolated reserves are predicted to do the least to conserve species, in part because of edge effects (see Chapter 23). Generally, the island biogeographic principles as applied to the design of nature reserves include the following (Diamond and May 1976):

- All things being equal, large reserves are more effective at supporting biodiversity than a group of small reserves of the same total area.
- Multiple reserves that are closer together are more effective than those that are farther apart.
- Multiple reserves that are linked by corridors (thin strips of natural habitat) are more effective than those that are not linked.

A focus of the debate centered on what came to be known as the "SLOSS" controversy: if we have finite resources available, should we use those to set aside a "single large [piece] or several small" pieces of habitat? Many scientific papers were written on the topic over a decade, but in the end it was recognized that the answer to the question was (as it is for so many ecological questions): "It depends." The choice of whether to have one large or several small reserves depends on the extremely variable, case-specific factors for the reserve in question, such as its habitat heterogeneity or the number of microclimates.

Human uses of land affect these decisions as well. One approach recognizes that habitat preservation is essential for species conservation and that the challenge of planning is to devise strategies to minimize disturbance of sensitive habitat areas, while simultaneously providing alternative sites for human uses. Exclusion of all human activity is usually impossible in conservation areas; in fact, areas where local people have been included and have a vested interest in preservation (e.g., Santa Rosa National Park in Costa Rica) have fared much better than "game preserves" such

as those in central Africa, where exclusion of villagers has only promoted poaching and resource exploitation. Adversarial relationships are counterproductive in the effort for habitat and species conservation.

As concerns of species diversity came to be recognized in the last three or four decades, the attention of conservationists in many countries was focused primarily on single-species projects such as "save the elephant," "save the panda," "save the tiger," and "save the condor," although no species can be long saved unless the ecosystem of which it is a part is saved first. For example, large mammals have a territory that they defend and a home range in which they hunt to obtain their food and mates. Thus, it is most appropriate that ecologists have emphasized the functioning of ecosystems alongside their structural and organismal components.

Endemism and Extinction Risk

As described previously, entire regions are sometimes considered to be at risk for extinction because they are "hotspots" of biodiversity due to high levels of endemism. Areas of significant endemism are good examples why species conservation is not just matter of "saving" a specific plant or animal from extinction. Most species cannot survive without a required range of environmental conditions specific to their evolution. Therefore, ecosystem conservation rather than just saving a short list of individual species is imperative, as is now widely recognized. Tropical rain forests have become the poster children for ecosystem conservation, and with the rate of forest loss, the statistics are alarming. However, it is impossible to minimize the effects of widespread habitat destruction in the tropical forests, if only because the conversion is so rapid and so recent in human history. When habitats are reduced in area, healthy populations of many species are affected, albeit slowly. With time, however, secondary effects such as increased inbreeding, genetic drift, or loss of genetic variation reduce species viability to the point of extirpation.

Legislative Efforts

The 1973 passage of the **Endangered Species Act** (**ESA**) closed a long battle by advocates for legislation to protect imperiled species. The act, and its subsequent amendments, require that species in decline be listed as either threatened (species at risk of becoming endangered in the foreseeable future) or endangered (those under immediate threat of extinction) and that a population recovery plan be designed and implemented for each species listed. Species with dedicated recovery plans have been found to show more significant signs of recovery than those for which plans have not been developed. These plans take an ecosystem approach, providing for the protection of the habitat essential for the listed species.

Although it has been successful in some cases, the ESA has been criticized for being too "reactive": listing species only when population numbers are at a crisis level, with very few individuals remaining. The ESA makes few provisions for being "proactive." A perceived bias in the species selected for listing has also been noted; for example, vertebrates are much more likely to receive attention and funding than are plants or invertebrates (Noss and Cooperrider 1994). The bald eagle was the first species to be listed. In 1978, there were only 791 nesting pairs in the lower 48 states. By 2006 when the species was taken off the list, or delisted, there were at least 9,780 nesting pairs.

Since the ESA was passed, the number of listed species has increased steadily and very few of the more than 1,300 endangered species on the list have been delisted. Most recently, the polar bear was listed under the ESA as a threatened species due to climate change and the threat of decreasing sea ice. This is the first species to be identified as threatened because of the accumulation of atmospheric greenhouse gases and associated climate change. The recently-realized effects of climate change have become superimposed on such pre-existing stresses on species as habitat loss, invasive species, pollution effects, over-exploitation, and disease. Lawler, Aukema, Grant et al. (2006) provide an assessment of 20 years of research studies aimed specifically at conservation science (Figure 25.9).

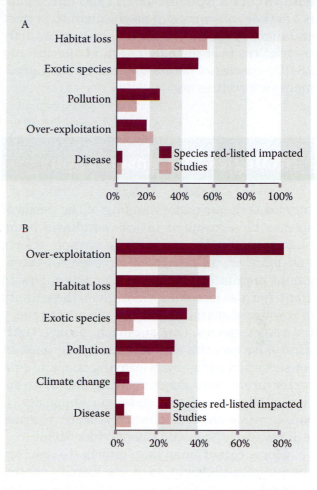

Figure 25.9 A comparison of the prevalence of different risks in biodiversity and the degree to which they are reported in the literature: (a) in all systems, and (b) in marine systems. The prevalence of threats to species in all systems was based on Wilcove et al. 1998; the prevalence of threats to marine systems based on Kappel 2005. (From Lawler, J. J., J. E. Aukema, and 3 others. 2006. Conservation science: a 20-year report card. *Frontiers in Ecology and Environment* 4: 473–480.)

The primary international treaty for biodiversity protection is the **Convention on Biological Diversity,** adopted in 1992 at the United Nations Conference on Environment and Development. To date, 188 countries have ratified the treaty; the United States is not one of them. The convention has three goals: (1) to conserve biodiversity, (2) to use biodiversity in a sustainable way, and (3) to share the genetic benefits of diversity fairly (i.e., to include countries from which species were collected for biotechnological exploration in any profits made). The **Convention on International Trade in**

Endangered Species (or CITES) agreement is another international mechanism to protect endangered species, as signatories voluntarily restrict trade of the 5,000 animal species and 28,000 species of plants covered by the agreement (www.cities.org).

Strategies for Conservation

Increases in our understanding of the needs of species have resulted in less emphasis being placed on single-species conservation efforts and more on an ecosystem or even an ecoregional approach. This calls to question how to identify areas for conservation given the scarcity of resources available and the sense of urgency to protect species. One approach, named **GAP analysis,** has the goal of identifying gaps in conservation areas by overlaying maps of protected areas onto maps showing the distribution of threatened and endangered species. If conservation has been effective, these two areas will coincide (Noss and Cooperrider 1994).

More recent mapping efforts to portray the complex distribution of the Earth's species have resulted in global maps of both terrestrial (Olson et al. 2001) and aquatic (Abell et al. 1999) ecoregions (Figure 25.10). **Terrestrial ecoregions** have been defined as "relatively large units of land containing a distinct assemblage of natural communities and species" (Olson et al. 2001); the **freshwater ecoregions** are based on freshwater biogeography, particularly the distribution of freshwater fishes (Abell et al. 1999). The ecoregions are of sufficient detail and resolution to allow areas of high diversity as well as representative communities to be identified for conservation priority. As Olson et al. (2001) point out, conservation strategies designed around ecoregions are particularly well suited to identify a full range of representative areas that act to preserve known and unknown species of regions and of all the associated ecological processes.

Finally, The Nature Conservancy has developed a seven-step framework to aid in regional conservation planning based on the principles of conservation biology and ecology.

This framework is now being used to prioritize areas for conservation (Groves et al. 2002):

1. Identify conservation targets, considering communities, ecosystems, abiotic features, and species of special importance (threatened, endangered, keystone, endemic, etc.).
2. Collect information and identify any information gaps, using sources such as published information, rapid ecological assessments, biotic inventories, and local experts.
3. Establish conservation goals that address both representation (i.e., protect areas that are good representatives of their ecosystem or habitat types) and habitat quality.
4. Assess existing conservation areas to determine if there are gaps in the protection that is being offered to the conservation targets identified in step 1.
5. Evaluate the ability of conservation target areas to maintain viable populations using tools such as GIS-based habitat suitability models.
6. Assemble a "portfolio" of conservation areas for preservation.
7. Identify priority areas for conservation, drawn from the portfolio developed in step 6.

Conservation of Natural Resources

There are long-standing assumptions that important, nonrenewable natural resources such as oil, coal, and metal ores will soon be exhausted. However, technological advances have made continued extraction of resources possible, beyond what was originally predicted, due to the ability to use lower-grade ores or extract more remote resource deposits cost effectively. As with biodiversity, a goal of modern conservation efforts is to develop the means to use natural resources sustainably. This is the topic of our next chapter; for now, we summarize broadly applicable principles

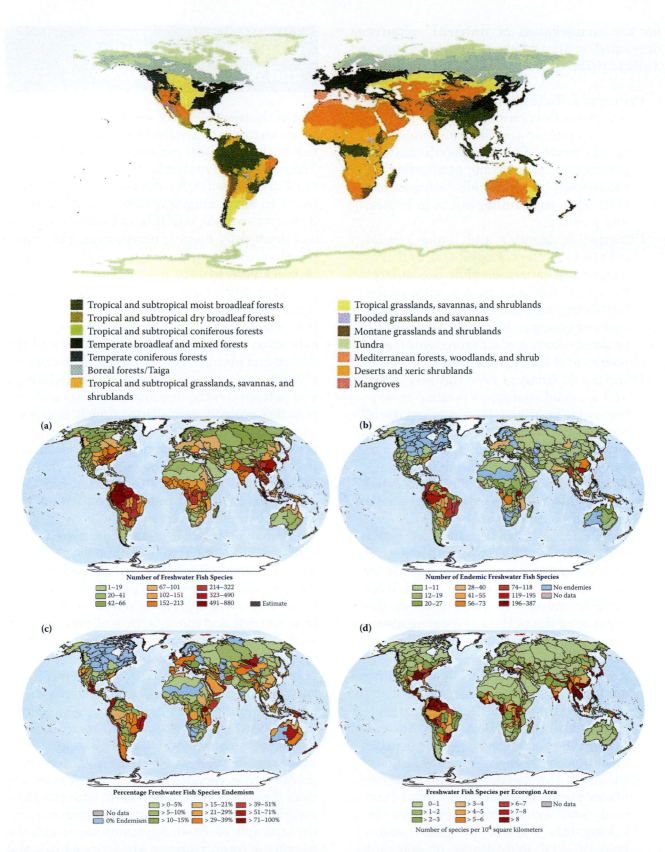

Figure 25.10 Aggregation of 867 terrestrial ecoregions into 14 biomes for conservation planning (top panel based on Olson et al. 2001) and bottom panel, freshwater ecoregions based on freshwater fish data: (a) species richness, (b) number of endemic species, (c) percent endemism, and (d) species per region. (From Abell, R. A. et al. 1999. *Freshwater Ecoregions of North America. A Conservation Assessment.* Washington, D.C.: Island Press.)

for the conservation of **natural resources** developed by Mangel et al. (1996) and Holt and Talbot (1978):

Principle 1. Take appropriate actions that recognize that human impacts on natural resources are the consequence of rapid growth of human population and per-capita consumption; applications of ecologically incompatible technologies, land uses, and management practices; and poverty.

Principle 2. Identify and make special efforts to protect ecosystems and life-supporting processes that are subject to critical limits as well as vital to the well-being and integration of humans and ecosystems, at scales appropriate to the dynamics of environmental, economic, and social processes.

Principle 3. Analyze how the ecosystem and its relationship with other ecosystems may be affected by proposed use of natural resources, determine how sensitive the results of the analysis are to changes in assumptions and uncertainties, and be prepared for unexpected events due to complex and stochastic behaviors of the natural systems.

Principle 4. Ensure that proposed use and management of land are within actual and potential biophysical capabilities and limits of ecosystems, avoid irreversible changes in ecosystems at the expense of the welfare of future generations, and require those most likely to benefit from use of the resource to pay the costs of associated environmental damage when damage is reversible.

Principle 5. Invoke the full range of relevant natural and social disciplines at the earliest possible stage of scientific investigation of environmental issues.

Principle 6. Develop agreed-upon mechanisms for making decisions about land uses and management practices by taking into account all users, stakeholders, and institutions at relevant community levels consistent with the temporal and spatial scales of the ecosystems involved.

Case Study: Recycling and Natural Resources

One of the most effective ways to conserve natural resources is to use the products derived from the stock of natural resources more than once. The recycling of materials benefits the environment in many ways: It conserves energy, reduces the need for mining activities, and reduces the flow of wastes into landfills and waterways that can drastically disturb ecosystems. For example, steel, aluminum, and copper produced from primary ores require more energy than those produced from secondary, recycled metal (or scrap) (Wernick and Themelis 1998). Reducing the amount of waste involves altering the extraction and processing of minerals and the design and manufacture of products on the supply side and changing consumption behaviors and patterns on the demand side. This strategy is called **source reduction**, and it conserves energy and minerals, prevents the degradation of natural resources, and protects public health. It should therefore not be surprising that one category of exports from the United States to China has increased from $1.4 billion in 2001 to $5 billion in 2005: trash!

Dematerialization of manufactured goods leads to products that consume less of each basic material. Recycling largely reduces the energy consumption that is needed to produce these materials. Recycling refers to the collection, sorting, and manufacture of some wastes, such as paper, glass, plastic, and metals, which are sold as new products in order to divert them from the waste stream. Recycling items such as paper, glass, plastics, and metals therefore has multiple benefits. There is a reduction of the amount of waste sent to landfills, and we preserve nonrenewable energy supplies and natural resources. In addition, manufacturing products from recycled material uses less energy than if raw materials were used, with the associated benefits of lower carbon emissions into the atmosphere. From a systems viewpoint, the benefits of resource reuse are far reaching, and it appears that China has realized this more than any other country, given the magnitude of its import as mentioned earlier.

Questions

1. Why is the participation of all concerned citizens important when handling environmental issues?
2. How can you compare the recycling of materials with the rehabilitation of damaged ecosystems?
3. What are the roles of renewable energy technologies, energy conservation, and energy efficiency in stabilizing global climate?
4. What are utilitarian- and values-based rationales for protecting biodiversity?
5. What institutional and personal barriers could prevent manufacturing and retail sectors from adopting waste minimization technology, making more with less (dematerialization), reusing, recycling, and designing products for longer life?

Sustainable Development

26

Approaches to Achieving Sustainability

Every treatise, long and short, must have a summation. It is great perseverance that has brought readers of the book to this point! Armed with the knowledge of fundamental ecological principles (Section A), contemporary environmental problems (Section B), and some relatively recent mitigation strategies (Chapters 21–25), it is only fair to ask: Is there a central rallying theme that should bring together scientists and professionals of all disciplines and the concerned general citizenry? The answer to the question is in the affirmative. The appropriate rallying theme, in our opinion, is *sustainable development*. Even a cursory search (as of February 5, 2008) of this two-word phrase shows 7,648 hits in the Library of Congress, 4,243 (2,216 books; 2,027 journals articles) in the National Agricultural Library, and 47,349 hits in the WorldCat. Either the phrase is inviting enough to inspire authors to write such an extraordinary number of articles, research papers, and books, or, simply, its importance is profound. In this case, it is both. Because of its central place in managing environmental issues, we must examine the concept of sustainable development and its ecological, economic, social, and cultural implications.

The Concept of Sustainable Development

The concept of sustainable development has its roots in the distant past, stretching back to the work of Thomas Malthus on the dangers of population growth in the late 1700s. Concerns about environmental degradation have come from poets and writers (see Wali 1999b). In the United States, the most powerful indictment on human actions came from George Perkins Marsh (1864): "[M]an is everywhere a disturbing agent. Wherever he plants his foot, the harmonies of nature are turned to discords....[O]f all organic beings, man alone is to be regarded as essentially a destructive power." Marcus Hall (2004) notes that Marsh's "key warnings about degradation were characteristically American—having been interpreted, produced, and packaged by an American for Americans."

It was the decade of the 1970s, however, as the environmental movement was growing, that we saw the concept gain standing and momentum. Debates over the consequences of human use of the environment, fears about the increasing rate of human population growth, and the tensions between economic development and environmental quality led to a series of influential writings focused on these issues. A few influential examples from the time include works such as those of Paul Ehrlich, Barbara Ward, and Rene Dubois, and E. F. Schumacher.[1] The 1970s and thereafter also witnessed the ushering in of several organizations that have become established as excellent resources for the synthesis and communication of environmental data. Of importance and endurance has been the establishment of the Worldwatch Institute by Lester Brown in 1974 and, since 1984, that organization's publication each year of the premier volume, *State of the World*. This widely perused annual assessment of global environmental issues is available in more than 40 countries and in almost as many languages. In his foreword to the first book (1984), Brown noted: "The yardstick by which we measure progress is sustainability—the extent to which our economic and social systems are successfully adjusting to changes in the underlying natural resource base." The quest continues.[2]

Two important components of sustainable development have been recognized. One has centered on the disparity in wealth that is evident if one looks across different segments of the world's population and the associated desire to help the poor. The second emphasized that economic development should not take place without consideration of its long-term environmental, social, and cultural consequences (Barbier 1987). This view emerged as a result of the growing awareness that the availability of natural resources and their capacity to provide ecosystem goods and services and to accommodate wastes is finite. It recognized the environmental constraints on human activities and acknowledged the need for fairness in terms of maintaining natural resource availability across generations.

The thinking for sustainable development began with the 1974 United Nations Cocoyoc Declaration, which addressed the use of natural resources in a sustainable way by promoting natural resource use only to the extent that it is adequate to meet human needs without

outstripping the Earth's capacity to meet those needs. A subsequent policy summit brought together world leaders to discuss the environmental and socioeconomic issues that are part of sustainable development. This group laid out the world conservation strategy (IUCN et al. 1980), which included one of the earliest attempts to define the term:

> For development to be sustainable, it must take account of social and ecological factors, as well as economic ones; of the living and non-living resource base; and of the long-term as well as the short-term advantages and disadvantages of alternative action.

The group went on to define sustainability from an ecological viewpoint, with the goal of maintaining the "essential ecological processes and life support systems, the preservation of genetic diversity, and the sustainable use of species and ecosystems."

Realizing both its conceptual and practical importance, in 1982 the United Nations established the World Commission on Environment and Development (WCED), headed by Gro Harlem Brundtland, first the environment minister and then prime minister of Norway. When the commission's report, *Our Common Future* (also known as the Brundtland Report), was published in 1987, sustainable development received worldwide attention and was established as an international policy issue. In what has come to be the most widely used definition, sustainable development was said to be "development that meets the needs of the current generation without undermining the ability of future generations to meet their own needs" (WCED 1987).

As noted earlier, this concept has been the subject of numerous studies and, not unexpectedly, the definitions that have been developed for it are equally numerous. Among many, we have a preference for the definition provided by the Philanthropic Foundation of Canada, which states (see www.pfc.ca/cms_en/page1112.cfm):

> [Sustainable development is] the ability to meet the needs of today's people and environment without compromising that of subsequent generations. When a program seeks to create sustainability, it aims to create an environment that can renew itself without damage to future stakeholders.

More recently, the United Nations sponsored the U.N. Conference on Environment and Development (UNCED), known as the Earth Summit, in 1992 in Rio de Janeiro. The conference delegates approved five agreements designed to foster the different aspects of sustainability (UNCED 1992) (from Dalal-Clayton 2003):

- **Agenda 21:** a global plan of action for sustainable development, defining more than 100 program areas ranging from trade and environmen, through agriculture and desertification and technology transfer;
- **the Rio Declaration on Environment and Development:** a listing of 27 key principles to guide the integration of environment and development policies, including some adopted by the United States, including the "polluter pays" principle and pollution prevention programs;
- **the Statement of Principles on Forests:** the first global consensus on the management, conservation, and sustainable development of the world's forests;
- **the Framework Convention on Climate Change:** a legally binding agreement to limit greenhouse gas emissions so as to stabilize global climate change; and
- **the Convention on Biological Diversity:** another legally binding agreement, with the goal of conserving the world's genetic, species, and ecosystem diversity and ensuring that the benefits of its use are shared in a fair and equitable way.

Ecological Controls on Sustainable Development

We emphasized the importance of ecological principles in managing our resources, both natural and made by humans. A sufficient number of case studies from around the world are now available that allow us, with a few measurements, to predict the environmental impacts of agricultural and industrial developments and to use a cost-benefit approach to make decisions. For example, given the state of water availability, water-intensive crop growth in California does not appear sustainable, as was shown by Reisner (1989). His major findings: agriculture accounts for 85% of all water use. Between 1972 and 1988, as the California economy grew from $220 to $550 billion, agriculture was static at $14 billion. Irrigated pasture, grass, and hay used 5.3 million acre-feet of water—equivalent to the consumption of water by the entire population of 27 million in the state. Alfalfa used 3.9 million acre-feet (the same amount as used by Los Angeles and the San Francisco Bay areas combined), cotton used 3.0 million acre-feet (enough for 15 times the population of Nevada), and rice used enough water for 10 million people. New Mexico's agricultural returns were 18% of the state's income, but used 92% of water (Reisner 1989). The principles of sustainability dictate water management in a more efficient and equitable manner.

Since the Brundtland Report, sustainable development programs have come to be distinguished as promoting a either "weak" or "strong." Proponents of weak sustainable development are in favor of continued economic growth, placing faith in the ability of new technologies, technological fixes, and our ability to substitute available resources for those that are limiting in order to overcome the biophysical constraints of the environment. The goal in this approach is to guide the current system in a more sustainable direction. On the other hand, the strong sustainable development school argues that indefinite economic growth is impossible on a finite planet. If we are to maintain the capacity of the ecological life-supporting systems for both present and future generations, then substantive changes must be made. This approach calls for a restructuring of our policies, practices, and economic system.

Are today's short-term financial profits the only measure of economic success? If so, then there is no future for sustainable development. If long-term, nonspeculative income and stability are the goal, then we must preserve our natural capital (i.e., our natural resources) and live only on the interest of our environmental investments. Gardner (2006) suggests three basic criteria for maintenance of natural capital and ecological sustainability:

- For renewable resources, the rate of harvest should not exceed the rate of regeneration (sustainable yield).
- The rates of waste generation from projects should not exceed the assimilative capacity of the environment (sustainable waste disposal).
- The depletion of nonrenewable resources should require comparable development of renewable substitutes for that resource.

Environmental constraints on long-term human population growth arise from these issues—particularly the rate-limited capacities of renewable natural resources to provide ecosystem services such as waste assimilation, production of food and fiber, regulation of climate, and provision of clean air and water. The limited availability of and increasing demands on nonrenewable natural resources such as metals and fossil fuels, coupled with the exponential growth of human population and consumption, will also impose limits. Ultimately, we must depend on the capacity of human beings and their institutions to respond to population pressures and to function within the carrying capacity of the Earth.

Revolutionary Thinking in Economics Is Needed

Conventional economic systems currently in vogue are built on fossil fuels. As human populations continue to increase in almost every part of the globe, their economic well-being is largely dependent on the continuous growth and expansion of fossil-fuel-based markets, goods, and services. However, as natural resources are depleted at ever-increasing rates, planners and economists are struggling to define and implement models for sustainable development. If we are to transition to a

sustainable economy, a key assumption of our current economic system must be addressed: namely, that "growth" (making an economy bigger) is the primary goal (Gardner and Prugh 2008). Despite the increasing threat to our natural capital, the consumer culture defines economic progress and has made economic growth the centerpiece. In no ecological or economic system can growth continue indefinitely; to do so will destroy the biosphere.

A contrasting economic goal to growth is "development," which can be thought of as making things better for the people who are part of the economic system. This entails improving human well-being by meeting the needs for food, shelter, clean water, good health, and the opportunity to work to one's potential. Incorporating this goal is a necessary part of transitioning to an economic system based on sustainable thinking. This means economic growth that does not deplete resources or manufactured inputs more quickly than they can be replaced by natural systems—with or without the assistance of human technology.

Unfortunately, in most cases, a long-term sustainable growth rate does not show the same level of short-term profits by which we currently define economic health. Stock market indicators show only today's gains or losses and are not tempered by future conditions. Many indicators are used to signal economic growth, foremost among them the gross domestic product (GDP). However, this index fails as a realistic measure of social welfare or human well-being because it gives no indication of sustainability (Talberth 2008). It does measure consumption (i.e., money is exchanged) but says nothing about quality of life. For instance, the crash of the *Exxon Valdez* and the ensuing oil cleanup added substantially to the GDP of the United States in that year; money was being spent (a lot of it), so GDP was increased, but there was no evidence of the enhancement of human or environmental well-being.

Without the ability to plan and regulate resource use in the long term, however, there is no hope for "sustainability." Our political system is similarly short-sighted: Decisions and plans are made and unmade within the 2- to 4-year terms of our elected representatives, and rarely has any policy on resource

Table 26.1
Economic Indicators Designed to Be Responsive to Progress in Sustainable Development

Economic Objective	Sample Indicators and Desired Direction of Effect	Description
Genuine human program	Genuine progress indicator (+)	Aggregate index of sustainable economic welfare
	Happy planet index (+)	Aggregate index of well-being based on life satisfaction, life expectancy, and ecological footprint
	Well-being index (+)	Aggregate index of well-being based on health, wealth, knowledge, community, and equality
	Human development index (+)	Aggregate index of well-being based on income, life expectancy, and education
Renewable energy platform	Carbon footprint (−)	Provides spatial and intensity measures of life cycle carbon emission
	Energy return on investment (+)	Ratio between energy a resource provides and the amount of energy required to produce it
	Energy intensity (−)	Energy used per unit of economic output
Social equity	Index of representational equity (−)	Measures consistency between ethnic composition of elected officials and that of the general population; zero indicates "perfect" consistency
	GINI coefficient (−)	Measures extent to which an income distribution deviates from an equitable distribution, zero indicating "perfect" equity
	Legal rights index (+)	Measures degree to which collateral and bankruptcy laws protect rights of borrowers and lenders, scale of 0–10
	Access to improved water and sanitation (+)	Percent of population with access to improved water and sanitation services
Protect and restore natural capital	Ecological footprint (−)	Ecologically productive land and ocean area appropriated by consumption activities
	Genuine savings (+)	Net investment in human-bulk and natural capital stocks adjusted for environmental equality changes
	Environmental sustainability index (=)	Weighted average of 21 separate environmental sustainability indicators
Economic localization	Local employment and income multiplier effect (+)	Direct, indirect, and induced local economic activity generated by a given expenditure
	Ogilvie index of economic diversity (−)	Measures how well actual industrial structure matches an ideal structure; zero indicates "perfect" diversity
	Miles to market (−)	Average distance a group of products travels before final sale

Source: From Talberth, J. 2008. In *State of the World 2008: Innovations for a Sustainable Economy,* 19–31. New York: W. W. Norton & Company.

use, regulation, or subsidy survived a change in the dominant political party. Talberth (2008) presents a sample of alternative macroeconomic indicators that respond to the challenges of sustainable development (Table 26.1). These are designed to address one of the following economic objectives featured in the framework of sustainable development:

- promote progress based on multiple dimensions of human well-being;
- foster a transition to the use of renewable energy sources;
- encourage the fair distribution of resources and opportunity;
- protect and restore natural capital; and
- promote a local economy.

Conservation, reuse, and recycling of natural resources are ideal ways to prolong supplies. This is in contrast to the view that humans are not facing any limitation in resources and that our only real problem is that of global distribution of goods and services and the technology to extract them. Some argue that there are enough food and technological luxury for every person on the Earth, if we can only figure out how to deliver them. But can every person on the Earth realistically have a suburban house, two sport utility vehicles, disposable income, and all the other material benefits enjoyed in the United States, Europe, Australia, Korea, and Japan? As it stands now, the economic disparity between the developed and developing countries continues to expand. Increasing natural resource limitations will exacerbate this trend.

Addressing poverty is a necessary part of sustainable development. People who are hungry, landless, or homeless do not have the luxury of making balanced decisions about resource use, and chronic poverty limits the usefulness of short-term efforts to improve living conditions. What good does it do to provide food for a month or two for people who have no land or investment capital to produce their own sustenance? The lure of employment has drawn rural populations to cities, with increasingly dire results for many. Sao Paolo, Mexico City, Singapore, and Calcutta are only the largest examples of a global trend in urbanization. Without a global effort to curb birth rates and to provide education to all populations, we can expect more disease, famine, and warfare in many areas.

Cases in Point: Food Production and Renewable Energy

Sustainable Food Production

Large areas of land worldwide have been converted for agricultural production. The increasing use of technology, fuels, and chemicals has dramatically increased crop production, but has also created the chronic agricultural impacts detailed in Chapter 12. Sustainable agriculture combines the goals of environmental and economic health and social equity. We provide one example of how sustainable development of one of our most important human resources can be ensured. In order to ensure sustainable productivity of food, we reiterate the following basic set of objectives and approaches for achieving sustainability (based on Edwards and Wali 1993; Miller and Wali 1996):

- Land remains the paramount resource base for all societies (Cook 1992) and is one of the universal forms of wealth. The cultural landscape that develops in any location strongly reflects the character and quality of this land.
- Land and quality soils are a finite resource; the overuse of soils is a long-term form of degradation, and recovery takes long time periods. No matter what the circumstance, land will continue to provide the primary base from which human sustenance is derived.
- The prime agricultural lands have already been developed in virtually all countries. The recent rate of land conversion has been intense; more land has been converted to cropland since 1945 than in the eighteenth and nineteenth centuries combined. The remaining land may support agriculture, but will not be as productive.
- The development of additional lands for agricultural purposes will require substantial investments to increase soil fertility, make water available, drain excess water, irrigate, and control erosion.
- Climate change will, in ways unknown, alter the extent and distribution of land suitable for agricultural production, requiring shifting patterns of crop production.
- Traditional agricultural systems, some of which were sustainable, are disappearing. In the United States, this is manifested in the sharp decrease in the number of small family farms. These are being replaced by farming systems that are more intensive and/or dependent on finite fossil fuels and off-farm resources.

- Non-fossil-fuel inputs (e.g., water, phosphorus) for industrial agriculture are also finite and should be managed with care.
- The trend of global economic development is toward an increasing global interdependence in food and energy; markets for food are global, making it routine for food commodities to be transported large distances relative to where they were grown.
- Lack of money coupled with high population density encourages exploitation of land and increases pressure to produce exports, thereby creating unsustainable conditions.
- Global indebtedness in the developed countries can reduce capital investments, development assistance, and technical aid to developing countries. Despite this, the long-standing links between developed and developing countries, in terms of aid and open markets, mean that approaches to global sustainable development will have to be multinational.
- The trend toward specialization and fragmentation of knowledge is a deterrent to the development of holistic agricultural systems and of policies for sustainability. A systems perspective is necessary for true sustainability. Systems thinking takes into account multiple scales, from the local farm to the local ecosystem and community and beyond to national and international food policies and markets. At the local level, farmers are stewards of the environment who need a rational farm policy that encourages sustainability to survive.

Sustainable farming practices also view the land as a system, incorporating soil type, climate, invasive species and other pests, topography, and climate into an understanding that allows site-specific decisions that are the best fit (Jackson and Jackson 2002). The following general principles can be used as a guide when selecting appropriate management practices (University of California Sustainable Agriculture Research and Education Program, http://www.sarep.ucdavis.edu):

- The selection of species and varieties well suited to a farm site will match crops to soil and foster better growth and pest resistance.
- The diversification of crops (including livestock) and cultural practices to enhance the biological and economic stability make the systems-diversified farm ecologically resilient and support more biodiversity. Optimal levels of diversity may be achieved by integrating crops and livestock.
- Soil management to enhance and protect its quality recognizes soils as living systems; a healthy soil is the key to productivity over the long term.
- For the efficient and humane use of inputs in moving to sustainable farming practices, it is not enough to substitute a synthetic product for a natural one. It also means substituting a more complete scientific understanding for those synthetic or conventional inputs.

Renewable Energy

Some resource uses are, by definition, unsustainable; our dependence on fossil fuels is a good example. Fossil fuels are the predominant energy source for humans, and their use has increased exponentially over the past 60 years (Wolfson 2008). Even if more fuel-efficient machines are developed and marketed, the sheer numbers of existing vehicles, tools, and buildings being produced would still require ever-increasing use of the remaining fuel. Fossil fuels (petroleum, coal, and natural gas) are finite resources, no longer being created by natural processes at a rate that would render them usable by humans. Therefore, the important question is not whether we will run out, but rather when or how soon.

With innovation, it is possible to substitute one resource for another and, in many cases, to develop entirely new ways to meet our needs. The use of timber for fuel gave way to coal and oil only about 150 years ago, and in the last two generations we have developed nuclear energy capability. These fuel sources, however, have heavy waste output, and the effects of pollution on humans and ecosystems have almost never been included in the equations of profit.

Most recently, the prospect of solar and wind energy, both infinitely renewable, holds promise for energy with minimal pollution. Geothermal and tidal energy are also potentially infinite and nonpolluting. However, as long as traditional fuel resources remain profitable, even if only because their production is subsidized by our government, there is little hope for a rapid conversion to cleaner or renewable fuel sources.

The widespread use of solar energy will be the ultimate solution to the energy problem. The sun is the source of essentially all energy on the planet, in whatever form it takes (coal, oil, wind, wave, etc.). Nuclear power may be the exception, but its use is limited by the amount of recoverable uranium needed to run nuclear power plants. The Earth receives far more solar energy each day than we have ever harnessed; it can cleanly and sustainably meet our energy needs. The total amount of energy arriving at any one spot is a function of latitude, season, cloud cover (climate conditions), and day length. In most regions of the world, solar energy is enough to power photovoltaic cells, which can be used to generate electricity effectively. To maximize energy capture, solar panels are deployed, then tilted and rotated during the day to track the sun, much like sunflowers do. This adds expense to the system but greatly increases its efficiency.

The peak average solar insolation in the United States is in Arizona. Solar energy here is so powerful that it would take solar panels covering only 17% of Arizona's land area, or 22,000 km^2, to provide enough electricity for the entire United States (Wolfson 2008). Although the technology to use solar energy is available and improving, it has not yet reached the point of cost-effective mass production. As the cost of fossil fuels continues to increase, alternatives like solar energy will become more widely available. This process may be accelerated by the rapid industrialization of China and India and the energy needs that their growing prosperity will require. Carbon dioxide emissions from the use of fossil fuels in China are rising at 10% per year and, in 2006, their total emissions were only 12%

below those of the United States (although the per-capita emissions are much higher in the United States) (Flavin 2008).

In the short term, reducing our individual use of energy to make our own lifestyle more sustainable (i.e., decreasing our ecological footprint) can save substantial amounts of energy. For instance, producing an aluminum can from recycled aluminum requires 95% less energy than generating a new one. Recycling just one can saves enough energy to run a 100-W incandescent light bulb for 4 hours. Likewise, recycling a plastic bottle preserves enough energy to power a light for 6 hours (see www.cancentral.com, www.cepc.net).

Carbon Markets

The Kyoto Protocol to the U.N. Framework Convention on Climate Change has established an international "carbon market." Under this protocol, 38 industrialized countries agreed that by 2012 they would cut greenhouse gas emissions to an average of 5% below 1990 levels. Since 1990, however, emissions have increased by 15%. Although there is doubt that this level of cutback will be enough to limit the forces of climate change, the Kyoto Protocol has established what is known as a market for trading carbon "credits." In essence, the structure of the agreement is implemented through a **cap and trade** program. The protocol sets the cap, or the upper limit, on how much carbon can be emitted on a global basis. The sources of carbon (industries) trade among themselves based on which source actually emits how much.

If a source does not emit as much carbon as it is allowed, it can be seen as the "right to pollute"; its excess carbon "allocation" can be sold to other industries that will be emitting more carbon than they are allotted. This allows flexibility in emissions while maintaining the cap. There are currently three major established carbon markets where trading is done: the European Union Emissions Trading Scheme (the EU-ETS), the Australian New South Wales Market, and the Chicago Climate Exchange in the United States. To date, the only market that has really been active is the EU-ETS based in London, where companies buy and sell the right to emit CO_2. Most recently, the New Green Exchange, where carbon "futures" will be traded, opened in New York. This essentially allows investors to bet on how much it will cost to pollute with CO_2 in the future. The money involved in these markets is staggering: The carbon trading market is currently estimated at US$30 billion (Green Chip Review 2008).

Other options for carbon offsets and working to achieve a **carbon-neutral** lifestyle are through investment in alternative energy sources or by carbon sequestration through reforestation. Forest areas continue to shrink on a global scale, but reforestation in the developed world is now balancing some of the loss, contributing to the provision of ecosystem services, and serving as an important natural carbon sink. This method of sequestering carbon is largely voluntary, such as when individuals contribute to a reforestation fund to offset the carbon emitted during an airplane flight. Because carbon offsets are typically sold when the trees are planted—despite the fact that it might take decades for them to sequester the carbon they are credited with removing from the atmosphere—some environmentalists have been skeptical about their value (www.davidsuzuki.org). It should be noted that new plantation forests do not provide the same ecosystem services as old-growth forest communities, so their ancillary benefits may not rise to expectations.

The Eco-City Concept: Green Cities

A hundred years ago, there was widespread idealism in Europe and North America that technological advances would enable city dwellers to have the best of both worlds: urban convenience, employment, culture, and society, along with the benefits of vast green spaces, recreation, clean air and water, and the unlimited personal space associated with rural living.

Schemes for such "garden cities," as advanced by the influential architect Le Corbusier in his "radiant city" plan for central France, are almost entirely based on a single concept of optimal land use that reflects a preindustrial landscape. This includes not only concentrating business and industry in dense urban areas, but also avoiding the vast suburban sprawl areas in which residential development leapfrogs indefinitely onto farmland and natural areas. These concepts were completely predicated on a network of efficient public transportation, mostly rail systems. During the last century, however, the once extensive rail lines that served every municipality in the United States have been reduced or abandoned in favor of private vehicle ownership and highway and road construction. For most Americans, it is impossible to conceive of walking or bicycling to work every day instead of commuting by car. As residential suburbs and malls spread across former woodlands, deserts, farms, and orchards, the hope of containing cities to a defined "footprint" has been abandoned. Formerly small concentric towns have merged into megacities on all three U.S. coasts, and urban sprawl is consuming once productive local farmland.

One glimmer of hope for American cities is the increasing realization that brownfield redevelopment may be more desirable and even more profitable than the cost of infrastructure expansion into "green fields." The premise of brownfield projects is that blighted urban areas, whether abandoned industrial sites or decaying tenements, are not only cheaper to buy than prime farmland at the margins, but also are already adjacent to all the utilities and employment needed to serve the development. Brownfield programs offer incentives for any site cleanup or remediation that may be needed in order to rebuild on-site. It is to be hoped that reconstruction in the centers of cities may begin to counteract the effects of 50 years of outward flight from urban centers and will reduce the need for more services and ever widening roads into undeveloped land.

Certainly, the issues of transportation are key to this reawakening. However, we continue to subsidize the petroleum industry, highway development, and airport expansion at a far higher rate than rail or other public transportation. People who ride the bus are receiving proportionally less federal subsidy than those who ride a jet plane. Unfortunately, many

Americans do not perceive the irony of spending 2 or more hours a day commuting to work so that they can live in a "country" setting. In the United States, more than 75% of the population lives in suburbs, and fewer than 2% of Americans are farmers.

Comprehensive and nonpolitical land use planning is essential to the preservation of our rural areas, whether for agriculture, forestry, or conservation. All landscapes are complex mosaics with fragments of human and natural history and dimensions in time as well as in space (Forman and Godron 1986). Only when landscapes are viewed as complex, interdependent systems, rather than parcels of land under various ownerships or restrictions, will ecological principles really inform land use decisions. For instance, land use planning should occur in watershed units—areas that share a common drainage system (see Chapters 5 and 16). However, political and property ownership boundaries rarely follow watershed perimeters, thus making planning and regulation more difficult.

In developing countries, the trend in the last half-century has been toward accelerated urban growth (with a different, less attractive kind of "suburb") as rural populations flock to cities for employment and opportunities. The burden of these millions of landless migrants has overloaded the infrastructure in the world's largest cities, such as Sao Paulo, Mexico City, and Singapore. Ironically, a World Bank-subsidized attempt to relocate urban poor families in Brazil to create new agricultural communities has accelerated the destruction of the world's largest rain forest. Whatever physical form is taken by expanding human cities, the relationships between the natural and constructed systems must be addressed simultaneously for truly sustainable planning and development.

Ecotourism

Can natural areas preserve themselves? It has been argued that future value of natural areas (such as potential foods or medicines) is meaningless and that only the economics of the moment—immediate profit—will slow the destruction of remaining natural areas. But how do we get money from a nature preserve unless we are harvesting something tangible? In the last generation, a new form of vacation travel has evolved from the once rare and expensive safari experience, which meant traveling at great effort and expense to an exotic locale, such as Africa or central Asia, and killing wildlife for sport. Now, safaris tend to be based on photography or animal encounters, rather than bagging a trophy. The economic gain comes in the form of service and hospitality employment for a region—many jobs that are likely to be held by local residents. To this end, the International Ecotourism Society defines ecotourism as "responsible travel to natural areas, which conserves the environment and improves the welfare of local people."

Ecotourism in its ideal form exploits the beauty of natural biodiversity in a nondestructive form. However, impacts of even nonextractive human visitation can be severe. Once again, the example of killing the goose that lays the golden egg comes to mind. In a fragile ecosystem, the result of human presence can destroy the attraction. Examples include many mountain-climbing base areas in the Himalayas, now fields of frozen garbage and discarded equipment; the denuding of the tepui plateaus in South America, where litter and trampling are destroying rare endemic habitats; the destruction of travertine stream terraces in Mexico, Jamaica, and the United States, simply by walking on them; and the potential of tourism to disrupt migration and breeding behaviors of marine mammals, such as during whale watching.

Sustainability *Inter Alia*

Defining Questions and Sustainability Types

In a succinct narrative, Gale and Cordray (1994) posed four questions that lie at the base of resource issues of sustainability.

"What is sustained?" may range from a single population to several species-populations interdependent on each other, a forest, a prairie, a wetland, or a large ecological system embracing a whole watershed.[3]

"Why sustain it?" has multiple reasons, the most obvious of which is economic. Although biological and aesthetic reasons apply, sustainability of food, fiber, and water occupies the basic tier. (Remember Maslow's hierarchy in Chapter 1?)

"How is sustainability measured?" continues to be the most difficult problem, given the entrenchment of current economic metrics.

"Because the accounting prices of environmental resources are generally not available [or not easily assessable], they are frequently regarded as having no value. As a result, depreciation of natural capital is regarded as inconsequential" (Dasgupta et al. 2000).

"What are the politics?" depends upon the competing parties in a given resource issue. Gale and Cordray provide a matrix of nine "sustainability types," which includes a good synthesis of the economic, the ecological, and the socio-cultural (Table 26.2).

Science of Sustainability

The urgency of problems attendant on many environmental resources was expressed by the Ecological Society of America in its Sustainable Biosphere Initiative (Lubchenco et al. 1991). Three research priorities—global change, biological diversity, and sustainable ecological systems—were recognized on the basis of two criteria: "(1) the potential to contribute to fundamental ecological knowledge, and (2) the potential to respond to major human concerns about the biosphere." Such research efforts are continuing at other important levels as well. For example, the U.S. National Academy of Science appears to have taken an active role in "sustainability science" (see Kates et al. 2001), and a number of papers have appeared on such

environmental resources as air, food, water, and biofuels.[4]

Millennium Development Goals (MDGs)

The United Nations has adopted the following eight MDGs to be achieved by 2015:

goal 1: eradicate extreme poverty and hunger;

goal 2: achieve universal primary education;

goal 3: promote gender equality and empower women;

goal 4: reduce child mortality;

goal 5: improve maternal health;

goal 6: combat HIV/AIDS, malaria, and other diseases;

goal 7: ensure environmental sustainability; and

goal 8: develop a global partnership for development.

Adopted by 189 nations, environmental sustainability is listed as one of the eight goals. However, progress made in the other seven goals will have an appreciable impact on sustainability. The progress toward realizing these goals is to be achieved through "**21 quantifiable targets** that are measured by 60 **indicators**."[5]

All these initiatives are laudable. Whatever the steps in achieving sustainability, we must avoid having biological, economic, social, and other discipline-oriented definitions; there must be one, fully integrated, common definition (Wali 1992). Discussions on sustainable development are numerous (see, for example, Sneddon, Howarth, and Norgaard 2006). But to achieve tangible results, we need cohesion.

Environmental Awareness Is Here to Stay

We invoked the words of Alexis de Tocqueville earlier (Chapter 21) when he wrote of the

Table 26.2
Four Defining Questions for Nine Types of Sustainability

	What Is Sustained?	Why Sustain It?	How Is Sustainability Measured?	What Are the Politics?
Dominant product	Yield of high-valued products	Economic efficiency	Quantity produced	Maintain flow of narrow resource-specific resources versus broad, diverse resource production
Dependent social system	Social systems (communities, families, occupations)	Lifestyle values	Social system persistence	Local, targeted resource-dependent social systems versus broadly distributed use or preservation
Human benefit	Diverse human benefits	Human rights to resource abundance	Range of ecosystem products and uses	Broadly distributed multiple uses versus ecocentrism or resource specialization
Global niche preservation	Globally unique ecological systems	Global human–ecosystem interdependence	Ecosystem health	"Spaceship earth" versus which niches to maintain
Global product	Globally important high-value products	Human need for products even if few areas produce them	Price and supply fit of local products into international market	International comparative advantage versus global exploitation and resource nationalism
Ecosystem identity	General types of ecosystems or resource uses	Commitment to ecosystem autonomy and naturalness	Persistence of global ecosystem diversity	Worth of general ecological characteristics versus market-driven ecosystem conversion
Self-sufficiency	Ecosystem integrity	Commitment to ecosystem autonomy and naturalness	Ecosystem integrity without external input	Ecosystem rights and values versus human values and needs
Ecosystem insurance	Ecosystem diversity	Insure against ecological disaster and diversity loss	Vitality and amount of insured resources, resistance to ecological crises	General need for reserved areas versus questions of future need and technological optimism
Ecosystem benefit	Undisturbed ecosystems	Respect rights inherent in natural ecosystems	Ecosystem continuity, natural evolution	Restorative intervention or ecocentric autonomy versus human dominance and use

Source: From Gale, R. P., and S. M. Cordray. 1994. *Rural Sociology* 59 (2): 311–332.

power of citizen groups in bringing about a change in the United States. Such indeed has been the case in the environmental movement. Although, as authors of this book, we strove to present the science of the environment without appending opinions, we must recognize the fact that welfare of the environment is everybody's business. "Corporate, conservative, and grassroots activism provided the main impetus toward changing environmental discourse in the last three decades of the twentieth century," wrote Buell (2003). Thus, the role of citizen groups, largely responsible for our advances in contemporary environmental policy and law, must be recognized with gratitude.

The increasing involvement of religious communities in environmental concerns has also been complementary. When Aldo Leopold introduced his land ethic nearly 60 years ago, he wrote that

"no important change in ethics was ever accomplished without an internal change in our intellectual emphasis, loyalties, affections, and convictions. The proof that conservation has not yet touched these foundations of conduct lies in the fact that philosophy and religion have not yet heard of it" (1949, p. 209).

Times have changed. Theologians began to engage ecological questions significantly in the 1960s, and by 1990 a group of Nobel laureate scientists famously called upon religious leaders to join them in the cause of addressing environmental problems.[6] Today, as the fields of environmental philosophy and ecotheology attest,[7] religion and philosophy have certainly "heard of" and are making important contributions to environmental thinking. Every major religious denomination in the United States now has an established environmental statement or program, and local religious communities increasingly engage in environmental education and action.[8]

Some recent examples highlight the character of this engagement:

The National Association of Evangelicals in the United States has named "creation care" as one of its seven social priorities.

The National Council of Churches claims the environmental crisis is *the* moral crisis of the age.

The Vatican recently included "environmental pollution" as one of seven new "deadly sins" recognized by the Church.

Environmental actions of Buddhist groups around the world have been documented.

Partnerships between religious and conservation groups have multiplied worldwide.[9]

Consider, for instance, the U.S. evangelical Christian leaders' recent call to action and their "four simple but urgent claims":

Claim 1. Human-induced climate change is real. Claim 2. The consequences of climate change will be significant and will hit the poor the hardest. Claim 3. Christian moral convictions demand our response to the climate change problem. Claim 4. The need to act now is urgent. Governments, businesses, churches, and individuals all have a role to play in addressing climate change—starting now.[10]

Clear and forthright statements like these are most helpful for discourse and action in a civil society.

The Road Ahead

As was done in the Millennium Ecosystem Assessment (MEA 2005), the valuation of ecosystem goods and services can be used as an effective strategy to impress upon decision-makers that humans are nested within the environment and depend on it for the basics of survival. The collective management of ecosystem goods and services (such as biogeochemical cycles, global climatic regulation, protective stratospheric ozone layer, biodiversity, and atmospheric processes), which by definition cannot belong to any individual as "property," is imperative. This means that the sustainable yield of renewable natural resources cannot exceed their rate of regeneration. Likewise, ecosystem health should be maintained by ensuring that the rate of environmental degradation does not exceed the rate of rehabilitation or restoration of damaged ecosystems.

Market failures and policy failures (such as distorted subsidies; powerful pressure groups; inadequate use of transparency, public participation, and power; bureaucracy; corruption; lack of coordination and information; and distributive injustice of wealth and power) have been identified as two of the major causes of the undervaluation of goods and services generated by environmental processes. Ultimately, our goal must be to right this imbalance so as to maintain the global stock of natural capital. Meeting human needs today and in the future, while preserving the Earth's life-support system, is at the center of our move toward sustainability.

Having traversed a multitude of environmental topics, we close our discourse with three observations. First, we live in an age of accessibility and availability of information undreamed of in human history. When we combine that with the active involvement of individual scientists in nearly all disciplines, professionals in economic and social sciences, governmental and nongovernmental agencies, and numerous citizen groups, the tremendously enthusiastic ferment in environmental aspects becomes obvious. Thus, the enrichment of knowledge in basic and applied ecology is most welcome.

Second, given the preceding observations, claiming that any book will capture the whole story on the totality of all environmental issues and their solutions would be more than audacious. Ours is one view of choosing the basics of ecology, a selection of major environmental problems, and some approaches to mitigate what has been degraded or destroyed and save what is left. Readers had perhaps heard of many key words before they decided to read this book; we now hope that they can visualize those topics in a more systematic, interconnected, and interdependent way. We have included a rich sampling of references, synthesis reports, and Web sites for students to expand their knowledge on topics of their choosing. There is no substitute for an environmentally literate, well-informed citizenry.

Third, we are optimistic that many environmental problems ultimately will find redress and that that which is threatened will be afforded protection. What we do not know is whether these measures will come soon enough to save any of the 1 billion hungry people from starvation; we surely hope they will.

Let us stay alert, engaged, and informed.

Notes

1. Dalal-Clayton (2003) notes the influence of the following work: *How to Be a Survivor: A Plan to Save Spaceship Earth,* by Paul Ehrlich (1971); "A Blueprint for Survival," by Edward Goldsmith and colleagues (1972); *Only One Earth,* by Barbara Ward and Rene Dubois (1972) for the U.N. Conference on the Human Environment in Stockholm; and *Small Is Beautiful,* by E. F. Schumacher (1973).

2. Lester Brown's 1984 foreword continued: "The intent merely is not to describe how things are, but to indicate whether they are getting better or worse. The primary focus in *State of the World—1984* is on the interplay between the changing resource base and economic system" (p. xv). The foreword concluded: "The tone of this report is not intended to be optimistic or pessimistic. Neither unfounded optimism nor undue pessimism provides a solid foundation for policymaking. *Only realism will do*" (p. xvii) (emphasis added). Since the founding of the Worldwatch Institute, several organizations have also become established that report on the state of the environment; these include the United Nations Environment Program, World Wildlife Fund, World Resources Institute, and the H. John Heinz III Center for Science, Economics, and the Environment.

3. Gordon Orians (1990) suggested the selection of a limited number of "valued ecosystem components (VECs)" for explicit environmental assessment and management. He added,

"A VEC can be a single species of economic (deer) or aesthetic (California condor) value, systems of interacting species (bees with plants they pollinate), or an entire ecosystem (a wetland or rain forest). Within a given ecosystem, a number of VECs may be present. Certain uses of an ecological system may be sustainable with respect to one VEC but not another" (p. 13).

4. The *Proceedings of the National Academy of Sciences* has carried a number of studies; see, for example, Clark (2007) and papers in *PNAS* volumes 100 and 104.
5. Millennium Development Goals, http://www.un.org/millenniumgoals/.
6. Preserving and Cherishing the Earth: An Appeal for Joint Commitment in Science and Religion, available at http://fore.research.yale.edu/publications/statements/preserve.html.
7. More than 2,000 sources in these fields are found in Wesley Wildman's *Bibliography in Ecological Ethics and Eco-Theology* at http://people.bu.edu/wwildman/WeirdWildWeb/proj_bibs_ecoethics_00.htm.
8. For further discussion on these topics and a review of social-science research examining religious affiliations and environmental attitudes and behaviors, see Hitzhusen (2006, 2007).
9. "For the Health of the Nation: An Evangelical Call to Civic Responsibility" (2005), http://www.nae.net/images/civic_responsibility2.pdf; "God's Earth Is Sacred" (National Council of Churches ecumenical environmental statement (2004) http://www.timesonline.co.uk/tol/comment/faith/article3517050.ece; "Buddhist Engaged Projects," http://fore.research.yale.edu/religion/buddhism/projects/ index.html; Applied Research Center; The National Religions Partnership for the Environment, www.arc.org; www.nrpe.org.
10. See www.christiansandclimate.org. In its preamble to "Climate Change—An Evangelical Call to Action," it says, "As American evangelical Christian leaders, we recognize both our opportunity and our responsibility to offer a biblically based moral witness that can help shape public policy in the most powerful nation on earth, and therefore contribute to the well-being of the entire world. Whether we will enter the public square and offer our witness there is no longer an open question. We are in that square and we will not withdraw."

Questions

1. What are the main premises and principles of sustainable development?
2. Give examples of indicators of economic development that also measure progress in sustainability.
3. Why is there a need to rethink the current economic patterns of consumption, waste generation, and distributive justice of income?
4. What are the main elements of sustainable food production?
5. Discuss the possibility for future economic growth in light of the inherent biophysical limits of the Earth.

References

Numbers at the ends of references refer to the chapters in which the work has been cited.

Abell, R. A., D. M. Olson, et al. 1999. *Freshwater ecoregions of North America. A conservation assessment*. Washington, D.C.: Island Press. [5, 25]

Abramovitz, J. N. 1996. *Imperiled waters, impoverished future: The decline of freshwater ecosystems*. Washington, D.C.: Worldwatch Institute. [16]

Abrol, I. P., J. S. P. Yadav, J.S.P., and F. I. Massoud. 1988. *Salt-affected soils and their management*. FAO Soils Bulletin 39, Rome, Italy. [12]

Acharya, A. 1995. The fate of the boreal forests. *World Watch* 8 (3): 20–29 (she quotes Myers for the comparative estimates on Russia and Brazil). [B, 24]

ACIA (Arctic Climate Impact Assessment). 2004. *Impacts of a warming Arctic: Arctic climate impact assessment*. New York: Cambridge University Press. [19]

Addiscott, T. M. 1995. Entropy and sustainability. *European Journal of Soil Science* 46:161–168. [7]

Alaback, P. B. 1996. Biodiversity patterns in relation to climate. In *High-latitude rainforests and associated ecosystems of the west coast of the Americas,* ed. R. G. Lawford, P. B. Alaback, and E. Fuentes, 105–133. *Ecological studies 116*. New York: Springer, 409 pp. [10]

Alberti, M. 2005. The effects of urban patterns on ecosystem function. *International Regional Science Review* 28:168–192. [15]

Alcamo, J., G. J. J. Kreileman, et al. 1994. Modeling the global society–biosphere–climate system. Part 1: Model description and testing. *Water, Air, and Soil Pollution* 76:1–35.

Allen, M. F. 1991. *The ecology of mycorrhizae*. New York: Cambridge University Press. [4]

Alward, R. D., J. K. Detling, and D. G. Milchunas. 1999. Grassland vegetation changes and nocturnal global warming. *Science* 283:229–231. [19]

American Forest and Paper Association. 1993. *Sustainable forestry principles and implementation guidelines*. Washington, D.C.: American Forest and Paper Association. [23]

American Heritage Dictionary of the English Language. 1973. Boston: The American Heritage Publishing Company, Inc., and Houghton–Miflin Company.

Anderson D. M. 2000. The harmful algae page. National Office for Marine Biotoxins and Harmful Algal Blooms, Woods Hole Oceanographic Institution, Woods Hole, MA. [16]

Anderson, J. R. 1999. Health effects of air pollution episodes. In *Air pollution and health*, S. T. Brown, J. Holgate, M. Samet, et al., ed. 461–482. San Diego, CA: Academic Press. [18]

Anderson, H. R. 2009. Air pollution and mortality: a history. *Atmospheric Environment* 43: 142–152. [18]

Anderson, J. E., D. W. Brady, et al. 1984. *Public policy and politics in America,* 2nd ed. Monterey, CA: Brooks/Cole Publishing Company. [21]

Anderson, J. M. 1981. *Ecology for environmental sciences: Biosphere, ecosystems and man*. New York: John Wiley & Sons. [11]

Andreae, M. O. 1995. Climatic effects of changing atmospheric aerosol levels. In *Future climates of the world: A modeling perspective,* ed. A. Henderson-Sellers, 347–389. Amsterdam: Elsevier Science B.V. [15, 18]

Andrewartha, H. G., and L. Birch. 1954. *The distribution and abundance of animals.* Chicago, IL: University of Chicago Press.

Antonovics, J., A. D. Bradshaw, and R. G. Turner. 1977. Heavy metal tolerance in plants. *Advances in Ecological Research* 7:1–85. [4]

Archibold, O. W. 1995. *Ecology of world vegetation.* London: Chapman & Hall. [10]

Arendt, A. A., K. A. Echelmeyer, W. D. Harrison, C. S. Lingle, and V. B. Valentine. 2002. Rapid wastage of Alaska glaciers and their contribution to rising sea level. *Science* 297:382–386. [19]

Arthur, W. B. 1990. Positive feedbacks in the economy. *Scientific American* 262:92–95, 98–99. [22]

Associated Press. 2006. Ozone hole matches record size. October 3, 2006. [18]

Ayensu, E., D. R. van Claaasen, et al. 1999. International ecosystem assessment. *Science* 286:685–686. [11, 19]

Bach, W. 1972. *Atmospheric pollution.* New York: McGraw–Hill Book Company. [18]

Bailey, A. W. 1976. Alberta's range land resources. *Rangeman's Journal* 3:44–46. [10]

Baker, H. G. 1988. Patterns of plant invasion in North America. In *Ecology of biological invasions of North America and Hawaii,* ed. H. A. Mooney and J. A. Drake. New York: Springer–Verlag. [20]

Barbault, R., and S. Sastrapradja. 1995. United Nations environmental program. In *Global Biodiversity Assessment,* ed. V. H. Heywood, 193–274. Cambridge, England: Cambridge University Press. [25]

Barbier E. B. 1987. The concept of sustainable economic development. *Environmental Conservation* 14:101–110. [26]

Barbour, M. G., and W. D. Billings. 1988. *North American terrestrial vegetation.* New York: Cambridge University Press. [10]

Barbour, M. G., J. H. Burk, W. D. Pitts, F. S. Gilliam, and M. W. Schwartz 1999. *Terrestrial plant ecology,* 3rd ed., p. 270, Figure 11-1. San Francisco, CA: Benjamin/Cummings. [9]

Bardgett, R. D., and D. A. Wardle. 2003. Herbivore-mediated linkages between aboveground and belowground communities. *Ecology* 84:2258–2268. [9]

Barnosky, A. D., and B. P. Kraatz. 2007. The role of climatic change in the evolution of mammals. *BioScience* 57:523–532 (Figure 1, p. 525). [1]

Barrett N. E., and J. P. Barrett. 1997. Reserve design and new conservation theory. In *The ecological basis of conservation: Heterogeneity, ecosystems, and biodiversity,* ed. S. T. A. Pickett, R. S. Ostfeld, M. Shachak, and G. E. Likens, 236–251. New York: Chapman & Hall. [25]

Barrow, C. J. 1995. *Developing the environment: Problems and management.* New York: John Wiley & Sons. [12]

Bartlett, A. A. 1978. Forgotten fundamentals of the energy crisis. *American Journal of Physics* 46:876–888. [11]

Baumol, W. J., and W. E. Oates. 1988. *The theory of environmental policy.* Cambridge, England: Cambridge University Press. [22]

Bazzaz, F. A. 1983. Characteristics of populations in relation to disturbance in natural and man-modified ecosystems. In *Disturbance and ecosystems: Components of response,* ed. H. A. Mooney and M. Godron, 259–275. Ecological Studies 44. New York: Springer–Verlag. [B, 24]

Bazzaz, F. A., and W. G. Sombroek. 1996. Global climatic change and agricultural production: An assessment of current knowledge and critical gaps. In *Global climate change and agricultural production,* ed. F. Bazzaz and W. Sombroek, 319–330. New York: Food and Agriculture Organization of the United Nations and John Wiley & Sons. [9]

Benda, L., K. Andras, D. Miller, and P. Bigelow. 2004. Confluence effects in rivers: Interactions of basin scale, network geometry, and disturbance regimes. *Water Resources Research* 40 (5): Article #W05402. [5]

Benedick, R. 1991. *Ozone diplomacy.* Cambridge, MA: Harvard University Press. [18]

Bernabo, C. J. 1989. Statement of Dr. J. Christopher Bernabo. Hearing on Global Change Research Act of 1989 before House Committee on Space, Science, and Technology, SubCommittee on Natural Resources, Agriculture, Research, and Environment, and SubCommittee on International Scientific Cooperation, 27 July 1989 (Document #104-74). Washington, D.C.: U.S. Government Printing Office. [21]

Biswas, A. K. 1998. Deafness to global water crisis: Causes and risks. *Ambio* 27 (6): 492–493. [11, 16]

Blackman, F. F. 1909. Optima and limiting factors. *Annals of Botany* 19:281–295. [2]

Blaney, B. 2008. EPA refuses to reduce ethanol requirements. Associated Press, August 8, 2008.

Bliss, L. C. 2000. Arctic tundra and polar desert biome. In *North American terrestrial vegetation,* ed. M. G. Barbour and W. D. Billings, 1–40. New York: Cambridge University Press. [10]

Bonner, J. T. 1965. *Size and cycle.* Princeton, NJ: Princeton University Press. [8]

Borchert, J. R. 1950. The climate of the central North American grassland. *Annals of the Association of American Geographers* 40:1–39. [10]

Borlaug, N. 1970. The Green Revolution, peace, and humanity. The Nobel Peace Prize, December 11, 1970, The Nobel Foundation, Stockholm, Sweden.

Botkin, D. B. 1990. *Discordant harmonies: A new ecology for the 21st century.* New York: Oxford University Press. [1]

Botsford, L. W., J. C. Castilla, and C. H. Peterson. 1997. The management of fisheries and marine ecosystems. *Science* 277:509–515. [16]

Boulding, K. E. 1966a. The economics of the coming spaceship Earth. In *Environmental quality in a growing economy,* ed. H. Jarrett, 3–14. Baltimore, MD: The John Hopkins Press. [22]

———. 1966b. Economics and ecology. In *Future environments of North America,* ed. F. F. Darling and J. P. Milton, 225–234. New York: The Natural History Press. [22]

Boyle, R. H. 1993. The killing fields—Toxic drainwater from irrigated farmland in California and other western states has created an environmental calamity. *Sports Illustrated* March 22:62–64, 66–69. [1]

Bradshaw, A. D. 1983. The reconstruction of ecosystems. *Journal of Applied Ecology* 20:1–17. [24]

———. 1987. Restoration: An acid test for ecology. In *Restoration ecology—A synthetic approach to ecological research,* ed. W. R. Jordan III, M. E. Gilpin, and J. D. Aber, 23–29. New York: Cambridge University Press. [24]

Brady, N. C., and R. F. Weil. 2008. *The nature and properties of soils,* 14th ed. Upper Saddle River, NJ: Prentice Hall. [4]

Braun, E. L. 1950. *Deciduous forests of eastern North America.* New York: Hafner Publishing Company. [10]

Braun-Blanquet, J. 1932. (1928). *Plant sociology: The study of plant communities.* (English translation by G. D. Fuller and H. S. Conard.) New York: McGraw–Hill. [9]

Brimblecombe P., and A. Y. Lein, eds. 1989. *Evolution of the global biogeochemical sulphur cycle.* New York: John Wiley & Sons. [14]

BP (British Petroleum). 1996. BP statistical review of world energy. Technical report, British Petroleum Co., London. [17].

———. 2008. BP statistical review of energy. June 2008 (http://www.bp.com/productlanding.do?categoryId=6929&contentId=7044622)

Broecker, W. 1991. The great ocean conveyor. *Oceanography* 4:79–89. [19]

Brönmark, C., and H. Lars-Anders. 1998. *The biology of lakes and ponds.* New York: Oxford University Press (2nd ed. In 2005). [5]

Brown, D., R. G. Hallman, C. R. Lee, J. G. Skogerboe, K. Eskew, R. A. Price, N. R. Page, M. Clar, R. Kort, and H. Hopkins. 1986. *Reclamation and vegetative restoration of problem soils and disturbed land.* Park Ridge, NJ: Noyes Data Corporation. 560 pp. [14]

Brown, L. R. 2004. *Outgrowing the earth.* New York: W. W. Norton. [12]

———. 2005. *Outgrowing the Earth: The food security challenge in an age of falling water tables and rising temperatures.* New York: W. W. Norton & Company. [11, 16]

———. 2008. Plan 3.0—Mobilizing to save civilization. New York: W. W. Norton & Company. [11, 16]

Brown, L. R., G. Gardner, and B. Halweil. 1999. Impacts of growth. *The Futurist* (February): 36–41. [11]

Brown, W. M., III, G. R. Matos, and D. E. Sullivan. 1998. Materials and energy flows in the earth science century. A summary of a workshop held by the USGS in November 1998. U.S. Geological Survey Circular 1194.

Browne, J. 2001. The role of corporate leadership. In *Managing the Earth—The Linacre lectures 2001,* ed. J. C. Briden and T. E. Downing, 101–109. Oxford, England: Oxford University Press. [22]

Bruno, J. F., K. E. Boyer, J. E. Duffy, S. C. Lee, and J. S. Kertesz. 2005. Effects of macroalgal species identity and richness on primary production in benthic marine communities. *Ecology Letters* 8:1165–1174. [8]

Bruno, J. F., J. J. Stachowics, and M. D. Bertness. 2003. Inclusion of facilitation into ecological theory. *Trends in Ecology and Evolution* 18:119–125. [8]

Bryant, D., D. Nielsen, and L. Tangley. 1997. *The last frontier forests: Ecosystems and economies on the edge.* Washington, D.C.: Forest Frontier Initiative, World Resources Institute. [13]

Bryson, R. A., and F. K. Hare. 1974. *Climates of North America. World survey of climatology*, vol. 11. New York: Elsevier Scientific Publishing Company. [18]

Buell, F. 2003. *From apocalypse to way of life: Environmental crisis in the American century.* New York: Routledge. [26]

Buol, S. W., F. D. Hole, R. J. McCracken, and R. J. Southard. 1997. *Soil genesis and classification,* 4th ed. Ames: Iowa State University Press. [10]

Burnett, D. 2005. Special report: America's cleanest (and dirtiest) cities. *Reader's Digest*, July, 80–87. The data for rating on the first four criteria were based on the U.S. Environmental Protection Agency, and for the fifth on the U.S. Bureau of Labor Statistics. [1]

Caldwell, L. K. 1990. *International environmental policy: Emergence and dimensions,* 2nd ed., pp. 49–50. Durham, NC: Duke University Press. (Stockholm Conference began 5 June 5, 1972; the quotes on UN resolution 2398 are from Caldwell.) [1]

Callaway, R. M. 2007. *Positive Interactions and Interdependence in Plant Communities.* New York: Springer. [9]

Callaway, R. M., and B. E. Mahall. 2007. Family roots. *Nature* 448:145–147. [9]

Capone, D. G., R. Popa, et al. 2006. Follow the nitrogen. *Science* 312:708–709. [6]

CDIAC (Carbon Dioxide Information Analysis Center). 1999. Oak Ridge National Laboratory, Oak Ridge, TN. [19]

Carpenter, S. R., and J. F. Kitchell, eds. 1993. *The trophic cascade in lakes.* New York: Cambridge University Press. [7]

Carson, R. 1962. *Silent spring.* Boston: Houghton, Mifflin Company; see also Woodwell, G. M. 1984. Broken eggshells. *Science83*. Washington, D.C.: American Association for the Advancement of Science. [1, 12, 21]

Caruccio, F. T. 1979. The nature of acid mine drainage reactions and their relation to overburden analysis. In *Ecology and coal resource development,* vol. 2, ed. M. K. Wali, 775–781. New York: Pergamon Press. [14]

Chapin, F. S., III, and G. R. Shaver. 1985. Arctic. In *Physiological ecology of North American plant communities,* ed. B. F. Chabot and H. A. Mooney, 16–40. New York: Chapman & Hall. [10]

Chapin, F. S. III, E. S. Zavaleta, V. T. Eviver, et al. 2000. Consequences of changing biodiversity. *Nature* 405:234–242. [20]

Cherrett, J. M. 1989. Key concepts: The results of a survey of our members' opinions. In *Ecological concepts: The contribution of ecology to an understanding of the natural world,* ed. J. M. Cherrett, 1–16. Oxford, England: Blackwell Scientific Publications. [2, 9]

Chew, S. C. 2001. *World ecological degradation—Accumulation, urbanization, and deforestation 3000 B.C.–A.D. 2000.* Lanham, MD: Rowan & Littlefield Publishers. [B, 24]

Choi, Y. D. 2004. Theories for ecological restoration in changing environment: Toward 'futuristic' restoration. *Ecological Research* 19:75–81. [24]

Choi, Y. D., V. M. Teperton, et al. 2008. Ecological restoration for future sustainability in a changing environment. *Ecoscience* 15:53–64. [24]

Choi, Y. D., and M. K. Wali. 1995. The role of switch grass (*Panicum virgatum* L.) in the revegetation of iron mine tailings in northern New York. *Restoration Ecology* 3:123–132. [17, 24]

Christensen, N. L., A. Bartuska, et al. 1996. The scientific basis for ecosystem management. *Ecological Applications* 6:665–691. [23]

Ciriacy-Wantrup, S. V. 1952. *Resource conservation: Economics and policies.* Berkeley: University of California Press. [22]

Clark, F. E. 1975. Viewing the invisible prairie. In *Prairie: A multiple view,* ed. M. K. Wali, 181–197. Grand Forks: University of North Dakota Press. [4]

Clark, W. C. 2007. Sustainability science: A room of its own. *Proceedings of the National Academy of Sciences* 104:1737–1738. [26]

Clausen, J., D. D. Heck, and W. M. Hiesey. 1948. Experimental studies on the nature of species III. Environmental responses of climatic races of *Achillea. Carnegie Institution of Washington Publication* 581:1–129. [3]

Clay, J. 2004. *World agriculture and the environment—A commodity-by-commodity guide to impacts and practices.* Washington, D.C.: Island Press. [12]

Clements, F. E. 1916. *Plant succession: An analysis of the development of vegetation.* Washington, D.C.: Carnegie Institution of Washington Publication 512. [9]

———. 1928. *Plant succession and indicators.* New York: The H. W. Wilson Company. [9]

Cleugh, H. 1995. Urban climates. In *Future climates of the world; A modeling perspective,* ed. A. Henderson-Sellers, 477–509. Amsterdam: Elsevier Science B.V. [15]

Cloud, P. 1969. *Resources and man.* A study and recommendations by the Committee on Resources and Man of the Division of Earth Sciences, National Academy of Sciences, National Research Council, with the cooperation of the Division of Biology and Agriculture. San Francisco, CA: W. H. Freeman. [11].

Coase, R. 1960. The problem of social cost. *Journal of Law and Economics* 3:1–44. [21]

Cobb, C., T. Halstead, and J. B. Cobb, Jr. 1995. If the GDP is up, why is America down? *The Atlantic Monthly* 276 (4): 59–78 (October). [22]

Cohen, J. E. 1995. *How many people can the Earth support?* New York: W. W. Norton & Company. [11]

Coleman, D. C., and D. A. Crossley, Jr. 1995. *Fundamentals of soil biology.* New York: John Wiley & Sons. [4]

Colinvaux, P. A. 1993. *Ecology 2,* 371–372. New York: John Wiley & Sons. [10]

Columbus *Dispatch,* The. 1997. Shuttle mission: Orbiting junk just misses satellite. August 13, p. 10A. [1]

Conant, J. B. 1951. *Science and common sense.* New Haven, CT: Yale University Press. [1]

Connell, J. H., and R. O. Slatyer. 1977. Mechanisms of succession in natural communities and their role in community stability and organization. *American Naturalist* 111:1119–1144. [9, 24]

Costanza, R. 2000. Social goals and the valuation of ecosystem services. *Ecosystems* 3: 4–10. [22]

Costanza, R., S. Farber, B. Castaneda, and M. Grasso. 2001. Green national accounting: Goals and methods. In *The economics of nature and the nature of economics,* ed. C. J. Cleveland, D. Stern, and R. Costanza, 262–282. Cheltenham, England: Edward Elgar.

Costanza, R., and Martinez-Alier, J. 1996. *Getting down to Earth—Practical applications of ecological economics.* Washington, D.C.: Island Press. [22]

Costanza, R. et al. 1997. *An introduction to ecological economics.* Boca Raton, FL: St. Lucie Press. [22]

Coupland, R. T. 1958. The effects of fluctuations in weather upon the grasslands of the Great Plains. *Botanical Review* 24:273–317. [10]

Cowles, H. C. 1899. The ecological relations of vegetation on the sand dunes of Lake Michigan. *Botanical Gazette* 27:95–117, 167–202, 281–308, 361–39. [9]

Craig, J. R., D. V. Vaughan, and B. J. Skinner. 1996. *Resources of the earth,* 2nd ed. Upper Saddle River, NJ: Prentice Hall. [5, 14]

Craul, P. J. 2000. *Urban soils.* New York: John Wiley & Sons. [15]

Crawley, M. J. 1997. The structure of plant communities. In *Plant ecology,* ed. M. J. Crawley, 475–531. Oxford, England: Blackwell Science. [9]

Critchfield, H. J. 1966. *General climatology.* Englewood Cliffs, NJ: Prentice Hall. [10]

Cubbage, F. W., J. O'Laughlin, and C. S. Bullock III. 1993. *Forest resource policy.* New York: John Wiley & Sons. [21]

Cunningham, S. 2002. *The restoration economy—The greatest new growth factor.* San Francisco: Berrett–Koehler Publishers. [24]

Currie, C. R., A. N. M. Bot, and J. J. Boomsma. 2003. Experimental evidence of a tripartite mutualism: Bacteria protect ant fungus gardens from specialized parasites. *Oikos* 101:91–102. [8]

Curtis, J. T., and R. P. McIntosh. 1951. An upland forest continuum in the prairie–forest border region in Wisconsin. *Ecology* 32:476–496. [9]

Czech, B., and H. E. Daly. 2004. In my opinion: The steady state economy—What it is, entails, and connotes. *Wildlife Society Bulletin* 32 (2): 598–605. [22]

Dalal-Clayton, B. 2003. *What is sustainable development?* London: International Institute for Environment and Development (IIED). [26]

Dale, V. H., S. Brown, et al. 2000. Ecological principles and guidelines for managing the use of land. *Ecological Applications* 10 (3): 639–670. [23]

Daly, H. 1990. Toward some operational principles of sustainable development. *Ecological Economics* 2:1–6. [22, 26]

Daly, H. E. 1991. *Steady-state economics,* 2nd ed. Washington, D.C.: Island Press. [22]

Daly, H. E., and J. Farley. 2004. *Ecological economics—Principles and applications.* Washington, D.C.: Island Press. [22]

Dansereau, P. 1957. *Biogeography: An ecological perspective.* New York: Ronald Press Company. [3]

Dasgupta, P. 2004. How best to face the coming storm. *Science* 305:1761. [Sect. C]

Daubenmire, R. F. 1968. *Plant communities: A textbook of plant synecology.* New York: Harper & Row. [9]

———. 1976. The use of vegetation in assessing the productivity of forest lands. *Botanical Review* 42:115–143.

Dauncey, G., with P. Mazza. 2001. *Stormy weather: 101 solutions to global climate change.* Gabiola Island, BC, Canada: New Society Publishers. [19]

David Suzuki Foundation. What is a carbon offset? www.davidsuzuki.org/Climate_Change/What_You_Can_Do/carbon_offsets.asp [26]

Davis, M. B., C. Douglas, et al. 2000. Holocene climate in the western Great Lakes National Parks and lakeshores: Implications for future climate change. *Conservation Biology* 14: 968–983. [19]

de Bilj, H. J., and P. O. Muller. 1996. *Physical geography of the global environment.* New York: John Wiley & Sons. [3]

de Duve, C. 1999. Letters: Skepticism and relativism. *Science* 285:200. [1]

Deevey, E. S., Jr. 1960. The human population. *Scientific American* 203 (5): 195–204. [11]

Deffeyes, K. S. 2001. *Hubbert's peak: The impending world oil shortage.* Princeton, NJ: Princeton University Press. [17]

———. 2005. *Beyond oil—The view from Hubbert's peak.* New York: Hill and Wang. [17]

DeFries, R. S., and J. R. G. Townshend. 1994. NDVI-derived land cover classification at global scales. *International Journal of Remote Sensing* 15:3567–3586. [23]

Denman, K. L., G. Brasseur, A. Chidthaisong, et al. 2007. Couplings between changes in the climate system and biogeochemistry. In *Climate Change 2007 – The Physical Science Basis. Contribution of Working Group I to the Fourth Assessment Report of the Intergovernmental Panel on Climate Change,* Ch. 7, 499–587. Cambridge, England: Cambridge University Press. [7, 19]

DePinto, J. V., and R. Narayanan. 1997. What other ecosystem changes have zebra mussels caused in Lake Erie? Potential bioavailability of PCBs. *Great Lakes Research Review* 3:1–9. [16]

de Ruiter, P. C., V. Wolters, J. C. Moore, and K. O. Winemiller. 2005. Food web ecology: Playing Jenga and beyond. *Science* 309:68–71. [23.]

de Tocqueville, A. 1835/1840. *Democracy in America,* ed. J. P. Mayer, trans. G. Lawrence. Garden City, NY: Doubleday, 1969.

Dickinson, R. E., P. J. Kennedy, and A. Henderson-Sellers. 1993. Biosphere–atmosphere transfer scheme (BATS) version 1e as coupled to the NCAR community climate model. NCAR technical note, NCAR/TN-387.

Dikhanov, Y. 2005. Trends in global income distribution, 1970–2015. Background note for *Human Development Report 2005.* New York: United Nations. [11]

Dinerstein, E., D. J. Graham, and D. M. Olson. 1995. *A conservation assessment of the terrestrial ecoregions of Latin America and the Caribbean.* Washington, D.C.: The World Bank. [10]

Dittmer, H. J. 1937. A quantitative study of the roots and root hairs of a winter rye plant (*Secale cereale*). *American Journal of Botany* 24:417–420. [2]

Dix, R. L. 1975. Colonialism in the Great Plains. In *Prairie: A multiple view,* ed. M. K. Wali, 15–24. Grand Forks: The University of North Dakota Press. [14, 17]

Dodds, W. K. 2002. *Freshwater ecology: Concepts & environmental applications.* San Diego, CA: Academic Press. [5].

Donahue, R. L., J. C. Shickluna, and L. S. Robertson. 1971. *An introduction to soils and plant growth.* Englewood Cliffs, NJ: Prentice Hall. [2]

Döös, B. R. 1994. Environmental degradation, global food production, and risk for large-scale migrations. *Ambio* 23 (2): 124–130. [11]

Dregne, H. E., and N. T. Chou. 1992. Global desertification dimensions and costs. In *Degradation and restoration of arid lands,* ed. H. E. Dregne, 249–282. Lubbock, TX: Texas Tech University. [B, 24]

Drollette, D. 1996. Australia fends off critic of plan to eradicate rabbits. *Science* 272:191–192. [20]

———. 1997. Wide use of rabbit virus is good news for native species. *Science* 275:154. [20]

Durant, W. 1935. *The story of civilization.* Volume 1. *Our oriental heritage.* New York: Simon and Schuster. [1]

Durnil, G. K. 1995. *The making of a conservative environmentalist.* Bloomington: Indiana University Press. [16, 21, 22]

Dyksterhuis, E.J. 1949. Condition and management of rangeland based on quantitative ecology. *Journal of Range Management* 2:104–115. [9]

Eavenson, H. N. 1935. *Coal through the ages.* New York: American Institute of Mining and Metallurgical Engineering. [17]

Edwards, C. A., D. E. Reichle, and D. A. Crossley, Jr. 1970. The role of soil invertebrates in turnover of organic matter and nutrients. In *Analysis of temperate forest ecosystems,* ed. D. E. Reichle, 147–172. New York: Springer–Verlag. [4]

Edwards, C. A., and M. K. Wali. 1993. The global need for sustainability in agriculture and natural resources. In *Agriculture and the environment,* ed. C. A. Edwards, M. K. Wali, D. J. Horn, and F. P. Miller, vii–xxv. New York: Elsevier. [26]

Edwards, K. J., P. L. Bond, T. M. Gihring, and J. F. Banfield. 2000. An archaeal iron-oxidizing extreme acidophile important in acid mine drainage. *Science* 287:1796–1799. [14]

Edwards, R. 1999. Dear Santa: Please bring me sulfur dioxide for Christmas—Environmental groups are pushing pollution allowances as the ultimate gift. Columbus *Dispatch,* December 19, 1999, p. 4D. [1]

Ehrlich, P. 1968. *The population bomb.* New York: Ballantine Books. [11]

———. 1971. *How to be a survivor: A plan to save spaceship Earth.* London: Ballantine Books. [26]

———. 1986. Which animal will invade? In *Ecology of biological invasions of North America and Hawaii,* ed. H. A. Mooney and J. A. Drake, 79–95. New York: Springer–Verlag. [20]

Einstein, A. 1940. Considerations considering the fundamentals of theoretical physics. *Science* 91:487–492. [1]

El-Hinnawi, E. 1985. *Environmental refugees.* United Nations Environment Program (UNEP), UNEP Office, Nairobi, Kenya, p. 40. [11]

Elton, C. S. 1927. *Animal ecology.* London: Sidgewick & Jackson. [2]

———. 1958. *The ecology of invasions by animals and plants.* London: Methuen. [20]

Emanuel, K. A. 2005. Increasing destructiveness of tropical cyclones over the past 30 years. *Nature* 436:686–688. [19]

Enger, E., and B. F. Smith. 1995. *Environmental science—A study of interrelationships.* Dubuque, IA: W. C. Brown. [5]

Environment Canada. 1997–1998. 1997 *Canadian acid rain assessment.* Volume one: *Summary of results.* Volume two: *Atmospheric science assessment report.* Volume three: *The effects on Canada's lakes, rivers and wetlands,* ed. D. S. Jeffries. Volume four: *The effects on Canada's forests,* P. Hall, W. Bowers, H. Hirvonen, G. Hogan, N. Forster, I. Morrison, K. Percy, and R. Cox. Volume five: *The effects on human health,* ed. L. Liu. Ottawa, Ontario, Canada. [18]

EHC (Environmental Health Center). 1998. Coastal challenges: A guide to coastal and marine issues. Division of the National Safety Council, Washington, D.C. [16]

Evans, F. C. 1975. The natural history of a Michigan field. In *Prairie: A multiple view,* ed. M. K. Wali, 27–51. Grand Forks: University of North Dakota Press. [9]

Evans, L. T. 1998. *Feeding the ten billion: Plants and population growth.* New York: Cambridge University Press. [11]

Evenson, R. E., and D. Collin. 2003. Assessing the impact of the Green Revolution, 1960–2000. *Science* 300:758–762. [12]

Evrendilek, F., and M. K. Wali. 2001. Modeling long-term C dynamics in croplands in the context of climate change: A case study of Ohio. *Environmental Modeling and Software* 16 (4): 361–375. [12]

———. 2004. Changing global climate: Historical carbon and nitrogen budgets and projected responses of Ohio's cropland systems. *Ecosystems* 7:381–392. [12, 19]

Faiz, A. 1993. Automotive emissions in developing countries: relative implications for global warming, acidification and urban air quality. *Transportation Research* 27A (3): 167–186. [15, 18]

Falkenmark, M. 1997. Meeting water requirements of an expanding world population. *Philosophical Transactions of the Royal Society of London Series B* 352:929–936. [11]

Farnsworth, N. R. 1988. Screening plants for new medicines. In *Biodiversity,* ed. E. O. Wilson, 83–97. Washington, D.C.: National Academy Press. 521 pp. [20]

Feshbach, M., and A. Friendly. 1992. *Ecocide in the USSR: Health and nature under siege.* New York: Basic Books. [17]

Fischer, G., and G. K. Heilig. 1997. Population momentum and the demand on land and water resources. *Philosophical Transactions of the Royal Society of London Series B* 352:869–889. [16]

Fisher, D. A., C. H. Hales, et al. 1990a. Model calculations of the relative effects of CFCs and their replacements on stratospheric ozone. *Nature* 344:508–512. [18]

———. 1990b. Model calculations of the relative effects of CFCs and their replacements on global warming. *Nature* 344:513–516. [18]

FISWRG (Federal Interagency Stream Restoration Working Group). 2001. *Stream corridor restoration: Principles, processes, and practices.* Washington, D.C.: Government Printing Office, Item No. 0120-A. Chapter 2, p. 27, Figure 2.9. [4]; p. 36, Figure 2.21. [6]

Fitter, A. H., and R. K. M. Hay. 1987. *Environmental physiology of plants,* 2nd ed. London: Academic Press. [3]

Flavin, C. 2008. Building a low-carbon economy. In *State of the world 2008: Innovations for a sustainable economy,* 75–90. New York: W. W. Norton & Company. [26]

Foley, J. A. et al. 2005. Global consequences of land use. *Science* 309:570–574 (Figure 1, p. 571). [B, 24]

FAO (Food and Agriculture Organization of the United Nations). 1985. Guidelines: Land evaluation for irrigated agriculture. *Soils Bulletin 55.* Food and Agriculture Organization of the United Nations, Rome, Italy. [23]

———. 1990. *FAO yearbook 1989. Production.* FAO Statistical Series no. 94, vol. 43, FAO, Rome, Italy. [11]

———. 1998. FAOSTAT. Rome, Italy. [11, 12]

———. 2005. Global forest resources assessment, progress towards sustainable forest management. FAO forestry paper 147. Rome, Italy. [13]

———. 2006. Livestock's long shadow—Environmental issues and options. Rome, Italy. [12]

Food Quality Protection Act. 1998. Food Quality Protection Act of 1996, P.L. 104–170, Title II, Section 303, enacted August 3, 1996. Codified in Title 7, U.S. Code, Section 136r-1. Integrated Pest Management.

Forcier, L. K. 1975. Reproductive strategies and the co-occurrence of climax tree species. *Science* 189:808–810. [9]

Forman, R. T. T., and M. Godron. 1986. *Landscape ecology.* New York: John Wiley & Sons. [26]

Fortescue, J. A. C. 1980. *Environmental geochemistry: A holistic approach. Ecological Studies 35.* New York: Springer–Verlag. [4, 14]

Franks, Jeff. 2005. Shuttle commander sees wide environmental damage. *Reuters,* August 5. [1]

Fritts, T. H., and G. H. Rodda. 1998. The role of introduced species in the degradation of island ecosystems: A case history of Guam. *Annual Review of Ecology Systems* 29:113–140. [20]

Fukasawa, M., H. Freeland, R. Perkin, T. Watanabe, H. Uchida, and A. Nishina. 2004. Bottom water warming in the North Pacific Ocean. *Nature* 427:825–827. [19]

Fullerton, D., and R. Stavins. 1998. How economists see the environment. *Nature* 395:433–434. [22]

Gale, R. P., and S. M. Cordray. 1994. Making sense of sustainability: nine answers to "What should be sustained?" *Rural Sociology* 59:311–332. [26]

Galloway, J. N. 1988. Effects of acid deposition on tropical aquatic ecosystems. In *Acidification in tropical countries,* ed. H. Rodhe and R. Herrera, 141–166. Chichester, England: John Wiley & Sons. [18]

Galloway, J. N., and E. B. Cowling. 1978. The effects of precipitation on aquatic and terrestrial ecosystems: A proposed precipitation chemistry network. *Journal of Air Pollution Control Association* 28:229–235. [18]

Gankenheimer, R., L. T. Molina, et al. 2002. The MCMA transportation systems: Mobility and air pollution. In *Air quality in the Mexico megacity,* ed. L. T. Molina and M. J. Molina, 213–284. Dordrecht, the Netherlands: Kluwer Academic Publishers. [15]

Gankin, R., and J. Major. 1964. *Arctostaphylos myrtifolia,* its biology and relationship to the problem of endemism. *Ecology* 45:792–808. [4]

Garcia, S. M., and R. Grainger. 1996. Fisheries. FAO fisheries technical paper 359. [16]

Gardner, G. 2006. *Inspiring progress: Religions' contributions to sustainable development.* Washington, D.C.: Worldwatch Institute. [26]

Gardner, G., and T. Prugh. 2008. Seeding the sustainable economy. In *State of the world 2008: Innovations for a sustainable economy,* 2–17. New York: W. W. Norton & Company. [26]

Gardner, G. T., and P. Sampat. 1998. Mind over matter: Recasting the role of materials in our lives. Worldwatch paper 144. Washington, D.C.: World Watch Institute. 60 pp. [14]

Gause, G. F. 1934. *The struggle for existence.* Baltimore, MD: Williams and Wilkins. [8]

Gellings, C. W., and K. E. Parmenter. 2004. Energy efficiency in fertilizer production and use. 15 pp. In *Efficient Use and Conservation of Energy*, C. W. Gellings and K. Blok, ed. *Encyclopedia of life support systems*. Oxford: Eolss Publishers (http://www.eolss.net). [12]

Ghassemi, F., A. J. Jakeman, and H. A. Nix. 1995. *Salinization of land and water resources*. Sydney, Australia: University of New South Wales Press. [12]

Gilpin, A. 2000. *Environmental economics—A critical overview*. New York: John Wiley & Sons. [22]

Girardet, H. 1999. *Creating sustainable cities*. Schumacher briefing no. 2. Devon, England: Green Books Ltd. [15]

Gnidovec, D. M. 2000. Utah mine king of copper, gold. Columbus *Dispatch* February 13, p. 7B. Dale Gnidovec has been contributing short, informative geological articles to the Sunday edition of the *Dispatch* since 1995. [14]

Godin, E., ed. 1991. *Canadian minerals yearbook: Review and outlook*. Ottawa, ON: Canada Ministry of Energy, Mines and Resources. [14]

Goldsmith, E., R. Allen, M. Allaby, J. Davoll, and S. Lawrence. 1972. A blueprint for survival. *The Ecologist* 2 (1). [26]

Golley, F. B. 1993. *A history of the ecosystem concept in ecology*. New Haven, CT: Yale University Press. [2]

Gorham, E. 1954. An early view of the relation between plant distribution and environmental factors. *Ecology* 35:97–98. [2]

———. 1998. Acid deposition and its ecological effects: A brief history of research. *Environmental Science & Policy* 1:153–166. [18]

Gosz, J. R. 1993. Ecotone hierarchies. *Ecological Applications* 3:369–376. [10]

Gotelli, N. J. 2001. *A primer of ecology*. Sunderland, MA: Sinauer Associates. [8]

Goudie, A., and J. C. Wilkinson. 1977. *The warm desert environment*. New York: Cambridge University Press. [10]

Gould, J. L., W. T. Keeton, and C. G. Gould. 1996. *Biological science,* 6th ed. New York: W. W. Norton & Company. [3]

Gould, S. J. 2001. Introduction. In *Evolution: The triumph of an idea,* C. Zimmer, pp. ix–xiv. New York: HarperCollins Publishers. [1]

Grant, V. 1963. *The origin of adaptations*. New York: Columbia University Press. [1]

Green Chip Review. 2008. http://www.greenchipstocks.com/ [26]

Green, T. G. A., B. Schroeter, and L. G. Sancho. 1999. Plant life in Antarctica. In *Handbook of functional plant ecology,* ed. F. I. Pugnaire and F. Valladares, 496–543. New York: Marcel Dekker, Inc. [2]

Greep, R. O. 1998. Whither the global population problem? *Biochemical Pharmacology* 55:383–386. [11]

Greller, A. M. 1988. Deciduous forest. In *North American terrestrial vegetation,* ed. M. G. Barbour and W. D. Billings, 287–316. New York: Cambridge University Press. [10]

Grime, J. P. 1979. *Plant strategies and vegetation processes*. New York: John Wiley & Sons. [9]

Grimm, N. B., J. M. Grove, et al. 2000. Integrated approaches to long-term studies of urban ecological systems. *BioScience* 50: 571–584 [15]

Grinnell, J. 1917. The niche relationships of the California thrasher. *Auk* 34:427–433. [2]

Groombridge, B., ed. 1992. *Global biodiversity: Status of the Earth's living resources*. London: Chapman & Hall. [20]

Grosholz, E. D., and G. M. Ruiz. 1996. Predicting the impact of introduced marine species: Lessons from the multiple invasions of the European green crab *Carcinus maenas. Biological Conservation* 78:59–66. [20]

Grumbine, R. E. 1994. What is ecosystem management? *Conservation Biology* 8:27–38. [23]

Guo, Q. 2005. Ecosystem maturity and performance. *Nature* 435:E6–E6.

Haberl, H., F. Krausmann, K-H. Erb, and N. B. Schulz. 2002. Human appropriation of net primary production. *Science* 296 (5575): 1968–1969. [11]

Haeckel, E. 1866. *Generelle Morphologie der Organismen: Allgemeine Grundzüge der organischen Formen-wissenschaft, mechanisch begründetdurch die von Charles Darwin reformirte Descendenz— Theorie.* 2 vols. Berlin: Reimer. [2]

Hall, C. A. S. 1992. Economic development or developing economics? What are our priorities? In *Ecosystem rehabilitation,* vol. 1, 101–126. *Policy issues,* ed. M. K. Wali. The Hague: SPB Academic Publishing. [17, 22]

Hall, M. 2004. The provincial nature of George Perkins Marsh. *Environment and History* 10 (2): 191–204.
[26]

Halle, F., R. A. A. Oldeman, and P. B. Tomlinson. 1978. *Tropical trees and forests.* Berlin: Springer–Verlag.
[10]

Hannah, L., J. L. Carr, and A. Lankerani. 1995. Human disturbance and natural habitat: A biome level
analysis of a global data set. *Biodiversity and Conservation* 4:128–155. [24]

Hansen, R. C., ed. 1999. *Taxus* and Taxol: A compilation of research findings. Special circular 150,
Ohio Agricultural Research and Development Center, Wooster, OH. 68 pp. [20]

Hanski, I. 1998. Metapopulation dynamics. *Nature* 396:41–49. [8]

Hardin, G. 1968. The tragedy of the commons. *Science* 162:1243–1248. [8]

Hare, F. K., and J. D. Ritchie. 1972. The boreal bioclimates. *The Geographical Review* 62:333–365. [10]

Harr, R. D. 1982. Fog drip in the Bull Run municipal watershed, Oregon. *Water Resources Bulletin*
18:785–789. [10]

Heaton, E. A., F. G. Dohleman, and S. P. Long. 2008. Meeting U.S. biofuel goals with less land: The potential
of *Miscanthus. Global Change Biology* 14:2000–2014. [17]

Hebblewhite, M., C. A. White, et al. 2005. Human activity mediates a trophic cascade caused by wolves.
Ecology 85:2135–2144 (Fig.1, p. 2138). [7]

Helms, J. A., ed. 1997. *Dictionary of forestry.* Bethesda, MD: Society of American Foresters. [24]

Hillel, D. 1994. *Rivers of Eden—The struggle for water and quest for peace in the Middle East.* New York:
Oxford University Press. [4, 16]

———. 1998. *Environmental soil physics.* New York: Elsevier Science Publishers. [12]

Hillgard, E.W. 1911. *Soils.* New York: Macmillan Publishing Company. [4]

Hinrichsen, D. 1988. Acid rain and forest decline. In *The Earth report: The essential guide to global
ecological issues,* ed. E. Golsmith and N. Hildyard, 63–78. Los Angeles: Price Stern Sloan, Inc. [18]

Hobbs, R. J. 2000. Land-use changes and invasions. In *Invasive species in a changing world,*
ed. H. A. Mooney and R. J. Hobbs, 55–64. Washington, D.C.: Island Press. [20]

Hodgson, G., and J. A. Dixon. 1988. Logging versus fisheries and tourism in Palawan: An environmental
and economic analysis. EAPI occasional paper no. 7. East–West Center, Honolulu, HI. [13]

Hoekstra, A. Y., and A. K. Chapagain. 2007. Water footprint of nations: Water use by people as a function of
their consumption pattern. *Water Resources Management* 21:35–48. [12]

Hofman. M. 1992. Yeast biology, a new surprising phase. *Science* 255:1510–1511. [8]

Hogg, D. W. 1999. Letters: Science and "truth." *Science* 285:663. [1]

Holt, S. J., and L. M. Talbot. 1978. New principles for the conservation of wild living resources.
Wildlife Monographs 59:1–33. [25]

Hore-Lacy, I. 2008. Advanced nuclear power reactors. World Nuclear Association. http://www.eoearth.org/
article/Advanced_nuclear_power_reactors

Horne, A. J., and C. R. Goldman. 1994. *Limnology,* 2nd ed. New York: McGraw–Hill. [5]

Houghton, R. A. 2003. Revised estimates of the annual net flux of carbon to the atmosphere from changes in
land use and land management 1850–2000. *Tellus* 55B:378–390. [B]

———. 2007. Balancing the global carbon budget. *Annual Review of Earth and Planetary Sciences*
35:313–347. [B, 19]

Hoyningen-Huene, P. 1999. The nature of science. *Nature & Resources* 35 (4): 4–8. [1]

Hrudey, S. E., P. Payment, P. Huck, R. W. Gillham, and E. Hrudey. 2003. A fatal waterborne disease
epidemic in Walkerton Ontario: Comparison with other waterborne outbreaks in the developed world.
Water Science and Technology 47:7–14. [21]

Hubbert, M. K. 1973. Survey of world energy resources. *The Canadian Journal of Mining and
Metallurgical Bulletin* 66:37–54. [17]

Humboldt, A. von, and A. Bonpland. 1807. *Essai sur la geographie des plantes.* Paris: Librarie Lebrault
Schoell. [2]

Hutchinson, G. E. 1957a. *A treatise on limnology,* vol. I. *Geography, physics and chemistry.* New York: John
Wiley & Sons. [5]

———. 1957b. Concluding remarks. *Cold Spring Harbor Symposium of Quantitative Biology* 22:415–427.
[2]

Hutchinson, G. L. 1995. Biosphere–atmosphere exchange of gaseous N oxides. In *Soils and global change,*
ed. R. Lal, J. Kimble, E. Levine, and B. A. Stewart, 219–236. Boca Raton, FL: CRC Press. [18]

Hutchinson, T. C. 1980. Conclusions and recommendations. In *Effects of acid precipitation on terrestrial ecosystems,* ed. T. C. Hutchinson and M. Havas, 617–627. New York: Plenum Press. [18]

IFA (International Fertilizer Industry Association). 1998. *Nitrogen, phosphate and potash statistics.* Paris: IFA. [12]

IFA (International Fertilizer Industry Association), UNEP (United Nations Environment Program), and UNIDO (United Nations Industrial Development Organization). 1998. Mineral fertilizer production and the environment. Technical report no. 26 (revised ed.). Part 1. The fertilizer industry's manufacturing processes and environmental issues. United Nations Publication, 66 pp. [12]

Imes, A. C., and M. K. Wali. 1977. An ecological–legal assessment of mined land reclamation laws. *North Dakota Law Review* 53:359–399. [14, 21, 24]

———. 1978. Governmental regulation of reclamation in the western United States: An ecological perspective. *Reclamation Review* 1:75–88. [14, 21, 24]

Inouye, D. W., B. Barr, K. B. Armitage, and B. D. Inouye. 2000. Climate change is affecting altitudinal migrants and hibernating species. *Proceedings of the National Academy of Sciences* 97:1630–1633. [19]

———. 1996. *Climate change 1995: The science of climate change,* ed. J. T. Houghton, L. G. M. Filho, et al. Cambridge, England: Cambridge University Press. [16, 19]

———. 2001. *Climate change 2001: The scientific basis,* ed. J. T. Houghton, Y. Ding, et al. Cambridge, England: Cambridge University Press. [19]

———. 2007. *The physical science basis,* 93–127. Working Group I Report. Frequently asked questions. New York: Cambridge University Press. [7, 10, 19]

IPCC, 2007. *Climate Change 2007 – The Physical Science Basis. Contribution of Working Group I to the Fourth Assessment Report of the Intergovernmental Panel on Climate Change*, Solomon, S., D. Qin, M. Manning, Z. Chen, M. Marquis, K. B. Averyt, M. Tignor and H. L. Miller, eds. Cambridge, England: Cambridge University Press. [7, 19]

IUCN (International Union for Conservation of Nature). 1994. *Guidelines for management planning for protected areas.* IUCN Commission on National Parks and Protected Areas with the assistance of the World Conservation Monitoring Center. [23]

Iverson, L. R., A. M. Prasad, et al. 2008. Modeling emerald ash borer in Ohio—It's mainly about roads. [20]

Iverson, L. R., and M. K. Wali. 1982. Reclamation of coal-mined lands: The role of *Kochia scoparia* and other pioneers. *Reclamation & Revegetation Research* 1:123–160. [8, 9, 24]

———. 1992. Grassland rehabilitation after coal and mineral extraction in the western United States and Canada. In *Ecosystem rehabilitation: Preamble to sustainable development,* vol. 2, 85–129. *Ecosystem analysis and synthesis,* ed. M. K. Wali. The Hague: SPB Academic Publishing. [24]

Jackson, D., and L. Jackson. 2002. *The farm as natural habitat: Reconnecting food systems with ecosystems.* Covelo, CA: Island Press. [26]

Janzen, D. H. 1966. Coevolution of mutualism between ants and acacias in Central America. *Evolution* 20:249–275. [8]

Jasanoff, S. S. 1990. *The fifth branch: Science advisors as policy-makers.* Cambridge, MA: Harvard University Press. [21]

Jelinski, L. W., T. E. Graedel, et al. 1992. Industrial ecology: Concepts and approaches. *Proceedings of the National Academy of Sciences* 89:793–797. [14]

Jenny, H. 1958. Role of the plant factor in the pedogenic functions. *Ecology* 39:5–16. [24]

———. 1980. *The soil resource: Origin and behavior.* New York: Springer–Verlag. [4]

Johannessen, O. M., E. V. Shalina, and M. W. Miles. 1999. Satellite evidence of an Arctic sea ice cover in transformation. *Science* 286:1937–1939. [19]

Johnson, E. A., and K. Miyanishi. 2007. Disturbance and succession. In *Plant disturbance ecology: The process and the response,* 1–14. San Diego, CA: Academic Press. [24]

Johnson, L. B. 1967. *Preserving our national heritage.* Message to the 89th Congress, 2d Session, Washington, D.C. [1]

Johnstone, N., and Karousakis, K. 1999. Economic incentives to reduce pollution from road transport: The case for vehicle characteristics taxes. *Transport Policy* 6: 99–108. [15, 18]

Jorgensen, S. E., B. D. Fath, et al. 2007. *A new ecology—Systems perspective.* Amsterdam: Elsevier. [2]

Josephy, A. M. 1975. The prairie and its people: Yesterday's history and today's challenges. In *Prairie: A multiple view,* ed. M. K. Wali, 7–13. Grand Forks: The University of North Dakota. [14, 17].

Kaiser, J. 1999a. Battle over a dying sea. *Science* 284:28–30. [16]

———. 1999b. Bringing the Salton Sea back to life. *Science* 287:565. [16]

Kates, R. W. 1996. Ending hunger: Current status and future prospects. *Consequences* 2 (2): 1–11. [12]

Kaufman, D. G., and C. M. Franz. 2000. *Biosphere 2000: Protecting our global environment.* Dubuque, IA: Kendall/Hunt Publishing Company. [5, 15]

Kegley, S., S. Orme, and L. Neumeister. 2000. Hooked on poison: Pesticide use in California, 1991–1998. Pesticide Action Network available online.

Keiter, R. B. 1998. Ecosystems and the law: Toward an integrated approach. *Ecological Applications* 8:333–341. [21, 23]

Kennedy, P. 1993. *Preparing for the twentieth century.* New York: Random House. [1]

Keoleian, G. A., K. Kar, M. M. Manion, and J. W. Bulkley. 1997. *Industrial ecology of the automobile: a life cycle perspective.* Warrendale, PA: Society of Automotive Engineers. 148 pp. [14]

Kerr, R. A. 1999. Will the Arctic Ocean lose all its ice? *Science* 286:1828. [19]

Kessler, M. A., and B. T. Werner. 2003. Geomorphology: Self-organization of sorted patterned ground. *Science* 299:380–383. [10]

Khoshoo, T. N. 1992. Degraded lands for agroecosystems. In *Ecosystem rehabilitation: Preamble to sustainable development,* vol. 2. *Ecosystem analysis and synthesis,* ed. M. K. Wali, 3–17. The Hague: SPB Academic Publishing. [B, 24]

Khush, G. S. 2006. Green Revolution rice. Invited public lecture at The Ohio State University, Columbus, March 20, 2006. [12]

Kiehl, J. T., and K. E. Trenberth. 1997. Earth's annual global mean energy budget. *Bulletin of the American Meteorology Society* 78:197–208. [7]

Killingbeck, K. T., and M. K. Wali. 1978. Analysis of a North Dakota gallery forest: Nutrient, trace element and productivity relations. *Oikos* 30:29–60. [7]

Kimmins, J. P. 2004. *Forest ecology—A foundation for sustainable forest management and environmental ethics in forestry,* 3rd ed. Upper Saddle River, NJ: Prentice Hall. [2]

Klein, D. R. 1959. Saint Matthew Island reindeer-range study. Special Science Report Wildlife no. 43, Washington, D.C.: U.S. Department of the Interior, Fish and Wildlife Service.

Knievel, D. P., and D. A. Schmer. 1971. Preliminary results of growth characteristics of buffalograss, blue grama, and western wheatgrass, and methodology for translocation studies using ^{14}C as a tracer. Technical Report Number 86, Grassland Biome, U.S. International Biological Program. Fort Collins, CO: Colorado State University.

Koenig, R. 2000. Wildlife deaths are a grim wake-up call in Eastern Europe. *Science* 287:1737–1738. [14]

Kolber, Z. S. 2006. Getting a better picture of the ocean's nitrogen budget. *Science* 312:1479–1480. [6]

Kormondy, E. J. 1996. *Concepts of ecology,* 4th ed. Upper Saddle River, NJ: Prentice Hall. [8]

Korpela, S. A. 2006. Oil depletion in the world. 2006. *Current Science* 91:1148–1152. [17]

Krajick, K. 2002. Ice man: Lonnie Thompson scales the peaks for science. *Science* 298: 518–522. [19]

Kramer, P. J. 1983. *Water relations of plants.* New York: Academic Press. [8]

Krebs, C. J. 1964. The lemming cycle at Baker Lake, Northwest Territories, during 1959–1962. Technical paper no. 15, Arctic Institute of North America. [8]

———. 1985. *Ecology: The experimental analysis of distribution and abundance,* 3rd ed. New York: Harper & Row, Publishers. [7]

Krupa, S. V., and H-J. Jager. 1996. Adverse effects of elevated levels of ultraviolet (UV)-B radiation and ozone (O_3) on crop growth and productivity. In *Global climate change and agricultural production,* ed. F. Bazzaz and W. Sombroek, 141–169. New York: John Wiley & Sons. [18]

Kubasek, N. K., and G. S. Silveraman. 1994. *Environmental law.* Englewood Cliffs, NJ: Prentice Hall. [16, 21]

Kump, L. R., and J. E. Lovelock. 1995. The geophysiology of climate. In *Future climates of the world; A modeling perspective,* ed. A. Henserson-Sellers, 537–553. World Survey of Climatology (Vol. 16). Amsterdam: Elsevier Science B.V. [15]

Kuznet, S. 1948. National income. A new version (discussion of the new Department of Commerce income series). *Review of Economics and Statistics* 30 (3): 151–179. [22]

Lackey, R. T. 1998. Seven pillars of ecosystem management. *Landscape and Urban Planning* 40:21–30. [23]

Lahaye W. S., R. J. Gutiérrez, and H. R. Akçakaya. 1994. Spotted owl metapopulation dynamics in southern California. *Journal of Animal Ecology* 63:775–785. [8]

Landsberg, H. E. 1981. *The urban climate.* New York: Academic Press. [15]

Larcher, W. 1995. *Physiological plant ecology. Ecophysical and stress physiology of functional groups,* 3rd ed. Heidelberg, Germany: Springer–Verlag. [7]

Laurance, W. F., P. Delamonica, S. G. Laurance, H. L. Vasconcelos, and T. E. Lovejoy. 2000. Rain forest fragmentation kills big trees. *Nature* 404: 836. [13]

Lawler, J. J., J. E. Aukema, J. B. Grant, et al. 2006. Conservation science: a 20-year report card. *Frontiers in Ecology and Environment* 4: 473–480. [25]

Lawrence, G. B., and T. G. Huntington. 1999. Soil-calcium depletion linked to acid rain and forest growth in the eastern United States. WRIR 98-4267. Denver, CO: U.S. Geological Survey. 12 pp. [18]

Laws, E. A. 1997. *Aquatic pollution,* 2nd ed. New York: John Wiley & Sons. [16]

Laycock, W. A. 1991. Stable states and thresholds of range condition on North American rangelands: A viewpoint. *Journal of Range Management* 44 (5):427–433. [9]

Lazarus, R. J. 2004. *The making of environmental law.* Chicago, IL: University of Chicago Press. [21]

Leape, J. P. 2006. Are we winning the race to stop biodiversity loss? A guest commentary. *Environmental News Network,* March 23.

Lee, T. M., and W. Jetz. 2008. Future battlegrounds for conservation under global change. *Proceedings of the Royal Society B* 275: 1261–1270. [25]

Leopold, A. 1949. *A Sand County almanac: And sketches here and there,* p. 209. New York: Oxford University Press. [26]

Leopold, D. J., and M. K. Wali. 1992. The rehabilitation of forest ecosystems in the eastern United States and Canada. In *Ecosystem rehabilitation: Preamble to sustainable development,* vol. 2, 187–231. *Ecosystem analysis and synthesis,* ed. M. K. Wali. The Hague: SPB Academic Publishing. [24]

Leslie, M. 2007. Biodiversity—The ultimate life list. *Science* 316:818. [2]

Levins, R. 1969. Some demographic and genetic consequences of environmental heterogeneity for biological control. *Bulletin of the Entomological Society of America* 15:237–240. [8]

Lindeman, R. L. 1942. The trophic-dynamic aspect of ecology. *Ecology* 23:399–418. [2, 7]

Lipton, P. 2007. The world of science. *Science* 316:834. [1]

Longman, K. A., and J. Jeník. 1987. *Tropical forest and its environment.* Essex, England: Longman Scientific & Technical. [10]

Loveland, T. R., and A. S. Belward. 1997. The International Geosphere–Biosphere Program Data and Information System global land cover data set (DISCover). *Acta Astronautica* 41:681–689. [24]

Lund, H. G., and S. Iremonger. 2000. Omissions, commissions, and decisions: The need for integrated resource assessments. *Forest Ecology and Management* 128:3–10. [12]

Lutz, W., M. R. Testa, and D. J. Penn. 2006. Population density is a key factor in declining human fertility. *Population and Environment* 28:69–81. [11]

Lydersen, K. 2005. Mountaintop removal meets fresh resistance in Tennessee. *The New Standard,* November 15.

MacArthur, R. H. 1958. Population ecology of some warblers of northeaster coniferous forest. *Ecology* 39:599–619. [8]

Mack, R. N., D. Simberloff, W. M. Lonsdale, H. Evans, M. Clout, and F. Bazzaz. 2000. Biotic invasions: Causes, epidemiology, global consequences and control. *Ecological Applications* 10:689–710. [20]

MacKinnon, J., and K. MacKinnon. 1986. *Review of the protected areas system in the Indo-Malayan realm.* Gland, Switzerland, and Cambridge, UK: IUCN/UNEP publication.

MacNab, J. 1985. Carrying capacity and related slippery shibboleths. *Wildlife Society Bulletin* 13:403–410. [8]

Mahlman, J. D. 1997. Uncertainties in projections of human-caused climate warming. *Science* 278 (5342): 1416–1417. [19]

Mainguet, M., and R. Létolle. 1998. Human-made desertification in the Aral Sea Basin: Planning and management failures. In *The arid frontier: Interactive management of environment and development,* ed. H. J. Bruins and H. Lithwick, 129–142. Dordrecht, the Netherlands: Kluwer Academic Publishers. [16]

Mairet, P. 1975. *Pioneer of sociology: The life and letters of Patrick Geddes.* London: Lund Humphries. [2]

Major, J. 1974. Kinds and rates of changes in vegetation and chronofunctions. In *Handbook of vegetation science,* vol. 8, 392–412. *Vegetation dynamics.* The Hague: Dr. W. Junk, Publishers. [9]

Malakoff, D. 1999. Plan to import exotic beetle drives some scientists wild. *Science* 284:1255. [20]

Malanson, G. P. 1993. *Riparian landscapes.* New York: Cambridge University Press. [24]

Malthus, T. R. 1798/1970. *An essay on the principle of population,* ed. A. Flew. London: Penguin. [11, 22]

———. 1914. *An essay on population*, vols. 1 and 2. London: J. M. Dent and Sons. (The "Essay on the Principle of Population as It Affects the Future Improvement of Society" was published anonymously in 1798.) [11]

Mangel, M., L. M. Talbot, et al. 1996. Principles for the conservation of wild living resources. *Ecological Applications* 6 (2), 338–362. [25]

Manion, P. D. 1990. *Tree disease concepts*, 2nd ed. Upper Saddle River, NJ: Prentice Hall. [18]

Margalef, R. 1968. *Perspectives in ecological theory*. Chicago: University of Chicago Press. [2]

Margulis, I., and J. E. Lovelock. 1989. Gaia and geognosy. In *Global ecology: Towards a science of the biosphere*, ed. M. B. Rambler and L. Margulis, 1–29. San Diego, CA: Academic Press. [2]

Markewitz, D., D. D. Richter, H. L. Allen, and J. B. Urrego. 1998. Three decades of observed soil acidification in the Calhoun Experimental Forest: Has acid rain made a difference? *Soil Science Society of America Journal* 62:1428–1439. Also, ENN 3.212.99 on Markewitz research. [18]

Marsh, G. P. 1864. *Man and nature, or physical geography as modified by human action*. Cambridge, MA: Harvard University Press. [26]

Marsh, W. M. 1987. *Earthscape: A physical geography*, p. 99, Figure 6.8. New York: John Wiley & Sons. [3, 9]

Martinez-Alier, J. 1987. *Ecological economics: Energy, environment, and society*. Cambridge, MA: Blackwell. [22]

Marx, D. H. 1991. Forest applications of *Psilothus tinctorius*. Symposium at the Marcus Walleberg Prize ceremonies. September 26, 1991, Stockholm, Sweden. [4]

Maslow, A. 1954. *Motivation and personality*. New York: Harper and Row. [26]

Mason, H. L., and J. H. Langenheim. 1957. Language analysis and the concept "environment." *Ecology* 38:325–340. [1, 2]

Massey, A. B. 1925. Antagonism of the walnuts (*Juglans nigra* L. and *J. cineraria* L.) in certain plant associations. *Phytopathology* 15:773–784. [8]

Mathews, E. 1983. Global vegetation and land use: New high-resolution databases for climate studies. *Journal of Climate and Applied Meteorology* 22:474–487. [24]

Mathews, E., and A. Hammond. 1999. *Critical consumption trends and implications: Degrading Earth's ecosystems*, 7. Washington, D.C.: World Resources Institute. [13]

Matos, G., and L. Wagner. 1998. Consumption of materials in the United States, 1900–1995. *Annual Reviews of Energy and the Environment* 23:107–122.

Matthews, S. N., R. J. O'Connor, L. R. Iverson, and A. M. Prasad. 2004. *Atlas of climate change effects in 150 bird species of the Eastern United States*. U.S. Department of Agriculture, Forest Service, General Technical Report NE-18. Radnor, PA. [19]

McBride, J. R., and P. R. Miller. 1987. Reponses of American forests to photochemical oxidants. In *Effects of atmospheric pollutants on forests, wetlands, and agricultural ecosystems*, ed. T. C. Hutchinson and K. M. Meema, 217–228. New York: Springer–Verlag. [18]

McConnell, C. R., and S. L. Brue. 2002. *Economics—Principles, problems, and policies*, 15th ed. New York: McGraw–Hill, Inc. [22]

McCulley, M. E. and M. J. Canny. 1988. Pathways and processes of water and nutrient movement in roots. *Plant and Soil* 111:159–170. [4]

McHarg, I. L. 1969. *Design with nature*. Garden City, NY: The Natural History Press. [23]

McKillop, A., and S. Newman, eds. 2006. *Final energy crisis*. London: Pluto Press. [17]

McNeely, J. A. et al., ed. 2001. *Global strategy on invasive alien species*. Gland, Switzerland: IUCN. [2, 20, 25]

Meier, M. F., and M. B. Dyurgerov. 2002. How Alaska affects the world—Global sea level rise is caused mainly by ocean expansion due to warming, and by ocean mass increase due to melting of glaciers on land. *Science* 297:350–351. [19]

Melillo, J. M., R. A. Houghton, and A. D. McGuire. 1996. Tropical deforestation and the global carbon budget. *Annual Reviews of Energy and the Environment* 21:293–310. [13]

Mendelsohn, R., and M. Balick. 1995. Private property and rainforest conservation. *Conservation Biology* 9: 1322–1323. [13]

Merriam, C. H. 1894. Laws of temperature control the geographic distribution of animals and plants. *National Geographic Magazine* 6: 229–238. [10]

Meyer, W. B., and B. L. Turner, II. 1992. Human population growth and global land-use/cover change. *Annual Review of Ecology and Systematics* 23:39–61. [11]

Meyers, R. A., J. A. Hutchings, and N. J. Barrowman. 1997. Why do fish stocks collapse? The example of cod in Atlantic Canada. *Ecological Applications* 7:91–106. [16]

Micklin, P. 2007. The Aral Sea disaster. *Annual Review of Earth and Planetary Sciences* 35:47–72. [16]

Micklin, P., and N. V. Aladin. 2008. Reclaiming the Aral Sea. *Scientific American* 298 (4): 64–71. [16]

Middleton, N., and D. Thomas, eds. 1998. *World atlas of desertification,* 2nd ed. New York: Arnold, and John Wiley & Sons, Inc. [11]

MEA (Millennium Ecosystem Assessment). 2003. *Ecosystems and human well-being: A framework for assessment,* Washington, D.C.: Island Press. [1, 22]

Miller, F. P. 2000. Land grant colleges of agriculture: Preempting a postmortem—Requisites for renaissance. Invited lecture, School of Natural Resources, The Ohio State University, Columbus, April 6. 42 pp. [12]

Miller, F. P., and M. K. Wali. 1995. Soils, land use, and sustainable agriculture: A review. *Canadian Journal of Soil Science* 75:413–422. [12, 26]

Miller, P. 2007. Swarm theory. *National Geographic* 21. [8]

Mitsch, W. J., and J. G. Gosselink. 2007. *Wetlands,* 4th ed. New York: Van Nostrand Reinhold. [5]

Mitsch, W. J., and S. E. Jørgensen. 2004. *Ecological design principles. Ecological engineering and ecosystem restoration,* chap. 5. New York: John Wiley & Sons. [24]

Mittermeier, R. A., N. Myers, and C. G. Mittermeier. 1999. Hotspots: Earth's biologically richest and most endangered terrestrial ecoregions. CEMEX Conservation International. [20, 25]

Molles, M. C., Jr. 2008. *Ecology: Concepts, applications*. New York: MacGraw–Hill Higher Education. [6, 8]

Mooney, H. A., S. P. Hamburg, and J. A. Drake. 1986. The invasions of plants and animals into California. In *Ecology of biological invasions of North America and Hawaii,* ed. H. A. Mooney and J. A. Drake, 250–272. New York: Springer–Verlag. [20]

Morris, J. G., Jr. 1999. Harmful algal blooms: An emerging public health problem with possible links to human stress on the environment. *Annual Review of Energy and the Environment* 24:367–390. [16, 21]

Morrison, P., and P. Morrison. 1987. *The ring of truth: An inquiry into how we know what we know.* New York: Random House. [7]

Myers, N. 1995. *Environmental exodus: An emergent crisis in the global arena,* 214. Washington, D.C.: The Climate Institute. [11]

———. 1996a. Environmental services of biodiversity. *Proceedings of the National Academy of Sciences* 93:2764–2769. [20]

———. 1996b. The world's forests: Problems and potentials. *Environmental Conservation* 23 (2): 156–168. [13]

———. 1998. Lifting the veil on perverse subsidies. *Nature* 392:327–328. [22]

Myers, N., and J. Kent. 1998. *Perverse subsidies—Tax $s undercutting our economies and environments alike.* Winnipeg, Manitoba, Canada: International Institute for Sustainable Development. [22]

Myneni, R. B., C. D. Keeling, C. J. Tucker, G. Asrar, and R. R. Nemani. 1997. Increased plant growth in the northern high latitudes from 1981 to 1991. *Nature* 386:698–702. [19]

Naiman, R. J., H. Decamps, and M. E. McClain. 2005. *Riparia: Ecology, conservation and management of streamside communities.* Amsterdam: Elsevier Publishers. [24]

NAPAP (National Acid Precipitation Assessment Program). 1998. *National Acid Precipitation Assessment Program Report to Congress—An Integrated Assessment.* Washington, D.C.: Executive Office of the President. [18]

Narain, S. 2007. We don't smell the air. *Down to Earth*—Science and environment online, September 25

National Atmospheric Deposition Program (NRSP-3)/National Trends Network. 2000. NADP Program Office, Illinois State Water Survey, Champaign, IL.

NCASI (National Council for Air and Stream Improvement). 2000. Land management tools for the maintenance of biological diversity: An evaluation of existing forestland classification schemes. Technical bulletin no. 800 prepared by K. S. Pregitzer and P. C. Goebel.

NET (National Environmental Trust, Physicians for Social Responsibility, and Learning Disabilities Association of America). 2000. Polluting our future: Chemical pollution in the U.S. that affects child development and learning. Washington, D.C. 31 pp. [16, 18]

NHLA (National Hardwood Lumber Association). 1998. Memphis, TN. [13]

NRC (National Research Council). 1980. *Surface mining: Soil, coal, and society*. Washington, D.C.: National Academy Press. [24]

———. 1983. *Acid deposition: Atmospheric processes in eastern North America*. Washington, D.C.: National Academy Press. [18]

Naveh, Z., and P. Kutiel. 1990. Changes in the Mediterranean vegetation in response to human habitation and land use. In *The Earth in transition: Patterns and processes of biotic impoverishment*, ed. G. M. Woodwell, 259–299. New York: Cambridge University Press. [20]

Naylor, R. L. 1996. Energy and resource constraints on intensive agricultural production. *Annual Reviews: Energy and the Environment* 21:99–123. [11]

Newman, P., and J. Kenworthy. 1999. *Sustainability and cities: Overcoming automobile dependence*. Washington, D.C.: Island Press. [15]

Nihlgard, B. 1985. The ammonium hypothesis, an additional explanation of forest dieback in Europe. *Ambio* 14 (1): 2–8. [18]

Nordstrom, D. K., and C. N. Alpers. 1999a. Geochemistry of acid mine waters. In *The environmental geochemistry of mineral deposits,* ed. G. S. Plumlee and M. J. Logsdon, 133–160. Review of Economic Geology, vol. 6A. Littleton, CO: Society of Economic Geologists, Incorporated. [14]

———. 1999b. Negative pH, efflorescent mineralogy, and consequences for environmental restoration at the Iron Mountain Superfund site, California. *Proceedings of the National Academy of Sciences,* 96:3455–3462. [14]

Nordstrom, D. K., C. N. Alpers, C. J. Ptacek, and D. W. Blowes. 2000. Negative pH and extremely acidic mine waters from Iron Mountain, California. *Environmental Science and Technology* 34:254–258. [14]

Normile, D. 2000. Warmer waters more deadly to coral reefs than pollution. *Science* 290:682–683. [19]

Northcote, T. G., and G. F. Hartman, eds. 2004. *Forests and fishes—World watershed interactions and management.* Oxford, England: Blackwell Science. [13]

Odum, E. P. 1959. *Fundamentals of ecology,* 1st ed. (2nd ed. 1959, 3rd ed. 1971). Philadelphia, PA: Saunders Publishing. [2]

———. 1969. The strategy of ecosystem development. *Science* 164:262–270. [9]

———. 1971. *Fundamentals of ecology,* 3rd ed, 378. Philadelphia, PA: Saunders Publishing. [2, 8, 10]

Odum, E. P., and G. W. Barrett. 2005. *Fundamentals of ecology,* 5th ed. New York: Thompson/Brooks–Cole. [7, 8]

Odum, H. T. 1962. Man in the ecosystem. In *Lockwood conference on the suburban forest and ecology,* 57–75. Bulletin 652. Storrs, CT: Connecticut Agricultural Experiment Station. [24]

Oki, T., and S. Kanae. 2006. Global hydrological cycles and world water resources. *Science* 313:1068–1072. [6]

Oldeman, L. R. 1994. The global extent of soil degradation. In *Soil resilience and sustainable land use,* ed. D. J. Greenland and I. Szabolcs, 99–118. Wallingford, Oxford, England: CAB International Publishers. [B, 12, 24]

Oldeman, L. R., R. T. A. Hakkeling, and W. G. Sombroek. 1991. *World map of the status of human-induced land degradation: An explanatory note,* 2nd revised ed. Wageningen, the Netherlands: UNEP, ISRIC, and Nairobi. [12]

Oldeman, L. R., and G. W. J. van Lynden. 1998. Revisiting the GLASOD methodology. In *Methods for assessment of soil degradation,* ed. R. Lal, W. H. Blum, C. Valentine, and B. A. Stewart, 423–441. New York: CRC Press. [12]

Olson, J. S. 1958. Rates of succession and soil changes on southern Lake Michigan sand dunes. *Botanical Gazette* 119:125–170. [9]

O'Neill, R. V. 1996. Making ecological magic from sunbeams. *Ecology* 77: 2263. [1]

Orians, G. H. 1990. Ecological concepts of sustainability. *Environment* 32 (9): 10–15, 34–39. [26]

Orlove, B., J. Chiang, and M. Cane. 2002. Ethnoclimatology in the Andes. *American Scientist* 90:428–435. [3]

O'Toole, C. 1993. Diversity of native bees and agroecosystems. In *Hymenoptera and biodiversity,* ed. J. LaSalle and I. D. Gault. Wallingford, Oxfordshire, England: AB International. [11]

Overbay, J. C. 1992. Ecosystem management, 3–15. In *Taking an ecological approach to management.* United States Department of Agriculture Forest Service Publication Wo-WSA-3. [23]

Overpeck, J. et al. 2000. Arctic environmental change of the last four centuries. *Nature* 403:756–758. [19]

Owen-Smith, N. 1971. Territoriality in the white rhinoceros (*Ceratotherium simum*) Burchell. *Nature* 231:294–296. [8]

Paine, R. T. 1966. Food web complexity and species diversity. *American Naturalist* 100:66–75. [8]

Parmesan, C., and G. Yohe. 2003. A globally coherent fingerprint of climate change impacts across natural systems. *Nature* 421:37–42. [19]

Patel, C. K. N. 1992. Industrial ecology. *Proceedings of the National Academy of Sciences* 89:798–799. [14]

Patz, J. A., M. A. McGeehin, et al. 2000. The potential impacts of climate variability and change for the United States: Executive summary of the Report of the Health Sector of the U.S. National Assessment. *Environmental Health Perspectives* 108:367–376. [19]

Pauly, D., and V. Christensen. 1995. Primary production required to sustain global fisheries. *Nature* 374:255–257 (see errata corrections in *Nature* 376:279, 1995). [11, 16]

Pechan. 1999. Muskingum River—Ohio Power Co 1997 plant-level utility data.

Penning de Vries, F. W. T., R. Rabbinge, and J. J. R. Groot. 1997. Potential and attainable food production and food security in different regions. *Philosophical Transactions of the Royal Society of London Series B* 352:917–928. [11, 16]

Peñuelas, J., and L. Filella. 2001. Responses to a warming world. *Science* 294:793–795. [19]

Persinger, M. 1980. *The weather matrix and human behavior.* Boulder, CO: Praeger Publishers. [3]

Pianka, E. R. 1970. On *r* and *K* selection. *American Naturalist* 104:592–597. [8]

Pickering K. T., and L. A. Owen. 1994. *An introduction to global environmental issues.* New York: Routledge. [12, 16, 17]

Piel, G. 1997. The urbanization of poverty worldwide. *Challenge* (January–Februrary): 58–68. [15]

Pielke, R. A., Jr. 2007. *The honest broker—Making sense of science in policy and politics.* New York: Cambridge University Press. [21]

Pigou, A. C. 1920. *The economics of welfare.* London: Macmillan. [21]

Pilon-Smits, E. 2005. Phytoremediation. *Annual Review of Plant Biology* 56:15–39. [24]

Pimentel, D. 1986. Biological invasions of plants and animals in agriculture and forestry. In *Ecology of biological invasions of North America and Hawaii*, ed. H. A. Mooney and J. A. Drake, 149–162. New York: Springer–Verlag. [20]

———. 1995. Protecting crops. In *The literature of crop science,* ed. W. C. Olsen, 49–66. Ithaca, NY: Cornell University Press. [12]

———. 1997. *Techniques for reducing pesticides: Environmental and economic benefits.* Chichester, England: John Wiley & Sons. [11]

———. 1999. Human resource use, population growth, and environmental destruction. *Bulletin of Ecological Society of America* 80 (1): 88–91. [12]

———. 2005. Aquatic nuisance species in the New York State Canal and Hudson River Systems and the Great Lakes Basin: An economic and environmental assessment. *Environmental Management* 35:692–701. [20]

Pimentel, D., O. Bailey, et al. 1999. Will limits of the world's resources control human numbers? *Environment, Development and Sustainability* 1:19–39. [11]

Pimentel, D., W. Dazhong, and M. Gieampetro. 1990. Technological changes in energy use in U.S. agricultural production. In *Agroecology,* ed. S. R. Gliessman, 305–321. New York: Springer. [12]

Pimentel, D., A. Greiner, and T. Bashore. 1998. Economic and environmental costs of pesticide use. In *Environmental toxicology: Current developments,* ed. J. Rose, 121–150. Amsterdam: Gordon and Breach Science Publishers. [12]

Pimentel, D., J. Houser, et al. 1997. Water resources: Agriculture, the environment, and society: An assessment of the status of water resources. *BioScience* 47 (2): 97–106. [16]

Pimentel, D., L. McLaughlin, et al. 1991. Environmental and economic effects of reducing pesticide use. *BioScience* 41 (6): 402–409. [12]

Pimm, S. L., and T. M. Brooks. 1997. The sixth extinction: How large, where, and when? In *Nature and human society: The quest for a sustainable world,* ed. P. H. Raven, 46–62. Washington, D.C.: National Academy Press. [20]

Pimm, S. L., G. E. Davis, et al. 1994. Hurricane Andrew. *Bioscience* 44 (4): 224–229. [3]

Pinstrup-Andersen, P., and R. Pandya-Lorch. 1996. Food for all in 2020: Can the world be fed without damaging the environment? *Environmental Conservation* 23 (3): 226–234. [12]

Pitelka, L. F. 1993. Biodiversity and policy decisions. In *Biodiversity and ecosystem function,* ed. E-D. Schulze and H. A. Mooney, 481–493. New York: Springer–Verlag. [20]

Plater, Z. J. B., R. H. Abrams, and W. Goldfarb. 1992. *Environmental law and policy.* St. Paul, MN: West Publishing Company. [21]

Platt, J. R. 1965. Strong inference. *Science* 149:607–613. [1]

Plochl M., and W. Cramer. 1995. Coupling global models of vegetation structure and ecosystem processes—An example from arctic and boreal ecosystems. *Tellus Series B—Chemical and Physical Meteorology* 47:240–250.

Pockley, P. 2000. Global warming identified as main threat to coral reefs. *Nature* 407:932. [19]

Poehlman J. M., and D. A. Sleper. 1995. *Breeding field crops,* 4th ed. Ames: Iowa State University Press. [12, 16]

Pollack, H. 2000. *Nature.* [19]

PRB (Population Reference Bureau). 2007. World population highlights: Key findings from PRB's 2007 World Population Data Sheet. *Population Bulletin* 62 (3): 1–16. [11]

Postel, S. 1997. *Last oasis: Facing water scarcity.* New York: W. W. Norton & Company. [12, 16]

Postel, S. L. 1998. Water for food production: Will there be enough in 2025? *BioScience* 48 (8): 629–637. [11, 12, 16]

Postel, S. L., G. C. Daily, and P. R. Ehrlich. 1996. Human appropriation of renewable freshwater. *Science* 271:785–788. [11, 16]

Powell, J. A., and R. A. Mackie. 1966. Biological interrelationships of moths and *Yucca whipplei. University of California Publications in Entomology* 42:1–59. 1965. Butterflies and plants: A study in coevoluton. *Evolution* 18:586–608. [8]

Poynter, J. 2006. *The human experiment: Two years and twenty minutes inside BIOSPHERE 2.* New York: Thunder's Mouth Press. [2]

Prasad, A. M., L. R. Iverson, S. Matthews, and M. Peters. 2007–ongoing. *A climate change atlas for 134 forest tree species of the Eastern United States* [database]. http://www.nrs.fs.fed.us/atlas/tree, Northern Research Station, USDA Forest Service, Delaware, OH. [13, 19].

Prather, M. J., and R. T. Watson. 1990. Stratospheric ozone depletion and future levels of atmospheric chlorine and bromine. *Nature* 344:729–734. [18]

Quevauviller, Ph. 2005. Groundwater monitoring in the context of EU legislation: Reality and integration needs. *Journal of Environmental Monitoring* 7:89–102. [21]

Rabalais, N. N., R. E. Turner, and W. J. Wiseman, Jr. 2001. Hypoxia in the Gulf of Mexico. *Journal of Environmental Quality* 30:320–329. [16]

Rafiqul, I., C. Weber, B. Lehman, and A. Voss. 2005. Energy efficiency improvements in ammonia production—perspectives and uncertainties. *Energy* 30: 2487–2504. [12]

Ramakrishnan, P.S. 1992. *Shifting agriculture and sustainable development.* Paris: UNESCO. [4]

Ramlogan, R. 1996. Environmental refugees: A review. *Environmental Conservation* 23 (1): 81–88. [11]

Ray, D. L., and L. Guzzo. 1990. *Trashing the planet.* Washington, D.C.: Regnery Gateway. [17]

Reeburgh, W. S. 1997. Figures summarizing the global cycles of biogeochemically important elements. *Bulletin of the Ecology Society of America* 78 (4):260–267. [6] (see also Web site.)

Rees, W. E. 1992. Ecological footprints and appropriated carrying capacity: What urban economics leaves out. *Environment and Urbanization* 4 (2): 121–130. [15]

Reid, W. V., and K. R. Miller. 1989. *Keeping options alive: The scientific basis for conserving biodiversity.* Washington, D.C.: World Resources Institute. [11]

Reiners, W. A. 1986. Complementary models for ecosystems. *American Naturalist* 127:59–73. [6, 7]

Reisner, M. 1989. Competition for water. *Issues in Science and Technology.* [16, 26]

———. 2000. Unleash the rivers. *Time* (April–May): 66–71. [16]

Rice, E. L. 1984. *Allelopathy,* 2nd ed. San Diego, CA: Academic Press. [8, 24]

Richards, J. F. 1993. Land transformation: In *The Earth as Transformed by Human Action*, ed. B. L. Turner II, W. C. Clark, R. W. Kates et al. 163–178. New York: Cambridge University Press. {Sec. B]

Richards, P. W. 1996. *The tropical rain forest: An ecological study.* New York: Cambridge University Press. [10]

Ricketts, T. H., E. Dinerstein, D. M. Olson, C. J. Loucks, W. Eichbaum, D. DellaSala, K. Kavanaugh, P. Jedao, P. T. Hurley, K. M. Carney, et al. 1999. *Terrestrial ecoregions of North America: A conservation assessment.* Washington, D.C.: Island Press. [10]

Riley, C. V. 1892. The yucca moth and yucca pollination. Missouri Botanical Garden, third annual report, 181–226. [8]

Robbins, C. S., D. K. Dawson, and B. C. Dowell. 1989. Habitat area requirements of breeding forest birds of the Middle Atlantic States. *Wildlife Monographs* 103:1–34. [13]

Roberts, J. 2004. *Environmental policy.* New York: Taylor & Francis. [21]

Rodgers, V. L., K. A. Stinson, and A. C. Finzi. 2008. Ready or not, garlic mustard is moving: *Allaria petiolata* as a member of eastern North American forests. *BioScience* 58:426–436. [20]

Rodhe, H., E. Cowling, I. Galbally, J. Galloway, and R. Harrera. 1988. Acidification and regional air pollution in the tropics. In *Acidification in tropical countries,* ed. H. Rodhe and R. Herrera, 3–39. Chichester, England: John Wiley & Sons. [18]

Rogner, H. H. 1997. An assessment of world hydrocarbon resources. *Annual Review of Energy and the Environment* 22:217–262. [14, 17]

Ronderos, A. 2000. Where giants once stood: The demise of the American chestnut and efforts to bring it back. *Journal of Forestry* 98 (2): 10–11. [20]

Root, T. L. et al. 2003. Fingerprints of global warming on wild animals and plants. *Nature* 421:57–60 (143 studies). [19]

Rosenfeld, D. 2000. Suppression of rain and snow by urban and industrial air pollution. *Science* 287:1793–1796. [18, 19]

Rosenfeld, D., Y. Rudich, and R. Lahav. 2001. Desert dust suppressing precipitation: A possible desertification feedback loop. *Proceedings of the National Academy of Sciences* 98:5975–5980. [3]

Rossiter, D. G. 1996. A theoretical framework for land evaluation. *Geoderma* 72 (3–4): 165–190. [23]

Rowe, J. S. 1952. [10]

Ruess, R. W., and M. K. Wali. 1980. Daily fluctuations in water potential and associated ionic changes in *Atriplex canescens. Oecologia* 47:200–203. [4]

Ruhe, R. V. 1975. Climate geomorphology and fully developed slopes. *Catena* 2:309–320. [4]

Ruttner, F. 1953. *Fundamentals of limnology.* English translation by D. G. Frey and F. E. J. Frey. Toronto, Ontario, Canada: University of Toronto Press. [3]

Sachs, J. D. 2005a. *The end of poverty—Economic possibilities for our time.* New York: The Penguin Press. [11, 22]

———. 2005b. *Common wealth—Economics for a crowded planet.* New York: The Penguin Press. [22]

Sadik, N. 1991. World population continues to rise. *The Futurist* (March/April): 9–14. [11]

Safaya, N. M., and M. K. Wali. 1992. Applicability of some U.S. environmental laws in the developing countries: An analysis of ecological and regulatory concepts. In *Ecosystem rehabilitation,* vol. 1. *Policy issues,* ed. M. K. Wali, 143–155. The Hague: SPB Academic Publishing. [21]

Sagan, C. 1994. *Pale blue dot: A vision of the human future in space.* New York: Random House, p. 6. [1]

Sala, O., F. S. Chapin III, et al. 2000. Global biodiversity scenarios for the year 2100. *Science* 287:1770–1774. [20, 25]

Samuelson, P. A., and W. D. Nordhaus. 2001. *Economics,* 17th ed. New York: McGraw–Hill. [22]

Sarewitz, D., and R. Pielke, Jr. 2000. Breaking the global-warming gridlock. *The Atlantic Monthly* 286 (1): 55–64. [19]

Scheidel, W. 2003. Demography of ancient world. In *Encyclopedia of population,* vol. 1, ed. P. Demeny and J. McNicoll, 44–48. New York: Macmillan Reference. [11]

Schindler, D. W. 1990. Experimental perturbations of whole lakes as tests of hypotheses concerning ecosystem structure and function. *Oikos* 57:25–41. [18]

———. 1992. A view of NAPAP from north of the border. *Ecological Applications* 2:124–130.

———. 1997. Widespread effects of climatic warming on freshwater ecosystems in North America. *Hydrological Processes* 11:1043–1067. [19]

———. 1998. A dim future for boreal waters and landscape. *BioScience* 48:157–164. [B, 24]

Schlesinger, W. H. 1997. *Biogeochemistry—An analysis of global change,* 2nd ed. San Diego, CA: Academic Press. [6]

———. 1999. Carbon sequestration in soils. *Science* 284:2095. [12, 13]

Schmid, P. E., M. Tokeshi, and J. M. Schmid-Arya. 2000. Relation between population density and body size in stream communities. *Science* 289:1557–1560. [8]

Schoener, T. W. 1983. Field experiments on interspecific competition. *American Naturalist* 122:240–285. [8]

Schultz, R. C., T. M. Isenhart, et al. 2004. Riparian forest buffers in agroecosystems—Lessons learned from the Bear Creek Watershed, central Iowa, USA. *Agroforestry Systems* 61:35–50. [24]

Schumacher, E. F. 1973. *Small is beautiful: A study of economics as if people mattered.* London: Abacus. [26]

Shantz, H. L. 1954. The place of grasslands in the Earth's cover of vegetation. *Ecology* 35:143–145. [10]

Sharma, D. C. 2004. Himalayan glaciers vanishing. *Frontiers in Ecology and the Environment* 2 (3): 118. [19]

Shelford, V. 1911. Physiological animal geography. *Journal of Morphology* 22:551–618. [2]

Schellnhuber, H.-J., and H. Held. 2002. How fragile is the Earth system? In *Managing the Earth–the Linacre Lectures 2001*, J. Briden and T. Downing, eds. 5–34, Oxford: Oxford University Press. [19]

Shiklomanov, I. A. 1993. World fresh water resources. In *Water in crisis: A guide to the world's fresh water resources,* ed. P. H. Gleick. New York: Oxford University Press. [5]

———, ed. 1996. *Assessment of water resources and water availability in the world.* Background report to the comprehensive freshwater assessment. St. Petersburg, Russia: State Hydrological Institute. [11, 16]

Simberloff, D. 1997. Biogeographic approaches and the new conservation biology. In *The Ecological Basis of Conservation: Heterogeneity, Ecosystems, and Biodiversity,* ed. S. T. A. Pickett, R. S. Ostfeld, M. Shachak, and G. E. Likens, 274–284. New York: Chapman & Hall, International Thomson Publishing.

Simberloff, D., and B. von Holle. 1999. Positive interactions of nonindigenous species: Invasional melt-down? *Biological Invasions* 1 (1): 21–32. [20]

Sims, P. L., and P. G. Risser. 2000. Grasslands. In *North American terrestrial vegetation,* 2nd ed., ed. M. G. Barbour and W. D. Billings, 323–356 (Table 9.5, p. 346). New York: Cambridge University Press. [9, 10, B, 24]

Singer, P. C., and M. W. Stumm. 1970. Acid mine drainage: the rate determining step. *Science* 167: 1121–1123. [14]

Smith, R. L. 1990. *Ecology and field biology,* 4th ed. New York: Harper Collins. [5]

Smith, W. B., J. L. Faulkner, and D. S. Powell. 1994. Forest statistics of the United States, 1992. Metric units. USDA Forest Service, North Central Forest Experiment Station, St. Paul, MN. [13]

Smith, W. B., P. D. Miles, et al. 2004. Forest resources of the United States, 2002. General technical report NC-241, North-Central Research Station, USDA, Forest Service, St. Paul, MN. [13]

Sneddon, C., R. B. Howarth, and R. B. Norgaard. 2006. Sustainable development in a post-Brundtland world. *Ecological Economics* 57:253–268. [26]

Snow, C. P. 1959. *The two cultures.* New York: Cambridge University Press. [1]

SER (Society for Ecological Restoration, International Science & Policy Working Group). 2004. *The SER international primer on ecological restoration*, accessed July 2005 (see http://www.ser.org). [24]

Socolow, R. H. 1999. Nitrogen management and the future of food: Lessons from the management of energy and carbon. *Proceedings of National Academy of Sciences* 96:6001–6008. [11]

SSSA (Soil Science Society of America). 1997. *Glossary of soil science terms.* Madison, WI: SSSA. [4]

Solley W. B., R. R. Pierce, and H. A. Perlman. 1998. USGS (U.S. Geological Survey). Estimated use of water in the United States in 1995. U.S. Geological Survey Circular 1200, U.S. Government Printing Office. Available at http://water.usgs.gov/watuse/pdf1995/html/ [16]

Sorensen, J. 1979. *Renewable energy.* New York: Academic Press. [7]

Speidel, J. J., D. C. Weiss, et al. 2007. Family planning and reproductive health: The link to environmental preservation. *Population and Environment* 28:247–258. [11]

Spellerberg, I. F. 2002. *Ecological effects of roads.* Plymouth, England: Science Publishers. [15]

Squires, G. D. 2002. Urban sprawl and the uneven development of metropolitan America. In *Urban sprawl—Causes, consequences, and policy responses,* 1–22. Washington, D.C.: The Urban Institute Press. [15]

Staley, S. R. 2000. The vanishing farmland myth and smart growth agenda, reason. Public Policy Institute policy brief no. 12, January 2000, Los Angeles, CA. [15]

Stanturf, J. A., S. H. Schoeholtz, et al. 2001. Achieving restoration success: Myths in bottomland hardwood forests. *Restoration Ecology* 9:189–200. [24]

Starfinger U., and H. Sukopp. 1994. Assessment of urban biotopes for nature conservation. In *Landscape planning and ecological networks,* ed. E. A. Cook and H. N. van Lier. New York: Elsevier. [15]

Steffen, W., J. Jager, D. J. Carson, and C. Bradshaw, eds. 2002. *Challenges of a changing Earth.* New York: Springer. [19]

Steiner, R. 1998. Deforestation in Alaska's coastal rainforest: Causes and solutions. Available at http://www.wrm.org.uy/english/u_causes/regional/north_america/Alaska.html [10]

Sterner, R. W., and J. J. Elser. 2000. *Ecological stoichiometry: The biology of elements from molecules to the biosphere.* Princeton, NJ: Princeton University Press. [6]

Stone, R. 1999. Coming to grips with the Aral Sea's grim legacy. *Science* 284:30–31, 33. [16]

Stork, N. E. 1999. The magnitude of global biodiversity and its decline. In *The living planet in crisis: Biodiversity science and policy*, ed. J. Cracraft and F. T. Grifo, 3–33. New York: Columbia University Press. [20]

Strauss, W., and N. Howe. 1991. *Generations*. New York: William Morrow.

Streitfeld, D. 2008. Uprising against the ethanol mandate. *The New York Times*, July 23, 2008. [12]

Surface Mining Control and Reclamation Act, Public Law 97–85. 1977. [24]

Surowieck, J. 2004. The wisdom of crowds—*Why the many are smarter than the few and how collective wisdom shapes business, economics, societies, and nations*. New York: Doubleday. [8]

Swaminathan, M. S. 1996. *Sustainable agriculture: Toward food security*. Delhi, India: Konark Publishers Ltd. [12]

Szabolcs, I. 1989. *Salt-affected soils*. Boca Raton, FL: CRC Press. [12]

———. 1992. Salinization and desertification. *Acta Agronomica Hungarica* 41:137–148. [12]

Takahashi, T, W. S. Broecker, and S. Langer. 1985. Redfield ratio based on chemical data from isopycnal surfaces. *Journal of Geophysical Research* 90:6907–6924. [6]

Talberth, J. 2008. A new bottom line for progress. In *State of the world 2008: Innovations for a sustainable economy*, 19–31. New York: W. W. Norton & Company. [26]

Theophrastus. ca. 300 B. C. *An enquiry into plants*. Book IV. *Of the trees and plants special to particular districts and positions*. Sir Arthur Hort Edition 1916. London: Heinesmann. [2]

Thomas, J. A. et al. 2004. Comparative losses of British butterflies, birds, and plants and the global extinction crisis. Northward migration. *Science* 303:1879. [19]

Thomas, M. F. 1974. *Tropical geomorphology; a study of weathering and landform development in warm climates*. London: Macmillan Press Limited, 332 pp. [10]

Thompson, D. 2000. Asphalt jungle. *Time* (April–May): 50–51. [15]

Thompson, F. 2003. The mistress of vision. In *Poems of Francis Thompson (1913)*. Whitefish, MT: Kessinger Publishing Company. [2]

Thompson, L. G., E. Mosley-Thompson, et al. 1995. Late-glacial stage and Holocene tropical ice core records from Huascaran, Peru. *Science* 269:46–50. [19]

Tietenberg, T. 2003. *Environmental and natural resource economics*, 6th ed. New York: Addison–Wesley. [22]

Tilman, D. 1990. Mechanisms of plant competition for nutrients: The elements of a predictive theory of competition. In *Perspectives on plant competition*, ed. J. B. Grace and D. Tilman, 117–141. New York: Academic Press. [9]

———. 1999. Global environmental impacts of agricultural expansion: The need for sustainable and efficient practices. *Proceedings of the National Academy of Sciences* 96:5995–6000. [11]

———. 2000. Causes, consequences and ethics of biodiversity. *Nature* 405:208–211. [20]

Tilman, E. A., D. Tilman, M. J. Crawley, and A. E. Johnston. 1999. Biological weed control via nutrient competition: Potassium limitations of dandelions. *Ecological Applications* 9:103–111. [4]

Tiner, R. W. 1996. Wetland definitions and classifications in the United States. In *National Water Summary on Wetland Resources*, chap. 2, water supply paper 2425. Denver, CO: U.S. Geological Survey. [5]

Toon, O. B. 2000. Atmospheric science—How pollution suppresses rain. *Science* 287:1763–1765. [3]

Trewavas, A. 2002. Malthus foiled over and over again. *Nature* 418:668–670. [11]

Umali, D. L. 1993. *Irrigation-induced salinity*. Washington, D.C.: World Bank. [12]

United Kingdom All Party Parliamentary Group on Population, Development and Reproductive Health. 2007. *Return of the population growth factor: Its impact upon the millennium development goals*. London: House of Commons. 74 pp. [11]

United Nations. 2006. *Water: A shared responsibility*. The United Nations world water report 2. World Water Assessment Program. Paris: UNESCO. [11, 12, 16]

UNCED (UN Conference on Environment and Development). 1992. *Agenda 21*. United Nations Conference on Environment and Development, United Nations General Assembly, New York. [21, 26]

UNDP (UN Development Program). 2005. *Human development report 2005: International cooperation at a crossroads—Aid, trade and security in an unequal world*. New York: UNDP. [11]

UNEP (UN Environment Program). 1992. Nairobi, Kenya. [12]

———. 1997. *World atlas of desertification*. Nairobi, Kenya. [12]

———. 2005. Montreal *Protocol handbook CIESIN's thematic guide on ozone depletion and global environmental change*. Nairobi, Kenya. [18, 25]

UNEP/ISRIC. 1991. *Mapping of soil and terrain vulnerability to specified chemical compounds in Europe at a scale of 1:5 M.* Proceedings of International Workshop (Wageningen, March 20–23, 1991), ed. N. H. Batjes and E. M. Bridges. Chemical Time Bombs (CTB) Project and ISRIC, Wageningen, vi + 177. [12]

UN Population Fund. 2007. *State of the world population 2007: Unleashing the potential of urban growth.* New York: United Nations. (www.unfpa.org) [15]

USDA (U.S. Department of Agriculture). 1955. Yearbook. *Water.* Washington, D.C.: Government Printing Office. [4]

———. 1994. *Keys to soil classification,* 6th ed. Washington, D.C.: Soil Conservation Service, Government Printing Office. [4]

———. 1995. Natural resources inventory—1992. U.S. Department of Agriculture Soil Conservation Service, Washington, D.C. [15]

———. 1997. *World soil resources.* Washington, D.C.: Natural Resources Conservation Service, Soil Survey Division. [4]

U.S. DOE (Department of Energy). Energy Information Administration. 2008. Monthly energy review. DOE/EIA-0035 (2008/03). Washington, D.C. March 2008. [17]

———. May 2008. (http://www.eia.doe.gov/fuelrenewable.html). [17]

———. 1967. *Surface mining and our environment—A special report to the nation.* Washington, D.C.: U.S. Government Printing Office. [1, 24]

USDI-BLM (U.S. Department of the Interior, Bureau of Land Management). 1994. *Ecosystem management in the BLM: From concept to commitment,* 573–183. Washington, D.C.: U.S. Government Printing Office. [23]

U.S. EPA (U.S. Environmental Protection Agency). 1996a. Hazardous waste characteristics scoping study. Washington, D.C.: U.S. Environmental Protection Agency, Office of Solid Waste. [15, 18]

———. 1996b. *Municipal solid waste factbook* (ver. 3). Washington, D.C.: Office of Solid Waste. [15]

———. 1998. *Guidelines for ecological risk assessment.* U.S. Environmental Protection Agency, Risk Assessment Forum, Washington, D.C., EPA/630/R095/002F. Available at http://cfpub.epa.gov/ncea/cfm/recordisplay.cfm?deid=12460 [23]

———. 1999. The ecological condition of estuaries in the Gulf of Mexico. EPA-620-R-98-004, Washington, D.C. [16]

———. 2001. Planning for ecological risk assessment: Developing management objectives (external review draft). U.S. Environmental Protection Agency, Risk Assessment Forum, Washington, D.C., EPA/630/R-01/001A. [23]

U.S. GAO (U.S. General Accounting Office). 2000. Invasive species: Federal and selected state funding to address harmful, non-native species.

USGCRP (U.S. Global Change Research Program). 2001. *Our changing planet.* A report by the Climate Change Science Program and the Subcommittee on Global Change Research. The White House, Washington, D.C. [19]

———. 2008. *Our changing planet.* Climate change science program for fiscal 2008. The White House, Washington, D.C. [18, 19]

U.S. National Academy of Sciences. 1974. *Rehabilitation potential of western coal lands.* Cambridge, MA: Ballinger Publishing Co. [24]

U.S. National Science and Technology Council. 1998. National acid precipitation assessment program biennial report to Congress: An integrated assessment. Washington, D.C. [18]

U.S. OTA (U.S. Office of Technology Assessment). 1990. Coping with an oiled sea: An analysis of oil spill response technologies. Washington, D.C. [16]

NSIDC (U.S. National Snow and Ice Data Center). 2000. *State of the cryosphere.* [19]

van der Ploeg, R. R., W. Bohm, and M. B. Kirkham. 1999. On the origin of the theory of mineral nutrition in plants. *Soil Science Society of America Journal* 63:1055–1062. [2]

van der Valk, A. G., and C. B. Davis. 1978. The role of the seed bank in the vegetation dynamics of prairie glacial marshes. *Ecology* 59:322–335. [9]

Van Wambeke, A. 1991. *Soils of the tropics: Properties and appraisal.* New York: McGraw–Hill, Inc. [10]

Vannote, R. L., G. W. Minshall, K. W. Cummins, J. R. Sedell, and C. E. Cushing. 1980. The river continuum concept. *Canadian Journal of Fisheries and Aquatic Science* 37:130–137. [5]

Vernadsky, V. 1928. *The biosphere.* English translation. New York: Copernicus–Springer. [2]

Vertuno, J. 2008. Texas closer to saddling up wind energy—Panel gives preliminary OK for building transmission lines to turbine fields. Associated Press, July 18, 2008.

Vinikov, K. Y., A. Robock, R. J. Stouffer, J. E Walsh, C. L. Parkinson, D. J. Cavalieri, J. F. B Mitchell, D. Garrett, and V. F. Zakharov. 1999. Global warming and Northern Hemisphere ice extent. *Science* 286:1834–1937. [19]

Vitousek, P. M. 1986. Biological invasions and ecosystem properties: Can species make a difference? In *Ecology of biological invasions of North America and Hawaii,* ed. H. A. Mooney and J. A. Drake, 163–175. New York: Springer–Verlag. [20]

Vitousek, P. M., C. M. D'Antonio, et al. 1996. Biological invasions as global environmental change. *American Scientist* 84:468–478. [20]

Vitousek, P. M., H. A. Mooney, et al. 1997. Human domination of Earth's ecosystems. *Science* 277:494–499. [B, 11, 24]

Vogt, K, J. C. Gordon, et al. 1997. *Ecosystems—Balancing science with management.* New York: Springer–Verlag. [9]

von Bertalanffy, L. 1950. An outline of general system theory. *British Journal for the Philosophy of Science* 1:134–165. [7]

Wackernagel, M., and W. Rees. 1996. *Our ecological footprint: Reducing human impact on the Earth.* Gabriola Island, BC, Canada: New Society Publishers. [11, 22]

Wackernagel, M., N. B. Schulz, et al. 2002. Tracking the ecological overshoot of the human economy. *Proceedings of the National Academy of Sciences* 99 (14): 9266–9271. [22]

Wagner, W., J. Gawel, et al. 2002. Sustainable watershed management: An international multiwatershed case study. *AMBIO* 31 (1): 2–13. [23]

Wali, M. K. 1975. The problem of land reclamation viewed in a systems context. In *Practices and problems of land reclamation in western North America,* ed. M. K. Wali. 1–17. Grand Forks: The University of North Dakota Press. [12]

———. 1987. The structure, dynamics, and rehabilitation of drastically disturbed ecosystems. In *Perspectives in environmental management,* ed. T. N. Khoshoo, 163–183. New Delhi, India: IBH & Oxford Publications. [B, 24]

———. 1992. Ecology of the rehabilitation process. In *Ecosystem rehabilitation: Preamble to sustainable development,* Vol. 1. *Policy issues,* ed. M. K. Wali, 3–23. The Hague: SPB Academic Publishing. [2, B, 24]

———. 1999a. Ecology today: Beyond the bounds of science. *Nature & Resources* 35 (2): 38–50. [2]

———. 1999b. Ecological succession and rehabilitation of disturbed terrestrial ecosystems. *Plant and Soil* 213:195–220. [3, 4, 9, B, 24, 26]

Wali, M. K., and R. L. Burgess. 1985. The interface of ecology and law: Science, the legal obligation, and public policy. *Syracuse Journal of International Law and Commerce* 12:221–253. [21]

Wali, M. K., F. Evrendilek, et al. 1999. Assessing terrestrial ecosystem sustainability: Usefulness of carbon and nitrogen models. *Nature & Resources* 35 (4): 21–33. [2, 6, 24, 26]

Wali, M. K., and P. B. Kannowski. 1975. Prairie ant mound ecology: Interrelationships of microclimate, soils and vegetation. In *Prairie: A multiple view,* ed. M. K. Wali, 155–171. Grand Forks: University of North Dakota Press. [4]

Wali, M. K., and V. J. Krajina. 1973. Vegetation–environment relationships of some sub-boreal spruce zone ecosystems in British Columbia. *Vegetatio* 26:237–381. [4, 9, 10]

Wali, M. K., and F. P. Miller. 1998. The soil ecosystem. In *Encyclopedia of environmental analysis and remediation,* vol. 8, 4403–4462. New York: John Wiley & Sons. [4]

Wali, M. K., N. M. Safaya, and F. Evrendilek. 2002. Ecological rehabilitation and restoration in the Americas with special reference to the United States of America. In *Handbook of restoration ecology,* vol. 2, 3–31. *Restoration in practice,* ed. M. R. Perrow and A. J. Davy. Cambridge, England: Cambridge University Press. [24]

Wali, M. K., and F. M. Sandoval. 1975. Regional site factors and revegetation studies in North Dakota. In *Practices and problems of land reclamation in Western North America,* ed. M. K. Wali, 133–153. Grand Forks: University of North Dakota Press. [14]

Walker, B., and W. Steffen, eds. 1996. *Global change and terrestrial ecosystems.* International Geosphere–Biosphere Program Book Series 2. Cambridge, England: Cambridge University Press. [11]

Walker, L. R., and R. del Moral. 2003. *Primary succession and ecosystem rehabilitation.* Cambridge, England: Cambridge University Press. [9, 24]

Walsh, J. 1984. Hunger in West Africa: A crisis in development. *Technology Review* 87 (6): 22–23. [12]

Ward, B., and R. Dubois. 1972. *Only one Earth—The care and maintenance of a small planet.* London: Deutsch. [26]

Ware, G. E., H. Nigg, and D. R. Doerge, eds. 2004. *Reviews of environmental contamination and toxicology,* vol. 183. New York: Springer–Verlag. [21]

Waring, R. H., and B. D. Cleary. 1968. Plant moisture stress: Evaluation by pressure bomb. *Science* 155:1248–1254. [4]

Warming, E. 1905. *Oecology of plants: An introduction to the study of plant communities.* Oxford, England: Clarendon Press (originally published in 1895 as *Plantesamfund: Gründtrak af den Ökologiska Plantegeografi.* Philipsen, Copenhagen, Denmark). [2]

Warrick, J. 1998. "Moutaintop removal" shakes coal state. *Washington Post* August, 31, p. A01. Similar stories appeared in *U.S. News and World Report* ("Shear Madness") and on CBS's *60 Minutes.* [14]

Watson, R.T., J. A. Dixon, et al. 1998. *Protecting our planet—Securing our future: Linkages among global environmental issues and human needs.* United Nations Environment Program, U.S. National Aeronautics and Space Administration, The World Bank, Washington, D.C. [11, 13, 17, 19]

Watts, S. E. 2001. Determining forest productivity and carbon dynamics in southeastern Ohio from remotely sensed data. PhD dissertation. The Ohio State University, Columbus. [18]

Weart, S. R. 2003. *The discovery of global warming.* Cambridge, MA: Harvard University Press. [19]

Webster, P. J., G. J. Holland, J. A. Curry, and H. R. Chang. 2005. Changes in tropical cyclone number, duration, and intensity in a warming environment. *Science* 309:1844–1846. [19]

Wells, J. V., and M. E. Richmond. 1995. Populations, metapopulations, and species populations: What are they and who should care? *Wildlife Society Bulletin* 23:458–462. [8]

Wernick, I. K., and N. J. Themelis. 1998. Recycling metals for the environment. *Annual Reviews of Energy and the Environment* 23:465–497. [14, 25]

Westman, W. E. 1990. Detecting early signs of regional air pollution injury to coastal sage scrub. In *The Earth in transition: Patterns and processes of biotic impoverishment,* ed. G. M. Woodwell, 323–345. New York: Cambridge University Press. [20]

Westoby, M., B. Walker, and I. Noy-Meir. 1989. Opportunistic management for rangelands not at equilibrium. *Journal of Range Management* 42 (4): 266–274. [9]

Wetzel, R. G. 2001. *Limnology—Lake and river ecosystems,* 3rd ed. New York: Academic Press. [5]

White, P. S., and S. T. A. Pickett. 1985. Natural disturbance and patch dynamics: An introduction. In *The ecology of natural disturbance and patch dynamics,* ed. S. T. A. Pickett and P. S. White, 3–13. San Diego, CA: Academic Press. [B, 24]

Whitehead, A. N. 1929. *The function of reason.* Princeton, NJ: Princeton University Press. [1]

Whittaker, R. H. 1954. The ecology of serpentine soils. IV. The vegetation response to serpentine soils. *Ecology* 35:275–288. [4]

———. 1974. Climax concepts and recognition. In *Handbook of vegetation science,* vol. 8. *Vegetation dynamics,* ed. R. Knapp, 137–154. The Hague: Dr. W. Junk, Publishers. [9]

———. 1975. *Communities and ecosystems,* 2nd ed. New York: Macmillan Publishing Company. [3, 9, 10]

Whittaker, R. H., and G. E. Likens. 1973. Carbon in the biota. In *Carbon and the biosphere,* ed. G. M. Woodwell and E. V. Pecan, 281–302. Brookhaven Symposium of Biology. 24. Springfield, VA: National Technology Information Services. [7, 10]

Wielgolaski, F. E., ed. 1997. *Ecosystems of the world. 3. Polar and alpine tundra.* Amsterdam: Elsevier Science Publishers. [10]

Wielgolaski, F. E., L. C. Bliss, J. Svoboda, and G. J. Doyle. 1981. Primary production in tundra. In *Tundra ecosystems: A comparative analysis,* ed. L. C. Bliss, O. W. Heal, and J. J. Moore, 187–225. New York: Cambridge University Press. [10]

Wikramanayake, E., E. Dinerstein, C. J. Loucks, D. M. Olson, J. Morrison, J. Lamoreaux, M. McKnight, and P. Hedao. 2002. *Terrestrial ecoregions of the Indo-Pacific: A conservation assessment.* Washington, D.C.: Island Press.

Wikum, D. A., and M. K. Wali. 1974. Analysis of a North Dakota gallery forest: Vegetation in relation to topographic and soil gradients. *Ecological Monographs* 44:441–464 + appendix. [4, 10]

Wilding, L. P. 1994. Soil. *Agronomy News,* April 22, p. 4. [4]

Will, G. W. 2001. Wow! Or maybe just sort of. Have new information technologies "changed everything"? They do make a swell jukebox. *Newsweek* (April 16, 2001): 64.

Williams, M., ed. 1993. *Planet management.* New York: Oxford University Press. [12, 13, 15, 17]

Wilson, E. O. 2000a. A global biodiversity map. *Science* 289:2279. [2, 8, 12]

————. 2000b. Vanishing before our eyes. *Time* (April–May) 29–34. [20, 25]

Wolfson, R. 2008. *Energy, environment and climate.* New York: W. W. Norton & Co. [26]

Wood, C. A. 1994. Ecosystem management: Achieving the new land ethic. *Renewable Natural Resources Journal* 12:6–12. [23]

Woodmansee, R. G. 1978. Additions and losses of nitrogen in grassland ecosystems. *BioScience* 28:448–453. [9]

Woodwell, G. M. 1984. Broken eggshells. In 20 Discoveries that shaped our lives. *Science* November. [7, 12]

————, ed. 1990. *The Earth in transition: Patterns and processes of biotic impoverishment.* New York: Cambridge University Press. [20]

————. 1992. When succession fails ... In *Ecosystem rehabilitation: Preamble to sustainable development,* vol. 1. *Policy issues,* ed. M. K. Wali, 27–35. The Hague: SPB Academic Publishing. [24]

Woodwell, G. M., C. F. Wurster, and P. A. Isaacson. 1967. DDT residues in an East Coast estuary: A case of biological concentration of a persistent insecticide. *Science* 156 (3776): 821–824. [7]

WCED (World Commission on Environment and Development). 1987. *Our common future* [the Brundtland Report]. Oxford, England: Oxford University Press. [26]

WEC (World Energy Council). 1998. Survey of energy resources. [17]

WHO (World Health Organization). 1995. Community water supply and sanitation: Needs, challenges and health objectives. 48th World Health Assembly, A48/INEDOC./2,28 April, Geneva, Switzerland. [16]

WMO (World Meteorological Organization). 1997. *The world's water: Is there enough?* World Meteorological Organization. United Nations Educational, Scientific and Cultural Organization, WMO-No: 857. Paris, France. [12]

WRI (World Resources Institute). 1987. *Frontier forests of the world. World resources 1994–95.* New York: Oxford University Press. [15]

————. 1998. *World resources: A guide to the global environment.* New York: Oxford University Press. [11, 13]

————. 2000. World Resources Report. *People and ecosystems: The fraying web of life.* Washington, D.C.: the World Bank, the United Nations Environment Program, and the World Resources Institute. [5, 11, 20, 25]

Worldwatch Institute. 1984 *et seq. State of the world.* New York: W. W. Norton & Company. [26]

Wright, H. A., and A. W. Bailey. 1982. *Fire ecology.* New York: John Wiley & Sons. [10]

Wuerthner, G. 1998. High stakes: The legacy of mining. *National Parks* 72 (7–8): 22–25. [14] www.ipcc.ch

Yaalon, D. H. 1971. Soil-forming processes in time and space. In *Paleaopedology—Origin, nature and dating of Paleosols,* ed. D. H. Yaalon, 29–39. Jerusalem: Israel University Press. [4]

Yergin, D. 1991. *The prize: The epic quest for oil, money and power.* New York: Simon and Schuster. [17]

Zedler, J. B., and S. Kercher. 2005. Wetland resources: Status, trends, ecosystem services, and restorability. *Annual Review of Environment and Resources* 30:39–74. [24]

Zipperer, W. C., J. Wu, R. V. Pouyat, and S. T. A. Pickett. 2000. The application of ecological principles to urban and urbanizing landscapes. *Ecological Applications* 10 (3): 685–688. [23]

Zonneveld, I. S., and R. T. T. Forman, eds. 1990. *Changing landscapes: An ecological perspective.* New York: Springer–Verlag. [26]

Zwinger, A. H., and B. Willard. 1972. *Land above the trees, a guide to American alpine tundra.* New York: Harper & Row Publishers. [10]

The following web sites are a treasure-trove of environmental information—data, figures, tables, and excellent syntheses of relevant topics

BP (British Petroleum) is an authoritative source of data on total reserves of fossil fuels (coal, gas, oil, nuclear) and renewable sources; trends for energy use, movement of energy resources, price structure and future trends; and educational and environmental materials. http://www.bp.com

Climate Change and Water, a report by the Intergovernmental Panel of Climate Change (IPCC) on the predicted impacts of climate change on the provision of water to peoples the world over, (Technical Paper 6, with WMO and UNEP) http://www.ipcc.ch/pdf/technical-papers/climate-change-water-en.pdf

Conservation of International Trade in Endangered Species of Wild Fauna and Flora (CITES) is an international agreement to prevent trade in endangered species: http://www.cites.org/

Earth Policy Institute is "dedicated to building a sustainable future as well as providing a plan of how to get from here to there": http://www.earth-policy.org/

New Delhi, India—Center for Science and Environment, a public interest agency that lobbies for the environment, sustainable development, and social justice: http://www.cseindia.org/

International Energy Agency (IEA) serves as a policy advisor to its 28 member countries so they may more effectively advocate for affordable and clean energy: http://www.iea.org/

National Wildlife Federation, working to protect wildlife species: http://www.nwf.org/

Oak Ridge National Laboratory, the Carbon Dioxide Information Analysis Center, stores and analyzes climate change data, in its role as part of the U.S. Department of Energy: http://cdiac.ornl.gov/

Population Reference Bureau provides information about population growth, health, and the environment: http://www.prb.org/

Stream Corridor Restoration, A Partnership of U. S. Government Departments, is a clearinghouse for information on stream corridor restoration: http://www.nrcs.usda.gov/technical/stream_restoration/scrhgr1.htm

International Union for the Conservation of Nature (IUCN), works to find solutions to environmental problems and development challenges: http://www.iucn.org/

IUCN Red List of Threatened Species—the listing of internationally threatened and endangered species, this site contains taxonomic, conservation status, and distribution information: http://www.iucnredlist.org/

The White House Council on Environmental Quality (CEQ), coordinates U.S. federal environmental efforts and policy: http://www.whitehouse.gov/administration/eop/ceq/

United Nations Department of Economic and Social Affairs, Division for Sustainable Development promotes an international action plan for sustainable development: http://www.un.org/esa/sustdev/documents/agenda21/index.htm

Useful Web Resources

United Nations Development Program (UNDP), works to advance development though local capacity building: http://www.undp.org/

United Nations Environment Program (UNEP), an international effort of care for the environment and humans today without compromising the needs of future generations: http://www.unep.org/

United Nations Food and Agricultural Organization (FAO) has the goal of defeating hunger worldwide: http://www.fao.org/ and statistical data: http://faostat.fao.org/site/291/default.aspx

United Nations Intergovernmental Panel on Climate Change (IPCC) provides accurate scientific information about climate change to decision-makers worldwide: http://www.ipcc.ch/index.htm

United Nations Millennium Project, commissioned in 2002, developed a plan to address the poverty, hunger, and disease that affects billions of people: http://unmp.forumone.com/

United Nations Millennium Ecosystem Assessment (MEA). The MEA "assessed the consequences of ecosystem change for human well-being." Between 2001 to 2005, the MEA involved the work of more than 1,360 experts worldwide. All reports are available at: http://www.millenniumassessment.org/en/index.aspx

United States Climate Change Science Program/US Global Change Research Program works on issues associated with global change and its effects on ecosystems and society: http://www.usgcrp.gov/usgcrp/default.php

United States Department of Agriculture, Natural Resources Conservation Service is devoted to the conservation of soil, water, and other natural resources on agricultural lands: http://www.nrcs.usda.gov/

United States Department of Commerce, National Oceanic and Atmospheric Administration (NOAA) is the scientific agency focused on weather, climate monitoring, coastal zone management, and fisheries management: http://www.noaa.gov/

United States Department of Energy, Energy Information Administration (EIA) is the site of "official energy statistics from the U.S. Government": http://www.eia.doe.gov/

United States Department of the Interior, Fish, and Wildlife Service provides a host of information on species conservation in the United States: http://www.fws.gov/

United States Department of the Interior, Geological Survey (USGS) "serves the Nation by providing reliable scientific information to describe and understand the Earth; minimize loss of life and property from natural disasters; manages water, biological, energy, and mineral resources; and enhances and protects our quality of life": http://www.usgs.gov/

United States Environmental Protection Agency (EPA) is the lead U.S. agency tasked to protect environmental quality through a rigorous scientific research program, education, and policy development: http://www.epa.gov/

United States Man and Biosphere Program (MAB) is an effort by U.S. Government agencies to fund research on ecosystem sustainability, global change, and biodiversity: http://www.state.gov/www/global/oes/mab.html

Wetlands International works to conserve wetland species, particularly waterfowl and fish species: http://www.wetlands.org/Default.aspx

World Business Council for Sustainable Development (WBCSD) is a CEO-led, global association of some 200 companies dealing exclusively with business and sustainable development: http://www.wbcsd.org

World Energy Council (WEC) is a group of representatives from the energy industry drawn from over 90 countries who work to promote the sustainable use of energy: http://www.world-energy.org/

World Health Organization (WHO) is an agency within the United Nations providing research and guidance on issues associated with human health: http://www.who.int/en/

World Meteorological Organization (WMO) is an agency within the United Nations responsible for monitoring the atmosphere, weather, and climate: http://www.wmo.int/pages/index_en.html

World Resources Institute (WRI) is an "environmental think tank that goes beyond research to find practical ways to protect the earth and improve people's lives": http://www.wri.org/

World Wildlife Fund (WWF) works internationally to conserve nature through the protection of species and their habitats: http://www.worldwildlife.org/

Worldwatch Institute, an independent agency, conducts research on environmental issues to provide sound scientific information to decision-makers, with a focus on climate change, environmental degradation, human population, and poverty: http://www.worldwatch.org

Abundance: the number of individuals of animals and plants per unit area and time.

Acid mine drainage (AMD): the process by which acids and soluble metal compounds contained in ores and coal deposits are generated upon exposure to air and water and washed into surface waters, groundwater, and soils, thus degrading the quality of surface water, groundwater, and aquatic life.

Acid soils: soils with pH below 7.0.

Active solar thermal systems: use of solar collectors of the flat plate type, generally installed on roofs for heating, cooling, and the production of hot water for residential, commercial, and industrial uses.

Adaptive ecosystem management: an iterative management system that uses information from monitoring to improve decision-making over time. Feedback from ecological, social, and economic systems may be used, with an understanding of the uncertainties associated with them.

Adiabatic lapse rate: the rate of change in temperature of a mass of air as it moves upwards.

Aerobic conditions: conditions for growth or metabolism in which the organism is sufficiently supplied with oxygen.

Aerosols: a mixture of solid and liquid particles suspended in the atmosphere.

Afforestation: forests established by planting or seeding on land where the previous ecosystem was not a forest.

Agroforestry: a collective name for land-use systems and practices where trees are deliberately integrated with crops or animals on the same land management unit, yielding wood products and crops.

Albedo: the extent to which an object diffusely reflects light from the sun.

Algal bloom: the explosive population growth of algae in aquatic ecosystems, typically a result of nutrient enrichment.

Alkaline (sodic) soils: soils in which there is an accumulation of sodium carbonates and bicarbonates near the soil surface.

Allelopathy: the process by which the metabolic products of one species can inhibit others.

Allogenic succession: changes in species composition that are imposed by factors external to the biotic community.

Alpha diversity: the biodiversity within a particular area, community, or ecosystem, measured by counting the number of taxa within the ecosystem.

Ammonification: microbial conversion of organic-nitrogen into ammonia (NH_3) or ammonium (NH_4).

Anaerobic conditions: circumstances under which reductive conditions prevail due to the absence of oxygen.

Anion exchange capacity (AEC): the total amount of anions that a soil can adsorb.

Anions: negatively charged ions.

Aphotic zone: the portion of a lake or ocean where there is little or no sunlight.

Aquaculture: the farming of aquatic organisms, particularly fish, practiced in inland freshwaters, in estuaries, and near-shore oceans.

Assimilation efficiency: the ratio of the amount of energy assimilated to the amount of energy ingested.

Association: a plant community of definite floristic composition, uniform site conditions, and uniform appearance.

Atmospheric deposition: a term that refers to the combination of wet deposition (rain, snow, and mist droplets—wet fall) and dry deposition (dust, aerosols and gases—dry fall).

Atmospheric window: parts of the electromagnetic spectrum through which more than 70% of the radiation emitted from the Earth's surface escapes into space.

Autogenic succession: changes in species composition that are driven by changing conditions coming from within the community.

Beneficiation: the process of separating the mineral in ores from the waste.

Beta diversity: species diversity between ecosystems that involves comparing the number of taxa that are unique to each of the ecosystems.

Bioaccumulation: the biological sequestering of substances that enter the organism through respiration, food intake, skin contact with the substance, and/or other means.

Biochemical conversion technologies: a technology involving three steps: (1) converting biomass to sugar or other fermentation feedstock; (2) fermenting these biomass intermediates using microorganisms, including yeast and bacteria; and (3) processing the fermentation product to yield fuel-grade ethanol and other fuels, chemicals, heat and/or electricity.

Biocides (or pesticides): chemicals to kill organisms of all kinds that we consider undesirable (pests) in agricultural and urban-industrial ecosystems. Includes fungicides (fungus-killers), herbicides and weedicides (plant-killers), insecticides (insect-killers), and rodenticides (rat- and mouse-killers).

Bioconcentration: the accumulation of non-biodegradable compounds in an organism to higher concentrations than found in its surrounding environment; also called bioaccumulation or biomagnification.

Biodegradable: able to be broken down over time by the action of organisms, particularly microorganisms.

Biofuels: fuels derived from relatively recently dead biological material. See also biomass fuels.

Biogeochemical cycle: a circular pathway by which a chemical element or molecule moves through both biotic and abiotic compartments of an ecosystem.

Biological clock: a roughly 24-hour cycle in the biochemical, physiological, or behavioral processes of living beings. Another term for circadian rhythm.

Biological diversity or biodiversity: the variety of all of the Earth's life forms.

Biological environment: the component of the environment encompassing all the biological factors that affect the existence and survival of an organism or groups of organisms and their relationships to each other, including humans.

Biological invasion: refers to the human-driven introduction of invasive plants, animals, and microorganisms into new regions, or to the natural expansion of a species range, where they adversely affect the structure and function of native species and ecosystems.

Biological pumps: biologically-mediated processes that transport about a fifth of the atmospheric carbon fixed by algae in the euphotic zone from the surface layer to the ocean's interior.

Biomagnification: the sequence of processes that results in higher concentrations in organisms at higher levels in the food chain (at higher trophic levels).

Biomass fuels: energy obtained from plant materials and animal wastes.

Biomass: plant organic matter that accumulates as biological mass in plant organs over time.

Biophilia: love of living things.

Bioremediation: the use of natural or bioengineered microorganisms to break down wastes in order to clean up soil and water pollution.

Biosphere: the upper thin mantle on the Earth's crust enveloped by the atmosphere that supports life.

Biotic potential: the maximum growth rate of a given population under ideal conditions.

Bottom ash: solids remaining as residuals after incineration.

Brownfields: a real property, the expansion, redevelopment, or reuse of which may be complicated by the presence or potential presence of a hazardous substance, pollutant, or contaminant (U.S. EPA 2008).

Bucket-wheel excavators: crawler-mounted machines that continuously dig, transport, and deposit overburden during surface mining.

Burden of proof: the demand for proof of causation to prove something in a legal action.

C_3 plants: during the first steps in CO_2 assimilation, C_3 plants form a pair of three carbon-atom molecules.

C_4 plants: during the first steps in CO_2 assimilation, C_4 plants initially form four carbon-atom molecules.

Caldera: a volcanic feature formed by the collapse of land following a volcanic eruption.

Cap and trade program: a trading scheme where a limit (cap) is applied to permitted emissions and trading of permits is allowed in order to keep within the cap.

Carbon sequestration: the removal and storage of carbon from the atmosphere in carbon sinks (such as oceans, forests or soils) through biogeochemical processes.

Carrying capacity: the maximum population size that can be supported indefinitely by a given habitat.

Cation exchange capacity (CEC): the soil's ability to adsorb ions.

Cations: positively charged ions.

Change detection of land use/cover: monitoring landscape conversions from one land use/cover to another, or tracking modifications within a given land use/cover for a given time and space, through remotely sensed data and GIS analysis.

Channelization: a procedure of making stream channels deeper and straighter to increase the rate at which water is carried out of an area.

Chestnut blight (*Cryphonectria parasitica*): a fungal disease that has virtually eliminated American chestnut trees (*Castanea dentata*) from forest ecosystems in the eastern United States.

Chronosequence: time sequence.

Circadian rhythm: the behavior of many animals and plants following a diurnal–nocturnal pattern of approximately 24 hours.

Clear-cutting: the removal of all the trees from a forested area at one time.

Climate: average conditions derived from long-term periods of observation of the day-to-day weather changes.

Climax community: a relatively steady endpoint toward which successional stages tend to proceed.

Coevolution: reciprocal evolution of morphological and chemical traits that have arisen in two species to provide ecological advantages to both.

Cohort: the members of a population that are the same age.

Collective management: management of common ecosystem goods and services whose property rights cannot be defined, including biogeochemical cycles, global climatic regulation, protective stratospheric ozone layer, biodiversity, and atmospheric processes.

Combined sewer system (CSO): a sewer system in which sewage, industrial wastes, and urban storm water runoff are mixed together in an underground system of storm sewer pipes.

Command and control: environmental instruments also known as standards or regulations that require government intervention in stating that polluters must not exceed a certain level of environmental degradation (command) and in monitoring and enforcing the standards (control).

Commensalism: a type of relationship between two species in which one benefits while the other neither benefits nor is harmed.

Comminution: breaking down of the ore to liberate the ore mineral particles.

Common resource: a resource that is shared at the global, regional, or local levels whose property rights are impossible to assign; sometimes referred to simply as a "commons."

Community ecology: a subdiscipline of ecology which studies the distribution, abundance, demography, and interactions between coexisting populations.

Community: a group of coexisting populations of different species (plants, animals, microorganisms) living in the same place at the same time.

Community-type: a discrete plant community so recognized after field study of many sample plots and their synthesis, utilizing their inclusion of many of the same species (referred to as an *association* in the European ecological literature).

Compensation point: when the light intensity is just enough to produce energy by photosynthesis equal to the energy used in respiration

Competition: rivalry for limiting resources between populations of different species or between members of the same species.

Competitive exclusion principle: the process by which two species with identical ecological niches cannot coexist.

Composting: transformation of biodegradable solid wastes such as agricultural and yard waste into soil humus by microbial action.

Condensation: the change of the physical state of matter from a gaseous phase into a liquid phase.

Conduction: the transfer of heat that occurs when two objects are in physical contact.

Connectivity: the degree of spatial and temporal connectedness or continuity of ecosystems or habitat types.

Conservation buffers: relatively narrow strips of permanent vegetation that exist naturally or are planted strategically to intercept runoff or wind and play a significant role in the protection of water, soil, and air quality and the enhancement of wildlife.

Consumers: (heterotrophs) organisms that cannot make their own food; they obtain food by consuming other organisms.

Consumption efficiency: a measure of how much of the total energy available is actually consumed by the next higher trophic level.

Contingent valuation: a questionnaire-based method in which non-use economic values of non-marketed goods and services are estimated contingent upon the nature of the constructed (hypothetical) market and of their description in the survey scenario.

Convection: the mass transfer of heat from a solid object to a fluid, either air or water.

Conventional-tillage farming: using a plow followed by a disk, so that the upper 15 cm of the soil is inverted and broken up with previous crop residues and any cover vegetation plowed under.

Coriolis effect: the Earth's rotation from west to east causing a large-scale deflection of the direction of wind and water to the right in the northern hemisphere and to the left in the southern hemisphere.

Cost-benefit analysis (CBA): a comparative analysis of the consequences of alternative courses of action in terms of both their costs and benefits to individuals or society as a whole. This assists in the process of decision-making about the allocation of resources.

Crassulacean acid metabolism (CAM): a photosynthetic pathway that occurs mainly in succulent plants belonging to the family Crassulaceae. Unlike the chemical reaction of the carbon dioxide accumulation by C_4 plants, carbon dioxide fixation and its assimilation by CAM plants are not separated spatially, but temporally. This allows the uptake of carbon dioxide during the night.

Criteria pollutants: the six most common air pollutants for which the U.S. EPA has set national air quality standards: carbon monoxide (CO), lead (Pb), particulate matter (PM), nitrogen dioxide (NO_2), sulfur dioxide (SO_2), and ozone (O_3).

Crop rotation: successive planting on the same field of different crops that have different nutrient needs, diseases, pests, and root growth patterns.

Cryosphere: a collective term for glaciers, ice caps, sea ice, and permafrost.

Decision-making: the daily process of identifying and choosing among alternatives based on information, values, and preferences available at the time.

Decomposers: (saprotrophs) organisms that break down and consume dead organisms as their food supply. Bacteria and fungi are the primary decomposer groups and are fundamental to the cycling of nutrients.

Decomposition: breakdown of coarser biotic residues into finer ones.

Deep-well disposal: the belowground disposal of hazardous liquid wastes by injection into geologic formations that are isolated from groundwater or other avenues of migration.

Deforestation: the removal of a forest so that the land is converted to other land uses and covers.

Deltas: deposition of the smaller rocks and sediments in rivers, lakes, or the oceans. These are transported from the steeper slopes upstream to areas where water movement slows in downstream regions that have shallower slopes.

Demand: the amount of an economic good or service that people are willing and able to buy at various possible prices.

Demand-side management: conservation and efficiency-oriented measures on the side of consumers.

Demographic water scarcity: shortage of water when human population growth puts pressure on the total amount of available water in aquifers and watercourses.

Demography: the scientific study of the size, composition, and rate of change of populations.

Demophoric growth: the growth of human and per capita technological energy consumption.

Dendrochronology: the method of dating past events using tree rings.

Denitrification: the process by which some bacteria, able to use nitrate as the terminal electron acceptor in respiration in place of oxygen ("N-breathing"), obtain energy by using organic matter in respiration with the associated reduction of nitrate to nitrous oxide (N_2O) and then to dinitrogen gas, N_2. Occurs under anaerobic (lack of oxygen) conditions.

Depletion of ozone layer: see *ozone layer*.

Deposit-refund scheme: a front-end payment (deposit) for a potential polluting activity and a guarantee of a return of the payment (refund) when it is shown that the polluting activity did not occur.

Derelict lands: destruction of landforms and vegetation at the local and regional scales because of mining or other industrial processes, with consequent modification of microclimates. Derelict lands cannot be used without rehabilitation/restoration.

Desertification: a human-induced process that permanently diminishes biological productivity, diversity and self-regenerative capacity of ecosystems, especially in arid, semiarid, and dry subhumid regions (drylands) of the globe.

Detoxification: conversion of wastes into less hazardous or non-hazardous materials.

Detritivore: saprophytic components of the decomposer community, dominated by microbes that break down dead, organic matter.

Development: the processes of growing or progressing.

Dew point: the temperature to which a given parcel of air must be cooled for water vapor to condense into water under constant barometric pressure.

Diffuse sources: non-point sources of pollution.

Dimictic lake: lakes that mix from top to bottom during two mixing periods each year.

Diploid: having two copies of each chromosome in cells.

Distributive justice of power: disparity in access to the processes of decision- and policy-making of the poor versus the rich.

Distributive justice of wealth: the degree of unequal economic growth (or the gap between poverty and overconsumption).

Disturbance of ecosystems: an event or a series of events that alters the environmental conditions of the environment and thus the relationships of organisms, both spatially and temporally. This often involves the destruction of biomass.

Dominance: the exertion of a major controlling influence of one or more species upon all others by virtue of their number, size, productivity, or related activities.

Doubling time: the period of time required for a quantity to double in size or value.

Dragline: In surface mining, this is a method to remove overburden using a massive excavation machine with a boom that operates a huge bucket to scoop up and move the overburden.

Dredging: the removal of unconsolidated material from rivers, streams, lakes, and seas.

Drilling: the extraction of liquid and gaseous mineral resources from below the Earth's surface.

Dry deposition: the deposition of particles to the Earth's surface between precipitation events.

Dutch elm disease: a fungal disease caused by one of three species of microfungus, *Ophiostoma* and spread to elm trees by native North American elm bark beetle, *Hylurgopinus rufipes*, and European elm bark beetle, *Scolytus multistriatus*.

Dynamic equilibrium: the process of change from one state to another that continually occurs with a balance, since changes are occurring in opposite directions at the same rate.

Dynamic global vegetation models (DGVMs):   simulation of changes in different vegetation communities at the global scale.

Dynamics: spatial and temporal changes in ecosystems.

Ecesis: the establishment of the new arrivals (pioneers) at a site.

Ecological economics: an emerging field which aims at integrating economics and ecology under one umbrella. It recognizes the economy as a subsystem of the environment that must operate within the limits of the biosphere.

Ecological engineering: reshaping and redesign of any ecosystem that requires engineering skills.

Ecological footprint (EF):   an accounting of how much ecologically productive land and water are needed to sustain one person's lifestyle.

Ecological pyramid of energy: a graphical representation with the available energy left at the top being considerably less than at the starting point as one goes up the food chain.

Ecological risk assessment: a process to evaluate potential cause-and--effect relationships between multiple stressors and thier impacts on ecosystem components.

Ecological succession: gradual (progressive or retrogressive) changes in vegetation and its biophysical environment over time.

Ecology: the scientific study of an organisms' adaptations to their environment over time, how and how many organisms live in a given space at a given time, and how they interact with each other and their biophysical environment.

Economic disincentives: taxes or fines on the depletion and degradation of natural resources, with the purpose of discouraging environmentally destructive activities as well as promoting more efficient use of natural resources.

Economic law of diminishing returns: each additional unit of variable input beyond some point yields less and less additional output in a production system with fixed and variable inputs.

Economic minerals: minerals that are central to the economy of a nation.

Economic value: contribution of goods and services to monetary wealth. Economic value is revealed and measured by individual's willingness to pay (WTP) for a gain (or to avoid a loss) and willingness to accept (WTA) compensation to tolerate a loss or forego a benefit.

Economics: the study of the allocation of resources to produce marketable (commercially priced) goods and services available for consumption based on people's willingness to pay on individual business (micro) or societal (macro) levels.

Ecoregions: The U.S. Geological Surveys classification systems of areas delineated based on similar ecosystem types and environmental factors such as climate, soils, and topography.

Ecosystem: an interacting and open system of all plants, animals (including humans), and micro-organisms (biotic factors) and all the non-living physical (abiotic) factors in a given time and space.

Ecosystem classification: quantification of the spatial and volumetric structure of landscape mosaics that can be used to support analysis and modeling of processes and decisions about use and management of natural resources.

Ecosystem diversity: the variety of natural ecosystems that occur within a larger landscape.

Ecosystem goods and services: the benefits and products provided by natural ecosystems. Services are categorized into *provisioning*, such as the production of clean water or food; *regulating*, such as the control of climate; *supporting*, including pollination and nutrient cycles; and *cultural*, such as recreational benefits.

Ecosystem management: ecologically compatible resource governance system that mediates the biophysical, social, and economic relationships or trade-offs among societal needs, economic development, and sustenance of life-supporting ecological processes through Cooperation, Coordination, Collaboration, and Consensus (the 4 Cs).

Ecosystem pools or stocks: those phases of a cycle where a material is held in varying quantities in an ecosystem compartment such as the atmosphere, water, soil, animals, or vegetation.

Ecosystem resilience: the rate of ecosystem recovery after human-induced or natural disturbances.

Ecosystem resistance: the ability of an ecosystem to persist, in terms of its structure and function, despite human-induced or natural disturbances.

Ecosystem rehabilitation: a process designed to assist the recovery of an ecosystem that has been degraded or destroyed by human actions.

Ecosystem restoration: a process designed to recreate what was there before or to mimic natural succession as much as possible.

Ecosystem stability: the self-regulating capability of a natural system which allows the return to a steady state in the face of a disturbance.

Ecotones: transition zones between major communities, or transitional communities.

Ecotourism: an ideal form of tourism that enjoys the beauty of natural biodiversity in a non-destructive form.

Ecotypes: a subspecies or race that is especially adapted to a particular set of environmental conditions.

Edaphic specialization: population and community differences at several levels due to the requirements of plants for or their tolerance of different micronutrients.

Edge effects: the boundary areas between forest and non-forest ecosystems caused by fragmentation (breaking up contiguous forests into patches) that have harsher microclimate conditions and more invasive and non-native species.

Effluent: in water quality terms, water flowing out of a plant or factory after it has been treated to remove pollutants.

El Niño/Southern Oscillation (ENSO): a large-scale disturbance in the circulation patterns of the atmosphere and the Pacific Ocean, affecting coastal upwellings in the eastern Pacific and weather patterns in many parts of the world.

Electrical conductivity (EC): a measure of a solution's ability to conduct electric current, used in soils to measure dissolved solutes.

Electromagnetic radiation: a self-propagating wave in space produced by any electric charge which accelerates or any changing magnetic field and classified by wavelength (in order of increasing frequency) into electrical energy, radio, microwave, infrared, the visible region we perceive as light, ultraviolet, X-rays, and gamma rays.

Emergent properties: properties that are apparent when the group is viewed as a whole, which are not apparent, or even predictable, when looking at the individual species.

Emigration: the migration of organisms (and people) out of an area.

Empirical models: models relying only on data to quantify the behavior of a system as a function of explanatory variables that do not explain the underlying processes and structures that produced that system's behavior.

Emulsification: a blend of immiscible liquids (i.e., those not capable of being mixed), such as oil and water.

Endangered species: species under immediate threat of extinction.

Endemism: the biological uniqueness of an area including species with ranges restricted to a particular region.

Endocrine-disrupting chemicals (EDCs): a group of chemicals that are able to imitate or modify the action of natural hormones of organisms (also known as hormone disruptors or xenoestrogens) such as alkylphenols, bisphenol A, polychlorinated biphenyls (PCBs), phthalates, and pesticides such as DDT and dioxin.

Energy crops: crops that are used for energy production (e.g., biofuels).

Energy flux: the amount of energy that flows through a unit area per unit time.

Energy intensity: the ratio of energy used to gross domestic product (GDP).

Entropy: the measure of relative disorder, which is a gauge of the inability of energy to do work.

Environment: the specific physical and biological conditions in a given space (spatial) and time (temporal) that support the development, growth, and reproduction of a particular species or group of species.

Environmental accounting: recognition and inclusion of the true costs of resource exploitation and the value of intangible ecosystem benefits such as clean air and water, peace, and stability.

Environmental economics: the application of the principles of (neoclassical) economics to the study of management of environment and natural resources incorporating some ecological principles and methodology, with the focus of assigning appropriate values for environment and natural resources to assess the full costs and benefits of economic activity on the environment.

Environmental ethics: the recognition and consideration of intrinsic values in our natural landscape in the process of decision-making.

Environmental gradients: ecological conditions that vary spatially and that affect the distribution and abundance of organisms.

Environmental justice: the disproportionate distribution of environmental costs and benefits over time and space within the present generation and between the present and future generations.

Environmental legislation or law: a body of state and federal statutes intended to protect the environment and mitigate environmental damages that give individuals and groups the right to bring legal actions to enforce the protections or demand revisions of private and public activities which may have detrimental effects on the environment.

Environmental policy: an agreed-upon course of actions to solve environmental problems established through integration of the activities and powers of the legislative, judicial, and executive branches of government.

Environmental refugees: both externally and internally displaced people because of environmental disturbances that jeopardized their existence and the quality of their lives.

Environmental science: the scientific study of the dynamic relationships and processes that occur among the biological and physical components of nature and their impact on economic and social systems at multiple spatial and temporal scales.

Epilimnion: the uppermost layer of water in direct contact with the atmosphere.

Erodibility: the vulnerability of soil to erosion.

Erosivity: the capacity of erosive agent to cause erosion in given circumstances.

Estivation: summer dormancy.

Eutrophic: nutrient-rich.

Eutrophication: conditions including lack of oxygen and severe reductions in water quality created by biological oxygen demand of extensive algal growth and decay in response to addition of nitrogen and phosphorus.

Evaporation: the process by which molecules in a liquid state spontaneously become gaseous.

Evapotranspiration: the combined processes of evaporation and transpiration.

Even-aged forest structure: forest with all of the trees being essentially the same age.

Experiment: a systematic method of investigations performed to falsify a hypothesis or identify a causal relationship between phenomena.

Experimental studies: scientific investigations designed to determine causal relationships and extract conclusions using experiments that manipulate a single factor or multiple factors about the hypothesis being tested.

Exponential growth: a system behavior that is exhibited when a component of a system feeds

Export production: the amount of organic matter produced in the ocean by primary production that is not recycled (re-mineralized) before it sinks into the aphotic zone.

***Ex situ* conservation:** methods for the conservation, protection, or banking of organisms and their genetic materials which involve moving individuals or groups from their natural habitats into captivity such as aquaria, botanical gardens, zoological gardens, arboreta, and seed banks.

Externalities: variables that are not accounted for, or are external to, the economic system and that impose positive or negative costs on the community. Negative externalities include costs associated with cleanup or health effects that are not borne by the polluters during the use of water, soil, or air for the discharge of industrial wastes. Positive externalities include the free use of the environment to get rid of unwanted waste.

Facilitation: a process by which an initial group of organisms inhabiting a given area modifies it so much that, over time, the habitat becomes more hospitable for the establishment of other succeeding species.

Falsification: the process of using an empirical method to refute or disprove a scientific hypothesis.

Feedback loops: inherent features of ecological systems referring to a process whereby an output from the system returns to itself as an input.

Field capacity: water remaining in soil after gravity-driven drainage.

First law of thermodynamics: conservation of energy/matter in which energy/matter can be converted from one form to another, but they can neither be destroyed nor created during the process.

First-generation pesticides: naturally occurring elements and compounds used to deter the consumption of/damage to crops by pests until the 1940s.

Fission: nuclear reaction by which the nucleus is split.

Floodplains: areas bordering rivers and streams at extreme risk of flooding.

Floods: the spread of water onto the floodplain and land areas beyond.

Fluvial: all topics related to flowing water such as rivers, streams, through flow, overland flow and percolation.

Fluxes: transfer rates of material through a reservoir during a given period, flow of energy and minerals.

Fly ash: one of the residues generated in the combustion of coal.

Food security: production of adequate and nutritious food that is accessible to everyone, especially the poor, by securing the natural resource base of agricultural productivity in the long term.

Forestland: an area of at least 0.4 ha (= 1 acre) with 10 percent forest cover.

Fossil fuels: finite energy resources that take millions of years to be produced, such as petroleum, coal, natural gas, and nuclear fuels.

Fragmentation: the breaking of forests into smaller patches through human activities (e.g., transportation or utility corridors, urbanization, and large clear-cuts), thereby reducing area, integrity, connectivity, and population sizes of habitats, and altering all or part of their microclimates.

Frequency: a measure of the number of occurrences of a repeating event per unit time.

Fuel assembly: pellets of ceramic uranium dioxide sealed in metal rods bundled together.

Functional diversity: variation in ecological processes.

Fundamental niche: the full range of environmental conditions under which an organism can exist.

Fusion: nuclear reaction by which the nucleus is fused.

Gamma diversity: taxonomic diversity of a region with several ecosystems.

Gamma rays: a form of electromagnetic radiation or light emission with the highest frequency and energy (the shortest wavelength) within the electromagnetic spectrum, produced by sub-atomic particle interactions.

Gangue: undesirable minerals such as carbonates or silicates interwoven with economically valuable metal-bearing minerals.

Gasification: the conversion of coal to a gaseous form in a reaction with steam, oxygen, and carbon dioxide.

General circulation models (GCMs):   models constructed to predict the overall impact of GHGs on climate change including temperature and other climactic variables.

Genetic diversity: a level of biodiversity that refers to the total number of genetic characteristics in the genetic makeup of a species.

Geochemically abundant metals: metals individually constituting 0.1 percent or more of the Earth's crust by weight, including iron, aluminum, silicon, manganese, magnesium, and titanium.

Geochemically scarce metals: metals individually constituting less than 0.1 percent of the Earth's crust by weight, including copper, lead, zinc, molybdenum, mercury, silver, and gold.

Geographical information systems: a spatial analysis tool of entering, storing, retrieving, transforming, measuring, combining, classifying, and displaying geographically referenced data that has been registered to a common coordinate system.

Geothermal energy: extracting thermal energy from hot water found underground at depths ranging from a few hundred meters to a few kilometers beneath the Earth's surface.

Glacial moraines: any glacially formed accumulation of unconsolidated debris which can occur in currently and formerly glaciated regions.

Glaciations or ice ages: cold periods that Earth has experienced in which glaciers expanded to cover an extensive portion of the Earth's surface. There have been four major periods of glaciation dating back about 2.5 billion years.

Global warming potential (GWP):   the ratio of both direct and indirect radiative forcing from one unit mass of a greenhouse gas to that of one unit mass of the reference gas (CO_2) over a period of time.

Graben lake: a type of lake that forms in the depression of the earth's crust between two parallel faults.

Grade: the amount of metal-bearing rock per given volume.

Green Revolution: the development and spread of high-yielding varieties and more nutrient rich crops during the 1960s.

Greenhouse effect: the natural process in which certain gases in the atmosphere (water vapor, carbon dioxide, nitrous oxide, and methane, for example) trap energy by absorbing long-wave radiation emitted from the Earth's surface, thus warming the Earth's surface and atmosphere.

Greenhouse gases (GHGs): gases such as carbon dioxide (CO_2), water vapor (H_2O), methane (CH_4), nitrous oxide (N_2O), tropospheric ozone (O_3), and chlorofluorocarbons (CFxClx) present in the atmosphere that reduce the loss of heat into space and therefore contribute to global temperatures through the greenhouse effect.

Gross domestic product (GDP): the total market value of all goods and services produced within a country in a given period of time.

Gross primary productivity (GPP): the photosynthetic fixation rate of atmospheric carbon dioxide.

Growing season: the length of time when climatic conditions are suitable for photosynthetic activity of plants.

Growth: increase in size, number, or value.

Gully erosion: the more severe form of rill erosion in which huge rivulets of fast-flowing water carve deep, wide channels in the soil, ultimately forming ditches and gullies.

Gyres: any manner of swirling vortex caused by the Coriolis effect driving the cell-like circulation pattern of ocean currents.

Habitat: the place where an organism lives and its living and non-living surroundings.

Harmful algal blooms (HABs): algae that grow quickly in the water, often producing toxins that can poison the food web, thereby adversely affecting and even killing zooplankton, shellfish, fish, birds, marine mammals, and humans.

Hazardous waste: wastes that exhibit one of the following features: ignitability (spontaneously combustible), corrosivity, reactivity (explosiveness and toxicity when mixed with water), and toxicity (harmful or fatal when ingested or absorbed).

Haze: an atmospheric phenomenon where dust, smoke and other dry particles obscure the clarity of the sky.

Heap leaching (hydrometallurgy): recovering the pure metal from impurities using chemicals.

Heat: a form of energy linked with the movement of atoms or molecules of a substance.

Hibernation: winter dormancy.

Hierarchy of human needs: a theory in psychology, proposed by Abraham Maslow (1943), in which a hierarchy of universal human needs is represented as a pyramid.

High-grading: timber harvesting where all or almost all of the trees of economic value are removed from the stand.

Histosols: soils comprised of poorly decomposed organic matter that are found in poorly-drained areas throughout the taiga.

Home range: a given space that animals tend to inhabit.

Homeotherms: warm-blooded animals who keep their body temperatures at approximately a constant level.

Hotspots: areas of high biological diversity that are at a high risk of extinction. Biodiversity hotspots must support at least 1,500 species of endemic plants and have lost a minimum of 70% of their original habitat area.

Human appropriation of net primary production: the fraction of the Earth's primary production used by humans, making it unavailable for other species.

Human carrying capacity: the controls over human populations imposed by local and global biophysical limitations.

Human development index (HDI): a composite index that measures the average achievements of a society as a provider of basic opportunity for the members of that society, based on the premise that the overall well-being of a society is not really represented by a measure that considers only the production of goods and services; this index was developed by The United Nations Development Programme.

Human resources (human capital): labor that fuels and controls socioeconomic systems as well as the mental energy that goes into business and information systems (e.g., labor, human ingenuity, and social values).

Humidity: the amount of water vapor in a given volume of air.

Humus: the fraction of slowly decomposable soil organic material.

Hydrarch succession: succession that starts primarily at the land–water interface.

Hydric habitat: aquatic habitats.

Hydrophytes: plant species growing in aquatic habitats.

Hydrosphere: the area of the earth covered by water.

Hypothesis: an explanation suggested for a phenomenon or an educated guess for a possible relationship between multiple phenomena.

Hypoxia: low oxygen concentrations in aquatic ecosystems, typically defined as less than 2 mg dissolved oxygen per liter. Caused by the microbial decay of plants, which consumes the oxygen dissolved in surface waters.

Igneous rocks: rocks formed from molten magma.

Immigration: the migration of organisms into a population from another area.

***In situ* conservation:** the practice of conservation in the context of "natural" systems or habitats.

Inceptisols: soils without well-developed organic layers.

Incident solar energy: a measure of the rate of solar energy arriving at the Earth's surface from the sun.

Incineration: the burning of material such as municipal solid waste.

Indicator plants: plants whose occurrence and abundance may signal the quantity, quality, and identity of mineral ores.

Industrialized agriculture: an agricultural production system that relies on high inputs of fossil fuel energy, water, fertilizers, and pesticides to produce high yields of single crops (monoculture) per unit area of cropland.

Infant mortality rate: the number of babies per 1,000 newborns each year who die before their first birthday.

Inference: the process of deriving a conclusion based solely on what one knows.

Infrared (IR): electromagnetic radiation whose wavelength is longer than that of visible light (longer than red), with wavelengths between 750 nm and 1 mm.

Inhibition: a process by which modifications of the environment by pioneer species inhibit subsequent recruitment of other species until the pioneer species eventually die.

Insolation (incoming solar radiation): a measure of incident solar radiation energy received on a given surface area in a given time.

Integrated pest management (IPM): an approach to managing agricultural pests by combining biological, cultural, physical, and chemical tools in a way that minimizes economic, health, and environmental risks through pesticide use.

Intercropping: biodiversity-enhancing practice of planting two or more mutually beneficial crops near one another on the same field either simultaneously with/without a row arrangement or in alternate strips of uniform width.

Interspecific interactions: the interactions between organisms belonging to different species.

Intertropical convergence zone (ITCZ): a low-pressure belt ringing the Earth at the equator formed by the upward movement of warm, moist air from the latitudes just above and below the equator.

Intuition: the act or faculty of knowing without the use of rational processes, a faculty of guessing accurately.

Invasional meltdown: a process by which the successful introduction and establishment of one non-indigenous species facilitates the establishment of other species from their home communities.

Invasive species: non-native, nonindigenous, or exotic species that invade an area and persist to the detriment of native species. It can also refer to invasive native species, i.e., those that grow to dominate an area and displace other species.

Ion exchange: reversible chemical reaction between an insoluble solid and a solution during which ions may be interchanged.

Jet streams: extremely fast-moving, narrow currents of air.

K-selected species: species characterized by large body size, long life expectancy, and the production of fewer offspring that require extensive parental care until they mature.

Keeling Curve: the most-cited and recognizable example of atmospheric CO_2 change that illustrates (a) that CO_2 concentrations have been rising since Keeling started his measurements in 1958, and (b) that in the summer when photosynthesis north of the equator is at its maximum, the atmospheric CO_2 is low, and then increases in the winter (annual oscillations).

Keystone predator: predator whose presence or absence determines the outcome of competition and, eventually determines the diversity and taxonomic composition of the entire community.

Kinetic energy: any form of energy associated with motion.

La Niña/Southern Oscillation: conditions opposite to those of El Niño and characterized by strong easterly trade winds and cold upwelling and sea surface temperatures (as much as $-14°C$ below normal).

Land cover: the biophysical appearance and the vegetation type present on the land surface, such as forest, agriculture, grassland, barren lands, deciduous or coniferous forests, water, ice, and snow.

Land management practices: a variety of methods to manage land use developed to meet human economic interests in response to changing environmental conditions.

Land suitability evaluation: an assessment of potential land performance and compatibility when the land is used for specified human purposes, e.g., the potential of a given piece of land to grow crops.

Land use conflicts: expanding and competing human uses of land and water resources that adversely affect one another or the structure and function of neighboring ecosystems.

Land use/cover dynamics: changes in the patterns of land uses/covers over time and space.

Land use: the way in which the land is used for human purposes such as nature conservation area, recreational area, agricultural land, rangeland, and urban-industrial area.

Land use/cover change: conversion of land between one land cover or use to another.

Law of demand and supply: describes the relationship between the buyers and sellers of a good: if demand exceeds supply, then the price rises, and the reverse is true if supply exceeds demand.

Leachate: liquid that seeps through the solid waste.

Leaching: the movement of nutrients into the ground water below the root zone.

Lentic (calm) systems: standing water bodies.

Life cycle assessment: the totality of environmental impacts resulting from a product during all stages of its life cycle, including its production, transportation, usage, and disposal stages ("cradle-to-grave" of a product).

Life expectancy: the average number of years that a newborn infant can be expected to live.

Light saturation: the light intensity above which light is no longer the factor limiting the overall rate of photosynthesis.

Limits to growth: negative feedbacks to prevent the exponential growth of a system from going on forever.

Limnetic zone: the well-lit, open surface waters in a lake, away from the shore.

Liquefaction: for coal, the distillation of coal to make liquid hydrocarbons yielding such fuels as gasoline, diesel oil, and alcohols.

Lithosphere: the bedrock and other rocky material on which soils have developed, including the plant root zone.

Littoral zone: the zone that extends from the high water mark, which is rarely inundated, to shoreline areas that are permanently submerged.

Logistic growth: the sigmoid (S-shaped) curve of growth during which the initial stage of growth is approximately exponential; then, as saturation begins, the growth slows, and at maturity, growth stops.

Lotic: aquatic ecosystems characterized by flowing water (rivers, streams, etc.).

Macronutrients: nutrients needed in relatively large quantities by plants and animals.

Malthusian: postulates an exponential increase in population numbers, but a linear growth in available resources.

Market and policy failures: market- and policy-induced failures as the major cause of under-valuation of goods and services generated by environmental processes or ecosystems.

Marketable (tradable) permits: actions for abating below the standard or pollution allowances for emitting a certain amount of pollution.

Market-based instruments: economic incentives for abating environmental degradation.

Megacities: those with more than 10 million residents.

Mesic habitat: habitats with intermediate moisture conditions.

Mesophytes: terrestrial plants which are adapted to neither a particularly dry nor particularly wet environment.

Mesosphere: a layer of the Earth's atmosphere that extends from approximately 45 to 80 km above the Earth's surface.

Metamorphic rocks: the product of the transformation of a pre-existing rock type subjected to heat and pressure (temperatures greater than 150 to 200 °C and pressures of 1500 bars) causing profound physical and/or chemical change, in a process called metamorphism, which means "change in form".

Metapopulation: sets of populations of the same species that are separated spatially and that are linked by immigration and emigration.

Microclimate: the climate of a small area that differs from the area around it. Differences may be due to factors such as temperature or moisture availability.

Microhabitat: a small, localized habitat with specialized conditions.

Micronutrients: nutrients needed in very small quantities.

Migration: the arrival of seeds, germinules and disseminules at a site.

Mine-mouth operations: locating the electricity-generating plants close to the mines when it is uneconomical to transport coal large distances.

Mineral prospecting: the initial step of locating ores in large-enough quantities to make it economically profitable to mine and process.

Mineral reserve: an economically, environmentally and legally extractable mineral resource.

Mineral resource: a concentration of minerals, in or on the Earth's crust, in such a form and amount that extraction is currently or potentially economically feasible.

Mineralization: conversion of fine organic materials into inorganic ones.

Minerals: naturally occurring inorganic substances with a characteristic chemical composition.

Minimum viable area: the minimum amount of area needed for the survival of plant and animal species and their dispersal and recolonization from one site to another.

Mitigation: approaches to compensate for the adverse effects of environmental degradation.

Monomictic lake: lakes that mix from top to bottom during one mixing period each year.

Mountaintop removal: the practice of physically decapitating the tops of mountains and placing them in adjacent valleys to expose coal seams.

Municipal sewage: liquid waste that is typically 98 percent water containing a variety of organic and toxic compounds that is usually discharged, whether treated or untreated, into bodies of water near urban-industrial ecosystems with the assumption that solution to pollution is dilution.

Municipal solid waste (MSW): solid materials disposed of by residential, commercial, and institutional sectors.

Mutualism: the type of relationship between two species in which both receive benefits from the other.

Natural capital: all the stocks of ecosystem goods and all the flows of ecosystem services upon which the survival and health of plants, animals, and humans depend.

Net primary productivity (NPP): the net rate of organic carbon gain after respiration losses.

Niche: the totality of environmental conditions under which a species can survive, grow, and reproduce.

Nitrate leaching: the movement of nitrate into the ground water below the root zone, where it can no longer be utilized by terrestrial plants, or its movement into surface waters, where it can contribute to downstream eutrophication of aquatic systems.

Nitrification: process performed only by specialized microbes (nitrifying bacteria such as *Nitrosomonas*) which oxidize ammonium (NH_4^+) to nitrite (NO_2^-) and then other genera (e.g. *Nitrobacter*) oxidize the nitrite to nitrate (NO_3^-).

Nitrogen fixation: the conversion of gaseous nitrogen to more widely bioavailable forms such as ammonia, NH_3 or nitrate, NO_3^-.

Nitrogen fixers: certain types of bacteria (heterotrophic and photosynthetic types) and fungi that have the capacity to use energy to drive a complex series of reactions (using the enzyme nitrogenase) to break the triple bonds of atmospheric N_2 and chemically reduce the nitrogen to NH^4 (ammonium), which they use for their own growth.

Nomadic herding: moving animals in a continuous search for fodder.

Non-biodegradable: compounds that are resistant to breaking down over time.

Non-hazardous wastes: urban-industrial and agricultural wastes that do not meet the definitions of hazardous wastes.

Non-municipal waste: agricultural waste, mining waste, and industrial waste.

Non-point sources of pollution: pollutants from agriculture, uncollected sewage, and urban stormwater that come from large land areas and are difficult to regulate or reduce.

Non-renewable resources: natural resources that have very slow renewal rates and thus cannot replenish themselves after they are consumed.

Normalized difference vegetation index (NDVI): a non-linear transformation of the visible (red) and near-infrared bands of satellite information defined as the difference between the visible (red) and near-infrared (nir) bands, over their sum that measures vegetation amount and condition.

Not In My BackYard (NIMBY) and Locally Unwanted Land Uses (LULU): describes public opposition to projects, such as waste disposal sites, planned for their neighborhoods.

Nuclear reactor core: fuel assemblies of uranium bundled together in a heavy vessel.

Nudation: creation of a bare or partially bare substrate by a disturbance event.

Nutrient cycles: cycling of only those macro- and microelements that are essential for biotic growth.

Nutrient spirals: concept invoked to explain the cycling of nutrients while they are carried downstream. The nutrient cycle in a lake becomes a nutrient spiral in a river as the flow moves nutrients downstream.

Nutrients: chemical elements and compounds found in the environment that plants and animals need to grow and survive.

Observation: the recording of data using scientific instruments.

Observational approaches: identifying, describing, and following the possible causes, effects, and controls associated with events and/or organisms for a given period of time in a given area without manipulating the system.

Occluded phosphorus: much of the phosphorus in the environment that is precipitated in iron, aluminum, calcium, and organic compounds is essentially unavailable to plants.

Ocean enclosure: a policy tool adopted to prevent overfishing by putting the management of marine organisms within 320 km of the shore under the jurisdiction of the country bordering the ocean.

Oil slick: a film of oil on the surface of the water.

Old-growth forests (late-successional forest, or frontier forests):   mature forests that have reached the final successional stage and have been left intact for several hundreds of years.

Oligotrophic: nutrient-poor.

Open dumping: disposal of waste into aquatic and terrestrial ecosystems, leading to their degradation.

Open pit mining: strip mining; process of extracting minerals from open pits.

Ore: the mineral deposit when the amount of metal-bearing rock per given volume (grade) is high enough to make mining economically feasible.

Orographic precipitation: precipitation on the windward side of mountains that is due to the rising of moist air across the mountain ridge. This results in adiabatic cooling, condensation of water vapor, and precipitation.

Overburden: the material overlying a mineral or coal deposit.

Overfishing: the harvest rate of fishes that exceeds their reproductive capacities to replace themselves.

Oxidation: the loss of electrons by a molecule, atom or ion.

Oxisols: highly-weathered soils with clay mineralogy.

Ozone depletion substances (ODS):   chlorine-containing compounds and nitrogen oxide (N_2O) that accelerate the breakdown rate of O_3 into molecular (O_2) and atomic (O) oxygen under the influence of energetic UV radiation.

Ozone layer: a layer in the Earth's atmosphere with a high concentration of ozone (O_3) that is vital to life because it absorbs much of the sun's damaging ultraviolet radiation.

Parasitism: a type of predation in which the host (prey) is not killed in the short-term.

Parent materials: the immediate products of weathering of rocks.

Passive solar thermal systems: the exploitation of solar energy for heating and cooling buildings by design, and not by using solar collectors.

Pathogens: biological agents that cause disease or illness to their host.

Persistent organic pollutants: organic chemicals that persist in the environment because they are slowly biodegradable or non-biodegradable. This confers high potentials for bioaccumulation in the food web.

Pesticides: a substance or mixture of substances used to kill organisms which have characteristics that are regarded by humans as injurious or unwanted.

Phenology: the study of the timing of the events in a life cycle, including morphological and physiological changes displayed by plants and animals.

Photic zone: the depth of the water in a lake or ocean that is exposed to sufficient sunlight for photosynthesis to occur.

Photon: the elementary particle responsible for electromagnetic radiation of all wavelengths.

Photoperiod: The duration of an organism's daily exposure to light; it is an important factor with regard to the effect of light exposure on growth and development.

Photoperiodism: in plants, this refers to their ability to respond to the length of period of light. Certain species (short-day plants) stop flowering as soon as the day length has passed a critical value, while long-day plants begin to flower only after such a value has been passed.

Photosynthesis: the conversion of light energy into chemical energy by green plants and photosynthetic bacteria.

Photosynthetic pathways: three photosynthetic pathways in plants known as C_3, C_4, and CAM (*crassulacean acid metabolism*). At some point during the photosynthetic process, all three use the enzyme rubisco to fix carbon dioxide in the Calvin cycle. In C_3 photosynthesis (most trees and shrubs), the first product of photosynthesis is a 3-carbon molecule. In C_4 photosynthesis (e.g., corn and sugarcane), the initial photosynthetic product is a 4-carbon molecule. In CAM photosynthesis CO_2 is fixed at night; this occurs in some epiphytes and succulents in very arid regions.

Photosynthetically active radiation (PAR): the visible wavelength range (400–700 nm) of solar radiation (about 45–50 percent of the incoming solar radiation), which plants absorb for use in photosynthesis.

Photovoltaic systems (PV): a system of solar cells in which incident solar radiation is converted directly into electrical energy for lighting and other purposes.

Phreatophytes: a deep rooted plant that obtains its water from a ground supply or the water table.

Physical environment: the component of the environment comprising the incoming radiant energy of the sun, air, water, and soil without which life would not be possible.

Phytoplankton: microscopic plants (the autotrophic component of plankton) that live in the ocean at or near the surface.

Phytoremediation: restoring or remediating degraded land areas or water bodies by the cultivation and growth of appropriately selected plants; the use of plants to absorb and accumulate toxic materials from the soil, after which plants are harvested and disposed of in a hazardous waste landfill.

Pigouvian tax: the idea first proposed in 1920 by the British economist, Arthur Cecil Pigou, referring to a tax levied on an agent causing an environmental externality as an incentive to avert or mitigate such damage.

Pioneer plants: first arrivals in succession.

Plant sociology: the study of plant communities and their interactions, and the social organization of plants.

Plantation: a large farm, often devoted to growing trees.

Poikilotherms: also known as ectotherms or cold-blooded animals; these species' body temperatures vary with the environment; its body heat is derived from the environment.

Point-sources of pollution: most municipal and industrial pollutants (or wastes) that are relatively easy to identify or detect, typically discharging from a pipe.

Polar easterlies: dense, cold air masses flowing out of large cells of high pressure centered over each pole (called polar highs) toward the equator.

Policy agenda: description of a problem under consideration for discussion as an issue of concern.

Policy formation: the totality of the creation, development, and adoption of a policy.

Policy formulation: the crafting of proposed alternatives or options for solving a problem.

Policy-making: a method and framework to enact laws, develop rules, and implement and enforce both.

Pollutant: any material that is put into the air, water, or soil that is likely to produce a harmful effect.

Polluter pays principle: assertion that the polluter should bear the cost of any abatement taken to maintain an acceptable level of environmental quality.

Polyculture: an agricultural system in which several mutually nurturing crops (e.g., nitrogen-fixing plants) or those that show mutually compatible uses of resources (e.g., the root systems) are grown simultaneously.

Polymictic lake: lakes that are too shallow to develop thermal stratification; thus, their waters can mix from top to bottom throughout the ice-free period.

Polyploidy: having more than two copies of each chromosome.

Population: the interbreeding members of a species found in a given habitat that collectively express genetic, morphological, physiological, and behavioral variations.

Population density: the number of individuals in a specific population per unit area at a given time.

Population dynamics: the rates at which individual organisms are added and lost from a given population due to births, deaths, emigration and immigration, the age structure of the population, and variations in population size over time.

Population momentum: a high proportion of individuals of reproductive age in a population that causes a time lag in the stabilization of population growth.

Positive feedback loop: a feedback loop in which the system responds to a change in the same direction as the change.

Potential energy: energy which is not associated with motion and gets stored in an object.

Precautionary principle: shifting the burden of proof from demonstrating the presence of risk to demonstrating the absence of risk, thus helping to avoid possible future harm associated with suspected, rather than conclusive, environmental risks.

Precipitation scavenging: attachment of a pollutant to water in the atmosphere.

Precipitation: all forms (rain, snow, hail, sleet, and fog) of water that reach the surface of the earth.

Precision farming: the recognition of spatially and temporally varying field conditions by farmers to apply the right quantities of fertilizers, chemicals, and water at the right places at the right times.

Predation: a form of exploitive species interaction where one organism benefits, while the other is harmed or killed.

Prediction: a scientific statement forecasting what is most likely to happen in the future under specific conditions.

Preventive measures: ecologically long-term and economically cost-effective solutions to environmental problems.

Primary aerosols: aerosols produced by the direct injection of particles into the atmosphere.

Primary air pollutants: air pollutants most commonly found in urban areas, including carbon monoxide, nitrogen oxides, sulfur oxides, hydrocarbons, and particulate matter (solid and liquid).

Primary consumers: herbivores.

Primary producers: autotrophs (literally, self-nourishing organisms), including photosynthetic plants, photosynthetic Cyanobacteria, and eukaryotic algae, as well as chemosynthetic organisms (e.g., sulfur-oxidizing bacteria) that oxidize reduced inorganic compounds to obtain energy.

Primary succession: succession that begins in an area where no plants have grown before and that has had no biological activity or accumulation of soil organic matter.

Principle of successive approximations: the three-step process of identification, feasibility, and development in mining ore.

Process-based ecosystem models: models used to mimic past, present, and future changes, based on the representation of components (structure) and their interactions (function) of a delineated system.

Producers: also known as autotroph (green plants); the organisms in an ecosystem that produce food from inorganic compounds and light (common) or inorganic chemicals (less common).

Production efficiency: the ratio of net secondary production to the amount of energy assimilated at a given trophic level.

Progression: change from a less to a more complex community.

Proven mineral reserves: an estimate of minerals that that can be extracted profitably for human use with currently available technology.

Pull factors: conditions that attract people to move to a new location.

Push factors: conditions perceived by people to be detrimental to their well-being that induce them to migrate.

Pyramid of biomass: the amount of dry matter found at each trophic level.

Pyramid of numbers: the number of individuals at each successive trophic level.

Pyrolysis: a form of incineration that chemically decomposes organic materials by heat in the absence of oxygen.

Quarrying: a form of an open pit mine from which ore, coal, stone, or gravel is extracted.

***R*-selected species:** species characterized by an early age of reproduction, high reproductive rates, and low survival rates.

Radiant energy: energy that comes from a light source, such as from the nuclear reactions within the core of the sun; energy traveling as electromagnetic waves.

Rain shadow: an area of reduced moisture (precipitation) on the lee sides of mountains.

Rate variables: a modeling term for variables that describe rates of processes.

Realized niche: a niche that is a subset of its fundamental niche, which a given population is forced to occupy as a result of interactions with other species, particularly due to competition with other organisms.

Recoverable reserves: estimates of mineral resources that are likely to be available for future human use.

Red tide: a bloom of large numbers of certain algal species (phytoplankton) with reddish pigments that make water appear red.

Reduction: the gain of electrons by a molecule, atom, or ion.

Reforestation: forests established by planting or artificial seeding following a human-induced or natural disturbance.

Regional heat island: the dome of heat over cities that traps suspended and gaseous pollutants, increasing urban air pollution by up to 1,000 times.

Relative humidity: the amount of water vapor that exists in a gaseous mixture of air and water relative to the total amount of water the air can hold at a given temperature.

Remote sensing: the science of acquiring information about the Earth's surface without actually being in contact with it by sensing, recording, processing, and analyzing reflected or emitted energy.

Renewable energy sources: natural resources such as sunlight, wind, rain, tides, and geothermal heat, which are naturally replenished.

Renewable resources: natural resources that regenerate at a rate as fast as or faster than their use by humans.

Repetition: the act of performing scientific experiments or observations again to test reliability and validity of results.

Replacement-level fertility: the number of children that a couple must have to replace

Reproductive revolution: the expanding access of women to education and reproductive health services as well as the development of family planning methods during the second half of the twentieth century.

Residence time: the average time a substance spends within a specified pool of an ecosystem, that is, how fast something moves through a system in equilibrium.

Residue management: an agricultural practice of keeping previous crop residues as a protective cover on the soil surface of a field to maintain soil organic matter, reduce evaporation by shading the soil surface, control both water and wind erosion, and improve infiltration of water into the soil.

Resilience: a measure of how rapidly an ecosystem recovers from disturbances.

Resistance: a measure of ecosystem capacity to absorb disturbances before a change in the state of an ecosystem occurs.

Retrogression: change from a more to a less complex community.

Rhizosphere: the zone of and around the roots of plants.

Rill erosion: the kind of erosion that occurs when the surface water forms fast-flowing stream-lets that cause shallow, narrow channels to erode in the soil.

Riparian zone: the border of vegetation that runs alongside the stream channel and is influenced by seasonal flooding and saturated soils.

River continuum concept: a model of changes that might take place as water travels from headwater streams to larger rivers.

Rotation length: the number of years required to establish and grow trees to a specified size or condition of maturity.

Saline soils: soil in which sodium chloride and sodium sulfates have accumulated to the extent that plant growth is negatively affected.

Salinization: Human-induced accumulation of salts (such as chlorides and sulfates of sodium, calcium, magnesium, and potassium) in the soil over time.

Saltwater intrusion: the movement of saltwater into aquifers due to excessive extraction of freshwater in coastal areas.

Scientific law: a statement explaining a phenomenon of nature, derived from a well-documented history of successful replication that can be generalized across a set of conditions.

Second-generation pesticides: pesticides such as DDT (dichloro-diphenyl-trichloroethane) that are formulated chemically (as opposed to being derived from plants).

Second-growth forests: areas where forests have grown back through secondary succession after timber cutting or major natural disturbances.

Second law of thermodynamics: the entropy law that states that while quantity remains the same (First Law), the quality of matter/energy deteriorates gradually over time. When energy is transferred or transformed, part of the energy always becomes unusable as heat waste.

Secondary aerosols: aerosols produced by the conversion of gaseous precursors into liquid or solid particles.

Secondary air pollutants: those that are formed by chemical reactions of other substances in the atmosphere.

Secondary consumers: organisms in the food chain which feed on primary consumers.

Secondary productivity: the accumulation rate of biomass by consumers or heterotrophic organisms, including herbivores, carnivores, omnivores, and detritivores.

Secondary salinization: human-induced accumulation of dissolved (soluble) salts in the soil.

Secondary succession: succession that begins in areas from which the original biotic community has been removed due to a disturbance.

Secure landfill/sanitary landfill: burial of solid waste in places lined with impermeable layers of clay and plastic to minimize the pollution of surface water and groundwater.

Sediment: any particulate matter that can be transported by fluid flow and is deposited eventually as a layer of solid particles on the bottom of a water body.

Sedimentary rocks: rocks formed from the deposition and recementing of weathered products of other rocks modified by the action of climatic variables and water.

Sedimentation (or siltation): the deposition of sediment (such as soil particles, organic matter) by wind or water.

Seedtree: a modified clear-cut, in which a few widely spaced residual trees are left throughout the cut area to provide seed for the purpose of establishing the next forest.

Selective cutting: the periodical removal of individual trees or small groups of trees from a stand based on their size, species, quality, condition, and spacing.

Serendipity: making discoveries by accident.

Seres or seral stages: each succeeding community of organisms in succession.

Shantytowns: in urban areas of developing countries, collections of shelters built illegally by people who have migrated, on land they do not own.

Sheet erosion: the loss of relatively uniform layers of topsoil by a wide flow of surface water down a slope or across a field.

Shelterwood: a method of harvesting trees that consists of a series of two or more partial cuts spaced over several years (commonly 5 to 15 years, depending on species and site) so that new trees can grow from seeds of the remaining trees. This method typically shows limited success.

Shifting agriculture: (also called slash-and-burn agriculture) a traditional agricultural system common in tropical forest areas that involves the cutting and burning of forests, cropping the land for a few years until the soil fertility is exhausted, and then abandoning the area and moving on to clear another forest patch for agriculture.

Smelting: using heat to recover metal.

Soil: the unconsolidated mineral and organic material on the surface of the earth that serves as a medium for the growth of land plants.

Soil classification: the systematic categorization of soils based on distinguishing characteristics.

Soil structure: the aggregation of soil particles under field conditions into primary and secondary structural units which may be described as platy, blocky, granular, columnar, or prismatic.

Soil texture: particle size distributionof a soil defined by the relative percentages of sand, silt, and clay.

Solar energy: the energy reaching the Earth from sunlight.

Solid wastes (trash or refuse): any mixture of solid materials that have been discarded.

Solifluction: the surface layers of the soil and ice melt during the summer, resulting in saturated soil that flows down slope over the soil that remains frozen underneath.

Source reduction: reducing the amount of waste that involves altering the extraction and processing of minerals and the design and manufacture of products on the supply side and changing consumption behaviors and patterns on the demand side.

Species diversity: the number and distribution of species in one location.

Species extinction: the complete elimination of a species from the earth.

Spent nuclear fuel: nuclear fuel that has been used in the operation of nuclear reactors to the point that it can no longer generate energy efficiently.

Spodosols: relatively infertile soils with low pH, a leached (eluviated) surface (A horizon) layer, and a subsoil zone characterized by iron and aluminum accumulation.

Spoil banks: in surface mining, the deposition of overburden in large piles or in a series of ridges.

State-and-transition model: description of vegetation dynamics in a given area by a set of discrete states, with defined transitions between states.

State variables: a modeling term for variables that describe the accumulation, stock, pool, or level of things.

Stemflow: the process by which the remainder of intercepted precipitation after direct absorbtion by the leaves, and throughfall, contributes to the soil water by running down stems to the ground.

Strategic minerals: minerals that are of importance to a nation's military defense.

Stratification: different vegetation layers in a community including trees, shrubs, herbs (non-woody green plants, including vines), mosses, and epiphytes.

Stratosphere: the atmospheric region that extends from approximately 10 to 45 km above the Earth's surface.

Stratospheric ozone: a concentrated layer of ozone (O_3) in the stratosphere that absorbs incoming UV radiation, thereby reducing its harmful effects on organisms.

Strip mining: extractions of flat-lying layers of minerals just beneath the surface by first removing the overlying soil and rock.

Structural diversity: the spatial arrangement of physical units.

Subsidence: the sinking of the surface level of the Earth relative to sea level.

Subsistence agriculture: the production of enough food to feed the farming families with little left over to sell for income or hold in reserve for hard times.

Subtropical highs: an area of high pressure that occurs at latitudes of 30°N and 30°S.

Sudden oak death: a disease of native oaks, tanoaks, and a variety of shrubs.

Surface mining: see strip mining.

Survivorship curves: graphical representation of patterns of age-specific mortality on the survivorship of cohort members.

Sustainability-oriented incentives/disincentives: actions to induce ecologically compatible changes in technology, land uses, management practices, institutional regimes, and individual habits of consumption.

Sustainable development: the ability to meet the needs of today's people and environment without compromising that of subsequent generations.

Sustainable yield: the level of harvesting timber at a rate that will not reduce the amount and quality of renewable natural resources in the future.

Symbiosis: a close ecological relationship between the individuals of two (or more) different species.

Systematics: the study of the diversity of past and present life on the earth, and the relationships among living things through time.

Tailings: huge piles of waste rock.

Taxonomy: branch of biology that covers the describing, identifying, classifying, and naming of different species.

Technical water scarcity: limited freshwater resources due in part to a lack of infrastructure.

Temperature: the average kinetic energy of the atoms or molecules of a substance.

Territory: the defended portion of the home range.

Tertiary consumers: carnivores that eat other carnivorous (or secondary) consumers which is usually the top predator in an ecosystem and/or food chain.

Theory: an explanation for a natural phenomenon that is so well established that new data are not likely to alter it.

Thermal inversions: deviations from the normal decrease in temperature with altitude in the troposhere, resulting in an extremely stagnant pocket of air at the Earth's surface. This occurs more frequently in valleys and low-lying areas, where cool air is trapped by warm air above.

Thermal stratification of lake: a change in the temperature at different depths in the lake.

Thermodynamics: the science of energy transfer and transformation.

Thermosphere: a layer of the Earth's atmosphere that extends from approximately 80 to 500 km above the Earth's surface with very high temperatures that may exceed 900°C.

Thinning frequency: number of years between two successive thinnings.

Thinning intensity: percentage removal of basal area.

Third law of thermodynamics: the law that states that all kinetic motion of molecules ceases at a temperature called absolute zero (−273°C).

Threatened species: species at risk of becoming endangered in the foreseeable future.

Throughfall: process by which the remainder of intercepted precipitation after the leaves directly absorb contributes to the soil water by dripping to the ground.

Tolerance: a process by which more tolerant, late-succession species invade and are able to mature in the presence of pioneer species, making conditions less favorable for subsequent recruitment of the pioneers.

Toposequence: natural gradients of variability in soil properties.

Total fertility rate: the average number of children born to each woman in a population during her lifetime.

Tradable or marketable permits: the permits or licenses issued by governments to businesses to release certain amounts based on safe minimum standards of pollution into the air and water, which can be bought and sold in the marketplace.

Trade-offs: an exchange of one thing in return for another to achieve a desired result.

Trade winds: blow out of the high-pressure area around the subtropical latitudes (30°N and 30°S) toward the low-pressure area around the equator

Traditional agricultural systems: labor-intensive agricultural production systems such as polyculture, subsistence agriculture, shifting agriculture, and nomadic herding.

Traffic congestion: overshooting the available capacity of the system due to clogged urban roads and limited parking spaces.

Tragedy of the Commons: refers to how individuals acting in their own best interest collectively can threaten their own livelihood and the sustainability of publicly owned resources/ecosystem services. This concept was made popular by an article by Garrett Hardin published in 1968.

Transaction costs: expenses in time and resources for interested parties to negotiate and bargain with one another in an effort to overcome problems.

Transboundary issues: issues whose scope crosses existing national boundaries.

Transpiration: process by which plants take up water from the soil, transport it through their stems, and lose it to the atmosphere, primarily through the stomata on the leaves.

Transportation corridors: the physical infrastructure such as roads, railroads, or canals that enables the exchange of people and economic goods and services.

Transuranic waste: waste of radioactive elements with atomic numbers and weights greater than that of uranium.

Trophic structures: the relationship of an organism to other organisms in the context of a food web.

Troposphere: the lowest layer of the Earth's atmosphere that extends from the Earth's surface to a height of about 10 km.

Tropospheric ozone: unnaturally high levels of ozone in the troposphere, formed by the reaction of VOCs and NO_x in the presence of heat and sunlight; a pollutant.

Turbidity: the haziness of a fluid caused by individual suspended solids.

Turnover time: the ratio of the mass of a reservoir to the total rate of removal from the reservoir.

Ultisols: soils that have developed on igneous and metamorphic substrates.

Ultraviolet light: electromagnetic radiation with wavelengths between 380 and 10 nm, which is shorter than those of violet light (beyond violet).

Underground mining: excavating deposits that lie deep below the Earth's surface.

Uneven-aged forest structure: forest with trees of various age classes.

Upwelling: involves the rising of cold, nutrient-rich bottom water to the surface.

Urban heat islands: elevated air temperatures in cities relative to surrounding rural areas.

Urban–industrial ecosystems: the centers of trade and commerce, political governance, and consumption, production, and the distribution of economic goods and services, cultural–political enrichment, and technological innovation.

Urban sprawl: a process by which uncontrolled expansion of urban land, coupled with rapidly growing urban populations, leads to the increase of residential, commercial, industrial, and transportation land uses into what were once natural or farmed areas.

Urban vegetation: naturally growing and planted trees, shrubs, and grasses in open spaces of urban areas that range from remnants of natural habitats, municipal recreation areas, greenbelts, greenways, and gardens to cemeteries and abandoned vacant lots.

Urbanization: the process of transformation from a rural-agrarian system to an urban-industrial system in which a citys' population increases along with related economic, social, and political changes.

Validation: the process of establishing, with a high degree of assurance, that a system accomplishes its intended requirements.

Verification: a process used to evaluate internal consistence for a system to work properly.

Vertisols: self-churning soils that have a high proportion of montmorillinitic clays that shrink and swell during wet-dry cycles.

Volatilization: removal of toxic compounds from soil or water by transforming these into volatile compounds (used commonly in remediation of sites by plants).

Waste disposal: any waste management operation that covers the following main operations of final treatment: (1) incineration without energy recovery (on land; at sea), and (2) biological, physical, chemical treatment resulting in products or residues that are discarded; and of final disposal: (1)deposit into or onto land (e.g. landfill), including specially engineered landfill, (2) deep injection, (3) surface impoundment, (4) release into water bodies, and (5) permanent storage.

Waste generation: the weight or volume of materials and products that enter the waste stream in the form of gases, solids, and liquids before recycling, composting, landfilling, or combustion take place.

Waste stream: the total solid, liquid, and gaseous waste generated by all contributors.

Water consumption: the use of water in a manner that is not immediately replaced as runoff because of evaporation, etc., and so is not available for use.

Water footprint: the volume of water needed for the production of goods and services consumed by the inhabitants of a given area.

Water mining: a progressive drop of the groundwater table due to withdrawal in excess of the recharge rate of an aquifer.

Water pollutants: the materials that alter the biological and chemical composition of water so as to harm the health, survival, or activities of aquatic life and human populations.

Water scarcity: shortage of water when its availability drops below 1,700 m^3 per person per year.

Water stress: shortage of water when its availability drops below 1,000 m^3 per person per year.

Water use efficiency: the carbon gain per unit of water lost by plants through transpiration.

Watershed or catchment area: the ecological unit that includes a river system and the area of land that catches the precipitation that drains to that river.

Weather: refers to the short-term conditions of the atmosphere at any time.

Weathering: a complex group of processes that brings about a breakdown and disintegration of any rock type over time.

Westerlies: the winds flowing out of the subtropical highs that dominate between 30 and 60 degrees latitude.

Wet deposition: the deposition of particles during precipitation events.

Willingness to accept (WTA): economic value of a good or a service to a person as revealed and measured by what an individual is willing to accept as a compensation to tolerate a loss or forego a benefit.

Willingness to pay (WTP): economic value of a good or a service to a person as revealed and measured by what an individual is willing to sacrifice for a gain (or to avoid a loss).

Wilting point: a condition in which soil is too dry to allow further water extraction by plants.

Withdrawals, off-stream uses or abstractions (water demand or use): water removed from rivers, lakes, and subterranean aquifers.

X-rays: electromagnetic radiation whose wavelengths are very short, between 0.01 and 10 nm.

Xerarch: succession that starts primarily in dry habitats.

Xeric: low moisture conditions.

Xerophytes: plants inhabiting arid areas.

Zero population growth: a condition where the birth rate of a population equals the death rate.

Zooplankton: microscopic animals that eat phytoplankton and float freely with oceanic currents and in other bodies of water.

A